RLC ELEMENTS

	Resistor	Inductor	Capacitor
Circuit symbol:	$i(t)$ $+$ $v(t)$ $-$	$i(t)$ $+$ $v(t)$ $-$	$i(t)$ $+$ $v(t)$ $-$
Element laws in time domain:	$v(t) = Ri(t)$	$v(t) = L\dfrac{d}{dt}i(t)$	$v(t) = \dfrac{1}{C}\displaystyle\int_{t_0}^{t} i(\tau)d\tau + v(t_0)$
Phasor impedance Z	R	$j\omega L$	$\dfrac{1}{j\omega C}$
Laplace impedance $Z(s)$	R	sL	$\dfrac{1}{sC}$

LAPLACE TRANSFORM PROPERTIES

	$f(t)$	$\mathbf{F}(s)$
1. Linearity	$c_1 f_1(t) + c_2 f_2(t)$	$c_1 \mathbf{F}_1(s) + c_2 \mathbf{F}_2(s)$
2. Differentiation	$d/dt\; f(t)$	$s\mathbf{F}(s) - f(0-)$
3. n-fold differentiation	$d^n/dt^n f(t)$	$s^n \mathbf{F}(s) - s^{n-1} f(0-) - s^{n-2} f'(0-)$ $- \ldots - s f^{(n-2)}(0-) - f^{(n-1)}(0-)$
3. Integration	$\int_0^t f(\tau)d\tau$	$\mathbf{F}(s)/s$
4. Time shift	$f(t - t_0)u(t - t_0), t_0 > 0$	$e^{-st_0}\mathbf{F}(s)$
5. Frequency shift	$e^{-s_0 t} f(t)$	$\mathbf{F}(s + s_0)$
6. Time-frequency scaling	$f(ct), c > 0$	$1/c\; \mathbf{F}(s/c)$
7. t-multiplication	$t\; f(t)$	$-d/ds\; \mathbf{F}(s)$
8. n-fold t-multiplication	$t^n f(t)$	$(-1)^n d^n/ds^n \mathbf{F}(s)$
9. Impulse response and transfer function are a LT pair	$h(t)$	$\mathbf{H}(s)$
10. Initial rate theorem		$f(0+) = \lim\limits_{s\to\infty} sF(s)$
11. Final value theorem		$f(\infty) = \lim\limits_{s\to 0} sF(s)$

Third Edition

ELECTRIC CIRCUIT ANALYSIS

David E. Johnson
Birmingham-Southern College

Johnny R. Johnson
University of North Alabama

John L. Hilburn
President, Microcomputer Systems Inc.

Peter D. Scott
State University of New York—Buffalo

Prentice Hall, Upper Saddle River, New Jersey 07458

Library of Congress Cataloging-in-Publication Data

Electric circuit analysis / David E. Johnson . . . [et al]. — 3rd ed.
    p.   cm.
    Rev. ed. of: Electric circuit analysis / David E. Johnson, Johnny
R. Johnson. John L. Hilburn. 2nd ed. c1992.
    Includes bibliographical references and index.
    ISBN 0–13–252479–1
    1. Electric circuit analysis. I. Johnson, David E.   II. Johnson,
David E.   Electric circuit analysis.
TK454.E57   1997
621.319′2—dc20   96-27572
                                                   CIP

MicroSim, PSpice, PLSyn, and Polaris are registered trademarks of
MicroSim Corporation. All other trademarks are the property of their
respective owners.

ACQUISITIONS EDITOR: *Eric Svendsen*
EDITOR-IN-CHIEF: *Marcia Horton*
PRODUCTION EDITOR: *Rose Kernan*
DESIGNERS: *Christine Wolf/Maureen Eide*
COVER DESIGNER: *Christine Wolf*
MANUFACTURING BUYER: *Donna Sullivan*
EDITORIAL ASSISTANT: *Andrea Au*
DEVELOPMENT EDITOR: *Sondra Chavez*

© 1997, 1992, 1989 by Prentice-Hall, Inc.
Simon & Schuster/A Viacom Company
Upper Saddle River, New Jersey 07458

The author and publisher of this book have used their best efforts in preparing this book.
These efforts include the development, research, and testing of the theories and programs to
determine their effectiveness. The author and publisher make no warranty of any kind,
expressed or implied, with regard to these programs or the documentation contained in this
book. The author and publisher shall not be liable in any event for incidental or
consequential damages in connection with, or arising out of, the furnishing, performance, or
use of these programs.

Printed in the United States of America

10 9 8 7 6 5 4 3 2 1

ISBN 0-13-252479-1

Prentice-Hall International (UK) Limited, *London*
Prentice-Hall of Australia Pty. Limited, *Sydney*
Prentice-Hall Canada Inc., *Toronto*
Prentice-Hall Hispanoamericana, S.A., *Mexico*
Prentice-Hall of India Private Limited, *New Delhi*
Prentice-Hall of Japan, Inc., *Tokyo*
Simon & Schuster Asia Pte. Ltd., *Singapore*
Editora Prentice-Hall do Brasil, Ltda., *Rio de Janeiro*

Contents

7 SECOND-ORDER CIRCUITS   271

8 SINUSOIDAL SOURCES AND PHASORS   315

9 AC STEADY-STATE ANALYSIS   359

Preface

In 1978 the first edition of "Basic Electric Circuit Analysis" was published. It was well received by students and instructors, and after a decade of widespread adoption and four editions, a somewhat expanded version entitled "Electric Circuit Analysis" became available in 1989. The book you hold is the third edition of "Electric Circuit Analysis." It is somewhat unusual for an engineering text project to survive and prosper over a duration approaching two decades, and for the loyalty of those who use and have come to rely on it we are grateful. We hope you will find its traditional virtues securely woven into the present edition, accompanied perhaps with a few enhancements in the bargain.

The continued acceptance of this book is, we believe, the consequence of two essential ingredients: a sound initial design, and an ongoing commitment to upgrade and adapt, to regularly reappraise both content and approach in light of actual experience. To do this effectively requires awareness of the needs of instructors and students, not in the abstract but "on the ground." The four authors listed on the spine of this book have among us over one century of circuits teaching experience, and we remain in communication with the larger community of instructors and students beyond our campuses through extensive regular surveys conducted by Prentice Hall. Changes both large and small can be traced to comments made by many insightful colleagues and students, and, yes, even to complaints about specific examples or sections of narrative which could be written more clearly, and, we hope, as a consequence have been. To these individuals, too numerous to list but not to remember, we owe a great debt.

SCOPE AND PURPOSE

This book is intended for a two semester or three quarter introductory course in linear lumped circuit analysis for the sophomore post-secondary year. The first portion of the book may also be used for a one semester or two quarter terminal course for non-majors. The development is calculus-based, as are the physical laws upon which the methods of circuit analysis stand. The exposure to differential and integral calculus routinely acquired by the engineering or physical science student in the freshman year is adequate background. Other mathematical topics are raised in a limited and self-contained manner, and are not assumed or required as prerequisite background. This includes the algebraic manipulation of complex numbers, vector-matrix formulation and solution of linear systems of equations, singularity functions, and most particularly, the solution of linear time-invariant differential equations. The first two are the subjects of appendices and numerous examples and exercises, the latter two are developed in the chapter in which they are needed.

Care has been taken to fashion the selection and order of material to be of use to the electrical engineering baccalaureate student, but also to students of other engineering and technical disciplines at the two and four year post-secondary level. The mathematical ideas and methods of electrical circuit *analysis* are carefully separated from physics-based issues of electrical *science*, such as Maxwell's Laws, which are neither covered here nor assumed as pre- or corequisites. Mathematical rigor has not been sacrificed where it is justified and can be built on. But in the best engineering spirit, where a conclusion is obvious, it is declared to be just that, without further ado.

Sufficient space has been allocated for judicious repetition, reinforcement and learning by example, but we have resisted expanding the length of the book to the extremes sometimes found in current practice. We have found that great length daunts many students, and impedes the instructor crafting a one semester or two trimester course. It may dictate a crazy-quilt approach, skipping sections and whole chapters in the interests of time. This leaves lingering doubts whether some needed concept may have been missed, and certainly does not help develop a sense of continuity. The first nine chapters may be comfortably covered in a single semester exposure to AC and DC circuits, with more ambitious instructors adding the AC power chapters 10 and 11, or an introduction to the use of Laplace transforms, chapters 12 and 13. It is our practice to cover all 18 chapters in a two semester circuits course.

COMPLETELY REWRITTEN THIRD EDITION

The third edition incorporates considerably more changes than its predecessor. The principal ways in which it differs from the second edition are:

Narrative

The text narrative has been extensively rewritten. More than half the paragraphs are new, and many others have been reworked. The goal has been to clarify the flow of ideas and accommodate the changes in order and emphasis, while keeping the book to its relatively compact size. We do not believe that great length is a virtue in textbooks, either for the student or instructor.

Organization

Sequence and order have been moderately reworked. Chapters on Independence of Equations and Network Theorems have disappeared, the first to an appendix and the second scattered among several chapters in which the theorems arise more naturally in context. Laplace transforms are introduced somewhat earlier, after phasors and AC circuits, but in time to be used to simplify development of transfer functions and input-output analysis without resort to the contrivance of complex frequency. Fourier series and transform are treated in a unified single chapter.

New examples and exercises

Most examples and exercises are new. Previous editions had an unswerving commitment to easy numerics, like $C = 1$ Farad. Recognizing their motivational value, some more practical examples and exercises have been added to the mix. Sharp price reduction for scientific calculators over the last decade makes them now universally available. Their use lowers the computational pain level for exercises with answers like $i_1(t) = 0.544 \cos(377t + 27.4°)$ A to easily tolerable levels.

New problems

Most end-of-chapter problems are new. In response to requests from adopters, effort has been made to expand the range of difficulty, from drill to very challenging. The most challenging problems have been set off as a separate group. In addition, those problems intended to be solved with aid of a digital computer are set off as a second separate group within the end-of-chapter problem sets containing them. Other problems not requiring computer support but for which an electronic calculator would be of significant benefit are marked with a calculator icon. Finally, just as several new examples and exercises have been added which emphasize practical values, for instance resistance in the kilohms range and current in the milliampere range, many new end-of-chapter problems are of this practical type as well.

New coverage

Circuit design. The payoff for studying circuit analysis is directing this skill towards the design of novel and useful electric circuits. Few practitioners view circuit analysis, the calculation of currents and voltages in a specified circuit, as a fully satisfying end in itself. Engineers and applied scientists are characterized by an interest in building things, a desire to blend creativity and imagination with mathematical analysis to produce things that did not exist before, to create practical devices that satisfy design requirements and specifications. The universal availability of the circuit simulators SPICE and MicroSim® PSpice® make it cost-effective (and safe) to challenge even the first-year circuits student with design problems. In our experience, starting from a blank sheet of paper and seeing one's own design emerge, ultimately demonstrating that it works, meets all design specs, is one of the most gratifying and motivating elements of circuits work. To this end, sections entitled "Design of ..." have been added to five chapters, and a new chapter (Chapter 18) entirely devoted to design. This design content is self-contained, each instructor may avail her- or himself of as much of this design exposure as judgement and time availability dictate.

Some topics have been expanded and new topics added. Op amps have been presented through a series of increasingly refined models, and the virtual short circuit principle justified in context of negative feedback circuits. Feedback, loading, and the concept of modular "building blocks," all essential to understanding modern linear op amp circuits, are introduced. The fundamental requirement of stability, sometimes overlooked in introductory circuits courses, and its link to frequency response and AC steady state is addressed. Design of active circuits using SPICE, an activity students find exciting and satisfying and which can stimulate interest in the more routine analysis tasks, is given space. Bode gain plots are developed, and their use related to the engineering design cycle for active filters and other active circuits exploited.

A SPICE command reference appendix is included. The public domain computer-aided circuit analysis and simulation software program SPICE has been covered tutorially in the body of the book, in sections entitled "SPICE and . . ." at the end of many chapters. In addition, a concise listing of each of the SPICE commands used in this text, their rules and formats, is included as an appendix. While tutorial discussions and step by step examples are useful to learn SPICE efficiently, it is convenient not to have to thumb through many pages before reminding oneself how a resistor is entered into a SPICE input file, for instance.

DESIGN AND ORGANIZATION OF THIS BOOK

The book contains eighteen chapters and five appendices. These logically divide into time domain, phasor and transform domain sets of chapters. The first seven chapters introduce passive (RLC) and active (source, op amp) circuit elements, and the time domain analysis of AC and DC circuits which are formed when these elements are interconnected. Chapters 8–11 cover phasor analysis of ac steady state circuits, including complex power and three-phase circuits. The final set of chapters, Chapters 12–18, develops the transform analysis methods of Laplace and Fourier, and applies them in framing various input-output descriptions and design methods.

Chapters 10 and 11, on AC steady state power and three phase circuits, may be skipped without disservice to the following material. Likewise, the final chapter, Chapter 18, on filter design may be considered optional for an introductory circuits course. The remaining 15 chapters are designed to be covered in order, and it is not recommended that any of these be skipped without leaving troublesome gaps in the student's preparation for subsequent material. Single-semester courses typically cover Chapters 1–9, rewarding the student who will not continue with a second circuits course with a useful familiarity with AC and DC analysis techniques.

CHAPTER PEDAGOGY

Each section of each chapter has numerous step-by-step solved examples and ends with answered exercises. The end-of-chapter problems are designed to range over the topics of all the sections at all levels of difficulty. Those requiring an electronic calculator are marked with a calculator icon. Boxed equations are particularly important and should not be passed over without clear comprehension as much depends on them. Italicized areas in the text are highlighted for emphasis, they are the narrative equivalents of boxed equations.

The appendices contain background material included to make the book more self-contained without interrupting the basic flow of the narrative. The use of the material from Appendix A on linear algebra begins in Chapter 4 and continues regularly thereafter. Appendix B on complex numbers should be reviewed before studying Chapter 8, which relies heavily on complex algebra and the complex exponential time function. Appendix C on circuit topology is perhaps less necessary to solve simple problems, but clarifies important logical issues related to forming and solving simultaneous algebraic equations of the type which are needed for nodal and mesh analysis. Appendix D is a reference guide for the use of SPICE, and Appendix E lists answers to selected odd-numbered problems.

USE OF SPICE

SPICE has become the de facto standard for computer-aided circuit simulation and analysis in the classroom. Its use as a limited but helpful tool for the study of electrical circuits is warmly embraced in this book. There are two domains in which we have found SPICE particularly valuable, in gaining experience with circuits too big to justify the often catastrophic level of computational effort required to extract just a few simple responses, and in design. In the first case, a five-mesh circuit is essentially as simple as a one- or two-mesh circuit, and the experience with multiple resonances, parasitic oscillations and other behaviors of complex circuits can be extremely interesting. In design, SPICE allows students to try their own ideas on how real circuits solving real problems can be configured, and in our experience this often forms the high point of the course from the student's viewpoint. It is not always easy to sustain motivation at the sophomore level, when basic science and mathematics are center stage and students interested in building things and exercising individual creativity are beginning to wonder if they are in the right place. They are indeed in the right place, design is what engineering is all about, and SPICE is quite useful in making this point when it may be most needed.

Because not all instructors agree concerning the best use of SPICE within introductory circuits courses, care has been taken to segregate SPICE-related material into final sections recognizable by their titles "SPICE and ..." in 8 separate chapters. These sections form a bare-bones but self-contained tutorial on the commands and syntax of SPICE. These can all be skipped without loss of continuity if SPICE is not available, or the instructor chooses to forego computer support. End-of-chapter problems requiring the use of SPICE are the final problems offered in each chapter containing SPICE material, and these problems are clearly marked.

INSTRUCTIONAL AIDS

In addition to the bound text, the following ancillary material is available to support the instructional use of the book.

Solutions manual. A complete set of clear, concise step-by-step solutions to all of the end-of-chapter problems.

Student problem set with solutions. This large format manual contains an additional set of approximately 500 problems with detailed step by step solutions covering all chapters and topics. This supplement may be purchased separately by the student seeking further guided practice beyond the solved examples and the answered exercises and odd numbered end-of-chapter problems.

Transparency masters. A set of approximately 200 selected figures and tables from the text are available in the medium of overhead transparency masters.

CONTINUED FEEDBACK IS ESSENTIAL

We have acknowledged our debt to the community of users whose comments were instrumental in the genesis of the present edition, and eagerly anticipate continued dialog with instructors and students in the future. This is how we learn and move the project closer to the center of your needs and concerns. We have established an electronic mail address which will be read regularly by the authors, and to which we urge and invite you to post reactions, questions, and yes, even gripes. All will be carefully considered as the next edition, or next new book, comes round. The Internet email address is: *peter@eng.buffalo.edu.*

In particular we would like to offer this challenge to students. If you find an error in a problem or exercise, let us know. If we agree, we will send you a Student Problem Set with Solution free of charge, and acknowledge your help in the next edition of this manual.

ACKNOWLEDGMENTS

This work has been greatly enriched by the efforts of those who have contributed to earlier editions, whose voice can clearly be heard here, as well as those active in preparing the current edition. Of those in the latter category, special thanks to Profs. Dennis Tyner, Indira Chatterjee, Keith A. Ross and Bruce Wollenberg for their careful reading and many useful suggestions, to Sondra Chavez, Alan Apt and Eric Svendsen for their invaluable editorial help. The last-named author shouldered the heaviest load in the current revision cycle, and he would like to save the final acknowledgment for his family, who somehow understood.

David E. Johnson
John L. Hilburn
Johnny R. Johnson
Peter D. Scott

Introduction

Alessandro Volta
1745–1827

This endless circulation of the electric fluid may appear paradoxical, but it is no less true and real, and you may feel it with your hands.

Alessandro Volta

Electric circuit theory had its real beginning on March 20, 1800, when the Italian physicist Alessandro Volta announced his invention of the electric battery. This magnificent device allowed Volta to produce *current electricity*, a steady, continuous flow of electricity, as opposed to *static electricity*, produced in bursts by previous electrical machines such as the Leyden jar and Volta's own *electrophorus*.

Volta was born in the Italian city of Como, then a part of the Austrian Empire, and at age 18 he was performing electrical experiments and corresponding with well-known European electrical investigators. In 1782 he became professor of physics at the University of Padua, where he became involved in controversy with another well-known electrical pioneer, Luigi Galvani, professor of anatomy at Bologna. Galvani's experiments with frogs had led him to believe that current electricity was *animal electricity* caused by the organisms themselves. Volta, on the other hand, maintained that current electricity was *metallic electricity*, the source of which was the dissimilar metal probes attached to the frogs' legs. Both men were right. There is an animal electricity, and Galvani became famous as a founder of nerve physiology. Volta's great invention, however, revolutionized the use of electricity and gave the world one of its greatest benefits, the electric current. Volta was showered with honors during his lifetime. Napoleon made him a senator and later a count in the French Empire. After Napoleon's defeat, the Austrians allowed Volta to return to his Italian estate as a citizen in good stead. Volta was rewarded 54 years after his death when the unit of electromotive force was officially named the *volt*.

Chapter Contents

Electric circuit analysis is the portal through which students of electric phenomena begin their journey. It is the first course taken in their majors by most electrical engineering and electrical technology students. It is the primary exposure to electrical engineering, sometimes the only exposure, for students in many related disciplines, such as computer, mechanical, and biomedical engineering. Virtually all electrical engineering specialty areas, including electronics, power systems, communications, and digital design, rely heavily on circuit analysis. The only study within the electrical disciplines that is arguably more fundamental than circuits is electromagnetic (EM) field theory, which forms the scientific foundation upon which circuit analysis stands. However, even the primacy of EM field theory over circuits is incomplete, since many EM field theory problems turn out to be best solved by resort to circuit analysis techniques. It is no exaggeration to say that the ideas and methods of this first circuits course are, for many, the most important to be encountered in their entire undergraduate curriculum. To begin our study of electric circuits, we need to consider what an electric circuit is and what we mean by its analysis and its design. Since this is a quantitative study, we must identify the quantities associated with it, in what units these quantities are measured, and the basic definitions and conventions to be used throughout. These are the topics we consider in this chapter.

1.1   DEFINITIONS AND UNITS

FIGURE 1.1 General two-terminal electrical element.

An electric *circuit*, or electric *network*, is a collection of electrical elements interconnected in some way. Later we shall define the electrical elements in a formal manner, but for the present we shall be content to represent a general *two-terminal element* as shown in Fig. 1.1. The terminals a and b are accessible for connection with other elements. Familiar examples of two-terminal elements are resistors, inductors, capacitors, and sources, each of which will be studied in later sections.

More general circuit elements may have more than two terminals. Transistors and operational amplifiers are common examples. Also, a number of simple elements may be combined by interconnecting their terminals to form a single package having any number of accessible terminals. We consider some multiterminal elements later, but our main concern will be simple two-terminal elements. An example of an electric circuit with six such elements is shown in Fig. 1.2.

To understand the behavior of a circuit element, we need to consider certain quantities associated with it, such as *current* and *voltage*. These quantities and others, when they arise, must be carefully defined. This can be done only if we have a standard system of units so that when a quantity is described by measuring it, we can all agree on what the measurement means. Fortunately, there is such a standard system of units that is used today by virtually all the professional engineering societies and the authors of most modern engineering textbooks. This system, which we use throughout the book, is the *International System of Units* (abbreviated SI), adopted in 1960 by the General Conference on Weights and Measures.

There are six basic units in the SI, and all other units are derived from them. Four of the basic units—the meter, kilogram, second, and coulomb—are important in the study of circuits, and we shall examine them in some detail. The other two are the kelvin and the candela, which are not essential to our study and thus will not be considered further.

The SI units are very precisely defined in terms of permanent and reproducible quantities. However, the complete definitions are both lengthy and esoteric.* Therefore, we shall simply name the basic units and then relate them to the very familiar *British System of Units*, which includes inches, feet, pounds, and so on. Note, however, that the SI incorporates the *metric system*, favored in all but a few countries in the world, rather than the more cumbersome British System. Prefixes for powers of 10 in the SI system are shown, along with their standard abbreviations, in Table 1.1.

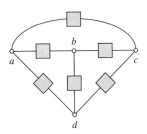

FIGURE 1.2 Electric circuit containing six elements.

Table 1.1	Prefixes in the SI	
Multiple	Prefix	Symbol
10^{12}	tera	T
10^{9}	giga	G
10^{6}	mega	M
10^{3}	kilo	k
10^{-3}	milli	m
10^{-6}	micro	μ
10^{-9}	nano	n
10^{-12}	pico	p
10^{-15}	femto	f

The basic unit of length in the SI is the *meter*, abbreviated m, which is related to the British unit of inches by the ratio 0.0254001 m/in. (to six significant digits). The basic

*Complete definitions of the basic units may be found, for example, in "IEEE, Recommended Practice for Units in Published Scientific and Technical Work," *IEEE Standard 268-1973*, IEEE, New York, 1973; also see updates 268-1978 and 268-1982 from the same source.

unit of mass is the *kilogram* (kg), related to the British pound mass by 0.453593 kg/lb, and the basic unit of time in both systems is the *second* (s).

The fourth basic unit in the SI is the *coulomb* (C), which is the basic unit used to measure electric charge. We defer formal definition of this unit to the next section, where we consider charge and *current* in more detail. The name *coulomb* was chosen to honor the French scientist, inventor, and army engineer Charles Augustin de Coulomb (1736–1806), who was an early pioneer in the fields of friction, electricity, and magnetism. We should note that all SI units named for famous people have abbreviations that are capitalized; otherwise, lowercase abbreviations are used.

The quantities we shall encounter most commonly in our study of circuits are current, voltage, and power. The SI units for these quantities are derived from the four basic units (meter, kilogram, second, and coulomb) introduced above. Units of current, voltage, and power are described next, along with related units of force and energy. Together, these five derived units, along with the four basic units, are all we shall need to describe the fundamental quantitative behavior of electric circuits.*

The fundamental unit of force is the *newton* (N), which is the force required to accelerate a 1-kg mass by 1 meter per second per second (1 m/s^2). Thus 1 N = 1 kg-m/s^2. The newton is named for the great English scientist, astronomer, and mathematician Sir Isaac Newton (1642–1727). Newton's prolific accomplishments include the law of universal gravitation, the wave nature of light, and many other fundamental contributions.

The fundamental unit of work or energy is the *joule* (J), named for the British physicist James P. Joule (1818–1889), who shared in the discovery of the law of conservation of energy and helped establish the idea that heat is a form of energy. A joule is the work done by a constant 1-N force applied through a 1-m distance. Thus 1 J = 1 N-m.

The fundamental unit of current is the *ampere*, named in honor of André Marie Ampère (1775–1836), who discovered the relationship between electric current and magnetic induction. One ampere or amp is the current that flows when 1 coulomb of charge passes each second (1 A = 1 C/s). This topic will be taken up in more detail in the next section.

The fundamental unit of electrical potential is the *volt*, named after Alessandro Volta, whose thumbnail biography is given on page 1. If a charge of 1 coulomb may be moved between two points in space with the expenditure of 1 joule of work, 1 volt is said to be the potential difference existing between these points (1 V = 1 J/C).

The last derived unit we shall need is the *watt* (W), which is the fundamental unit of power, the rate at which work is done or energy expended. The watt is defined as 1 joule per second (1 J/s) and is named in honor of James Watt, the Scottish engineer whose engine design first made steam power practical and triggered the Industrial Revolution.

Example 1.1	Let us convert the speed limit of 55 miles per hour into kilometers per hour.

$$1 \text{ inch} = 0.0254 \text{ meter}$$

so

---

*Other measurable quantities in electric circuits, such as those associated with magnetic and electrooptic phenomena, will not be needed here. These quantities and their units will therefore not be formally considered.

$$1 \text{ mile} = (0.0254 \text{ m/in.}) \ (12 \text{ in./ft}) \ (5280 \text{ ft/mi}) \ (1/1000 \text{ km/m})$$

or

$$1 \text{ mi} = 1.609 \text{ km}$$

Thus

$$55 \text{ mi/h} = (55 \text{ mi/h}) \ (1.609 \text{ km/mi}) = 88.5 \text{ km/h}$$

We note that the units cancel at each step, which may be used to indicate which factors are needed in the conversion process.

EXERCISES

1.1.1. Find the number of picoseconds in (a) 20 μs, (b) 2 ms, and (c) 100 ns.

Answer (a) 2×10^7; (b) 2×10^9; (c) 1×10^5

1.1.2. If a microprocessor can perform 4 million instructions per second, how many nanoseconds are required to execute an operation requiring three instructions?

Answer 750 ns

1.1.3. A 10,000-lb mass spacecraft in deep space is being accelerated at 1.5 g (1 g = 32 ft/s^2). How many newtons of force is the engine applying?

Answer 6.67×10^4 N

1.1.4. The fastest recorded speed run by a human being for 10 m or more is 26.32 mph, set jointly by two runners in 1987. If this speed were sustained over 100 m, how long would it take to cover that distance? (The 100-m dash record at that time was 9.83 s.)

Answer 8.50 s

1.2 CHARGE AND CURRENT

We are familiar with the gravitational force of attraction between bodies, which is responsible for holding us on the earth and which causes an apple dislodged from a tree to fall to the ground rather than soar into the sky. There also exist much stronger forces, far out of proportion to the masses of the bodies, which are both attractive and repulsive. These forces cannot be gravitational in nature.

One of the most important of these forces is electrical and is caused by the presence of *electrical charges.** We explain the existence of electrical forces of both attraction and repulsion by postulating that there are two kinds of charges, positive and negative, and that unlike charges attract and like charges repel.

According to the standard theory, which has been verified by careful and repeated testing, matter is made up of atoms composed of a number of particles. The most

---

*Others include magnetic force and the so-called weak and strong fundamental forces acting between subatomic particles.

important of these particles are protons (positive charges) and neutrons (neutral, no net charge) found in the nucleus of the atom and electrons (negative charges) moving in orbit about the nucleus. Normally, the atom is electrically neutral; the negative charge of the electrons balances the positive charge of the protons. Atoms may become positively charged by losing electrons and become negatively charged by gaining electrons from other atoms.

As an example, we may transfer negative charge to a balloon by rubbing it against our hair. The balloon will then stick to a wall or ceiling, evidence of the attractive force between the balloon's excess electrons and the wall, which becomes positively charged when its surface electrons drift toward its interior due to repulsion by the net negative charge of the balloon.

We now define the *coulomb* (C), introduced in Section 1.1, by stating that an electron has a negative charge of 1.6021×10^{-19} coulomb. Putting it another way, a coulomb is the aggregate charge of about 6.24×10^{18} electrons. These are, of course, mind-boggling numbers, but their sizes enable us to use more manageable numbers, such as 2 C, in the work to come.

The symbol for charge will be taken as Q or q, the capital letter usually denoting constant charges, such as $Q = 4$ C, and the lowercase letter indicating a time-varying charge. In the latter case we may emphasize the time dependency by writing $q(t)$. This practice will be carried over to other electrical quantities as well.

The primary purpose of an electric circuit is to move charges at desired rates along specified paths. The motion of charges constitutes an *electric current*, denoted by the letters i or I, taken from the French word *intensité*. By definition, current is the time rate of change of charge, or

$$i = \frac{dq}{dt} \tag{1.1}$$

As introduced in Section 1.1, the basic unit of current is the ampere. One ampere of current is said to flow past a given cross section if the net rate of charge movement crossing it is 1 coulomb per second (1 C/s).

In circuit theory, current is generally specified as the movement of positive charges. This convention was proposed by the great American scientist, inventor, and diplomat Benjamin Franklin (1706–1790), who guessed that electricity traveled from positive to negative. While this is indeed true in some physical media, we now know that in the important case of metallic conductors so common in real-world circuits, the opposite is the case. In metals and most other conductors, free electrons with negative charge are the current carriers, rather than Franklin's postulated positive charge flow. Fortunately from a circuit theory point of view, there is no difference between positive charge motion in one direction and equal but opposite negative charges moving in the other. Thus we shall cling to the traditional convention of positive charge motion by defining *conventional current* and using this to measure current regardless of the true identity of the charge carrier. Conventional current is the *equivalent flow of positive charges* in a given conductor. Unless otherwise specified, all currents are to be understood as conventional currents.

Current flow along a lead or through an element will be specified by two indicators: an arrow, which establishes a *current reference direction*, and a value (variable or fixed), which quantifies the current flow in the reference direction. Figure 1.3(a) depicts a current i_1 flowing left to right through a lead, and Fig. 1.3(b) a current of 7 A flowing from right to left through an element. In the latter case, 7 A of conventional (positive) charge flows right to left, which may physically consist of an equal rate of negative charges flowing left to right. Returning to Fig. 1.3(a), note that if the variable i_1 is positive conventional current flows left to right, whereas if it is negative, current is said to flow from right to left. The arrow does not indicate the *actual* direction of current flow but rather, *the direction of conventional current flow if the algebraic sign of the current value is positive.* The arrow points opposite to the direction of conventional current flow if the sign of the current value is negative.

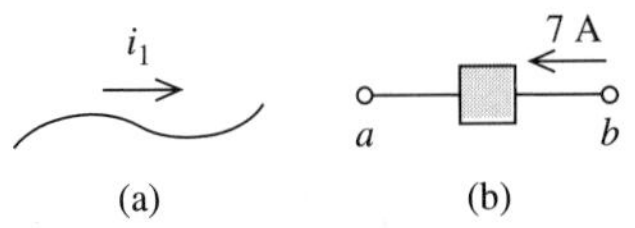

(a)   (b)

FIGURE 1.3

As another example, suppose that the current in the wire of Fig. 1.4(a) is 3 A. That is, 3 C/s passes some specific point in the wire in the left-to-right direction. In Fig. 1.4(b), −3 A passes right to left. These two cases result in exactly the same charge transfer and are thus equivalent ways to denote precisely the same current flow. *We may always reverse both the algebraic sign and the reference direction of a current without changing either its value or direction.* This corresponds to reversing the sign of a number twice, thus leaving it unchanged. Note that without knowing its reference direction, a current's numerical value alone is not enough to specify the current. Both value and reference direction are needed.

3 A   −3 A

(a)   (b)

FIGURE 1.4   Two representations of the same current.

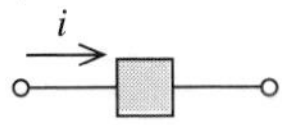

FIGURE 1.5   Current flowing in a general element.

Figure 1.5 represents a general circuit element with a current i flowing from the left toward the right terminal. The total charge entering the element between time t_0 and t is found by integrating (1.1). The result is

$$q_T = q(t) - q(t_0) = \int_{t_0}^{t} i \, d\tau \tag{1.2}$$

We should note at this point that we are considering the network elements to be *electrically neutral*. That is, no net positive or negative charge can accumulate in the element. Any positive charges entering must be accompanied by equal positive charges leaving (or, equivalently, an equal negative charge entering). This property is a consequence of Kirchhoff's current law, which is discussed in Section 1.3. Thus the current shown entering the left terminal in Fig. 1.5 must leave the right terminal. Note that the arrow could as well have been drawn pointing out the right side of the element as into the left with no change in current reference direction.

Example 1.2

As an example, suppose that the current entering a terminal of an element is $i = 4$ A. The total charge entering the terminal between $t = 0$ and $t = 3$ s is given by

$$q = \int_0^3 4dt = 12 \text{ C}$$

Several types of current are frequently encountered, some of which are shown in Fig. 1.6. A constant current, as shown in Fig. 1.6(a), will be termed a *direct current*, or dc. An *alternating current*, or ac, is by definition a sinusoidal current, such as that of Fig. 1.6(b). Sinusoids are discussed in some detail in Chapter 8. Figure 1.6(c) and (d) illustrate, respectively, an *exponential current* and a *sawtooth current*.

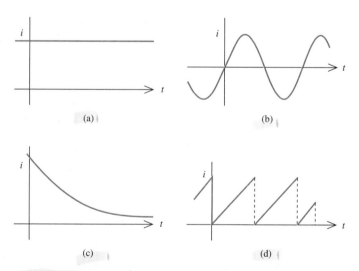

FIGURE 1.6 (a) Dc; (b) ac; (c) exponential; (d) sawtooth current.

There are many commercial uses for dc, such as in flashlights and power supplies for electronic circuits, and ac is the common household current found all over the world. Exponential currents appear quite often (whether we want them or not!) when a switch is actuated to close a path in an energized circuit. Sawtooth waves are useful in devices such as cathode ray tubes (CRTs) used for displaying electrical waveforms visually.

EXERCISES

1.2.1. How many electrons are represented by a charge of 0.32042 pC?

Answer 2 million

1.2.2. The total charge entering a terminal of an element is given by $q = 4t^3 - 5t$ mC. Find the current i at $t = 0$ and at $t = 2$ s.

Answer −5 mA; 43 mA

1.2.3. The current entering a terminal is given by $i = 1 + \pi \sin 2\pi t$ A. Find the total charge entering the terminal between $t = 0$ and $t = 1.5$ s.

Answer 2.5 C

1.3 VOLTAGE, ENERGY, AND POWER

Charges in a conductor, exemplified by free electrons, may move in a random manner. However, if we want some concerted motion on their part, such as is the case with an electric current, we must apply an external or *electromotive force (emf)*. Thus work is done on the charges. In Section 1.2 we defined voltage across an element as the work done in moving a unit charge (+1 C) through the element from one terminal to the other.

Voltage across an element will be designated by two indicators: a plus–minus sign pair, which establishes a *voltage reference direction*, and a *value* (variable or fixed), which quantifies the voltage across the element in the specified reference direction. In Fig. 1.7(a) we see a voltage (or potential difference) of value v_6 volts across the element, measured as potential at the left end of the element minus potential at the right end. Thus, if $v_6 > 0$ V, the left end is at the higher potential, and if $v_6 < 0$ V, the right end is at the higher potential. Note that specifying a reference direction does not by itself indicate which end of the element is at the higher potential. We need not try to guess the direction of potential drop when assigning a voltage reference direction for a voltage variable. In Fig. 1.7(b) the right end of the element is +23 V higher than the left, so there is a +23-V drop right to left across the element or, equally well, a +23-V rise left to right across the element.

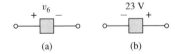

(a)   (b)

FIGURE 1.7 Specifying voltage value and reference direction.

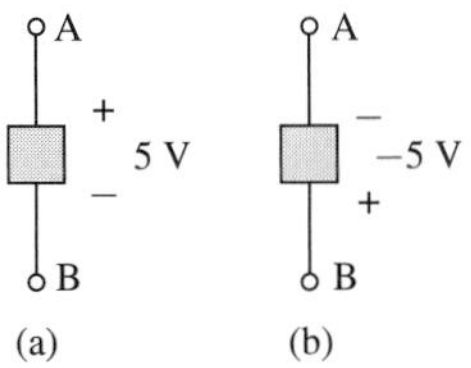

As examples, Fig. 1.8(a) and (b) are two versions of exactly the same voltage. In (a), terminal A is +5 V above terminal B, and in (b), terminal B is −5 V above A (or +5 V below A).

We may also use a *double-subscript* notation, v_{ab}, to denote the potential of point a with respect to point b. In this case we have, in general, $v_{ab} = -v_{ba}$. Thus in Fig. 1.8(a), $v_{AB} = 5$ V and $v_{BA} = -5$ V.

A

+

5 V

−

B

(a)

A

−

−5 V

+

B

(b)

FIGURE 1.8 Two equivalent voltage representations.

In transferring charge through an element, work is being done, as we have said. Or, putting it another way, energy is being supplied. To know whether energy is being supplied *to* the element or *by* the element to the rest of the circuit, we must know both the polarity of the voltage across the element and the direction of the current through the element. If a positive current enters the positive terminal, an external force must be driving the current and is thus supplying or delivering energy to the element. The element is *absorbing* energy in this case. If, on the other hand, a positive current leaves the positive terminal (enters the negative terminal), the element is delivering energy to the external circuit.

As examples, in Fig. 1.9(a) the element is absorbing energy. A positive current enters the positive terminal. This is also the case in Fig. 1.9(b). In Fig. 1.9(c) and (d), a positive current enters the negative terminal, and therefore the element is delivering energy in both cases.

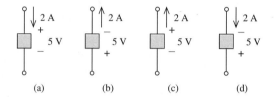

(a)   (b)   (c)   (d)

FIGURE 1.9 Various voltage–current relationships.

Let us consider now the *rate* at which energy is being delivered to or by a circuit element. If the voltage across the element is v and a small charge Δq is moved through the element from the positive to the negative terminal, the energy absorbed by the element, say Δw, is given by

$$\Delta w = v \, \Delta q$$

If the time involved is Δt, then the rate at which the work is being done, or that the energy w is being expended, is given, as Δt gets smaller and smaller, by

$$\lim_{\Delta t \to 0} \frac{\Delta w}{\Delta t} = \lim_{\Delta t \to 0} v \frac{\Delta q}{\Delta t}$$

or

$$\frac{dw}{dt} = v \frac{dq}{dt} = vi \tag{1.3}$$

Since by definition the rate at which energy is expended is power, denoted by p, we have

$$p = \frac{dw}{dt} = vi \tag{1.4}$$

We might observe that (1.4) is dimensionally correct, since the units of vi are (J/C)(C/s) or J/s, which is watts (W), defined earlier.

The quantities v and i are generally functions of time, which we may also denote by $v(t)$ and $i(t)$. Therefore, p given by (1.4) is a time-varying quantity. It is sometimes

FIGURE 1.10 Element whose terminal variables satisfy the passive sign convention.

called the *instantaneous power* because its value is the power at the instant of time at which v and i are measured.

Note carefully the relationship between the current and voltage reference directions of the element of Fig. 1.10. *If the current reference direction arrow points into the positive end of the voltage reference direction, the current and voltage so defined are said to satisfy the passive sign convention.* In this case, $vi > 0$ means the element is *absorbing* power; $vi < 0$ means it is *delivering* power at that instant of time. More will be said about this in Section 1.4.

Example 1.3

As examples, in Fig. 1.9(a) and (b) the passive sign convention is satisfied (arrow points into the positive end or equivalently out of the negative end). The element is absorbing power: $p = (5)(2) = 10$ W. In Fig. 1.9(c) and (d) it is *delivering* 10 W to the external circuit, since $vi = +10$ W, but these variables violate the passive sign convention.

Before ending our discussion of power and energy, let us solve (1.4) for the energy w delivered to an element between time t_0 and t. We have, upon integrating both sides between t_0 and t,

$$w(t) - w(t_0) = \int_{t_0}^{t} vi \, d\tau \qquad (1.5)$$

Example 1.4

For example, if, in Fig. 1.10, $i = 2t$ A and $v = 6$ V, the energy delivered to the element between $t = 0$ and $t = 2$ s is given by

$$w(2) - w(0) = \int_{0}^{2} (6)(2\tau) \, d\tau = 24 \text{ J}$$

Since the left member of (1.5) represents the energy delivered to the element between t_0 and t, we may interpret $w(t)$ as the energy delivered to the element between the beginning of time and t and $w(t_0)$ as the energy between the beginning of time and t_0. In the beginning of time, which let us say is $t = -\infty$, the energy delivered by that time to the element was zero; that is,

$$w(-\infty) = 0$$

If $t_0 = -\infty$ in (1.5), we shall have the energy delivered to the element from the beginning up to t, given by

$$w(t) = \int_{-\infty}^{t} vi \, dt \qquad (1.6)$$

This is consistent with (1.5), since

$$w(t) = \int_{-\infty}^{t} vi \, dt = \int_{-\infty}^{t_0} vi \, dt + \int_{t_0}^{t} vi \, dt$$

By (1.6), this may be written

$$w(t) = w(t_0) + \int_{t_0}^{t} vi\, dt$$

which is (1.5).

EXERCISES

1.3.1. Find v if $i = 6$ mA and the element is (a) absorbing power of $p = 18$ mW and (b) delivering to the external circuit a power $p = 12$ mW.
Answer (a) 3 V; (b) -2 V

EXERCISE 1.3.1

1.3.2. If $i = 3$ A and $v = 6$ V in Exercise 1.3.1, find (a) the power absorbed by the element and (b) the energy delivered to the element between 2 and 4 s.
Answer (a) 18 W; (b) 36 J

1.3.3. A two-terminal element absorbs w millijoules of energy as shown. If the current entering the positive terminal is $i = 100\cos 1000\pi t$ mA, find the element voltage at $t = 1$ ms and $t = 4$ ms.
Answer -50 V; 5 V

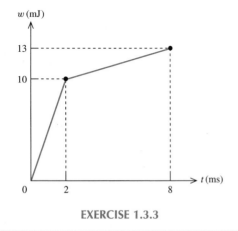

EXERCISE 1.3.3

1.4 PASSIVE AND ACTIVE ELEMENTS

We may classify circuit elements into two broad categories, passive elements and active elements, by considering the energy delivered to or by them. A circuit element is said to be *passive* if it cannot deliver more energy than has previously been supplied to it by the rest of the circuit. That is, referring to (1.6), at each t the net energy absorbed by a passive element up to t must be nonnegative:

$$w(t) = \int_{-\infty}^{t} p(\tau)\, d\tau = \int_{-\infty}^{t} vi\, d\tau \geq 0 \qquad (1.7)$$

The reference directions of v and i in this equation are assumed to satisfy the passive sign convention introduced in Section 1.3 (the arrow for current reference points into the positive end of the voltage reference direction). Only if the passive sign convention is satisfied will this integral be guaranteed nonnegative for passive elements (which explains the choice of the name "passive sign convention"). As we shall explore later, examples of passive elements are resistors, capacitors, and inductors.

An *active* element is any element that is not passive. That is, (1.7) does not hold at each time t. Examples of active elements are generators, batteries, and electronic devices that require power supplies.

We are not ready at this stage to begin a formal discussion of the various passive elements. This will be done in later chapters. In this section we give a brief discussion of two very important active elements, the independent voltage source and the independent current source.

An independent voltage source is a two-terminal element, such as a battery or a generator, that maintains a specified voltage between its terminals regardless of the rest of the circuit it is inserted into. The voltage is completely independent of the current through the element.

The symbol for a voltage source is shown in Fig. 1.11(a). The value v_s is the *source function* and the plus–minus signs within the source give the *source function reference direction*. The source function v_s, which may be constant or a function of time $v_s(t)$, is assumed known. In addition, we will always define a terminal voltage variable across every element in a circuit, including voltage sources. v is the terminal voltage in Fig. 1.11(a). By definition of an independent voltage source, its terminal voltage simply equals the specified value:

$$v = v_s \tag{1.8a}$$

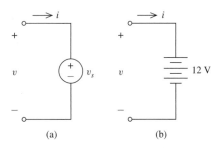

(a)   (b)

FIGURE 1.11   Independent voltage source.

In other words, *the voltage across an independent voltage source always equals its specified source function, regardless of the current through it.* Note that, if the terminal voltage reference direction and the source function reference direction are opposite one another, a minus sign must be inserted in the element law, $v = -v_s$, since the terminal voltage is defined in this case as the negative of the source function.

Equation (1.8a) is an example of an *element law*, an equation involving the element's terminal voltage and/or current, which describes the behavior of an element. Each element has a distinct element law, as we shall see for current sources, resistors, and each other type of element as it is introduced.

Another symbol that is often used for a constant voltage source, such as a battery with 12 V across its terminals, is shown in Fig. 1.11(b). In the case of constant sources, we shall use Fig. 1.11(a) and (b) interchangeably.

An independent current source is a two-terminal element through which a specified current flows. The value of this current is given by its source function (i_s in Fig. 1.12)

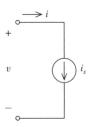

FIGURE 1.12
Independent current source.

and its source function reference direction by the arrow inside the source. As with the independent voltage source, the source function i_s [or $i_s(t)$ in the time-varying case] is assumed known and is related to this element's terminal current variable (i in Fig. 1.12) by the element law for an independent current source:

$$i = i_s \qquad (1.8b)$$

If the source function i_s and terminal current i reference directions are opposed rather than in agreement as shown in Fig. 1.13, a minus sign must be introduced into the terminal law, $i = -i_s$.

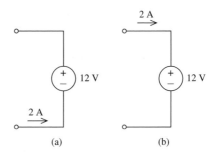

FIGURE 1.13   (a) Source delivering power; (b) source absorbing power.

Example 1.5

To show that independent sources are, in general, active elements, we need only demonstrate that there exist circumstances under which they may be made to supply rather than absorb net energy. For a current source with source function $i_s(t)$ A and element law $i(t) = i_s(t)$, suppose that we arrange for the voltage across the source to be $v(t) = -i_s(t)$ V, with reference directions as shown in Fig. 1.12. Then, by (1.5), $p(t) = v(t)i(t) = -i_s^2(t)$, and the net energy $w(t)$ absorbed by the source must be nonpositive for each t. If $i_s(t)$ is not identically equal to zero for all t, $w(t)$ must be not just nonpositive but, in fact, negative for some t. This violates the requirement (1.7) that passive elements never absorb negative net energy, in other words never supply net energy, under any circumstances. A similar argument can be used to show that a voltage source is also an active element. Note that this result assumes that the source function $i_s(t)$ is not identically equal to zero for all time t, in which case $w(t)$ would also be zero for all t. Sources with zero source functions are passive elements; all other sources are capable of producing net energy and are thus active elements.

Independent sources are usually meant to deliver power to the external circuit, although in a particular circuit they may be either net suppliers or absorbers of power. In Fig. 1.13(a), $v = +12$ V is the terminal voltage across the source and $i = +2$ A its terminal current. Since the reference directions of these terminal variables do not satisfy the passive sign convention (the arrow points into the minus end), $p = vi = 24$ W > 0 implies that the source is *supplying* 24 W of power to the rest of the circuit. In Fig. 1.13(b), the reference directions do satisfy the passive sign convention, so $p = vi = 24$ W > 0 signifies that the battery is *absorbing* 24 W, as would be the case when it is being charged by a more powerful source somewhere in the circuit.

The sources that we have discussed here, as well as the circuit elements to be considered later, are *ideal elements*. That is, they are mathematical models that approximate the actual or physical elements only under certain conditions. For example, an ideal automobile battery supplies a constant 12 V, no matter what external circuit is connected to it. Since its current is completely arbitrary, it could theoretically deliver an infinite amount of power. This, of course, is not possible in the case of a physical device. A real 12-V automobile battery supplies roughly constant voltage only as long as the current it is required to deliver is low. When the current exceeds a few hundred amperes, the voltage drops appreciably from 12 V.

EXERCISES

1.4.1. Find the power being supplied by the sources shown.
Answer (a) 36 W; (b) 20 W; (c) −24 W; (d) −45 W

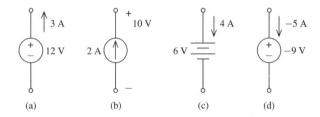

EXERCISE 1.4.1

1.4.2. The terminal voltage of a voltage source is $v = 6 \sin 2t$ V. If the charge leaving the positive terminal is $q = -2 \cos 2t$ mC, find (a) the power supplied by the source at any time and (b) the energy supplied by the source between 0 and t seconds.
Answer (a) $24 \sin^2 2t$ mW; (b) $12t - 3 \sin 4t$ mJ

1.5 CIRCUIT ANALYSIS AND DESIGN

Let us now turn to the words *circuit analysis*, which are contained in the title of the book, and the related idea of circuit design, and consider their meaning. Generally, if an electric circuit is subjected to an *input* or excitation in the form of a voltage or a current provided by an independent source, then an *output* or response is elicited. The input and output are voltages or currents associated with particular elements in the circuit. There may be, of course, more than one input and more than one output.

There are two main branches of circuit theory, and they are closely linked to the fundamental concepts of input, circuit, and output. The first branch is *circuit analysis*, which, given an input and a circuit, is the process of determining the output. The other branch is *circuit design*, which, given a set of inputs and their desired outputs, is the process of creating a circuit that realizes this desired input–output relationship (often subject to additional side constraints such as size, cost, etc).

The balance of emphasis in this book tilts strongly toward analysis. While circuit design is the subject of several chapter sections entitled "Design of ..." and is the sole focus of the final chapter, the balance of this study is devoted to analysis. This may seem a somewhat puzzling choice. After all, design, the creation of something new and useful, is for most of us more satisfying and motivating than analyzing a given circuit by solving for its currents and voltages. The emphasis on analysis perhaps merits some comment.

First consider the process of circuit design, which may be viewed as a cyclic process. Given the desired inputs and outputs, an initial guess of a circuit is made. This circuit is then analyzed to determine its actual outputs for the desired inputs, and the result is compared to the desired input–output relationship. A judgment is made as to whether the design goals have been met satisfactorily. If not, modifications in the circuit are made and this iterative-improvement design cycle is then repeated.

Conceiving a first-cut or initial-guess circuit and selecting the specific modifications to be made during each redesign cycle are seldom matters on which two circuit designers would agree exactly. These choices depend heavily on individual skill, experience, and intuition. Matters largely of judgment and creativity, they constitute the "art" component of circuit design. They cannot be codified into explicit, mechanistic rules any more than the process by which a sculptor works can. *Circuit design is a creative human activity,* like the sculptor's process in designing a statue.

Circuit analysis constitutes the complementary "science" component embedded within the circuit design cycle. Without analysis, the input–output behavior of even the initial-guess circuit could not be determined; thus not a single improvement in the first-cut circuit can be made with confidence. We cannot expect to improve steadily what we don't clearly understand. Thus, while circuit analysis can be performed without reference to circuit design, the opposite is not true.

Since circuit analysis forms the necessary foundation for circuit design, it must be studied before meaningful design is possible. Just as a jazz musician must master the musical scales before creating really interesting and original music, the elements of circuit analysis are our scales and must be mastered if we are to accept the challenge of designing useful and interesting circuits.

Finally, we note that circuit design may be approached on two levels. *Schematic design* is the solution of a design problem by specifying a circuit diagram, while *physical design* is the further conversion of the schematic to actual hardware. In this study we content ourselves with schematic design. Unlike physical design, schematic design does not require a parts inventory, measurement instruments, or access to an electronics laboratory. The elements of schematic design are more abstract than physical elements such as resistors, capacitors, and the like: equations to model the idealized behavior of the elements contained in the circuit diagram, and the mathematical methods to exercise the circuit analytically to predict its idealized input–output behavior. In addition, computer-aided circuit design software tools such as SPICE or PSpice will prove vital to all but the simplest schematic design exercises, speeding the many calculations that must be undertaken at each repetition of the design cycle. Without these simulation tools we could not tackle designs of circuits containing many loops and nodes, that is, we would limit ourselves to a narrow selection of "plain vanilla" design exercises. With these tools in hand, the scope of accessible schematic designs will prove satisfyingly broad and practical. Our goal in this study is to present the systematic methods of analysis upon which design stands, and to offer a first exposure to the challenges and satisfactions of circuit design.

SUMMARY

We begin our study of electrical circuits by defining those electrical quantities that will be of interest, primarily current, voltage, and power, and the selection of a consistent system of units for their measurement. The SI system chosen incorporates the metric system (meters, kilograms, seconds) and uses the coulomb, a measure of charge, as a fundamental unit. SI units of current, voltage, and power are amperes, volts, and watts, respectively.

- Current has both magnitude and direction. Its direction is determined by its current reference direction arrow and the sign of its algebraic value. A positive value indicates equivalent positive current flow in the direction of the arrow, and negative current indicates the opposite direction of flow.

- Voltage, or potential difference, also has magnitude and direction. Its direction is determined by its voltage reference direction plus–minus signs and its algebraic sign. A positive voltage indicated voltage drop in the plus-to-minus direction (higher electric potential at the plus sign), and negative value indicates the opposite direction of voltage drop.

- Passive elements cannot supply more energy on balance than they receive from other elements. Active elements, such as independent sources, can.

- The instantaneous power absorbed by a two-terminal subnetwork equals its current–voltage product if the reference directions satisfy the passive sign convention (current arrow points into voltage plus sign or out of minus sign). Negative power absorbed is equivalent to positive power supplied.

- Circuit analysis is the process of determining output currents and voltages for a specified circuit. Circuit design is the discovery of a circuit with some required input–output behavior.

PROBLEMS

1.1. Einstein's formula for conversion of mass to energy is $E = mc^2$, where E is energy, m is mass, and $c = 3.02 \times 10^8$ m/s is the speed of light. How many joules of energy are there in 5 micrograms of matter?

1.2. Taking the speed of light to be $c = 3.02 \times 10^8$ m/s and a year to contain exactly $365\frac{1}{4}$ days, how many kilometers are there in a light-year?

1.3. The fastest recorded speed for any ground vehicle was 6121 mph (an unmanned rocket sled in 1982). For many years, the official air speed record for manned flight (a SR-71 "Blackbird" aircraft in 1964) was 980.4 m/s. Compare these speeds in kilometers per hour.

1.4. $f(t)$ is the charge in coulombs passing a certain cross section along a wire per unit time. Find (a) the total charge passing the cross section, (b) the time at which exactly 1 C has passed, (c) the current at $t = 5$ s, and (d) the current at $t = 8$ s.

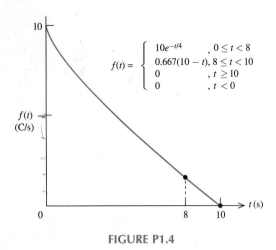

$$f(t) = \begin{cases} 10e^{-t/4} & , 0 \le t < 8 \\ 0.667(10 - t), & 8 \le t < 10 \\ 0 & , t \ge 10 \\ 0 & , t < 0 \end{cases}$$

FIGURE P1.4

1.5. How long would it take a force of 10 N to accelerate a 1000-lb mass object to a speed of 55 mph?

1.6. If the function $f(t)$ is the current in amperes entering the positive terminal of an element at time t (seconds), find (a) the total charge that has entered between 4 and 9 s, (b) the charge entering at 8 s, and (c) the current at 1, 5, and 8 s. Take the charge to be 0 at $t = 0$.

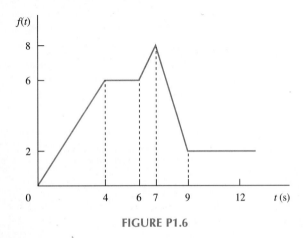

FIGURE P1.6

1.7. If $f(t)$ in Problem 1.6 is the charge entering the element in coulombs and t is in seconds, find the charge entering the element between 4 and 9 s, the current at 6.5 s, and the current at 8 s.

1.8. If 10 J of work is required to move $\frac{1}{4}$ C of negative charge from point a to point b, find v_{ab} and v_{ba}.

1.9. If the voltage across an element is 8 V and the current i entering the positive terminal is as shown, find the power delivered to the element at $t = 7$ ms and the total charge and total energy delivered to the element between 0 and 10 ms.

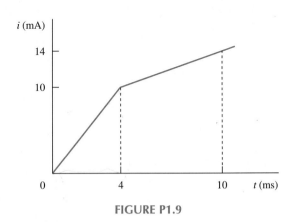

FIGURE P1.9

1.10. If the function graphed in Problem 1.9 is the voltage v (volts) across an element versus the time (ms), and the current entering the positive terminal is 3 mA, find the power delivered to the element at 2 ms and at 6 ms.

1.11. Find the power delivered to an element at $t = 2$ ms if the charge entering the positive terminal is

$$q = 10 \cos 125\pi t \ \text{mC}$$

and the voltage is

$$v = 6 \sin 125\pi t \ \text{V}$$

1.12. If $v(t) = 4t$ V and $i(t)$ is as shown, find the power $p(t)$ dissipated by the element and total energy supplied to the element between times $t = 0$ and $t = 10$.

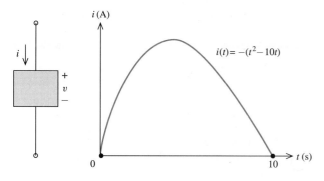

$i(t) = -(t^2 - 10t)$

FIGURE P1.12

1.13. Repeat Problem 1.12 for $v(t) = 4(t - 5)$ V.

1.14. The power delivered to an element is $p = 10t \cos^2 t$ W and the voltage is $v(t) = 2 \cos t$ V. Find the current entering the positive terminal and the charge between times 0 and 2π s.

1.15. An element has $v(t) = 4e^{-t}$ V for $t \geq 0$ s. If we wish this element to supply net energy of $2t$ J for $t \geq 0$, what must $i(t)$ be?

$v(t)$

$i(t)$

FIGURE P1.15

1.16. Show that by definition of a passive element, any element satisfying $v = mi$, where $m \geq 0$ and the element variables i and v satisfy the passive sign convention, is passive.

1.17. The element shown satisfies $v = 2i + 1$. Is this element active or passive? Find the net energy between $t = -\infty$ and $t = 1$ s if $v = 0$ for $t < 0$ and $v = \frac{1}{2}$ V for $t \geq 0$.

i

v

FIGURE P1.17

1.18. An element has constant current and voltage, $v = +100$ V and $i = +10$ A, with i and v satisfying the passive sign convention. How much energy is required to move each coulomb through the element? How long does it take to move 1 C through? Repeat if $i = -10$ A.

1.19. If $v_{s1} = 100$ V and $i_{s2} = -10$ A, which source is supplying energy and which is dissipating energy? Holding i_{s2} fixed at -10 A, what would v_{s1} be so that 1000 J/h was being supplied by the current source to the voltage source?

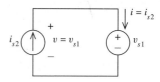

$i = i_{s2}$

i_{s2} $v = v_{s1}$ v_{s1}

FIGURE P1.19

1.20. If a current $i = 0.4$ A is entering the positive terminal of a battery with terminal voltage $v = 12$ V, the battery is in the process of being charged. (It is absorbing rather than delivering power.) Find (a) the energy supplied to the battery and (b) the charge delivered to the battery in 2 h (hours). Note the consistency of the units 1 V = 1 J/C.

1.21. Suppose that the voltage in Problem 1.20 varies linearly from 6 to 18 V as t varies from 0 to 10 min. If $i = 2$ A during this time, find (a) the total energy supplied and (b) the total charge delivered to the battery.

1.22. If a current source and a voltage source are connected back to back as in Problem 1.19, are there values v_{s1} and i_{s2} for which both sources are supplying equal nonzero power to one another? If so, what values? If not, show why not.

1.23. The power absorbed by a circuit element is shown. At what time is the net energy a maximum? What is the maximum value of the net energy?

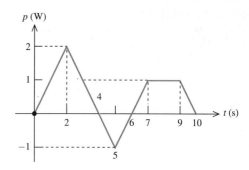

FIGURE P1.23

1.24. The net energy $f(t)$ in joules absorbed by an element is shown in the figure. [$f(t)$ equals net energy from $t = -\infty$ to time t.] $f(t) = 6 \sin(\pi/2)t$ for $0 \leq t \leq 3$ and is piecewise linear thereafter as shown. What is the peak power supplied to the element? What is the peak power dissipated by the element? Is this a passive or active element?

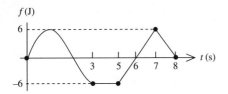

FIGURE P1.24

1.25. Suppose that an element obeys $v = Ki^p$, where i and v satisfy the passive sign convention, $K > 0$, and p is a positive integer. Find those values of p for which the element is passive.

1.26. An element has a power rating of 1 kW. If the element satisfies the equation $v = 2i + di/dt$, what is the maximum voltage $v(t)$ we may have across the element if $i(t) = 4\cos 2t$ A?

1.27. A circuit breaker is set to trip if the current through the voltage source reaches ± 20 A (that is, if $|i| > 20$). If the net energy delivered by the source from $t = -\infty$ to time t is 0 for $t < 1$ s and $10(\cos \pi t + 1)$ J for $t \geq 1$ s, will the circuit breaker ever trip? If so, when will it first trip?

FIGURE P1.27

1.28. A certain element with i and v that satisfy the passive sign convention obeys the equation $i(t) = \int_{t_0}^{t} v(\tau)\, d\tau + i(t_0)$. We wish to supply this element 5 J of energy using a constant voltage between $t = 0$ and $t = 1$ s. What should be the value of this voltage? What is the peak power we must be able to supply?

1.29. When fully charged, a car battery's stored charge depends on the temperature. If $q_0(T) = \frac{3}{4}T + 15$, where $q_0(T)$ is the stored charge (units are kilocoulombs) and T the Fahrenheit temperature, how many times can we attempt to start the car in $100°$F weather? In $32°$F weather? Assume that each attempt lasts 10 s and draws 90 A of current from the battery.

More Challenging Problems

1.30. The electrical potential at a distance r away from a certain electrical charge is $20/r$ V (r in meters). How much work is required to move a $+1$-C charge from 10 m away to 1 m away? If the movement were made at uniform velocity over 9 s, what would be the maximum power required?

1.31. The element shown has the relationship $v(t) = |i(t)|$ between its current and voltage variables. Sketch the power $p(t)$ if $i(t) = 10\cos 2\pi t$. Is this element active or passive?

FIGURE P1.31

1.32. The power delivered by an independent current source is $p(t) = 15e^{-2t}\sin t$ W for $t \geq 0$. If the voltage across the source is $v(t) = 30e^{-t}$ V for $t \geq 0$, find the source function $i_s(t)$ and determine the total energy delivered by the source for $0 \leq t < \infty$.

FIGURE P1.32

1.33. Determine if the rest of the circuit to which the source described in Problem 1.32 is connected is passive or active. Justify your answer.

2

Resistive Circuits

André Marie Ampère
1775–1836

I shall call the first "electric tension" [voltage] and the second "electric current."

André Marie Ampère

On September 11, 1820, the exciting announcement was read to the French Academy of Sciences of the discovery by the Danish physicist Hans Christian Oersted that an electric current produces a magnetic effect. One member of the academy, André Marie Ampère, a French mathematics professor, was highly impressed and within a week had repeated Oersted's experiment, given a mathematical explanation of it, and—in addition—discovered that electric currents in parallel wires exert a magnetic force on each other.

Ampère was born in Lyons, France, and at an early age had read all the great works in his father's library. At age 12 he was introduced to the Lyons library and because many of its best mathematical works were in Latin, he mastered that language in a few weeks. Despite two crushing personal tragedies—at age 18 he witnessed his father's execution on the guillotine by the French Revolutionaries, and later his young, beloved wife died suddenly after only four years of marriage—Ampère was a brilliant and prolific scientist. He formulated many of the laws of electricity and magnetism and was the father of electrodynamics. The unit of electric current, the *ampere*, was chosen in his honor in 1881.

Chapter Contents

The simplest and most commonly used circuit element is the resistor. All electrical conductors exhibit properties that are characteristic of resistors. When currents flow in metals, for instance, electrons that make up the current collide with the orderly lattice of atoms of the metal. This impedes or resists the motion of the electrons. The larger the number of collisions, the greater the resistance of the conductor. In other materials the charge carriers and their surrounding milieu may differ from free electrons flowing through crystal lattices, but the principle that movement of charge is resisted remains the same. We shall consider a resistor to be any device that exhibits solely a resistance. Materials that are commonly used in fabricating resistors include metallic alloys and carbon compounds and, in the case of resistors that are deposited onto integrated-circuit chip substrates, semiconductors.

In this chapter we first introduce two general laws governing the flow of current and the pattern of voltages in any circuit, called Kirchhoff's current law and Kirchhoff's voltage law. We next introduce the basic relationship linking current and voltage in resistors, Ohm's law. Armed with these tools, we begin our study of systematic circuit analysis. The fundamental concept of equivalent circuits is introduced. Simplifications available through the use of equivalent circuits are then discovered in the context of series–parallel and Thevenin–Norton equivalents.

2.1   KIRCHHOFF'S LAWS

For purposes of circuit analysis, an electric circuit is fully characterized by specifying just two of its characteristics: the elements it contains and how those elements are interconnected. Nothing more need be known to determine the resulting currents and voltages.

In this section we consider two simple laws first stated in 1847 by the German physicist Gustav Kirchhoff (1824–1887), which together define how the interconnection of elements constrains their possible currents and voltages. These laws are valid for circuits containing elements of all types: resistors, inductors, capacitors, sources, and others. They are called Kirchhoff's current law and Kirchhoff's voltage law. These two interconnection laws, together with element laws describing the behavior of each individual element in the circuit, are all the equations we will need for systematic circuit analysis. The element law for a resistor, Ohm's law, will be taken up in Section 2.2, and systematic circuit analysis methods are developed in Chapter 4.

Our development of Kirchhoff's laws will not include rigorous proofs. These are best supplied in the context of the study of electromagnetic fields, the theory upon which Kirchhoff's laws are based. For the present circuit theory purposes, we content ourselves with defining the laws and justifying their plausability.

A *circuit* consists of two or more circuit elements connected by means of perfect conductors. Perfect conductors are idealized leads or wires that allow current to flow without impediment (no charge accumulations or voltage drops along the leads, no power or energy dissipation). For circuits so defined, the energy can be considered to reside, or be lumped, entirely within each circuit element, and thus the circuit is called a *lumped-parameter circuit*.

A point of connection of two or more circuit elements, together with all the connecting wire* in unbroken contact with this point, is called a *node*. An example of a circuit with three nodes is shown in Fig. 2.1(a). Nodes are indicated in diagrams in two ways: by a dashed line enclosing the node, as with nodes 1 and 3, or by marking a typical point within the node, as with node 2 of this figure. Whichever is used, it is important to remember that a node is all the wire in direct contact with a given point, and thus any two points that can be traversed by moving exclusively along connecting leads are both part of the same node. The points marked a and b in Fig. 2.1(a) are part of the same node, designated node 1. We are now ready to discuss the all-important laws of Kirchhoff.

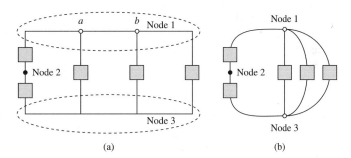

FIGURE 2.1   (a) Three-node circuit; (b) same circuit redrawn.

Kirchhoff's current law (KCL) states that

> The algebraic sum of the currents entering any node is zero.

---

*The leads that connect elements will frequently be called "wires" even though they may take on quite different physical form. For instance, integrated-circuit chips leads are narrow, shallow conducting channels called *traces* deposited onto the chip substrate.

For example, the currents entering the node of Fig. 2.2 are i_1, i_2, $-i_3$, and i_4 (since i_3 leaving is $-i_3$ entering). Therefore, KCL for this case is

$$i_1 + i_2 + (-i_3) + i_4 = 0$$

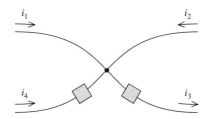

FIGURE 2.2   Current flowing into a node.

For the sake of argument let us suppose that the sum were not zero. In such a case we would have

$$i_1 + i_2 - i_3 + i_4 = \Psi \neq 0$$

where Ψ has units of coulombs per second and must be the rate at which charges are accumulating in the node. However, a node consists of perfect conductors and cannot accumulate charges. In addition, a basic principle of physics states that charges can be neither created nor destroyed (conservation of charge). Therefore, our assumption is not valid, and Ψ must be zero, demonstrating the plausibility of KCL. For every charge carrier entering a node, one is immediately forced out, leaving the net current entering the node at each instant of time exactly equal to zero.

There are several ways that KCL may be stated, all logically equivalent but each phrased slightly differently. Multiplying both sides of the KCL equation by -1, we obtain

$$(-i_1) + (-i_2) + i_3 + (-i_4) = 0$$

Examining this equation, each term on the left is a current *exiting* the node. Generalizing, we arrive at a second form of KCL.

> The algebraic sum of the currents leaving any node is zero.

Next, this same equation may be rearranged in the form

$$i_1 + i_2 + i_4 = i_3$$

where currents whose reference directions point into the node are gathered on one side and those exiting the node on the other. All minus signs disappear, and we have the third form of KCL.

> The sum of the currents entering any node equals the sum of the currents leaving the node.

Depending on the reference directions assigned to currents impinging on a node, one of these three forms may be more convenient to use when KCL is required. In general, KCL may be expressed mathematically as

$$\sum_{n=1}^{N} i_n = 0 \qquad (2.1)$$

where i_n is the nth current entering (or leaving) the node and N is the number of node currents.

Example 2.1

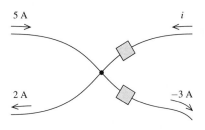

FIGURE 2.3 Example of KCL.

As an example of KCL, let us find the current i in Fig. 2.3. Summing the currents entering the node, we have

$$5 + i - (-3) - 2 = 0$$

or
$$i = -6 \text{ A}$$

We note that -6 A entering the node is equivalent to 6 A leaving the node. Therefore, it is not necessary to guess the correct current direction when assigning reference directions to current variables. We still arrive at the correct answer in the end.

We may find the current i slightly more conveniently by noting that since i flows into the node, it must equal the sum of all the other currents if they are defined to be flowing out (by reversing their reference directions if necessary), yielding

$$i = -3 + 2 + (-5) = -6 \text{ A}$$

which agrees with the previous answer.

We now move to Kirchhoff's voltage law, which states that

> The algebraic sum of voltage drops around any closed path is zero.

As an illustration, application of this principle to the closed path *abcda* of Fig. 2.4 gives

$$-v_1 + v_2 - v_3 = 0$$

where the algebraic sign for each voltage has been taken as positive when going from $+$ to $-$ (higher to lower potential) and negative when going from $-$ to $+$ (lower to higher potential) in traversing the element.

As with KCL, temporarily suppose the contrary, that the sum is not zero. Then

$$-v_1 + v_2 - v_3 = \Phi \neq 0$$

where Φ has units of volts and must equal the potential difference between the initial point and endpoint of the traversal. But this is a closed path, the two points are one

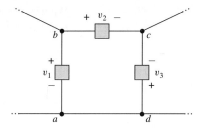

FIGURE 2.4 Voltages around a closed path.

and the same, and a node can have no potential difference with itself. The supposition is invalid, and the plausibility of KVL is demonstrated. KVL is in fact an instance of an even more general physical principle that potential differences along a closed path in any conservative energy field must sum to zero. The electric fields produced in lumped-parameter circuits are such fields, and so (for instance) is the field of gravity. If we hike along a path with uphill and downhill sections, the sum of all the altitude changes around any closed path must be zero, for we will have returned to the altitude (gravitational potential) at which we started.

As with KCL, there are two additional equivalent statements of KVL worth noting. Multiplying each term in the *sum of drops* form of KVL stated above by -1 yields the *sum of rises* form.

> The algebraic sum of voltage rises along any closed path equals zero.

The final form of KVL arises when voltages encountered plus-to-minus in the direction of motion around the closed path are summed on one side of the equation and minus-to-plus on the other, as in the loop of Fig. 2.4, yielding

$$v_1 + v_3 = v_2$$

> The sum of voltage drops equals the sum of voltage rises along any closed path.

In general, KVL may be expressed mathematically as

$$\sum_{n=1}^{N} v_n = 0 \tag{2.2}$$

where v_n is the nth voltage drop (or rise) in a loop containing N voltages. The signs of the terms are as noted in the illustrations, positive for plus-to-minus traversal in the sum of drops form or for minus-to-plus traversal using the sum of rises form, and negative if the traversal is in the opposite sense.

Example 2.2

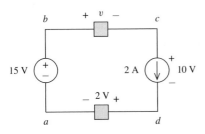

FIGURE 2.5   Circuit to illustrate KVL.

We will use KVL to find v in Fig. 2.5. Traversing the loop clockwise beginning at point a, using the sum of drops form of KVL, yields

$$-15 + v + 10 + 2 = 0$$

or $v = 3$ V. Had we chosen to move counterclockwise from b, we would have

$$+15 - 2 - 10 - v = 0$$

yielding the same result. The sum of rises form yields exactly these same two equations, the first starting from the point b and moving counterclockwise, the second from point a moving clockwise. Finally, equating drops and rises clockwise, we have

$$v + 10 + 2 = 15$$

Clearly, none of these equations is more informative than the others. Their choice is left to convenience or some arbitrary convention. All other things being equal, we will adopt the convention of using the sum of drops form and clockwise traversal.

There is a common situation, however, where a special form of the sum of drops will be useful. Suppose that there is a single unknown voltage in a loop, such as v in Fig. 2.5, which we wish to solve for. Traversing the loop in the plus-to-minus direction of the unknown voltage (clockwise in Fig. 2.5) and summing the drops to zero and then moving all terms but the unknown to the other side of the equation, we have

$$v = +15 - 2 - 10$$

from which the value of v may be immediately computed. Examining this equation, the left-hand side is the voltage drop from the point b to the point c, while the right-hand side is the negative sum of voltage drops from c to back to b, in other words, the sum of voltage drops from b to c by this alternative (counterclockwise) route $badc$. Indeed, the voltage drop between two nodes must be the same regardless of the path used to compute it, in the present case

$$v = v_{bc} = v_{ba} + v_{ad} + v_{dc}$$

or

$$v = 15 - 2 - 10 = 3 \text{ V}$$

Thus, by using the principle that *the voltage drop between two nodes equals the sum of the voltage drops between the same nodes along any path*, a loop containing a single unknown voltage can be conveniently described and the unknown solved.

By use of similar reasoning, it is easily demonstrated that *a desired current into a node equals the sum of all other currents at that node, counting outward flows as positive*. This may be used to solve for a single unknown current at a node and in fact was employed to solve the final example in the earlier discussion of KCL in this section.

Example 2.3

We will use KCL and KVL to find all unknown currents and voltages in Fig. 2.6. Summing voltage drops around the right loop clockwise gives

$$v_1 + 5 + 3 + 2 = 0$$

or
$$v_1 = -10 \text{ V}$$

Repeating for the left loop, we have

$$11 - v_2 - 9 - (-10) = 0$$

or
$$v_2 = 12 \text{ V}$$

In the left loop equation we have used the fact that $v_1 = -10$ V. We could have solved for v_2 without knowing v_1 by writing the outer loop equation:

$$v_2 = -9 + 5 + 3 + 2 + 11 = 12 \text{ V}$$

Here we equated the voltage drop across an element to the sum of voltage drops between the same two points along another path.

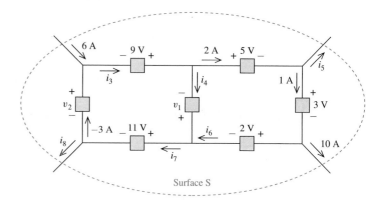

FIGURE 2.6   Circuit for Example 2.3.

Next we determine the unknown currents. At the upper-left node, KCL in the sum in equals-sum-out form yields

$$i_3 = 6 + (-3) = 3 \text{ A}$$

At the node immediately to the right,

$$i_4 = 3 + (-2) = 1 \text{ A}$$

where i_4 flows out, and 3 A and −2 A both flow in (+2 A out equals −2 A in). At the upper-right node,

$$i_5 = 2 + (-1) = 1 \text{ A}$$

while at the lower right,

$$i_6 = 1 + (-10) = -9 \text{ A}$$

Finally,

$$i_7 = i_4 + i_6 = -8 \text{ A}$$

$$i_8 = -(-3) + i7 = -5 \text{ A}$$

Note that Kirchhoff's laws are quite indifferent to the *identities* of the elements populating a circuit; they are sensitive only to the circuit *topology*, how the elements are interconnected.

Before concluding our discussion of Kirchhoff's laws, note that if the sum of currents entering any closed region in a circuit, such as that bounded by the surface S in Fig. 2.6, were not zero, charge would be accumulating somewhere in that region. This would imply a violation of KCL at one or more nodes in the region. This illustrates the generalized form of KCL.

> The algebraic sum of currents entering any closed surface is zero.*

Applying generalized KCL to the circuit of Fig. 2.6, we may sum the currents into the surface S to zero, or

$$6 + (-i_5) + (-10) + (-i_8) = 0$$

In Example 2.3, the values of i_5 and i_8 were computed using the nodal forms of KCL to be $i_5 = 1$ A and $i_8 = -5$ A, which is consistent with the foregoing result.

EXERCISES

2.1.1. Find i_1 and i_2 for the circuit shown.
Answer $i_1 = 5$ A; $i_2 = -2$ A

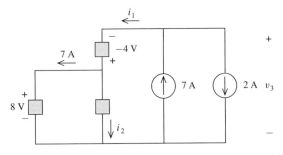

EXERCISE 2.1.1

2.1.2. In the circuit for Exercise 2.1.1, find v_3.
Answer 12 V

---

*The surface cannot pass through an element, which is considered to be concentrated at a point in lumped-parameter networks.

2.1.3. Sketch a loop satisfying the equation $v_1 - v_2 + v_3 - v_4 - 7 = 0$. Assume that this equation is KVL in the sum-of-drops form with clockwise traversal.

Answer

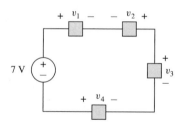

EXERCISE 2.1.3

2.2 OHM'S LAW

Georg Simon Ohm (1787–1854), a German physicist, is credited with formulating the current–voltage relationship for a resistor based on experiments performed in 1826. In 1827 he published the results in a paper titled "The Galvanic Chain, Mathematically Treated." As a result of his work, the unit of resistance is called the *ohm*. It is ironic, however, that Henry Cavendish (1731–1810), a British chemist, discovered the same result 46 years earlier. Had he not failed to publish his findings, the unit of resistance might well be known as the *caven*.

Ohm's law states that the voltage across a resistor is directly proportional to the current flowing through the resistor. The constant of proportionality is the resistance value of the resistor in ohms. The circuit symbol for the resistor is shown in Fig. 2.7. For element current i and voltage v defined to satisfy the passive sign convention introduced in Section 1.3 (current reference direction arrow points into the plus end of the voltage reference direction), Ohm's law is

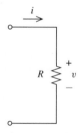

FIGURE 2.7 Circuit symbol for the resistor.

$$v = Ri \qquad (2.3)$$

where $R \geq 0$ is the resistance in ohms.

The symbol used to represent the ohm is the capital Greek letter *omega* (Ω). By (2.3), we have $R = v/i$ so, dimensionally,

$$1 \ \Omega = 1 \ \text{V/A}$$

Example 2.4

If $R = 3 \ \Omega$ and $v = 6$ V in Fig. 2.7, the current is

$$i = \frac{v}{R} = \frac{6 \ \text{V}}{3 \ \Omega} = 2 \ \text{A}$$

If R is changed to 1 kΩ (using the power of 10 prefixes of Table 1.1), then with the same voltage the current becomes

$$i = \frac{6 \ \text{V}}{1 \ \text{k}\Omega} = 6 \ \text{mA}$$

The process is streamlined by noting that $V/k\Omega = mA$, $V/M\Omega = \mu A$, and so on.

Since R is constant, (2.3) describes a linear current–voltage (i–v) graph, as shown in Fig. 2.8(a). For this reason, a resistor obeying Ohm's law is called a *linear resistor*. Resistors with different i–v laws are called *nonlinear resistors*. In nonlinear resistors, the resistance varies with the current flowing through it. An example is an ordinary incandescent lamp, whose typical i–v graph is shown in Fig. 2.8(b). Since sets of linear equations are far easier to solve than sets containing even a single nonlinear equation, the incorporation of a nonlinear resistor into a network complicates its analysis dramatically.

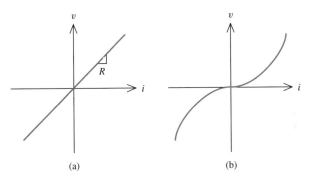

(a)   (b)

FIGURE 2.8   (a) i–v graph for an example linear resistor; (b) i–v graph for an example nonlinear resistor.

In reality, all practical resistors exhibit nonlinear behavior because the resistance of any resistor is affected by conditions such as temperature, which are in turn linked to past and present current flow through the element (current flow causes heating). Many materials, however, closely approximate an ideal linear resistor over a limited range of currents and environmental conditions. In this book our interest will be exclusively on linear elements, including linear resistors (which we will hereafter simply refer to as "resistors"). Implied by the choice to concentrate on linear elements is the result that our equations and techniques will work only over a limited operating range. A 1-kΩ resistor pulled out of a lab drawer will reliably draw 1 mA of current when placed across a 1-V source, but do not expect this same 1-kΩ resistor to conduct 1 kA of current if you should find a 1-MV power supply with which to excite it. In all likelihood you will measure zero current, because after a brief transient you will be left with just a wisp of smoke and a bad smell in the room (or worse).

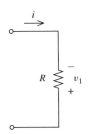

FIGURE 2.9   Resistor with terminal variables i and v_1 not satisfying the passive sign convention.

It was stated that Ohm's law takes on the familiar form $v = Ri$ if the terminal variables satisfy the passive sign convention. If we choose to state this law using variables that violate the convention, the form taken on by Ohm's law changes accordingly. In Fig. 2.9, suppose that we use the variables i and v_1, rather than i and v as used in Fig. 2.7, as the element current and voltage variables, where $v_1 = -v$. Then, substituting v_1 for v in $v = Ri$, Ohm's law becomes $v_1 = -Ri$. Both $v = Ri$ and this last equation are valid expressions of Ohm's law. To avoid the minus sign, we will almost always choose to define resistor current and voltage variables that satisfy the passive sign convention.

In Section 1.4, passive elements were defined to be those incapable of supplying positive net energy. For resistors the instantaneous power is

$$p = vi = Ri^2 = \frac{v^2}{R} \tag{2.4}$$

where v and i have been assumed to satisfy the passive sign convention. Note that since p is always positive for a resistor ($R \geq 0$, and the squares of i and v are also nonnegative) the net energy dissipated by this element, the integral of its instantaneous power p, must be nonnegative as well. Thus a resistor does indeed satisfy the definition of a passive element. Resistors dissipate, but under no circumstances supply, electrical power or energy.

Another important quantity very useful in circuit analysis is known as *conductance*, defined by

$$G = \frac{1}{R} \tag{2.5}$$

The SI unit for conductance is the *siemens*, symbolized by S and named for the brothers Werner and William Siemens, two noted German engineers of the late nineteenth century. Thus 1 S = 1 A/V. (Another unit of conductance, widely used in the United States, is the *mho*, which is *ohm* spelled backward. The symbol for the mho is an inverted omega.) Combining (2.3) to (2.5), we see that Ohm's law and the formulas for instantaneous power may be expressed in terms of conductance as

$$i = Gv \tag{2.6}$$

and

$$p = \frac{i^2}{G} = Gv^2 \tag{2.7}$$

The important concepts of short circuit and open circuit may be defined in terms of resistance. A *short circuit* is a resistor of zero ohms resistance, in other words a perfect conductor capable of carrying any amount of current without sustaining a voltage drop across it. An *open circuit* is a resistor of zero siemens conductance, in other words a perfect insulator capable of supporting any voltage without permitting any current flow through it. In light of (2.5), an open circuit is equivalent to an infinite resistance, or a break in the wire. Two points may be short-circuited by joining them with a wire, which we always assume to be a perfect conductor. They may be open-circuited by removing all current pathways between them.

Example 2.5

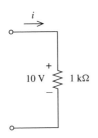

FIGURE 2.10   Circuit for Example 2.5.

We wish to determine the current i and the power p absorbed by the 1-kΩ resistor in Fig. 2.10 and the energy it dissipates each half-hour.

Since v and i together satisfy the passive sign convention, we may use Ohm's law in the $v = Ri$ form. Then $i = 10$ V/1000 Ω = 10 mA and $p = vi = (10$ V$)(0.01$ A$)$, or 100 mW. Since this power is constant over the half-hour, the total energy absorbed during the period t_0 to $t_1 = t_0 + (30$ min$)(60$ s/min$)$ s is, by (1.7),

$$w(t_1) - w(t_0) = \int_{t_0}^{t_1} p \, dt$$

$$= (0.1 \text{ W})(1800 \text{ s}) = 180 \text{ J}.$$

EXERCISES

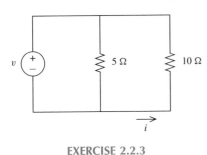

EXERCISE 2.2.3

2.2.1.   The terminal voltage of a 10-kΩ resistor is 5 V. Find (a) the conductance, (b) the terminal current, and (c) the power dissipated.

Answer (a) 0.1 mS; (b) 0.50 mA; (c) 2.5 mW

2.2.2.   A long cable carries 3 A of current while dissipating 72 W. What are the resistance of the cable and the voltage drop across its length?

Answer $R = 8$ Ω; $v = 24$ V

2.2.3.   Find (a) the current i and (b) the power delivered to the resistors if $v = -100$ V.

Answer (a) +10 A; (b) 1 kW to the 10-Ω resistor, 2 kW to the 5-Ω resistor

2.3   EQUIVALENT SUBCIRCUITS

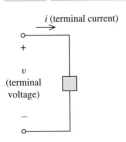

FIGURE 2.11
Two-terminal subcircuit. The shaded region may contain any intercon-nection of elements.

A generally useful strategy in analyzing electric circuits is to *simplify wherever possible.* We shall see how to remove a part of a circuit, replacing it with a simpler subcircuit containing fewer elements, without altering any current or voltage outside that region. The simpler circuit can then be analyzed, and the results will apply equally to the original, more complex, circuit. Such a beneficial trade is possible only when the original and replacement subcircuits are equivalent to one another in a specific sense to be defined.

A *subcircuit* is any part of a circuit. A subcircuit containing any number of in-terconnected elements, but with only two accessible terminals, is called a *two-terminal subcircuit.* Examples include two-terminal elements, as illustrated in Fig. 1.10, and any arbitrary circuit in which a pair of leads is made accessible for measurement or inter-connection with other elements. The voltage across and current into these terminals are called the *terminal voltage* and *terminal current* of the two-terminal subcircuit, as in Fig. 2.11.

The behavior of a two-terminal subcircuit, what it "does" to any circuit containing it, is completely described by its *terminal law*. The terminal law is a function of the form $v = f(i)$ or $i = g(v)$, where v and i are the terminal variables.

Two-terminal subcircuits containing just a single element were introduced in Section 1.1. For these simplest of subcircuits, called *two-terminal elements*, the terminal law is also referred to as the *element law*. The element law for a resistor is Ohm's law, $v = f(i) = Ri$ [or $i = g(v) = Gv$]. The element law for the independent voltage source of Fig. 2.12(a) is, by KVL, $v = v_s$. This element law is of the form $v = f(i)$, where the function $f(i)$ is the constant* equal to v_s. The independence of the terminal voltage v from the current i flowing through the source is what characterizes this element as an ideal voltage source. The element law for the independent current source shown in Fig. 2.12(b) is $i = i_s$ by KCL. An independent current source maintains its terminal current i at the specified source function value i_s regardless of its terminal voltage.

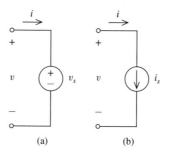

(a)   (b)

FIGURE 2.12 Independent source elements. v_s, i_s are the source functions and v, i the terminal variables.

The terminal laws for ideal voltage and current sources are specified in terms of only one terminal variable. All other elements, such as the resistor, as well as subcircuits with several elements, have terminal laws that are equations linking their two terminal variables. The terminal law is what is needed to understand the action of elements and subcircuits.

We come to the central point of this discussion. *Two two-terminal subcircuits are said to be equivalent if they have the same terminal law.* A subcircuit with terminal law $v = f_1(i)$ and a second with terminal law $v = f_2(i)$ are equivalent if $f_1(i) = f_2(i)$ for each i, or if $g_1(v) = g_2(v)$ for every v.

Example 2.6

Let's find the terminal laws for the two subcircuits of Fig. 2.13. That for Fig. 2.13(a) is clearly $v = 2i$. In (b), by Ohm's law applied to each resistor, $v_{21} = i$, $v_{22} = i$, and by KVL, $v = v_{21} + v_{22}$ or $v = i + i = 2i$. Since the terminal laws are identical, the subcircuits are indeed equivalent.

---

*By constant we mean not dependent on i. The source functions v_s and i_s might in fact vary with time, $v_s = v_s(t)$ and $i_s = i_s(t)$.

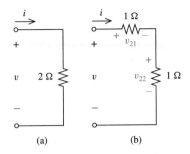

(a)   (b)

FIGURE 2.13   Equivalent subcircuits.

The significance of the notion of equivalence is that *equivalent subcircuits may be freely exchanged without altering any external current or voltage.* Suppose that a two-terminal subcircuit A is removed from a circuit and an equivalent subcircuit B inserted in its place as in Fig. 2.14. For any equation that could have been written for circuit (a), the identical equation will be true of circuit (b). For instance, around the outer loop we will have in either case

$$v_2 + v - v_3 - v_1 = 0$$

and rewriting this equation in terms of currents by applying the terminal laws yields

$$5i_2 + f_1(i) - 5 - 4i_1 = 0$$

for (a) and

$$5i_2 + f_2(i) - 5 - 4i_1 = 0$$

for (b).

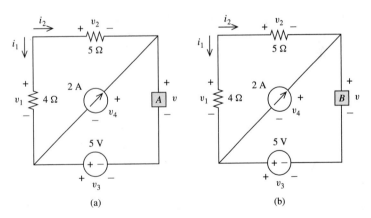

(a)   (b)

FIGURE 2.14   Exchanging equivalent subcircuits.

Clearly, these equations are identical if and only if the circuits are equivalent. Since corresponding equations for the two equivalents are all identical, so must be their solutions. We conclude that equivalent circuits result in equal currents and voltages in any circuit to which they are connected.

The notion of equivalent circuits will prove very helpful in simplifying circuit problems. We will proceed by replacing more complex subcircuits by simpler ones. Classes of equivalents to be explored in this chapter include series–parallel and Thevenin–Norton equivalents. The test for equivalence will always be the identity of the terminal laws of the two subcircuits.

While the currents and voltages *external* to equivalent subcircuits will not be changed when one is exchanged for the other in any circuit, the *internal* behavior of the equivalents may be quite different. For instance, in Example 2.6 there is no element in Fig. 2.14(b) with the same terminal voltage as the element in Fig. 2.14(a). All that is guaranteed by their equivalence is that they can be freely interchanged in any circuit without affecting any currents or voltages measurable outside their own boundaries. Based on their effects (resulting currents and voltages) on the rest of the circuit, there is no way to tell which of two equivalents is in the circuit.

Finally, while there is a useful distinction between the terms *subcircuit* and *circuit*, that of part versus whole, it is often clear enough which we are talking about from the context. Wherever this is the case, we will freely use the term *circuit* to refer to either a complete circuit or a subcircuit; thus we will use terms such as *equivalent circuit* and *two-terminal circuit*. Only where it is thought necessary to emphasize the incompleteness of a part of a circuit will the term *subcircuit* be used.

EXERCISES

2.3.1. Find the terminal laws for the two circuits sketched.
Answer (a) $v = 9i$; (b) $v = (2 + R)i$

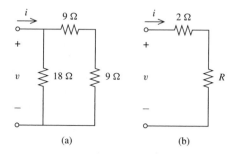

(a)   (b)

EXERCISE 2.3.1

2.3.2. Find a value for R for which the two circuits of Exercise 2.3.1 are equivalent.
Answer $R = 7\ \Omega$

2.3.3. Find an equivalent to the two-terminal circuit shown in Exercise 2.3.1(a) containing just a single element.
Answer 7-Ω resistor

Now that the laws of Kirchhoff and Ohm and the notion of equivalent circuits have been introduced, we are prepared to analyze resistive circuits. We begin with a *simple circuit*, which we define as one that can be completely described by a single equation. One type, which we consider in this section, is a circuit consisting of a single closed path, or loop, of elements. By KCL, each element has a common current, say i. Then Ohm's law and KVL applied around the loop yield a single equation in i that describes the circuit completely.

Two adjacent elements are said to be connected in *series* if they share a common node that has no other currents entering it. Nonadjacent elements are in series if they are each in series with the same element. So chains of series elements may be formed of any length. For instance, in Fig. 2.15(a) the resistors R_1 and R_2 are in series since they share exclusive access to a common node; similarly, R_2 and R_3 are in series. Since R_1 and R_3 are both in series with the same element R_2, they are in series with one another as well. Note that, by KCL, elements connected in series must all carry the same current.

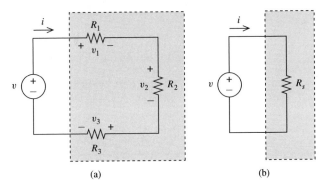

FIGURE 2.15   (a) Single-loop circuit; (b) equivalent circuit.

The single-loop networks of this section consist entirely of elements connected in series. Consider first the circuit of Fig. 2.15(a). The first step in the analysis of this or any circuit is the assignment of current and voltage variables, both names and reference directions, to each element. Since all elements in the loop are in series, they carry a common current, which we will note as i and assign a clockwise reference direction. We could have assigned its reference direction counterclockwise had we preferred; in the end we would draw consistent conclusions. It is never necessary somehow to guess the "right" reference direction in order to arrive at a correct answer. The opposite choice of reference direction for i will just result in the opposite sign when we compute its value. Since, for any number I, I amperes clockwise is the same as $-I$ amperes counterclockwise, the two results must be identical.

We next make the voltage variable assignments v_1, v_2, and v_3 for the resistors R_1, R_2, and R_3, respectively. Note that the voltage reference directions have been chosen to satisfy the passive sign convention jointly with the current reference direction already chosen for their common current i. This will permit us to use Ohm's law for the resistors without a minus sign. Applying KVL gives

$$v = v_1 + v_2 + v_3$$

Applying Ohm's law to each term on the right side of this equation and gathering the common factor i yields

$$v = (R_1 + R_2 + R_3)i \qquad (2.8)$$

Then, knowing the source function v, we may solve for the loop current i:

$$i = \frac{v}{R_1 + R_2 + R_3} \qquad (2.9)$$

Equation (2.8) or (2.9) also serves as the terminal law for the two-terminal subcircuit shaded in blue in Fig. 2.15(a), since it links its terminal current i and terminal voltage v (by KVL, v is both the source function of the voltage source and the terminal voltage into the shaded region). Examining Fig. 2.15(b), the terminal law of its shaded region is

$$v = R_s i$$

Comparing this result with (2.8), we see that the terminal laws are identical; hence the two shaded circuits of Fig. 2.15 are equivalent if R_s is the sum of the three resistances:

$$R_s = R_1 + R_2 + R_3$$

Generalizing in the obvious way, for N resistors connected in series,

$$R_s = \sum_{i=1}^{N} r_i \qquad (2.10)$$

A chain of series resistors is equivalent to a single resistor whose resistance is the sum of the series resistances. Less formally, series resistance adds. This equivalent resistor can be freely substituted in any circuit without changing any current or voltage external to the equivalent circuits themselves.

Continuing our analysis of Fig. 2.15(a), we substitute the current from (2.9) into Ohm's law $v_1 = R_1 i$ for the first resistor, yielding

$$v_1 = \frac{R_1}{R_s} v$$

Similarly,

$$v_2 = \frac{R_2}{R_s} v$$

$$v_3 = \frac{R_3}{R_s} v$$

Note that these voltages are in proportion to their resistances. This is the *principle of voltage division: the voltage across series resistors divides up in direct proportion to their resistances.* For this reason, the circuit of Fig. 2.15(a) is called a *voltage divider.* Larger resistances have larger voltage drops in a voltage divider.

Example 2.7

We seek values for R_1 and R_2 in Fig. 2.16 so that the voltage across R_2 is 8 V. With $v_2 = 8$, by KVL $v_1 = 4$. Since R_1 and R_2 form a voltage divider, their voltages are in proportion to their resistance. Then a voltage ratio of $v_2/v_1 = 2$ requires a resistance ratio $R_2/R_1 = 2$. Any R_1 together with $R_2 = 2R_1$, such as $R_1 = 1 \ \Omega$ and $R_2 = 2 \ \Omega$, will do.

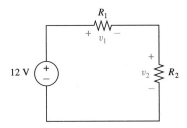

FIGURE 2.16 Voltage-divider example.

Example 2.8

As another example, suppose that $v = 120 \sin t$ V in Fig. 2.17. Let us determine the loop current, resistive voltages, and the element power deliveries or dissipations. Replacing the three series resistors by a single 60-Ω equivalent, the loop current is, by Ohm's law,

$$i = \frac{120 \sin t}{60} = 2 \sin t \ \text{A}$$

and the voltages are

$$v_1 = R_1 i = 60 \sin t \ \text{V}$$

$$v_2 = R_2 i = 40 \sin t \ \text{V}$$

$$v_3 = R_3 i = 20 \sin t \ \text{V}$$

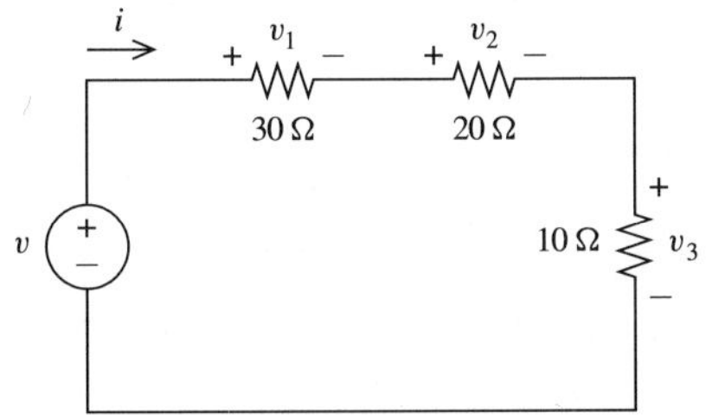

Although we did not explicitly use the voltage-divider principle, note that these three voltages are in proportion to their resistances.

We next compute the powers. Since v_1, v_2, and v_3 each satisfies the passive sign convention relative to i, the power dissipated by these elements is just their vi products, or

$$p_1 = v_1 i = 120 \sin^2 t \ \text{W}$$

$$p_2 = v_2 i = 80 \sin^2 t \ \text{W}$$

$$p_3 = v_3 i = 40 \sin^2 t \ \text{W}$$

Note that the power dissipations are also in proportion to the resistances. This follows from the voltage-divider principle and the fact that series elements carry a common current. Finally, v and i together violate the passive sign convention for the source (the current reference direction points out of, not into, the plus end of the voltage reference direction). Thus a positive vi product means instantaneous power *supplied*, not dissipated, by this element. Then

FIGURE 2.17 Single-loop circuit.

the power delivered by the source is

$$vi = 240 \sin^2 t \ \text{W}$$

Thus the power delivered in this circuit equals, at each instant of time, the total power dissipated. This is an illustration of *Tellegen's theorem*, or conservation of electrical power, which we take up in Chapter 10.

We next consider the series connection of N voltage sources as in Fig. 2.18(a). By KVL, the total voltage v is the algebraic sum of the source functions:

$$v = v_{s1} + v_{s2} + \cdots + v_{sN}$$

This is also the terminal law for the two-terminal circuit. Comparing the terminal law $v = v_s$ for Fig. 2.18(b), we conclude that the two are equivalent if

$$v_s = v_{s1} + v_{s2} + \cdots + v_{sN}$$

A chain of series voltage sources is equivalent to a single voltage source whose source function is the algebraic sum of the series source functions. More concisely, series voltage sources add.

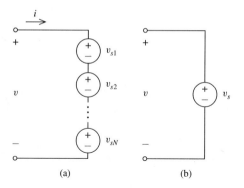

FIGURE 2.18 (a) Series voltage sources; (b) equivalent circuit.

Example 2.9

Find i in Fig. 2.19(a).

Since the three voltage sources are in series, we may replace them by a single equivalent source. With reference directions as indicated in Fig. 2.19,

$$v_s = v_{ba} = -7 + 18 + 5 = 16 \ \text{V}$$

Then $v_R = 16 = 4i$, so $i = 4$ A. Note that we could have assigned the opposite reference direction for the source function v_s, in which case we would have had $v_s = -16$ V and $v_R + v_s = 0$, so $v_R = +16$ V and once again $i = 4$ A. It is not necessary to guess the best reference direction for an equivalent source function; either will do.

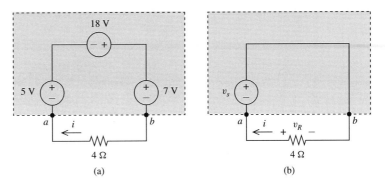

FIGURE 2.19 (a) Single-loop circuit; (b) equivalent circuit.

Finally, consider the series interconnection of the remaining type of element found in resistive circuits, the current source. From Fig. 2.20(a) we have $i_{s1} = i_{s2} = \cdots = i_{sN}$, and at the top node i equals this common value $i = i_{s1}$. From Fig. 2.20(b) we see that $i = i_s$, and the circuits are equivalent if the single source function i_s equals the common value of the series current source functions.

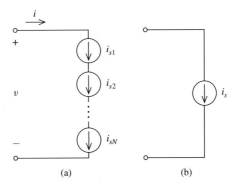

FIGURE 2.20 (a) Series current sources; (b) equivalent circuit.

But what if the series current source functions we are given are not all equal? Apparently, charge will be accumulating at the node connecting two unequal sources and KCL will be violated. Yet we know this cannot happen, since we have agreed that KCL will not be violated by any lumped-parameter circuit. This is a case of *inconsistent mathematical assumptions*. Either the given element laws are incorrect or KCL is violated. Once a mathematical inconsistency is discovered in a problem statement, there is nothing more to be done than to note this fact and stop. Inconsistent equations cannot be solved. In the real world, if we were to series-connect two current sources with unequal source functions, they would not behave like ideal current sources whose currents were independent of their voltages. The "give" would be on the part of the element law, bringing the assumptions back to consistency.

We conclude that *a chain of current sources in series must have equal source functions, in which case the chain is equivalent to any one of them. If the given source functions are not all equal, there is a mathematical inconsistency and the circuit cannot be analyzed.*

2.4.1. Reduce the circuit to two equivalent elements and find i and the the instantaneous power dissipated by the 50-V source.

Answer $2 - 4\cos 2t$ A; $200\cos 2t - 100$ W

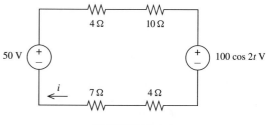

EXERCISE 2.4.1

2.4.2. Find R in the circuit shown.

Answer 6 Ω

2.4.3. Find values for i_{s1} and v_{s2} so that $v = 4$ V.

Answer -6 A; 0 V

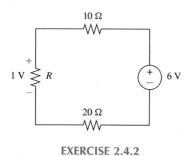

EXERCISE 2.4.2

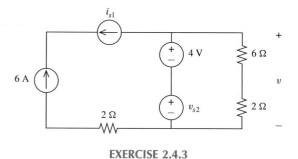

EXERCISE 2.4.3

2.5 PARALLEL EQUIVALENTS AND CURRENT DIVISION

Another important simple circuit is the single-node-pair resistive circuit. Just as single-loop circuits lead naturally to series interconnections and voltage dividers, single-node-pair circuits will be shown in this section to lead to the equally useful ideas of parallel interconnection and current division.

Two elements are connected in *parallel* if together they form a loop containing no other elements. By KVL, elements in parallel all have the same voltage across them. The single-node-pair circuit of Fig. 2.21 contains three resistors and a current source all in parallel, since any pair of the four elements forms a loop containing no other elements.

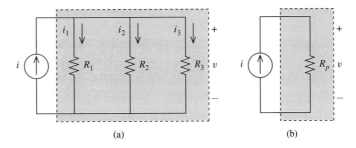

FIGURE 2.21   (a) Single-node-pair circuit; (b) equivalent circuit.

Designate the common voltage of these elements v and define resistive currents to satisfy the passive sign convention relative to v. Applying KCL at the upper node, we obtain

$$i = i_1 + i_2 + i_3$$

Using Ohm's law in the form $i = Gv$ for each term on the right and gathering the common voltage factor yields

$$i = (G_1 + G_2 + G_3)v \qquad (2.11)$$

where each conductance G_i is the inverse of the corresponding resistance R_i. Then, knowing the current source function i, we may solve for v:

$$v = \frac{i}{G_1 + G_2 + G_3} \qquad (2.12)$$

Equation (2.11) or (2.12) also serves as the terminal law for the two-terminal circuit shaded in blue in Fig. 2.21(a), since it links terminal current i and voltage v. The terminal law for the shaded circuit in Fig. 2.21(b) in terms of the conductance $G_p = 1/R_p$ is

$$i = G_p v$$

Comparing this with (2.11), the two shaded circuits of Fig. 2.21 are equivalent if G_p is the sum of the three conductances:

$$G_p = G_1 + G_2 + G_3$$

Generalizing in the obvious way for N resistors connected in parallel, *a set of parallel resistors is equivalent to a single resistor whose conductance is the sum of the parallel conductances.* Informally, parallel conductance adds. By denoting the conductance of this equivalent resistor as

$$G_p = \sum_{i=1}^{N} G_i \qquad (2.13)$$

where $R_p = 1/G_p$ and remembering that $R_i = 1/G_i$ for each i, we have (2.13) in terms of the resistances:

$$\frac{1}{R_p} = \sum_{i=1}^{N} \frac{1}{R_i} \qquad (2.14)$$

A set of parallel resistors is equivalent to a single resistor whose resistance is the inverse of the sum of the inverses of the parallel resistances.
This *inverse–inverse* law has an even simpler form when $N = 2$. In this case (2.14) becomes

$$\frac{1}{R_p} = \frac{1}{R_1} + \frac{1}{R_2}$$

or $$R_p = \frac{R_1 R_2}{R_1 + R_2} \qquad (2.15)$$

The equivalent resistance of two resistors in parallel is the product of their resistances divided by the sum. Unfortunately, this *product-by-sum rule* does not directly apply to sets of more than two resistors in parallel. In this case, (2.13) or (2.14) should be used.
Note by (2.13) that adding parallel resistors can only increase the equivalent conductance, in other words decrease the equivalent resistance. Putting resistors in parallel reduces the overall resistance below that of any of them individually. For N equal R-Ω resistors in parallel, by (2.14), $R_p = R/N$; in other words, putting N equal resistors in parallel reduces the resistance by a factor of N.
Continuing our analysis of Fig. 2.21(a), we substitute the voltage from (2.12) into Ohm's law in the form $i_1 = G_1 v$ for the first resistor, yielding

$$i_1 = \frac{G_1}{G_p} i \qquad (2.16a)$$

Similarly,

$$i_2 = \frac{G_2}{G_p} i \qquad (2.16b)$$

$$i_3 = \frac{G_3}{G_p} i \qquad (2.16c)$$

Note that these currents are all in proportion to their conductances. This is the *principle of current division: the current through parallel resistors divides up in direct proportion to their conductances.* For this reason, the circuit of Fig. 2.18(a) is called a *current divider.* Smaller resistances (larger conductances) have larger current flows in a current divider.

For two resistors in parallel, $G_p = G_1 + G_2$ and (2.16a) and (2.16b) can be rewritten in terms of the resistances $R_1 = 1/G_1$, $R_2 = 1/G_2$, and $R_p = 1/G_p$ as

$$i_1 = \frac{R_2}{R_1 + R_2} i \qquad (2.17a)$$

$$i_2 = \frac{R_1}{R_1 + R_2} i \qquad (2.17b)$$

For two resistors in parallel, current divides inversely with their resistances. Note that the same caution holds with this as with the product-by-sum rule: it works only in the case of exactly two resistors in parallel.

Example 2.10

Let us determine how the 18 mA delivered by the source divides among the four resistors in Fig. 2.22. Given their resistances, we note their conductances to be $G_1 = 4$ mS, $G_2 = G_3 = 2$ mS, and $G_4 = 1$ mS. Then the currents will be, by the current-divider principle, in the proportions $4 : 2 : 2 : 1$, with R_1 having the largest current and R_4 the smallest. The equivalent parallel conductance is the sum

$$G_p = 4 + 2 + 2 + 1 = 9 \text{ mS}$$

and by (2.16), for the case of four parallel resistors,

$$i_1 = \frac{G_1}{G_p} i = \frac{4}{9} (18 \text{ mA}) = 8 \text{ mA}$$

$$i_2 = \frac{G_2}{G_p} i = \frac{2}{9} (18 \text{ mA}) = 4 \text{ mA}$$

$$i_3 = \frac{G_3}{G_p} i = \frac{2}{9} (18 \text{ mA}) = 4 \text{ mA}$$

$$i_4 = \frac{G_4}{G_p} i = \frac{1}{9} (18 \text{ mA}) = 2 \text{ mA}$$

Note that the currents are in the predicted proportions $(4 : 2 : 2 : 1)$.

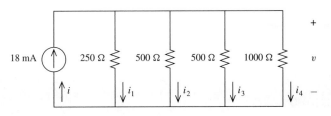

FIGURE 2.22 Current-divider example.

Continuing the example, let us see how the power divides. The voltage v is found from the equivalent

$$v = \frac{i}{G_p} = \frac{18 \text{ mA}}{9 \text{ mS}} = 2 \text{ V}$$

or by dividing any of the resistive currents computed above by the conductance of that resistor. Then noting that v and that each resistive current in Fig. 2.22 satisfies the passive sign convention, the dissipated power in R_1 is

$$p_1 = vi_1 = (2 \text{ V})(8 \text{ mA}) = 16 \text{ mW}$$

Similarly, p_2 and p_3 are found to be 8 mW and p_4 is 4 mW. Turning to the source, since v and i together violate the passive sign convention for the source, the power supplied by this element (rather than dissipated by) is given by the vi product,

$$vi = (2 \text{ V})(18 \text{ mA}) = 36 \text{ mW}$$

Thus the 36 mW produced by the source is divided among the resistors comprising the current divider in the same ratio as the currents themselves, and the power supplied equals the power dissipated over the entire circuit.

We next consider current sources connected in parallel. Examining Fig. 2.23(a),

$$i = i_{s1} + i_{s2} + \cdots + i_{sN}$$

while, for Fig. 2.23(b), $i = i_s$. The terminal laws are identical if

$$i_s = i_{s1} + i_{s2} + \cdots + i_{sN}$$

A set of parallel current sources is equivalent to a single current source whose source function is the sum of the parallel source functions. Concisely, parallel current sources add.

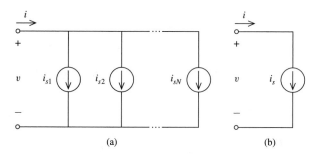

(a)   (b)

FIGURE 2.23   (a) Parallel current sources; (b) equivalent circuit.

Example 2.11

Find v and i_1 in Fig. 2.24.

Since the two current sources are in parallel, they may be replaced by an equivalent current source as shown. Note that with

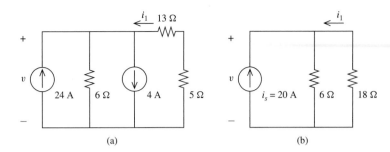

FIGURE 2.24 (a) Circuit; (b) its equivalent.

the chosen reference direction for the equivalent source, $i_s = +24 - 4 = +20$ A. The two series resistors are equivalent to a single 18-Ω resistor, which is then in parallel with the 6-Ω resistor, and by the product-by-sum rule for two parallel resistors,

$$R_{eq} = \frac{(18)(6)}{18 + 6} = \frac{9}{2}\Omega$$

Then $v = R_{eq}i_s = (9/2)(20) = 90$ V and $i_1 = -v/18 = -5$ A. Note the minus sign in Ohm's law here, since i_1 flows out of the positive end of v and this pair of variables violates the passive sign convention.

Finally, voltage sources in parallel are vulnerable to the mathematical inconsistency problem that we noted in Section 2.4 with current sources in series. If two or more voltage sources in parallel are all equal, they can be replaced by an equivalent single source of the common value, and if they are not all equal, there is an inconsistency between KVL and the element laws for these sources. *A set of voltage sources in parallel must have equal source functions, in which case any one of them is equivalent to the entire set. If the given source functions are not all equal, there is a mathematical inconsistency and the circuit cannot be analyzed.*

EXERCISES

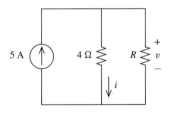

EXERCISE 2.5.1

2.5.1. Find R and v if $i = 1$ A.
Answer 1 Ω; 4 V

2.5.2. Determine the equivalent resistance of the circuit shown.
Answer 3 Ω

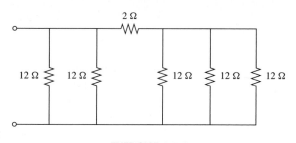

EXERCISE 2.5.2

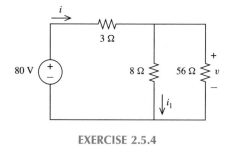

2.5.3. We wish to construct a $\frac{6}{5}$-Ω equivalent resistance by putting some number N_1 of 100-Ω resistors and some number N_2 of 3-Ω resistors in parallel. How many of each are needed?

Answer $N_1 = 50$; $N_2 = 1$

2.5.4. Find the equivalent resistance seen by the source and use this to find i, i_1, and v.

Answer 10 Ω; 8 A; 7 A; 56 V

EXERCISE 2.5.4

2.6 THEVENIN AND NORTON EQUIVALENTS

The series and parallel equivalents described thus far are limited to elements of the same type: resistors in series with other resistors, current source in parallel with current sources, and so on. In this section we develop a pair of equivalents, called Thevenin and Norton forms, containing both resistors and sources. They will prove to be of great utility in simplifying many circuit analysis problems.

Consider the pair of two-terminal circuits shown in Fig. 2.25. Applying KVL to (a) gives

$$v = -R_T i + v_T \qquad (2.18)$$

while KCL at the upper node of (b) yields

$$i_N = i + \frac{v}{R_N} \qquad (2.19a)$$

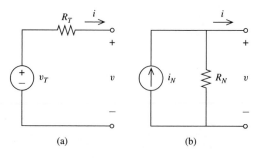

FIGURE 2.25 (a) Thevenin and (b) Norton subcircuits.

Solving the last for v yields

$$v = -R_N i + R_N i_N \qquad (2.19b)$$

Comparing the terminal law for the circuit (a) given in (2.18) with that for (b) in (2.19b), they are identical, and thus the two circuits are equivalent if $R_T = R_N$ and $v_T = R_N i_N$. The circuit of Fig. 2.25(a) is called the *Thevenin form*, a series combination of *Thevenin equivalent voltage source* v_s and *Thevenin equivalent resistance* R_T, and that of Fig. 2.25(b) the *Norton form*, a parallel combination of *Norton equivalent current source* i_N and *Norton equivalent resistance* R_N.

The Thevenin form with voltage source v_T and series resistance R_T is equivalent to the Norton form with current source i_N and parallel resistance R_N, if

$$\text{(a)} \quad R_T = R_N \quad \text{and}$$

$$\text{(b)} \quad v_T = R_N i_N \qquad (2.20)$$

From our previous discussion of equivalents in Section 2.3, we know that two equivalent subcircuits can be freely exchanged in any circuit without altering any current or voltage external to the subcircuits. If it is to our advantage, we may replace a Thevenin form by a Norton form (or vice versa) before computing a desired current or voltage. The following example shows how it may be to our advantage to do so.

Example 2.12

Find i_1 in Fig. 2.26.

First, replace the Norton form to the right of the dashed line by its Thevenin equivalent. Since $R_N = 4\ \Omega$ and $i_N = 1\ \text{A}$, to be equivalent we require that $R_T = 4\ \Omega$ and $v_T = (4)(1) = 4\ \text{V}$. Performing this replacement, we have the circuit of Fig. 2.27(a). This simple single-loop network may be further simplified by combining series resistors and series sources, yielding the circuit of Fig. 2.27(b). From this circuit, $i_1 = 12/6 = 2\ \text{A}$. Note that these additional series simplifications were not possible before the Thevenin–Norton transformation.

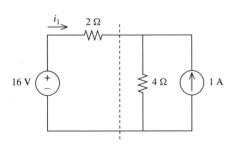

FIGURE 2.26   Circuit for Example 2.12.

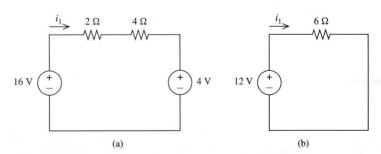

(a)   (b)

FIGURE 2.27   (a) Circuit of Fig. 2.26 after Thevenin–Norton transformation; (b) combining series elements.

Next consider an arbitrary two-terminal resistive circuit with terminal law $v = f(i)$ as shown in Fig. 2.28(a). Comparing the terminal law for the Thevenin form, (2.18), the two circuits of Fig. 2.28(a) and (b) are equivalent if

$$f(i) = -R_T i + v_T \tag{2.21}$$

Given $f(i)$, we can find slope $(-R_T)$ and intercept (v_T) values to satisfy this requirement as long as the terminal law $v = f(i)$ also describes a straight line. This will always be the case with resistive circuits of interest here, which contain only linear resistors and independent sources. This follows from the linearity of Kirchhoff's laws and the element laws and the fact that combinations of linear equations remain linear.

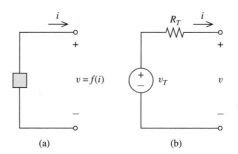

(a)   (b)

FIGURE 2.28   (a) Two-terminal circuit; (b) Thevenin form.

Equation (2.21) is satisfied by selecting R_T and v_T so that the slope of $f(i)$ equals $-R_T$ and the intercept is v_T. The resulting Thevenin form, Fig. 2.26(b), is then equivalent to the given circuit, Fig. 2.26(a). The Norton equivalent may be found by first finding the Thevenin equivalent and then performing the Thevenin–Norton transformation described by (2.20). Or the Norton equivalent may be found directly by matching the slope of $f(i)$ to $-R_N$ and the intercept to $R_N i_N$.

Example 2.13

The voltage v_0 in Fig. 2.29(a) is required. We first simplify the circuit by replacing the circuit to the right of the dashed line by its Norton equivalent. To do so, we will compute its terminal law. By KCL,

$$i_1 = 10 + i - 6 = i + 4$$

and by KVL

$$v = -4i - 8i_1 - 2i + 4$$

Combining, we have the terminal law

$$v = -14i - 28$$

The equivalent Thevenin form then has $R_T = 14\ \Omega$ and $v_T = -28$ V, and the equivalent Norton form has $R_N = R_T = 14\ \Omega$ and $i_N = v_T/R_N = -2$ A. Choosing the Norton form, the circuit after replacement is shown in Fig. 2.29(b). This is a simple circuit of the

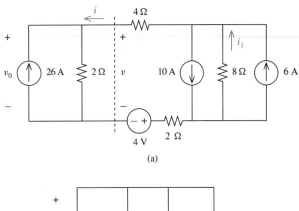

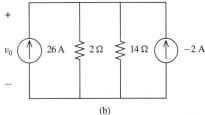

FIGURE 2.29 Circuit for Example 2.13.

single-node-pair type, with KCL

$$26 - 2 = \frac{v_0}{2} + \frac{v_0}{14}$$

or $v_0 = 42$ V.

As Example 2.13 illustrates, a circuit with many elements will have Thevenin and Norton equivalents with only two. Replacing the original by its two-element equivalent has the desired effect of simplifying the circuit.

Given the terminal law for a circuit, it is straightforward to find its Thevenin and Norton equivalents. In the previous examples, we derived the terminal law from the circuit diagram by applying Kirchhoff's laws and the element laws. It would be more convenient to determine the equivalents directly from the circuit diagram. Setting $i = 0$ in (2.21) yields

$$f(0) = v_T \tag{2.22}$$

With terminal law $v = f(i)$, $f(0)$ must be the terminal voltage of the given circuit when its terminal current i is zero. Since no current can flow if the terminals of the circuit are open circuited, $f(0)$ is called the *open-circuit voltage* $v = v_{oc}$ of the circuit, and we see by (2.22) that the Thevenin equivalent voltage v_T is equal to the open-circuit voltage of the circuit. If instead of open circuiting the terminals, we short circuited them, a *short-circuit current* $i = i_{sc}$ would flow. Since $v = f(i)$, and across a short circuit there is no voltage $v = 0$, (2.21) in the short-circuit case becomes

$$0 = -R_T i_{sc} + v_T \tag{2.23}$$

which may be solved for R_T. Combining these two results, we can state:

Given a subcircuit with open-circuit voltage v_{oc} and short-circuit current i_{sc}, its Thevenin equivalent may be found from

$$\text{(a)} \quad v_T = v_{oc} \quad \text{and}$$
$$\text{(b)} \quad R_T = v_{oc}/i_{sc}$$

$$(2.24a)$$

Use of Eq. (2.20) permits translation of these results to the Norton form.

Given a subcircuit with open-circuit voltage v_{oc} and short-circuit current i_{sc}, its Norton equivalent may be found from

$$\text{(a)} \quad i_N = i_{sc} \quad \text{and}$$
$$\text{(b)} \quad R_N = v_{oc}/i_{sc}$$

$$(2.24b)$$

Example 2.14

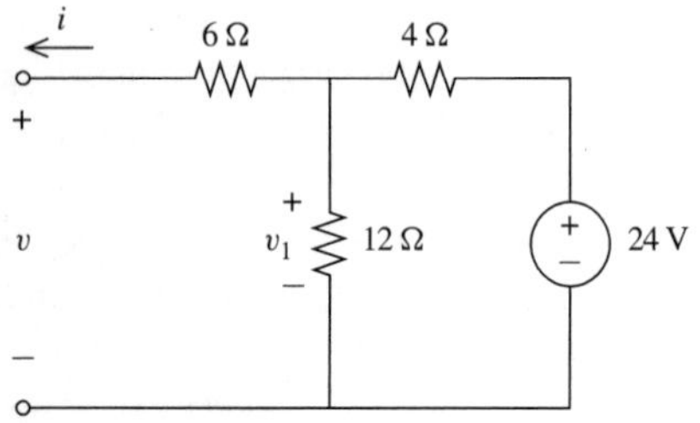

FIGURE 2.30   Circuit for Example 2.14.

Find Thevenin and Norton equivalents for the circuit of Fig. 2.30.

To find v_{oc}, we assume that the terminals are open circuited; in other words, $i = 0$. Then, with no current in the 6-Ω resistor, by voltage division

$$v_1 = \left(\frac{12}{16}\right) 24 = 18 \text{ V}$$

and by KVL around the left loop $v = v_{oc} = 0 + 18 = 18$ V. So the Thevenin equivalent voltage source is $v_T = v_{oc} = 18$ V. To find i_{sc}, we assume that the terminals are short circuited, $v = 0$. Then the two leftmost resistors are in parallel (they form a loop containing no other voltage drops), and by the product-by-sum rule,

$$R_{eq} = \frac{(12)(6)}{12 + 6} = 4 \ \Omega$$

In this case v_1 is, by voltage division of 24 V between the two 4-Ω resistors,

$$v_1 = \frac{4}{8} 24 = 12 \text{ V}$$

Since v_1 is also the voltage across the 6-Ω resistor under these short-circuit conditions, by Ohm's law $v_1 = 6i$. Then, solving for $i = i_{sc}$ gives

$$i_{sc} = \frac{v_1}{6} = \frac{12}{6} = 2 \text{ A}$$

Hence $R_T = v_{oc}/i_{sc} = 18/2 = 9 \ \Omega$. The Thevenin form and corresponding Norton form ($R_N = R_T = 9 \ \Omega$, $v_T = v_{oc} = 18$ V, $i_N = i_{sc} = \frac{1}{2}$ A) are shown in Fig. 2.31.

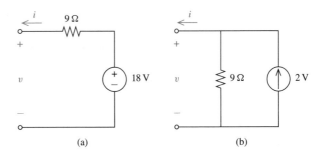

(a)   (b)

FIGURE 2.31   (a) Thevenin equivalent; (b) Norton equivalent.

There is another way to find the common resistance $R_T = R_N$ in the Thevenin and Norton forms, which is sometimes simpler than computing the ratio of open-circuit voltage v_{oc} to short-circuit current i_{sc}. Returning to (2.20) and remembering that $f(i)$ for linear resistive circuits describes a straight-line relationship, we note that if all independent sources internal to the subcircuit were set to zero (or "killed," to use a more vivid terminology), this would set the intercept of this straight line to zero; that is, the terminal law $v = -R_T i + v_T$ becomes simply $v = -R_T i$. This follows from the fact that in the absence of internal sources, the open-circuit voltage v_{oc} must be zero, since there is no source of power remaining to create any currents or voltages anywhere in the subcircuit. Then, with all internal sources killed and defining $i' = -i$, we have $v = -R_T/(-i')$, or

$$R_T = \left. \frac{v}{i'} \right|_{\text{sources killed}} \tag{2.25}$$

Equation (2.25) suggests that R_T (or R_N) *can be computed as the resistance looking into the terminals of the subnetwork when all internal sources have been killed.* The sign change $i' = -i$ was necessary to create a pair of terminal variables (i', v) that together satisfy the passive sign convention so that Ohm's law will hold without a minus sign.

Note that killing a source means setting its source function to zero. Since a 0-V voltage source is simply a short circuit, *independent voltage sources are killed by replacing them by short circuits.* Similarly, since a 0-A current source is an open circuit, *kill independent current sources by replacing them with open circuits.*

Example 2.15

Consider the circuit of Fig. 2.30. After killing internal sources, we have the circuit shown in Fig. 2.32. The resistance looking into its terminals is found by combining the parallel 12- and 4-Ω resistors

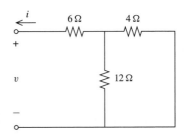

FIGURE 2.32   Circuit for Example 2.15.

into a single equivalent of 3 Ω and adding this to the 6-Ω resistance that this equivalent is in series with, yielding $R_T = R_N = 9\ \Omega$. This agrees with our calculation in Example 2.14.

Finding Thevenin–Norton equivalents for a given circuit is no more or less difficult than finding any two of its open-circuit voltage, short-circuit current, and look-in resistance with internal sources killed. If the circuit contains many loops and nodes, it may not yet be clear how to find these quantities. Systematic methods for finding any current or voltage in any circuit are developed in Chapter 4. For now we content ourselves with being aware of the power of Thevenin and Norton equivalents to simplify circuit analysis and being able to compute and use these equivalents in simple circuits.

EXERCISES

2.6.1. Find the Thevenin equivalent of the circuit to the right of line cd.
Answer $v_T = 6$ V; $R_T = 1.5$ kΩ

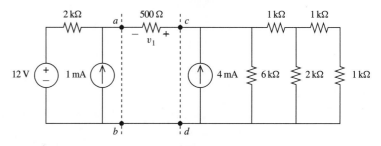

EXERCISE 2.6.1

2.6.2. Repeat Exercise 2.6.1 for the circuit to the left of line ab.
Answer $v_T = 14$ V; $R_T = 2$ kΩ

2.6.3. Use the results of Exercises 2.6.1 and 2.6.2 to find v_1.
Answer -1 V

2.7 PRACTICAL SOURCES AND RESISTORS

In Chapter 1 an independent voltage source was defined as an ideal element whose terminal voltage is equal to its source function, $v = v_s$ in Fig. 1.11, regardless of the current flowing through it. Similarly, an ideal current source always carries current exactly equal to its source function, $i = i_s$ in Fig. 1.12, no matter what terminal voltage this requires. One clear sign that these ideal sources are not physically realizable is that such devices must be capable of producing infinite power. This can be seen by connecting an R_L-Ω load resistor across the terminals of an ideal voltage source and considering what happens as R_L goes to zero. Since the voltage across the resistor remains fixed at v_s regardless of R_L, the power v_s^2/R_L goes to infinity as R_L goes to zero. An ideal current source with R_L Ω across its terminals is similarly required to produce infinite power $i^2 R_L$ as R_L goes to infinity, since its terminal current i is likewise fixed.

Practical sources, such as the familiar dry cells found in flashlights and calculators or dc power supplies in electronics equipment, are not capable of such infinitely powerful behavior. How then are we to model practical, physically realizable independent voltage and current sources? In Section 2.6 it was shown that there exist Thevenin and Norton equivalent circuits for all linear two-terminal subcircuits. We will use this universal modeling capability to represent practical sources by their Thevenin or Norton equivalent circuits. The presence of a resistor in both circuits limits their current, voltage, and power to finite values and makes them useful models for physically realizable sources. Practical sources connected to load resistors are shown in Fig. 2.33.

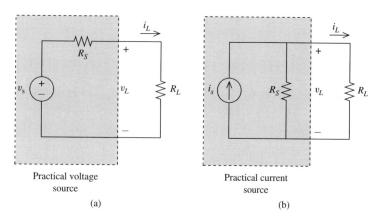

FIGURE 2.33 Practical sources connected to a load resistor: (a) voltage source; (b) current source.

The effect of the source resistance R_S in the practical voltage source model of Fig. 2.33(a) may be seen by plotting load voltage v_L and current i_L versus load resistance R_L as shown in Fig. 2.34. For large values of R_L the source is lightly loaded (little current is being drawn), and the source produces values closely approximating the ideal voltage source $R_S = 0$, as shown. As R_L converges toward zero, v_L decreases from its ideal value, as does i_L. The power dissipated by the load is

$$P = \frac{v_L^2}{R_L} = \left(\frac{R_L}{R_L + R_S} v_s\right)^2 \frac{1}{R_L} = \frac{R_L v_S^2}{(R_L + R_S)^2} \tag{2.26}$$

In the case of an ideal voltage source we have no source resistance ($R_S = 0$) and, examining (2.26) with $R_S = 0$, the power supplied to the load by the ideal source goes to infinity as the load resistance R_L goes to zero. For a practical source we have $R_S > 0$, and by (2.26) the power goes to zero, not infinity, as R_L goes to zero. The source resistance R_S serves to limit the maximum or short-circuit current the source may supply, thereby forcing the power to zero as R_L becomes small.

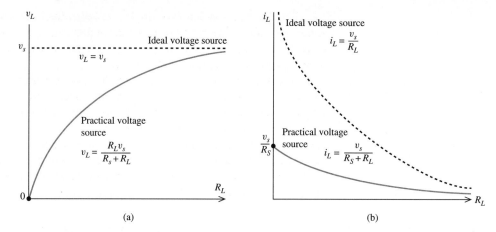

FIGURE 2.34 Comparison of practical and ideal voltage sources: (a) v_L versus R_L; (b) i_L versus R_L.

Similarly, the load current and voltage comparisons for ideal and practical current sources are shown in Fig. 2.35. The practical values are reduced for current sources with R_L large. The power to the load is, noting the current divider in Fig. 2.33(b),

$$P = i_L^2 R_L = \left(\frac{R_S}{R_L + R_S} i_S \right)^2 R_L \tag{2.27}$$

For an ideal current source with R_S infinite, P goes to infinity with R_L. For a practical current source with finite source resistance $R_S > 0$, as R_L goes to infinity in (2.27), P goes to zero. The source resistance R_S serves to limit the open-circuit voltage the practical current source may produce, and thus its power to the load. The comparisons for ideal and practical current sources are shown in Fig. 2.35.

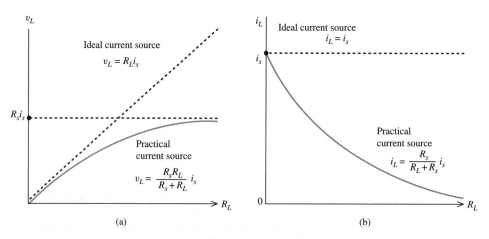

FIGURE 2.35 Comparison of practical and ideal current sources: (a) v_L versus R_L; (b) i_L versus R_L.

Thus by the addition of a single resistor to the ideal source we have a model that accurately reflects the behavior of practical sources. The Thevenin and Norton sources v_S and i_S in Fig. 2.33 may be constants for the dc case, or more general sources for signal generators or other non-dc sources.

The Thevenin or Norton equivalent for a given source may found by measuring any two of the open-circuit voltage, short-circuit current, and look-in resistance with internal sources killed, as discussed in Section 2.6. The last measurement is often difficult to arrange when it is a source that is being modeled. Also, short-circuiting a source to measure the short-circuit current may result in excessive load current, possibly damaging the source. Fortunately, it is not necessary to make these specific measurements. It is possible to determine the Thevenin or Norton parameters from *any* two i–v pairs, as we see in the following example.

Example 2.16

The open-circuit voltage across the terminals of a fresh C cell is measured to be 1.55 V. When a 29-Ω resistor is placed across it, its terminal voltage drops to 1.45 V. Find the Thevenin and Norton equivalents of the C cell, and the maximum current that can be drawn from this battery.

The Thevenin equivalent voltage $v_T = v_{oc}$, its open-circuit voltage. This is 1.55 V as shown in Fig. 2.36(a). Noting that R_T and 29 Ω form a voltage divider, we find that

$$\left(\frac{29}{29 + R_T}\right)(1.55) = 1.45$$

or
$$R_T = 2\ \Omega$$

The Thevenin equivalent of the C cell is shown in Fig. 2.36(b). Maximum current will flow if the device is short-circuited, $v_{ab} = 0$. In this case

$$i_{\max} = i_{sc} = \frac{1.55}{2} = 775\ \text{mA}$$

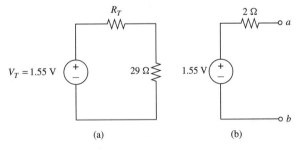

FIGURE 2.36   (a) Circuit for the example; (b) practical source.

Note that while ideal sources can supply power to a circuit without themselves dissipating any fraction, this is not the case for practical sources. For instance, in Example 2.16, if the C cell is supplying 100 mA, there will be $P = 0.1^2(2)$ or 20 mW power dissipated by the internal source resistance of the cell. Sources feel warm to the touch when they are operating because they always drop some of the power they produce internally across their source resistance. This power is not available to supply to other elements.

Example 2.17

A practical current source with a 1-A short-circuit current and $R_s = 100\ \Omega$ look-in resistance is to be used to deliver $\frac{1}{12}$ A to a 100-Ω load R_L. Determine an external resistance R_{SH} that can be placed in "shunt," or across the source terminals, to achieve this result. Then combine R_{SH} with the original current source in a Norton source model, and determine the maximum power dissipation internal to this new practical source under all possible load conditions.

We first use current division to determine R_{SH}. Let $G_{SH} = 1/R_{SH}$ be the conductance. By current division, to achieve the prescribed load current

$$i_L = \left(\frac{0.01}{0.01 + 0.01 + G_{SH}} \right)(1) = \frac{1}{12}$$

and solving for G_{SH}, we obtain

$$G_{SH} = (12)(0.01) - 0.02 = 0.1\ \text{S}$$

So $R_{SH} = 10\ \Omega$. Then the Norton equivalent of the left of ab in Fig. 2.37(a) has Norton equivalent resistance

$$R_N = \frac{(10)(100)}{10 + 100} = 9.09\ \Omega$$

and short-circuit current 1 A. The new source model, including the shunt resistor, is shown in Fig. 2.37(b). Maximum power dissipation will occur when the load is an open circuit, for then the entire 1 A will flow through the source resistance 9.09 Ω:

$$P = (1^2)(9.09) = 9.09\ \text{W}$$

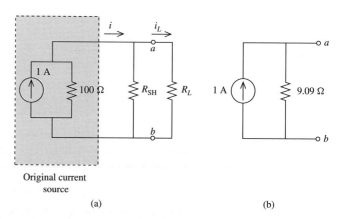

FIGURE 2.37   (a) Circuit for Example 2.17; (b) Norton source model.

We next turn to practical resistors. Resistors are manufactured from a variety of materials and are available in many sizes and values. Their characteristics include a nominal resistance value, an accuracy with which the actual resistance approximates the nominal value (known as *tolerance*), a power dissipation, and a stability as a function of temperature, humidity, and other environmental factors. They may be produced as discrete elements or deposited onto substrates as part of an integrated circuit containing other elements such as transistors and capacitors.

The most common type of discrete resistor found in electrical circuits is the carbon composition or carbon film resistor. The composition type is made of hot-pressed carbon granules. The carbon film device consists of carbon powder that is deposited on an insulating substrate. A typical resistor of this type is shown in Fig. 2.38. Multicolored bands, shown as a, b, c, and % tolerance, are painted on the resistor body to indicate the nominal value of the resistance. The color code for the bands is given in Table 2.1.

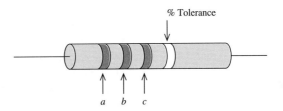

FIGURE 2.38 Carbon resistor.

Table 2.1	Color Code for Carbon Resistors		
	Bands *a, b,* and *c*		
Color	Value	Color	Value
Silver[a]	−2	Yellow	4
Gold[a]	−1	Green	5
Black	0	Blue	6
Brown	1	Violet	7
Red	2	Gray	8
Orange	3	White	9
	% Tolerance Band		
	Gold	± 5%	
	Silver	± 10%	

[a]These colors apply to band *c* only.

Bands a, b, and c give the nominal resistance of the resistor, and the tolerance band gives the percentage by which the resistance may deviate from its nominal value. Referring to Fig. 2.38, the resistance is

$$R = (10a + b)10^c \pm \% \text{ tolerance} \tag{2.28}$$

by which we mean that the % tolerance of the nominal resistance is to be added or subtracted to give the range in which the resistance lies.

Example 2.18

As an example, suppose that we have a resistor with band colors of yellow, violet, red, and silver. The resistor will have a value given by

$$R = (4 \times 10 + 7) \times 10^2 \pm 10\%$$

$$= 4700 \pm 470 \ \Omega$$

Therefore, the resistance value lies between 4230 and 5170 Ω.

Values of carbon resistors range from 2.7 to 2.2×10^7 Ω, with wattages from $\frac{1}{8}$ to 2 W. For resistance values less than 10 Ω, we see from (2.28) that the third band must be gold or silver. Carbon resistors are inexpensive but have the disadvantage of a relatively high variation of resistance with temperature (typically increasing 0.1% per degree Celsius).

Another resistor type that is commonly used in applications requiring a high power dissipation is the wire-wound resistor. This device consists of a metallic wire, usually a nickel–chromium alloy, wound on a ceramic core. Low-temperature-coefficient wire permits the fabrication of resistors that are very precise and stable, having accuracy and stability on the order of ± 1 to $\pm 0.001\%$.

The metal film resistor is another useful discrete resistor type. These resistors are made by vacuum deposition of a thin metal layer on a low-thermal-expansion substrate. The resistance is then adjusted by etching or grinding a pattern through the film. Accuracy and stability for these resistors approach that of wire-wound types, and high resistance values are much easier to attain.

Integrated circuits are rapidly superceding discrete element circuits for most low-power applications. An integrated circuit is a single monolithic chip of *semiconductor* (material with conducting properties between those of a conductor and an insulator) in which active and passive elements are fabricated by diffusion and deposition processes. Integrated circuits containing hundreds of thousands of transistors, resistors, and other elements are available on chips less than 1 in. square. An integrated-circuit resistor has the typical structure shown in Fig. 2.39. The resistive layer is very thin, typically a fraction of a micrometer (10^{-6} m), and its resistance is set by its length, width, and doping. It consists of the same doped semiconductor material used to form transistors and capacitors on the chip.

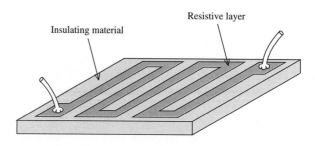

FIGURE 2.39 Integrated-circuit resistor.

EXERCISES

2.7.1. A practical source supplies $i_L = 2$ A dc to a 10-Ω load and $i_L = 1$ A dc to a 50-Ω load. Determine its Thevenin and Norton parameters V_T, R_T, I_N, and R_N.

Answer 80 V; 30 Ω; $\frac{8}{3}$ A; 30 Ω

2.7.2. A 500-mA fuse is located in the wire connecting a power supply to its load. If the power supply has an open-circuit voltage of $+100$ V dc and a source resistance of 30 Ω, what is the range of acceptable load resistances that will not cause the fuse to blow?

Answer $R_L > 170$ Ω

2.7.3. An automobile battery in a poor state of charge has an open-circuit voltage of $+11.5$ V dc and when loaded by the starter motor, which we model as a 0.8-Ω resistor, supplies 10 V across the starter motor terminals. Determine the power delivered to the load and the power being dissipated internally in the source resistance under these conditions.

Answer 125 W; 18.75 W

2.7.4. Find the resistance range of carbon resistors having color bands of (a) brown, black, red, silver; (b) red, violet, yellow, silver; and (c) blue, gray, gold, gold.

Answer (a) 900 to 1100 Ω; (b) 243 to 297 kΩ; (c) 6.46 to 7.14 Ω

2.8 AMMETERS, VOLTMETERS, AND OHMMETERS

In the real world of physical design and use of electric circuits, those quantities of interest, such as current, voltage, and resistance, must be measured by instruments. Ideally, these measurement devices do not disturb the circuits they are assessing. An *ideal ammeter* measures the current flowing through its terminals and has zero voltage across its terminals. In contrast, an *ideal voltmeter* measures the voltage across its terminals and has a terminal current of zero. An *ideal ohmmeter* measures the resistance connected between its terminals and delivers zero power to the resistance.

Practical measuring instruments only approximate these ideal devices. Practical ammeters, for instance, will not have zero terminal voltages. Similarly, voltmeters will not have zero terminal currents, and ohmmeters will deliver nonzero power to the load they are measuring.

The most common type of ammeter consists of a mechanical movement known as a d'Arsonval meter. This device is constructed by suspending an electrical coil between the poles of a permanent magnet. A dc current passing through the coil causes a rotation of the coil, as a result of magnetic forces, that is proportional to the current. A pointer is attached to the coil so that the rotation, or meter deflection, can be observed visually. d'Arsonval meters are characterized by their *full-scale current*, which is the current that will cause the meter to read its greatest value. Meter movements having full-scale currents from 10 μA to 10 mA are common.

An equivalent circuit for the d'Arsonval meter consists of an ideal ammeter in series with a resistance R_M, as shown in Fig. 2.40. In this circuit, R_M represents the

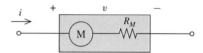

FIGURE 2.40 Equivalent circuit for a d'Arsonval meter.

resistance of the electrical coil. Clearly, a voltage appears across the ammeter terminals as a result of the current i flowing through R_M. R_M is usually a few ohms, and the terminal voltage for a full-scale current is typically from 20 to 200 mV.

The d'Arsonval meter is an ammeter suitable for measuring dc currents not greater than the full-scale current I_{FS}. Suppose that we wish, however, to measure a current that exceeds I_{FS}. We must not allow a current greater than I_{FS} to flow through the device without risking damage to the delicate device. A circuit to accomplish this is shown in Fig. 2.41(a), where R_p is a parallel resistance which reduces the current flowing through the meter coil.

From current division we see that

$$I_{FS} = \frac{R_p}{R_M + R_p} i_{FS}$$

where i_{FS} is the current that produces I_{FS} in the d'Arsonval meter. (Clearly, i_{FS} is the maximum current the ammeter can measure.) Solving for R_p, we have

$$R_p = \frac{R_M I_{FS}}{i_{FS} - I_{FS}}$$

A dc voltmeter can be constructed using the basic d'Arsonval meter by placing a large resistance R_s in series with the device, as shown in Fig. 2.41(b). It is obvious that the full-scale voltage, $v = v_{FS}$, occurs when the meter current is I_{FS}. Therefore, from KVL,

$$-v_{FS} + R_s I_{FS} + R_M I_{FS} = 0$$

from which

$$R_s = \frac{v_{FS}}{I_{FS}} - R_M$$

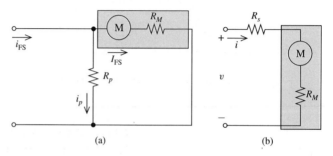

(a) (b)

FIGURE 2.41 (a) Ammeter circuit; (b) voltmeter circuit.

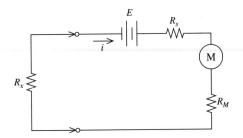

FIGURE 2.42   Ohmmeter circuit.

The *current sensitivity* of a voltmeter, expressed in ohms per volt, is the value obtained by dividing the resistance of the voltmeter by its full-scale voltage. Therefore,

$$\Omega/\text{V rating} = \frac{R_s + R_M}{v_{\text{FS}}} \approx \frac{R_s}{v_{\text{FS}}}$$

(*Note:* "≈" means approximately equal to.)

A simple ohmmeter circuit employing a d'Arsonval meter for measuring an unknown resistance R_x is shown in Fig. 2.42. In this circuit, the battery E causes a current i to flow when R_x is connected into the circuit. Applying KVL, we have

$$-E + (R_s + R_M + R_x)i = 0$$

from which

$$R_x = \frac{E}{i} - (R_s + R_M)$$

We select E and R_s such that for $R_x = 0$, $i = I_{\text{FS}}$. Therefore,

$$I_{\text{FS}} = \frac{E}{R_s + R_M}$$

Combining the last two equations, we find that

$$R_x = \left(\frac{I_{\text{FS}}}{i} - 1\right)(R_s + R_M)$$

A very popular general-purpose meter which combines the three circuits described previously is the *VOM* (voltmeter–ohmmeter–milliammeter). In the VOM, provisions are made for changing R_p and R_s so that a wide dynamic range of operation is provided. Scales may then be laid out on the face of the instrument to correspond with the resistance values R_x given in the preceding equation for each available R_s.

Electronic volt-ohm-ammeters employ active circuits, such as precision op amp stages, instead of the passive d'Arsonval movements. They can be made to measure with precision unavailable using the d'Arsonval meter, and have the added advantage of coupling naturally with digital readouts.

EXERCISES

2.8.1.   A d'Arsonval meter has $I_{\text{FS}} = 1$ mA and $R_m = 50\ \Omega$. Determine R_p in Fig. 2.41 so that i_{FS} is (a) 1.0 mA, (b) 10 mA, and (c) 100 mA.
Answer (a) Infinite; (b) 5.556 Ω; (c) 0.505 Ω

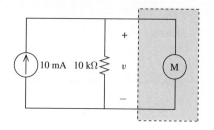

EXERCISE 2.8.3

2.9 DESIGN OF RESISTIVE CIRCUITS

Limited to resistors and independent sources, resistive circuits cannot be expected to have the versatility of those containing other types of circuit elements, such as dependent sources, inductors, and capacitors. These elements will be introduced in the next several chapters, and the wider range of design opportunities they afford will be explored in stages throughout the balance of our study. Resistive circuits are not without important practical applications, however. In this section we present three design examples using resistive circuits.

Series–parallel transformations may be used to reduce a set of interconnected resistors to a single resistor in order to simplify their analysis. They may also be used to expand a single resistor to many for purposes of design. This typically occurs when the single resistor is not of an available resistance, or when its power rating is insufficient, as in the first example.

Design Example 2.1

For our first design example, we shall use any number of available resistors, which are of the specific component values 470 Ω and 6.8 Ω, to synthesize a desired 50-Ω load resistor ($\pm$1%). Each of the available resistors is related at $\frac{1}{4}$ W maximum power dissipation, and our 50-Ω load resistor must dissipate at least 2 W.

As with most design problems, there is no unique circuit that will satisfy the design specifications, and we begin by seeking a circuit layout that appears capable of doing the job, being prepared to modify as necessary. If we put N identical R-Ω resistors in parallel, adding their conductances yields $N(1/R)$, so the equivalent resistance is R/N. If we put $N = 10$ of the available 470-Ω resistors in parallel, this yields an equivalent of 47 Ω, as shown in Fig. 2.43(a). Then if R_{e2} is put in series, we would need

$$R_{e2} = 50 - 47 = 3\ \Omega$$

How can we synthesize a 3-Ω equivalent resistance from the available ones? We cannot have two or more in series, since even the smallest available resistance exceeds 3 Ω and series resistance adds. They must be in parallel once again. The required conductance is

$$G_{e2} = \frac{1}{R_{e2}} = 333.3\ \text{mS}$$

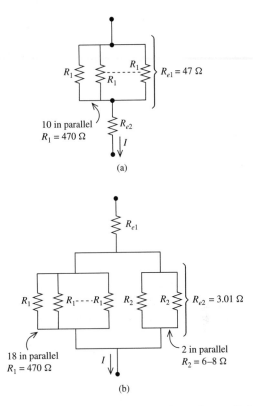

FIGURE 2.43 (a) First step: ten 470-W resistors in parallel; (b) completing the design of a 50-W equivalent load.

and the two available conductances are

$$G_1 = \frac{1}{R_1} = \frac{1}{470} = 2.13 \text{ mS}$$

$$G_2 = \frac{1}{R_2} = \frac{1}{6.8} = 147 \text{ mS}$$

Since parallel conductances add, we seek integers n_1 and n_2 such that to a good approximation, $n_1 G_1 + n_2 G_2$ yields the required conductance G_{e2}. Setting $n_2 = 2$ and $n_1 = 18$ gives

$$G_{e2} = (2)(147) + (18)(2.13) = 332.3 \text{ mS}$$

This yields

$$R_{e2} = \frac{1}{G_{e2}} = \frac{1}{0.3323} = 3.01 \ \Omega$$

so that the series combination

$$R_{e1} + R_{e2} = 47 + 3.01 = 50.01 \ \Omega$$

is well within the ±1% requirement. This configuration is shown in Fig. 2.43(b).

We need to determine if this resistor array satisfies the power requirement as well. In order that the circuit dissipate 2 W of power, we must have

$$I^2(50.01) = 2$$

or $I = 0.20$ A. The total current I passing through the circuit divides equally among the 10 resistors in R_{e1}, so for $I = 0.20$ A, the power dissipated in each is

$$P = \left(\frac{0.2}{10}\right)^2 (470) = 0.188 \text{ W}$$

well below the $\frac{1}{4}$ W permitted. The power dissipated in R_{e2} divides proportionally with the conductance of the parallel elements, so each of the two 147-mS conductances dissipate more power than the other conductances. By current division, the power dissipated in each 147-mS (6.8 Ω) element is

$$P = \left[\left(\frac{147}{332.3}\right)(0.2)\right]^2 (6.8) = 0.053 \text{ W}$$

again below the $\frac{1}{4}$ W maximum. Thus we have a solution.

Note that this solution is hardly unique. We could have begun by placing seven 6.8-Ω resistors in series rather than ten 470-Ω resistors in parallel. This may seem to save resistors, but minimizing the number of resistors used was not an explicit design specification. Among distinct solutions to a design problem it is common sense to reduce the component count where possible, and this may seem to favor the series alternative. Note, however, that had we done so, the entire current $I = 0.2$ A needed to meet the 2-W power requirement would flow through each 6.8-Ω resistor, yielding

$$P = (0.2^2)(6.8) = 0.272 \text{ W}$$

which would exceed the $\frac{1}{4}$-W limit per resistor. We could work around this problem by substituting a pair of parallel 6.8-Ω resistors in series with another such pair for each of the seven proposed 6.8-Ω resistors, since each four-resistor array would have the same equivalent resistance of 6.8 Ω, but each resistor would dissipate only one-fourth as much power as before. But in this case the component count advantage would be lost.

Design Example 2.2

Design a practical electrical source (by specifying its Thevenin equivalent components) which delivers at least 16 V dc to an 8-Ω load but is current limited to 3 A.

We will determine the Norton source, then transform to get the desired Thevenin values. Connecting the source and load as shown in Fig. 2.44, we wish I_L to be less than or equal to 3 A regardless of

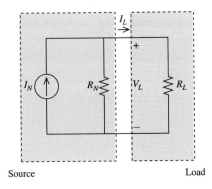

Source

Load

FIGURE 2.44

R_L. Setting $R_L = 0$ clearly achieves the greatest value for I_L, which is I_N itself . Thus the Norton equivalent current source $I_N = 3$ A. When $R_L = 8$ Ω, the load voltage should be 16 V,

$$V_L = 16 = \frac{I_N}{G_N + G_L}$$

where G_N and G_L are the conductances. Setting $G_L = \frac{1}{8}$ S and solving for G_N yields

$$G_N = \tfrac{3}{16} - \tfrac{1}{8} = \tfrac{1}{16} \text{ S}$$

or $R_N = 16$ Ω. The Norton source elements are $I_N = 3$ A and $R_N = 16$ Ω, so the Thevenin source has $V_T = (3)(16) = 48$ V and $R_T = 16$ Ω. This source will deliver 16 V to an 8-Ω load and never produce more than 3 A, as required.

One very practical class of resistive circuits are those that supply relatively fixed currents or voltages to variable loads. Our final design example is a circuit of this type.

Design Example 2.3

Our goal is to design a battery charger circuit that will deliver controlled levels of dc current to charge batteries of different sizes. We wish to deliver 10 mA of current to larger batteries, which will be connected between terminals a and a', and 1 mA to smaller batteries being charged at terminals b–b'. At any terminal voltage with batteries from zero to full charge 1.5 V, and any battery internal resistance from zero up to 30 Ω, the charging currents must remain within ±5% of the target values.

We begin the design by noting that the circuit shown in Fig. 2.45(a) will deliver

$$I = \frac{V_S - V_B}{R_S + R_B} \qquad (2.29)$$

where V_s and R_s are the source voltage and resistance of the charger circuit, V_B the battery voltage, and R_B its internal resistance. Since I must be relatively constant as V_B and R_B vary, we will set $V_S \gg V_B$

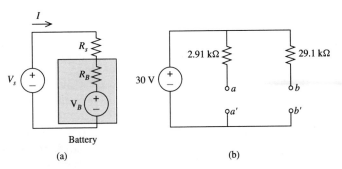

FIGURE 2.45 (a) Source and battery; (b) final circuit.

and $R_S \gg R_B$ so, to a good approximation, $I \simeq V_S/R_S$. Now the numerator in (2.29) will vary by 5% as V_B varies from 0 to 1.5 V if $V_S = (20)(1.5) = 30$ V. Then setting $V_s/R_S = 1$ mA yields

$$0.001 = \frac{30}{R_S} \quad \text{or} \quad R_S = 30 \text{ k}\Omega \quad\quad (2.30)$$

By (2.29), the maximum charging current occurs with $V_B = R_B = 0$, and by (2.29), $I_{max} = 1$ mA. I_{min} occurs with $V_B = 1.5$ V, $R_B = 30$ Ω, or

$$I_{min} = \frac{30 - 1.5}{30,000 + 30} = 0.946 \text{ mA}$$

The worst-case error occurring at I_{min} is -5%, which is slightly beyond our tolerance. This can easily be remedied by "tuning" this first-cut try.

Since the range of errors is $+0$ to -5%, let's increase the maximum current by, say, 3%. Then retaining $V_S = 30$ V for convenience gives

$$I_{max} = 0.00103 = \frac{30}{R_S}$$

or $R_S = 29.1$ kΩ. The minimum current is then increased to

$$I_{min} = \frac{30 - 1.5}{29,100 + 30} = 0.978 \text{ mA}$$

The errors now range between $+3$ and -2.2%, well within the specified $\pm 5\%$ tolerance.

Using the same 30-V source to produce the other desired charging current of 10 mA, note from Fig. 2.45(a) that if we scale R_S by $\frac{1}{10}$, I will roughly scale by 10 as desired, since R_S still dominates R_B and V_S dominates V_B. The maximum and minimum currents are

$$I_{max} = \frac{30}{2910} = 10.3 \text{ mA}, \quad\quad I_{min} = \frac{30 - 1.5}{2910 + 30} = 9.7 \text{ mA}$$

Once again the range of errors ($\pm 3\%$) is well within the $\pm 5\%$ tolerance. The final design is shown in Fig. 2.45(b).

The process of design begins with a complete statement of design specifications and constraints. A first-cut circuit layout is conceived and first-cut element values are selected. Circuit analysis methods are then used to determine the performance of the proposed circuit. In the examples above, the circuit analysis methods invoked included series–parallel and Thevenin–Norton transformations, also current–voltage division. If the resulting element values do not satisfy the specifications and constraints, changes are made and the cycle repeated. Even if a solution has been found, it may often be improved further, unless the optimal solution has been discovered. If desired, another cycle may then be executed to improve performance, as in Design Example 2.3.

In very simple designs it may be possible to solve for solution values of the circuit elements exactly rather than through repeated cut-and-try cycles. This is the case with the examples above. But as the designs get more complex, it is usually preferable to improve iteratively by cut-and-try rather than to attempt to solve a complex system of coupled equations which are often nonlinear in the circuit parameters being sought. This will be the case for most of the design examples in subsequent chapters.

DESIGN EXERCISES

2.9.1. Design a circuit that supplies the following dc voltages from a single voltage source: 1 V, 2 V, 3 V, 4 V, 6 V, all ±5% into resistive loads of 100 Ω or more. Use as few resistors and as low a voltage source as possible. The voltages supplied need not have a common ground terminal.

One of many possible solutions

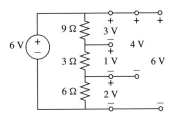

EXERCISE 2.9.1

2.9.2. Design a buffer circuit to go between a 100-V dc source and a variable resistive load with nominal value R Ω, which will protect the load from both overvoltage and overcurrent conditions. The load current should never exceed 1 A or the load voltage exceed 60 V, even as the load resistance varies unpredictably from its nominal value during operation. At its nominal value of R Ω the load should dissipate at least 10 W. Specify both the buffer circuit and an acceptable nominal value for the load.

One of many possible solutions

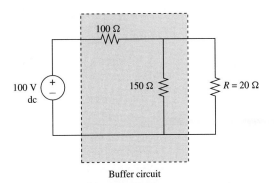

EXERCISE 2.9.2.

The behavior of an electrical circuit is a consequence of two factors: the individual behaviors of each element populating the circuit, and the manner in which they are interconnected. There are two interconnection laws, the Kirchhoff current and voltage laws. Each type of element has its characteristic element (or i–v) law, an equation involving its terminal variables. In this chapter we introduce two elements, resistors and independent sources, which when interconnected form resistive circuits. Others are introduced in later chapters.

- KCL asserts that the net current entering any node or region is zero.

- KVL asserts that the net voltage drop around any closed loop is zero.

- The i–v law for resistors, Ohm's law, is $v = Ri$, with R the resistance in ohms, and is valid as long as i and v satisfy the passive sign convention.

- Resistors are passive elements; that is, they cannot supply power.

- Two subcircuits are equivalent if they have the same i–v law. Equivalent subcircuits may be freely interchanged without changing any external variable. Examples include series–parallel and Thevenin–Norton equivalents.

- Series resistance adds, parallel conductance adds.

- Parallel resistance obeys the reciprocal–reciprocal law: the equivalent resistance is the reciprocal of the sum of the reciprocals of the parallel resistances.

- Series voltage divides in proportion to resistance, parallel current divides in proportion to conductance (or inversely proportional to resistance).

- A Thevenin form is a two-terminal series interconnection of voltage source and resistor, a Norton form is a parallel interconnection of current source and resistor.

- Any two-terminal subnetwork has a Thevenin equivalent and a Norton equivalent form. These may be found by determining the open-circuit voltage, short-circuit current, and resistance looking into the terminal pair.

- Thevenin and Norton transformations are useful in replacing a selected region of a circuit to be analyzed by a simpler one which will not result in any changed currents or voltages outside its own boundaries.

- Practical voltage and current sources may be modeled accurately by their Thevenin and Norton forms.

- Ammeters, voltmeters, and ohmmeters based upon the d'Arsonval movement are in wide use. The force between a fixed and variable magnetic field moves a pointer.

While KCL, KVL, i–v laws, and equivalent subcircuits are introduced here in the context of resistive circuits studied in the time domain, these fundamental ideas will apply equally well as we broaden our scope in later chapters to include other types of elements and other domains of analysis.

PROBLEMS

2.1. Write KCL at each node.

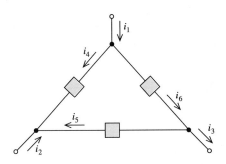

FIGURE P2.1

2.2. Write KCL for this node:
(a) In sum-entering-equals zero form.
(b) In sum-leaving-equals zero form.
(c) In sum-entering-equals-sum-leaving form.

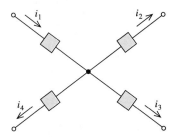

FIGURE P2.2

2.3. KCL equations for a certain three-node circuit are:

Node 1: $i_1 + i_2 - i_3 = 0$
Node 2: $-i_1 - i_2 - i_4 = 0$
Node 3: $i_3 + i_4 = 0$

Sketch the circuit, indicating current reference directions.

2.4. Write KVL around each of the three loops.

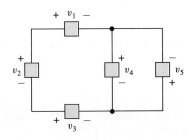

FIGURE P2.4

2.5. Write KVL for this loop:
(a) In sum-of-drops-equals-zero form.
(b) In sum-of-rises-equals-zero form.
(c) In sum-of-drops-equals-sum-of-rises form.

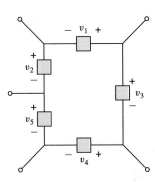

FIGURE P2.5

2.6. Assign names and reference directions for all voltages, then write KVL for each of the seven loops in this circuit.

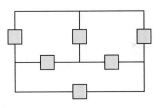

FIGURE P2.6

2.7. Supply reference directions so that these KCL equations are all valid and the passive sign convention is satisfied everywhere.

$$i_1 + i_4 - i_2 = 0$$
$$i_7 + i_8 - i_5 = 0$$
$$i_1 + i_3 - i_5 = 0$$
$$i_6 - i_4 - i_8 = 0$$
$$i_2 + i_3 - i_6 - i_7 = 0$$

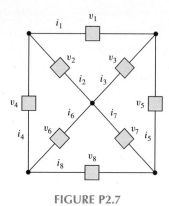

FIGURE P2.7

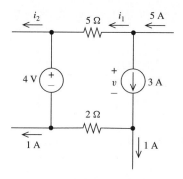

FIGURE P2.12

2.13. Assign voltage reference directions for each resistor so that the passive sign convention is satisfied. Determine each voltage if $i_2 = 2$ A and $i_4 = -4$ A.

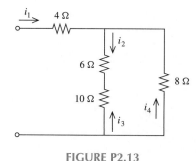

FIGURE P2.13

2.8. In a given circuit, suppose that we write KCL at each node in the sum-entering-equals-zero form.
(a) In how many equations will a given element current appear? With what signs? Justify.
(b) If all these equations were added together, what would we get?
(c) Use part (b) to show that the last node equation is just the negative sum of all the previous ones.

2.9. Assign names and reference directions; then write KCL at each node and KVL around each loop.

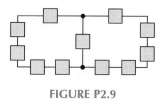

FIGURE P2.9

2.14. Assign voltage reference directions for all five elements so that the voltage drop across each element is positive in the clockwise direction; then assign current reference directions that satisfy the passive sign convention.

2.10. A toaster is essentially a resistor that becomes hot when it carries a current. If a toaster is dissipating 960 W at a voltage of 120 V, find its current and its resistance.

2.11. Find the energy used by a toaster with a resistance of 12 Ω, which is operated at 120 V for 10 s.

2.12. Find i_1, i_2, and v.

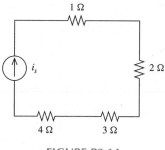

FIGURE P2.14

2.15. Find v for each subcircuit. In each case, $R = 1\ \text{k}\Omega$.

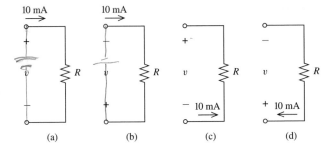

FIGURE P2.15

2.16. Find v_1, i_2, and the power produced (or dissipated) by the 7-V voltage source.

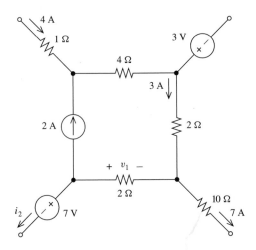

FIGURE P2.16

2.17. Find all currents and voltages.

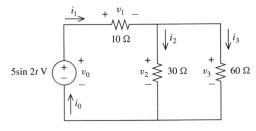

FIGURE P2.17

2.18. Determine the power supplied by, or dissipated by, each of the four elements in the circuit of Problem 2.17.

2.19. By what factor would we increase the source function in the circuit of Problem 2.17 if we wished to double the power it supplies to the rest of the circuit? Justify.

2.20. Find the terminal law $v = f(i)$. Rewrite as $i = g(v)$.

$$V = 10 - 4\left(2+i\right)$$

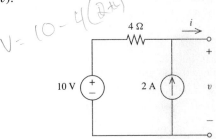

FIGURE P2.20

2.21. Find the terminal law $v = f(i)$. Rewrite as $i = g(v)$.

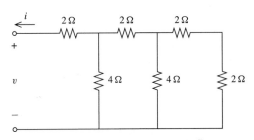

FIGURE P2.21

2.22. Find the terminal law for this circuit. Note that only one of the two forms $v = f(i)$ and $i = g(v)$ exists in this case.

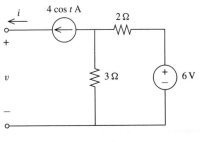

FIGURE P2.22

2.23. (a) Find the equivalent resistance looking in terminals a–b if terminals c–d are open, and if terminals c–d are shorted together. (b) Find the equivalent resistance looking in terminals c–d if terminals a–b are open, and if terminals a–b are shorted together.

2.24. Design a voltage divider that delivers +3, +6, +12, and +24 V, all with a common negative terminal. All resistors must be integer multiples of 1 kΩ, and the voltage source should be as low voltage as possible.

2.25. A 50-V source and two resistors, R_1 and R_2, are connected in series. If $R_2 = 4R_1$, find the voltages across the two resistors.

2.26. A 12-V source in series with a resistive load R carries a current of 60 mA. If a resistor R_1 is added in series with the source and load, find R_1 so that the voltage across it is 8 V.

2.27. If $v_x = 10$ V and $v_y = 20$ V, find R_x and R_y.

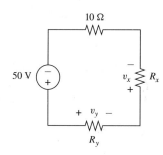

FIGURE P2.27

2.28. If all resistors in a voltage divider are scaled by the same factor α, how does this affect the voltages across them? The currents through them? The power? Use this result to design a voltage divider using a +5-V source with outputs of +4.0, +3.0, and +1.5 V (all with respect to a common negative terminal) and a power dissipation of 1 mW. Start by taking R_1 to be 1 Ω; compute the other R's and scale.

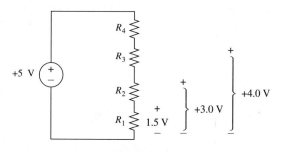

FIGURE P2.28

2.29. A voltage divider is to be constructed using a +24-V battery and 1-kΩ resistors. Design a voltage divider with a +18-V output using as few resistors as possible. Design another one with a +18.1-V output.

2.30. What must R be to set the equivalent resistance of this circuit to 300 Ω?

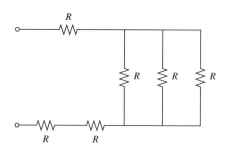

FIGURE P2.30

2.31. Find i and the total power dissipated by this circuit if $i_2 = 1$ A.

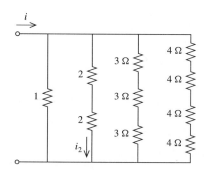

FIGURE P2.31

2.32. Determine the equivalent resistance of the circuit of Problem 2.31.

2.33. If the conductances of all resistors in a current divider are scaled by the same factor α, how does this affect the currents through them? The voltages across them? The power? Use this result to design a current divider using a +1-A source with outputs of 0.4, 0.3, and 0.1 A and a power dissipation of 1 mW. Start by taking G_1 to be 1 S; compute the other G's and scale.

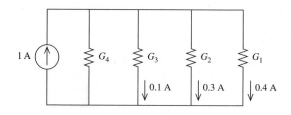

FIGURE P2.33

2.34. Find i and the power delivered to the 4-Ω resistor.

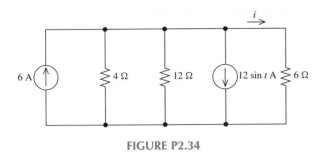

FIGURE P2.34

2.35. A current divider consists of the parallel connection of a 20-, a 40-, a 60-, and a 120-kΩ resistor. Find the equivalent resistance of the divider, and if the total current entering the divider is 120 mA, find the current in the 20-kΩ resistor.

2.36. Find i_1 and i_2.

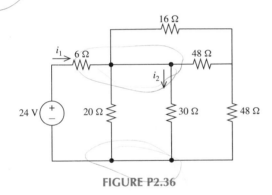

FIGURE P2.36

2.37. A current divider is to be constructed using a +2-mA constant current source and 1-MΩ resistors. Design a current divider with a +1.5-mA output. Design another with a +1.51-mA output.

2.38. Find the Thevenin equivalent circuit.

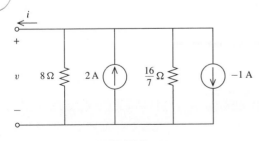

FIGURE P2.38

2.39. Suppose that the terminal laws for N_1 and N_2 are $v = -2i + 1$ and $v = -4i - 7$. Find the Thevenin equivalent if N_1 and N_2 are placed in series.

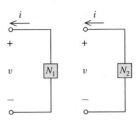

FIGURE P2.39

2.40. Ten identical copies of a subcircuit N with terminal law $i = g(v) = -\frac{1}{3}v + 2$ are put in series. Find the terminal law of the new subcircuit and its Norton equivalent.

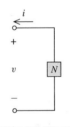

FIGURE P2.40

2.41. Use a series of Thevenin–Norton and series–parallel transformations to reduce to a simple two-node circuit the circuit shown. Which variables from the original circuit remain in the simple circuit?

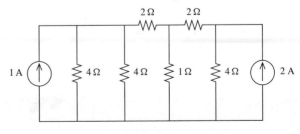

FIGURE P2.41

2.42. Find the Thevenin equivalent of everything except the R-Ω resistor. Then find R so that $i = 1$ A.

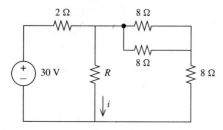

FIGURE P2.42

2.43. Find the Thevenin equivalent of this subcircuit. Use this to show that resistors in parallel with voltage sources have no effect on the rest of the circuit. Find a similar relation involving resistors and current sources.

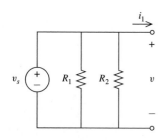

FIGURE P2.43

2.44. Find v. Use a Thevenin–Norton transformation to help.

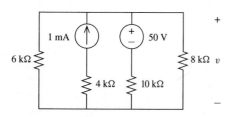

FIGURE P2.44

2.45. Find a value for R such that $i = 1$ mA.

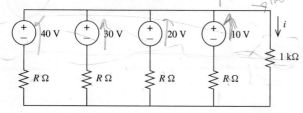

FIGURE P2.45

2.46. Find the Thevenin equivalent of this circuit by first finding and using the Thevenin equivalent of the current source and its parallel resistors.

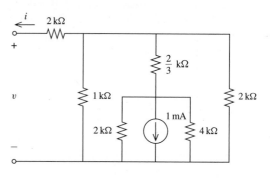

FIGURE P2.46

2.47. Find the Norton equivalent of the subcircuit of Problem 2.46.

2.48. Find the Norton equivalent to the left of ab, then the Thevenin equivalent to the left of cd, then the Norton equivalent to the left of ef. Having simplified to a two-node circuit, finally find v.

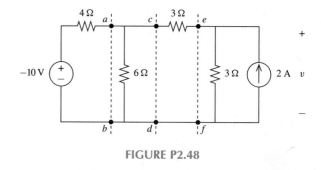

FIGURE P2.48

2.49. N satisfies the terminal law $v = -7i + 4$. Find the Thevenin and Norton equivalents for the given circuit.

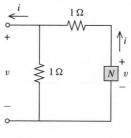

FIGURE P2.49

2.50. Under what circumstances will the Thevenin equivalent of a given subcircuit exist but not the Norton equivalent? Give an example.

2.51. Under what circumstances will the Norton equivalent of a given subcircuit exist but not the Thevenin equivalent? Give an example.

2.52. Find i and v.

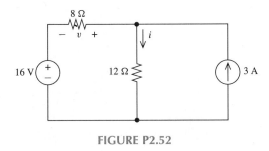

FIGURE P2.52

2.53. Using any number of 10 Ω resistors, design a subcircuit whose equivalent resistance is 144.7 Ω.

Design Problems

2.54. Use any number of 2.2-kΩ $\frac{1}{4}$-W and 56-Ω $\frac{1}{8}$-W resistors, together with 1.5-V batteries, to produce a practical source supplying 25 mA of short-circuit current and 15 mA of current to a 50-Ω load (each $\pm$5%).

2.55. Design a dc power supply circuit with four pairs of load terminals marked 1–1' through 4–4'. When a load resistor is connected between 1 and 1', it should get 1 A dc; between 2 and 2', 500 mA dc; 3 and 3', 250 mA dc; and 4 and 4', 50 mA dc. The load resistor can be any $R_L \leq 50$ Ω. Use one independent current source of 2 A dc and as few internal resistors as possible. Tolerance on the load currents is $\pm$10%.

2.56. An alternator with an open-circuit voltage of 22 V dc and a short-circuit current of 10 A dc is to be used to charge automobile batteries safely with open-circuit voltage in the range 0 to 12 V and Thevenin equivalent impedance in the range 1 to 10 Ω. Design a resistive circuit limiting the power supplied to any battery in these ranges to 10 W while permitting the maximum number of batteries to be charged at the same time.

More Challenging Problems

2.57. Find i and R.

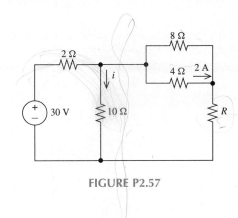

FIGURE P2.57

2.58. Find Thevenin and Norton equivalents.

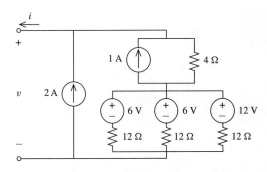

FIGURE P2.58

2.59. A certain circuit has seven elements and obeys the KCL equations

$$i_1 = i_2$$
$$i_2 = i_3 - i_4$$
$$i_7 = i_1 + i_5$$
$$i_3 + i_5 + i_6 = 0$$

(a) Sketch the circuit, labeling currents $i_1, i_2, \ldots, i_7$ and voltages $v_1, \ldots, v_7$. Define the voltages for each element to satisfy the passive sign convention.
(b) Write three distinct KVL equations for this circuit.

2.60. (a) Find the i–v law $v_2 = f_2(i_2)$ for the circuit shown in (b). (b) If $v_1 = 3i_1 - \sin t$, nodes a–a' are connected together, and nodes b–b' are connected together, find i_1 and v_2.

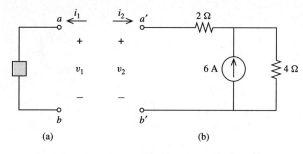

(a)   (b)

FIGURE P2.60

2.61. A total current of 20 A divides among three resistors as shown. If the largest is $R_1 = 1$ kΩ, find R_2 and R_3 so that 12 A flows through R_3 and $R_2 = 2R_3$.

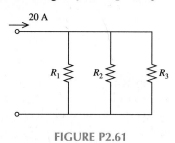

FIGURE P2.61

2.62. Sketch the Thevenin and Norton equivalents to the circuit shown.

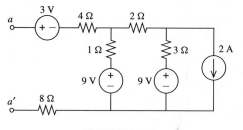

FIGURE P2.62

Dependent Sources and Op Amps

Georg Simon Ohm
1787–1854

I herewith present to the public a theory of galvanic electricity [Ohm's law].

Georg Simon Ohm

The most basic and most widely used of all the laws of electricity, Ohm's law, was published in 1827 by the German physicist Georg Simon Ohm in his great work, *The Galvanic Chain, Mathematically Treated*. Without Ohm's law we could not analyze the simplest galvanic chain (electric circuit), but at the time of its publication, Ohm's work was denounced by critics as "a web of naked fancies," the "sole effort" of which was "to detract from the dignity of nature."

Ohm was born in Erlangen, Bavaria, the oldest of seven children in a middle-class-to-poor family. He was an early dropout at the University of Erlangen but returned in 1811 and earned his doctorate and the first of several modest, low-paying mathematics teaching positions. To improve his lot, he threw himself into his electrical research at every opportunity allowed by his heavy teaching duties, and his efforts culminated in his famous law. Despite the misplaced criticisms of his work, during his lifetime Ohm received the fame that was due him. The Royal Society of London awarded him the Copley Medal in 1841, and the University of Munich gave him its Professor of Physics chair in 1849. He was also honored after his death when the *ohm* was chosen as the unit of electrical resistance.

Chapter Contents

The voltage and current sources of Chapters 1 and 2 are independent sources, as defined in Section 1.4. We may also have *dependent sources*, which are very important in circuit theory, particularly for understanding electronic circuits (circuits containing elements such as transistors or vacuum tubes that need to be coupled to power supplies). In this chapter we shall define dependent sources and consider an additional circuit element, the operational amplifier or *op amp*. Dependent sources are central to the design of electronic amplifiers and to a wide variety of other circuits of great practical interest. Op amps behave like dependent sources, and op amps are a convenient way to fill the requirement for dependent sources in these circuits, and the design of simple op amp circuits.

We shall first analyze a few simple circuits containing resistors and sources, both independent and dependent. As we shall see, the analysis is very similar to that performed in Chapter 2. We shall then turn our attention to the op amp and its behavior as a linear circuit element. Ideal and nonideal op amp models will be presented and the uses of these models in analyzing op amp circuits explored. A set of op amp building block circuits will be introduced, easily analyzed circuits from which complex circuit behaviors can be built by combination. The chapter concludes with a look at practical op amps in the real world of physical packages, finite limitations, and environmental constraints, and the design of simple op amp circuits.

3.1 DEFINITIONS

A *dependent or controlled voltage source* is a voltage source whose terminal voltage depends on, or is controlled by, a voltage or a current defined at some other location in the circuit. Controlled voltage sources are categorized by the type of controlling variable. A *voltage-controlled voltage source (VCVS)* is controlled by a voltage and a *current-controlled voltage source (CCVS)* by a current. The symbol for a dependent voltage source with source function v_s is shown in Fig. 3.1(a).

A *dependent or controlled current source* is a current source whose current depends on, or is controlled by, a voltage or a current defined at some other location in the circuit.

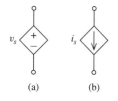

FIGURE 3.1
(a) Dependent voltage source; (b) dependent current source.

A *voltage-controlled current source (VCCS)* is controlled by a voltage and a *current-controlled current source (CCCS)* by a current. The symbol for a dependent current source with value (source function) i_s is shown in Fig. 3.1(b).

Figure 3.2 illustrates the four types of *linear* controlled sources and shows the voltage or current on which they are dependent. The quantities μ and β are dimensionless constants, commonly referred to as the *voltage gain* and *current gain*, respectively. The constants r and g have dimensions of ohms and siemens, units of resistance and conductance, and are thus called the *transresistance* and *transconductance*, respectively. The prefix *trans* is to remind us that the current and voltage are not measured at the same location. These four parameters each scale the ratio of the controlled variable produced per unit controlling variable.

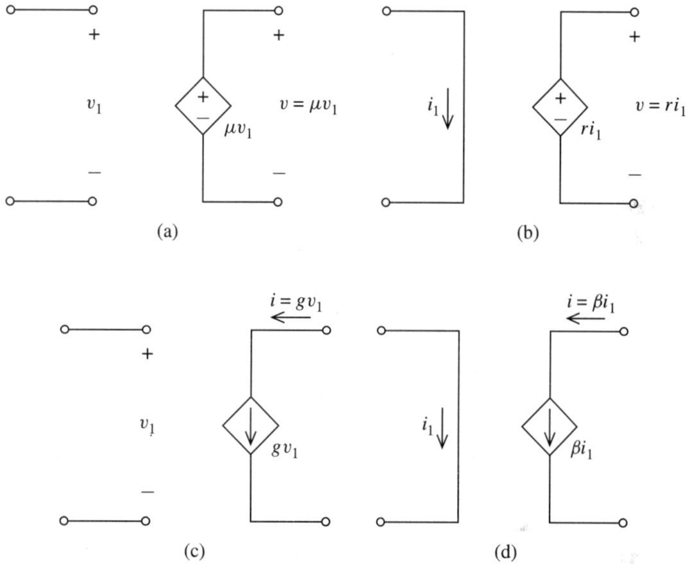

FIGURE 3.2   (a) VCVS; (b) CCVS; (c) VCCS; (d) CCCS.

Nonlinear dependent sources are those in which the controlled variable is not simply proportional to the controlling variable as in the *linear dependent sources* above. In the present book we are interested only in studying linear circuits, and thus nonlinear controlled sources will not be considered further.

As an example of a circuit containing a dependent source, in Fig. 3.3 we have an independent source, a dependent source, and two resistors. The dependent source is a CCVS with controlling current i_1 and transresistance $r = 0.5\ \Omega$. Note that if i_1 were redefined to be the current flowing downward through the rightmost arm of this circuit rather than the middle arm, the controlled source would then be an element whose voltage was 0.5 times its own current, and thus it would be identical to an ordinary 0.5-Ω resistor. This illustrates that only if the controlling variable is located *elsewhere* in the circuit does the element truly have to be represented as a controlled source.

Dependent sources are essential for producing *amplifiers*, circuits that produce outputs more powerful than their inputs. They are also integral to active filters (to be studied in Chapter 14) and to electronic circuits of all kinds. Among their many important uses,

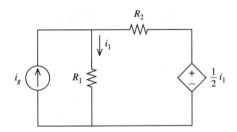

FIGURE 3.3   Circuit containing a dependent source.

dependent sources are used to prevent loading, isolate one part of a circuit from another, and provide exotic behaviors such as negative resistance. As we saw in Chapter 2, the resistor is a passive element with positive resistance. However, by means of dependent sources we may create circuits exhibiting negative resistance, as we shall see (Exercise 3.2.3).

3.2   CIRCUITS WITH DEPENDENT SOURCES

Circuits containing dependent sources are analyzed in the same manner as those without dependent sources. All the tools introduced in Chapter 2 may be applied: the Kirchhoff current and voltage laws, Ohm's law for resistors, series–parallel equivalents, Thevenin–Norton equivalents, and voltage and current division. In Chapter 4 we develop systematic strategies for using these tools, following a step-by-step plan that will permit us to solve for any variable in any circuit. For now, we are content to use the tools of circuit analysis on an ad hoc basis and limit our treatment to relatively small circuits.

Example 3.1

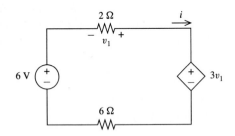

FIGURE 3.4   Dependent source example.

Find the current i in the circuit of Fig. 3.4. The dependent source is a VCVS with voltage gain of $+3$ controlled by the voltage v_1.

Applying Kirchhoff's voltage law around the circuit, we have

$$-v_1 + 3v_1 + 6i = 6 \qquad (3.1)$$

and by Ohm's law

$$v_1 = 2(-i) = -2i \qquad (3.2)$$

Using (3.2), we may eliminate v_1 from (3.1), resulting in

$$2(-2i) + 6i = 6$$

or $i = 3$ A. Thus the dependent source has complicated matters only to the extent of requiring the extra equation (3.2).

Let's use this example to consider further the significance of the voltage gain μ. With voltage gain of 3, we have determined the current i to be 3 A. If this gain were doubled, we would just replace $3v_1$ by $6v_1$ in (3.1) and repeat the subsequent steps to get $i = -1.5$ A. It was stated previously that the controlled source parameter (μ, β, r, or g) scales the size of the controlled variable per unit controlling variable. This does not imply that the circuit currents and voltages scale by the same ratio. Doubling the voltage gain in this example changed i from 3 A to -1.5 A, while the controlled voltage (VCVS terminal voltage) was reversed in sign but unchanged in magnitude. What has doubled is only the *ratio* of the controlled voltage per unit controlling voltage, not any particular current or voltage.

EXERCISES

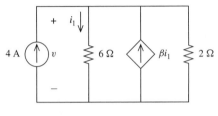

EXERCISE 3.2.2

3.2.1. In the circuit of Fig. 3.4, add an 18-Ω resistor in parallel with the given 6-Ω resistor and find i, v_1, and the resistance seen by the independent source (i.e., the resistance that when attached across the independent source in place of the rest of the circuit would yield the same response i).
Answer 12 A; -24 V; 0.5 Ω

3.2.2. Find the current gain β that results in $v = 8$ V.
Answer 1

3.2.3. In the circuit of Exercise 3.2.2, change the 2-Ω resistor to 12 Ω and find the resistance seen by the independent source for $\beta = 2$ (see Exercise 3.2.1 for a definition of "resistance seen"). Note that a negative resistance is possible only when a dependent source is present.
Answer -12 Ω

3.3 OPERATIONAL AMPLIFIERS

A dependent source has the important capability of generating a current or voltage that is an *amplified* version of its controlling variable. One practical circuit element embodying this capacity is the *operational amp (op amp)*, defined as *an electronic device that under proper circumstances behaves like a voltage-controlled voltage source (VCVS) with very high gain*. The circuit symbol for the op amp is shown in Fig. 3.5(a). The terminal

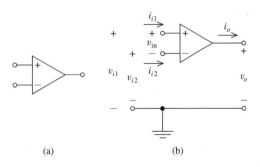

(a)   (b)

FIGURE 3.5 Operational amplifier (op amp): (a) circuit symbol; (b) currents and voltages.

marked by a plus sign in the circuit symbol is called the *noninverting input*, that by the minus sign the *inverting input*, and the unmarked terminal is the *output*. The voltages at these three nodes, v_{i1}, v_{i2}, and v_o, respectively, are all defined relative to *ground*, a node that is usually omitted from the op amp circuit symbol but is part of every op amp circuit. The op amp input voltage, or voltage across the op amp input terminals, is

$$v_{\text{in}} = v_{i1} - v_{i2}$$

Sometimes a fourth wire is added to the circuit symbol as shown in Fig. 3.6 to show the connection to the ground node explicitly. We will adopt the three-wire convention of Fig. 3.5(a) when drawing op amps in circuits.

Since the op amp is an electronic device, a power supply of the right ratings must be correctly connected to the op amp in order to energize it. This is one of the "proper circumstances" required in the definition to achieve useful op amp behavior, and these considerations will be briefly discussed in Section 3.7. Although well-behaved op amp circuits cannot be physically built and operated without due consideration of how these proper circumstances can be assured, the behavior of op amps as linear circuit elements may be analyzed assuming that this has been done. A satisfactory discussion of these requirements and how to meet them must be deferred to an electronics course, which often follows basic circuits courses such as this. For the present purpose of understanding how op amps behave in linear circuits, *we will assume that the proper environment for this behavior has been established in all op amp networks we will analyze.* In particular, since power supplies help establish these circumstances but otherwise do not influence linear op amp behavior, they will generally be omitted from op amp circuit diagrams.

Under the proper operating conditions, an op amp presumably behaves like a VCVS with high gain. The simplest circuit model for an op amp, called the *ideal voltage amplifier* model and shown in Fig. 3.7, consists of nothing more than a VCVS and some connecting wire. Note that this model shows that v_o, the op amp output voltage, is controlled by v_{in}, the difference between the op amp input terminal voltages. The VCVS voltage gain A dictates the degree of amplification of the input voltage in passage through the op amp (ratio of v_o to v_{in}). A is frequently referred to as the *open-loop gain* of the op amp. This nomenclature is related to the concept of feedback discussed in the next section. As we shall see, op amps are frequently used in configurations in which the op amp output is connected back to its input. Such circuits are called *closed loop* and

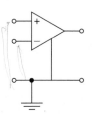

FIGURE 3.6 Alternative circuit symbol for the op amp.

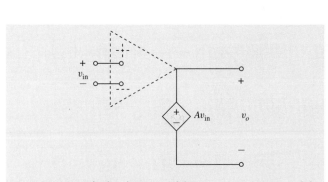

FIGURE 3.7 Ideal voltage amplifier op amp circuit model.

in general have different gain (ratio of output to input) than the open-loop op amp (no feedback), whose open-loop gain is simply the VCVS voltage gain A.

To use this model, we simply replace the op amp circuit symbol by the model and apply the methods of Section 3.2 to analyze the resulting circuit.

Example 3.2

Suppose that we connect an independent source across the input terminals of an op amp, and a load resistor between output and ground. The source has a Thevenin equivalent voltage of 1 mV and Thevenin equivalent resistance of 10 kΩ. The load resistance is 1 kΩ, and the op amp has an open-loop gain of 100,000. Find the source current i_s and load voltage v_1.

Examining Fig. 3.8(b), there can be no current ($i_s = 0$ A) through the 10-kΩ resistor since it is in series with an open circuit. Then, by KVL, $v_{in} = 0.001$ V. Thus the voltage of the VCVS is $100,000 \times 0.001 = 100$ V, which by KVL is also v_1. We see that the source voltage has been amplified by the open-loop gain and transferred across to the load. Note also that the power delivered to the load did not come from the independent source, which produces no power since $i_s = 0$ A. It must have come from the op amp or, more specifically, from the power supply energizing the op amp, even though this necessary source of power is not shown explicitly.

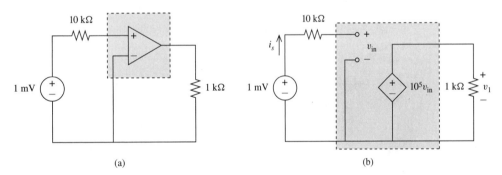

FIGURE 3.8   Circuit for Example 3.2: (a) circuit modules; (b) ideal voltage amp model substitution.

Reflection upon the ideal voltage amplifier model of Fig. 3.7 reveals some apparent shortcomings. First, the model predicts that an op amp draws no current at all from the input source (the input terminals are open circuited) and thus no power, yet performs its job of amplification. But basic physics suggests that any amplifier must draw at least a little power from its source in order to sense the signal it is supposed to amplify. Communication always requires some minimal expenditure of power by the source in order to transmit the message, yet the model predicts that no power is needed. Second, an equally questionable prediction is apparent on the output side of the model. If a load resistor is applied across the output, by decreasing its resistance further and further the current and thus the power delivered to the load by the op amp may be made arbitrarily large (the load voltage is fixed by the input voltage and open-loop gain, while power to the load is v_1^2/R). Yet every device in the real world has some finite maximum power capability.

An *improved op amp model*, which remedies these shortcomings by the addition of two resistors to the ideal voltage amplifier model, is shown in Fig. 3.9. R_i is called the *input resistance* and R_o the *output resistance*. The ideal voltage amplifier model may be seen as a special case of the improved model with $R_i =$ infinity and $R_o = 0$. By requiring these values to be both nonzero and finite in the improved model, we guarantee that some power will be expended by the source and that a finite maximum current and power will be developed across the load, no matter how small the load resistance is. Typical values for common op amps are $R_i = 1$ MΩ and $R_o = 30$ Ω. Just as with the previous model, the improved model is used by substituting it for the op amp circuit symbol in the circuit to be analyzed.

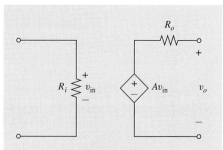

FIGURE 3.9 Improved op amp model.

Example 3.3

We will use the improved op amp model with $R_i = 1$ MΩ, $R_o = 30$ Ω, and $A = 10^5$ to determine v_2 in the circuit shown in Fig. 3.10(a). After replacement, the circuit is as shown in Fig. 3.10(b). To simplify analysis, everything to the left of the dashed line is first converted to its Thevenin equivalent. The open-circuit voltage at terminals a and b is, by voltage division,

$$v_{oc} = v_T = \frac{60 \times 10^3}{90 \times 10^3} v_1 = \tfrac{2}{3} v_1$$

and the resistance looking into terminals ab with source v_1 killed is the parallel equivalent of the 60-kΩ and 30-kΩ resistors:

$$R_T = \frac{(30 \times 10^3)(60 \times 10^3)}{90 \times 10^3} = 20 \text{ k}\Omega$$

Substituting the Thevenin equivalent circuit, we have the circuit of Fig. 3.11. Calling the loop current i, by KVL around the loop

$$2.02 \times 10^6 \, i + 10^5 v_{in} - \tfrac{2}{3} v_1 = 0$$

But since $i = v_{in}/10^6$, this may be solved for v_{in}:

$$v_{in} = \tfrac{2}{3} \times 10^{-5} v_1$$

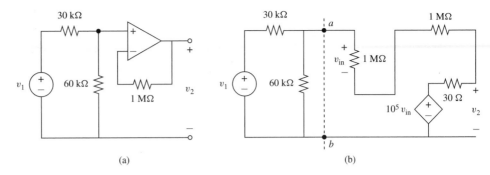

FIGURE 3.10 (a) Circuit for Example 3.3; (b) after replacement of the op amp by the improved op amp model.

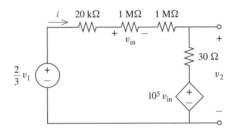

FIGURE 3.11 Circuit of Fig. 3.10 after Thevenin equivalent substitution.

Finally, since

$$v_2 = 10^5 v_{in} + 30i$$

we have the desired result:

$$v_2 = \tfrac{2}{3}v_1$$

Which op amp model should be used in a given problem? As in any model-based analysis, there is always some distance between physical objects and each of their models. An unending sequence of ever more accurate and more complex models may be invoked to represent a given object. The best model to use is not necessarily the most accurate, but rather the simplest one with the accuracy required for a given purpose. For instance, Ohm's law itself is only a simple, low-order model for the current–voltage behavior of a typical physical resistor. No physical resistor obeys Ohm's law exactly, and thus the predictions of Ohm's laws can be improved by using more complex models. More accurate models are available over specific ranges of current and voltage for specific types of resistors, but are unnecessary for most circuit analysis purposes and certainly are more difficult to use. We use Ohm's law because it is simple, yet for many physical resistors is quite accurate over a relatively broad range of currents, voltages, and temperatures.

The simpler ideal voltage amplifier model should be used in those settings where its predictions do not differ meaningfully from those of the the improved model. In Example 3.2, had we used the improved model with $R_i = 1$ MΩ, $R_o = 30$ Ω and the same $A = 100,000$ as before, by a voltage divider calculation in the input loop the input voltage is

$$v_{\text{in}} = \frac{0.001 R_i}{R_i + 10,000} = 0.00099 \text{ V} \tag{3.3}$$

compared to 0.001 V previously, while a second voltage divider applied to the output loop shows that

$$v_1 = \frac{(99.0)50000}{50,000 + R_o} = 98.9 \text{ V} \tag{3.4}$$

No current or voltage has changed by more than a fraction of 1%, which for most purposes means that the simple model is accurate enough. As we progress in our knowledge of op amp circuit analysis, we will develop guidelines for model selection. For now, the model to use will always be specified in each problem requiring analysis of an op amp circuit.

The two models introduced in this section are simple to use and in many circumstances predict the actual behavior of physical op amps quite accurately. We should never forget, though, that a model is different from the device it is modeling. A physical op amp is constructed out of many transistors, resistors, and other elements. Figure 3.12 is

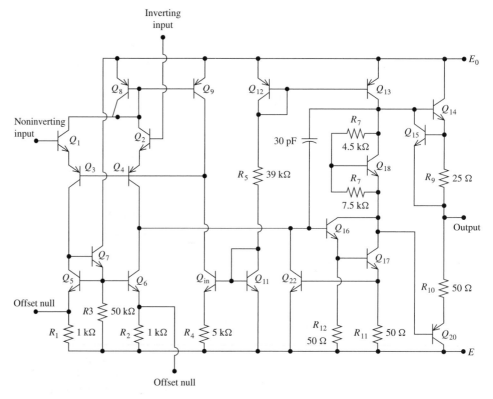

FIGURE 3.12 741 op amp. (Courtesy of Fairchild Semiconductor Corp.)

the circuit diagram for the 741 op amp, one of the most popular inexpensive low-power op amps. It is shown to illustrate the distance between a simple model, such as Figs. 3.7 and 3.9, and the circuit being modeled. This circuit diagram is complicated because it is not a simple matter to create a device that under the proper circumstances behaves like a voltage-controlled voltage source of high gain. Considerable circuit complexity is required to design a device calculated to behave so simply.

EXERCISES

3.3.1. Determine the voltages v_a and v_b. Use the ideal voltage amplifier model with open-loop gain $A = 100{,}000$.

Answer 0.200 V; 2.00 V

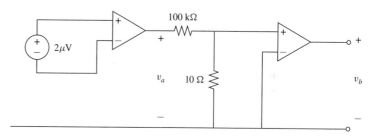

EXERCISE 3.3.1

3.3.2. Repeat Exercise 3.3.1 with the improved op amp model using $R_i = 1$ MΩ, $R_o = 30$ Ω. Are the voltages within 1% of Exercise 3.3.1? Within 0.1%?

Answer 0.200 V; 2.00 V; yes; yes

3.3.3. Using the circuit shown in Exercise 3.31 and the improved op amp model of Exercise 3.3.2, suppose that a load resistor R_1 was connected across v_b. For what value of R_1 would the current delivered to R_1 be a maximum? What would that maximum current be?

Answer 0 Ω; $\frac{1}{15}$ A

3.3.4. Redo Example 3.3 using the ideal voltage amplifier model for the op amp. Do the results agree with the example in which the improved op amp model was used?

Answer $v_2 = \frac{2}{3}v_1$; yes

3.4 ROLE OF NEGATIVE FEEDBACK

Typical op amps may have open-loop gains in the range of 10^5 to 10^6 or more, yet relatively few circuits require such large voltage amplification. Why is the op amp required to behave like a VCVS of very high gain, since this gain is seldom used to create such extreme voltage amplification? To understand how open-loop gain can be converted to other benefits no less important to successful circuit design than raw amplification, we will consider the concept of negative feedback.

Negative feedback is said to exist within an electrical circuit, or dynamic system of any kind, if *any change of the output has the algebraically opposite effect on the input.* In a negative feedback system, an increase in the algebraic value of the output (more positive or less negative) yields a decrease in the algebraic value of the input (less positive or more negative), and vice versa. Consider an athlete running a marathon. Define as input the concentration of oxygen in the blood and as output the athlete's resulting running speed. If at some point it should be decided to pick up the pace (increased output), oxygen will be consumed more rapidly and blood oxygen levels will decrease in turn (decreased input). If, instead, the pace is slowed, less oxygen is burned per unit time and its blood concentration will increase (increased input) due to the decreased pace (decreased output). This is an example of a negative feedback system. Changes in the output of a negative feedback system lead to opposite input changes, which tend to return the output toward its initial value. By this process, negative feedback systems tend to resist change and thereby to *stabilize* their own outputs.

Negative feedback may be achieved in an op amp circuit by supplying a current path between the output terminal and the inverting input terminal. Consider the circuit of Fig. 3.13(a), called a *voltage follower circuit* for reasons that will soon be apparent. Replacing the op amp by its ideal voltage amplifier model as in Fig. 3.8(b), KVL around the left loop yields

$$-v_1 + v_{\text{in}} + Av_{\text{in}} = 0 \tag{3.5}$$

or $v_{\text{in}} = v_1/(A+1)$. The right loop shows that $v_2 = Av_{\text{in}}$. Eliminating v_{in} from these last two equations and solving for v_2 yields the result

$$v_2 = \frac{A}{A+1}v_1 \tag{3.6}$$

First let us verify that this is truly a negative feedback circuit. If the op amp output voltage v_2 were somehow to increase slightly, the voltage at the inverting input terminal v_{i2} would be increased by the same amount, while the voltage at the noninverting input

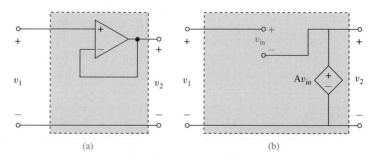

(a)   (b)

FIGURE 3.13 Voltage follower: (a) circuit diagram; (b) using ideal voltage amplifier model.

would be unaffected (it is determined solely by v_1). Since $v_{in} = v_{i1} - v_{i2}$, the increase in the op amp output v_2 will indeed result in an algebraic decrease in the op amp input v_{in}, the definitive sign of negative feedback.

Next we compare the behavior summarized in (3.6) with the op amp circuits of Section 3.3 in which negative feedback was not applied. When the voltage v_1 was applied directly across the op amp input terminals with no feedback, the op amp output voltage v_2 was Av_1. With negative feedback the ratio of output v_2 to circuit input v_1, or the *voltage transfer ratio*, has dropped from A to $A/(A+1)$. For $A = 100,000$, this is a drop from a voltage transfer ratio of 100,000 to just a bit less than unity. Indeed, for any reasonably large value of A, the voltage transfer ratio is quite close to unity, whence the name *voltage follower* (the output v_2 "follows" or matches the input v_1 without amplification or any other changes).

By making the negative feedback connection, it appears that we have thrown away the $A = 100,000$ open-loop op amp gain, instead settling for a voltage transfer ratio (with feedback) of approximately unity. What have we gained in compensation for the tremendous reduction in available amplification? Suppose that over time, as the circuit ages, the open-loop gain of our op amp declines 10%, from 100,000 to 90,000. In the circuit lacking negative feedback, the voltage transfer ratio will change between the same two values, or change 1 part in 10. In our voltage follower, $A/(A+1)$ will go from 0.999990 to 0.999989 as A declines from 100,000 to 90,000, a change of 1 part in 10^6. Thus, at the cost of a five-order-of-magnitude reduction in amplification, we have gained a five-order-of-magnitude increase in the constancy (decrease in the sensitivity) of the circuit's amplification factor.

Constancy of gain is essential in many applications and useful in most others. Using modern integrated-circuit technology, it is easy to build an op amp with high open-loop gain and do so inexpensively, but very difficult to regulate the value of the gain with high precision. Suppose that two op amps coming off the production line have open-loop gains that differ by 10%. If otherwise identical circuits in which these op amps are used have outputs that differ by 10% also, the performance of the resulting electronic end product could not be guaranteed to any degree of precision. By using op amps only in stabilizing negative feedback configurations, the copy-to-copy end circuit variations can be kept very low even when the copy-to-copy op amp variations are rather high. Other gain variations compensated by negative feedback include those due to temperature, aging of other circuit elements connected to the op amp, and variable load resistance. Negative feedback is also essential to decrease undesirable variations of gain with frequency, that is, increase the stage bandwidth, as discussed in Chapter 14.

In summary, *using negative feedback trades raw gain for constancy of gain.* The resulting circuits are highly insensitive to the numerical value of the open-loop op amp gain A, provided only that A can be counted on to be reasonably large. The voltage transfer ratios across these circuits may then be set by selecting the circuit type and the values of any external elements (e.g., resistors as in Fig. 3.15), matters much more easily controlled by the circuit designer than the open-loop op amp gains. Negative feedback configurations are sometimes referred to as *closed-loop circuits*, since the feedforward path through the op amp and the feedback path together form a closed loop. Because of its extreme usefulness in creating reliable circuit designs, all op amp circuits analyzed in the remainder of this book are of the negative feedback type.

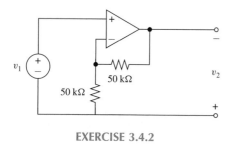

3.4.1. Determine the voltage transfer ratio v_2/v_1 of a voltage follower using the improved op amp model with $A = 100{,}000$, $R_i = 1$ MΩ, and $R_o = 30$ Ω and with a load resistor of $R_1 = 100$ Ω connected across v_2. Is it within 1% of that computed using the ideal voltage amplifier model?
    Answer 1.00; yes

3.4.2. Determine the voltage transfer ratio v_2/v_1. Use the ideal voltage amplifier model.
    Answer $2A/(A + 2)$, or in the high open-loop gain limit, $v_2/v_1 = 2$

3.5   OP AMP BUILDING BLOCK CIRCUITS

In this section we present several simple op amp circuits that are frequently used as modules or *building blocks* in the design of more complex circuits. Each building block performs one basic operation, such as the amplification of its input or the addition of two or more inputs. By interconnecting these building blocks, more complex results may be built up, in the same way that an arbitrarily complicated mathematical expression can be evaluated by the simple step-by-step application of add, subtract, multiply, and divide operations or a wall built by properly stacking building blocks made of brick or stone.

The first building block circuit is the *voltage follower* shown in Fig. 3.14. For this and the other building blocks to be described in this section, what the circuit "does" is summarized in its voltage transfer equation. *The voltage transfer equation specifies the output voltage in terms of the input voltage (or voltages) when the output port is open circuited.* The voltage-follower circuit was analyzed in Section 3.4, where the voltage transfer equation was found to be simply $v_2 = v_1$. Thus what the voltage follower does is produce an output voltage that equals (follows) its input voltage. This may seem like a rather modest achievement, but as we will see in Section 3.6, voltage followers are of the utmost practical benefit. Note that the input voltage to this circuit, v_1, should not be confused with the input voltage to the op amp, v_{in}. The voltage transfer equation and corresponding voltage transfer ratio (ratio of output voltage to input voltage, v_2/v_1

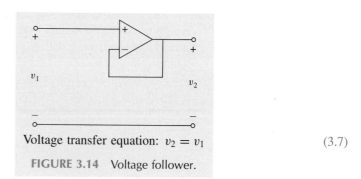

Voltage transfer equation: $v_2 = v_1$ (3.7)

FIGURE 3.14   Voltage follower.

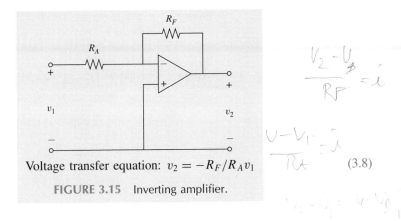

Voltage transfer equation: $v_2 = -R_F/R_A v_1$ (3.8)

FIGURE 3.15 Inverting amplifier.

$\dfrac{V_2 - V_B}{R_F} = i$

$\dfrac{V - V_1}{R_A} = i$

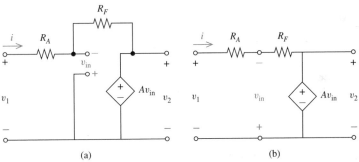

(a)

(b)

FIGURE 3.16 Analyzing the inverting amplifier: (a) using the ideal voltage amplifier model; (b) redrawn for clarity.

in Fig. 3.14) are defined in terms of the input to and output from the circuit, not the input to the op amp that is contained in the circuit.

The voltage transfer equation for the voltage follower was derived using the ideal voltage amplifier model. The same result may also be derived using the improved op amp model of Fig. 3.9, as shown in Exercise 3.5.1. For each building block, use of the ideal voltage amplifier model and use of the improved op amp model leads to the same voltage transfer equation in the limit as A goes to infinity (which we use to define the voltage transfer function since we know that A will be very high for practical op amps). Because analysis is a bit more laborious using the improved op amp model with its two additional resistors, but the results do not differ in the high-op-amp-gain limit from those derived using the simpler ideal voltage amplifier model, we will use the latter to study each building block circuit.

The second building block is the *inverting amplifier* of Fig. 3.15. Note that this circuit incorporates negative feedback, as evidenced by the current path from the output back to the inverting input. Indeed, all the building blocks will incorporate negative feedback. Replacing the op amp by its ideal voltage amplifier model results in the circuit of Fig. 3.16(a).

Examining Fig. 3.16(b), we see that

$$v_{\text{in}} = -A v_{\text{in}} - R_F i$$

Solving for i and using the fact that $v_2 = Av_{in}$ yields

$$i = \frac{-(A+1)}{R_F} v_{in} = \frac{-(A+1)}{R_F} \frac{v_2}{A} \tag{3.9}$$

Applying KVL around the outer loop gives

$$v_2 = v_1 - (R_A + R_F)i \tag{3.10}$$

Substituting (3.9) into (3.10) to eliminate i yields

$$v_2 = v_1 + \frac{R_A + R_F}{R_F} \frac{A+1}{A} v_2 \tag{3.11}$$

Gathering v_2 terms and solving for v_2 gives the result

$$v_2 = \frac{-R_F}{R_A + 1/A(R_A + R_F)} v_1 \tag{3.12}$$

In the high-op-amp-gain limit as A becomes arbitrarily large, we have the voltage transfer equation for the inverting amplifier:

$$v_2 = \frac{-R_F}{R_A} v_1 \tag{3.13}$$

From (3.13) we see that the voltage transfer ratio is always negative for the inverting amplifier, whence its name. This building block circuit functions to create an inverted, scaled copy of the input with the scale (amplification) factor set by the designer's choice of external resistances R_F and R_A. Just as with all other building blocks, the voltage transfer equation shows no dependence on open-loop gain A, a consequence of our insistence on negative feedback.

The next building block circuit is the *inverting summer* of Fig. 3.17. It differs from the inverting amplifier by having more than one input. To analyze this circuit, we first convert it to an equivalent inverting amplifier and then use the results just derived for this circuit. The Thevenin equivalent of the circuit to the left of the dashed line in Fig. 3.18(a) is conveniently evaluated by first replacing each series combination of voltage source and resistor by its Norton equivalent and then adding current sources and combining parallel resistors. The result is shown in Fig. 3.18(b).

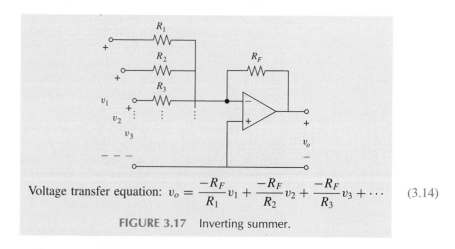

Voltage transfer equation: $v_o = \dfrac{-R_F}{R_1} v_1 + \dfrac{-R_F}{R_2} v_2 + \dfrac{-R_F}{R_3} v_3 + \cdots$ (3.14)

FIGURE 3.17   Inverting summer.

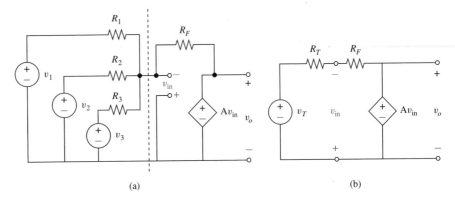

FIGURE 3.18 Analyzing the inverting summer: (a) using ideal voltage amplifier model; (b) after Thevenin transformation.

In Fig. 3.18(b),

$$R_T = R_1 || R_2 || R_3 = \frac{1}{1/R_1 + 1/R_2 + 1/R_3} \tag{3.15}$$

$$v_T = R_T \left(\frac{v_1}{R_1} + \frac{v_2}{R_2} + \frac{v_3}{R_3} \right) \tag{3.16}$$

To fix ideas, exactly three inputs have been shown in Fig. 3.18(a) and used in the calculations above. Figure 3.18(b) contains a single-input inverting amplifier with input voltage v_T and voltage transfer ratio $-R_F/R_T$. Applying voltage transfer equation (3.13) for this circuit with input $v_1 = v_T$ and input resistance $R_A = R_T$ yields the voltage transfer equation for the three-input case, or in general (by adding the ellipsis, "...") the voltage transfer equation (3.14). The inverting summer output is the sum of its inverted and scaled inputs.

If it is desired to amplify without sign change, the *noninverting amplifier* shown in Fig. 3.19 may be used. Replacing the op amp with its ideal voltage amplifier model results in Fig. 3.20. Examining the output circuit, R_F and R_A are in series across v_2, or

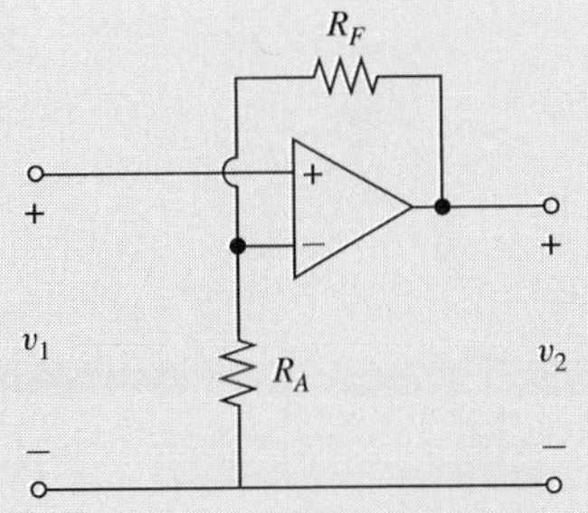

Voltage transfer equation: $v_2 = \left(1 + \dfrac{R_F}{R_A} \right) v_1$ (3.17)

FIGURE 3.19 Noninverting amplifier.

by voltage division

$$v_{R_A} = \frac{R_A}{R_A + R_F} v_2 \tag{3.18}$$

KVL around the left loop yields

$$v_1 = v_{\text{in}} + v_{R_A} \tag{3.19a}$$

Substituting v_2/A for v_{in} and (3.18) for v_{R1} gives

$$v_1 = \frac{v_2}{A} + \frac{R_A}{R_A + R_F} v_2 \tag{3.19b}$$

The first term goes to zero in the high-op-amp-gain limit, leaving the voltage transfer equation (3.17). For instance, $R_F = 10$ kΩ, $R_A = 5$ kΩ results in a voltage transfer ratio of $+3$. Note that whereas the inverting amplifier has no restriction on voltage transfer ratio magnitude, the noninverting voltage transfer ratio must be greater that unity (for passive resistors with nonnegative resistance). To achieve values less than unity, a voltage should first be divided (see Exercise 3.5.2) before being input to a noninverting amplifier.

The *noninverting summer* of Fig. 3.21 may be analyzed exactly as the previous (inverting) summer. First convert to an equivalent single-input circuit by Thevenin

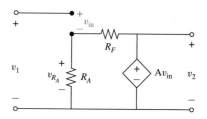

FIGURE 3.20   Substituting the ideal voltage amplifier model.

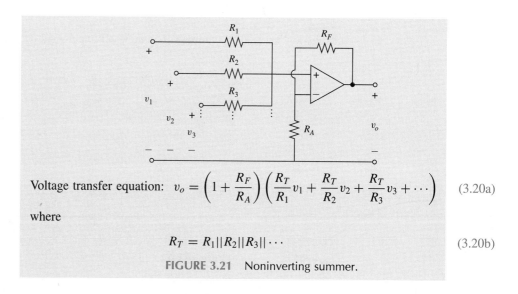

Voltage transfer equation: $v_o = \left(1 + \dfrac{R_F}{R_A}\right)\left(\dfrac{R_T}{R_1} v_1 + \dfrac{R_T}{R_2} v_2 + \dfrac{R_T}{R_3} v_3 + \cdots\right)$ (3.20a)

where

$$R_T = R_1 || R_2 || R_3 || \cdots \tag{3.20b}$$

FIGURE 3.21   Noninverting summer.

transformation and then apply the voltage transfer equation for the single-input case already determined. By (3.17), the voltage transfer equation for the noninverting summer is

$$v_o = \left(1 + \frac{R_F}{R_A}\right)v_T \qquad (3.21)$$

with v_T given by (3.16). While (3.16) is specific for three inputs, the ellipsis $\cdots$ in (3.20) once again indicates the immediate generalization to any number of inputs v_1, v_2, $\ldots$. This building block has the effect of scaling and summing its inputs (without sign inversion).

These five circuits constitute the bulk of the building blocks we will use. One additional one will be introduced later. It is somewhat surprising, given the simplicity of these circuits, that they are quite sufficient to build up many circuits of considerable practical significance. We need only consider how they may be interconnected, as in Section 3.6, and allow inductors and capacitors in the building blocks to be ready to understand and to design many interesting circuits such as filters and analog computers.

EXERCISES

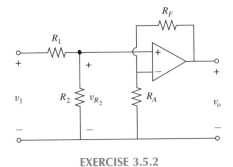

EXERCISE 3.5.2

3.5.1. Using the improved op amp model (Fig. 3.9) with $R_i = 1\ M\Omega$ and $R_o = 30\ \Omega$, first determine the output v_2 of the voltage follower (Fig. 3.14) in terms of op amp gain A and then in the high-op-amp-gain limit. Does the voltage transfer equation agree with that calculated in (3.7)?

$$Answer \quad \frac{A + 0.00003}{A + 1.00003}v_1;\ v_1;\ \text{yes}$$

3.5.2. This exercise shows how to get a positive voltage transfer ratio of less than 1. Using the ideal op amp model, determine the voltage v_{R2}. Then determine v_o, noting that v_{R2} is the input to the noninverting amplifier. Specify values for the resistances yielding an overall voltage transfer ratio of $\frac{1}{3}$ (answer not unique).

$$Answer \quad \frac{R_2}{R_1 + R_2}v_1;\ \frac{R_2}{R_1 + R_2}\left(1 + \frac{R_F}{R_A}\right)v_1;\ R_F = R_A;\ R_1 = 5R_2$$

3.5.3. Design a noninverting summer whose voltage transfer equation is $v_o = 2v_1 + 5v_2$.

Answer In Fig. 3.21, pick any R_A and R_2; then set $R_F = 6R_A$ and $R_1 = 2.5R_2$.

3.6 INTERCONNECTING OP AMP BUILDING BLOCKS

Op amp building blocks are simple modules that may be connected to produce more complex circuit behaviors. In this section we consider one particularly useful way to interconnect these modules and how to predict the behavior of the interconnected circuit from that of the modules that are its parts.

A *port* in a circuit is a pair of wires to which another subcircuit may be attached. Op amp building block circuits each have one output port and at least one input port. Circuits containing a single input and a single output port, for example the voltage follower and inverting and noninverting amplifiers of Section 3.5, are called *two-port circuits*. Those with additional ports, such as the inverting and noninverting summers, are *n-port circuits*.

The most direct way to connect a pair of two-port circuits is to tie the output port of one to the input port of the other. This is called the *cascade interconnection* of two-ports. Cascading two (or more) two-ports results in an interconnected circuit that is also a two-port, as shown in Fig. 3.22. Here v_1 is the input voltage to the overall cascaded circuit and v_3 the output voltage.

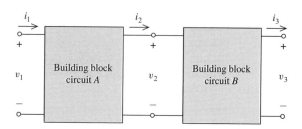

FIGURE 3.22 Cascade interconnection of two op amp building block units.

Every op amp building block is characterized by its voltage transfer equation, which specifies the output voltage as a function of the input voltage or voltages under the condition that the output port is open circuited. When op amp building blocks are interconnected, it is the voltage transfer function across the entire interconnected circuit that describes its behavior. Thus for two-port op amp building blocks in cascade, we will determine the overall voltage transfer equation.

Referring to Fig. 3.22, take the output port (v_3) to be open circuited as required by definition of the voltage transfer equation. Applying the voltage transfer equation for building block B, $v_3 = K_B v_2$, where K_B is the voltage transfer ratio. Then, to find the voltage transfer equation $v_3 = K_C v_1$ for the overall cascade, we need only express v_2 in terms of v_1. If $i_2 = 0$, the output of building block A is also open circuited, and v_2 is found by applying A's voltage transfer equation $v_2 = K_A v_1$. The overall result is then $v_3 = K_A K_B v_1$; that is, $K_C = K_A K_B$.

Example 3.4

Find the voltage transfer equation (input v_1, output v_3) for the circuit of Fig. 3.23. Use the ideal voltage amplifier model for the op amps.

The rightmost building block circuit is a noninverting amplifier with $v_3 = (1 + R_F/R_A)v_2$ or $v_3 = 3v_2$. $i_2 = 0$ since the the ideal voltage amplifier model for its op amp has an open-circuited input. The leftmost building block is an inverting amplifier with voltage transfer ratio $-R_F/R_A = -4$. Since its output port is open circuited ($i_2 = 0$), v_2 may be exactly evaluated using the voltage transfer equation for this building block, $v_2 = -4v_1$. Then, combining, $v_3 = 3(-4v_1) = -12v_1$. The voltage transfer ratio for the cascade is the product of the individual voltage transfer ratios.

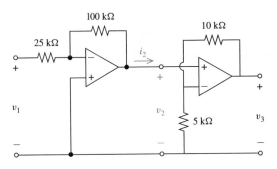

FIGURE 3.23   Circuit for Example 3.4.

Suppose that i_2 is not equal to zero in Fig. 3.22. The ideal voltage amplifier model of the op amp predicts that the output voltage of each building block must be independent of i_2 since it is just the voltage across a VCVS (see Fig. 3.7), whose source function does not depend on the output current i_2. So even though open-circuit output is assumed in its definition, the voltage transfer equation continues to apply in computing the output voltage for a building block circuit even when the output is not open circuited. This analysis using the ideal voltage amplifier model suggests that connecting a load, such as another op amp building block, to the output port of a previous building block will not affect its output voltage.

Thus *for building blocks in cascade, the overall voltage transfer ratio is simply the product of the individual voltage transfer ratios*, provided only that the ideal voltage amplifier model of the op amps may be accurately used. The same principle applies to n-port op amp building blocks as well, as illustrated in the next example.

Example 3.5

Find the voltage transfer equation for the circuit of Fig. 3.24. Use the ideal voltage amplifier model for the op amps.

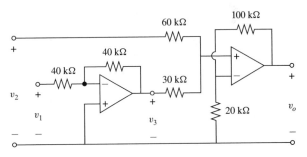

FIGURE 3.24   Circuit for Example 3.5.

The rightmost building block is a two-input noninverting summer with input voltages v_2 and v_3 and output v_o. Using the voltage transfer equation (3.17) for this building block, we have

$$v_o = \left(1 + \frac{100\text{K}}{20\text{K}}\right)(60\text{K}||30\text{K})\left(\frac{v_2}{60\text{K}} + \frac{v_3}{30\text{K}}\right)$$

where K stands for multiplication by 10^3 and $||$ is the parallel equivalent operator.

$$60K||30K = \frac{1}{1/60K + 1/30K} = 20K$$

Then

$$v_o = 2v_2 + 4v_3 \tag{3.22}$$

The remaining building block is an inverting amplifier with

$$v_3 = \frac{-40K}{40K}v_1 = -v_1$$

Combining these two results, the overall voltage transfer equation is

$$v_o = 2v_2 + 4(-v_1) = 2v_2 - 4v_1 \tag{3.23}$$

By combining building blocks of different types, we may achieve results beyond what would be possible using any one alone. This example shows how a *difference*, as opposed to an inverting or noninverting sum, of voltages may be generated by combining inverting and noninverting building blocks.

The main result of this section, that when op amp building blocks are cascaded the overall voltage transfer equation may be found simply by combining the voltage transfer equations of each individual building block, was justified using the ideal voltage amplifier model for each op amp. The result followed from the lack of dependence of the output voltage on the output current. The improved op amp model of Fig. 3.9 contains an additional resistor R_o in its output loop. The current produced by its VCVS, which supplies both the output current (i_2 in Fig. 3.22) and the current to the feedback path present in every op amp building block, must all pass through R_o. By KVL, the improved op amp model thus predicts that the output voltage will be reduced as the output current increases (their sum equals Av_{in}, which does not depend on this current). The two models seem to disagree as to whether the output voltage will remain constant, as the ideal voltage amplifier model predicts, or will in fact decrease as the output current increases as the improved model suggests.

Which model gives the right prediction? As might be anticipated in calling it an "improved" op amp model, direct measurement verifies that as the output current of any building block circuit increases, its output voltage declines as predicted by the improved model. The decline of load (output) voltage with load current is called *loading* and is quite a general phenomenon, not restricted to op amp building blocks. As explored in Exercise 3.6.1, loading is the consequence of voltage division between internal resistance and load resistance. Some degree of loading must occur whenever a source has nonzero Thevenin equivalent resistance and is connected across a finite load resistance.

Must we then abandon our main results, derived using the ideal voltage amplifier model, which has no internal resistance and hence manifests no loading? For sufficiently low levels of output current the degree of loading, that is, reduction of output voltage, is negligible and our result holds with excellent accuracy. Current levels may be kept low by keeping the resistances in the building block circuits high, which is why values of 5 kΩ and above are used in the examples and problems.

In summary, to a good approximation a cascade of op amp building blocks has a voltage transfer equation that may be computed simply by combining the individual voltage transfer equations. *For two-ports in cascade, the overall voltage transfer ratio is the product of the constituent two-port voltage transfer ratios.* Using this combining rule, circuits involving multiple building blocks may be designed to accomplish relatively complex goals and yet be analyzed easily by multiplying voltage transfer ratios. The range and versatility of the building block approach to op amp circuit design, suggested by the results of this section, will be fully realized with the introduction of capacitors and inductors into the building blocks in Chapter 7.

EXERCISES

3.6.1. Consider a source with Thevenin equivalent resistance R_T and Thevenin equivalent voltage v_T. If a load of R_L ohms is placed across the source terminals as in part (a) of the figure, (a) what is the open-circuit load voltage (voltage across R_L when R_L is infinite)? (b) Find the value for resistance R_L at which loading has reduced v_L to half of its maximum (open circuit) value. (c) If a voltage follower is inserted between source and load as shown in part (b) of the figure, show that loading is eliminated. *Buffering* of a source from a load to eliminate loading is one of the main uses of a voltage follower.

Answer v_T, R_T, V_L (no buffer) $= (R_L + V_T)/(R_L + R_T)$, but V_L (with buffer) $= V_T$. Loading reduces the output voltage by a factor equal to the voltage-divider ratio.

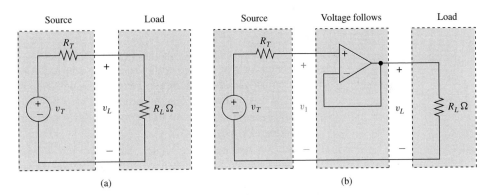

(a)   (b)

EXERCISE 3.6.1 Loading with and without a buffer between source and load: (a) direct connection; (b) buffered connection using a voltage follower.

3.6.2. Design a circuit with the voltage transfer equation $v_o = 8v_1 - 2v_2 - v_3$. Use only inverting building blocks and select all resistances to be in the range 5 to 500 kΩ.

Answer Express v_o as $v_o = -2(-4v_1) - 2v_2 - v_3$. Get the term in parentheses as the output of an inverting amplifier with input v_1 and voltage transfer ratio -4. Then route this voltage, along with the voltages

v_2 and v_3, to an inverting summer with appropriate voltage transfer ratios $(-2, -2, \text{ and } -1)$.

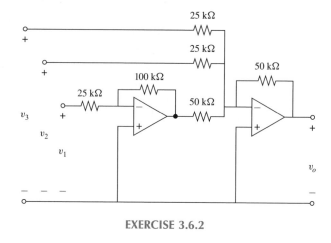

EXERCISE 3.6.2

3.6.3. Determine the voltage transfer ratio v_2/v_1.

$$Answer \quad \frac{R_2}{R_1 + R_2}\left(1 + \frac{R_F}{R_A}\right)\frac{R_4}{R_3 + R_4}$$

EXERCISE 3.6.3

3.7 PRACTICAL OP AMPS

Op amps are not merely schematic diagrams to which the laws of circuit analysis may be applied but physical devices with size and weight requiring power supplies and attention to other practical considerations. Here the issues of physical packaging and limits of performance and environment are introduced very briefly. These topics are taken up at a more satisfying level of detail in courses in electronics circuits, which typically follow a first course in circuit analysis.

Physical Packaging

The first few generations of commercial op amps, which made their appearance more than 50 years ago, were large, heavy devices fabricated from vacuum tubes and discrete RLC components. Size, cost, and appetite for power limited their use to the most critical applications, particularly the performance of key mathematical operations in predigital computer fire-control units for air defense batteries (hence the name *operational amplifier*). In 1962, Fairchild Electronics introduced the $\mu A709$, the first monolithic op amp, miniaturized onto an eggshell-thin chip of silicon just a fraction of an inch square. The sharp reductions in size and cost created by this use of integrated-circuit technology opened the door to the extensive use of op amps in a broad range of industrial and consumer electronics. Successive generations of op amps have taken advantage of advances in microelectronics, resulting in steadily improved integration, performance, and reliability, all at prices less than that of a postage stamp. Today's op amp is truly a wonder of invention and refinement and is one of the prime workhorses of modern electronic design.

Op amps are currently available in a variety of convenient packages and ratings. The popular general-purpose μA741 op amp in an 8-pin dual-in-line (DIP) package is shown in Fig. 3.25. Pins 4 and 7 are assigned to be connected to the power supply. Without the proper dc bias currents supplied by the power supply at the rated voltage levels, the transistors contained in the op amp circuit will not act to produce the VCVS with high-gain behavior that characterizes all functional op amps.

FIGURE 3.25   Op amp (the μA741 in an 8-pin DIP package). Millimeter scale.

Besides their packaging as separate chips, op amps are frequently deployed as modules contained in other integrated-circuit chips. Voltage followers are commonly used as buffers at the input and output pins of integrated circuits to prevent loading. Op amps form essential parts of analog-to-digital (AD) and digital-to-analog (DA) converters, instrumentation amplifiers, active filters, and numerous other chip-level electronic modules. It is not uncommon for dozens of op amp modules to be designed into a single very large scale integrated (VLSI) circuit chip.

Performance and Environmental Limits

The op amp was defined in Section 3.3 as an electronic device that under certain circumstances behaves like a VCVS with high gain. The circumstances that need to be established include proper environmental conditions and operation within a certain "performance envelope" or permissible range of the circuit variables. With careful attention to these constraints, the op amp, despite all its physical complexity, will indeed behave very much as predicted by its simple ideal voltage amplifier model.

Environmental conditions that need to be maintained include use of a power supply developing the right voltages and free from spikes and surges, adequate shielding from other electronic and magnetic signals, and operation in the proper temperature range. Microminiaturization, with all its benefits, puts many elements on a very small piece of real estate and may result in overheating and circuit failure. Adequate cooling of densely populated circuit boards is an important consideration in designing modern electronic systems. For very dense circuits, such as the central processing unit of supercomputers, elaborate refrigeration systems are required to carry off the heat produced.

Performance limitations include voltage, current, power, and frequency. All voltages and currents must be kept within the linear regime of operation. Typically, this means that voltages cannot exceed the power supply values; and currents must be strictly limited by rated values. If called upon to produce too much output voltage, the op amp will *clip* near its power supply values; that is, the output will be limited in amplitude by the positive and negative power supply values. If too much power is demanded, the circuit will fail to conform to its predicted linear behavior or, more ominously, may overheat and ultimately burn out. Indeed, current-limiting circuits protecting against this are built into most popular op amp designs. Op amps are also limited in the frequency band over which they may act. While more broadband specialty op amps are available, most op amp circuits are designed to operate over frequencies well below 1 MHz. The frequency dependence of op amps will be taken up in Chapter 14.

The experienced op amp circuit designer will take other practical factors into consideration as well: shock and vibration, offset currents and voltages, and slew rate limits (maximum rate of change in volts/second, for example), and it is not surprising that an electronic device such as an op amp can be made to perform best in the hands of a skilled circuit designer. What is perhaps more surprising is that well-behaved op amp circuits can be made rather easily by those with less than expert status. Using today's highly integrated, miniaturized op amps, a few inexpensive resistors and batteries are all that is really required to construct an op amp circuit that behaves with a high degree of accuracy as the ideal voltage amplifier model of the op amp predicts.

3.8 DESIGN OF SIMPLE OP AMP CIRCUITS

In Chapter 2 we considered strictly resistive circuits, those containing only resistors and independent sources. In the design section of that chapter it was noted that the versatility of resistive circuits, their range of useful applications, is limited. The addition of just one more element, the op amp, greatly expands the range of potential applications, as we shall see below. The examples presented here point to the design of electrocardiogram amplifiers for biomedical application, analog-to-digital converters in computer engineering, and an averaging circuit for signal processing applications. Many more examples could be offered as well, but it is hoped that these are persuasive of the range and versatility of the op amp in practical applications.

The design examples here illustrate uses of op amp circuits constructed using the straightforward building block cascade approach described in Sections 3.5 and 3.6. Other interconnections of op amp stages will be considered in later chapters, and elements other than op amps and resistors, which will further enhance the variety of electronic circuits that can be designed.

Design Example 3.1

The electrical potential difference $v_a(t) - v_b(t)$ measured between two electrodes secured to points a and b on the body's wall contains information on the state of health of the heart. Each time the heart beats, its mechanical contraction is triggered by electrical currents flowing through the heart. The strength and sequence of these crucial cardiac currents are encoded in the electrocardiogram (ECG) signal $v_o(t)$. We wish to design an ECG amplifier that will amplify this signal from its body surface value of approximately 2 mV peak to peak to the level of 1 V peak to peak needed to drive a stripchart recorder. The accompanying common-mode signal $v_{\text{cm}}(t) = \frac{1}{2}[v_a(t) + v_b(t)]$, which contains only noise and irrelevant signals from other body sources, should be small or nonexistent in the output.

We begin our design by noting that the desired overall voltage transfer equation for our circuit is $v_o = 500(v_a - v_b)$. A circuit realizing this voltage transfer equation will scale a 2-mV difference signal to 1 V, as required. Moreover, the addition of a constant to both v_a and v_b does not change the output of this circuit, which is sensitive only to their potential difference. Thus the common-mode signal $v_{\text{cm}}(t)$ is effectively rejected by a circuit with this difference-voltage transfer equation.

We will use a two-stage design incorporating an inverting amplifier and an inverting summer as shown in Fig. 3.26. First an inverting amplifier will create

$$v_c = \mu_1 v_a$$

where $\mu_1 = R_{F1}/R_{A1}$ is the voltage gain for this stage. Then this output, along with v_a, will be input to an inverting summer whose voltage transfer equation will be

$$v_o = -\mu_{2c} v_c - \mu_{2b} v_b$$

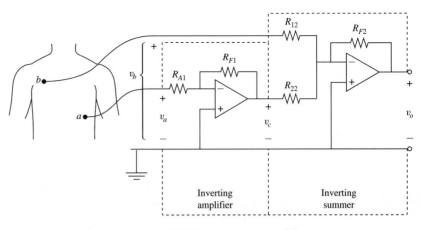

FIGURE 3.26 ECG amplifier.

where $\mu_{2b} = R_{F2}/R_{12}$ and $\mu_{2c} = R_{F2}/R_{22}$. Combining these two equations gives

$$v_o = \mu_1\mu_{2c}v_a - \mu_{2b}v_b$$

Thus if we can set $\mu_1\mu_{2c} = \mu_{2b} = 500$, we will have our design. First setting

$$\mu_{2b} = \frac{R_{F2}}{R_{12}} = 500$$

we select $R_{12} = 1$ kΩ and $R_{F2} = 500$ kΩ. Now we may divide the desired gain $\mu_1\mu_{2c} = 500$ between the two stages any way we wish. It is helpful to balance the gains so we set $\mu_1 = \mu_{2c} = \sqrt{500}$. Then since

$$\mu_{2c} = \sqrt{500} = \frac{R_{F2}}{R_{22}} = \frac{5 \times 10^5}{R_{22}}$$

we have that $R_{22} = 22.4$ kΩ. Finally, $\mu_1 = R_{F1}/R_{A1} = \sqrt{500}$, so selecting $R_{F1} = 100$ kΩ, we have fixed the value of the final resistance:

$$R_{A1} = \frac{R_{F1}}{\sqrt{500}} = 4.47 \text{ k}\Omega$$

which completes the design.

As with almost all designs, this solution is not unique. Indeed, more attractive solutions can be found, solutions that use only a single op amp, have less sensitivity to parameters, are easier to adjust, or possess a broader range of operating frequencies. But the circuit shown in Fig. 3.26 is quite capable of satisfying the simple ECG amplifier design specifications as stated.

Design Example 3.2

Our goal is to design a digital-to-analog converter (DAC). Assume that we have an 8-bit digital input signal with transistor-to-transistor (TTL) logic levels $+5$ V and 0 V, and that the full-scale analog output should range from -10 V when the digital input is all zeros (0 V at each input pin, digital value 00000000_2) to 0 V for the highest input value ($+5$ V at each input pin, digital value 11111111_2).

Since each pin represents a power of 2, the formula by which a digital value is converted to its analog value will be

$$v_a = S\left(\sum_{k=0}^{7} 2^k v_k\right) + v_B \tag{3.24}$$

where v_B is a bias voltage that sets the zero output value and S is a scale factor. In our case we wish $v_0 = v_1 = \cdots = v_7 = 0$ V to map into $v_a = -10$ V, or

$$-10 = v_B$$

so $v_B = -10$ V. We also want $v_0 = v_1 = \cdots = v_7 = +5$ V to map into $v_a = +10$ V, or

$$0 = S(255)(5) + v_B$$

which yields $S = 2/255$. Thus our voltage transfer equation is, by (3.24),

$$v_a = \left(\sum_{k=0}^{7} \frac{2^{k+1}}{255} v_k \right) - 10 \qquad (3.25)$$

This can be realized with a nine-input noninverting summer as shown in Fig. 3.27. Examining the form of the voltage transfer equation for this building block which is given in Fig. 3.21, in order that the gains on the eight successive single-bit inputs $v_7, v_6, \ldots, v_0$ each be in the ratio of 2, we must have $R_6 = 2R_7$, $R_5 = 2R_6 = 4R_7$, and so on. Then with $G_T = 1/R_T$,

$$G_T = G_7 + G_6 + \cdots + G_0 + G_B$$

$$= G_7(1 + \tfrac{1}{2} + \tfrac{1}{4} + \cdots + \tfrac{1}{128}) + G_B$$

$$= \frac{255 G_7}{128} + G_B$$

Now to keep the numbers simple let us select $G_B = 129/128\, G_7$ so that

$$G_T = 3G_7$$

Then $R_T = \tfrac{1}{3} R_7$, and with the resistive ratios already defined we have, by the voltage transfer equation in Fig. 3.21,

$$v_a = \left(1 + \frac{R_F}{R_A}\right)\left(\frac{1}{3}\right)\left(\frac{v_0}{128} + \frac{v_1}{64} + \cdots + \frac{v_7}{1} + \frac{129 v_B}{128}\right) \qquad (3.26)$$

Comparing (3.26) and (3.25), the v_0 terms will be identical if

$$\left(1 + \frac{R_F}{R_A}\right)\left(\frac{1}{3}\right)\left(\frac{1}{128}\right) = \frac{2}{255}$$

Solving the last for R_F yields

$$R_F = 2.01 R_A$$

Thus in Fig. 3.27 we set $R_A = 10$ kΩ, $R_F = 20.1$ kΩ. Since the v_0 terms in (3.25) and (3.26) are now identical and the ratios of these terms to the $v_1, \ldots, v_7$ terms in the two equations are identical, all eight of these terms must agree in the two equations. In Fig. 3.27 we show the bias voltage being produced by voltage division of a -15-V source (this is a common power supply voltage). Finally, each of the conductances $G_0, G_1, \ldots, G_6$, and G_B have been determined only as ratios compared to G_7. We may choose any convenient value for G_7 (or R_7) and the rest follow. If we select $R_7 = 1$ kΩ, then $R_6 = 2$ kΩ, $R_5 = 4$ kΩ, $\ldots$, $R_0 = 128$ kΩ, and $R_B = 128/129\, R_7 = 992\ \Omega$.

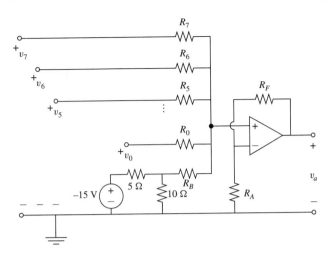

FIGURE 3.27   Eight-bit DAC.

The resulting design will convert TTL digital inputs to analog outputs in the range ± 10 V. DACs are used wherever computers must be interfaced to nondigital circuitry, for instance in computer displays such as cathode ray tubes and computer-controlled electromechanical devices such as industrial robots.

Design Example 3.3

For our final design example, we wish a circuit that will produce an output voltage which is the average of the outputs of several input voltages. Signal averagers are widely used in signal processing circuits; for instance, at the outputs of noisy transducers averaging is used to minimize the effects of random noise on repeated samples of the same input signal.

For N inputs $v_1, v_2, \ldots, v_N$, the average is

$$v_0 = \frac{1}{N}(v_1 + v_2 + \cdots + v_N)$$

This is the voltage transfer equation for a noninverting summer, each of whose gains is $1/N$. Referring to the circuit diagram for this building block shown in Fig. 3.21, in order for the gains to be all equal we require that $R_1 = R_2 = \cdots = R_N = R$, so

$$R_T = R_1 || R_2 || \cdots || R_N = \frac{R}{N}$$

Then from that figure,

$$v_o = \left(1 + \frac{R_F}{R_A}\right) \frac{1}{N}(v_1 + v_2 + \cdots + v_N)$$

Note that for any finite values for R_F and R_A, the gains on each input signal will be too large, each by the same factor. An easy way to remedy this problem is to set R_F and R_A equal, so that

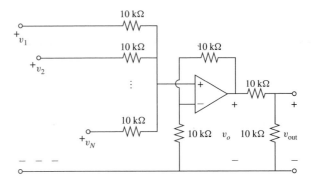

FIGURE 3.28 Signal averager circuit.

$R_F/R_A = 1$. Then v_o is too large by a factor of 2, so we voltage divide the output v_o by 2. The resulting circuit with all resistors set to a common value of 10 kΩ is shown in Fig. 3.28.

One limitation of this solution is that the output impedance of this circuit is larger than most of our op amp circuits due to the output voltage divider. If we are using the output of this circuit to drive a low-input-impedance circuit, such as a stripchart recorder, it would probably be necessary to buffer the connection with, say, a voltage follower. There are many other signal averaging circuits with features beyond those offered by our solution, but Fig. 3.28 does the job of signal averaging simply and accurately.

DESIGN EXERCISES

3.8.1. A certain photoresistor has resistance that is greater than 1 kΩ in the dark, and drops below 1 kΩ in the presence of moderate light. Design a light alarm using this photoresistor as the sensor. As the indicator of an alarm condition, use a buzzer that sounds when the voltage across it exceeds 10 V dc.

One of many possible solutions

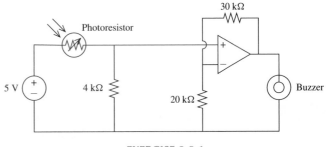

EXERCISE 3.8.1

3.8.2. Produce a DAC circuit with two outputs, each the analog-converted value of one "nibble," the top or bottom 4 bits of the 8-bit input. Assume TTL values and scale each of the two analog outputs to the inverted single-ended range 0 V (all 4 bits 0 V) to -75 V (all 4 bits $+5$ V).

One of many possible solutions

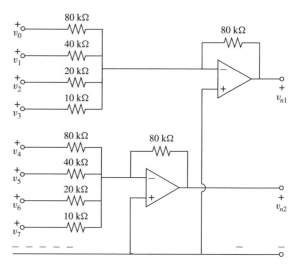

EXERCISE 3.8.2 v_O is the least significant bit, v is the most significant bit of the input byte.

SUMMARY

Dependent sources, those whose source functions depend on the value of a current or voltage elsewhere in the circuit, are used to model the behavior of active devices such as transistors and op amps. An op amp is a device which, under proper conditions, behaves like a voltage-controlled voltage source with very high gain. Op amps are among the most versatile of active devices and will be treated as a basic circuit element for the remainder of this book.

- Circuits containing dependent sources may be analyzed by treating the sources as if they were independent, then substituting the controlling variable into the dependent source function.

- The ideal voltage amplifier model for the op amp consists in an open-circuit input and voltage-controlled voltage source output.

- The improved op amp model replaces the open-circuit input with a high but finite (e.g., 1 MΩ) input resistance, and adds a low (e.g., 30 Ω) resistance in series in the output circuit. This model is more accurate but contains more elements.

- An op amp is in a negative feedback configuration when an incremental increase in its output voltage leads to a decrease in its input voltage.

- Negative feedback is necessary for useful op amp behavior in all linear circuit applications, such as the op amp building block circuits: inverting and noninverting amplifiers and summers, and the voltage follower.

- The voltage transfer equation is a linear equation relating the input voltage or voltages of an op amp circuit to its output voltage. The output-to-input voltage ratio is called the voltage transfer ratio.

- The voltage transfer ratio of two op amp building blocks connected in cascade is the product of their individual voltage transfer ratios.

The circuit diagrams for physical op amps are complicated, containing many transistors and other elements. But when properly configured, op amps can be modeled very simply and are an essential element in modern circuit design.

PROBLEMS

3.1. Draw the circuit symbols for two distinct controlled sources, each of which is equivalent to a 1-kΩ resistor.

3.2. Give examples of an active controlled source and a passive controlled source (see Problem 3.1).

3.3. For what value of transconductance of the dependent source in Problem 3.8 does a mathematical inconsistency arise?

3.4. Find v_1 and the power delivered to the 10-Ω resistor.

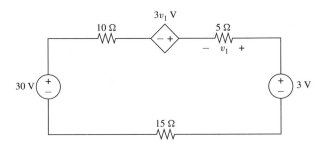

FIGURE P3.4

3.5. Find i_1.

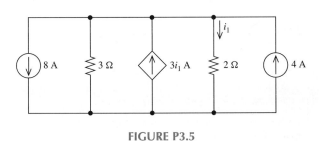

FIGURE P3.5

3.6. Find i.

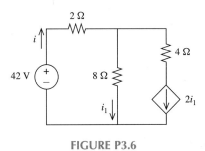

FIGURE P3.6

3.7. Find v_1.

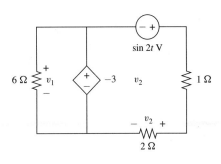

FIGURE P3.7

3.8. Which supplies more power, the independent source or the dependent source?

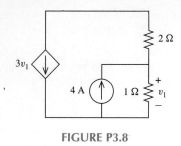

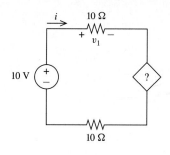

FIGURE P3.8

FIGURE P3.11

3.9. Find the Thevenin equivalent circuit.

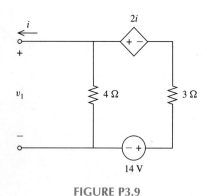

FIGURE P3.9

3.12. We wish to make $i = -1$ A. Find two different kinds of dependent sources that can do this, specifying their source functions.

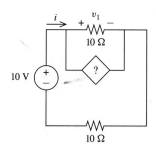

FIGURE P3.12

3.10. What single resistor is this circuit equivalent to?

3.13. Find v_1 and i_1.

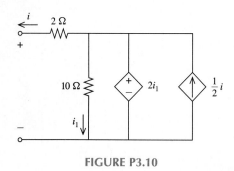

FIGURE P3.10

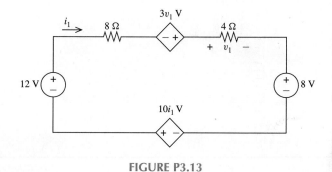

FIGURE P3.13

3.11. We wish to make $i = 1$ A. Find two different kinds of dependent sources that can do this, specifying their source functions.

3.14. Find i_1 and v if (a) $R = 4$ Ω and (b) $R = 12$ Ω.

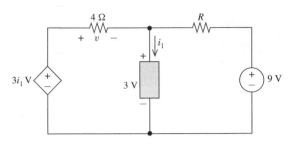

FIGURE P3.14

3.15. Find i.

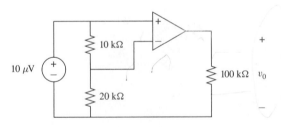

FIGURE P3.15

3.16. Find v_o. Use the ideal op amp model with $A = 100,000$.

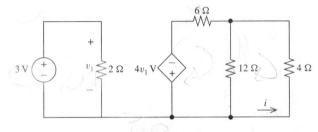

FIGURE P3.16

3.17. Repeat Problem 3.16, but use the improved op amp model with parameters $A = 100,000$, $R_i = 1$ MΩ, and $R_o = 30$ Ω.

3.18. Show that the improved op amp model of Fig. 3.9 can be redrawn using a CCCS rather than a VCVS.

3.19. Using an inverting amplifier, design a circuit with $v_2 = -9v_1$. Keep resistances in the range 5 to 500 kΩ.

3.20. Using an inverting amplifier, design a circuit with $v_2 = -\frac{1}{9}v_1$. Keep resistances in the range 5 to 500 kΩ.

3.21. Using a noninverting amplifier, design a circuit with $v_2 = 7v_1$. Keep all resistors in the range 5 to 500 kΩ.

3.22. If all the external resistors in any one of the building blocks were scaled by the same constant, what effect would

this have on the voltage transfer function? On the currents through the external resistor?

3.23. Compare the power delivered to the 20-kΩ resistor in this buffered circuit with that if the voltage follower were removed (a and b shorted together).

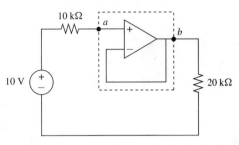

FIGURE P3.23

3.24. If we wish to supply $\frac{1}{2}$ mW of power to the 15-kΩ load, $R_F = ?$

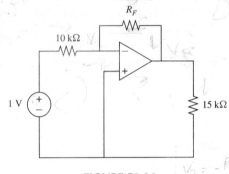

FIGURE P3.24

3.25. Using a voltage divider followed by a noninverting amplifier, design a circuit with $v_2 = +\frac{1}{7}v_1$. Can we reverse the order of these two subcircuits and get the same result?

3.26. Find k in the voltage transfer function $v_2 = kv_1$.

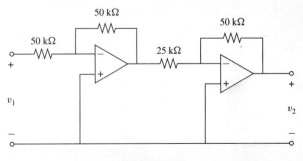

FIGURE P3.26

3.27. Find k in the voltage transfer function $v_2 = kv_1$.

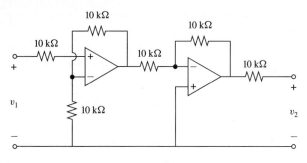

FIGURE P3.27

3.28. Show how you can create a circuit with $v_2 = 1000v_1$ by cascading three *identical* amplifier stages. Keep all R's in the range 5 to 500 kΩ.

3.29. Design a circuit with $v_3 = -2v_1 - 5v_2$. Keep all resistors in the range 5 to 500 kΩ.

3.30. Design a circuit with $v_5 = -v_1 - 2v_2 - 3v_3 + 4v_4$. Keep all resistors in the range 5 to 500 kΩ.

3.31. Find the voltage transfer function $v_3 = k_1v_1 + k_2v_2$.

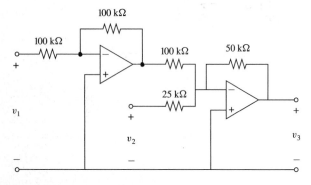

FIGURE P3.31

3.32. Design a circuit with $v_3 = 3v_1 + 2v_2$ using one noninverting summer.

3.33. Repeat Problem 3.32 using two inverting stages instead of one noninverting.

3.34. Find the voltage transfer equation (inputs v_1, v_2; output v_3).

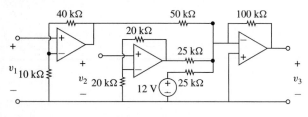

FIGURE P3.34

3.35. Design a circuit with $v_3 = 5v_1 + 4v_2$ two ways. (1) Use a single op amp, and (2) use op amps in inverting building blocks only.

Design Problems

3.36. An intensive care unit patient monitoring system has three output voltages: $v_1(t)$ measures the pulse rate, $v_2(t)$ the average blood pressure, and $v_3(t)$ the respiration rate. The relationship between these voltages and the quantities they measure are shown below.

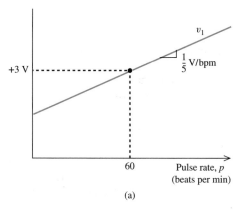

(a)

FIGURE P3.36a

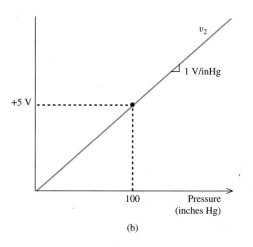

(b)

FIGURE P3.36b

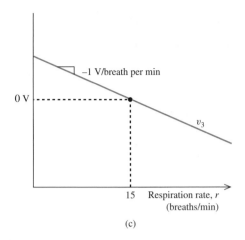

-1 V/breath per min

0 V

v_3

15   Respiration rate, r
(breaths/min)

(c)

FIGURE P3.36c

Design a circuit whose output is the alarm voltage $v_A = 2\Delta v_1 + 3\Delta v_2 + \Delta v_3$, where Δv_i is the fractional deviation of v_i from its nominal value shown. For instance, if the patient's pulse rate is 90 bpm, $\Delta v_1 = (90-60)/60 = +0.5$. The inputs to this circuit can be v_1, v_2, v_3, and ± 15-V dc sources.

3.37. A photoresistor has a dark resistance of 1 kΩ, and its resistance drops 20 Ω/lux (a unit of light intensity). Design a circuit employing two such photoresistors that will output a +5-V signal when the light incident on one, the test photoresistor, is 1 lux brighter than the other or reference photoresistor, and −5 V when the reference photoresistor is 2 lux brighter.

3.38. Design a circuit whose voltage transfer equation is $v_{out} = 2500v_1 - 150v_2$. The voltage gain on any input to any single op amp stage should not exceed 8. Use as few op amps as you can.

More Challenging Problems

3.39. Find the Thevenin and Norton equivalents.

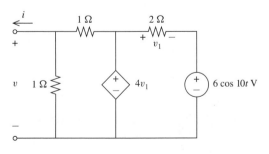

FIGURE P3.39

3.40. Find the voltage transfer function $v_3 = k_1 v_1 + k_2 v_2$ when $R = 100$ kΩ. Repeat when $R = 10$ kΩ.

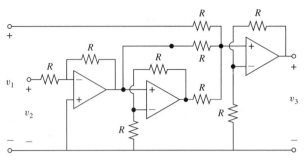

FIGURE P3.40

3.41. Design a circuit satisfying $v_5 = v_1 - v_2 + v_3 - v_4$. Keep all R's in the range 5 to 500 kΩ.

3.42. For what range of resistance R_1 do the values of v_2 computed using the ideal voltage amplifier model of Fig 3.7 and the improved model of Fig. 3.9 differ by at least 1%? Use $R_i = 1$ MΩ, $R_o = 30$ Ω.

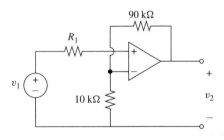

FIGURE P3.42

3.43. For what range of resistance R_L do the values of v_2 computed using the ideal voltage amplifier model of Fig. 3.7 and the improved model of Fig. 3.9 differ by at least 1%? Use $R_i = 1$ MΩ, $R_o = 30$ Ω.

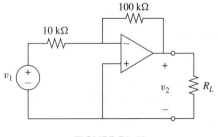

FIGURE P3.43

Gustav Robert Kirchhoff
1824–1887

There must be a fundamental story here [on his research with Bunsen].

Gustav Robert Kirchhoff

4

Analysis Methods

Ohm's law is fundamental to electric circuits, but to analyze even the simplest circuit requires two additional laws formulated in 1847 by the German physicist Gustav Robert Kirchhoff. These laws—Kirchhoff's current law and Kirchhoff's voltage law—are the more remarkable when we consider that Kirchhoff's principal interest was in his pioneering work in spectroscopy with the noted German chemist Robert Bunsen, to whom we owe the Bunsen burner. In that field there is another law of Kirchhoff: Kirchhoff's law of radiation.

Kirchhoff was born in Königsberg, East Prussia, the son of a lawyer. He entered the University of Königsberg at age 18 and graduated with his doctorate five years later. Upon graduation in 1847 he married the daughter of Friedrich Richelot, one of his famous mathematics teachers, and at the same time received a rare travel grant for further study in Paris. The political unrest that led to the 1848 wave of revolutions in Europe forced him to change his plans, and he became a lecturer in Berlin. Two years later he met Bunsen and the two began their famous collaboration. Kirchhoff's great success in spectroscopy drew attention away from his contributions in other branches of physics, but without his electrical laws there would be no circuit theory.

Chapter Contents

In Chapter 2 we considered methods of analyzing simple circuits, which we recall are those that may be described by a single equation involving a single unknown current or voltage. More general circuits must be described by a set of simultaneous equations involving several unknown circuit variables.

In this chapter we meet the challenge of these more general circuits by developing systematic ways of formulating and solving sets of equations that permit *complete analysis of any linear circuit*. We shall consider two methods, one based primarily on Kirchhoff's current law (the method called nodal analysis) and one on Kirchhoff's voltage law (mesh analysis).

Each of these methods results in a set of equations that contains only a small subset of all the currents and voltages in the circuit. It should be evident from our work in previous chapters that a complete analysis of a circuit can be performed by first *breaking the circuit*, that is, solving for the values of a few key circuit variables. Once these are known, the rest then follow easily. For example, in a simple circuit consisting of a single loop, the key variable is the loop current. Once we have broken the circuit by solving for this current, we may find every voltage in the circuit by applying the individual element laws with known current (and of course, every element current is just equal to the loop current).

We begin in Sections 4.1 and 4.2 by showing how the principles of proportionality and superposition can be used to divide a linear circuit problem involving several sources into component problems each involving a single variable or single source. This may sometimes be used to make more general circuits simple and thus solvable by those methods introduced previously, which are generally applicable only to simple circuits. We next develop the methods of nodal analysis and mesh analysis, a pair of techniques that proves indispensable in the practice of circuit analysis and design. The virtual short and open principles presented in Section 4.8 facilitate analysis of circuits containing op amps. The chapter concludes with a first look at the computer-based circuit simulation and analysis program called SPICE. We will make frequent use of SPICE in the course of our study to check our numerical results and to gain hands-on experience with a wider range of circuits than could be comfortably solved without the level of computational support provided by a digital computer. Solving, for instance, 10 simultaneous circuit analysis equations in 10 unknowns definitely loses its charm after the first few hours of hand calculation, and SPICE relieves us of that task.

4.1 LINEARITY AND PROPORTIONALITY

In Chapter 2 we defined a linear resistor as one that satisfied Ohm's law:

$$v = Ri$$

and we considered circuits that were made up of linear resistors and independent sources. We defined dependent sources in Chapter 3 and analyzed circuits containing both independent and dependent sources. The dependent sources that we considered all had source functions of the form

$$y = kx \tag{4.1}$$

where k is a constant and the variables x and y were circuit variables (voltages or currents). Clearly, Ohm's law is a special case of (4.1). In (4.1) the variable y is proportional to the variable x, and the graph of y versus x is a straight line passing through the origin. For this reason, some authors refer to elements that are characterized by (4.1) as *linear elements*. For our purposes, we shall define a linear element in a more general way, which includes (4.1) as a special case. If x and y are circuit variables associated with a two-terminal element, we shall say that the element is *linear* if multiplying x by a constant K results in the multiplication of y by the same constant K. This is called the *proportionality property* and evidently holds for all elements obeying (4.1) since, after multiplying both sides by K,

$$(Ky) = k(Kx)$$

Thus an element described by (4.1) is linear. In addition, elements with terminal laws of the forms

$$y = k\frac{dx}{dt}, \qquad y = k \int x(t)\, dt \tag{4.2}$$

are also linear, since these laws also imply that

$$(Ky) = k\left(\frac{d}{dt}Kx\right), \qquad (Ky) = k \int (Kx)\, dt$$

The op amp is a multiterminal element and is described by more than one equation. However, we shall use the op amp only in a feedback mode, as stated in Chapter 3, and in this case the op amp model circuits that result consist of linear elements. In this case the models introduced in Chapter 3, consisting of linear dependent sources and linear resistors, may be used in place of the op amp. Therefore, we may add the op amp to our list of linear elements.

We shall define a *linear circuit* as any circuit containing nothing but linear elements and independent sources. As examples, almost all the circuits we have considered thus far are linear circuits. Indeed, our attention throughout this book is limited exclusively to linear circuits. The reasons for the focus on linear circuits are twofold: first, a great many interesting circuits, such as amplifiers and filters, are linear, and second, the techniques for analyzing linear circuits are very powerful and form the basis for all circuit analysis, whether linear or nonlinear.

The analysis equations of a linear circuit are obtained by applying KVL, KCL, and the element laws for each element in the circuit. The equations that result can be generally expressed as

$$a_1 x_1 + a_2 x_2 + \cdots + a_n x_n = y \qquad (4.3a)$$

where the x_i's are circuit variables (currents and/or voltages) and y is a net algebraic sum of independent sources. For instance, if (4.3a) is a KVL equation around a loop consisting of resistors and independent voltage sources, each x_i may be a current, each a_i (plus or minus) a resistance value, and y the algebraic sum of voltage source functions. If (4.3a) is a KCL equation, the x_i's may be currents, a_i's plus or minus conductances, and y the net current into the node. All circuit variables in a given equation need not be of the same type, some of the unknown x_i's might in fact be currents and others voltages.

A sum of scaled variables, as on the left side of (4.3a), is referred to as a *linear combination* of x_1 through x_n. Note that linear combinations of linear elements also satisfy the *proportionality property*; that is, *scaling each source by K will scale each circuit variable by K*, since (4.3a) implies that

$$a_1(Kx_1) + a_2(Kx_2) + \cdots + a_n(Kx_n) = Ky \qquad (4.3b)$$

Thus if $x_1, \ldots, x_n$ were the circuit variable values when y was the source, $Kx_1, \ldots, Kx_n$ will be their values with the net source y scaled by K.

Example 4.1

We seek i_1 and i_2 in Fig. 4.1. Applying KVL clockwise around the left loop,

$$-2i_2 + 4i_1 = v_{g1} \qquad (4.4a)$$

we see that the source v_{g1} equals a linear combination of i_1 and i_2. By KCL at the top right node,

$$i_1 + i_2 = i_{g2} \qquad (4.4b)$$

which, like all KCL and KVL equations in a linear circuit, also equates a linear combination of circuit variables to a net source value. Solving the last for i_2 and using this to eliminate i_2 from (4.4a), the current i_1 is found to be

$$i_1 = \tfrac{1}{6} v_{g1} + \tfrac{1}{3} i_{g2} \qquad (4.5a)$$

and substituting this value for i_1 into (4.4b),

$$i_2 = -\tfrac{1}{6} v_{g1} + \tfrac{2}{3} i_{g2} \qquad (4.5b)$$

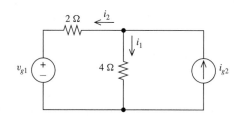

FIGURE 4.1   Linear circuit with two sources.

The solution expresses the circuit variables as linear combinations of the sources. By the proportionality principle, scaling each source by 2 should double all circuit variables. Examining (4.5), if v_{g1} and

i_{g2} are doubled, the right sides of these equations are doubled and thus so are i_1 and i_2.

Example 4.2

We shall use the proportionality principle to assist in calculation of v_1 in the circuit of Fig. 4.2. Such a circuit is sometimes called a *ladder network*, due to its ladderlike configuration. Note that the unknown is far away from the only circuit variable we initially know, the value of the source. Thus several equations involving several unknowns can be expected if we begin writing equations at the source end of the ladder.

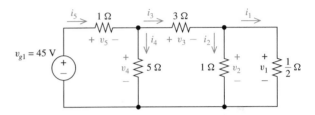

FIGURE 4.2   Ladder network example.

Instead, let us *guess a solution*, $v_1 = 1$ V, and see where this assumption leads. If $v_1 = 1$ V, surely $i_1 = 2$ A by Ohm's law, and $v_2 = 1$ V since parallel elements have the same voltage drop. Thus $i_2 = 1$ A, and at the top right node, $i_3 = i_2 + i_1 = 3$ A. Since i_3 flows through a 3-Ω resistor, $v_3 = 9$ V. Then, by KVL, $v_4 = v_3 + v_2 = 10$ V. Continuing right to left in this manner, $i_4 = 2$ A, $i_5 = i_3 + i_4 = 5$ A, $v_5 = 5$ V, and finally, if our guess ($v_1 = 1$ V) were right, we would have $v_{g1} = v_4 + v_5 = 15$ V.

This is not correct, since actually $v_{g1} = 45$ V. But after all, this was no more than a guess to get us started. By the proportionality relation, if a 15-V source gives an output $v_1 = 1$ V, as we have discovered, our 45-V source will give three times as much, so the correct answer must be

$$v_1 = \frac{45}{15}(1) = 3 \text{ V}$$

This method of assuming an answer for the output, working backward to obtain the corresponding input and finally adjusting the assumed output to be consistent with the actual input by means of the proportionality relation, is particularly easy to apply to the ladder network, but may be applied to other circuits as well.

A nonlinear circuit is, of course, one that has at least one element whose terminal relation is not of the form (4.1) or (4.2). An example is given in Exercise 4.1.4, for which it is seen that the proportionality property does not apply.

EXERCISES

EXERCISE 4.1.1

4.1.1. Find i_1 and v_2, with (a) the source values as shown, (b) the source values divided by 2, and (c) the source values multiplied by -2. Note how the principle of proportionality applies in (b) and (c).
Answer (a) 2 A, 36 V; (b) 1 A, 18 V; (c) -4 A, -72 V

4.1.2. Suppose that $v_{g1} = 90$ V in Fig. 4.2. Find v_1, i_1, v_2, i_2, v_3, i_3, v_4, i_4, v_5, and i_5.
Answer 6 V; 12 A; 6 V; 6 A; 54 V; 18 A; 60 V; 12 A; 30 V; 30 A

4.1.3. Find v and i using the principle of proportionality.
Answer 8 V; 3 A

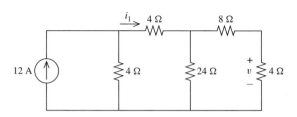

EXERCISE 4.1.3

EXERCISE 4.1.4

4.1.4. In this circuit R_{nl} is a nonlinear resistor with terminal law $v = 2i^3$. Find the source function v_g that supports $i = 1$ A and then that for $i = 2$ A, and show that the principle of proportionality is *not* satisfied. Why?
Answer 5 V; 28 V. Violation of the principle of proportionality is that these answers are not in the proportion $1 : 2$ (this is a nonlinear circuit).

4.2   SUPERPOSITION

In this section we consider linear circuits with more than one independent source. The linearity property makes it possible, as we shall see, to analyze these circuits by adding together the responses due to each source separately.

Recall from (4.3) that each circuit equation may be expressed as

$$a_1 x_1 + a_2 x_2 + \cdots + a_n x_n = y$$

where the x_i's are circuit variables and y the net source value in this KCL or KVL equation. To be specific, suppose that there are exactly two sources, y_a and y_b:

$$a_1 x_1 + a_2 x_2 + \cdots + a_n x_n = y_a + y_b \tag{4.6}$$

Now suppose that we "kill" the source b; that is, we set its source function y_b to zero but otherwise leave the circuit unchanged. Then, in this new circuit, we would have an equation corresponding to the one above:

$$a_1 x_1^a + a_2 x_2^a + \cdots + a_n x_n^a = y_a \tag{4.7a}$$

In this equation, the left-hand side would be the same linear combination as before (the a_i's are unchanged), since only the value of source y_b was changed, not any of the linear elements or their placement in the circuit. The values of the circuit variables will be different due to the presence of just the single source a alone and are designated with an a superscript. x_i^a is thus the response to source a alone. If, instead, we kill the other source, source a, the circuit equation (4.6) becomes

$$a_1 x_1^b + a_2 x_2^b + \cdots + a_n x_n^b = y_b \tag{4.7b}$$

As before, the a_i's in (4.7b) will be identical to those in (4.6), and the x_i^b's are the response to source b alone. Adding (4.7a) and (4.7b) together gives

$$a_1(x_1^a + x_1^b) + a_2(x_2^a + x_2^b) + \cdots + a_n(x_n^a + x_n^b) = y_a + y_b$$

Comparing (4.6) with the preceding equation, we arrive at our central result. Thinking of any of the x_i circuit variables as the response of this circuit to the sources, we have

$$x_i = x_i^a + x_i^b$$

or *the overall response of a circuit containing several sources is the sum of the responses to each individual source with the other sources killed.* This is the *principle of superposition.*

Our justification above assumed exactly two sources, but clearly can be applied, after slight generalization, to the case of several sources. Remember from Chapter 2 that current sources are killed, or zeroed out, by replacing them by open circuits and voltage sources by short circuits. Note also that *superposition only holds in general for linear circuits*, those for which (4.6) may be assumed, just as the proportionality principle is guaranteed only for linear circuits.

Example 4.3

First consider the circuit of Fig. 4.1, which was analyzed in Section 4.1. The circuit variables i_1 and i_2 were found in (4.5). Let us now use superposition to find these responses. First we kill the current source, resulting in the modified circuit of Fig. 4.3(a), and determine i_1^a and i_2^a. This is the component a problem, and i_1^a and i_2^a are the components of the responses i_1 and i_2 due to source a (the voltage source). This is a single-loop circuit, and

$$i_1^a = \frac{v_{g1}}{6}$$

$$i_2^a = -\frac{v_{g1}}{6}$$

Next we kill instead the voltage source, resulting in the component b problem shown in Fig. 4.3(b). By current division,

$$i_1^b = \tfrac{1}{3}i_{g2}$$

$$i_2^b = \tfrac{2}{3}i_{g2}$$

By the superposition principle, each response is the sum of its component responses, or

$$i_1 = i_1^a + i_1^b = \tfrac{1}{6}v_{g1} + \tfrac{1}{3}i_{g2}$$

$$i_2 = i_2^a + i_2^b = -\tfrac{1}{6}v_{g1} + \tfrac{2}{3}i_{g2}$$

These results indeed agree with our previous calculations (4.5).

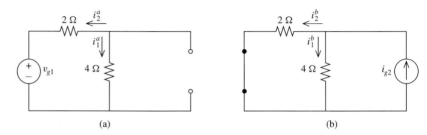

(a)   (b)

FIGURE 4.3   (a) Component a problem; (b) component b problem.

Example 4.3 demonstrates that a trade-off exists in the use of superposition. On the one hand, there are two (or more) component problems to solve instead of just the one we were given originally. On the other hand, each component problem is simpler than the original, both because it contains only one source and because killing sources reduces the number of nodes and/or loops in the circuit. Indeed, both component problems in Example 4.3 involve simple circuits (single loop in component a, single pair of nodes in component b), whereas the original circuit was not simple and could not be analyzed without solving simultaneous equations in two unknowns [(4.4a) and (4.4b)].

Example 4.4

As a second example, let us find the voltage v in the circuit with three independent sources shown in Fig. 4.4. Using superposition, we will find the components of v due to each source separately, referring to the 6-V source as source a, the 18-V source as source b, and the 2-A source as source c.

The component a problem is shown in Fig. 4.5(a), in which the b and c sources have been killed. Killing the 18-V source shorts nodes b and d together into a single node. The circuit has been drawn to show this. The parallel equivalent of the 2- and 6-Ω resistors is $\frac{3}{2}$ Ω, and v^a can be found by voltage division:

$$v^a = \frac{3(6)}{3 + \frac{3}{2}} = 4 \text{ V}$$

The component b problem is shown in Fig. 4.5(b). Killing the 6-V source has drawn nodes a and c together as shown. Once again, after replacing the 2- and 3-Ω resistors by their parallel equivalent $\frac{6}{5}$ Ω, v^b can be found by voltage division:

$$v^b = \frac{\frac{6}{5}}{\frac{6}{5} + 6}(18) = 3 \text{ V}$$

The component c problem [Fig. 4.5(c)] has three resistors in parallel:

$$v^c = -\frac{1}{\frac{1}{6} + \frac{1}{2} + \frac{1}{3}}(2) = -2 \text{ V}$$

By superposition, the value of v is the sum of its components:

$$v = v^a + v^b + v^c = 4 + 3 - 2 = 5 \text{ V}$$

Note that we have solved a set of simple circuit problems in place of an original problem that, if not for superposition, would have required simultaneous equations.

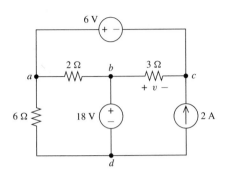

FIGURE 4.4   Example with three independent sources.

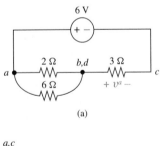

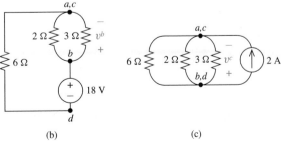

FIGURE 4.5   (a) Component a; (b) component b; (c) component c.

Superposition is a powerful tool that allows us to solve for currents and voltages in *linear* circuits containing several sources by adding (superposing) component currents and voltages. In general, superposition will not work in nonlinear circuits (those that contain one or more nonlinear elements). Its validity is also limited, even in linear circuits, to calculation of currents and voltages. In particular, the total power dissipated by an element in a linear circuit is *not* in general the sum of separate powers due to each source. This is because power is a *nonlinear* function of the circuit variables current and voltage. The use of superposition is limited to calculating currents and voltages in linear circuits.

Example 4.5

This example illustrates the proper use of superposition when power is to be calculated and also its use when there is a dependent source present. We seek the voltage v and the power dissipated by the 3-Ω resistor in the circuit of Fig. 4.6. To find v, we shall use superposition. Only *independent sources* generate components, so we decompose into two component problems a and b, as shown in Fig. 4.7. For component a, we kill the current source. Figure 4.7(a) is a single-loop circuit with the KVL equation

$$-12 + 3i_1^a + 2i_1^a + i_1^a = 0$$

or $i_1^a = 2$ A, and

$$v^a = 3i_1^a = 6 \text{ V}$$

In the component b problem, with the voltage source killed, KCL at the top node yields a current downward through the middle branch of $6 + i_1^b$. KVL around the left loop is then

$$i_1^b + 3(6 + i_1^b) + 2i^b = 0$$

or $i_1^b = -3$ A, and

$$v^b = 3(6 + i_1^b) = 9 \text{ V}$$

Then, by superposition of component voltages,

$$v = v^a + v^b = 6 + 9 = 15 \text{ V}$$

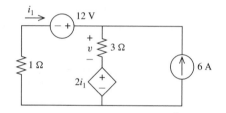

FIGURE 4.6   Circuit with a dependent source.

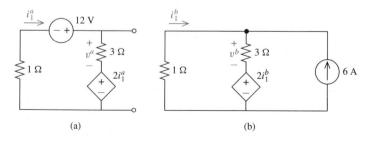

(a)   (b)

FIGURE 4.7   (a) Component a; (b) component b.

We also need the power through the 3-Ω resistor. Since we know its voltage v,

$$p = \frac{v^2}{R} = \frac{15^2}{3} = 75 \text{ W}$$

Note that we found v by superposition and then used the total voltage v *after* superposition of voltage components to compute the power. Had we computed the power through the resistor in the component problems separately and tried to superpose them, this would have given us a different and erroneous result, since the sum of component powers, $(v^{a2} + v^{b2})/R$, is not the same as the power due to the sum of components, $v^2/R = (v^a + v^b)^2/R$. Even in linear circuits, power does not superpose; only voltage and current do.

EXERCISES

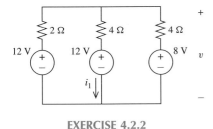

EXERCISE 4.2.2

4.2.1. Solve Exercise 4.1.1 using superposition.

4.2.2. Find v and i_1 by superposition. Check by Thevenin–Norton transformations.
    Answer 11 V; $-\frac{1}{4}$ A

4.2.3. Replace the 8 V by $4i_1$ in Exercise 4.2.2, converting an independent source to a dependent source. Find v and i_1 by superposition.
    Answer 8 V; -1 A

4.3   NODAL ANALYSIS

In this section we develop a general method of circuit analysis in which voltages are the unknowns to be found. A convenient choice of voltages for many networks is the set of *node voltages*. Since a voltage is defined as existing between two nodes, it is convenient to select one node in the network to be a *reference node* and then associate a voltage at each of the other nodes. *The voltage of each of the nonreference nodes with respect to the reference node is defined to be a node voltage.* It is common practice to select reference directions for these voltages so that the plus ends are all at the nonreference nodes and the minus ends all at the reference node. For a circuit containing N nodes, there will be $N-1$ nonreference nodes and thus $N-1$ node voltages. Nodal analysis is a method in which we will break the circuit, that is, solve for a key set of circuit variables, by finding the node voltages themselves. Any other current or voltage will follow easily once the circuit is broken.

The reference node is often chosen to be the node to which the largest number of branches are connected. Many practical circuits are built on a metallic base or chassis, and usually a number of elements are connected to the chassis, which becomes a logical choice for the reference node. In many cases, such as in electric power systems, the chassis is shorted to the earth itself, becoming part of a single chassis–earth node. For

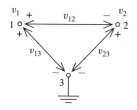

FIGURE 4.8 Reference and nonreference nodes.

this reason, the reference node is frequently referred to as *ground* or the *ground node*. The reference node is thus at ground potential or zero potential, and each other node may be considered to be at some potential above or below zero specified by the value of its node voltage.

The equations of nodal analysis are obtained by applying KCL at the nodes. Recall that each term in a KCL equation is an element current. For a resistor, this current is proportional to its voltage. This voltage, like any element voltage, is equal to a node voltage (if one end of the element is tied to the reference node) or the difference of two node voltages (if both ends are tied to nonreference nodes). For example, in Fig. 4.8 the reference node is node 3 with zero or ground potential. The symbol shown attached to node 3 is the standard symbol for ground, as noted in Chapter 3. The nonreference nodes 1 and 2 have node voltages v_1 and v_2. Thus the element voltage v_{12} with the polarity shown is

$$v_{12} = v_1 - v_2$$

The other element voltages shown are

$$v_{13} = v_1 - 0 = v_1$$
$$v_{23} = v_2 - 0 = v_2$$

These equations may be established by applying KVL around the loops (real or imagined). Evidently, if we know all the node voltages, we may find all the element voltages and thence all the element currents.

The application of KCL at a node, expressing each unknown current in terms of the node voltages, results in a *node equation*. Clearly, simplification in writing the resulting equations is possible when the reference node is chosen to be a node with a large number of elements connected to it. As we shall see, however, this is not the only criterion for selecting the reference node, although it is frequently the overriding one. Since we are going to apply KCL systematically at circuit nodes, the most straightforward case to consider is that of circuits whose only sources are independent current sources. We begin with examples of this type.

In the network shown in Fig. 4.9(a), there are three nodes, dashed and numbered as shown. [This may be easier to see in the redrawn version of Fig. 4.9(b).] Since there are four elements connected to node 3, we select it as the reference node, identifying it by the ground symbol shown.

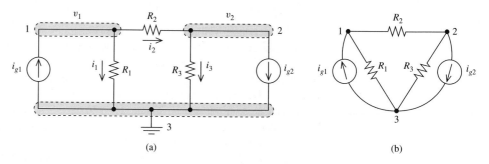

FIGURE 4.9 Circuit containing independent current sources.

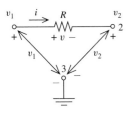

FIGURE 4.10 Single element.

Before writing the node equations, consider the element shown in Fig. 4.10, where v_1 and v_2 are node voltages. The element voltage v is given by

$$v = v_1 - v_2$$

and thus by Ohm's law we have

$$i = \frac{v}{R} = \frac{v_1 - v_2}{R}$$

or

$$i = G(v_1 - v_2)$$

where $G = 1/R$ is the conductance. That is, *the current from node 1 to node 2 through a resistor is the difference of the node voltage at node 1 and the node voltage at node 2 divided by the resistance R, or multiplied by the conductance G.* This relation will allow us to write the node equations rapidly by inspection directly in terms of the node voltages.

Now returning to the circuit of Fig. 4.9, the sum of the currents leaving node 1 must be zero, or

$$i_1 + i_2 - i_{g1} = 0$$

In terms of the node voltages, this equation becomes

$$G_1 v_1 + G_2(v_1 - v_2) - i_{g1} = 0$$

We could have obtained this equation directly using the procedure of the preceding paragraph. Applying KCL at node 2 in a similar manner, we obtain

$$-i_2 + i_3 + i_{g2} = 0$$

or

$$G_2(v_2 - v_1) + G_3 v_2 + i_{g2} = 0$$

Instead of summing currents leaving the node to zero, we could have used the form of KCL that equates the sum of currents leaving the node to the sum of currents entering the node. Had we done so, the terms i_{g1} and i_{g2} would have appeared on the right-hand side:

$$G_1 v_1 + G_2(v_1 - v_2) = i_{g1}$$

$$G_2(v_2 - v_1) + G_3 v_2 = -i_{g2}$$

Rearranging these two equations results in

$$(G_1 + G_2)v_1 - G_2 v_2 = i_{g1} \tag{4.8a}$$

$$-G_2 v_1 + (G_2 + G_3)v_2 = -i_{g2} \tag{4.8b}$$

These equations exhibit a symmetry that may be used to write the equations in the rearranged form (4.8) directly by inspection of the circuit diagram. In (4.8a) the coefficient of v_1 is the sum of conductances of the elements connected to node 1, while the coefficient of v_2 is the negative of the conductance of the element connecting node 1 to node 2. The same statement holds for (4.8b) if the numbers 1 and 2 are interchanged. Thus node 2 plays the role in (4.8b) of node 1 in (4.8a). That is, it is the node at which KCL is applied. In each equation the right-hand side is the current from the current sources that enters the corresponding node.

In general, in networks containing only conductances and current sources, KCL applied at the kth node, with node voltage v_k, may be written as follows. *On the left* side of the node k equation, the coefficient of the *k*th-node voltage is the sum of the conductances connected to node *k*, and the coefficients of the other node voltages are the negatives of the conductances between those nodes and node *k*. The right side of this equation consists of the net current flowing into node *k* due to current sources.

This predictable pattern makes it easy to write down the node equations. Note that the signs, positive on the left-hand side for v_k terms and negative for other node voltage terms, and positive on the right-hand side for current sources flowing into node *k*, are a consequence of the form of KCL chosen. While other forms could be used quite as correctly, we advocate sticking to the form recommended, with the payoff that the terms will always fall in this pattern. It helps to make the pattern of signs fixed and predictable, so we can focus our attention on the larger issues when analyzing a circuit.

Nodal analysis consists in writing KCL node equations described above at all non-reference nodes in the circuit. This yields $N - 1$ linear equations in a similar number of unknowns (the node voltages). As discussed in Appendix C, these equations are linearly independent and thus are guaranteed to possess a unique solution. The node voltages may be found by a variety of means, including Gauss elimination, Cramer's rule, and matrix inversion.

Example 4.6

Consider the circuit of Fig. 4.11. The bottom node has been selected as the reference node since so many elements connect to it. The resistors are labeled according to their conductances.

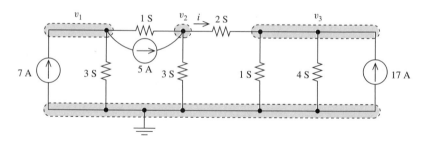

FIGURE 4.11   Circuit for Example 4.6.

Since there are three nonreference nodes, there will be three equations in three unknown node voltages. At node v_1, we note that the sum of conductances is $3 + 1 = 4$, the negative of the conductance connecting node v_2 to node v_1 is -1, and the net source current entering node 1 is $7 - 5 = 2$. Thus the first node equation is

$$4v_1 - v_2 = 2 \tag{4.9}$$

Similarly, at nodes v_2 and v_3, we have

$$-v_1 + 6v_2 - 2v_3 = 5 \tag{4.10a}$$

$$-2v_2 + 7v_3 = 17 \tag{4.10b}$$

We may solve (4.9) and (4.10) for the node voltages using any one of a variety of methods for solving simultaneous equations. Three such methods are matrix inversion, Cramer's rule, and

Gaussian elimination. For the reader who is not familiar with these methods, a discussion is given in Appendix A. Selecting Cramer's rule, first find the determinant of the coefficient matrix, given by

$$\Delta = \begin{vmatrix} 4 & -1 & 0 \\ -1 & 6 & -2 \\ 0 & -2 & 7 \end{vmatrix} = 145 \tag{4.11}$$

To determine v_1, we replace the first column of the coefficient matrix by the vector of constants on the right-hand side of (4.9)–(4.10), compute its determinant, and divide by the determinant of the coefficient matrix already found.

$$v_1 = \frac{\begin{vmatrix} 2 & -1 & 0 \\ 5 & 6 & -2 \\ 17 & -2 & 7 \end{vmatrix}}{\Delta} = 1 \text{ V}$$

v_2 is found by replacing the second and v_3 the third column of the coefficient matrix and calculating as above, yielding $v_2 = 2$ V and $v_3 = 3$ V.

Now that we have broken the circuit by finding the node voltages, we may easily find any other voltage or current. For example, if we want the current i in the 2-S element, it is given by

$$i = 2(v_2 - v_3) = 2(2 - 3) = -2 \text{ A}$$

Note that the coefficient matrix shown in (4.11) is symmetric [the (i, j) and (j, i) elements are equal]. This follows from the fact that the conductance between nodes i and j is that between nodes j and i. Symmetry further simplifies writing the node equations. While symmetry will hold as a general rule for all circuits not containing dependent sources, symmetry of the coefficient matrix cannot be counted on in that case, as we shall see in the next example.

Example 4.7

Consider the circuit of Fig. 4.12, which contains dependent current sources. We will begin by writing the node equations exactly as if the sources were independent. At node 1,

$$(1)(v_1) + (1)(v_1) + (2)(v_1 - v_2) = 5 - 5i$$

and at node 2,

$$\tfrac{1}{2}(v_2) + (2)(v_2 - v_1) = 5i + 2v$$

We next express the controlling variables for the dependent sources, i and v in these equations, in terms of the node voltages. By Ohm's law,

$$i = v_1$$

and by inspection

$$v = v_1 - v_2$$

Substituting the last two equations into the preceding two,

$$(1)(v_1) + (1)(v_1) + (2)(v_1 - v_2) = 5 - 5v_1$$

$$\tfrac{1}{2}(v_2) + (2)(v_2 - v_1) = 5v_1 + 2(v_1 - v_2)$$

These two equations in two unknowns can be solved by Cramer's rule, matrix inversion, or Gauss elimination, as desired. Selecting matrix inversion, we first rewrite as

$$\begin{bmatrix} 9 & -2 \\ -9 & \tfrac{9}{2} \end{bmatrix} \begin{bmatrix} v_1 \\ v_2 \end{bmatrix} = \begin{bmatrix} 5 \\ 0 \end{bmatrix} \tag{4.12}$$

The determinant of the coefficient matrix is $(9)\left(\tfrac{9}{2}\right) - (-9)(-2) = 45/2$ and the inverse is

$$\frac{2}{45} \begin{bmatrix} \tfrac{9}{2} & 2 \\ 9 & 9 \end{bmatrix} = \begin{bmatrix} \tfrac{1}{5} & \tfrac{4}{45} \\ \tfrac{2}{5} & \tfrac{2}{5} \end{bmatrix}$$

Then

$$\begin{bmatrix} v_1 \\ v_2 \end{bmatrix} = \begin{bmatrix} \tfrac{1}{5} & \tfrac{4}{45} \\ \tfrac{2}{5} & \tfrac{2}{5} \end{bmatrix} \begin{bmatrix} 5 \\ 0 \end{bmatrix} = \begin{bmatrix} 1 \\ 2 \end{bmatrix}$$

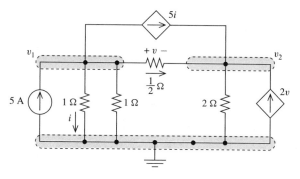

FIGURE 4.12 Circuit containing dependent sources.

From Example 4.7 we see that the presence of dependent sources destroys the symmetry in the coefficient matrix [see (4.12)] and that in such circuits the elements of this matrix may no longer simply be interpreted as sums of conductances, since the dependent sources also contribute. On the other hand, the presence of dependent sources has not significantly complicated nodal analysis, requiring only an additional substitution step, replacing controlling variables by node voltages.

EXERCISES

4.3.1. Take all resistors in Fig. 4.9 to be $1\ \Omega$ and both current source functions to be 1 A. Using nodal analysis, find the node voltages and the three labeled currents.

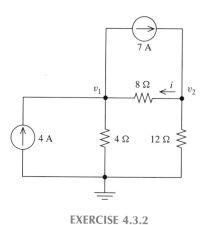

EXERCISE 4.3.2

Answer $v_1 = 1$ V; $v_2 = -1$ V; $i_1 = 1$ A; $i_2 = 2$ A; $i_3 = -1$ A

4.3.2. Using nodal analysis, find v_1, v_2, and i.
Answer 4 V; 36 V; 4 A

4.3.3. Write the nodal equations directly in vector-matrix form. Do not solve.

Answer

$$
\begin{bmatrix}
4 & -3 & -1 & 0 & 0 \\
-3 & 6 & 0 & -1 & -2 \\
-1 & 0 & 4 & -1 & 0 \\
0 & -1 & -1 & 6 & -1 \\
0 & -2 & 0 & -1 & 5
\end{bmatrix}
\begin{bmatrix}
v_1 \\
v_2 \\
v_3 \\
v_4 \\
v_5
\end{bmatrix}
=
\begin{bmatrix}
3 \\
3 \\
1 \\
3 \\
-2
\end{bmatrix}
$$

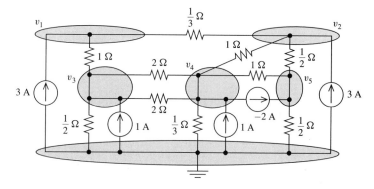

EXERCISE 4.3.3

4.4   CIRCUITS CONTAINING VOLTAGE SOURCES

At first glance it may seem that the presence of voltage sources in a circuit complicates nodal analysis. We can no longer write the KCL node equations, since there is no way to express the currents through these circuit elements in terms of their node voltages. As discussed in Chapter 2, the element law for a voltage source does not relate its current to its voltage, so we cannot use it to replace a current unknown by a voltage unknown in the node equation.

However, as we shall see, nodal analysis in the presence of voltage sources proves no more complicated, requiring only a small modification to the basic method for writing the equations of nodal analysis presented in Section 4.3. In fact, we will come to welcome voltage sources, since they reduce the number of simultaneous node equations that must be solved, yielding one less equation per voltage source.

Example 4.8

To illustrate the procedure, let us consider the circuit of Fig. 4.13. For convenience we have labeled the resistors by their conductances. Note that we have enclosed voltage sources in separate regions indicated by dashed lines. Recalling that the generalized form of KCL states that all currents entering a closed region must sum to zero

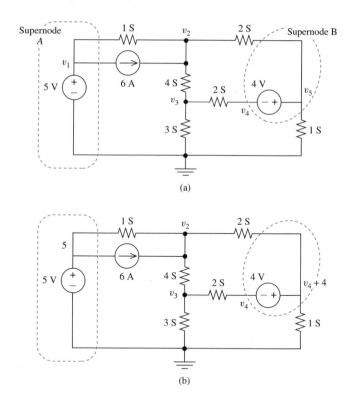

FIGURE 4.13 (a) Circuit containing voltage sources; (b) node voltages in supernodes relabeled.

(just as they do for ordinary nodes), we will refer to each of these regions as a *supernode*.

Each supernode contains two nodes, one a nonreference node and another node that may be a second nonreference node (e.g., supernode B in Fig. 4.13) or the reference node (e.g., supernode A). Supernodes containing the reference node have one node voltage variable, such as v_1 in Fig. 4.13(a). But since there is an element with known voltage drop connecting v_1 to a known node voltage (zero since it is the reference node), v_1 is not in fact an unknown and may be determined immediately:

$$v_1 = 5 + 0 = 5 \text{ V}$$

Similarly, supernodes not containing the reference node have two node voltage variables [v_4 and v_5 in Fig. 4.13(a)]. However, since their difference is known,

$$v_5 - v_4 = 4$$

we may consider just one of these node voltages as an unknown, say v_4, and the other known in terms of v_4 as

$$v_5 = v_4 + 4$$

The number of node voltage unknowns is thus reduced by the number of independent voltage sources. This is shown in Fig. 4.13(b) by relabeling each supernode so that one of its node voltages is expressed as the algebraic sum of the node voltage at the other plus the known voltage drop across the voltage source.

To complete the formulation of the nodal analysis equations, let us *apply KCL to all supernodes not containing the reference node and to all other nonreference nodes.* This has the effect of reducing the number of node equations by one per voltage source. *The presence of voltage sources thus reduces the number of equations and the number of unknowns in nodal analysis by one per voltage source.* For instance, in Fig. 4.13(a) there are five nonreference nodes, and if the voltage sources were all current sources, five equations in five unknowns would be required. Because there are two voltage sources, however, we may break the circuit by using just three equations in three unknowns, as we now demonstrate.

We do not require a KCL equation for supernode A because it contains the reference node. Referring to Fig. 4.13(b), equating currents out of node 2 through resistors to net current into node 2 via current sources in the usual way,

$$(1)(v_2 - 5) + (4)(v_2 - v_3) + (2)[v_2 - (v_4 + 4)] = 6$$

while at node 3

$$(4)(v_3 - v_2) + (3)(v_3 - 0) + (2)(v_3 - v_4) = 0$$

and at supernode B,

$$(2)(v_4 - v_3) + (2)[(v_4 + 4) - v_2] + (1)[(v_4 + 4) - 0] = 0$$

Rearranging these equations in vector-matrix form yields

$$\begin{bmatrix} 7 & -4 & -2 \\ -4 & 9 & -2 \\ -2 & -2 & 5 \end{bmatrix} \begin{bmatrix} v_2 \\ v_3 \\ v_4 \end{bmatrix} = \begin{bmatrix} 3 \\ 0 \\ -12 \end{bmatrix}$$

The coefficient matrix has all the same properties it did in the absence of voltage sources: the on-diagonal elements are the sum of all conductances into the node or supernode at which that equation was written, the (i, j) element is the negative sum of conductances connecting the ith equation's node or supernode with the jth equation's, and the matrix is symmetric. The source vector on the right-hand side, however, is no longer simply the net current in due to current sources, since it now contains terms due to the voltage sources as well.

Example 4.9

As a second example, let us find v and i in the circuit of Fig. 4.14. Assigning node voltages of 0 V and $0 + 20 = 20$ V to one supernode and v and $v + 3$ to the other, there is one supernode not containing the reference node, and there we must write a KCL equation. Summing currents into this region in the usual way gives

$$\frac{1}{6 \times 10^3}[(v + 3) - 20] + \frac{1}{2 \times 10^3}[(v + 3) - 0] + \frac{1}{4 \times 10^3}(v - 0)$$
$$= 6 \times 10^{-3}$$

Solving the equation yields $v = 8$ V. Having broken the circuit, we may now easily find any other circuit variable, such as i. This current is

$$i = \frac{1}{6 \times 10^3}[(v + 3) - 20] = \frac{1}{6 \times 10^3}(-9) = -\frac{3}{2} \text{ mA}$$

Note that while there are three nonreference nodes in this problem, we had to solve only a single equation in one unknown, not three in three, because of the welcome presence of the two voltage sources.

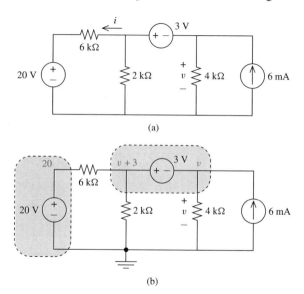

(a)

(b)

FIGURE 4.14 (a) Circuit for Example 4.9; (b) redrawn with supernodes and node voltages labeled.

Example 4.10

Next consider an example containing a dependent voltage source as shown in Fig. 4.15. In Example 4.7 we initially treated dependent current sources as if they were independent, for the purpose of writing the nodal equations, and then replaced their controlling variables by node voltages. We proceed in the same fashion here.

Equations are needed at nodes v_1 and v_2, but not at either supernode since they both contain the reference node. The two equations required are

$$4v_1 + 2(v_1 - v_2) = 10 \qquad (4.13a)$$

$$2(v_2 - v_1) + 9(v_2 - 10v) = -6 \qquad (4.13b)$$

In writing these equations, we proceeded as if the dependent voltage source were an independent source of value $10v$. Next we replace the controlling variable by node voltages. Since v is the element voltage between nodes with voltages given as $10v$ and v_1,

$$v = 10v - v_1$$

or $v = \frac{1}{9}v_1$. Substituting this into (4.13b), (4.13a) becomes

$$6v_1 - 2v_2 = 10$$
$$-12v_1 + 11v_2 = -6$$

Solving by Gaussian elimination, we multiply the second equation by $\frac{1}{2}$ and add it to the first, yielding

$$v_2 = 2 \text{ V}$$

Substituting this back into the first equation yields $v_1 = \frac{7}{3}$ V. With the circuit broken, analysis may be completed by inspection. For instance, the current i is

$$i = 2(v_2 - v_1) = 2(2 - \frac{7}{3}) = \frac{4}{3} \text{ A}$$

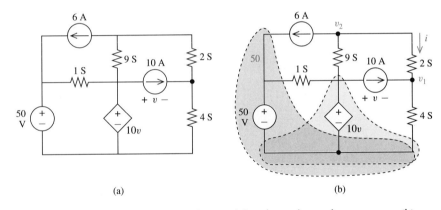

(a)   (b)

FIGURE 4.15   (a) Circuit containing dependent voltage sources; (b) re-drawn to show supernodes.

EXERCISES

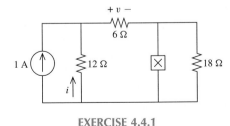

EXERCISE 4.4.1

4.4.1.   Using nodal analysis, find v if element x is a 3-V independent voltage source with the positive terminal at the top.
Answer 3 V

4.4.2.   Find v in Exercise 4.4.1 if element x is a 2-A independent current source directed downward.
Answer 8 V

4.4.3.   Find v in Exercise 4.4.1 if element x is a dependent voltage source of $12i$ V with the positive terminal at the bottom.
Answer 12 V

4.5 MESH ANALYSIS

In the nodal analysis of the preceding sections we applied KCL at the nonreference nodes of the circuit. We now consider a related method known as *mesh analysis*, in which KVL is applied around certain loops in the circuit. As we shall see, in this case the unknowns in the resulting equations will be currents.

We restrict attention in this section to *planar circuits*, circuits that can be drawn on a plane surface (like a piece of paper) so that no elements or connecting wires cross. In this case the plane is divided by the circuit diagram into distinct areas in the same fashion that the solid framework in a window separates and outlines each individual windowpane. The closed boundary of each such windowpane area is called a *mesh* of the circuit. Thus a mesh is a special case of a *loop*, which we consider to be any closed path of elements in the circuit passing through no node or element more than once. In other words, a mesh is a loop that contains no elements within it.

Example 4.11

The circuit of Fig. 4.16 is planar and contains three meshes identified by the arrows. Mesh 1 contains the elements R_1, R_2, R_3, and v_{g1}; mesh 2 contains R_2, R_4, v_{g2}, and R_5; and mesh 3 contains R_5, v_{g2}, R_6, and R_3.

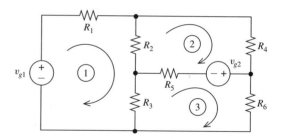

FIGURE 4.16 Planar circuit with three meshes.

In the case of *nonplanar circuits*, we cannot define meshes and mesh analysis cannot be performed. Mesh analysis is thus not as general as nodal analysis, which had no such topological restriction. A KVL-based technique may still be employed, however, even for nonplanar circuits. The procedure in this case is similar to mesh analysis, but the equations are not as easily formulated. This technique, *generalized loop analysis*, is summarized briefly in Appendix C. Fortunately, the large majority of electric circuits we shall have need to analyze in practice are planar, and for those we may choose nodal or mesh analysis. For nonplanar circuits, the only general method developed in this book is nodal analysis.

We define a *mesh current* as the current that circulates around a mesh. If an element is located on a single mesh such as v_{g1} or R_2 in Fig. 4.16, it carries an element current just equal to this mesh current (mesh current i_1 for v_{g1}, i_2 for R_4). If an element is located on the boundary between two meshes, such as R_2 in the same figure, its element current is the algebraic sum of the two mesh currents circulating through it ($i_1 - i_2$ for element R_2). Labeling element currents with capital letters and mesh currents with lowercase letters, both R_1 and v_{g1} have the same element current I_1, where

$$I_1 = i_1$$

The element current I_3 defined positive downward through R_3 is

$$I_3 = i_1 - i_2$$

Mesh analysis consists in writing KVL around each mesh in the circuit, using mesh currents as the unknowns. The resulting system of equations is guaranteed to be linearly independent, as shown in Appendix C, and thus possesses an unique solution. As with nodal equations, the solution may be found by any convenient method such as Gauss elimination, Cramer's rule, or matrix inversion.

Example 4.12

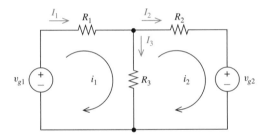

FIGURE 4.17   Circuit with two meshes.

Consider the circuit of Fig. 4.17. Writing KVL clockwise around the first mesh,

$$R_1 i_1 + R_3(i_1 - i_2) = v_{g1}$$

and clockwise around the second mesh,

$$R_2 i_2 + R_3(i_2 - i_1) = -v_{g2}$$

Note that in the second equation we have expressed the element current through R_3 as $i_2 - i_1$, so its reference direction arrow points in the direction in which we are traversing the loop (clockwise). This way the voltage drop is positive in this KVL equation. Had we used $i_1 - i_2$ as the element current, as we did in the first equation, we would have had a minus sign in front of the corresponding term, $-R_3(i_1 - i_2)$ in the second equation. We will always use implied element current reference directions that point in the direction in which we are traversing the mesh, thus keeping the signs for all resistive voltage drop terms positive. This is for convenience, of course; it is never incorrect to replace a term like $+R_3(i_2 - i_1)$ in any equation by $-R_3(i_1 - i_2)$.

There is also a *shortcut method* of writing mesh equations that is similar to the shortcut nodal method of Section 4.2. Rearranging the two equations of Example 4.12,

$$(R_1 + R_3)i_1 - R_3 i_2 = v_{g1} \qquad (4.14\text{a})$$

$$-R_3 i_1 + (R_2 + R_3)i_2 = -v_{g2} \qquad (4.14\text{b})$$

we note that in the first equation, corresponding to the first mesh, the coefficient of the first current is the sum of the resistances associated with the first mesh, and the coefficient of the other mesh current is the negative of the resistance common to that mesh and the first mesh. The right side of the first equation is the net voltage rise in the direction of traversal due to voltage sources in the first mesh. The same pattern applies to each equation. *On the left side of the mesh k equation, the coefficient of the mesh k current is the sum of the resistances of mesh k, and the coefficients of the other mesh currents are the negatives of the resistances on the boundary between their meshes and mesh k. The right side of the mesh k equation consists of the net voltage rise in this mesh due to voltage sources.* This shortcut procedure is a consequence of selecting all the mesh currents in the same direction (clockwise in Fig. 4.17) and writing KVL as each mesh is traversed in the direction of its mesh current. Of course, the shortcut method applies only when no sources are present except independent voltage sources.

Example 4.13

In Fig. 4.16 define i_1, i_2, and i_3 as the mesh currents circulating clockwise in meshes 1, 2, and 3, respectively. Applying the shortcut method to mesh 1, we have

$$(R_1 + R_2 + R_3)i_1 - R_2i_2 - R_3i_3 = v_{g1} \qquad (4.15a)$$

This result may be checked by applying KVL to mesh 1, resulting in

$$R_1i_1 + R_2(i_1 - i_2) + R_3(i_1 - i_3) = v_{g1}$$

The two results are evidently the same.

Applying KVL to meshes 2 and 3 yields, in the same manner,

$$-R_2i_1 + (R_2 + R_4 + R_5)i_2 - R_5i_3 = -v_{g2} \qquad (4.15b)$$

$$-R_3i_1 - R_5i_2 + (R_3 + R_5 + R_6)i_3 = v_{g2} \qquad (4.15c)$$

The analysis is completed by solving these three mesh equations for the three unknown mesh currents.

The same symmetry is present in the mesh equations as was noted in the nodal equations. Rewriting the three mesh equations (4.15) as a single vector matrix equation, we have

$$\begin{bmatrix} R_1 + R_2 + R_3 & -R_2 & -R_3 \\ -R_2 & R_2 + R_4 + R_5 & -R_5 \\ -R_3 & -R_5 & R_3 + R_5 + R_6 \end{bmatrix} \begin{bmatrix} i_1 \\ i_2 \\ i_3 \end{bmatrix}$$
$$= \begin{bmatrix} v_{g1} \\ -v_{g2} \\ v_{g2} \end{bmatrix}$$

The diagonal elements are the sums of the resistances in the meshes, and the off-diagonal elements are the negatives of the resistances common to the meshes corresponding to the row and column of the element. That is, $-R_2$ in row 1, column 2 or in row 2, column 1 is the negative of the resistance common to meshes 1 and 2, and so on. Thus the matrix is symmetric. As was the case with nodal analysis, this symmetry is not preserved if there are dependent sources present.

Example 4.14

We seek the current i in Fig. 4.18. This circuit contains a dependent voltage source. As with nodal analysis, we work with dependent sources by first writing the analysis equations as if the source were independent; we then replace the controlling variable by the desired variables of the analysis method (in this case the mesh currents). Using the shortcut method yields

$$6i_1 - 3i_2 = -2v \qquad (4.16a)$$

$$-3i_1 + 7i_2 = 30 \qquad (4.16b)$$

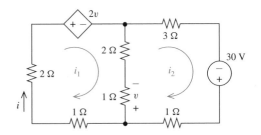

FIGURE 4.18 Example with a dependent source.

In (4.16), 6 Ω is the resistance around mesh 1, 7 Ω is the resistance around mesh 2, -3 Ω is the negative of the net resistance on the border between the meshes, $-2v$ is the clockwise voltage rise due to voltage sources in mesh 1, and 30 is the rise due to sources in mesh 2. Replacing the controlling variable v by $1(i_2 - i_1)$ via Ohm's law gives us

$$\begin{bmatrix} 4 & -1 \\ -3 & 7 \end{bmatrix} \begin{bmatrix} i_1 \\ i_2 \end{bmatrix} = \begin{bmatrix} 0 \\ 30 \end{bmatrix}$$

We will solve for i_1, which is the desired current i, using Cramer's rule. The determinant of the coefficient matrix is

$$\Delta = (4)(7) - (-1)(-3) = 25$$

Replacing the first column of the coefficient matrix by the vector of right-hand sides and calculating the determinant yields

$$i_1 = \frac{\left| \begin{bmatrix} 0 & -1 \\ 30 & 7 \end{bmatrix} \right|}{\Delta} = \frac{6}{5} \text{ A}$$

EXERCISES

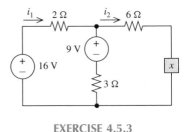

EXERCISE 4.5.3

4.5.1. Using mesh analysis, find i_1 and i_2 in Fig. 4.17 if $R_1 = 2$ Ω, $R_2 = 4$ Ω, $R_3 = 3$ Ω, $v_{g1} = 9$ V, and $v_{g2} = -5$ V.
Answer 3 A; 2 A

4.5.2. Repeat Exercise 4.5.1 with $R_1 = 1$ Ω, $R_2 = 2$ Ω, $R_3 = 4$ Ω, $v_{g1} = 21$ V, and $v_{g2} = 0$. Check by using equivalent resistance and current division.
Answer 9 A; 6 A

4.5.3. Using mesh analysis, find i_1 and i_2 if element x is a 6-V independent voltage source with the positive terminal at the top.
Answer 2 A; 1 A

4.5.4. Repeat Exercise 4.5.3 if element x is a dependent voltage source of $6i_1$ V with the positive terminal at the bottom.
Answer 5 A; 6 A

As in the case of nodal analysis with voltage sources, mesh analysis results in fewer equations if current sources are present. To illustrate this point, let us consider the circuit of Fig. 4.19(a), which has two current sources and a voltage source.

Circuits with three meshes will, in the absence of current sources, have three unknown mesh currents. Note, however, that each current source bordering only one mesh, such as i_{g1} in Fig. 4.19(a), results in one less *unknown* mesh current, since the mesh current must agree with the given current source function. Thus we relabel the mesh current i_3 in Fig. 4.19(b) by its known value,

$$i_3 = -i_{g1}$$

Similarly, for each current source on the boundary between two meshes, one of the two mesh currents may be expressed in terms of the other. The current source between meshes 1 and 2 permits us to relabel i_2 in terms of i_1, since by examination of Fig. 4.19(a),

$$i_{g2} = i_2 - i_1$$

or
$$i_2 = i_{g2} + i_1$$

as shown in Fig. 4.19(b). *The presence of current sources thus reduces the number of unknowns in mesh analysis by one per current source.*

We next consider the selection of a set of KVL equations to match this reduced number of unknown currents. Imagine the circuit that results if we were to kill all current sources by replacing them by open circuits. Some of its meshes correspond to meshes of the original circuit, others to loops in the original circuit that are not meshes in the original circuit since they are not empty (they had current sources inside them). These are called *supermeshes*. This new circuit has one less mesh per current source than the original. *KVL around loops in the circuit defined as the set of meshes that result when all current sources are removed yields the complete set of mesh analysis equations for circuits containing current sources.* In Fig. 4.19(b), if the current sources were killed, there would remain only a single mesh traversing the three resistors and the voltage source. This is not a mesh in the original circuit, since it encloses a current source. Writing KVL around

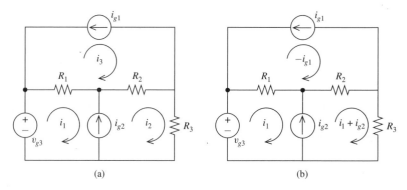

(a)   (b)

FIGURE 4.19 (a) Circuit with current sources; (b) mesh currents relabeled.

this supermesh of the original circuit gives

$$R_1(i_1 + i_{g1}) + R_2(i_1 + i_{g2} + i_{g1}) + R_3(i_1 + i_{g2}) = v_{g3}$$

This is a single equation in a single unknown mesh current i_1. Solving yields

$$i_1 = \frac{v_{g3} - R_1 i_{g1} - R_2(i_{g1} + i_{g2}) - R_3 i_{g2}}{R_1 + R_2 + R_3}$$

This solution is considerably easier than the three equations in three unknown mesh currents we would be left with if the current sources were voltage sources. We welcome current sources in mesh analysis for this reason.

Example 4.15

Let us apply mesh analysis to the circuit of Fig. 4.20(a). Examining this figure, the current sources are related to the mesh currents by

$$9 = i_1 - i_3 \tag{4.17a}$$

$$2v_x = i_4 - i_3 \tag{4.17b}$$

Solving (4.17a) for i_3, we have

$$i_3 = i_1 - 9$$

and (4.17b) for i_4 gives

$$i_4 = 2v_x + i_3 = 2v_x + i_1 - 9$$

These equations permit us to relabel the four mesh currents in terms of only two unknown mesh currents, i_1 and i_2. This is shown in Fig. 4.20(b).

The two equations required to find these two unknown mesh currents are found by imagining the current sources killed. In this modified circuit only two meshes would exist, one corresponding to mesh 2 of the original circuit and the other to a supermesh consisting of the union of the old meshes 1, 3, and 4 from Fig. 4.20(a). Note that this supermesh is a loop but *not* a mesh in our actual circuit of

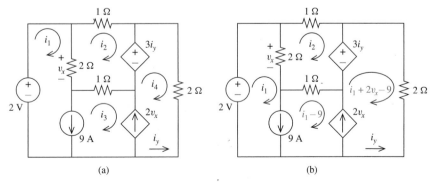

(a)   (b)

FIGURE 4.20 (a) More complicated circuit; (b) mesh currents relabeled.

Fig. 4.20(a) or (b), since the current sources lie inside its boundary (hence its designation as a supermesh). Writing KVL around mesh 2 in Fig. 4.20(b) gives

$$1i_2 + 3i_y + 1(i_2 - i_1 + 9) + 2(i_2 - i_1) = 0 \qquad (4.18a)$$

and around the supermesh,

$$2(i_1 - i_2) + 1(i_1 - 9 - i_2) - 3i_y + 2(i_1 + 2v_x - 9) - 2 = 0 \quad (4.18b)$$

Gathering common terms yields

$$-3i_1 + 4i_2 = -9 - 3i_y \qquad (4.19a)$$

$$5i_1 - 3i_2 = 29 - 4v_x + 3i_y \qquad (4.19b)$$

Thus far we have ignored the fact that two of the sources are, in fact, dependent. This is the same two-step strategy we employed with nodal analysis in the presence of dependent sources. To complete the analysis, the controlling variables v_x and i_y will be replaced by mesh currents. Inspecting the circuit diagram Fig. 4.20(b), this is readily accomplished.

$$v_x = 2(i_1 - i_2) \qquad (4.20a)$$

and
$$i_y = -(i_1 + 2v_x - 9)$$

$$= -(i_1 + 2[2(i_1 - i_2)]) - 9$$

or
$$i_y = -5i_1 + 4i_2 + 9 \qquad (4.20b)$$

Using (4.20) to eliminate v_x and i_y from (4.19) yields

$$-18i_1 + 16i_2 = -36$$

$$28i_1 - 23i_2 = 56$$

Solving, we find that $i_1 = 2$ A and $i_2 = 0$ A. This result may be confirmed by use of any of our linear algebraic methods (Gauss, Cramer, or matrix inversion) or by back substitution of i_1 and i_2 into these two equations.

With the circuit now broken, any other variable needed may be easily recovered. The other two mesh currents may be found from their labels in Fig. 4.20(b) in terms of i_1 and i_2. All resistive voltages follow from Ohm's law.

In general, before analyzing any circuit, we should note how many equations are required in nodal analysis and in mesh analysis and choose the method with fewer equations. The number of nodal equations is the number of nodes (less one, the reference node) minus the number of voltage sources. The number of mesh equations is the number of meshes minus the number of current sources. It is further assumed that series and parallel equivalents have been substituted where possible before performing either nodal or mesh analysis. Series elements multiply the number of nodes unnecessarily; similarly, parallel elements cause meshes to proliferate.

EXERCISES

4.6.1. Using mesh analysis, find i.

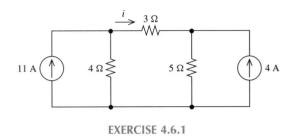

EXERCISE 4.6.1

Answer 2 A

4.6.2. Using mesh analysis, find v in Fig. 4.6. This was found in Example 4.5 using superposition.

Answer 15 V

4.6.3. In Fig. 4.19(a), let $R_1 = 4\ \Omega$, $R_2 = 6\ \Omega$, $R_3 = 2\ \Omega$, $i_{g1} = 4A$, $i_{g2} = 6A$, and $v_{g3} = 52$ V. Find the power delivered to R_3.

Answer 18 W

4.7 DUALITY

There is an interesting symmetry of opposites exhibited by pairs of network equations which we have encountered. For example, Ohm's law may be stated in v form as

$$v = Ri \tag{4.21a}$$

or in i form as

$$i = Gv \tag{4.21b}$$

Equation (4.21b) may be derived from (4.21a) by solving for i and replacing $\frac{1}{R}$ by G, of course. But note that either equation may be formed from the other by *replacing each symbol by its paired opposite: i by v, v by i, R by G, G by R.*

Similarly, in the case of series resistances, $R_1, R_2, \ldots, R_n$, the equivalent resistance was shown in Section 2.3 to follow the "series resistance adds" rule:

$$R_s = R_1 + R_2 + \cdots + R_n$$

Replacing R by G and *series* by *parallel*,

$$G_p = G_1 + G_2 + \cdots + G_n$$

which is the "parallel conductances add" rule of Section 2.4. It is clear that one of these equations may be obtained from the other by interchanging resistances and conductances and the subscripts s and p (i.e., series and parallel).

There is a systematic symmetry of opposites, or *duality*, between resistance and conductance, current and voltage, and series and parallel. We acknowledge this by

Table 4.1	Some Dual Quantities		
Dual Quantities		**Dual Quantities**	
v	i	Series	Parallel
R	G	Mesh	Node
Open circuit	Short circuit	KVL	KCL
Ind. voltage source	Ind. current source	v form	i form
VCVS	CCCS	Thevenin form	Norton form
CCVS	VCCS	Voltage divider	Current divider

defining these quantities as *duals* of each other. That is, R is the dual of G, i is the dual of v, series is the dual of parallel, and vice versa in each case.

Another simple case of dual equations is

$$v = 0$$

and its dual

$$i = 0$$

In the general case, an element described by $v = 0$ is a *short circuit*, and one described by $i = 0$ is an *open circuit*. Thus short circuits and open circuits are duals. Table 4.1 is a table of duals.

Note that *dual* quantities are not *equivalents*; they do not even fit in the same circuit. For instance, R and G are duals, but an 8-Ω resistor and 8-S conductance are not equivalent. Similarly, the dual to a Thevenin form (series voltage source and resistor) is a Norton form (parallel current source and conductance), but they are not equivalent since they do not in general satisfy $V_T = R_T I_N$ or $R_N = R_T$ as required for Thevenin-Norton equivalents.

Example 4.16

Consider the circuit of Fig. 4.21(a). The mesh equations are given by

$$(R_1 + R_2)i_1 - R_2 i_2 = v_g$$
$$-R_2 i_1 + (R_2 + R_3)i_2 = 0 \tag{4.22}$$

To obtain the dual of (4.22), we simply replace the R's by G's, the i's by v's, and v by i. The result is

$$(G_1 + G_2)v_1 - G_2 v_2 = i_g$$
$$-G_2 v_1 + (G_2 + G_3)v_2 = 0 \tag{4.23}$$

These are the nodal equations of a circuit having two nonreference node voltages, v_1 and v_2, three conductances, and an independent current source i_g. From our shortcut procedure for nodal equations we see that G_1 and G_2 are connected to the first node, G_2 is common to the two nodes, G_2 and G_3 are connected to the second node, and i_g enters the first node. Since G_1 and i_g are not connected to the second node and G_3 is not connected to the first node, these elements

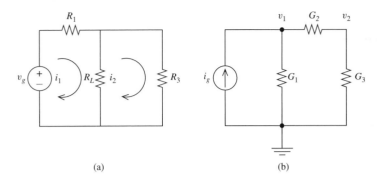

FIGURE 4.21 (a) Circuit; (b) dual circuit.

are connected to the reference node. The primal (original) circuit and its dual are shown in Fig. 4.21.

Figure 4.21(a) may be described as v_g in series with R_1 and the parallel combination of R_2 and R_3. Replacing the quantities in this statement by their duals, we see correctly that Fig. 4.21(b), its dual circuit, is i_g in parallel with G_1 and the series combination of G_2 and G_3.

Next consider systematic construction of the dual circuit from the circuit diagram of the original, or *primal, circuit*. As illustrated in Example 4.16, meshes in the original (primal) circuit are dual to nodes in the dual circuit. Each mesh current in the primal became a node voltage in the dual, thus each mesh in the primal corresponds to a node in the dual. This suggests that we begin constructing the dual circuit diagram by placing a node in the center of each mesh of the primal as shown in Fig. 4.22(a). One additional node is needed in the dual circuit, the reference node, for which there is no node voltage variable or corresponding mesh in the primal circuit. Elements on the outer periphery of the primal circuit have only one mesh current passing through them, so they will be dual to elements with only one node voltage variable across them. These must be elements connected to the reference node in the dual circuit, since all others will have two node voltages across them. Thus we add one last node to our dual circuit, the reference node, placing it somewhere outside the primal circuit as shown in Fig. 4.22(a).

To complete the dual circuit we must add its elements. Note that all terms in the first primal mesh equation will be dual to terms in the first dual node equation. So we attach one element at node 1 of the dual through each of the primal circuit elements surrounding this node. The element we attach will be the dual element, of course. The completed dual circuit is shown in Fig. 4.22(b). The dual circuit has the same number of elements as the primal and one more node than the primal has meshes. Note that the controlled source, a current-controlled current source (CCCS), has gone over to its dual controlled source, a voltage-controlled voltage source (VCVS). The controlling current $i_c = i_3 - i_2$ is replaced by its dual, the controlling voltage $v_c = v_3 - v_2$. Since the reference arrows for the primal source and its mesh current pointed in the same direction, the reference $+/-$ signs for the dual source agree with the dual variable, v_3 (plus at v_3).

Duality has many uses in circuit theory. It may be used to generalize a theoretical result immediately to its dual without further proof. For instance, we could have invoked

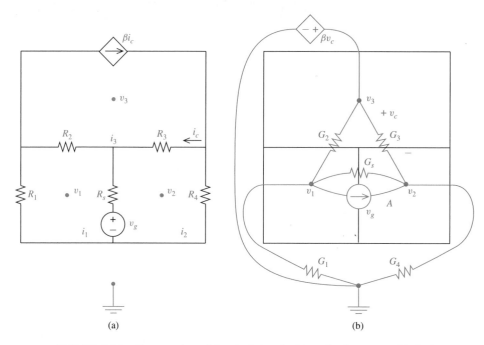

FIGURE 4.22 Construction of the dual circuit: (a) node placement; (b) dual circuit.

duality to write down the current-divider law for conductances in parallel from its dual, the voltage-divider law for resistances in series. Duality can also be used to produce circuits whose *currents* have certain desirable properties from other circuits whose *voltages* have those properties, or vice versa. Since nodal analysis on the dual is identical to mesh analysis on the primal, duality further permits us to use just one general method, say nodal analysis, whichever way is most numerically efficient. For instance, a circuit with fewer meshes than node voltages can be dualized and then nodal analysis done with the same benefits as mesh analysis on the primal. This is particularly useful in the production of computer-based tools, which by this duality-based approach only need to "know" one method, either nodal or mesh analysis, both need not be programmed.

Example 4.17

Suppose we are required to use mesh analysis to find the current i_L of Fig. 4.23(a). If we were to apply mesh analysis to this circuit directly, there would be three simultaneous equations in three unknown mesh currents. The dual circuit is shown in Fig. 4.23(b), and while there are four meshes, the presence of three current sources permits the circuit to be broken by solving a single equation in one unknown. Around the supermesh formed by the resistances,

$$(1)i_1 + \tfrac{1}{2}i_2 + \tfrac{1}{2}i_3 + \tfrac{1}{4}i_4 = 0 \qquad (4.24)$$

We may eliminate $i_2, i_3,$ and i_4 by noting from the circuit diagram that the currents through the current sources are $i_4 - i_1 = 6, i_4 - i_2 = 2,$ and $i_4 - i_3 = 3,$ so that in terms of $i_4, i_1 = i_4 - 6, i_2 = i_4 - 2,$

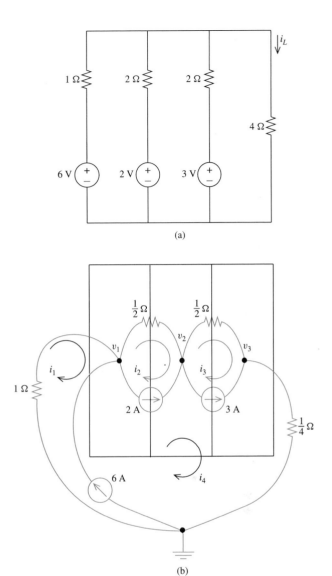

FIGURE 4.23 (a) Circuit for Example 4.17; (b) dual circuit.

and $i_3 = i_4 - 3$. Substituting these into (4.24) gives us

$$(1)(i_4 - 6) + \tfrac{1}{2}(i_4 - 2) + \tfrac{1}{2}(i_4 - 3) + \tfrac{1}{4}i_4 = 0$$

Solving, we have $i_4 = \frac{34}{9}$ A. The desired current i_L is the mesh current in the third mesh in the primal circuit. This current is dual to the node voltage v_3 in the dual circuit, which is

$$v_3 = \frac{i_4}{4} = \frac{17}{18} \text{ V}$$

so $i_L = \frac{17}{18}$ A.

EXERCISES

4.7.1. A circuit satisfies the equations

$$3x_1 - x_2 = -6$$

$$-x_1 + 4x_2 = 16$$

where x_1 and x_2 are both node voltages or both mesh currents. Sketch the primal and dual circuits.

Answer

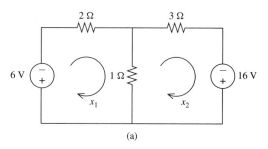

(a)

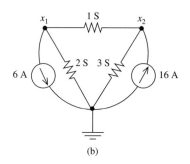

(b)

EXERCISE 4.7.1

4.7.2. Find the dual of Fig. 4.15.

Answer

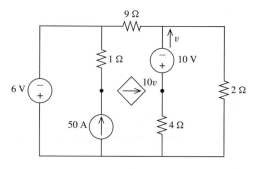

EXERCISE 4.7.2

4.7.3. Find a circuit that is *self-dual*; that is, its dual is identical to itself.

Answer There are many self-dual circuits. Here is one.

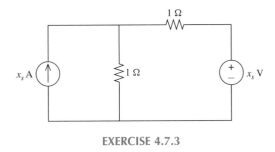

EXERCISE 4.7.3

4.8 VIRTUAL SHORT PRINCIPLE FOR OP AMPS

A general approach to the analysis of circuits containing op amps was developed in Chapter 3. Each op amp in the circuit diagram is replaced by a suitable model, for instance, the ideal voltage amplifier model of Fig. 3.7 or the improved model of Fig. 3.9. Since these models contain only resistors and dependent sources, the resulting circuit may be analyzed just as any other linear circuit would be.

There is another approach to analyzing op amp circuits that, where applicable, is far simpler to carry out. It does not require replacing op amps by models but is applied to the original circuit diagram directly. For each op amp in the circuit (Fig. 4.24), two rules are used:

1. The current into the op amp input terminals is zero: $i_{i1} = i_{i2} = 0$.
2. The voltage across the op amp input terminals is zero: $v_{in} = 0$.

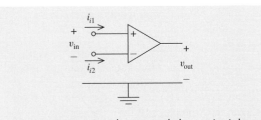

FIGURE 4.24   Virtual open and short principles.

The first of these rules asserts that the op amp input behaves like an open circuit. Since the ideal voltage amplifier model of Fig. 3.7 employs an open-circuited input, this rule, the *virtual open principle*, is not hard to justify. The second rule asserts that the op amp input also behaves like a short circuit (no voltage drop across its terminals). This rule, the *virtual short principle*, requires a bit more reasoning. We first give examples of the use of these rules and then consider the justification of the virtual short principle and the key assumption under which it is valid.

Example 4.18

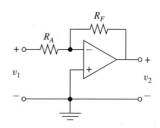

FIGURE 4.25   Inverting amplifier.

In Section 3.5 the inverting amplifier circuit shown in Fig. 4.25 was analyzed. Replacing the op amp by its ideal voltage amplifier model with open-loop gain A, it was found that the output voltage v_2 is given by

$$v_2 = \frac{-R_F}{R_A + \frac{1}{A}(R_A + R_F)} v_1 \qquad (4.25a)$$

In the high-op-amp-gain limit (as A becomes arbitrarily large), we have the voltage transfer relation for the inverting amplifier (3.8) repeated here:

$$v_2 = -\frac{R_F}{R_A} v_1 \qquad (4.25b)$$

We now show how (4.25b) can be derived from the virtual open and virtual short principles *without resort to models or passing to limits*. Since the noninverting input terminal in Fig. 4.25 is at ground, by the virtual short principle, so is the inverting input terminal. Then the node connecting the two resistors is at ground potential, implying that the full voltage v_1 is across R_A and v_2 is across R_F. Summing the currents into this node gives

$$\frac{v_1}{R_A} + \frac{v_2}{R_F} = 0 \qquad (4.26)$$

In this equation the current into the op amp at this node is zero by the virtual open principle. Rearranging (4.26) slightly yields the voltage transfer relation (4.25b).

Example 4.19

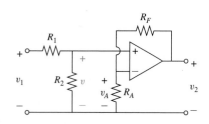

FIGURE 4.26   Circuit for Example 4.19.

Find v_2 in Fig. 4.26.

Since no current flows into the op amp at the inverting input terminal (the virtual open principle), R_F and R_A are in series, and thus they divide the voltage v_2, with

$$v_A = \frac{R_A}{R_A + R_F} v_2$$

Since no voltage is dropped across the op amp input (the virtual short principle), this is also the voltage at the noninverting input terminal, or

$$v = v_A$$

Looking at the left loop, R_1 and R_2 are in series across v_1 (again, no current into the op amp due to the virtual open principle). Thus by voltage division of v_1,

$$v = \frac{R_2}{R_1 + R_2} v_1$$

Combining these last three equations, it must be that

$$\frac{R_A}{R_A + R_F} v_2 = \frac{R_2}{R_1 + R_2} v_1$$

$$\text{or} \qquad v_2 = \left(1 + \frac{R_F}{R_A}\right) \frac{R_2}{R_1 + R_2} v_1$$

To justify the virtual short principle, we reason as follows. Assume, in a given op amp circuit, that the op amp output voltage remains finite in the high-op-amp-gain limit as the op amp gain A becomes arbitrarily large. Since the output voltage is A times the input, *as A becomes arbitrarily large the op amp output voltage will remain finite only if the input voltage becomes arbitrarily small.* Passing to the high-op-amp-gain limit, the input must be exactly zero, or we would have the product of a nonzero input times an infinite gain producing an infinite output voltage, contradicting the assumption that the output remain finite.

Thus, for circuits whose op amp output voltages remain finite in the high-op-amp-gain limit, the virtual short principle follows. What class of circuits is this? Certainly, the circuit described by (4.25a) is in this class. Note that as A gets larger and larger, the op amp output voltage v_2 tends not to infinity, but rather, to the finite value $(-R_F/R_A)v_1$ specified in (4.25b).

Indeed, any circuit with negative feedback, such as all the op amp building block circuits described in Chapter 3, will be "well behaved" enough to tend to a finite output in the high-op-amp-gain limit. Thus *we may use the virtual short and open principles (4.25) to analyze any op amp circuit that employs negative feedback.* As has been noted previously, only such circuits are practical linear op amp circuits anyway, since the rest will have op amp output voltages that "blow up," exceeding their physical output voltage ratings when plugged in. This violation of safe limits will result in behavior certainly nonlinear and perhaps rapidly destructive. So our virtual open and short rules will be valid for any useful linear op amp circuit.

Example 4.20

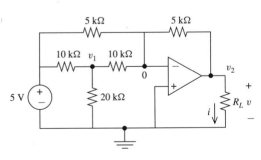

FIGURE 4.27   More complicated op amp circuit.

The op amp circuits analyzed so far have been of the simple type, that is, reducible to a single equation. This example demonstrates how to to solve a general op amp circuit problem. We will use nodal analysis together with the virtual short and open principles. We seek the load current i and the voltage v.

The circuit of Fig. 4.27 has two unknown node voltages, labeled v_1 and v_2. By the virtual short principle, the inverting input terminal is at ground potential, and we have labeled its node voltage 0 V. Thus we need two node equations. For circuits not containing op amps, we write KCL at each node or supernode containing an unknown node voltage. Where op amps are present, we modify that strategy by *writing KCL at the inverting input node rather than the op amp output node.* At nodes 1 and the noninverting input node, respectively,

$$\frac{1}{10 \times 10^3} + \frac{1}{20 \times 10^3} v_1 + \frac{1}{10 \times 10^3}(v_1 - 5) = 0 \quad (4.27a)$$

$$\frac{1}{10 \times 10^3}(0 - v_1) + \frac{1}{5 \times 10^3}(0 - 5) + \frac{1}{5 \times 10^3}(0 - v_2) = 0 \quad (4.27b)$$

The virtual open principle was used in (4.27b). Solving (4.27a) for v_1 yields

$$v_1 = 2 \text{ V}$$

and substituting into (4.27b) gives

$$v_2 = \frac{1.2}{-0.2} = -6 \text{ V}$$

Then the desired unknowns are $i = -6/1000 = -6$ mA and $v = 5 - v_1 = 3$ V.

We will consistently choose not to write node equations at op amp output nodes to avoid complications due to the incomplete nature of the op amp circuit diagram. As mentioned in Chapter 3, there are several connections to the op amp not shown in the circuit diagram. In particular, the lead connecting the op amp output to ground is usually omitted (see Fig. 3.6). Indeed, drawing the op amp as a three-terminal device appears to violate the generalized KCL, since each of the two input currents is zero (the virtual open principle), while the op amp output current in general is not zero. The missing output lead to ground makes it difficult to determine the op amp output current directly and thus hard to write a node equation there. Instead, we choose to write KCL at the inverting input node, where all currents can be easily determined. Once the circuit is broken, the op amp output current can be found easily if so desired.

The virtual short and open principles rules are designated "virtual" because in practice the input current and input voltage to physical op amps being operated in negative feedback circuits are small enough to be virtually, although not exactly, zero. This is exactly what we would expect if the op amp input had high but not infinite resistance, and the op amp gain A were large but finite (conditions characteristic of physical op amps). The virtual open and short principles are excellent approximations and may be used to simplify analysis wherever op amps are operated in the linear, negative feedback mode. Note, however, that use of the virtual short principle applies exactly in the high-op-amp-gain limit only. If the circuit variables for a specific finite gain A are desired, the less convenient model-substitution method developed in Chapter 3 should be employed.

EXERCISES

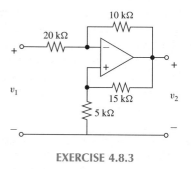

EXERCISE 4.8.3

4.8.1. Use the virtual short and open (* 11 12) principles to find the voltage transfer ratio v_2/v_1 for the noninverting amplifier of Fig. 3.19.
Answer $1 + R_F/R_A$

4.8.2. Use the virtual short and open principles to find the voltage transfer ratio v_2/v_1 for the voltage follower of Fig. 3.14.
Answer 1

4.8.3. Use nodal analysis and the virtual short and open principles to find the voltage transfer ratio v_2/v_1.
Answer $-4/5$

We have seen in this chapter that general linear resistive circuits can be analyzed efficiently by writing and then solving a set of simultaneous nodal or mesh equations. The computations required for hand solution of a simple circuit are minimal, and one with just two or three unknowns (node voltages or mesh currents) is still reasonable. But solving perhaps six equations in six unknowns or 20 equations in 20 unknowns by hand takes quite a long time. Moreover, the great number of steps makes the hand solution process error prone, demanding even more time for careful step-by-step checking. Most of this extensive work time is occupied doing repetitive calculations, with little payoff in terms of learning or new insights.

For this reason, generations of students of electrical engineering had to content themselves with working with a limited number of circuits each of very modest size. Valuable experience that could be gained through wider-ranging analysis exploring somewhat bigger circuits was precluded by the daunting prospect of the price demanded, hours of repetitive hand "cranking." Fortunately, this is no longer the case. Modern technology offers the digital computer, a splendid tool capable of relieving us of the tedious and inefficient use of time inherent in lengthy hand calculations. Over the last four decades the computer has revolutionized how every kind of engineering is done, whether the design of a space station, the improvement of automobile emission control devices, or the analysis of electrical circuits. With computer support for repetitive calculations, larger and more realistic problems can be solved and more experience with more complex systems can be gained. Both the student and the engineering practitioner benefit by the opportunity for more advanced, more realistic problem-solving experience afforded by the assistance of a computer equipped with the right software programs. While many useful computer programs are available to aid in the analysis of electrical circuits, we shall settle on one that is widely available on a variety of computer platforms. This software program is called SPICE (Simulation Program with Integrated Circuit Emphasis).

We shall make frequent use of this tool. One common use will be to check hand results (many of the problems in subsequent chapters ask the student to solve a circuit problem by hand and then "check using SPICE"). A second will be to solve problems involving many nodes and/or meshes. A third important use of SPICE will be to explore the effect of changing the value of a single circuit element, such as a specific resistance within a circuit. This is especially important for "cut-and-try" design, where we may try several values of a circuit element before settling on one resulting in the best circuit performance. Typically, the entire numerical analysis needs to be repeated for each value tried, and the effort involved effectively eliminates hand calculation as a realistic possibility in many design studies. The alternative to repeated trials is to write down an equation explicitly describing the functional dependence of a selected measure of circuit performance on the parameter in question and then somehow solve this equation for the optimal value. With complex circuits this is seldom a practical alternative to the repetitive process of cut-and-try.

SPICE will prove helpful in each of these ways, but it remains only a computational tool. One way in which we cannot use SPICE is to substitute for the intellectually challenging, sometimes painful process of acquiring a mastery of circuits concepts and methods. Problem-solving skills deepen only through concentrated thought and repeated experience. Just as ownership of a good desk calculator does not make one a skilled accountant or access to a digital music synthesizer does not make one a noted composer, so in electrical circuit work the computer can really only save us a bit of time. But since time is a precious commodity to every busy student or professional, we shall take maximum advantage of this time-saving tool where we can.

SPICE, developed with the support of the National Science Foundation by students and faculty at the University of California at Berkeley, is a program that has become the de facto standard for analog circuit simulation. SPICE's success is due to a combination of virtues: the program is powerful, easy to operate, and inexpensive, is supported on most computers of engineering significance, and is well maintained. This is not to suggest that SPICE is the absolute best product available for circuit analysis; indeed, there are more powerful and more comprehensive software packages available commercially. But SPICE will do what is needed for our purposes and is more widely and inexpensively available than other choices.

PSpice, a version of SPICE rewritten for personal computers such as the IBM PC family, the Apple Macintosh, and many other smaller computer platforms, is often used for illustration in this book. Although they differ in some particulars, where possible we shall stick to features common to both the SPICE and PSpice "dialects." Indeed, when we refer to SPICE, we will generically mean both SPICE, as written at the University of California at Berkeley for mainframe computers, and PSpice, as written by the MicroSim Corporation for various personal computers and workstations. Where necessary to distinguish them, we will explicitly do so. Other variants of SPICE have been developed, such as HSPICE and SPICE-Plus, which bear considerable similarity as well.

The student unfamiliar with SPICE is encouraged to review Appendix D. The remainder of this section is devoted to an introduction to those SPICE formats and commands necessary for the class of circuit analysis problems we have studied thus far, linear resistive circuits with constant (dc) independent sources.

Before the SPICE program can be run, the first step is the creation of a SPICE *input file*. The purpose of the SPICE input file is threefold: to specify completely the circuit we wish to analyze, to alert the SPICE program to which type of analysis we wish done, and to identify the desired output variables. The SPICE input file is formatted as an ordinary text file (ASCII file) and is typically created with a utility software program called a *text editor* available on the computer system being used.*

The SPICE input file is organized as a set of statements, each of which terminates with a carriage return. The first statement is a *title statement*, identifying the problem under study. There is no particular format to this statement, except that the first column must not be blank. The last statement is the *.END statement*. All other statements come between these two.

The circuit is identified by a set of *element statements*. Each statement specifies the element type, name, location in the circuit, and its element value. The format for a

---

*It is assumed the reader is familiar with the operating system, a text editor, and a command to print a text file to screen or hard copy on the computer system on which SPICE will be run.

resistor element statement is

```
RXXXXX          N1      N2      VALUE
```

where **R** indicates it is a resistor, **XXXX** is its name, **N1** and **N2** are the two nodes at its terminals, and **VALUE** its resistance in ohms. Note that symbols from SPICE input or output files will always be distinguished by the use of the boldface fixed-width font shown.

The format for an independent dc voltage source is

```
VXXXXX          N1    N2    DC    VALUE
```

and the independent dc current source is

```
IXXXXX          N1    N2    DC    VALUE
```

Although the order in which the nodes are listed does not matter for a resistor, for independent sources order is used to specify the reference direction. The plus sign of the voltage source reference direction is taken to be at the node **N1**, and the arrow of the current source reference direction is taken to point from **N1** toward **N2**. **VALUE** is the source function in volts (for the voltage source) or amperes (for the current source).

The SPICE input file will also contain *control statements* that specify the type of analysis to be done and the outputs to be viewed. The default mode of analysis is dc; thus, in the absence of statements specifying other modes of analysis (such as **.AC** or **.TRAN** statements used in later chapters), we will get the dc solution we are presently interested in generating. The output may be a list of numbers, or it may be plotted. For dc circuits, whose circuit variables do not vary over time, only the numerical output needs to be considered. For outputting a list of numbers, the control statement is

```
.PRINT     DC CVLIST
```

CVLIST is a list of the circuit variables we wish printed out. Each voltage in the list is formatted **V(I,J)**, the node voltage at node I minus that at node J (if $J = 0$, the reference node, the second argument may be omitted). Each current in **CVLIST** is formatted **I(NXXXX)**. This is the current I through element **NXXXX**, with the reference arrow for I pointing from the first node to the second node appearing in the element statement for **NXXXX**. For instance, **V(2,5)** is the voltage at node 2 minus the voltage at node 5, **V(3)** is the voltage at node 3 relative to node 0 (ground), and **I(VREF)** is the current through the voltage source **VREF**, with current reference direction pointing from the plus end of **VREF** (the first node on the **VREF** element statement) toward the minus end.

As the first SPICE example, consider the circuit of Fig. 4.28, where the node designations have been circled for clarity. Note that the reference node is designated as node 0. The SPICE input file that

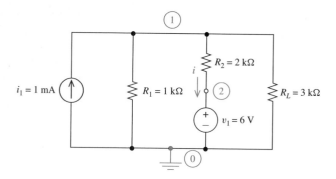

FIGURE 4.28 Circuit for first SPICE example.

will result in a dc analysis and output the values of node voltage 1 and the current i through the 6 V source, is given next.

```
SPICE input file for first example
*That was the title   statement.
*Here are the element   statements:
IS1   0   1   DC     1MA
VS1   2   0   DC     6
R1    1   0   1KILOHM
R2    1   2   2K
RL    1   0   3K
*Here are the control   statements:
.PRINT    DC    V(1)    I(VS1)
*Type of analysis defaults to dc
.END
```

Note that the element statements fully describe the circuit (which elements are where and what the source reference directions are). *The element statements comprise a complete representation of the circuit diagram.*

The default units for the element values are volts, amperes, and ohms unless otherwise specified. Thus the value of 6 in the voltage source element statement means 6 V. The units may be scaled by a power of 10 if an appropriate suffix is attached to the numerical value: M means 10^{-3}, K means 10^{+3}, and so on (see Appendix D for the complete list). Other suffix letters are ignored (such as the A in the IS1 element statement added only for clarity). Any line beginning with an asterisk * is a comment statement and is ignored by the SPICE program.

Suppose that this SPICE input file is typed into the computer and stored under the filename EX4-21.CIR. The SPICE circuit analysis program itself may now be run, using a command syntax that will vary with the variant of SPICE and operating system being used. Upon invoking the SPICE program, the user specifies EX4-

$\tt 21.CIR$ as the input file to be analyzed and declares a name for the output file to be created (typically, $\tt EX4-21.OUT$). After execution is complete, examination of this output file on the terminal screen or as hard copy will reveal (among other material, such as an echo of the SPICE input file) the lines

```
NODE VOLTAGE       NODE VOLTAGE
  (1)5.1429           (2)6.0000

VOLTAGE SOURCE CURRENTS
NAME CURRENT
VS1 -7.143E-04
```

Thus the values for node voltage 1 and current i are 5.1429 V and -0.7143 mA, respectively.

A somewhat peculiar limitation in some versions of SPICE is that only currents through independent voltage sources may be included in the output statement as part of $\tt CVLIST$. This limitation is found, for instance, in SPICE versions 1 and 2 and also in SPICE version 3 when using "batch mode." To output currents that do not happen to flow through independent voltage sources, we must add a *dummy voltage source*, a $+0$ V voltage source in series with the desired current. This source will have no other effect on the circuit, since no current or voltage will be changed by its inclusion. Most SPICE variants, such as PSpice, permit the more general form $\tt I(NXXXX)$ in $\tt CVLIST$, and dummy voltage sources are not needed to $\tt .PRINT$ currents. The documentation for the specific variant and version of SPICE should be consulted to determine whether only currents through independent voltage sources are permitted.

Along with the resistors and independent sources discussed above, resistive circuits may also contain dependent sources. These come in the four "flavors" illustrated in Fig. 3.1. In SPICE, the four types of dependent sources have element statements with the following formats:

VCVS:	EXXXXX	N1	N2	NC1	NC2	MU
VCCS:	GXXXXX	N1	N2	NC1	NC2	G
CCCS:	FXXXXX	N1	N2	VCONT		BETA
CCVS:	HXXXXX	N1	N2	VCONT		R

The conventions for these elements are illustrated in Fig. 4.29. In each case, $\tt N1$ and $\tt N2$ are the source nodes. For the voltage sources (VCVS and CCVS), $\tt N1$ is at the positive end and $\tt N2$ at the negative end of the voltage reference direction, just as was the convention for independent voltage source element statements. For the current sources (VCCS and CCCS), the arrow, indicating current reference direction, points from $\tt N1$ to $\tt N2$, once again consistent with the case of independent sources. $\tt NC1$ and $\tt NC2$ are the nodes of the controlling voltage, with $\tt NC1$ at its plus end. $\tt VCONT$ is the name of an independent voltage source through which the controlling current flows. This current is defined to have its reference direction arrow point from the positive node of $\tt VCONT$ to the negative. If we wish to use a controlling current that does not flow through an independent voltage source, we need to insert a dummy voltage source in series with this current.

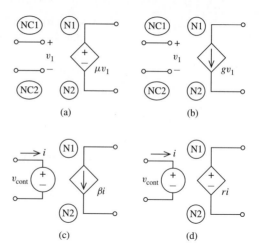

(a)

(b)

(c)

(d)

FIGURE 4.29   Dependent source conventions in SPICE.

The final field in each of these statements is the constant that multiplies the controlling variable to give the dependent source function. These were defined in Chapter 3 as voltage gain μ for the VCVS, transconductance g for the VCCS, current gain β for the CCCS, and transresistance r for the CCVS.

SPICE Example 4.22

This example contains two controlled sources, a CCCS and a VCVS. Since the controlling current i does not flow through a voltage source, we must insert a dummy voltage source in series to permit this current to be used. We wish to determine the voltage v and also the power dissipated by the 2-Ω resistor (see Fig. 4.30). SPICE always outputs the power delivered by each independent source in a dc problem without even being asked. Power delivered by or dissipated by other elements, though, must be computed by hand from current and voltage. So we will print out v and the current through the 4-V source, which happens to be in series and thus carries the

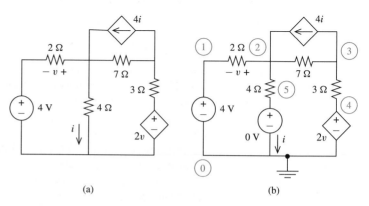

(a)   (b)

FIGURE 4.30   (a) Circuit; (b) prepared for input as SPICE input file.

same current as the desired current through the 2-Ω resistor. The SPICE input file follows.

```
SPICE input file for the second example
*
V1        1        0        DC        4
R1        1        2        2
R2        2        5        4
*Here comes the dummy voltage source:
VD        5        0        0
R3        2        3        7
*Here is the CCCS with current gain 4:
F1        3        2        VD        4
R4        3        4        3
*And the VCVS with voltage gain 2:
E1        4        0        2        1        2
.PRINT    DC    V(2,1)    I(V1)
.END
```

The resulting SPICE output shows us that the 4-V source current is -1.4 A and the voltage v is -28.0 V. Then the power dissipated by the 2-Ω resistor is $(-1.4)(-28.0) = 39.2$ W.

SPICE has many additional element statements, modes of analysis, and other features that we consider as the need for them arises in later chapters. For the present, contenting ourselves with dc analysis of linear resistive circuits, we introduce just two additional capabilities of SPICE: the dc sweep and the subcircuit.

A dc sweep is performed when we wish to step through several values of a dc source, each time performing a dc analysis and printing the outputs. This mode of analysis is useful, for instance, in determining whether an output variable will exceed specified limits over an anticipated range of input (dc source) values. A dc sweep mode of analysis is triggered by including a .DC control statement in the SPICE input file:

```
.DC  SNAME  START   STOP  DELTA
```

SNAME is the name of the independent source to be swept (current or voltage), START and STOP are the initial and final source values desired, and DELTA is the increment between successive source values within the sweep. The next example will include a dc sweep.

A powerful feature of SPICE is the ability to define subcircuits, which can be typed once in a SPICE input file and then referenced as often as needed without repeating its element statements. Each time the subcircuit is needed, it is "called," similar to a subroutine call in FORTRAN or a function invocation in C. The subcircuit is defined by including a set of lines beginning with a .SUBCKT statement and ending with an

· ENDS (end subcircuit) statement. The format is

```
.SUBCKT   SUBNAME NA      NB       NC...
*Subcircuit element  statements go here
...
...
.ENDS
```

SUBNAME is a name given to the subcircuit. The remaining fields on the · SUBCKT statement are the numbers of the nodes of the subcircuit at which the subcircuit will be connected to the main circuit. The subcircuit is called from elsewhere in the SPICE input file by a statement of the form

```
XYYYYY  N1  N2  N3...      SUBNAME
```

Here X indicates a subcircuit call, YYYYY is an arbitrary identifier for this particular subcircuit invocation, N1, N2, N3, · · · are the node numbers within the main circuit at which the subroutine is connected, and SUBNAME is the name of the subcircuit to be used.

SPICE is informed which subcircuit node to connect to which main circuit node by the order in which nodes appear. *The first node on the* · SUBCKT *statement will be connected to the first node on the* X *subroutine call statement, the second to the second, and so on.* The node numbers within a subcircuit definition are independent of the node numbers in the main part of the SPICE input file (with the exception only of the reference node or common ground, node 0, which always refers to the same node wherever it may appear). This independent numbering of nodes corresponds to the independent use of "formal" variables in a FORTRAN subroutine or C function and "actual" variables in the main program. Independent node numbering is what permits us to insert the same subroutine in several different places in the main circuit without confusion.

SPICE Example 4.23

As a final example, let us consider the circuit containing two op amps shown in Fig. 4.31. We will replace each op amp by the im-

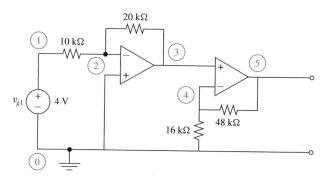

FIGURE 4.31   Circuit for Example 4.23.

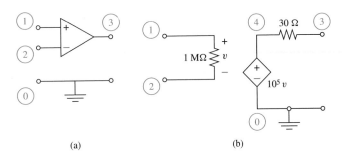

FIGURE 4.32   Subcircuit for Example 4.23.

proved circuit model of Fig. 3.9 (repeated in Fig. 4.32), defining this op amp model as a ·SUBCKT to avoid entering its element statements twice. The type of analysis we will do is a dc sweep of the independent dc source between the limits of −5 and +5 V, printing out the op amp output voltages. Note that same node numbers are used independently in the main circuit and the subcircuit. The SPICE input file used is:

```
Ex4-23: circuit with two op amps
*
*Main circuit:
VG1      1       0       DC              1
R1       1       2       10K
R2       2       3       20K
X1       0       2       3       OPAMP
*Order on nodes on above  statement matches
*order on .subckt  statement: + in, - in, out.
R3       4       0       16K
R4       4       5       48K
X2       3       4       5       OPAMP
*Here comes the op amp subcircuit:
.SUBCKT  OPAMP   1       2               3
*Node 1 is the +in, 2 the -in, and 3 the output node
RIN      1       2       1MEG
E1       4       0       1       2       100K
RO       4       3       30
.ENDS
*Finally the analysis mode and output statments:
.DC      VG1     -5      +5              1
.PRINT DC        V(3)    V(5)
.END
```

After running SPICE with this input file, the output contains the following lines:

VG1	V(3)	V(5)
-5.00E+00	1.00E+01	4.00E+01
-4.00E+00	8.00E+00	3.20E+01
-3.00E+00	6.00E+00	2.40E+01
-2.00E+00	4.00E+00	1.60E+01
-1.00E+00	2.00E+00	7.99E+00
0.00E+00	0.00E+00	0.00E+00
1.00E+00	-2.00E+00	-7.99E+00
2.00E+00	-4.00E+00	-1.60E+01
3.00E+00	-6.00E+00	-2.40E+01
4.00E+00	-8.00E+00	-3.20E+01
5.00E+00	-1.00E+01	-4.00E+01

These are the op amp outputs for input voltage VG1 swept from -5 to $+5$ V by increments of $+1$ V. Note that in each row V(3)$= -2$ VG1, confirming the observation that the first op amp stage is an inverting amplifier with gain of $-R_2/R_1 = -2$. Also, V(5)$= 4$ V(3), since V(3) is input into a noninverting amplifier with gain of $(1 + R_4/R_3) = +4$. Note also that the column of V(3) and of V(5) values both scale with VG1, a consequence of the principle of proportionality. For instance, the values of these variables for VG1 $= 2$ V are double those for VG1 $= 1$ V.

For all its power, SPICE is just a numerical algorithm and will have round-off errors. For instance, V(5) is exactly four times V(3), as expected, for almost all values of V(3). But the value V(3)$= 2.0000$ V yields V(5)$= 7.9999$ V, not 8.000 V. SPICE is not an algorithm endowed with artificial intelligence. It does strictly numerical calculations, not symbolic ones. With a hand calculator we may be surprised that division by a large number followed by multiplication by that same number sometimes will not return the original value to the display, so we cannot expect SPICE to give exact values, just close numerical approximations.

EXERCISES

4.9.1. Write a SPICE input file for the circuit shown in Exercise 4.1.1.

Answer

```
Exercise 4.9.1 SPICE input file
R11       0          1          10
R2        1          2          2
V1        2          0          DC         24
R3        2          3          3
I2        0          3          DC         4
.PRINT    DC    I(V1)    V(3,0)
.END
```

4.9.2. Write a SPICE input file for Fig. 4.6.
Answer

```
Exercise 4.9.2 SPICE input file
R1        1          0          1
V1        2          1          DC         12
R2        2          3          3
H1        3          0          V1         -2
I2        0          2          DC         6
.PRINT DC    I(V1)    V(2,3)
.END
```

The *negative* of the value I(V1) will correspond to the desired output variable *i* in the figure. This sign change could be avoided, as well as the minus sign on the transresistance of the controlled source (see the H statement), by putting a dummy voltage source in series with the 12-V source but with opposite reference direction as the 12-V source.

SUMMARY

In this chapter tools of great generality for the analysis of linear circuits are introduced: the principle of superposition, the methods of nodal and mesh analysis, and a computer program capable of simulating a very broad range of circuits called SPICE.

- Linear circuits satisfy the principle of proportionality: scaling all independent sources by the same factor scales all responses similarly.
- They also satisfy the principle of superposition: the total response is the superposition, or sum, of responses to each independent source with the others killed.
- In nodal analysis the key variables are the node voltages, and KCL is written at all nodes but the reference node.
- In nodal analysis, each voltage source defines a supernode containing only one unknown node voltage and KCL is applied to the supernode.
- In mesh analysis the key variables are the mesh currents, and KVL is written around all meshes.

- In mesh analysis each current source defines a supermesh containing only one unknown loop current and KVL is applied around the supermesh.

- Once a circuit is broken, it is an easy matter to express any desired current or voltage in terms of the known key variables (node voltages or mesh currents).

- Dual circuits are constructed by placing dual node i inside primal mesh i for each primal mesh, and a reference node outside. For each element in primal meshes i and j, its dual is connected between the nodes i and j of the dual circuit.

- The virtual short and virtual open principles for analyzing op amp circuits in negative feedback configurations state that the voltage across, and the current into, the input terminals of the op amp both equal zero.

- Nodal analysis is recommended for op amp circuits using the virtual short and open principles. Do not write nodal KCL equations at op amp output nodes.

- SPICE requires an input file which details the circuit, establishes the desired mode of operation and indicates the desired outputs variables and formats.

The examples in this chapter apply the principle of superposition, and the methods of mesh and nodal analysis, strictly to resistive circuits studied in the time domain. This is not a limitation of these methods but reflects the limited experience we have gained so far. Superposition, mesh and nodal analysis will continue to apply, to be our most basic tools, throughout the book: in the analysis of linear circuits containing storage elements (Chapters 5 to 7), those studied in the phasor domain (Chapters 8 to 11), the s-domain (Chapters 12 to 16), and the Fourier domain (Chapter 17).

PROBLEMS

4.1. By what factor must we scale *both* source functions in order to get $v = +1$ V across the 10-Ω resistor?

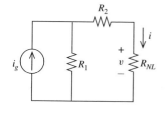

FIGURE P4.3

4.4. If $i = 1$ A, what must v_g be? Use this to determine i if $v_g = 1$ kV.

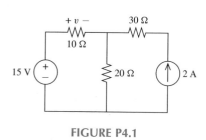

FIGURE P4.1

4.2. What does the principle of proportionality predict will happen to a linear circuit if all independent sources are scaled by zero?

4.3. R_1 and R_2 are linear resistors, but R_{NL} satisfies $i = e^{-v}$. Demonstrate that this circuit violates the proportionality principle.

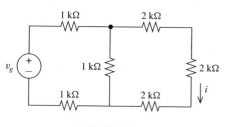

FIGURE P4.4

4.5. Use proportionality to find the source function v_g that makes $v = 6$ V.

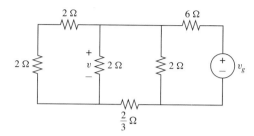

FIGURE P4.5

4.6. Find v using the proportionality principle. All resistors are 10 Ω.

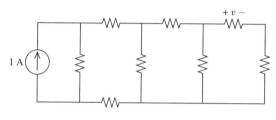

FIGURE P4.6

4.7. Find v using the proportionality principle.

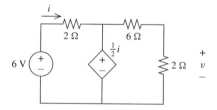

FIGURE P4.7

4.8. Find v by superposition. Check using Thevenin–Norton transformations.

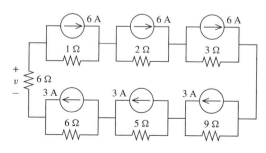

FIGURE P4.8

4.9. Solve for i by superposition.

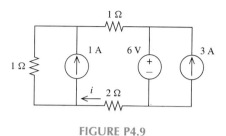

FIGURE P4.9

4.10. Use superposition to find v_{g1} and i.

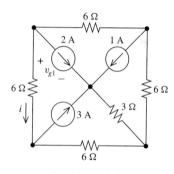

FIGURE P4.10

4.11. Use superposition to find the three resistive currents shown.

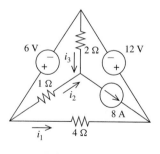

FIGURE P4.11

4.12. Use superposition to find v. Check by using Thevenin–Norton transformations.

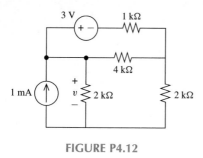

FIGURE P4.12

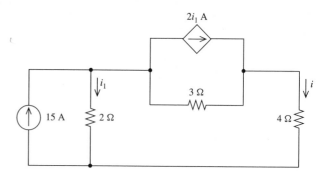

FIGURE P4.16

4.13. What should we replace the 6-A source function by in Problem 4.18 in order that $v_1 = 0$?

4.14. Using nodal analysis, find v_1 and v_2.

4.17. Use nodal analysis to find the indicated variables. Check using superposition.

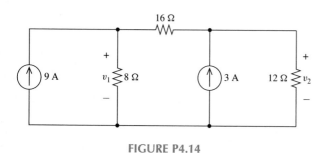

FIGURE P4.14

4.15. Using nodal analysis, find i.

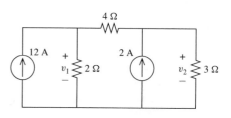

FIGURE P4.17

4.18. Repeat Problem 4.17 for the circuit shown.

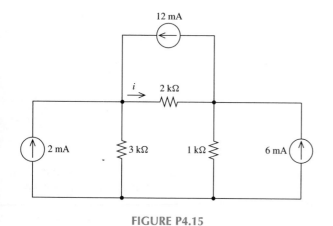

FIGURE P4.15

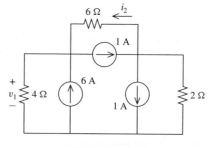

FIGURE P4.18

4.16. Using nodal analysis, find i and i_1.

4.19. Repeat Problem 4.17 for the circuit shown.

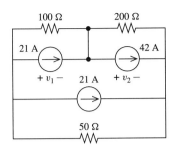

FIGURE P4.19

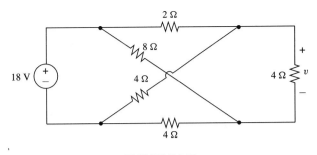

FIGURE P4.22

4.20. Find v and i using nodal analysis.

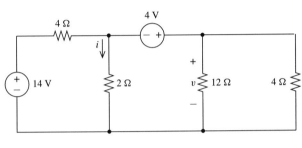

FIGURE P4.20

4.21. Using nodal analysis, find v.

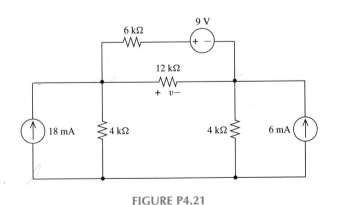

FIGURE P4.21

4.22. Using nodal analysis, find v.

4.23. Using nodal analysis, find i.

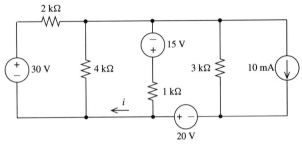

FIGURE P4.23

4.24. Repeat Problem 4.17 for the circuit shown.

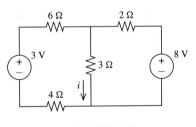

FIGURE P4.24

4.25. Replace the 8-V independent source in Problem 4.24 with an $8i$ CCVS (same reference direction as shown). Find i.

4.26. Show that by the very definition of node voltages, KVL around every loop is satisfied automatically.

4.27. Show that by the very definition of mesh currents, KCL at each node is satisfied automatically.

4.28. Use mesh analysis to find the indicated variables. Check using superposition.

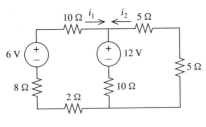

FIGURE P4.28

4.29. Repeat Problem 4.28 for the circuit given.

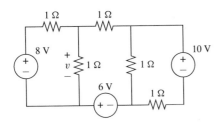

FIGURE P4.29

4.30. Use mesh analysis to find the indicated variables. Solve the mesh equations by Cramer's rule.

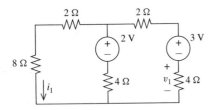

FIGURE P4.30

4.31. Repeat Problem 4.30 using matrix inversion.

4.32. Repeat Problem 4.30 using Gauss elimination.

4.33. Repeat Problem 4.30 if the 2-V independent source is replaced by a VCVS (same reference direction) with source function $-2v_1$.

4.34. Solve Problem 4.3 using mesh analysis.

4.35. Using mesh or loop analysis, find the power delivered to the 8-Ω resistor.

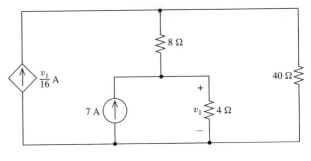

FIGURE P4.35

4.36. Solve Problem 4.15 using mesh analysis.

4.37. Find i using both nodal and loop analysis.

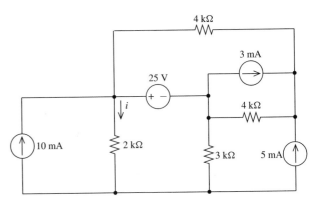

FIGURE P4.37

4.38. Find i_1 and i_2 using loop or mesh analysis.

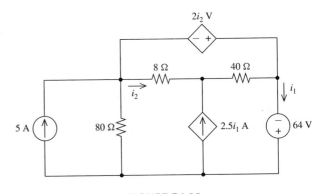

FIGURE P4.38

4.39. Find v_1.

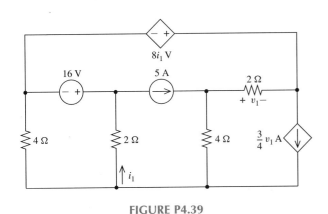

FIGURE P4.39

4.40. Solve Problem 4.56 by mesh analysis.

4.41. Solve Problem 4.23 using nodal analysis, and solve again using mesh analysis.

4.42. Using the virtual short and open principles, find v_2 in terms of v_{g1} and v_{g2}.

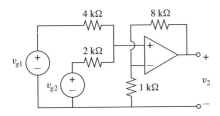

FIGURE P4.42

4.43. Repeat Problem 4.42 for the circuit shown.

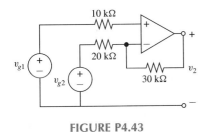

FIGURE P4.43

4.44. Use the virtual short and open principles to find the voltage transfer ratio v_2/v_1.

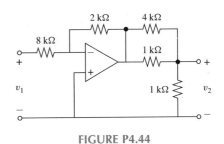

FIGURE P4.44

4.45. Repeat Problem 4.44 for this circuit. $R = 10 \text{ k}\Omega$.

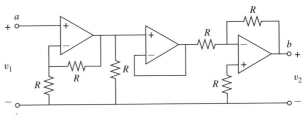

FIGURE P4.45

4.46. Repeat Problem 4.45, but add a 10-kΩ resistor between a and b.

SPICE Problems

4.47. Use SPICE to solve Problem 4.53.

4.48. Use SPICE to solve Problem 4.23.

4.49. Use SPICE to solve Problem 4.28. Do a dc sweep from 0 to 12 V on the 6-V source, with 1-V increments.

4.50. Use SPICE to solve Problem 4.44. Replace the op amp by the improved op amp model of Fig. 3.9, with $R_i = 1 \text{ M}\Omega$, $R_o = 30 \ \Omega$, and $A = 10^5$.

4.51. Use SPICE to solve Problem 4.46. Replace the op amps as directed in Problem 4.50.

4.52. This problem is intended to check how close the input current and voltage to an op amp are to zero (as the virtual open and short principles suggest). In the circuit of Problem 4.44, replace each op amp by its improved model (Fig. 3.9) with $R_i = 1 \text{ M}\Omega$, $R_o = 30 \ \Omega$, and $A = 10^5$. Let $v_1 = 1$ V. Use SPICE to determine the currents into each op amp and the voltages across these inputs.

More Challenging Problems

4.53. Use proportionality to find i_1 and i_2.

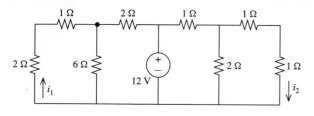

FIGURE P4.53

4.54. Find i using nodal analysis.

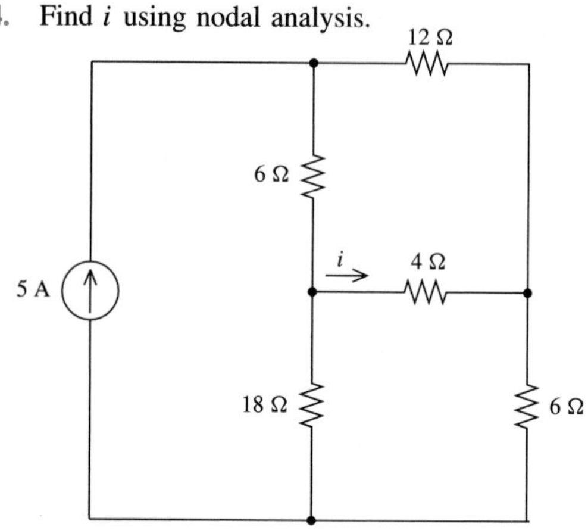

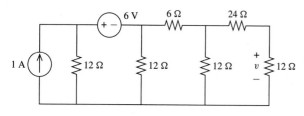

FIGURE P4.54

4.55. Use superposition to find v. For each component problem, use the proportionality principle; that is, assume that the component $v = 1$ V.

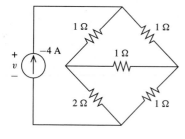

FIGURE P4.55

4.56. Use nodal analysis to find v for the circuit shown.

FIGURE P4.56

4.57. Solve by nodal analysis; then solve again using mesh analysis. Find all indicated variables.

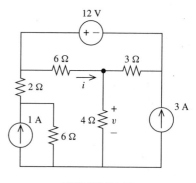

FIGURE P4.57

4.58. Use the virtual short and open principles to find the voltage transfer ratio v_2/v_1.

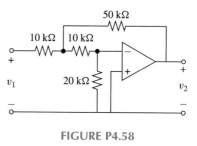

FIGURE P4.58

4.59. Use the virtual short and open principles to find the voltage transfer ratio v_2/v_1.

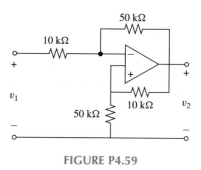

FIGURE P4.59

5

Energy-Storage Elements

Michael Faraday
1791–1867

My greatest discovery was
Michael Faraday.

Sir Humphry Davy

On August 29, 1831, Michael Faraday, the great English chemist and physicist, discovered electromagnetic induction, when he found that moving a magnet through a coil of copper wire caused an electric current to flow in the wire. Since the electric motor and generator are based on this principle, Faraday's discovery profoundly changed the course of world history. When asked by the British prime minister years later what use could be made of his discoveries, Faraday quipped, "Some day it might be possible to tax them."

Faraday, one of 10 children of a blacksmith, was born near London. He was first apprenticed to a bookbinder, but at age 22 he realized his boyhood dream by becoming assistant at the Royal Institution to his idol, the great chemist Sir Humphry Davy. He remained at the Institution for 54 years, taking over Davy's position when Davy retired. Faraday was perhaps the greatest experimentalist who ever lived, with achievements to his credit in nearly all the areas of physical science under investigation in his time. To describe the phenomena he investigated, he and a science-philosopher friend invented new words such as electrolysis, electrolyte, ion, anode, and cathode. To honor him, the unit of capacitance is named the *farad*.

Chapter Contents

Up to now we have considered only resistive circuits, that is, circuits containing resistors and sources (and op amps, which can be replaced by models consisting of resistors and sources). The terminal characteristics of these elements are simple algebraic equations that lead to circuit equations that are algebraic. In this chapter we introduce two important dynamic circuit elements, the capacitor and the inductor, whose terminal equations are integrodifferential equations (involve derivatives and/or integrals) rather than algebraic equations. These elements are referred to as *dynamic* because they store energy that can be retrieved at some later time. Another term used, for this reason, is *storage elements*. We first describe the property of capacitance and discuss the mathematical model of an ideal capacitor. The terminal characteristics and energy relations are then given, followed by equivalents for parallel and series connections of two or more capacitors. We then repeat this procedure for the inductor. The chapter concludes with a discussion of practical capacitors and inductors and their equivalent circuits.

5.1   CAPACITORS

A capacitor is a two-terminal device that consists of two conducting bodies that are separated by a nonconducting material. Such a nonconducting material is known as an *insulator* or a *dielectric*. Because of the dielectric, charges cannot move from one conducting body to the other within the device. They must therefore be transported between the conducting bodies via external circuitry connected to the terminals of the capacitor. One very simple type, the parallel-plate capacitor, is shown in Fig. 5.1. The conducting bodies are flat, rectangular conductors that are separated by the dielectric material.

Starting with a capacitor that is initially uncharged, suppose that by means of some external circuit we have transferred q coulombs of charge from one plate to the other. To determine the potential difference which results, we imagine a $+1$-C test charge subsequently being moved between points b and a in Fig. 5.1. Since the test charge is positive and there is $+q$ charge on the plate toward which we are moving, work must be done to move it. Equal work must be done on it due to attractive force of the $-q$ charge on the

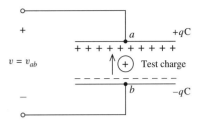

FIGURE 5.1 Parallel-plate capacitor.

plate we are leaving. Thus, from the definition of potential difference between two points in space in Chapter 2 as the net work done in transporting a +1-C test charge between the points, the plate with the $+q$ charge is at a higher potential than the other one; that is,

$$v_{ab} = f(q) \tag{5.1a}$$

where apparently the function f is an increasing one, since the larger the separated charge q, the greater the work done on the test charge.

For many physical capacitors, the general charge–voltage relationship (5.1a) is linear. A *linear capacitor* is defined as a two-terminal device whose charge–voltage relationship is a straight line intersecting the origin, $v_{ab} = f(q) = Kq$, or

$$q = Cv \tag{5.1b}$$

Here $v = v_{ab}$ is the voltage between the plates, and $C = 1/K$ is the slope of the graph plotting q as a function of v. C is known as the *capacitance* of the device, with the dimension of coulombs per volt. The unit of capacitance is known as the *farad* (F), named for the famous British physicist Michael Faraday (1791–1867). A 1-farad capacitor will separate 1 coulomb of charge for each volt of potential difference between its plates.

It is interesting to note in the example above that the net charge within the capacitor is always zero. Charges removed from one plate always appear on the other, so the total charge remains zero. We should also observe that charges leaving one terminal enter the other. This fact is consistent with the requirement that current entering one terminal must exit the other in a two-terminal device, a consequence of KCL. Since the current is defined as the time rate of change of charge, differentiating (5.1b), we find that

$$i = C\frac{dv}{dt} \tag{5.2}$$

which is the current–voltage relation or terminal law for a capacitor.

The circuit symbol for the capacitor and the current–voltage convention that satisfies (5.2) are shown in Fig. 5.2. It is apparent that moving a charge of $+q$ in Fig. 5.1 from the lower to the upper plate represents a current flowing into the upper terminal and thence onto the upper plate. The movement of this charge causes the upper terminal to become more positive than the lower one by an amount v. Hence the current–voltage convention of Fig. 5.2 is satisfied. Thus, in order that the terminal law (5.2) be satisfied, the current and voltage reference directions must be related as shown in Fig. 5.2, with

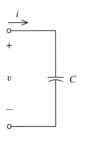

FIGURE 5.2 Circuit symbol for a capacitor.

the arrow pointing into the positive end of the voltage reference direction. This is the passive sign convention we required for use with resistors earlier. If the passive sign convention is violated by reversing either the voltage or the current reference direction, the current entering the positive terminal is $-i$ and (5.2) becomes

$$i = -C\frac{dv}{dt}$$

We recall that for this case a minus sign was also required in Ohm's law.

Example 5.1

As an example, suppose that the voltage on a 1-μF capacitor is $v = 6 \cos 2t$ V. Then, with terminal variables satisfying the passive sign convention as in Fig. 5.2,

$$i = C\frac{dv}{dt} = (10^{-6})(-12 \sin 2t)$$

$$= -12 \sin 2t \ \mu A$$

In (5.2) we see that if v is constant, the current i is zero. Therefore, *a capacitor acts like an open circuit to a dc voltage.* On the other hand, the more rapidly v changes, the larger the current flowing through its terminals.

Example 5.2

Consider a voltage that increases linearly from 0 to 1 V in a^{-1} s, given by

$$v = \begin{cases} 0, & t < 0 \\ at, & 0 \le t \le a^{-1} \\ 1, & t > a^{-1} \end{cases}$$

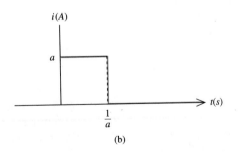

(a)

(b)

FIGURE 5.3 Voltage and current waveforms in a 1-F capacitor.

If this voltage is applied to the terminals of a 1-F capacitor (an unusually large value chosen for numerical convenience), the resulting current (in amperes) is

$$i = \begin{cases} 0, & t < 0 \\ a, & 0 \le t \le a^{-1} \\ 0, & t > a^{-1} \end{cases}$$

Plots of v and i are shown in Fig. 5.3. We see that i is zero when v is constant and that it is equal to a constant when v increases linearly. If the slope a is made larger, i must increase. It is apparent that if $a^{-1} = 0$ (a is infinite), v changes abruptly (in zero time) from 0 to 1 V.

In general, any abrupt or instantaneous changes in voltage, such as in the example above, require an infinite current flow through the capacitor. An infinite current, however, is a physical impossibility. Thus abrupt or instantaneous changes in the voltage across a capacitor are not possible, and *the voltage across a capacitor is always continuous* even though the current may be discontinuous. This is called the *continuity principle*. (Circuits that appear to violate this principle, but themselves prove to contain mathematical contradictions, are taken up in Section 5.9.)

If voltage is continuous in time, then, by (5.1b), charge must be also. Thus an alternative statement of the continuity principle, particularly useful in circuits with several capacitors, is that the *total charge cannot change instantaneously* (conservation of charge).

Let us now find $v(t)$ in terms of $i(t)$ by integrating both sides of (5.2) between times t_0 and t. The result is

$$v(t) = \frac{1}{C} \int_{t_0}^{t} i(\tau)\, d\tau + v(t_0) \qquad (5.3a)$$

where $v(t_0) = q(t_0)/C$ is the voltage on C at time t_0. In this equation, the integral term represents the voltage that accumulates on the capacitor in the interval from t_0 to t, whereas $v(t_0)$ is that which accumulates from $-\infty$ to t_0. The voltage $v(-\infty)$ is taken to be zero. Thus an alternative form of (5.3a) is

$$v(t) = \frac{1}{C} \int_{-\infty}^{t} i(\tau)\, d\tau \qquad (5.3b)$$

Example 5.3

In applying (5.3b), we are obtaining the area associated with a plot of i from $-\infty$ to t. In Fig. 5.3, for example, since $v(-\infty) = 0$ and $C = 1$ F, we have

$$v(t) = \frac{1}{1} \int_{-\infty}^{t} (0)\, d\tau + v(-\infty) = 0, \qquad t \leq 0$$

Therefore, $v(0) = 0$, and

$$v(t) = \int_{0}^{t} a\, d\tau + v(0) = at, \qquad 0 \leq t \leq a^{-1}$$

This last equation shows that $v(1/a) = 1$, so

$$v(t) = \frac{1}{1} \int_{1/a}^{t} (0)\, d\tau + v\frac{1}{a} = 1, \qquad t \geq a^{-1}$$

which agrees with $v(t)$ in Fig. 5.3. In this example we see that $v(t)$ and $i(t)$ do not necessarily have the same shape. Specifically, their maxima do not necessarily occur at the same time, nor their minima, unlike the case for the resistor. In fact, inspection of Fig. 5.3 reveals that the current may be discontinuous, even though the voltage must be continuous, as stated previously.

EXERCISES

5.1.1. A 100-nF capacitor has a voltage $v = 10 \sin 1000t$ V. Find its current.

Answer $\cos 1000t$ mA

5.1.2. A 400-μF capacitor has voltage as shown. It is piecewise linear except for $0 \le t \le 6$, $v(t) = (t)(6 - t)$. Find its current i at $t = -4, -1, 1, 5,$ and 9 s.

Answer $-0.5, 2, 1.6, -1.6,$ and -1.6 mA

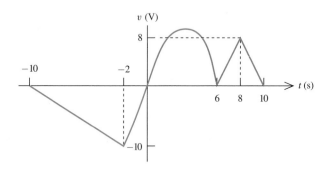

EXERCISE 5.1.2

5.1.3. A constant current of 10 mA is charging a 10-μF capacitor (entering its positive voltage terminal). If the capacitor was initially charged to 5 V, find the charge and voltage on it after 20 ms.

Answer 0.25 mC; 25 V

5.1.4. Let the graph of Exercise 5.1.2 be the graph of the current in milliamperes versus time in milliseconds in a 0.25-μF capacitor. Find the voltage at $t = -4, -1, 6,$ and 9 ms.

Answer $-90, -190, -56,$ and 0 V

5.2 ENERGY STORAGE IN CAPACITORS

The terminal voltage across a capacitor is accompanied by a separation of charges between the capacitor plates. These charges have electrical forces acting on them. An *electric field*, a basic quantity in electromagnetic theory, is defined as the position-dependent force acting on a unit positive charge. Thus the forces acting on the charges within the capacitor can be considered to result from an electric field. It is for this reason that the energy stored or accumulated in a capacitor is said to be stored in the electric field located between its plates. Quantitatively, the energy stored in a capacitor, from (1.6) and (5.2), is given by

$$w_C(t) = \int_{-\infty}^{t} vi \, d\tau = \int_{-\infty}^{t} v \left(C \frac{dv}{d\tau} \right) d\tau$$

$$w_C(t) = \int_{-\infty}^{t} Cv \, dv = \tfrac{1}{2}Cv^2(t) \qquad\qquad (5.4)$$

The last equality follows from $v(-\infty) = 0$.

From this result we see that $w_c(t) \geq 0$. Therefore, by (1.7), *the capacitor is a passive circuit element.* The only other passive element we have encountered thus far was the resistor. Resistors dissipate energy, as we have seen. Unlike the resistor, the ideal capacitor cannot dissipate any energy. The net energy $w_C(t)$ supplied to it by the external circuit is stored in the electric field within the device and can be fully recovered.

Consider, for instance, a 1-F capacitor that has a voltage of 10 V. The energy stored is

$$w_c = \tfrac{1}{2}Cv^2 = 50 \text{ J}$$

Suppose that the capacitor is disconnected from any external circuit so that no additional current can flow, and the charge, voltage, and energy all remain constant. If we subsequently connect a resistor across the capacitor, a current flows through it as the charges separated on the plates seek one another, until all the stored energy (50 J) has been dissipated in the form of heat by the resistor and the voltage across the RC combination settles to zero. Systematic analysis of such circuits is found in Chapter 6.

As was pointed out earlier, the voltage on a capacitor is a continuous function of time. Thus, by (5.4), we see that *the energy stored in a capacitor is also continuous.* This is not surprising, since otherwise, energy would have to be transported from one place to another in zero time. Since power is the rate of energy change per unit time, instantaneous change of energy demands infinite power, a physical impossibility.

To illustrate continuity of capacitor voltage, let us consider Fig. 5.4, which contains an *ideal switch* that is opened at $t = 0$, as indicated. An ideal switch is defined to be a two-terminal element that switches between an open circuit and a short circuit instantaneously. Unless otherwise specified, all switches used in this book are considered to be ideal.

To discuss the effect of the switching action, we first need to consider two times near $t = 0$. We shall denote as $t = 0-$ the time just before the switching action and $t = 0+$ as the time just after the switching action. Theoretically, no time has elapsed between $0-$ and $0+$, but the two times represent radically different states of the circuit. $v_c(0-)$ is the voltage on the capacitor just before the switch acts, and $v_c(0+)$ is the voltage immediately after switching. Similarly, $i_C(0-)$ and $i_C(0+)$ are the capacitive current just before and just after the switch changes state. More formally, $v_c(0-)$ is the limit of $v_c(t)$ as t approaches zero from the left [through negative ($t < 0$) values], and $v_c(0+)$ is the limit as t approaches zero from the right [through positive ($t > 0$) values]. Indeed, mathematically, the times $0-$ and $0+$ are convenient fictions without precise definition (when is the first instant greater than zero?), although the values of circuit variables such as $v_C(0-)$ and $v_C(0+)$ are defined precisely by the limits from the left and right.

Example 5.4

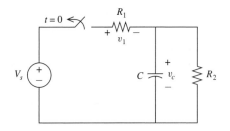

FIGURE 5.4 Circuit illustrating continuity of capacitor voltage.

Suppose that in Fig. 5.4 we have $V_s = 6$ V and $v_c(0) = 4$ V. Just prior to the switching action ($t = 0-$), we have $v_1(0-) = V_s - v_c(0-) = 2$ V. Immediately after the switch is opened, we have $v_1(0+) = 0$, since no current is flowing in R_1 (it is in series with an open circuit). However, since v_c, a capacitive voltage, must be continuous, we have

$$v_c(0+) = v_c(0-) = 4 \text{ V}$$

Thus the voltage on resistor R_1 has changed abruptly, but that on the capacitor has not. The voltage on R_2 is the same as the capacitor voltage and thus has not changed abruptly either.

Although apparently prohibited by the continuity principle, circuit diagrams can be drawn for which capacitor voltages seem forced to change abruptly. For example, consider two capacitors and an ideal switch forming a closed loop. If the capacitors have unequal voltages just as the switch closes, by continuity their values must remain unequal at $t = 0+$, while by KVL their voltages must be equal at $t = 0+$. Apparently, we must either concede an infinite power flow (by accepting violation of the continuity principle) or abandon KVL for these circuits. We shall consider such *singular circuits* in more detail in Section 5.9. Singular circuits are similar to the circuits discussed in Chapter 2 that contain unequal voltage sources in parallel or current sources in series. Such circuits lead to mathematical contradictions and cannot be analyzed consistently. In practice, the contradiction is removed by replacing ideal elements by more accurate physical models of them, including Thevenin equivalent resistance in the case of the sources of Chapter 2 and parallel leakage resistance in the present case of capacitors. In the real world, continuity and the Kirchhoff laws are both valid in lumped physical circuits.

EXERCISES

5.2.1. A 0.2-μF capacitor has a charge of 20 μC. Find its voltage and energy.
Answer 100 V; 1 mJ

5.2.2. If the energy stored in a 0.5-F capacitor is 25 J, find the voltage and charge.
Answer 10 V; 5 C

5.2.3. In Fig. 5.4, let $C = \frac{1}{4}$ F, $R_1 = R_2 = 4$ Ω, and $V = 20$ V. If the current in R_2 at $t = 0-$ is 2 A directed downward, find at $t = 0-$ and at $t = 0+$ (a) the charge on the capacitor, (b) the current in R_1 directed to the right, (c) the current in C directed downward, and (d) dv_C/dt.
Answer (a) 2, 2 C; (b) 3, 0 A; (c) 1, -2 A; (d) 4, -8 V/s

5.3 SERIES AND PARALLEL CAPACITORS

In this section we determine the equivalents for series and parallel connections of capacitors. Recall that two subcircuits are equivalent if they have the same terminal law. We shall match the terminal law for series-connected capacitors with that for a single capacitance, the series equivalent capacitance, and then repeat for the parallel case.

Let us first consider the series connection of N capacitors, as shown in Fig. 5.5(a). Applying KVL, we find

$$v = v_1 + v_2 + \cdots + v_N \tag{5.5}$$

Using the terminal law (5.3a) gives

$$v(t) = \left(\frac{1}{C_1} \int_{t_0}^{t} i \, d\tau + v_1(t_0) \right)$$
$$+ \left(\frac{1}{C_2} \int_{t_0}^{t} i \, d\tau + v_2(t_0) \right) + \cdots + \left(\frac{1}{C_N} \int_{t_0}^{t} i \, d\tau + v_N(t_0) \right)$$
$$= \left(\sum_{n=1}^{N} \frac{1}{C_n} \right) \int_{t_0}^{t} i \, d\tau + v(t_0)$$

where in the last we have used (5.5). This is the desired terminal law for the series-connected capacitors. The terminal law for Fig. 5.5(b) is simply

$$v(t) = \frac{1}{C_s} \int_{t}^{t} i \, d\tau + v(t_0)$$

The two terminal laws are identical, hence the circuits are equivalent, if

$$\frac{1}{C_s} = \sum_{n=1}^{N} \frac{1}{C_n} \tag{5.6}$$

The equivalent of a chain of series-connected capacitors is a single capacitor whose inverse capacitance is the sum of inverses of the series capacitances.

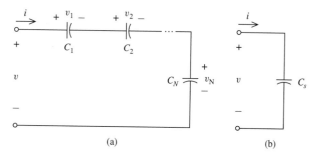

(a)   (b)

FIGURE 5.5 (a) Series capacitors; (b) equivalent circuit.

In the case of just two capacitors in series, this general result simplifies to

$$C_s = \frac{C_1 C_2}{C_1 + C_2}$$

For two capacitors in series, the equivalent capacitance is the product over the sum of the two individual capacitances. The product-by-sum equivalence rule was introduced in (2.15) for the case of two resistors in parallel. As with parallel resistors, this rule is limited to the special case of exactly two such elements. For three or more capacitors in series, the general result (5.6) should be used.

We next turn to the parallel connection of N capacitors, as shown in Fig. 5.6(a). KCL at the top node yields

$$i = C_1 \frac{dv}{dt} + C_2 \frac{dv}{dt} + \cdots + C_N \frac{dv}{dt}$$

or

$$i = \left(\sum_{n=1}^{N} C_n \right) \frac{dv}{dt}$$

The corresponding terminal law for Fig. 5.6(b) is

$$i = C_p \frac{dv}{dt}$$

Comparing the last two equations, the terminal laws are identical and the circuits equivalent if

$$C_p = \sum_{n=1}^{N} C_n \qquad (5.7)$$

The equivalent of parallel capacitors is a single capacitor whose capacitance is the sum of capacitances of the parallel capacitors. More simply put, parallel capacitance adds.

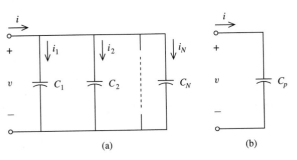

(a)   (b)

FIGURE 5.6   (a) Parallel capacitors; (b) equivalent circuit.

Example 5.5

Determine the equivalent capacitance C_{eq} of the circuit shown in Fig. 5.7(a).

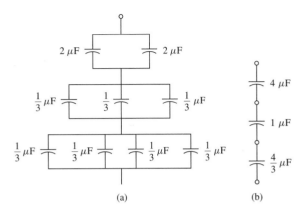

(a)   (b)

FIGURE 5.7   Circuit for Example 5.5.

Since parallel capacitances add, Fig. 5.7(a) may be immediately simplified to its equivalent, Fig. 5.7(b). Then, using the series equivalent rule (5.6), with μF units,

$$\frac{1}{C_{eq}} = \frac{1}{4} + \frac{1}{1} + \frac{1}{\frac{4}{3}} = 2$$

or $C_{eq} = \frac{1}{2}\mu F$.

Example 5.6

Determine the voltage $v_1(t)$ for $t \geq 0$ in Fig. 5.8(a) if $v_1(0) = 10$ V. The equivalent of the series-connected capacitors is shown in Fig. 5.8(b),

$$\frac{1}{C_s} = \frac{1}{1} + \frac{1}{2} + \frac{1}{2}$$

or $C_s = \frac{1}{2}$ F. Then

$$i = C_s \frac{d}{dt}(3\sin 2t) = \frac{1}{2}(6\cos 2t) = 3\cos 2t \text{ A}$$

The voltage across the 1-F capacitor is then given by (5.3), or

$$v_1(t) = \frac{1}{C_1}\int_{t_0}^{t} i(\tau)\,d\tau + v_1(t_0) = \int_{0}^{t} (3\cos 2\tau)\,d\tau + 10$$

which yields $v_1(t) = \frac{3}{2}\sin 2t + 10$ V. Note that evaluating this result at $t = 0$ gives the correct initial value for $v_1(t)$. We may also check that

$$i = C_1 \frac{dv_1}{dt} = 1\frac{d}{dt}\left(\frac{3}{2}\sin 2t + 10\right) = 3\cos 2t \text{ A}$$

which is consistent with our computed value for $i(t)$.

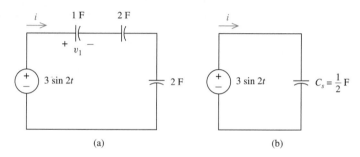

(a)
(b)

FIGURE 5.8   Circuit for Example 5.6.

It is interesting to note that the equivalent capacitance of series and parallel capacitors follows the same rules as equivalent conductance of series and parallel conductances. That is, for the parallel case the equivalent parameter (conductance G_P or capacitance C_P) is the sum of the individual parameters, while in series the equivalent parameter (G_S or C_S) is the inverse of the sum of inverses of the individual parameters. This correspondence follows from the fact that the terminal laws for the conductance and capacitance are of the same form: the parameter (G or C) multiplies the terminal voltage (or derivative of terminal voltage) to yield terminal current. This functional correspondence makes the foregoing derivations of capacitive equivalents correspond line for line to those for conductance equivalents presented earlier in Chapter 2.

EXERCISES

5.3.1. Find the maximum and minimum values of capacitance that can be obtained from ten 1-μF capacitors. How should they be connected?
Answer 10 μF (parallel); 0.1 μF (series connected)

5.3.2. Find the equivalent capacitance.
Answer 20 μF

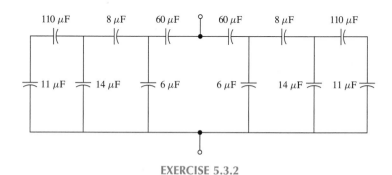

EXERCISE 5.3.2

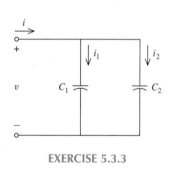

EXERCISE 5.3.3

5.3.3. Derive an equation for current division between two parallel capacitors by finding i_1 and i_2.
Answer $C_1 i/(C_1 + C_2)$; $C_2 i/(C_1 + C_2)$

5.3.4. Derive an equation for voltage division between two initially uncharged series capacitors by finding v_1 and v_2.

Answer $C_2 v/(C_1 + C_2)$; $C_1 v/(C_1 + C_2)$

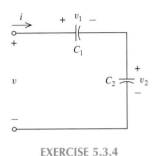

EXERCISE 5.3.4

5.4   INDUCTORS

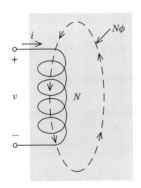

FIGURE 5.9 Simple model of an inductor.

In the preceding sections we found that the terminal characteristics of the capacitor are the result of forces that exist between electric charges due to their separation in space. Just as these charges exert position-dependent or *electrostatic* force upon one another, so moving charges, or currents, exert *electrodynamic* force.

The force that is exchanged by two neighboring current-carrying wires was determined experimentally by the French scientist André Marie Ampère (1775–1836) in the early nineteenth century. This force can be characterized by postulating the existence of a *magnetic field*. The magnetic field, in turn, can be thought of in terms of a *magnetic flux* that forms closed loops about electric currents. The cause of the flux is motion of charge in the wires, that is, the current. The study of magnetic fields, like that of electric fields mentioned in Section 5.3, comes in a course devoted to electromagnetic theory. Here we are interested primarily in the role of magnetic fields in linear lumped circuits, most prominently in the circuit element called an inductor.

An *inductor* is a two-terminal device that consists of a coiled conducting wire wound around a core. A current flowing through the device produces a magnetic flux ϕ that forms closed loops threading its coils, as shown in Fig. 5.9. Suppose that the coil contains N turns and that the flux ϕ passes through each turn. In this case the total flux linked by the N turns of the coil, denoted by λ, is

$$\lambda = N\phi$$

This total flux is commonly referred to as the *flux linkage*. The unit of magnetic flux is the weber (Wb), named for the German physicist Wilhelm Weber (1804–1891). In a linear inductor, the flux linkage is directly proportional to the current flowing through the device. For linear inductors, therefore, we may write

$$\lambda = Li \tag{5.8}$$

where L, the constant of proportionality, is the inductance in webers per ampere. The unit of 1 Wb/A is known as the *henry* (H), named for the American physicist Joseph Henry (1797–1878).

In (5.8) we see that an increase in i produces a corresponding increase in λ. This increase in λ induces a voltage in the N-turn coil. The fact that voltages are induced by time-varying magnetic flux was first discovered by Henry. Henry, however, repeating the mistake of Cavendish with the resistor, failed to publish his findings. As a result, Faraday is credited with discovering the law of electromagnetic induction. This law states that the induced voltage is equal to the time rate of change of the total magnetic flux. In mathematical form, the law is

$$v = \frac{d\lambda}{dt}$$

which with (5.8) yields the terminal law for an inductor:

$$v = L\frac{di}{dt} \tag{5.9}$$

Clearly, as i increases, a voltage is developed across the terminals of the inductor, the polarity of which is shown in Fig. 5.10. This voltage opposes any additional increase in i, for if this were not the case, that is, if the polarity were reversed, the induced voltage would "aid" the current. This physically cannot be true because the current and voltage would then both increase indefinitely.

The circuit symbol for the inductor whose terminal law is given in (5.9) and shown in Fig. 5.10. Just as in the cases of the resistor and the capacitor, the voltage and current reference directions are assumed to satisfy the passive sign convention. If either reference direction (but not both) is reversed, a negative sign must be introduced into the terminal law, yielding $v = -L(di/dt)$.

The terminal law of the inductor, (5.9), shows that if i is constant, the voltage v is zero. Therefore, *an inductor acts like a short circuit to a dc current.* On the other hand, the more rapidly i changes, the greater is the voltage that appears across its terminals. This is due to the more rapidly time-varying magnetic flux i produced, which is the cause of the induced voltage v across the terminals of the inductor.

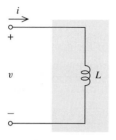

FIGURE 5.10   Circuit symbol for an inductor.

Example 5.7

Consider a current that increases linearly from 0 to 1 V in a^{-1} s, given by

$$i = \begin{cases} 0, & t < 0 \\ at, & 0 \le t \le a^{-1} \\ 1, & t > a^{-1} \end{cases}$$

If this current is applied to the terminals of a 1-H inductor, the resulting voltage is

$$v = \begin{cases} 0, & t < 0 \\ a, & 0 \le t \le a^{-1} \\ 0, & t > a^{-1} \end{cases}$$

Plots of i and v are shown in Fig. 5.11. We see that v is zero when i is constant and that it is equal to a when i increases linearly. If a

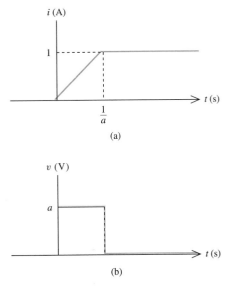

FIGURE 5.11 Current and voltage waveforms in a 1-H inductor.

is made larger, v changes more rapidly and v increases. It is apparent that if $a^{-1} = 0$ (a is infinite) i changes abruptly (in zero time) from 0 to 1 V.

In general, abrupt changes in the current through an inductor, such as in Example 5.7, require that an infinite voltage appear across the terminals of the inductor. As observed in the case of the capacitor, infinite current or voltage is a physical impossibility. Thus instantaneous changes in the current through an inductor are not possible. We observe that *the current through an inductor is always continuous.* This is the other half of the *continuity principle,* shown previously to apply to capacitive voltage. Inductive current is also continuous, even though the voltage across an inductor may change discontinuously, as may capacitive current.

Another interesting parallel to our previous discussion of the capacitor is revealed by comparing Example 5.7 with Example 5.2. We see that by reversing the roles of i and v the same terminal pair is produced in the inductor as in the capacitor. This is less surprising when we note that the terminal laws for these two elements, $i = C(dv/dt)$ and $v = L(di/dt)$, are of the same form with only the current and voltage interchanged. This *duality* of capacitor and inductor will make the remaining results for the inductor strongly reminiscent of the previous ones for the capacitor.

Let us find the current $i(t)$ in terms of the voltage $v(t)$ for the inductor. Integrating (5.9) from time t_0 to t and solving for $i(t)$, we have

$$i(t) = \frac{1}{L} \int_{t_0}^{t} v(\tau)\, d\tau + i(t_0) \tag{5.10}$$

In this equation the integral term represents the current buildup from time t_0 to t, whereas $i(t_0)$ is the current at t_0. Obviously, $i(t_0)$ is the current that accumulates from $t = -\infty$ to t_0, where $i(-\infty) = 0$. Thus an alternative to (5.10) is

$$i(t) = \frac{1}{L} \int_{-\infty}^{t} v(\tau) \, d\tau$$

Example 5.8

Consider a 1-H inductor. In the application of (5.10) we are obtaining the net area under the graph of v from $-\infty$ to t, since $i(t_0)$ represents the area from $-\infty$ to t_0. Setting $t_0 = 0$ in Fig. 5.11, for instance, we can compute the current at $t = 1/a$ to be

$$i\left(\frac{1}{a}\right) = \frac{1}{1} \int_{0}^{1/a} v(\tau) \, d\tau + 0 = \int_{0}^{1/a} a \, d\tau = 1 \text{ A}$$

If we prefer, we may set $t_0 = \frac{1}{2}a$, a time at which $i(t_0) = \frac{1}{2}$, and recompute

$$i\left(\frac{1}{a}\right) = \frac{1}{1} \int_{1/2a}^{1/a} v(\tau) \, d\tau + v\frac{1}{2a} = \int_{1/2a}^{1/a} a \, d\tau + \frac{1}{2} = 1 \text{ A}$$

In this example we see that v and i, just as in the case of the capacitor, do not necessarily have the same variation in time. Inspection of Fig. 5.11, for example, shows that the voltage can be discontinuous even though the current is always continuous.

EXERCISES

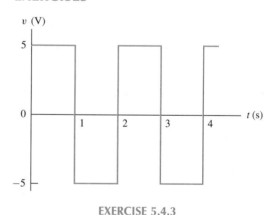

EXERCISE 5.4.3

5.4.1. A 10-mH inductor has a current of $50\cos 1000t$ mA. Find its voltage and its flux linkage. Assume that the passive sign convention is satisfied.

Answer $-0.5 \sin 1000t$ V; $0.5 \cos 1000t$ mWb

5.4.2. Find the current $i(t)$ for $t > 0$ in a 20-mH inductor having a voltage of $-5 \sin 50t$ V if $i(0) = 5$ A. Assume that the passive sign convention is satisfied.

Answer $5 \cos 50t$ A

5.4.3. Find the current in a 0.5-H inductor, for $0 \le t \le 2$ s, if $i(0) = 0$ and the voltage is as shown.

Answer $10t$ A, $0 \le t \le 1$; $10(2 - t)$ A, $1 < t \le 2$

5.5 ENERGY STORAGE IN INDUCTORS

A current i flowing through an inductor causes a total flux linkage λ to be produced, which passes through the turns of the inductor's coil. Just as work was performed in moving charges between the plates of a capacitor, work must be done by the external circuit to establish the flux ϕ in the inductor. This work or energy is said to be stored in the magnetic field. Just as was the case for energy stored in the electric field of a capacitor, it can be reclaimed later by the external circuit. Using (1.6) and (5.9), the energy stored in an inductor is given by

$$w_L(t) = \int_{-\infty}^{t} vi\, d\tau = \int_{-\infty}^{t} \left(L \frac{di}{d\tau} \right) i\, d\tau \qquad (5.11a)$$

$$w_L = \int_{i(-\infty)}^{i(t)} Li\, di = \tfrac{1}{2} Li^2(t) \qquad (5.11b)$$

Inspection of the result reveals that $w_L(t) \geq 0$. Therefore, by (1.7), we see that *the inductor is a passive circuit element*. Like the ideal capacitor, the ideal inductor does not dissipate any power. Consider, for example, a 2-H inductor that is carrying a current of 5 A. The energy stored is

$$w_L = \tfrac{1}{2} Li^2 = 25 \text{ J}$$

Suppose that this inductor is disconnected instantaneously from the external circuit that caused the 5-A current to flow and simultaneously is connected across a resistor. Current will flow through the inductor–resistor combination until all the energy previously stored in the inductor (25 J) has been dissipated by the resistor and the current settles to zero. Complete analysis of circuits of this type is found in Chapter 6.

Example 5.9

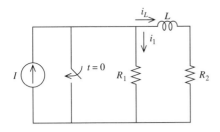

FIGURE 5.12   Circuit for Example 5.9.

Since inductive currents are continuous, it follows that the energy stored in an inductor, like that stored in a capacitor, is also continuous. To illustrate this, consider the circuit of Fig. 5.12, which contains a switch that is closed at $t = 0$, as indicated. Suppose that $i_L(0-) = 2$ A and $I = 3$ A. Then $i_1(0-) = 3 - 2 = 1$ A. Just after the switch is closed, we have $i_1(0+) = 0$ since a short circuit is placed across R_1. However, we have, by continuity,

$$i_L(0+) = i_L(0-) = 2 \text{ A}$$

Thus the resistor current has changed abruptly, but the inductive current has not.

An example of a singular circuit for which inductive currents appear to be discontinuous is given in Section 5.9. As in the case of singular capacitive circuits, the apparent discontinuity in the energy stored in the inductors cannot be accounted for in the ideal lumped-circuit model. Physical circuits having inductors contain associated resistance that does not permit the infinite inductor voltages that must accompany abrupt changes in inductor currents. We shall be concerned primarily with circuits of this type.

5.5.1. Derive an expression for the energy stored in an inductor in terms of the flux linkage λ and the inductance L.

Answer $\lambda^2/2L$

5.5.2. A 40-mH inductor has a current $i = 100 \cos 10\pi t$ mA. Find the flux linkage, voltage, and the energy at $t = 1/30$ s.

Answer 2 mWb; $-20\sqrt{3}\pi$ mV; 50 μJ

5.5.3. A 2-mH inductor has a voltage $v = 2 \cos 1000t$ V with $i(0) = 1.5$ A. (a) Find the energy stored in the inductor at $t = \pi/6$ ms. (b) What is the maximum energy stored and at what time(s)? (c) What is the minimum energy stored and at what time(s)?

Answer (a) 4 mJ; (b) $\frac{25}{4}$ mJ at $t = 2\pi n + \pi/2$; (c) $\frac{1}{4}$ mJ at $t = (2n + 1)\pi + \pi/2$, $n = 0, 1, \ldots$

5.5.4. In Fig. 5.12, let $I = 5$ A, $R_1 = 6$ Ω, $R_2 = 4$ Ω, $L = 2$ H, and $i_1(0-) = 2$ A. If the switch closes at time $t = 0$, find $i_L(0-)$, $i_L(0+)$, $i_1(0+)$, and $di_L(0+)/dt$.

Answer 3 A; 3 A; 0; -6 A/s

5.6 SERIES AND PARALLEL INDUCTORS

In this section we determine the equivalent inductance for series and parallel connections of inductors. Just as we found the series and parallel capacitive equivalents to follow the rules derived for conductances, we will now see that series and parallel inductors have equivalents whose values follow the same rules we derived for resistors.

First consider the series connection of N inductors shown in Fig. 5.13(a). Summing voltage drops around the loop and using the terminal law $v = L(di/dt)$ gives

$$v = L_1 \frac{di}{dt} + L_2 \frac{di}{dt} + \cdots + L_N \frac{di}{dt}$$

$$= \left(\sum_{n=1}^{N} L_N \right) \frac{di}{dt}$$

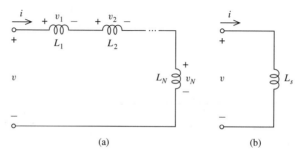

(a)   (b)

FIGURE 5.13 (a) Series inductors; (b) equivalent circuit.

Comparing the terminal law from Fig. 5.13(b), $v = L_S(di/dt)$, the two circuits are equivalent if

$$L_S = \sum_{n=1}^{N} L_N \qquad (5.12)$$

The equivalent of a chain of series inductors is a single inductor whose inductance is the sum of the individual inductances. Less formally, series inductance adds.

The parallel case is illustrated in Fig. 5.14(a). Summing currents at the top node and using the terminal law in the form (5.10),

$$i(t) = \left(\frac{1}{L_1} \int_{t_0}^{t} v \, d\tau + i_1(t_0) \right) + \left(\frac{1}{L_2} \int_{t_0}^{t} v \, d\tau + i_2(t_0) \right) + \cdots$$

$$+ \left(\frac{1}{L_N} \int_{t_0}^{t} v \, d\tau + i_N(t_0) \right)$$

Gathering like terms and noting that $i = i_1 + i_2 + \cdots + i_N$, the terminal law is

$$i(t) = \left(\sum_{n=1}^{N} \frac{1}{L_n} \right) \int_{t_0}^{t} v \, d\tau + i(t_0)$$

Comparing the terminal law of Fig. 5.14(b) given in (5.10), they match if

$$\frac{1}{L_P} = \sum_{n=1}^{N} \frac{1}{L_n} \qquad (5.13)$$

The equivalent of parallel inductors is a single inductor whose inverse inductance is the sum of the inverses of the parallel inductances.

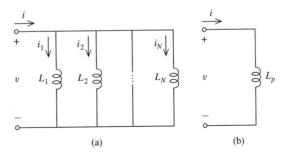

(a)   (b)

FIGURE 5.14   (a) Parallel inductors; (b) equivalent circuit.

In the case of two parallel inductors L_1 and L_2, (5.13) may be simplified to

$$L_P = \frac{L_1 L_2}{L_1 + L_2}$$

which is directly analogous to the equivalence of two parallel resistances (2.15). As mentioned previously in that case, the product-by-sum rule only holds for exactly two elements. For three or more, the general result (5.13) should be applied.

Example 5.10

We wish to find v in Fig. 5.15. We will first find an equivalent inductor. The two 1-H inductors in series are equivalent to a 2-H inductor, which is in parallel with the 2-H inductor shown in the figure. Their parallel equivalent is

$$L_{eq1} = \frac{(2)(2)}{2+2} = 1 \text{ H}$$

L_{eq1} is in turn in series with $\frac{2}{3}$ H, yielding a series equivalent of $\frac{5}{3}$ H. Finally, this is in parallel with the $\frac{5}{2}$-H inductor for an overall equivalent L_{eq}:

$$\frac{1}{L_{eq}} = \frac{1}{\frac{5}{3}} + \frac{1}{\frac{5}{2}} = \frac{3}{5} + \frac{2}{5} = 1$$

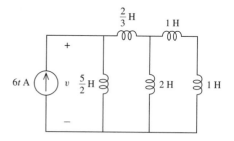

FIGURE 5.15 Circuit for Example 5.10.

and $L_{eq} = 1$ H. Then, for this circuit,

$$v = 1 \frac{d}{dt}(6t) = 6 \text{ V}$$

EXERCISES

5.6.1. Find the maximum and minimum values of inductance that can be obtained using five 20-mH inductors and ten 10-mH inductors.
Answer 200 mH; 800 μH

5.6.2. Find the equivalent inductance. Inductor values shown are in millihenries.
Answer 10 mH

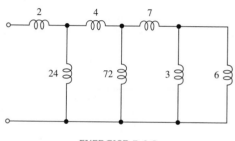

EXERCISE 5.6.2

5.6.3. Derive an equation for voltage division between two series inductors by finding v_1 and v_2.

Answer $L_1v/(L_1 + L_2)$; $L_2v/(L_1 + L_2)$

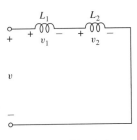

EXERCISE 5.6.3

5.6.4. Derive an equation for current division between two parallel inductors with no initial current by finding i_1 and i_2.

Answer $\dfrac{L_2 i}{L_1 + L_2}$; $\dfrac{L_1 i}{L_1 + L_2}$

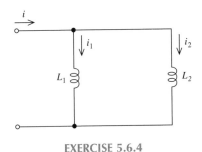

EXERCISE 5.6.4

5.7 DC STEADY STATE

Before introduction of capacitors and inductors, we focused on resistive circuits, those containing only resistors and sources. In resistive circuits whose independent sources are all dc (constant values), it is significant that all currents and voltages will also be dc. This follows from the fact that nodal or mesh analysis equations will contain no time-varying quantities. The solutions of these equations will clearly be constants, that is, dc values. Substituting these constant node voltages or mesh currents into Ohm's law to find the remaining circuit variables will produce only more dc currents and voltages.

A circuit is said to be in *dc steady state* if all currents and voltages in that circuit are constant. Clearly, only circuits whose independent sources are all dc can possibly attain dc steady state, for if some source function were not constant, that current or voltage would not be dc. From Section 5.6, resistive circuits containing only dc independent sources are always in dc steady state. However, the introduction of derivatives and integrals into the analysis equations by inductors and capacitors makes time-varying solutions a distinct possibility even when sources are constant. The relationship of dc steady state to circuits containing inductors and capacitors is considered next.

It was noted earlier in this chapter that since $i = C \, dv/dt$ is the terminal law for a capacitor, if v is a constant, then $i = 0$; that is, the capacitor is equivalent to an open circuit. Similarly, if i is a constant, $v = L(di/dt)$ suggests that the inductor is equivalent to a short circuit since its voltage is $v = 0$. Thus *in dc steady state, where all currents and voltages are constants, we may analyze the circuit by replacing inductors by short circuits and capacitors by open circuits.*

Example 5.11

Assume that the circuit of Fig. 5.16(a) is in dc steady state. Find i and v.

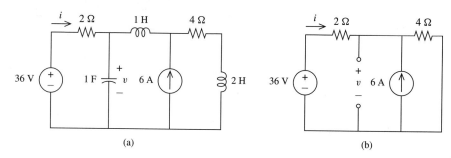

(a)   (b)

FIGURE 5.16   (a) Circuit; (b) dc steady-state equivalent.

In dc steady state we may replace inductors by shorts and capacitors by open circuits, as shown in Fig. 5.16(b). Applying mesh analysis, this is a two-mesh circuit. The presence of the current source reduces the analysis to the single equation around the super-mesh consisting of the outer loop,

$$2i + 4(i + 6) = 36$$

or $i = 2$ A. Then, by KVL,

$$v = 4(i + 6) = 4(8) = 32 \text{ V}$$

The assumption of dc steady state in *RLC circuits* (those with resistors, inductors, and capacitors), which contain dc sources only, thus leads to self-consistent results. Replacing inductors and capacitors by shorts and opens, we are left with a resistive circuit to analyze and are thus guaranteed a dc steady-state outcome of the analysis. Self-consistency, however, only proves that dc steady state is possible with such circuits, not that it is inevitable. As discussed in Chapter 6, most *RLC* circuits with dc sources eventually converge to dc steady state. But some never do; and even for those that do, the next example shows that they typically do not "start" in dc steady state.

Example 5.12

Suppose that before time $t = t_0$ the capacitor in Fig. 5.16 is charged up to $v = 100$ V and at t_0 is switched into the position shown in the circuit. Then, by continuity of capacitive voltages, its value at $t = t_0+$ will be 100 V as well. But the dc steady-state value for this voltage is 32 V, as shown in Example 5.11. We must conclude that this circuit is not in dc steady state at time $t = t_0+$, since for $t > t_0$ at least one current or voltage in this circuit will not be dc.

Dc steady state is an important concept that we will meet again at several points in our investigation of linear circuits. It is particularly helpful in determining initial conditions for many circuits with time-varying responses. Fortunately, performing a dc steady-state analysis on an RLC circuit is simplified by the replacement of inductors and capacitors by shorts and opens, thus becoming the straightforward job of solving a reduced resistive, rather than RLC, circuit.

EXERCISES

5.7.1. Find the dc steady-state value of i_1, i_2, and i_3.
Answer 0; 3 A; $\frac{3}{2}$ A

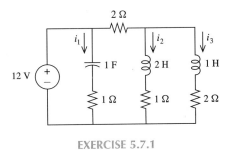

EXERCISE 5.7.1

5.7.2. Assume that this circuit is in dc steady state just before $t = 0$. Find $v(0-)$ and $v(0+)$. Is the circuit in dc steady state just after $t = 0$?
Answer 1 V; 0 V; no

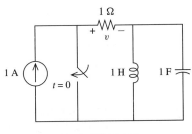

EXERCISE 5.7.2

5.7.3. Find a value for the source function v_g so that $i(0+) = 2$ A. For this v_g, find $v(0-)$ and $v(0+)$.

Answer -4 V; -4 V; -4 V

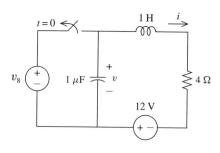

EXERCISE 5.7.3

5.8 PRACTICAL CAPACITORS AND INDUCTORS

Commercially important capacitors are produced either as discrete elements or are deposited onto integrated-circuit (IC) substrates during the IC manufacturing process. Discrete capacitors are available in a wide variety of types, values, and voltage ratings. The capacitor type is generally classified by the kind of dielectric used, and its capacitance is determined by the type of dielectric and the physical geometry of the device. The voltage rating, or working voltage, is the maximum voltage that can safely be applied to the capacitor. Voltages exceeding this value may permanently damage the device by destroying or breaking down the dielectric.

Simple discrete capacitors may be constructed using two strips of metal foil separated by a dielectric insulating material. The foil and dielectric are pressed together into sheets that are then rolled or folded into a compact package. Electrical conductors attached to each metal-foil sheet constitute the terminals of the capacitor.

Unlike ideal capacitors, practical capacitors dissipate small but nonzero amounts of power. This is due primarily to *leakage currents* that flow within the dielectric material in the device. Practical dielectrics have a nonzero conductance that allows charge to leak directly from one capacitor plate back to the other (internally, in addition to the nonleakage current moving through the external circuit path). Leakage current may be included in a circuit model for a practical capacitor by placing a resistance in parallel with an ideal capacitor, as shown in Fig. 5.17.

Common types of discrete capacitors, designated by their dielectric, include ceramic, Mylar, Teflon, and polystyrene. In addition, larger capacitive values are usually of the electrolytic type requiring polarized use (one terminal of the capacitor must be kept at a higher potential than the other). The materials and structures of integrated-circuit capacitors depend on the IC technology being used, frequently consisting of a thin layer of an insulating oxide dielectric between layers of doped semiconductor materials acting as the conducting plates.

Practical inductors are available only as discrete elements (or preassembled into electronic packages containing many discrete elements), not as constituents of integrated

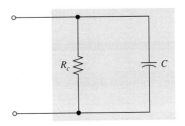

FIGURE 5.17   Circuit model for a practical capacitor.

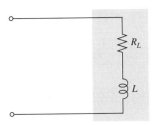

FIGURE 5.18   Circuit model for a practical inductor.

circuits. Like practical capacitors, they dissipate a small but nonzero amount of power. This dissipation results from losses associated with the fact that the wire making up the inductor coil has some resistance and also from core losses (losses due to currents induced in the core by the strong magnetic field present there). The magnetic flux is best focused in the inductor core by the use of ferromagnetic materials, but this also increases the core losses. These losses may be modeled by inserting a series resistor into the circuit model for a practical inductor as shown in Fig. 5.18.

The unavailability of IC inductors, a consequence of the difficulty of reproducing coils on flat surfaces and also of the poor magnetic properties of most IC materials, has limited the use of inductors in modern miniaturized circuit design. Much effort has been devoted to finding inductorless designs for IC applications. There are still many electronic applications, however, where inductors are commonplace, ranging from telephone circuits and radio receivers to power supplies and electric motors.

EXERCISES

5.8.1.   Suppose that we wish to construct a 1-F capacitor by putting 10-μF capacitors in parallel. How many do we need? If the leakage resistance is 10 MΩ for each 10-μF capacitor, what will the total leakage resistance and current be if the 1-F capacitor is put across a 100-V dc source?

Answer 100,000; 100 Ω; 1 A

5.8.2.   Suppose that we create a 50-mH inductor by winding 10,000 turns of thin wire around a 1-cm-diameter form. If the wire has a resistance of 0.01 Ω/cm, what power will be dissipated by the inductor if it carries a current of 20 mA dc? What would it be if this practical inductor was an ideal inductor?

Answer 40π mW; 0 mW

5.9   SINGULAR CIRCUITS

A circuit in which a switching action takes place that *appears* to violate the continuity principle by producing discontinuities in capacitor voltages or inductor currents is called a *singular circuit*. In this section we consider such circuits and show how the apparent violation may be consistently interpreted and resolved.

Example 5.13

Let us consider first Fig. 5.19(a), where the 1-F capacitors C_1 and C_2 have voltages of 1 V and 0 V, respectively, just prior to the closing of the switch. That is, $v_1(0-) = 1$ V and $v_2(0-) = 0$ V. Just after the switch closes, continuity requires that $v_1(0+) = 1$ and $v_2(0+) = 0$. But KVL applied at time 0+ demands that $v_1(0+) + v_2(0+) = 0$. Either continuity or KVL is apparently violated. Which rule should give way?

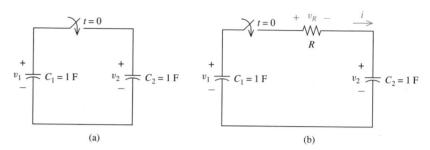

FIGURE 5.19   (a) Singular circuit with capacitors; (b) improved circuit model.

Recall that our justification of continuity was based on the assertion that infinite currents and voltages are not possible in *practical* circuits. In the realm of pure mathematics, there is nothing wrong with the concept of infinity. It serves us well in imagining what the function $f(x) = 1/x$ approaches as x goes to zero, or what the sum of all positive integers up to N goes to as N gets bigger and bigger.

What goes wrong in singular circuits is that they are not useful models for *practical* circuits, those that can be approximated in practice in the real world by physical means. They are similar to other impractical circuits we have identified previously, that is, parallel voltage sources, or series current sources, with different source functions. We earlier concluded that the terminal law for an ideal independent source, so useful in the context of most circuits, clashes irreconcilably with the Kirchhoff laws for these peculiar circuits. We agreed in that case to *resolve the apparent contradictions when they arise by use of a better model.* In the case of the sources, this was done by adding a finite, nonzero Thevenin or Norton equivalent resistance to the ideal source model, effectively eliminating the mathematical conflict. If we stubbornly insist on the original, inadequate source models in these circuits, we simply have to admit an inherent mathematical contradiction and refuse to analyze the self-contradictory circuit further.

In the case of singular circuits such as Fig. 5.19(a), the apparent contradiction between continuity and the Kirchhoff laws is also removed by using a better circuit model. One such model takes the nonzero resistance of any practical wire into account by adding a series resistance into the loop, as shown in Fig. 5.19(b). Then we may have continuity satisfied, that is, $v_1(0+) = 1$ and $v_2(0+) = 0$, and at the same time have KVL satisfied, since the difference of these voltages will now be accounted for as a voltage drop across the wire v_R, where $v_R(0+) = v_1(0+) - v_2(0+) = 1$ V. Note that if the wire resistance R is small, as we might expect in practice, then by Ohm's law the current in the wire, $i(0+) = v/R = 1/R$, will be quite large. Indeed, in the limit as R gets smaller and smaller, the current goes to infinity, the "impractical" value demanded by the circuit of Fig. 5.19(a). Normally, of course, we need not model the wire resistance since this is not required to limit the current to a finite value.

The second example of a singular circuit is shown in Fig. 5.20(a). Once again a contradiction between continuity and a Kirchhoff law is apparent at time $t = 0+$, the time continuity of inductive current i and Kirchhoff's current law applied at node a. Once again, resolution of the conflict requires use of a more practical model. Suppose that we were to build this circuit physically. What would happen when we threw the switch open? The power stored in the inductor's magnetic field would not instantaneously drop to zero, as it would be required to do if the current instantaneously dropped to zero. The current would seek another path. Indeed, we would probably be greeted with a quite visible spark arcing across the air gap produced as the switch is opened. A model for the conductivity of the air gap is included in Fig. 5.20(b). If conductance G is small, as expected in practice (air is a poor conductor), then with $i(0+) = 1$ A by continuity, we have $v(0+) = 1/G$, a large value. Indeed, this circuit suggests the design of automobile breaker point ignition systems, in which the goal is to produce a very high voltage by interrupting the current through the ignition coil or inductor with a pair of moving breaker points. The inevitable arcing predicted by our analysis is what causes these points to corrode and eventually fail.

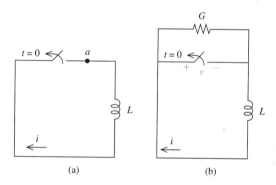

FIGURE 5.20 (a) Singular circuit with inductor; (b) improved circuit model.

EXERCISES

5.9.1. Suppose that $i = \frac{1}{2}$ A in Fig. 5.20(b). If the air gap conductance $G = 10$ μS between the switch terminals, how much voltage is produced across the air gap immediately after the switch is thrown open?
Answer 50 kV

5.9.2. If an ideal 100-μF capacitor is charged to 100 V and then suddenly shorted with an ideal connecting wire, assuming that KVL is not violated, how much energy must be instantaneously shed by the capacitor? In the same situation, using a practical model, we include 0.01-Ω

wire resistance in series with the ideal capacitor. What is the current flow through the wire just after it is connected? What is its dc steady-state value?

Answer 1/2 J; 10 kA; 0 A

SUMMARY

This chapter adds inductors and capacitors to the set of circuit elements we employ, thereby completing the list. Their i–v laws contain derivatives and integrals, thus bringing calculus, and most particularly differential equations, front and center. Although not capable of generating energy, these elements store energy drawn from the rest of the circuit and later return it. This produces more complex behavior than resistors are capable of, which is explored in Chapters 6 and 7.

- Capacitors obey the i–v law $i = C\,dv/dt$, where C is the capacitance in farads and i and v are assumed to satisfy the passive sign convention.
- Energy is stored by a capacitor in the electrical field between its plates. Its value is $\frac{1}{2}Cv^2$.
- Parallel capacitance adds, series capacitance obeys the reciprocal–reciprocal law.
- Inductors obey the i–v law $v = L\,di/dt$, where L is the inductance in henries and i and v are assumed to satisfy the passive sign convention.
- Energy is stored by an inductor in the magnetic field threading its turns. Its value is $\frac{1}{2}Li^2$.
- Series inductance adds, parallel inductance obeys the reciprocal–reciprocal law.
- In dc steady state $dv/dt = di/dt = 0$, so capacitors behave like open circuits, inductors like short circuits.
- The continuity principle states that inductive currents and capacitive voltages cannot jump discontinuously. To do so requires infinite power.
- Singular circuits are those in which either the Kirchhoff laws or the continuity principle must be violated. They can be avoided with the use of more realistic circuit models.

PROBLEMS

5.1. Find the current i through a 20-μF capacitor whose voltage is $v(t) = 2t \cos 6t$ V. Assume that the passive sign convention is satisfied.

5.2. Find the current i through a 1-F capacitor whose voltage v is shown, (a) assuming that i and v satisfy the passive sign conventions and (b) assuming that they do not.

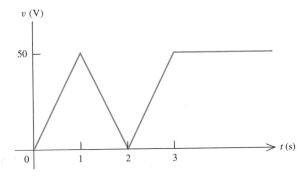

FIGURE P5.2

5.3. If this subcircuit has terminal law $i = 12(dv/dt)$ and i_1 is found to be $20\cos 2t$ A when v is $4\sin 2t$ V, find C_1 and C_2.

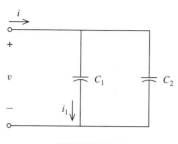

FIGURE P5.3

5.4. A 3-pF capacitor is in series with a current source of value $i = 5\sin 1000\pi t$ μA. If $v(0) = 0$ V, what is its voltage v at $t = 1$ ms? At $t = 1$ s? Assume that v and i satisfy the passive sign convention.

5.5. Find the voltage v across a 1-F capacitor whose current in milliamperes is given in the graph for Problem 5.2.

5.6. The voltage across a 0.2-μF capacitor is the triangular wave shown. Find the current and power for $0 < t < 3$ s.

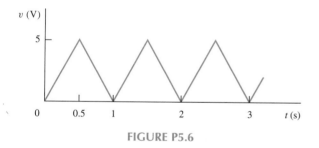

FIGURE P5.6

5.7. What constant current is required to deliver a charge of 40 μC to a 1-μF capacitor in 4 ms?

5.8. How long will it take for a constant 20-mA current to deliver a charge of 80 μC to a 1-μF capacitor? If the same charge resides on a 20-μF capacitor, what is the voltage?

5.9. Find v_g if $v_C = 25e^{-10t}$ V.

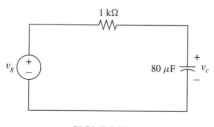

FIGURE P5.9

5.10. Find the energy stored in the capacitor in Problem 5.1 versus time.

5.11. A capacitor with an equal and opposite charge of 4 μC on each plate is found to have a voltage of 80 V. Find its capacitance and the energy stored.

5.12. A 100-μF capacitor is placed in parallel with an independent voltage source of te^{-t} V at $t = 0$. At what time instant will the power transferred by the source to the capacitor be maximum? What will this maximum power be?

5.13. Determine the energy stored in the capacitor and the energy dissipated by the resistor as functions of t for $t \geq 0$. As t gets larger and larger, which element is the larger net energy "customer" for the source?

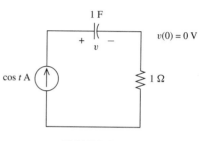

FIGURE P5.13

5.14. The current through a 0.01-F capacitor is $5\cos 25t$ A. Find the voltage v and the power p for $t > 0$ if $v(0) = 0$. Find the maximum value of p and the smallest value of t for which it occurs.

5.15. All capacitors are 1 μF. Find the equivalent capacitance at a–b.

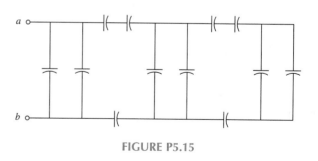

FIGURE P5.15

5.16. The capacitances are in picofarads. Find the equivalent capacitance at a–b.

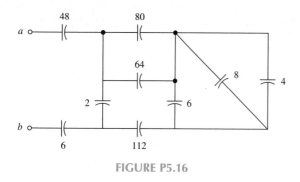

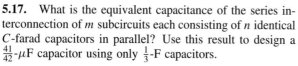

FIGURE P5.16

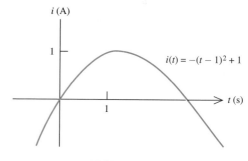

FIGURE P5.21

5.17. What is the equivalent capacitance of the series interconnection of m subcircuits each consisting of n identical C-farad capacitors in parallel? Use this result to design a $\frac{41}{42}$-μF capacitor using only $\frac{1}{3}$-F capacitors.

5.18. We wish to construct a 12.35-μF capacitor out of 10-μF capacitors. Show how this can be done. How many 10-μF capacitors do you need?

5.19. Find i and v for $t \geq 0$ if all the capacitors are initially discharged at $t = 0$.

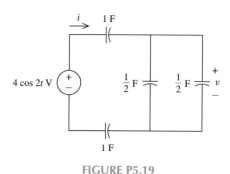

FIGURE P5.19

5.20. Find the voltage v across a 20-mH inductor whose current is $i(t) = 4 \sin 10\pi t + 1$ A. Assume that the passive sign convention is satisfied.

5.21. Find the voltage v across a 1-H inductor whose current i is shown. Assume that the passive sign convention is satisfied.

5.22. If an inductor with current $i(t) = 20 \cos 100t$ mA has a voltage $v(t) = 5 \sin 100t$ V: (a) What is its inductance? (b) Do i and v together satisfy the passive sign convention?

5.23. For what value of ω will the amplitude of the voltage across a 2-mH inductor with current $i = 18 \cos \omega t$ A be +12 V?

5.24. Solve Problem 5.2 if v is the voltage across a 50-mH inductor, and $i(0) = 0$.

5.25. What class of voltage waveforms will produce piecewise linear current waveforms through an inductor? Justify.

5.26. Let $v(t) = \sin \pi t$ V for $0 \leq t \leq 1$ and $2 \leq t \leq 3$, but otherwise $v(t) = 0$. If this is the voltage across an inductor, find the current through it for $t \geq 0$. Assume that $i(0) = 1$ A and that i and v together satisfy the passive sign convention.

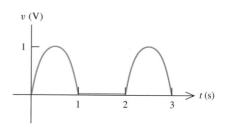

FIGURE P5.26

5.27. If $i(0) = 10$ A is the initial current in a 1-H inductor, find the currents for $t > 0$ for the two voltages shown. Assume that the passive sign convention is satisfied.

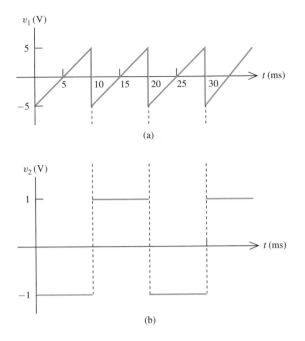

(a)

(b)

FIGURE P5.27

5.28. What type of current waveforms in inductors will produce piecewise exponential voltage waveforms such as those shown in (a)? Piecewise sinusoidal as in (b)?

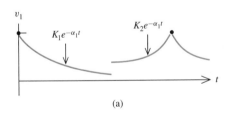

(a)

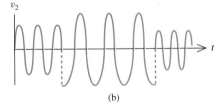

(b)

FIGURE P5.28

5.29. Find the terminal voltage of a 10-mH inductor if the current is (a) 2 A, (b) $20t$ A, (c) $10 \sin 100t$ A, and (d) $10(1 - e^{-5t})$ A.

5.30. For the circuit shown in (a), the source voltage is given in (b). Find the current i if $i_L(0) = -1$ A for (a) $0 < t < 1$ s and (b) $1 < t < 2$ s.

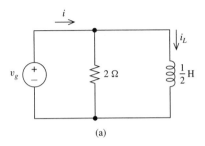

(a)

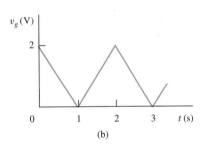

(b)

FIGURE P5.30

5.31. The current through a 0.1-H inductor is $i = 10 \cos 10t$ mA. Find (a) the terminal voltage, (b) the power, (c) the stored energy, and (d) the maximum value of the power being absorbed.

5.32. Find the energy stored versus t for the inductor of Problem 5.20.

5.33. Repeat Problem 5.32 for the inductor of Problem 5.21.

5.34. What is the maximum power that must be generated by a voltage source of $10 \cos 2t$ V that is placed in parallel with a 10-mH inductor at $t = 0$. Assume that the inductive current at $t = 0-$ is zero.

5.35. Find w_2/w_1, the ratio of energy stored in two series inductors if both start at $t = 0$ with $i(0) = 0$.

5.36. If $v(0-) = 9$ V, $i(0-) = 1$ A, and the switch is opened at $t = 0$, find $i(0+)$ and $di(0+)/dt$.

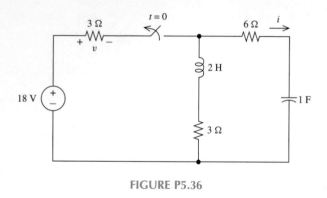

FIGURE P5.36

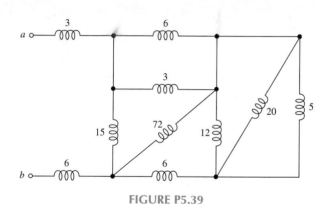

FIGURE P5.39

5.37. The inductances shown are all in mH. Find the equivalent inductance seen at terminals a–b.

5.40. This circuit is in dc steady state at $t = 0-$. Find $v(0-)$, $v(0+)$, $i(0-)$, and $i(0+)$.

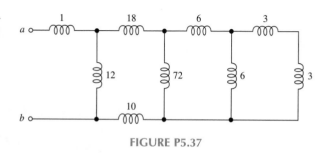

FIGURE P5.37

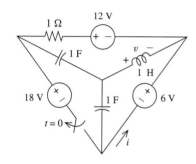

FIGURE P5.40

5.38. If $v(0-) = 72$ V, find i_1, i_2, i_c, v_c, and v at $t = 0-$ and at $t = 0+$.

5.41. Find the dc steady-state currents i_1 and i_2.

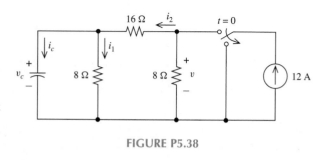

FIGURE P5.38

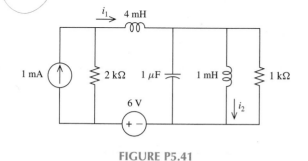

FIGURE P5.41

5.39. If inductances are all in mH, find the equivalent inductance seen at terminals a–b.

5.42. If the circuit is in dc steady state at $t = 0-$, find v_1 and v_2 at $t = 0-$ and $t = 0+$.

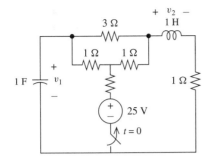

FIGURE P5.42

5.43. Show that this circuit does not have a dc steady state. Why?

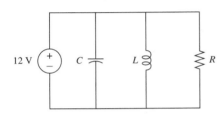

FIGURE P5.43

5.44. Find i, v_1, and v_2 in dc steady state.

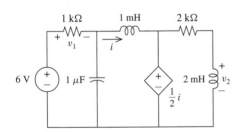

FIGURE P5.44

5.45. Repeat Problem 5.41 for this circuit.

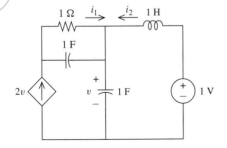

FIGURE P5.45

5.46. Repeat Problem 5.42 for this circuit. Note there are two switches, both acting at the same time $t = 0$.

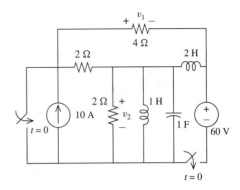

FIGURE P5.46

5.47. Assume that the circuit is in dc steady state at $t = 0-$. Find $i(0-)$ and $i(0+)$, in terms of R and C.

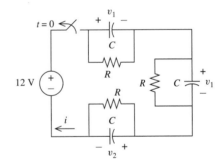

FIGURE P5.47

5.48. Repeat Problem 5.43 using practical inductors and capacitors in place of the ideal ones shown. Does the same conclusion hold?

More Challenging Problems

5.49. Sketch a subcircuit containing five capacitors and two marked terminals a–b such that none of the capacitors are in series or parallel. Is this possible with a four-capacitor subcircuit? If so, sketch.

5.50. Find C so that $i(t) = 12e^{-4t}$ A, $t > 0$. Assume that all capacitive voltages are zero at $t = 0-$.

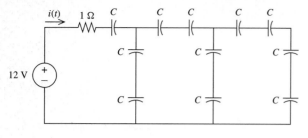

FIGURE P5.50

5.51. Find the dc steady-state power dissipated by each of the three resistors, and produced by each of the four sources.

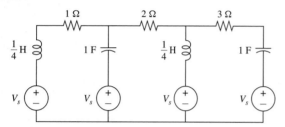

FIGURE P5.51

5.52. Write (do not solve) the mesh equations. The only unknowns should be mesh currents. Then repeat for nodal equations.

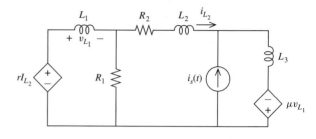

FIGURE P5.52

6

First-Order Circuits

Joseph Henry
1797–1878

Blot out these two names [Joseph Henry and Michael Faraday] and the civilization of the present world would become impossible.

H. S. Carhart

Michael Faraday's great discovery in 1831 of electromagnetic induction was being independently duplicated at about the same time by American physicist Joseph Henry, but Faraday was credited with the discovery because his results were published first. Henry became famous, however, as the discoverer of the inductance (called self-inductance) of a coil and as the developer of powerful electromagnets capable of lifting thousands of pounds of weight. He was also America's foremost nineteenth-century physicist and the first secretary of the newly formed Smithsonian Institution.

Henry was born near Albany, New York, and his early years were spent in poverty. His ambition was to become an actor until by chance at age 16 he happened upon a book of science, which caused him to devote his life to the acquisition of knowledge. He enrolled in the Albany Academy and upon graduation became a teacher there. In 1832 he joined the faculty of the College of New Jersey, now Princeton, and in 1846 joined the Smithsonian Institution. In his honor the unit of inductance was given the name *henry* 12 years after his death.

Chapter Contents

In Chapter 4 resistive circuits were analyzed, and in Chapter 5 the two energy storage elements, capacitors and inductors, were introduced. In this chapter we analyze the class of circuits that result when we add to resistive circuits a single storage element. Such a circuit is called a *first-order circuit* because, as we shall see, the storage element results in a first-order differential equation characterizing the circuit.

We concern ourselves first with simple RC and RL circuits that contain no independent sources. Lacking such sources, any circuit response can only be the result of initial capacitive or inductive energy stored within the circuit and is thus dictated by the nature of the circuit itself. For this reason the response is known as the *natural response* of the circuit. The natural response is characterized by a single time constant or rate of exponential decay.

Following our study of circuits without sources, we shall consider the total response of first-order circuits containing sources. We shall find that it consists of two additive parts, a natural response, identical in form to the response in the absence of sources, and a *forced response*, governed by the form of the forcing functions (source functions of the independent sources). The case in which all independent sources are dc, that is, constant, is considered in detail. Step and pulse sources and their responses are next discussed. The chapter concludes with a look at how SPICE can be used to determine responses for these circuits.

6.1   SIMPLE *RC* AND *RL* CIRCUITS WITHOUT SOURCES

We begin our study of first-order circuits by considering the simple one-loop circuit containing just a capacitor and a resistor, as shown in Fig. 6.1. Suppose that the capacitor is charged to a voltage of V_0 volts at an *initial time*, which we shall take as $t = 0$. By initial time we mean the beginning of the time period we are interested in, not the time of the circuit's construction (the circuit must have a history extending beyond our initial time, since the value of V_0 volts must have been established by external circuitry before this time).

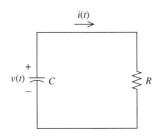

FIGURE 6.1 Source-free RC circuit.

Since there are no current or voltage sources in the network, the circuit response (v or i) is due entirely to the energy that is stored initially in the capacitor. The energy at the initial time $t = 0$ is, by (5.4),

$$w_C(0) = \tfrac{1}{2}CV_0^2 \tag{6.1}$$

We seek to determine $v(t)$ for $t > 0$. Applying KCL at the top node yields

$$C\frac{dv}{dt} + \frac{v}{R} = 0$$

or

$$\frac{dv}{dt} + \frac{1}{RC}v = 0 \tag{6.2}$$

which is a *first-order differential equation*. (The *order* of a differential equation is defined as the order of the highest-order unknown derivative contained in the equation.)

Several methods are available for solving first-order differential equations. One straightforward method is to rearrange the terms in the equation so as to separate the variables v and t. These variables may be separated by rewriting (6.2) as

$$\frac{dv}{v} = -\frac{1}{RC}dt$$

Then, taking the indefinite integral of each side, we have

$$\int \frac{dv}{v} = -\frac{1}{RC}\int dt \tag{6.3}$$

or

$$\ln(v) = -\frac{t}{RC} + k$$

with k the constant of integration. Exponentiating both sides, we have

$$v = Ke^{-(t/RC)} \tag{6.4}$$

where e is the base of the natural logarithms, $e = 2.71828\ldots$, and $K = e^{+k}$.

For this solution to be valid in the interval of interest $t > 0$, continuity requires that $v(0)$ must match the specified initial condition $v(0) = V_0$. Thus while $v(t)$ in (6.4) will satisfy the differential equation for any K, for only a single value of K will this solution satisfy both differential equation and initial condition. This is found by forcing $v(0)$ in (6.4) to the required value of V_0:

$$v(0) = Ke^0 = K = V_0$$

whence

$$v(t) = V_0 e^{-(t/RC)} \tag{6.5}$$

Note that all other circuit variables follow immediately from (6.5), although for the moment we will keep our focus on the capacitive voltage $v(t)$.

A graph of the circuit response $v(t)$ is shown in Fig. 6.2. The voltage begins at V_0, as required, and then decays exponentially to zero with time. The rate at which it decays to zero is given by the RC product of the circuit. Since this response is governed by the circuit elements themselves and not by some independent source "forcing" a different

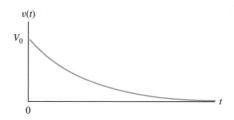

FIGURE 6.2   Voltage $v(t)$ in the circuit of Fig. 6.1.

behavior during $t \geq 0$, this response is called the *natural response* of the circuit. The natural response is synonymous with the response in the absence of independent sources.

At any time t, the energy stored in the capacitor $w_C(t)$ is given by $\frac{1}{2}Cv^2(t)$. The resistor stores no energy; thus the total energy stored in the circuit at time t is, by (6.5),

$$w_C(t) = \frac{1}{2}C\left[V_0 e^{-(t/RC)}\right]^2 = \frac{1}{2}CV_0^2 e^{-2t/RC}$$

Comparing with the initial stored energy given by (6.1) gives

$$w_C(t) = w_C(0)e^{-2t/RC} \tag{6.6}$$

Note that the stored energy is exponentially decaying to zero from its initial value $w(0)$. Where does this lost energy go? We know that the power dissipated by a resistor is

$$p_R(t) = \frac{v_R^2}{R} = \frac{\left[V_0 e^{-(t/RC)}\right]^2}{R}$$

Integrating from the initial time 0 to time t, we have the total energy dissipated by the resistor from time 0 to t:

$$w_R(t) = \frac{V_0^2}{R} \int_0^t e^{-2\tau/RC}\, d\tau$$

or
$$w_R(t) = \frac{1}{2}CV_0^2\left(1 - e^{-2t/RC}\right) = w_C(0)\left(1 - e^{-2t/RC}\right)$$

Comparing the last to (6.6), we see where the missing energy has gone.

$$w_C(0) - w_C(t) = w_R(t)$$

At each instant of time, the decrease in energy stored in the capacitor since $t = 0$ exactly equals the total energy dissipated by the resistor up to that time. As time increases, less energy remains stored in the circuit; and in the limit as t gets very large, all the initial stored energy $w_C(0)$ has been dissipated as heat in the resistor and no stored energy remains. As this occurs, $v(t)$ and all other voltages and currents in the circuit go to zero as well. Clearly, it is the internal energy stored at the initial time that is driving this "natural" response.

Example 6.1

Consider the circuit of Fig. 6.1 with $C = 1\ \mu\text{F}$, $R = 1\ \text{k}\Omega$, and an initial voltage of $v(0) = 5$ V across the capacitor. We wish to know the current $i(t)$ and the maximum power dissipated by the resistor. From (6.5), the capacitive voltage is

$$v(t) = V_0 e^{-t/RC} = 5e^{-1000t} \text{ V}$$

Then, since $v(t)$ is also the voltage across the resistor,

$$i(t) = \frac{v(t)}{R} = 0.005e^{-1000t} \text{ A}$$

The power dissipated by the resistor is

$$p_R(t) = \frac{v^2(t)}{R} = 0.025e^{-2000t} \text{ W}$$

Thus the maximum power of 25 mW is dissipated right at the initial time $t = 0$. This is confirmed by the graph in Fig. 6.2, which shows that the sharpest reduction in capacitive voltage, and thus stored power, occurs at the initial time.

Example 6.1 used elements with values commonly found in low-power circuits: resistance in kilohms and capacitance on the scale of microfarads. This is a departure from most other examples, which have used elements on the scale of 1 Ω and 1 F. While convenient for calculation, this scale of R and C values is not often seen in practice. Resistors of the scale of 1 Ω lead to high current and power dissipation at common voltage levels of 1.5 to 150 V, and capacitors on the scale of 1 F would be massively large and expensive.

For circuits with elements in the far more common kΩ–μF ranges, it is often convenient to use a *scaled* metric system to write equations. Carrying cumbersome powers of 10 through most calculations can be avoided. Suppose that we use the scaled units listed in Table 6.1. The first three are basic SI quantities, the rest are derived from them. Since $R = E/I$ (we are using E to avoid confusing voltages or electrical potential difference, with the unit of volts), then V/mA = kΩ and the scaled resistance unit is the kilohm. Similarly, the scaled unit of capacitance is mA–ms/V or μF. In this system inductance is measured as V–ms/mA = V–s/A, which is the basic SI unit of henries (H). *The scaled SI unit system V–mA–ms–kΩ–μF–H is consistent without the need to carry powers of 10.* Had we used this scaled system in Example 6.1 with $R = 1$ kΩ, $C = 1$ μF, and $V_o = 1$ V, applying (6.5) we would have written

$$v(t) = 5e^{-t/(1)(1)} = 5e^{-t} \text{ V}$$

Table 6.1	Standard and Scaled SI Unit Systems		
	Quantity	Standard SI Unit	Scaled SI Unit
Basic:	Voltage (E)	V	V
	Current (I)	A	mA
	Time (t)	s	ms
Derived:	Resistance (E/I)	Ω	kΩ
	Capacitance ($I - t/E$)	F	μF
	Inductance ($V - t/I$)	H	H
	Power ($E - I$)	W	mW

and the time constant is $\tau = 1$, with units of milliseconds. Continuing, we find that

$$i(t) = \frac{v(t)}{R} = \frac{5e^{-t}}{1} = 5e^{-t} \text{ mA}$$

$$P_R(t) = \frac{v^2(t)}{R} = \frac{(5e^{-t})^2}{1} = 25e^{-2t} \text{ mW}$$

Comparing Example 6.1, the scaled system results in more compact equations. The reader is encouraged to use this system on problems that fit these scaled units. Many such will be found in the exercises and in the end-of-chapter problems in the remainder of this book.

We next turn to the simple RL circuit shown in Fig. 6.3. As with the previous RC circuit, there are no external sources and the response will be driven by initial stored energy, in this case proportional to the square of the initial current $i(0) = I_0$ through the inductor. The energy stored at the initial time is, by (5.11),

$$w_L(0) = \tfrac{1}{2}LI_0^2$$

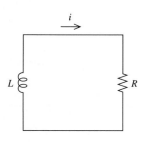

FIGURE 6.3 Source-free RL circuit.

Applying KVL yields

$$L\frac{di}{dt} + Ri = 0$$

or

$$\frac{di}{dt} + \frac{R}{L}i = 0 \qquad (6.7)$$

This equation is of the same form as that of (6.2) for the RC circuit. We may therefore solve it by the same method, separation of variables.

Instead, let us observe that since the present equation is of the same form as (6.2), we expect its solution will be of the same form as well. We know that the form of this previous solution, given in (6.5), is a constant times an exponential function of time. Guided by this experience, let us introduce a *trial natural solution*,

$$i(t) = Ke^{st} \qquad (6.8)$$

where K and s are constants to be determined. If the trial natural solution truly solves the differential equation, we must be able to substitute it successfully into this equation, or

$$\frac{d}{dt}(Ke^{st}) + \frac{R}{L}(Ke^{st}) = 0$$

from which we see that

$$\left(s + \frac{R}{L}\right)(Ke^{st}) = 0$$

If our trial natural solution is to work, this equation must be valid for all $t \geq 0$. However, the factor Ke^{st} cannot be zero for $t \geq 0$, since that would require $K = 0$; hence $i(t) = 0$ for all $t \geq 0$ in contradiction to the initial value $i(0) = I_0$. We conclude that if the trial natural solution is valid, it must be that the other factor is zero:

$$s + \frac{R}{L} = 0 \qquad (6.9)$$

This is called the *characteristic equation* for this differential equation, and its solution s specifies the exponent in the solution of the differential equation, in this case

$$s = -\frac{R}{L}$$

We conclude that a valid natural solution must be of the somewhat more specific form

$$i(t) = Ke^{-(R/L)t}$$

Any $i(t)$ of this form satisfies the differential equation (6.7). The only other test a solution must pass is that it satisfy the given initial condition $i(0) = I_0$. Applying this requirement,

$$i(0) = I_0 = Ke^0 = K$$

The only trial natural solution satisfying both the initial condition and the differential equation is the one with specified constants s and K, or

$$i(t) = I_0 e^{-(R/L)t} \qquad (6.10)$$

which is the desired solution. We may verify this solution by substituting $i(t)$ given by (6.10) into the original differential equation we set out to solve (6.7) and confirming that it matches the required initial condition as well. It passes these tests.

Let us review our reasoning. We guessed at an exponential form for the solution and then found values for its parameters s and K that made our guess demonstrably correct. There is no doubt the result is the desired solution, since it solves the differential equation and has the right initial value, and we ask no more of a solution than these two requirements.

Example 6.2

To demonstrate the method of characteristic equations in the case that the initial time is not zero, consider the circuit of Fig. 6.3 with $R = 2\ \Omega$, $L = 1$ H, and initial current $i(t_0) = 3$ A at time $t_0 = 10$ s. By (6.7) we have

$$\frac{di}{dt} + 2i = 0$$

The characteristic equation is $s + 2 = 0$ or $s = -2$. Then the trial natural solution is

$$i(t) = Ke^{-2t}$$

and evaluating at $t_0 = 10$ in order to match initial conditions,

$$i(t_0) = 3 = Ke^{-2(10)}$$

which yields $K = 3e^{+20}$. Thus the desired current is

$$i(t) = 3e^{20}e^{-2t} \text{ A}$$

or
$$i(t) = 3e^{-2(t-10)} \text{ A}$$

A plot of this current is shown in Fig. 6.4.

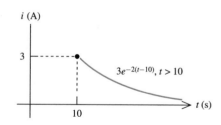

FIGURE 6.4 $i(t)$ in Example 6.2 for $t \geq 10$.

Examining (6.5) and (6.10), we see that the capacitive voltage and inductive current in these single-loop, source-free circuits are both decaying exponentials. Since the derivative of an exponential is also an exponential, the current through the capacitor $i = C(dv/dt)$ and the voltage across the inductor $v = L(di/dt)$ in these circuits are also decaying exponentials with the same exponents. The significance of the constants $1/RC$ and R/L in these exponentials will be taken up in Section 6.2.

EXERCISES

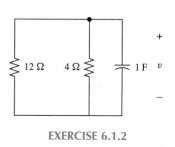

EXERCISE 6.1.2

6.1.1. In Fig. 6.1, let $t_0 = 0$, $V_0 = 10$ V, $R = 1$ kΩ, $C = 1$ μF. Find v and i at $t = 1$ ms, $t = 2$ ms and $t = 5$ ms.
Answer 3.68 V, 3.68 mA; 1.35 V, 1.35 mA; 0.067 V, 0.067 mA

6.1.2. If $v(0) = +10$ V, at what time will $v(t)$ equal $+1$ V?
Answer 6.9 s

6.1.3. If $w(t)$ is the energy stored in the capacitor in Exercise 6.1.2, find a formula for $w(t)$, $t > 0$. At what time does half the original stored energy remain?
Answer $50e^{-2t/3}$ J; 1.04 s

6.1.4. In Fig. 6.3, $R = 1$ kΩ. If the voltage across the inductor, defined to satisfy the passive sign convention together with i, is $v_L(t) = 10e^{-200t}$ for $t > 0$ s, find L and the initial current $i(0)$.
Answer 5 H; 10 mA

6.2 TIME CONSTANTS

In networks that contain energy-storage elements it is very useful to characterize with a single number the rate at which the natural response decays to zero. This role is played by a quantity called the time constant of the circuit.

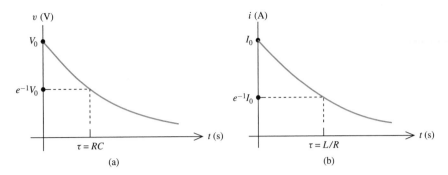

FIGURE 6.5　(a) *RC* circuit response; (b) *RL* circuit response.

Consider the graph shown in Fig. 6.5(a), which illustrates a natural response (the capacitive voltage) of the simple *RC* circuit discussed in Section 6.1. The time required for a natural response to decay by a factor of $1/e$ is defined as the *time constant* of the circuit, which we shall denote by τ. For the given *RC* circuit, the natural response is

$$v(t) = V_0 e^{-t/RC} \tag{6.11}$$

By definition of the time constant τ, we require that τ satisfy

$$\frac{v(t + \tau)}{v(t)} = \frac{1}{e}$$

or

$$v(t + \tau) = e^{-1} v(t)$$

Substituting for the voltage v, this equation becomes

$$V_0 e^{-(t+\tau)/RC} = e^{-1} V_0 e^{-t/RC}$$

or, after canceling common factors, we have

$$\tau = RC \tag{6.12}$$

The units of τ are $\Omega - F = (V/A)(C/V) = C/A = s$. In terms of the time constant τ, the natural response is

$$v(t) = V_0 e^{-t/\tau} \tag{6.13}$$

The time constant for the *RL* circuit is computed similarly from Fig. 6.5(b), with

$$i(t) = I_0 e^{-(R/L)t}$$

In this case we have

$$I_0 e^{-(R/L)(t+\tau)} = e^{-1} I_0 e^{-(R/L)t}$$

which, after cancellation, identifies the time constant τ as

$$\tau = \frac{L}{R} \tag{6.14}$$

and we may rewrite the response in terms of τ as

$$i(t) = I_0 e^{-t/\tau} \tag{6.15}$$

The time constant τ units are $(H/\Omega) = [V/(A/s)]/(V/A) = s$, once again.

In both RC and RL circuits, each τ seconds the natural response is reduced by a factor of $1/e$ relative to its value at the beginning of that one-time-constant interval. The response at the end of one time constant is reduced to $e^{-1} = 0.368$ of its initial value. At the end of two time constants, it is equal to $e^{-2} = 0.135$ of its initial value, and at the end of five time constants it has become $e^{-5} = 0.0067$ of its initial value. After four or five time constants, the response is essentially zero.

For the RC circuit of Fig. 6.1, we see in Fig. 6.2 that the capacitive voltage $v(t)$ is exponentially decaying to zero with time constant $\tau = RC$. Let us complete our analysis of this circuit by considering the other circuit variables. The resistive voltage just equals $v(t)$ by KVL. The mesh current can be found by dividing the resistive voltage by R and is thus also an exponential decay with $\tau = RC$. *All currents and voltages are exponential decays with the same time constant τ.* Examination of the RL circuit of Fig. 6.3 shows the same statement is true there as well. The behavior of all circuit responses is fixed by a single time constant, $\tau = RC$ in the RC circuit and $\tau = L/R$ in the RL circuit. This is why we refer to τ as the time constant of the *circuit*, rather than the time constant of any specific current or voltage within the circuit.

An interesting property of exponential functions is shown in Fig. 6.6. A tangent to the curve at $t = 0$ intersects the time axis at $t = \tau$. This is easily verified by considering the equation of a straight line tangent to the curve at $t = 0$, given by

$$v_1(t) = mt + V_0$$

where m is the slope of the line. Differentiating v, we have

$$\frac{dv}{dt} = \frac{-V_0}{\tau} e^{-(t/\tau)}$$

Thus the slope of v at $t = 0$ is $-V_0/\tau$. Then, with $m = -V_0/\tau$,

$$v_1(t) = \frac{-V_0}{\tau} t + V_0$$

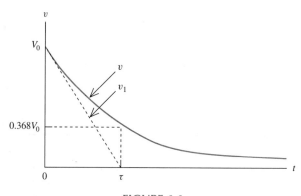

FIGURE 6.6

and

$$v_1(\tau) = 0$$

as asserted. Similarly, a tangent to the curve at a time t_1 intersects the time axis at $t_1 + \tau$ (see Problem 6.9). This fact is often useful in sketching the exponential function.

From Fig. 6.6 we see that an alternative definition for the time constant is the time required for the natural response to become zero if it decreased at a constant rate equal to the initial rate of decay. Of course, it does not decrease at a constant rate, since its decay rate is everywhere proportional to its value, which is steadily approaching zero.

Knowledge of the time constant allows us to predict the general form of the response, (6.13) or (6.15), but to complete the solution we must know the initial value V_0 or I_0. Frequently, this must be determined in a circuit with switches. Since capacitive voltages and inductive currents are continuous (as discussed in Chapter 5), the desired values just after the switch acts at 0+ can be determined from the state of the circuit just before, at $t = 0-$.

Example 6.3

To illustrate the procedure, let us find the capacitor voltage $v(t)$ in Fig. 6.7(a), given that the circuit was in dc steady state just before the opening of the switch. In dc steady state the capacitor may be replaced by (is equivalent to) an open circuit. The resistance to the left of this open circuit is

$$R_{eq} = 8 + \frac{(3)(6)}{3+6} = 10 \ \Omega$$

and, by voltage division,

$$v(0-) = \frac{100 R_{eq}}{R_{eq} + 15} = 40 \ V$$

At $t = 0$ the switch opens. By continuity of capacitive voltage, $v(0+) = v(0-) = 40$ V. For $t > 0$, we have a sourceless circuit, as shown in Fig. 6.7(b), where resistors to the left of the capacitor have been replaced by their equivalent $R_{eq} = 10 \ \Omega$. The time constant for this RC network is simply the product of the capacitance and the equivalent resistance, given by

$$\tau = R_{eq} C = (10)(1) = 10 \ s$$

Therefore, by (6.13) the voltage is

$$v(t) = 40 e^{-\frac{t}{10}} \ V$$

Every 10 s, the voltage will decline by a factor of $1/e$, or 0.368. After 50 s, the voltage has declined to $40 e^{-5} = 0.27$ V, illustrating that after 5 time constants the response has been reduced by more than two orders of magnitude. Thus after 10 time constants there is a reduction of more than four orders of magnitude, after 15 time constants a reduction of more than six orders of magnitude, and so on.

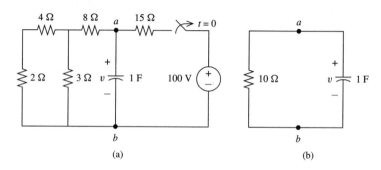

FIGURE 6.7 (a) Circuit for Example 6.3; (b) equivalent for $t \geq 0$.

EXERCISES

6.2.1. In an RC circuit, determine (a) τ for $R = 2$ kΩ and $C = 10$ μF, (b) C for $R = 10$ kΩ and $t = 20$ μs, and (c) R for $v(t)$ on a 2-μF capacitor to halve every 20 ms.
Answer (a) 20 ms; (b) 2 nF; (c) 14.43 kΩ

6.2.2. An RC circuit consists of a 20-kΩ resistor and a 0.05-μF capacitor. It is desired to decrease the current in the network by a factor of 5 without changing the capacitor voltage. Find the necessary values of R and C.
Answer 100 kΩ; 0.01 μF

6.2.3. In a single-loop RL circuit, the current is determined to be 2 mA at $t = 10$ ms and 100 μA at $t = 46$ ms. Find the time constant τ and the initial current at time $t = 0$.
Answer 12 ms; 4.6 mA

6.3 GENERAL FIRST-ORDER CIRCUITS WITHOUT SOURCES

The simple first-order RC and RL circuits examined in Section 6.2 were seen to have decaying responses governed by a single time constant. Generalizing, we must expect the same behavior from any circuit that is equivalent, whether it is simple (solvable from a single equation) or not.

Example 6.4

We wish to determine the time constant of the circuit of Fig. 6.8(a). We first find an equivalent circuit of the simple form of Fig. 6.8(b).

$$C_{eq} = \frac{(4)(12)}{4 + 12} = 3 \text{ F}$$

$$R_{eq} = \frac{(20)(5)}{20 + 5} = 4 \text{ }\Omega$$

The time constant τ in Fig. 6.8(b) is

$$\tau = R_{eq}C_{eq} = (4)(3) = 12 \text{ s}$$

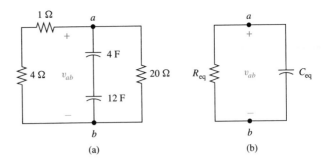

FIGURE 6.8 (a) RC circuit; (b) equivalent circuit.

Since Figs. 6.8(a) and (b) are equivalent circuits, the voltage $v_{ab}(t)$ must be the same. This voltage has time constant 12 s in (b), so it must also in (a). Indeed, all currents and voltages in both Figs. 6.8(a) and (b) decay exponentially to zero with $\tau = 12$ s.

From Example 6.4 we see that RC circuits containing a single equivalent R_{eq} and C_{eq} can be analyzed by reduction to these equivalents, with $\tau = R_{eq}C_{eq}$. Clearly, the same statement applies to RL circuits with single equivalent inductance and resistance, with $\tau = L_{eq}/R_{eq}$. Note that it may not always be possible to use series and parallel rules to find these equivalents, however. The next example shows the general approach in this case.

Example 6.5

We wish to find the time constant of the circuit of Fig. 6.9(a). The circuit to the right of the inductor is purely resistive, and we seek its equivalent resistance R_{eq}. Since none of the resistors is in series or parallel, we cannot use series–parallel equivalents. To find R_{eq}, we will force i to flow into the circuit, as shown in Fig. 6.9(b), and determine the resulting voltage v. Since the terminal law for the circuit is $v = R_{eq}i$, the ratio between v and i is the desired R_{eq}.

Figure 6.9(b) is a three-mesh circuit with one current source, so two mesh equations are required. The mesh equations are

$$\begin{bmatrix} 4 & -2 \\ -2 & 5 \end{bmatrix} \begin{bmatrix} i_1 \\ i_2 \end{bmatrix} = \begin{bmatrix} i \\ i \end{bmatrix}$$

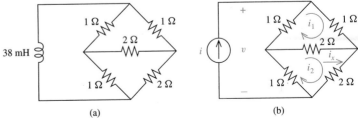

(a)   (b)

FIGURE 6.9 (a) RL circuit; (b) circuit to determine R_{eq}.

Inverting the matrix, the determinant is $(4)(5) - (-2)(-2) = 16$, and

$$\begin{bmatrix} i_1 \\ i_2 \end{bmatrix} = \frac{1}{16} \begin{bmatrix} 5 & 2 \\ 2 & 4 \end{bmatrix} = \frac{1}{16} \begin{bmatrix} 7i \\ 6i \end{bmatrix} \tag{6.16}$$

By KVL around the outer loop,

$$v = v_{ab} = 1i_1 + 2i_2$$

$$= \tfrac{1}{16}(7i + 12i)$$

$$= \tfrac{19}{16}i$$

Thus $R_{\text{eq}} = \frac{19}{16}\ \Omega$ and

$$\tau = \frac{L}{R_{\text{eq}}} = \frac{0.038}{19/16} = 0.032\ \text{s} = 32\ \text{ms}$$

Op amp circuits will also exhibit first-order behavior when they contain one storage element or when all storage elements can be combined into a single equivalent. They can be analyzed efficiently using nodal analysis and the virtual short- and open-circuit principles as discussed in Section 4.8.

Example 6.6

Suppose that at $t = 0$, the output in Fig. 6.10 is $v_2(0) = V_0$ and that $v_g(t) = 0$ for $t > 0$. We wish to find $v_2(t)$ for $t > 0$. By the virtual short principle, node 1 is at ground potential. Summing currents into this node gives

$$\frac{v_g - 0}{R_A} + \frac{v_2 - 0}{R_F} + C\frac{d(v_2 - 0)}{dt} = 0$$

or, with $v_g = 0$ for $t > 0$,

$$\frac{dv_2}{dt} + \frac{1}{R_F C}v_2 = 0$$

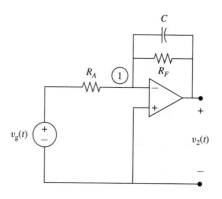

FIGURE 6.10   Circuit for Example 6.6.

This is our standard unforced first-order differential equation with the solution

$$v_2(t) = v_2(0)e^{-t/R_F C} = V_0 e^{-t/R_F C}$$

In summary, a sourceless circuit containing a single storage element, or the equivalent of a single storage element, will have all responses decay to zero with time constant $\tau = RC$ in the RC case or $\tau = L/R$ in the RL case. Note, however, that circuits with two or more storage elements may well not have a single equivalent storage element and thus will not be broken through the solution of a single first-order differential equation. These are not first-order circuits, and their responses cannot be characterized by a single time constant. Examples include any circuit with both capacitors and inductors and the RC circuit of Problem 6.16. Such circuits are taken up in Chapter 7.

EXERCISES

6.3.1. The circuit is in steady state at $t = 0-$ and the switch is moved from position 1 to position 2 at $t = 0$. Find the response $v(t)$ for $t > 0$. What is the time constant τ?

Answer $16e^{-4t}$ V; $\frac{1}{4}$ s

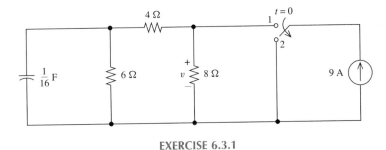

EXERCISE 6.3.1

6.3.2. Find i for $t > 0$ if the circuit is in steady state at $t = 0-$.

Answer $\frac{1}{4}e^{-2t}$ A

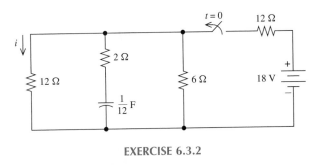

EXERCISE 6.3.2

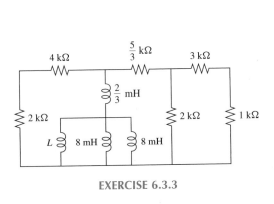

EXERCISE 6.3.3

6.3.3. For what value of L will the time constant τ be 1 μs?

Answer 2 mH

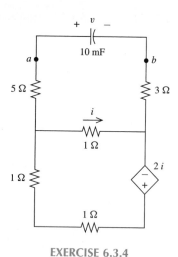

6.3.4. (a) Find the equivalent resistance of the circuit below a–b. (b) Use the result to find $v(t)$ for $t > -1$ if $v(-1) = 20$ V.

Answer (a) 10 Ω; (b) $20e^{-10(t+1)}$

EXERCISE 6.3.4

6.4 CIRCUITS WITH DC SOURCES

In the preceding sections we have considered source-free circuits whose responses have been the result of initial energies stored in capacitors and inductors. All independent current or voltage sources were removed or switched out of the circuits prior to finding the natural responses. It was shown that these responses, when arising in circuits containing a single equivalent capacitor or inductor, decay to zero exponentially over time.

In this section we examine circuits that in addition to having initial stored energies are driven by constant (dc) independent current or voltage sources. For these circuits we shall obtain solutions that are the result of both initial stored energy and that continually supplied by the sources. We shall find that the responses in this case consist of two parts, one of which is, like the sources themselves, a constant.

Let us begin by considering the circuit of Fig. 6.11. The network consists of the parallel connection of a constant current source and a resistor that is switched at time $t = 0$ across a capacitor having a voltage $v(0-) = V_0$ V. For $t > 0$, the switch is closed and a nodal equation at the upper node is given by

$$C\frac{dv}{dt} + \frac{v}{R} = I_0$$

or

$$\frac{dv}{dt} + \frac{1}{RC}v = \frac{I_0}{C} \tag{6.17}$$

This first-order differential equation is identical to that for the source-free RC case (6.2), except for the addition of a *forcing term* on the right-hand side. A forcing term in a differential equation is a term $f(t)$ independent of the unknown time functions.

As in the unforced case, this first-order differential equation may be solved either by separation of variables or by a trial solution approach. We choose the latter, since this method extends to higher-order circuits where separation of variables cannot be used.

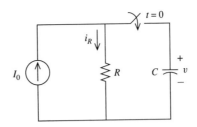

FIGURE 6.11   *RC* network with dc source.

Examining (6.17) we see that the left-hand side is a linear combination of v and its derivative, while the right-hand side is just a constant. Thus $v(t)$ must be some time function with the property that a linear combination of v and its derivative is constant for all $t > 0$. What kinds of functions have this property? Certainly, $v(t)$ itself equal to a constant is one such class of functions, for the derivative of a constant is zero; so every linear combination of a constant and its derivative is, as required, also just a constant.

Guided by this observation, let us use a *trial forced solution* of the form

$$v_f(t) = A$$

where A is a constant to be determined. As we did with the trial natural solution employed earlier, we will substitute it into the differential equation that it must satisfy to test its validity. Replacing v by $v_f(t) = A$ in (6.17) yields

$$\frac{d}{dt}A + \frac{A}{RC} = \frac{I_0}{C}$$

or
$$A = I_0 R$$

Thus we have found a (constant) time function $v_f(t)$ that satisfies the differential equation:

$$v_f(t) = I_0 R$$

Any such time function satisfying the forced differential equation is called a *forced solution*. We have found a forced solution to (6.17).

There is a second requirement that any proposed solution to our original problem must satisfy, however. It is not enough to satisfy the differential equation, as $v_f(t)$ has been shown to do. A valid solution must also satisfy the initial condition $v(0-) = v(0+) = V_0$ V (we have used continuity of capacitive voltages to project V_0 across the switch time $t = 0$). Evaluating our forced solution $v_f(t)$ at $t = 0+$, we have $v_f(0+) = I_0 R$. So the forced solution agrees with the given initial condition if and only if $V_0 = I_0 R$. Examining Fig. 6.11, V_0 and I_0 are clearly the result of quite separate causes; V_0 is an initial condition established before $t = 0$, and I_0 is an independent source function. Thus, while $V_0 = I_0 R$ may be satisfied for particular choices of initial conditions and source functions, there is no reason in general for this equation to be true.

Thus, if our forced solution $v_f(t)$ happens to satisfy the given initial condition, we have found the solution; it is $v(t) = v_f(t)$. In most cases, $v_f(t)$ will not satisfy the initial conditions. We are apparently stuck, for there are no additional free parameters in our forced solution $v_f(t)$ to use to satisfy the initial condition.

Let $v_n(t)$ denote any solution to the *unforced* version of our differential equation; that is, $v_n(t)$ satisfies

$$\frac{dv_n}{dt} + \frac{1}{RC}v_n = 0 \qquad (6.18)$$

The unforced version is formed by setting the forcing term to zero. From previous work we know that the trial natural solution for v_n is of the form

$$v_n = Ke^{st}$$

with characteristic equation

$$s + \frac{1}{RC} = 0$$

or $s = -1/RC$, so

$$v_n = Ke^{-(1/RC)t} \qquad (6.19)$$

We have thus far discovered a v_f that solves the forced differential equation (6.17), and a v_n that, for any value of its free parameter K, satisfies the associated unforced differential equation.

Consider finally a *sum of forced and natural solutions*, $v = v_n + v_f$. Substituting this sum for v on the left side of (6.17),

$$\frac{d}{dt}(v_n + v_f) + \frac{1}{RC}(v_n + v_f) = \left(\frac{dv_n}{dt} + \frac{1}{RC}v_n\right) + \left(\frac{dv_f}{dt} + \frac{1}{RC}v_f\right)$$

$$= 0 + \frac{I_0}{C}$$

where the first term in parentheses on the right evaluates to zero since v_n satisfies the unforced differential equation, and the second evaluates to I_0/C since v_f satisfies the forced version. Thus we see that $v = v_n + v_f$ satisfies the forced differential equation *regardless of the value of the free parameter K in the natural solution v_n*. Returning to the requirement that not only the forced differential equation, but also the initial condition, be satisfied,

$$v(t) = v_n(t) + v_f(t) = Ke^{-t/RC} + RI_0 \qquad (6.20)$$

Evaluating at $t = 0$, the initial condition is matched if

$$v(0+) = V_0 = K(1) + RI_0$$

which requires that

$$K = V_0 - RI_0$$

Inserting this into (6.20), we have found our desired solution to the forced differential equation (6.17) with given initial condition:

$$v(t) = (V_0 - RI_0)e^{-(1/RC)t} + RI_0$$

This solution may be tested by substitution into (6.17) and noting that evaluation at $t = 0$ does yield the specified initial value V_0.

Summarizing this method, the following steps solve the forced differential equation with constant forcing term and given initial conditions:

1. Using a trial forced solution equal to the unknown constant A, substitute into the forced differential equation to find the forced solution.
2. Using a trial natural solution equal to Ke^{st}, substitute into the unforced differential equation, solving the characteristic equation for s.
3. The total solution is the sum of natural and forced solutions. Evaluate the sum at the initial time and set equal to the required initial value to find K.

Example 6.7

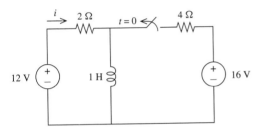

FIGURE 6.12   Circuit for Example 6.7.

We wish to find $i(t)$ for $t > 0$ in the circuit of Fig. 6.12. Before $t = 0$, the circuit is as shown in Fig. 6.13(a). Assuming that this circuit has had sufficient time to go to dc steady state prior to $t = 0$, the inductor acts like a short circuit and

$$i_L = i + i_1 = \frac{12}{2} + \frac{16}{4} = 10 \text{ A}$$

at $t = 0-$. At $t = 0$ the switch opens, and for $t > 0$ the circuit is redrawn as in Fig. 6.13(b). In this circuit, by KVL,

$$\frac{di}{dt} + 2i = 12 \tag{6.21}$$

and $i(0+) = i(0-) = 10$ A by continuity of inductive currents. The trial forced solution is $i_f = A$, and inserting this into the forced differential equation (6.21) gives

$$0 + 2A = 12$$

or $A = 6$. The unforced version of this equation is

$$\frac{di_n}{dt} + 2i_n = 0$$

with characteristic equation

$$s + 2 = 0$$

or $s = -2$. Putting these results together yields

$$i(t) = i_n(t) + i_f(t) = Ke^{-2t} + 6$$

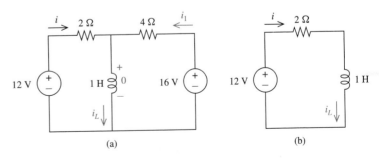

(a)   (b)

FIGURE 6.13   (a) Circuit before $t = 0$; (b) circuit after $t = 0$.

and matching initial conditions, we have

$$i(0+) = 10 = K(1) + 6$$

We conclude that $K = 4$ and the total solution is

$$i(t) = 4e^{-2t} + 6 \text{ A}$$

Example 6.8

This example, shown in Fig. 6.14, demonstrates application of these ideas to op amp circuits. The circuit is in dc steady state just before the switch acts at $t = 0$, and we wish to find $v_2(t)$ for all times t. With the op amp input shorted to ground for $t < 0$, v_1 and v_2 are zero for $t < 0$. For $t > 0$, the voltage at node 1 is 10 V, since the voltage at the noninverting input is 10 V by the virtual open principle, and v_1 equals this value by the virtual short principle. The node equation at the inverting input node is then

$$\left(\frac{1}{25 \times 10^3}\right)(10 - 0) + (50 \times 10^{-6})\frac{d}{dt}(10 - v_2)$$

$$+ \left(\frac{1}{20 \times 10^3}\right)(10 - v_2) = 0$$

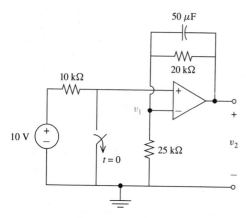

FIGURE 6.14   Circuit for Example 6.8.

or

$$\frac{dv_2}{dt} + v_2 = 18 \tag{6.22}$$

The trial forced solution is $v_{2f} = A$, and plugging it into (6.22), $A = 18$. The characteristic equation of the unforced version of (6.22) is $s + 1 = 0$, or

$$v_{2n}(t) = Ke^{-t}$$

Then the total solution is

$$v_2(t) = v_{2n}(t) + v_{2f}(t) = Ke^{-t} + 18$$

At $t = 0+$, there is no current flow through the 10-kΩ resistor, so the voltage at the noninverting input is 10 V. Since no voltage drop occurs across the op amp input terminals (the virtual short principle), $v_1(0+) = 10$ V also. Continuity of the capacitive voltage requires that

$$v_2(0+) - v_1(0+) = v_2(0-) - v_1(0-) = 0 - 0 = 0$$

Thus $v_2(0+) = 10$ V also, and evaluating the constant K, we obtain

$$v_2(0+) = K(1) + 18 = 10 \text{ V}$$

or $K = -8$. Thus

$$v_2(t) = 18 - 8e^{-t} \text{ V}, \qquad t \geq 0 \tag{6.23}$$

which, combined with the earlier observation that $v_2(t) = 0$ for $t < 0$, completes the solution for all t.

The presence of the op amp has not altered the fact that a single storage element will yield a single first-order differential equation (6.22). As recommended in Chapter 4, we used nodal analysis with op amp circuits and wrote our node equation at an op amp input node rather than the output node.

Example 6.9

Our final example, with circuit shown in Fig. 6.15, illustrates a general first-order circuit containing several independent and dependent sources. In this circuit we seek the current $i(t)$ through the inductor for $t > 0$, where $i(0-) = 2$ A.

Such problems can be approached several different ways. A basic approach would be to perform mesh or nodal analysis on the circuit. This would lead to coupled first-order differential equations, which could then be reduced to a single first-order differential equation by substitution. The reduction is not always straightforward, and here we recommend other routes to securing this single first-order differential equation. In this example we employ the strategy of simplification by Thevenin–Norton transformation. Another method, based on superposition, will be developed in Section 6.5.

We will find the Thevenin transform of the entire circuit less its storage element. Then, replacing all but the storage element by the two-element equivalent, we will be left with a simple one-loop circuit to analyze. To do so, we will find the open-circuit voltage v_{oc} and short-circuit current i_{sc} defined in Fig. 6.16 at the inductor terminals ab. v_{oc} is found from Fig. 6.16(a). By KVL,

$$1i_1 + 2i_1 + 2(i_1 + 10) = 80$$

or $i_1 = 12$. Then around the left loop, again by KVL,

$$1(12) + v_{oc} + 3(12) = 80$$

so $v_{oc} = 32$ V. To find i_{sc} in Fig. 6.16(b), we write the two required mesh analysis equations,

$$1i_1 = -3v + 80$$

$$2i_2 + 2(i_2 + 10) = 3v$$

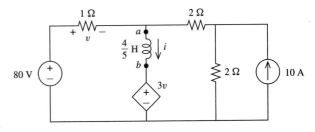

FIGURE 6.15   Circuit for Example 6.9.

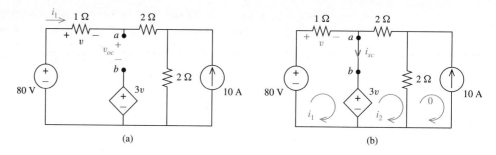

FIGURE 6.16   (a) Circuit for v_{oc}; (b) circuit for i_{sc}.

Substituting $v = i_1$, by the first mesh equation we have $i_1 = 20$. Using this in the second, we have

$$2i_2 + 2(i_2 + 10) = 3(20) = 60$$

or $i_2 = 10$. Then

$$i_{sc} = i_1 - i_2 = 20 - 10 = 10 \text{ A}$$

The Thevenin equivalent is shown in Fig. 6.17. We have reduced the problem to a familiar one. By KVL, around this simple RL circuit,

$$\frac{di}{dt} + 4i = 40 \tag{6.24}$$

with initial condition $i(0-) = i(0+) = 2$ A. We have solved this first-order differential equation several times before, and

$$i(t) = Ke^{-4t} + A = 10 - 8e^{-4t} \text{ A} \tag{6.25}$$

with $A = 10$ found by substitution of $i_f = A$ into (6.24) and K by then matching the initial condition.

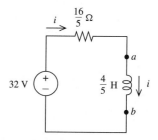

FIGURE 6.17   Circuit equivalent of Fig. 6.15.

The solutions that we have encountered so far in this chapter are often referred to in other more descriptive terms. Two such terms that are widely used are the *transient response* and the *steady-state response*. The transient response is the transitory portion of the total response, which approaches zero as time increases. The steady-state response, on the other hand, is that part of the total response that remains after the transient response has become zero. In the case of dc sources, the steady-state response is constant and is the dc steady state discussed in Section 5.7. In Example 6.9, for the response variable $i(t)$ given in (6.25), the transient response is $-8e^{-4t}$ A and the dc steady-state response is 10 A.

EXERCISES

6.4.1. Find v for $t > 0$ if the circuit is in steady state at $t = 0-$.
Answer $12 - 8e^{-t/8}$ V

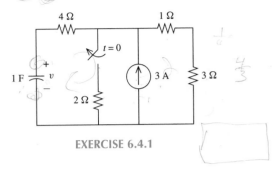

EXERCISE 6.4.1

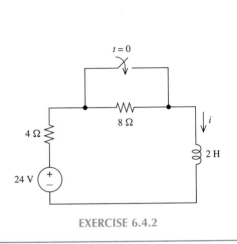

EXERCISE 6.4.2

6.4.2. The circuit is in steady state at $t = 0-$. Find i for $t > 0$.
Answer $6 - 4e^{-2t}$ A

6.4.3. Resolve Exercise 6.4.1 if the switch closes at $t = 0$ rather than opens.
Answer $4 + 8e^{-3/16t}$ V

6.5 SUPERPOSITION IN FIRST-ORDER CIRCUITS

In our discussion of resistive circuits, we saw that superposition could be applied to the analysis of circuits with more than one source. The overall response is computed as a superposition (i.e., a sum) of responses to each source individually, with all other sources killed.

To see how superposition applies in the case of first-order circuits, consider the general first-order differential equation governing all such circuits,

$$\frac{dx}{dt} + \frac{1}{\tau}x = f(t), \qquad x(t_0) = x_0 \tag{6.26}$$

In this equation, $x(t)$ is the circuit variable (current or voltage) and $f(t)$ the forcing function. If we kill all independent sources in the circuit, we will have the unforced equation, that is, $f(t)$ replaced by 0, but otherwise the equation is unchanged. Let $x_{ic}(t)$ solve the problem with independent sources killed; that is,

$$\frac{dx_{ic}}{dt} + \frac{1}{\tau}x_{ic} = 0, \qquad x_{ic}(t_0) = x_0$$

Here $x_{ic}(t)$ is the response due to the initial conditions. Now suppose we return the sources into the circuit, but instead, kill (zero out) the initial condition. The response in this case we label $x_s(t)$, the response due to the independent sources, satisfying

$$\frac{dx_s}{dt} + \frac{1}{\tau}x_s = f(t), \qquad x_s(t_0) = 0$$

It is easy to see that the superposition of these two responses $x(t) = x_{ic}(t) + x_s(t)$ satisfies the original forced differential equation, since

$$\frac{dx}{dt} + \frac{1}{\tau}x = \frac{d}{dt}(x_{ic} + x_s) + \frac{1}{\tau}(x_{ic} + x_s)$$

$$= \left(\frac{d}{dt}x_{ic} + \frac{1}{\tau}x_{ic}\right) + \left(\frac{d}{dt}x_s + \frac{1}{\tau}x_s\right) = 0 + f(t)$$

and that it satisfies the initial condition,

$$x(t_0) = x_{ic}(t_0) + x_s(t_0)$$

$$= x_0 + 0$$

This shows that *the total response is a superposition of the initial condition response with all independent sources killed and the response to independent sources with the initial condition killed.* The initial condition may be thought of as another source, along with the independent sources. This makes physical sense since the initial condition is indeed a source of energy for the natural response. $x_{ic}(t)$ is the natural response to this initial condition "source" with the other (independent) sources killed and must be superposed with the forced response $x_s(t)$ driven by those independent sources, for which the initial condition source is in turn killed.

In the case that there is more than one independent source, superposition may further be applied to compute $x_s(t)$, the forced response with initial conditions killed. That is, *the forced response $x_s(t)$ is the superposition of the responses to each of the independent sources with all others killed (including the initial conditions).*

Example 6.10

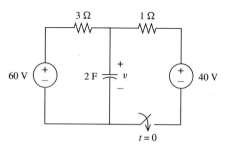

FIGURE 6.18   Circuit for Example 6.10.

Consider the circuit of Fig. 6.18, in which we wish to find v. Prior to $t = 0$ the switch is open, and, as shown in Fig. 6.19(a), in dc steady state $v(0-) = 60$ V. At $t = 0$ the switch closes, and for $t > 0$ we have the circuit of Fig. 6.19(b). We will compute v for $t > 0$ as the superposition of three components: forced components due to the two independent sources and a natural component due to the initial condition $v(0+) = v(0-) = 60$ V.

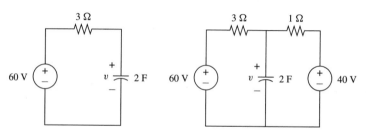

FIGURE 6.19   (a) Equivalent for $t < 0$; (b) for $t > 0$.

Denote the component due to the 60-V source as v_{1s}. Killing the other source, the node equation at the top center node is

$$\frac{1}{3}(v_{1s} - 60) + \frac{1}{1}(v_{1s} - 0) + \frac{2dv_{1s}}{dt} = 0$$

or

$$\frac{dv_{1s}}{dt} + \frac{2}{3}v_{1s} = 10 \qquad (6.27)$$

The initial condition imposed on v_{1s} is $v_{1s}(0+) = 0$, since we are required to kill the initial conditions when computing the forced components. The solution to (6.27) with the given initial conditions, using the method described in Section 6.4, may easily be verified to be

$$v_{1s}(t) = 15[1 - e^{-(2/3)t}]$$

The component due to the 40-V source, v_{2s} is found by killing the 60-V source (and the initial condition). The node equation becomes

$$\frac{1}{3}v_{2s} + \frac{1}{1}(v_{2s} - 40) + \frac{2dv_{2s}}{dt} = 0$$

or

$$\frac{dv_{2s}}{dt} + \frac{2}{3}v_{2s} = 20 \qquad (6.28)$$

with $v_{2s}(0+) = 0$, the solution is

$$v_{2s}(t) = 30[1 - e^{-(2/3)t}]$$

The final component is $v_{ic}(t)$, found by killing both independent sources and exciting a response by using the true initial condition. From Fig. 6.19(b), upon shorting the two voltage sources the two resistors are in parallel with an equivalent of $\frac{3}{4}$ Ω and $RC = (\frac{3}{4})(2) = \frac{3}{2}$ in this sourceless circuit, or

$$\frac{dv_{ic}}{dt} + \frac{2}{3}v_{ic} = 0 \qquad (6.29)$$

with initial condition $v_n(0+) = v_n(0-) = 60$. The solution is $v_{ic} = Ke^{-2/3t}$ with K computed to match the initial condition, or

$$v_{ic}(t) = 60e^{-(2/3)t} \text{ V}$$

We have found the responses driven by each source, the two independent sources, and the initial condition. It only remains to superpose the components:

$$v(t) = v_{ic}(t) + v_{1s}(t) + v_{2s}(t)$$
$$= 15e^{-(2/3)t} + 45 \text{ V} \qquad (6.30)$$

This is the solution for $t > 0$, with $v(t) = 60$ V the solution for $t < 0$.

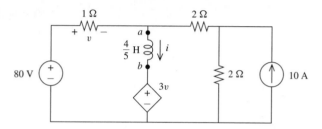

FIGURE 6.20 Circuit for Example 6.11.

Example 6.11

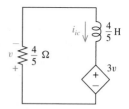

FIGURE 6.21 Circuit for component i_{ic}.

We shall redo Example 6.9 using superposition. The circuit diagram shown in Fig. 6.15 is repeated for convenience as Fig. 6.20. First we will compute i_{ic}, the component of i due to the initial conditions with all independent sources killed. In this case the resistors may be replaced by an equivalent of value $R_{eq} = \frac{4}{5}$, and we have a single-loop circuit as shown in Fig. 6.21. By KVL,

$$\frac{4}{5}\left(\frac{di_{ic}}{dt}\right) + 3v + \frac{4}{5}i_{ic} = 0 \tag{6.31}$$

Using Ohm's law, $v = \frac{4}{5}i_{ic}$, and dividing by $\frac{4}{5}$, the above is

$$\frac{di_{ic}}{dt} + 4i_{ic} = 0$$

The given initial value is $i(0-) = i(0+) = 2$ A, leading to the component

$$i_{ic}(t) = 2e^{-4t} \text{ A} \tag{6.32}$$

Next we need the forced components i_{f1} due to the independent voltage source and i_{f2} due to the current source. To determine i_{f1} we kill the initial conditions and all other independent sources, in this case the current source, as shown in Fig. 6.22(a). Here the branch containing the inductor has been drawn on the right for convenience.

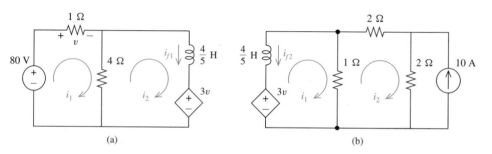

(a)   (b)

FIGURE 6.22 (a) Circuit for component i_{f1}; (b) circuit for i_{f2}.

Chapter 6 First-Order Circuits

The mesh equations for Fig. 6.22(a) are

$$i_1 + 4(i_1 - i_2) = 80$$

$$\frac{4}{5}\frac{d}{dt}i_2 + 3i_1 + 4(i_2 - i_1) = 0$$

Solving the first of these equations for i_1 and substituting into the second yields

$$\frac{d}{dt}i_2 + 4i_2 = 20, \qquad i_2(0+) = 0$$

with zero initial inductive current, since this is a forced response with initial conditions killed. The solution to the above is

$$i_2(t) = 5(1 - e^{-4t}) = i_{f1}(t) \qquad (6.33)$$

The third and final component, $i_{2f}(t)$, is computed with the voltage source killed and the current source returned to the circuit, as shown in Fig. 6.22(b) (the leftmost two parallel branches have been drawn in reverse order for convenience). The mesh equations are

$$\frac{4}{5}\left(\frac{di_1}{dt}\right) + (i_1 - i_2) - 3(i_2 - i_1) = 0$$

$$2i_2 + 2(i_2 + 10) + (i_2 - i_1) = 0$$

Solving the second mesh equation for i_2 and substituting into the first gives us

$$\frac{di_1}{dt} + 4i_1 = -20, \qquad i_1(0+) = 0$$

once again using the zero initial inductor current to derive initial conditions. This solution is

$$i_1(t) = -5(1 - e^{-4t}) \text{ A}$$

From Fig. 6.22(b) we see that

$$i_{f2}(t) = -i_1(t) = 5(1 - e^{-4t}) \text{ A} \qquad (6.34)$$

Using superposition, the total solution is the sum of components (6.32) to (6.34):

$$i(t) = i_{ic}(t) + i_{f1}(t) + i_{f2}(t)$$

$$= 2e^{-4t} + 10(1 - e^{-4t}) = 10 - 8e^{-4t} \text{ A}$$

which agrees with our result from Example 6.9.

6.5.1. Use superposition to find i for $t > 10$ s. Assume that the circuit is in dc steady state at time $t = 10-$ s.

$Answer$ $5 + 10[1 - e^{-500(t-10)}]$ A

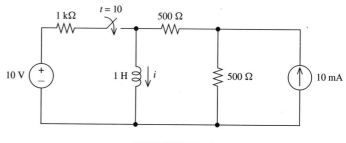

EXERCISE 6.5.1

6.5.2. Repeat Exercise 6.5.1 using the method of Thevenin transform.

6.5.3. Use superposition to find v_3 in terms of v_1 and v_2.

$Answer$ $v_3 = -R_F C \dfrac{dv_1}{dt} - \dfrac{R_F}{R_A} v_2$

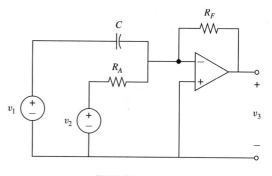

EXERCISE 6.5.3

6.6 UNIT STEP FUNCTION

In the preceding sections we analyzed circuits in which independent sources have been suddenly inserted into the networks. At the instant these sources are applied, the voltages or currents at the points of application change abruptly. Forcing functions whose values change in this manner are called *singularity functions*, since they have one time instant at which they exhibit singular or unusual behavior.

There are many singularity functions that are useful in circuit analysis (and we defer their full description to Chapter 13). One of the most important is the *unit step function*, so named by the English engineer Oliver Heaviside (1850–1925). The unit step

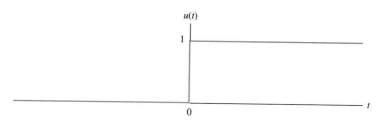

FIGURE 6.23   Graph of the unit step function $u(t)$.

function is the function that is equal to zero for all negative values of its argument and that is equal to one for all positive values of its argument. If we denote the unit step function by the symbol $u(t)$, a mathematical description is

$$u(t) = \begin{cases} 0, & t < 0 \\ 1, & t > 0 \end{cases} \tag{6.35}$$

From its graph shown in Fig. 6.23 we see that at $t = 0$, $u(t)$ changes abruptly from 0 to 1. Nowhere else does its value change except at this "singular" moment in its behavior. Some authors define $u(0)$ to be 1; others prefer $u(0) = \frac{1}{2}$; but we are leaving $u(t)$ undefined at $t = 0$.

A voltage that steps from 0 to V volts at $t = 0$ may be represented by the product $Vu(t)$. Clearly, this voltage is 0 for $t < 0$ and V volts for $t > 0$. A voltage step source of V volts is shown in Fig. 6.24(a). A circuit that is equivalent to this source is shown in Fig. 6.24(b). A short circuit exists for $t < 0$, and the voltage is zero. For $t > 0$, a voltage V appears at the terminals of the equivalent circuit.

Equivalent circuits for a current step source of I amperes are shown in Fig. 6.25. An open circuit exists for $t < 0$, and the current is zero. For $t > 0$, the switching action causes a terminal current of I amperes to flow at the terminals of the equivalent circuit.

Unit step sources of the type shown in Figs. 6.24 and 6.25 will be frequently used in circuits involving sources that instantaneously switch in or out. Note in these

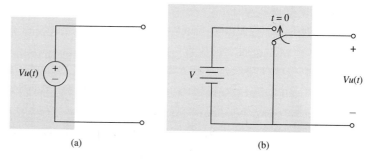

(a)   (b)

FIGURE 6.24   (a) V volt voltage step source; (b) equivalent circuit.

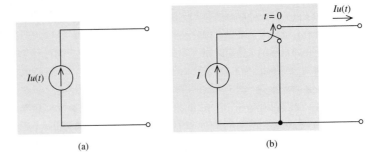

FIGURE 6.25   (a) I-ampere current step source; (b) equivalent circuit.

figures that the unit step forms Figs. 6.24(a) and 6.25(a) are simpler schematically than the combinations of dc sources and switches they replace in Figs. 6.24(b) and 6.25(b), resulting in simpler circuit diagrams.

This representation of switched variables by unit steps is not limited to the simple cases shown above. Just as we used the unit step to represent a voltage or current that jumps from 0 to another value at $t = 0$, we may use sums of scaled and time-shifted unit step functions to represent any "staircase" voltage or current, that is, any function with jump discontinuities that remains constant between jumps.

Returning to the definition of the unit step function given in (6.35), suppose that we replace t by $t - t_0$ in the three places that it occurs, which results in

$$u(t - t_0) = \begin{cases} 0, & t < t_0 \\ 1, & t > t_0 \end{cases} \qquad (6.36)$$

The graph of this function is shown in Fig. 6.26. $u(t - t_0)$ is the unit step time-shifted by t_0 *to the right*, that is, delayed by t_0 seconds.

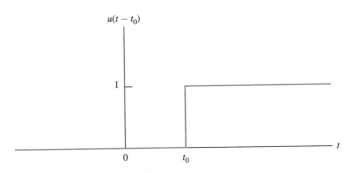

FIGURE 6.26   Time-shifted unit step.

Example 6.12

Consider the voltage $v(t)$ formed as a linear combination of a unit step with a second, time-shifted unit step:

$$v(t) = K_1 u(t) + K_2 u(t - t_0)$$

Suppose first that the second unit step is right-shifted (delayed); in other words, $t_0 > 0$. We will determine the value of $v(t)$ by use of (6.35) and (6.36). For $t < 0$, neither unit step has "switched on" yet

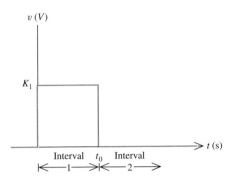
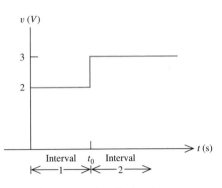

FIGURE 6.27 Two linear combinations of shifted unit steps.

and $v(t) = 0$. For t in the interval $0 < t < t_0$, the first unit step has switched on (its argument is positive), but the second has not yet (its argument is still negative), so $v(t) = K_1$. After t_0 they are both switched on and $v(t) = K_1 + K_2$. So

$$v(t) = \begin{cases} 0, & t < 0 \\ K_1, & 0 < t < t_0 \\ K_1 + K_2, & t > t_0 \end{cases} \qquad (6.37)$$

By proper selection of the scaling factors K_1 and K_2, we can create any desired level value in the intervals marked 1 and 2 in Fig. 6.27. For instance, with $K_2 = -K_1$ we have the rectangular voltage pulse shown in Fig. 6.27(a), and with $K_1 = 2$ and $K_2 = 1$ we have a staircase function with the level values 2 and 3 shown in Fig. 6.27(b).

We may also *time-reverse* a unit step. Replacing t by $-t$ in (6.35),

$$u(-t) = \begin{cases} 1, & t < 0 \\ 0, & t > 0 \end{cases}$$

Time-reversed unit steps are useful in describing sources with non-zero values until the switch time, after which they are switched out of the circuit.

Example 6.13

Consider the circuit of Fig. 6.28(a). For all $t < 0$ the voltage source has a value of 10 V, as shown in Fig. 6.28(b), leading to a dc steady-state current of $i(0-) = 2$ A. After $t = 0$ the source may be replaced by a short circuit [since $u(-t) = 0$ for $t > 0$], leading to the circuit of Fig. 6.28(c). From our previous analysis of the simple sourceless RL circuit, we know that (6.10)

$$i(t) = I_0 e^{-t/\tau} \text{ A}$$

With $\tau = L/R = 4$ and I_0 determined by the initial condition,

$$i(0+) = i(0-) = 2 = I_0(1)$$

whence the total solution

$$i(t) = 2e^{-t/4} \text{ A}$$

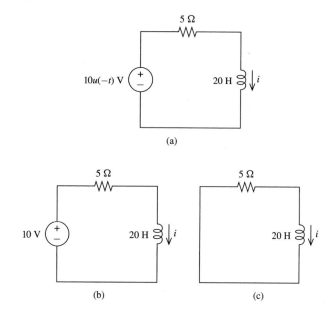

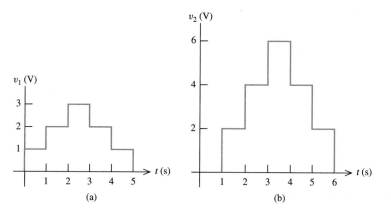

FIGURE 6.28 (a) Circuit with time-reversed step source; (b) equivalent for $t < 0$; (c) equivalent for $t > 0$.

EXERCISES

6.6.1. Express $v_1(t)$ as a sum of scaled and time-shifted unit steps. How is $v_2(t)$ related to $v_1(t)$? Use these two to express $v_2(t)$ as a sum of scaled and time-shifted unit steps.

Answer $u(t) + u(t-1) + u(t-2) - u(t-3) - u(t-4) - u(t-5)$; $v_2(t) = 2v_1(t-1)$; $2[u(t-1) + u(t-2) + u(t-3) - u(t-4) - u(t-5) - u(t-6)]$

EXERCISE 6.6.1

6.6.2. Express $i(t)$ as a sum of two time-reversed and one normal unit step.

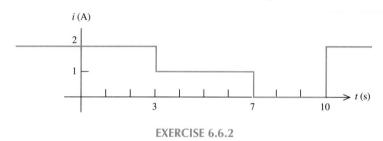

EXERCISE 6.6.2

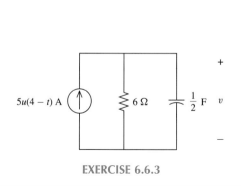

EXERCISE 6.6.3

6.6.3. Find $v(t)$ for $t > 4$. Assume that the circuit is in dc steady state just before the unit step jump time.
Answer $30e^{-(t-4)/3}$ V

6.7 STEP AND PULSE RESPONSES

The step response is defined as the response of a circuit having only one independent source, which is a unit step function. The response and step source can each be either a current or a voltage. Since the unit step source presents zero source function until time $t = 0-$, the dc steady-state response by the circuit at $t = 0-$ must be zero. There can be no initial energy stored in the storage elements at the singular time when the source switches on because there has been no source for $t < 0$ to energize the storage elements. Thus the step response is the response to a unit step input with no initial energy stored in the circuit.

Example 6.14

As an example, let us find the step response $v(t)$ in the simple RC circuit of Fig. 6.29(a) having an input of $v_g(t) = u(t)$ V. The equivalent circuit is shown in Fig. 6.29(b). For $t < 0$, $v_g(t) = 0$ and so the dc steady state at $t = 0-$ is $v(0-) = 0$ V. By continuity of capacitive voltages, $v(0+) = 0$ V as well. Applying KCL for $t > 0$ yields

$$\frac{dv}{dt} + \frac{1}{RC}v = \frac{1}{RC} \tag{6.38}$$

The characteristic equation is

$$s + \frac{1}{RC} = 0$$

leading to the natural response

$$v_n(t) = Ke^{-t/RC} \tag{6.39}$$

The trial forced solution is $v_f = A$, which when substituted into (6.38) yields $A = 1$. Combining this with the natural solution (6.39),

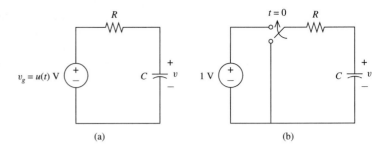

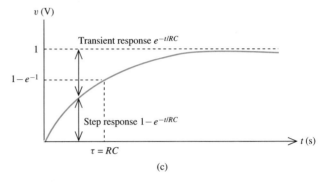

FIGURE 6.29 (a) RC circuit with a voltage step input; (b) equivalent circuit; (c) step response of RC circuit.

the total solution is

$$v(t) = Ke^{-t/RC} + 1$$

Finally, matching the initial condition $v(0+) = 0$ gives $K = -1$. The result is

$$v(t) = \begin{cases} 0, & t < 0 \\ 1 - e^{-t/RC}, & t > 0 \end{cases} \qquad (6.40)$$

where we have used the fact that before the singular time all currents and voltages were zero. This can be written somewhat more concisely as

$$v(t) = (1 - e^{-t/RC})u(t) \qquad (6.41)$$

avoiding the "laundry list" form of (6.40). The step response is shown in Fig. 6.29(c).

Once again it is the time constant that tells us the rate at which the transient response decays to zero. Each period of duration of one time constant, the transient reduces to e^{-1} times its value at the beginning of that period. Circuits with short time constants reach their steady-state value more quickly.

Note by (6.41) that multiplication of any other time function by a unit step forces the product to be zero prior to the singular time and to be equal to the other time function afterward. We will make frequent use of this convenient property of the unit step to simplify other equations arising in subsequent chapters.

Example 6.15

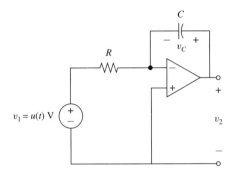

FIGURE 6.30   Circuit for Example 6.15.

As a second example, let us find $v_2(t)$ in the circuit of Fig. 6.30, consisting of a resistor, a capacitor, and an op amp, when the input is a unit step voltage source. The nodal equation at the inverting terminal of the op amp is, upon division by C,

$$\frac{dv_2}{dt} = -\frac{1}{RC}v_1$$

Integrating both sides of this equation between the limits of $0+$ and t, we have

$$v_2(t) = -\frac{1}{RC}\int_{0+}^{t} v_1(\tau)\,d\tau + v_2(0+) \qquad (6.42)$$

While our current interest is centered on the specific case $v_1(t) = u(t)$, note that by (6.42) the general effect of this circuit is to create an output voltage $v_2(t)$ that is proportional to the integral of the input $v_1(t)$ plus a constant. For this reason the circuit is called an *integrator*. Some authors prefer the term *inverting integrator* because of the negative sign in the voltage transfer equation (6.42).

In the case of a unit step input $v_1(t) = u(t)$, all currents and voltages in the circuit are zero at $0-$ and, by continuity, $v_C(0+) = 0$ V as well. The virtual short principle tells us that the inverting terminal is at ground potential; thus $v_C = v_2$ and the initial condition term in (6.42) is $v_2(0+) = 0$ V. For $t > 0$, the integrand $v_1(t) = u(t) = 1$; thus

$$v_2(t) = -\frac{1}{RC}\int_{0+}^{t} d\tau + 0 = \left(-\frac{1}{RC}\right)t$$

The step response of our integrator circuit is 0 for $t < 0$ and the above for $t > 0$, or, using multiplication by $u(t)$ to make this equation concise, for all t

$$v_2(t) = -\frac{1}{RC}tu(t)$$

The function $tu(t)$ is called the *unit ramp* singularity function and is shown in Fig. 6.31(a). Our step response $v_2(t)$ is a negative-going ramp function, scaled by $-1/RC$ and is shown in Fig. 6.31(b).

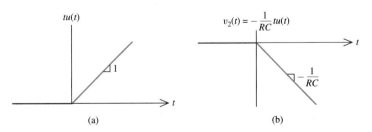

FIGURE 6.31   (a) Unit ramp function; (b) step response of the circuit of Fig. 6.30.

Closely related to the step response is the pulse response. We saw in Section 6.6 how a linear combination of two unit steps could be used to produce a rectangular pulse, for instance Fig. 6.27(a). We now consider the response of a first-order circuit to a pulse input. As before, we assume no other independent sources are present in the circuit.

Prior to the leftmost jump, or *leading edge* of the pulse, there are no currents or voltages for the same reason mentioned for the step response: the dc steady-state response to a zero-value source must be zero. Nothing has turned on yet. During the pulse, the response must be identical to a step response, since the pulse and equal-amplitude step inputs will only differ after the second jump time, which has not yet occurred. Finally, after this second jump time, or *trailing edge* of the pulse, the source is back to zero. We need only compute the response of a sourceless circuit with the proper initial condition (applied at the trailing edge of the pulse).

Example 6.16

To illustrate the pulse response of a first-order circuit, consider Fig. 6.32(a), where the source is the current pulse

$$i_g(t) = 6(u(t) - u(t-1)) \text{ A}$$

This source is zero before $t = 0$, supplies 6 A in the interval $0 < t < 1$, and then shuts down again. It is sketched in Fig. 6.33(a). Replacing the source and 5-Ω resistor by its Thevenin equivalent, we have the single-loop circuit of Fig. 6.32(b). KVL around the loop and division by 5 yield

$$\frac{di}{dt} + i = i_g(t) \qquad (6.43)$$

Before $t = 0$ all currents and voltages are zero, since there is no source of energy to stimulate nonzero responses. Thus $i(0-) = i(0+) = 0$ A. For $0 < t < 1$, $i_g(t) = 6$, and for this interval of time we have

$$\frac{di}{dt} + i = 6, \quad i(0+) = 0 \qquad (6.44)$$

The characteristic equation is $s + 1 = 0$, and the natural response is Ke^{-t}. The trial forced solution is A, and plugging in we see that

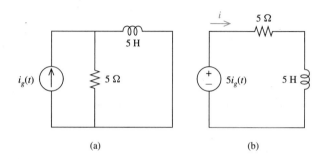

(a)   (b)

FIGURE 6.32   (a) Circuit for Example 6.16; (b) equivalent circuit.

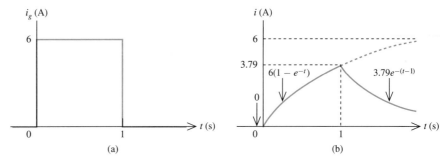

FIGURE 6.33   (a) Pulse current source; (b) pulse response.

$A = 6$. Matching the zero initial condition,

$$i(0+) = K(1) + 6$$

or $K = -6$. We have thus far that $i(t) = 0$ for $t < 0$, and

$$i(t) = 6(1 - e^{-t}) \text{ A}, \quad 0 < t < 1 \tag{6.45}$$

This portion of the response is seen in Fig. 6.33(b). The dashed line shows that, if the source did not switch back off at $t = 1$, the response would approach the value of 6, characteristic of the step response for (6.44). This is only reasonable, since if the source did not switch back off it would be a step source.

Returning to (6.43), for $t > 1$ we have $i_g(t) = 0$ or

$$\frac{di}{dt} + i = 0, \quad t > 1 \tag{6.46}$$

Thus in the interval $t > 1$ we need only solve an unforced first-order circuit. The time constant is $\tau = 1$, and the solution is

$$i(t) = Ke^{-t}, \quad t > 1 \tag{6.47}$$

Since (6.45) is valid for $t > 1$, its initial condition must be at $t = 1+$, just after the trailing edge of the pulse. We have already determined the current just before this time, since by (6.45)

$$i(1-) = 6(1 - e^{-1}) = 3.79 \text{ A}$$

By continuity of inductive currents, $i(1+) = 3.79$ as well. To simplify the initial condition matching, we will rewrite the natural solution (6.47) as

$$i(t) = K_1 e^{-(t-1)}, \quad t > 1 \tag{6.48}$$

These two forms are equivalent, with the new unknown K_1 replacing $K (K = K_1 e^1)$. Then, by (6.48) evaluated at time 1+, $K_1 = 3.79$ and

$$i(t) = 3.79 e^{-(t-1)} \text{ A}, \quad t > 1$$

which completes the full time history of $i(t)$. A graph of this pulse response is shown in Fig. 6.33(b).

Comparing the pulse input to pulse response in Fig. 6.33, we see that the effect of this first-order circuit has been to smooth the input and slow down its abrupt transitions. Although the pulse jumps instantaneously to its full value at $t = 0$, its pulse response grows smoothly for $0 < t < 1$. And although the pulse returns instantaneously to zero at $t = 1$, its response decays to zero continuously with time constant $\tau = 1$ s. This smoothing effect is characteristic of a class of circuits known as low-pass filters, which will be taken up in Chapter 14.

EXERCISES

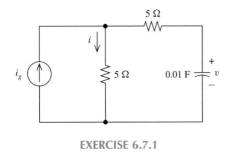

EXERCISE 6.7.1

6.7.1. Find the step responses i and v [step response implies $i_g = u(t)$ A].

Answer $(1 - \frac{1}{2}e^{-10t})u(t)$ A; $5(1 - e^{-10t})u(t)$ V

6.7.2. Find the response i to $v_g = 42u(t)$ V.

Answer $2(1 - e^{-7t})u(t)$ mA (t in ms)

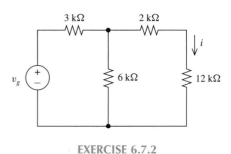

EXERCISE 6.7.2

6.7.3. Use the answer from Exercise 6.7.2 to find the response i to $v_g = 210u(t)$ V.

Answer By linearity, $10(1 - e^{-7t})u(t)$ A

6.8 SPICE AND THE TRANSIENT RESPONSE

SPICE was introduced earlier and used to analyze resistive circuits with dc sources. The program, of course, does much more than this. Here we discuss its use in obtaining transient responses.

The basic format of a SPICE input file and the element statements for resistors, dc sources, and linear controlled sources were introduced in Section 4.9. The format for storage element statements (capacitors and inductors) is

```
CXXXXX     N1     N2     VALUE     <IC = VOLTS>
LXXXXX     N1     N2     VALUE     <IC = AMPS>
```

Here C or L indicates capacitor or inductor, XXXXX is the name given this specific capacitor or inductor, N1 and N2 are the element nodes, and VALUE is the capacitance in farads or inductance in henrys. The final field is optional (angle brackets will be used to mark optional fields in SPICE statements), consisting of the initial conditions. The initial time at which these conditions are applied is always time zero, $t = 0$. The reference direction for the initial capacitive voltage of value VOLTS has its plus sign at the node N1, and that for the initial inductive current AMPS has its arrow pointing from N2 toward N1.

The control statement that will direct SPICE to perform a transient analysis is the .TRAN statement, whose basic format is

```
.TRAN       TSTEP       TSTOP       <UIC>
```

TSTEP is the time step between values to be printed or plotted (output statements for transient analysis will be described shortly). TSTOP is the end of the time interval, assumed to begin at time $t = 0$, during which analysis will be performed and output produced. Inclusion of the optional UIC field directs SPICE to use the initial conditions specified by the user on the storage element statements. If this field is absent, SPICE will compute the dc steady state at $t = 0-$ and use the steady-state inductor currents and capacitor voltages as the initial conditions.

The output of a transient analysis can either be printed as a list of numbers or plotted. The formats for these output control statements are

```
.PRINT    TRAN    CVLIST
.PLOT     TRAN    CVLIST
```

where CVLIST is the list of desired circuit variables.

SPICE Example 6.17

This example illustrates the use of SPICE both to determine initial conditions and the subsequent transient analysis. We seek the current $i_L(t)$ in Fig. 6.34. Since the voltage source is $12u(-t)$ V, its value for $t < 0$ is 12 V. The initial condition $i_L(0+) = i_L(0-)$ may be found by replacing the inductor with a short circuit and finding the dc value for i_L with the 12-V source in place.

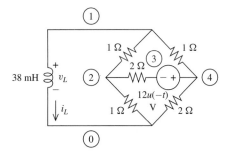

FIGURE 6.34   Circuit for Example 6.17.

```
Example 6.17 run 1: determining the IC
*PUT A DUMMY V SOURCE IN PLACE OF THE
*SHORTED L SO WE CAN PRINT IL=I(VD)
VD        1        0        DC    0
R1        1        2        1
R2        1        4        1
R3        2        3        2
*HERE IS THE 12 V SOURCE
VS        4        3        DC         12
R4        2        0        1
R5        4        0        2
.PRINT DC    I(VD)
.END
```

The printed output contains the lines

```
VOLTAGE SOURCE CURRENTS
NAME              CURRENT

VD                6.316E-01
VS               -3.789E+00
```

Thus the initial current is $i_L(0-) = i_L(0+) = 0.6316$ A. For $t > 0$ the circuit is source free, and the input file to plot $i_L(t)$ for $t \geq 0$ using this initial condition is

```
Example 6.17 run 2: the response for t>0
*ENTER THE IC CALCULATED IN RUN 1
L1        0        1        38M      IC=0.6316
R1        1        2        1
R2        1        4        1
R3        2        3        2
*THE FOLLOWING STATEMENT CAN BE OMITTED
*AND THE NODES RENUMBERED IF PREFERRED
VS        3        4        DC    0
R4        2        0        1
R5        4        0        2
.TRAN    0.2      10        UIC
*PUT IN A DUMMY V SOURCE IF YOUR VERSION
*OF SPICE DOESN'T PERMIT FORMS LIKE I(L1)
.PLOT         TRAN    I(L1)
.END
```

The output is shown in Fig. 6.35. The current $i_L(t)$ decays exponentially from its initial value, as any sourceless RC or RL circuit variable must. To determine τ, note that at the end of one time

```
**** 10/10/94 09:20:11 ********* Evaluation PSpice (January 1991) ************

 Ex 6.17 run 2: the response for t>0

 ****     TRANSIENT ANALYSIS                 TEMPERATURE =   27.000 DEG C

******************************************************************************


  TIME        I(L1)
(*)----------    0.0000E+00   2.0000E-01   4.0000E-01   6.0000E-01   8.0000E-01
                        - - - - - - - - - - - - - - - - - - - - - - - - - - -
  0.000E+00  6.316E-01  .            .            .            . *          .
  4.000E-03  5.575E-01  .            .            .         *  .            .
  8.000E-03  4.923E-01  .            .            .     *      .            .
  1.200E-02  4.344E-01  .            .            . *          .            .
  1.600E-02  3.830E-01  .            .           *.            .            .
  2.000E-02  3.382E-01  .            .        *   .            .            .
  2.400E-02  2.984E-01  .            .     *      .            .            .
  2.800E-02  2.631E-01  .            .   *        .            .            .
  3.200E-02  2.324E-01  .            . *          .            .            .
  3.600E-02  2.050E-01  .            *            .            .            .
  4.000E-02  1.808E-01  .           *.            .            .            .
  4.400E-02  1.597E-01  .         *  .            .            .            .
  4.800E-02  1.409E-01  .        *   .            .            .            .
  5.200E-02  1.242E-01  .       *    .            .            .            .
  5.600E-02  1.097E-01  .      *     .            .            .            .
  6.000E-02  9.678E-02  .     *      .            .            .            .
  6.400E-02  8.534E-02  .     *      .            .            .            .
  6.800E-02  7.536E-02  .    *       .            .            .            .
  7.200E-02  6.649E-02  .   *        .            .            .            .
  7.600E-02  5.863E-02  .   *        .            .            .            .
  8.000E-02  5.178E-02  .  *         .            .            .            .
  8.400E-02  4.568E-02  .  *         .            .            .            .
  8.800E-02  4.028E-02  .  *         .            .            .            .
  9.200E-02  3.557E-02  . *          .            .            .            .
  9.600E-02  3.138E-02  . *          .            .            .            .
  1.000E-01  2.767E-02  . *          .            .            .            .
  1.040E-01  2.444E-02  . *          .            .            .            .
  1.080E-01  2.156E-02  .*           .            .            .            .
  1.120E-01  1.901E-02  .*           .            .            .            .
  1.160E-01  1.679E-02  .*           .            .            .            .
  1.200E-01  1.481E-02  .*           .            .            .            .
  1.240E-01  1.306E-02  .*           .            .            .            .
  1.280E-01  1.154E-02  .*           .            .            .            .
  1.320E-01  1.018E-02  .*           .            .            .            .
  1.360E-01  8.974E-03  .*           .            .            .            .
  1.400E-01  7.925E-03  .*           .            .            .            .
  1.440E-01  6.992E-03  *            .            .            .            .
  1.480E-01  6.165E-03  *            .            .            .            .
  1.500E-01  5.791E-03  *            .            .            .            .
                        - - - - - - - - - - - - - - - - - - - - - - - - - - -
```

FIGURE 6.35 SPICE transient analysis output.

constant the response decays to $e^{-1}(0.6316) = 0.2324$ V. Examining Fig. 6.35, this occurs at 32 ms, so $\tau = 32$ ms. Note that this agrees with the time constant calculated in Example 6.5 using the same circuit, which was done without the help of SPICE.

SPICE may also be used to compute step and pulse responses. To determine the step response, we need only set all initial conditions to zero and include the independent step source as a dc source in the SPICE input file. Pulse responses are handled similarly, except we declare the source to be a *pulse source* through use of the element

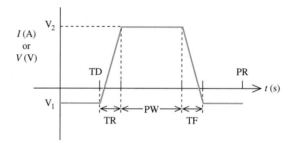

FIGURE 6.36 Parameters of a SPICE pulse source.

statements

```
VXXXXX   N1   N2   PULSE(V1 V2   TD   TR   TF   PW   PR)
IXXXXX   N1   N2   PULSE(V1 V2   TD   TR   TF   PW   PR)
```

The seven numbers within the parentheses specify the parameters of the pulse source, as shown in Fig. 6.36. The rise and fall times of the pulse TR and TF cannot be set to zero. To model instantaneous switching, these values should be set much smaller than the time constant of the circuit.

SPICE Example 6.18

We will compute the unit step and pulse response $v_2(t)$ for the op amp circuit shown in Fig. 6.37. We will use the improved op amp model (Fig. 3.9).

```
Example 6.18 (FIG. 6.38)
*FOR STEP RESPONSE WE'LL USE A DC SOURCE
VG        1        0          DC        1
RA        1        2          25K
RF        2        3          50K
CF        2        3          20U      IC=0
XOA       0        2          3        OPAMP
*HERE IS THE OP AMP SUBCIRCUIT
.SUBCKT OPAMP         1        2        3
*NODE 1 IS +IN, 2 -IN, 3 OUTPUT
RIN       1        2          1MEG
E1        4        0          1        2        100K
ROUT      4        3          30
.ENDS
*FINALLY, THE CONTROL STATEMENTS
.TRAN     0.1      10         UIC
.PLOT           TRAN         V(3)      V(1)
.END
```

The step response output is shown in Fig. 6.38. Based on Example 6.6, we would expect the step response to have a transient decaying to zero with time constant $\tau = R_F C_F$ or (50 kΩ)(20 μF)

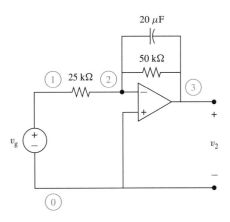

FIGURE 6.37 Circuit for Example 6.18.

$= 1000$ ms, that is, 1 s. Noting from Fig. 6.38 that the steady-state value is -2 V, the initial transient must be $+2$ V and the value after one time constant should be $e^{-1}(2) - 2 = -1.26$. This occurs at $t = 1.00$ s as expected.

Continuing the example, we will find the pulse response to an input pulse of amplitude 1 V and duration 1 s. Replacing the VG element statement in the SPICE input file by

```
VG     1      0          PULSE ( 0  1  0  1M  1M  1  10 )
```

and rerunning SPICE results in the pulse response output shown in Fig. 6.39. The step and pulse responses are identical until the pulse trailing edge ($t = 1$ s), after which the pulse response exhibits a natural decay to zero with time constant $\tau = 1$ s, while the step response continues toward the forced dc steady state of -2 V. From Fig. 6.37, we see that in dc steady state, with the capacitor equivalent to an open circuit, this is an inverting amplifier (as discussed in Chapter 3) with voltage transfer ratio of -2. Thus with a unit step input, we should closely approach this steady state after 5 to 10 time constants (5 to 10 s). The data in Fig. 6.38 confirm this.

SPICE includes several other transient sources as well. Each is of the form

```
VXXXXX     N1     N2     TSPEC
IXXXXX     N1     N2     TSPEC
```

where N1 and N2 are the positive and negative source nodes, and TSPEC is the transient source specification. Three of the most useful transient sources are the piecewise linear PWL, exponential EXP, and damped sinusoidal SIN sketched in Figs. 6.39 and 6.40.

In addition to the coarse, line-printer style of graph afforded by the .PLOT statement, PSpice permits terminals equipped with graphics adapters to display plotted output

```
**** 10/10/94 09:26:24 ********* Evaluation PSpice (January 1991) ************
EXAMPLE 6.18: STEP RESPONSE

****      TRANSIENT ANALYSIS              TEMPERATURE =   27.000 DEG C


********************************************************************************


  TIME        V(3)
 (*)----------   -2.0000E+00  -1.5000E+00  -1.0000E+00  -5.0000E-01   0.0000E+00
            - - - - - - - - - - - - - - - - - - - - - - - - - - - - - - - - - -
 0.000E+00 -7.956E-06 .            .            .            .            *
 1.000E-01 -1.894E-01 .            .            .            .         *  .
 2.000E-01 -3.594E-01 .            .            .            .  *         .
 3.000E-01 -5.140E-01 .            .            .            * .          .
 4.000E-01 -6.548E-01 .            .            .         *  .            .
 5.000E-01 -7.844E-01 .            .            .      *     .            .
 6.000E-01 -8.996E-01 .            .            .   *        .            .
 7.000E-01 -1.006E+00 .            .            * .          .            .
 8.000E-01 -1.100E+00 .            .         *  .            .            .
 9.000E-01 -1.187E+00 .            .      *     .            .            .
 1.000E+00 -1.264E+00 .            .   *        .            .            .
 1.100E+00 -1.335E+00 .            . *          .            .            .
 1.200E+00 -1.398E+00 .           * .           .            .            .
 1.300E+00 -1.456E+00 .         .*  .           .            .            .
 1.400E+00 -1.507E+00 .        *    .           .            .            .
 1.500E+00 -1.555E+00 .      *.     .           .            .            .
 1.600E+00 -1.597E+00 .      *      .           .            .            .
 1.700E+00 -1.636E+00 .    *        .           .            .            .
 1.800E+00 -1.670E+00 .    *        .           .            .            .
 1.900E+00 -1.702E+00 .   *         .           .            .            .
 2.000E+00 -1.730E+00 .  *          .           .            .            .
 2.100E+00 -1.756E+00 . *           .           .            .            .
 2.200E+00 -1.779E+00 . *           .           .            .            .
 2.300E+00 -1.801E+00 .*            .           .            .            .
 2.400E+00 -1.820E+00 .*            .           .            .            .
 2.500E+00 -1.837E+00 *             .           .            .            .
 2.600E+00 -1.852E+00 *             .           .            .            .
 2.700E+00 -1.867E+00 . *           .           .            .            .
 2.800E+00 -1.879E+00 . *           .           .            .            .
 2.900E+00 -1.891E+00 . *           .           .            .            .
 3.000E+00 -1.901E+00 . *           .           .            .            .
 3.100E+00 -1.911E+00 . *           .           .            .            .
 3.200E+00 -1.919E+00 . *           .           .            .            .
 3.300E+00 -1.927E+00 . *           .           .            .            .
 3.400E+00 -1.934E+00 . *           .           .            .            .
 3.500E+00 -1.940E+00 . *           .           .            .            .
 3.600E+00 -1.946E+00 .*            .           .            .            .
 3.700E+00 -1.951E+00 .*            .           .            .            .
 3.800E+00 -1.956E+00 .*            .           .            .            .
 3.900E+00 -1.960E+00 .*            .           .            .            .
 4.000E+00 -1.964E+00 .*            .           .            .            .
 4.100E+00 -1.967E+00 .*            .           .            .            .
 4.200E+00 -1.970E+00 .*            .           .            .            .
 4.300E+00 -1.973E+00 .*            .           .            .            .
 4.400E+00 -1.976E+00 .*            .           .            .            .
 4.500E+00 -1.978E+00 .*            .           .            .            .
 4.600E+00 -1.980E+00 .*            .           .            .            .
 4.700E+00 -1.982E+00 *             .           .            .            .
 4.800E+00 -1.984E+00 *             .           .            .            .
 4.900E+00 -1.985E+00 *             .           .            .            .
 5.000E+00 -1.987E+00 *             .           .            .            .
 5.100E+00 -1.988E+00 *             .           .            .            .
 5.200E+00 -1.989E+00 *             .           .            .            .
 5.300E+00 -1.990E+00 *             .           .            .            .
 5.400E+00 -1.991E+00 *             .           .            .            .
 5.500E+00 -1.992E+00 *             .           .            .            .
 5.600E+00 -1.993E+00 *             .           .            .            .
 5.700E+00 -1.993E+00 *             .           .            .            .
 5.800E+00 -1.994E+00 *             .           .            .            .
 5.900E+00 -1.995E+00 *             .           .            .            .
 6.000E+00 -1.995E+00 *             .           .            .            .
 6.100E+00 -1.996E+00 *             .           .            .            .
 6.200E+00 -1.996E+00 *             .           .            .            .
```

FIGURE 6.38 Step response for Example 6.18.

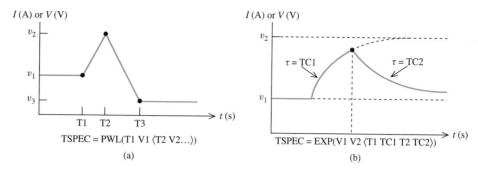

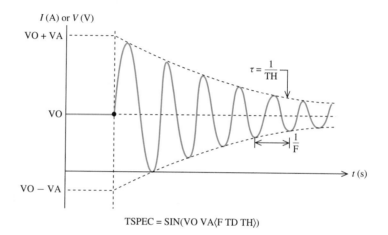

FIGURE 6.39   (a) Piecewise linear source; (b) exponential source.

TSPEC = SIN(VO VA⟨F TD TH⟩)

FIGURE 6.40   Damped sinusoidal source.

at the maximum available screen resolution. The control statement

.PROBE

when added to the input file results in the creation of an auxiliary data file that supports the operation of PROBE, a graphics postprocessor software program bundled with PSpice. Upon execution, PROBE permits the circuit variables to be plotted and scales for the axes and other display characteristics to be selected and changed interactively. A PROBE plot of the same pulse response .PLOTted in Fig. 6.41 is shown in Fig. 6.42.*

Finally, a circuit with nonzero or multiple switch times can be analyzed with SPICE by shifting the switch time (or times) to zero. For instance, suppose that we have a circuit in dc steady state at $t = 0-$, at which time a switch acts, followed by a second switch action at $t = t_0 > 0$. To determine the conditions at $t = 0-$, we first do a dc analysis of the circuit with the switches in their $t < 0$ positions. The resulting values are used as initial conditions on the storage element statements for a second SPICE run, this time a transient analysis of the circuit with its switches in their $0 < t < t_0$ positions.

---

*Users of SPICE version 3, DESIGN CENTER, and other SPICE variants or supersets may also have access to high-resolution graphics. Consult your user's guide.

EXAMPLE 6.18: PULSE RESPONSE

```
 ****      TRANSIENT ANALYSIS               TEMPERATURE =    27.000 DEG C


***************************************************************************************


   TIME        V(3)
   (*)----------    -1.5000E+00   -1.0000E+00   -5.0000E-01   0.0000E+00    5.0000E-01
                  - - - - - - - - - - - - - - - - - - - - - - - - - - - - - -
   0.000E+00  1.680E-08 .            .            .            *            .
   1.000E-01 -1.890E-01 .            .            .        *   .            .
   2.000E-01 -3.603E-01 .            .            .     *      .            .
   3.000E-01 -5.113E-01 .            .            .  *         .            .
   4.000E-01 -6.570E-01 .            .         *   .            .            .
   5.000E-01 -7.822E-01 .            .      *      .            .            .
   6.000E-01 -9.014E-01 .            .   *         .            .            .
   7.000E-01 -1.004E+00 .            . *           .            .            .
   8.000E-01 -1.101E+00 .          *  .            .            .            .
   9.000E-01 -1.186E+00 .       *     .            .            .            .
   1.000E+00 -1.266E+00 .     *       .            .            .            .
   1.100E+00 -1.149E+00 .        *    .            .            .            .
   1.200E+00 -1.041E+00 .           *..            .            .            .
   1.300E+00 -9.419E-01 .            . *           .            .            .
   1.400E+00 -8.541E-01 .            .    *        .            .            .
   1.500E+00 -7.710E-01 .            .      *      .            .            .
   1.600E+00 -6.986E-01 .            .         *   .            .            .
   1.700E+00 -6.307E-01 .            .           * .            .            .
   1.800E+00 -5.715E-01 .            .            .*           .            .
   1.900E+00 -5.159E-01 .            .            . *          .            .
   2.000E+00 -4.674E-01 .            .            .   *        .            .
   2.100E+00 -4.220E-01 .            .            .    *       .            .
   2.200E+00 -3.823E-01 .            .            .      *     .            .
   2.300E+00 -3.452E-01 .            .            .       *    .            .
   2.400E+00 -3.127E-01 .            .            .         *  .            .
   2.500E+00 -2.823E-01 .            .            .          * .            .
   2.600E+00 -2.558E-01 .            .            .           *.            .
   2.700E+00 -2.309E-01 .            .            .            *            .
   2.800E+00 -2.093E-01 .            .            .            *            .
   2.900E+00 -1.889E-01 .            .            .            *            .
   3.000E+00 -1.712E-01 .            .            .            .*           .
   3.100E+00 -1.545E-01 .            .            .            .*           .
   3.200E+00 -1.400E-01 .            .            .            .*           .
   3.300E+00 -1.264E-01 .            .            .            . *          .
   3.400E+00 -1.145E-01 .            .            .            . *          .
   3.500E+00 -1.034E-01 .            .            .            . *          .
   3.600E+00 -9.367E-02 .            .            .            .  *         .
   3.700E+00 -8.457E-02 .            .            .            .  *         .
   3.800E+00 -7.662E-02 .            .            .            .  *         .
   3.900E+00 -6.917E-02 .            .            .            .  *         .
   4.000E+00 -6.268E-02 .            .            .            .  *.        .
   4.100E+00 -5.658E-02 .            .            .            .   *.       .
   4.200E+00 -5.127E-02 .            .            .            .   *.       .
   4.300E+00 -4.628E-02 .            .            .            .   *.       .
   4.400E+00 -4.193E-02 .            .            .            .   *.       .
   4.500E+00 -3.786E-02 .            .            .            .   *.       .
   4.600E+00 -3.430E-02 .            .            .            .   *.       .
   4.700E+00 -3.097E-02 .            .            .            .   *.       .
   4.800E+00 -2.806E-02 .            .            .            .   *.       .
   4.900E+00 -2.533E-02 .            .            .            .   *.       .
   5.000E+00 -2.295E-02 .            .            . .          .   *.       .
   5.100E+00 -2.072E-02 .            .            .            .   *.       .
   5.200E+00 -1.877E-02 .            .            .            .    *       .
   5.300E+00 -1.695E-02 .            .            .            .    *       .
   5.400E+00 -1.536E-02 .            .            .            .    *       .
   5.500E+00 -1.386E-02 .            .            .            .    *       .
   5.600E+00 -1.256E-02 .            .            .            .    *       .
   5.700E+00 -1.134E-02 .            .            .            .    *       .
   5.800E+00 -1.027E-02 .            .            .            .    *       .
   5.900E+00 -9.275E-03 .            .            .            .    *       .
   6.000E+00 -8.404E-03 .            .            .            .    *       .
   6.100E+00 -7.587E-03 .            .            .            .    *       .
   6.200E+00 -6.874E-03 .            .            .            .    *       .
```

FIGURE 6.41 Pulse response for Example 6.18 (.PLOT output).

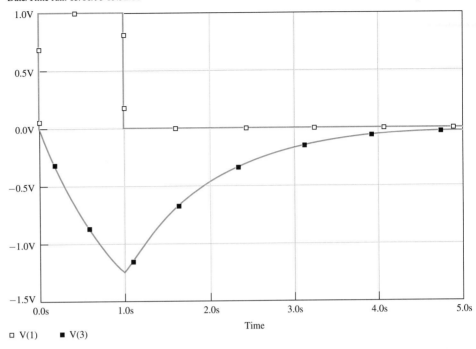

□ V(1) ■ V(3)

FIGURE 6.42 Pulse response for Example 6.18 (PROBE output).

We terminate this analysis at the time t_0, outputting the values of the capacitive voltages and inductive currents. These values are input as initial conditions for a final run, using a SPICE input file reflecting the circuit configuration that results by setting the switches in their $t > t_0$ positions. Since SPICE always begins a transient analysis at $t = 0$, we will remember to add t_0 to every time listed in the final output. An example with multiple switch times is given in Exercise 6.8.3.

EXERCISES

6.8.1. Using SPICE, find (a) $i_L(0-)$ and (b) i_L (30 μs).
    Answer (a) -2.475 mA; (b) $+0.285$ mA

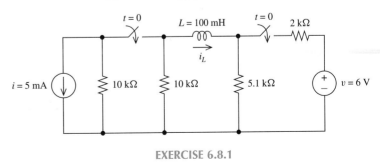

EXERCISE 6.8.1

6.8.2. Repeat Exercise 6.8.1 for the case in which the rightmost switch closes at time $t = 20 \ \mu s$ rather than $t = 0$.

Answer (a) -2.475 mA; (b) $+0.218$ mA

6.8.3. Plot the response $v(t)$ for $0 < t < 10$ ms if v_g is a pulse of amplitude 10 V and duration 2 ms, and the initial value of v is $+2$ V.

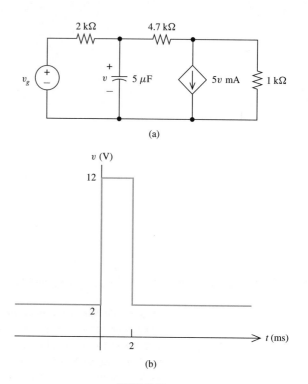

(a)

(b)

EXERCISE 6.8.3

6.8.4. Write a SPICE input file for this circuit using two subcircuits. Assume that all initial currents are zero. Omit the .TRAN or .DC statement and output control statements.

Answer

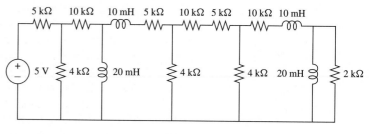

EXERCISE 6.8.4

```
Exercise 6.8.4
VS        1       0        DC       5
XR1       1       2        RSUBCKT
XL1       2       3        LSUBCKT
XR2       3       4        RSUBCKT
XR3       4       5        RSUBCKT
XL2       6       5        LSUBCKT
RL        6       0        2K
.SUBCKT           RSUBCKT           1       3
R1        1       2        5K
R2        2       3        10K
R3        2       0        4K
.ENDS
.SUBCKT           LSUBCKT   1       2
L1        1       2        10M      IC=0
L2        1       0        20M      IC=0
.ENDS
.END
```

6.9   DESIGN OF FIRST-ORDER CIRCUITS

First-order circuits have two parameters that may be assigned. These may be specified as the resistance R and inductance L, resistance R and capacitance C, or the time constant τ and the gain of the circuit. First-order circuits may be either passive, designed using only RLC elements, or active, incorporating op amps or other active devices. What they have in common is that they are described by first-order differential equations.

In Design Example 6.1 we demonstrate one common use of first-order circuits: to block or trap dc. Often, a dc bias voltage is produced by a source circuit and it is desired to prevent this dc value from appearing across the terminals of the load. In this case a first-order circuit can be interposed between source and load, serving as a buffer between them.

Design Example 6.1

Design a first-order buffer circuit B which blocks any dc source voltage component from appearing across the load R_L [see Fig. 6.43(a)]. In addition, the half-amplitude pulse width of the step response across the load should be no more than 1 ms for any load R_L.

The $i-v$ law for a capacitor is $i = C\,dv/dt$. If the voltage $v(t)$ is a constant (dc), then $dv/dt = 0$ and no dc current will flow through the capacitor. Thus we can place any capacitor C in series with the source voltage to block dc. If the buffer circuit B in Fig. 6.43(b) was only the capacitor, in other words if the resistor labeled R_B in this figure were omitted, the time constant of the resulting RC circuit would be $\tau = R_L C$ and could be arbitrarily long for large R_L regardless of our choice for C. So we put a resistor R_B

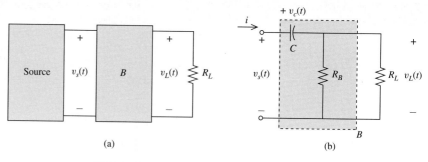

FIGURE 6.43   (a) Block diagram; (b) buffer circuit design.

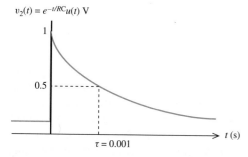

$v_2(t) = e^{-t/RC}u(t)$ V

FIGURE 6.44   Step response of the load voltage $v_L(t)$.

at the output of B as shown. Determination of appropriate values for C and R_B follow.

Let R denote the parallel equivalent of R_B and R_L. The voltage response $v_C(t)$ across the capacitor to a unit step input $v_S(t) = u(t)$ V is, repeating (6.41),

$$v_C(t) = (1 - e^{-t/RC})u(t)$$

and thus by KVL applied to Fig. 6.43(b),

$$v_L(t) = u(t) - (1 - e^{-t/RC})u(t) = e^{-t/RC}u(t)$$

This pulselike step response is sketched in Fig. 6.44. In order to have the response reduced to 0.5 V at $t = \tau_h = 0.001$ s, defined as the *half-amplitude pulse width*, we need

$$v_L(t)|_{t=0.001} = 0.5 = e^{-0.001/\tau}$$

where $\tau = RC$ is the time constant. Taking logarithms gives

$$\frac{-0.001}{\tau} = \log 0.5 = -0.693$$

or $\tau = 1.44$ ms. Thus in order that the half-amplitude pulse width $\tau_h = 1$ ms or less, it is necessary that the circuit time constant $\tau = RC$ be 1.44 ms or less. If we choose $R_B = 100$ kΩ, then since R is the parallel equivalent of R_B and another resistance (R_L), R cannot exceed 100 kΩ. Then choosing C so that

$$RC = 0.00144 = (10^5)C$$

or $C = 14.4$ nF, the design is complete. For open-circuit loads (R_L infinite) the half-amplitude pulse width will be 1 ms, and for finite loads leading to shorter time constants the half-amplitude pulse width will be less than 1 ms, as required. Note that any selection of R_B and C such that R_BC is less than or equal to 1.44 ms will also satisfy the design requirements.

If voltage gains of greater than unity are required in a first-order circuit, active design must be employed. An example based upon the inverting amplifier op amp building block follows.

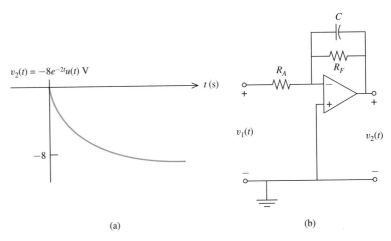

Design Example 6.2

Design a circuit whose unit step response is the voltage waveform shown in Fig. 6.45(a).

$v_2(t) = -8e^{-2t}u(t)$ V

(a)

(b)

FIGURE 6.45   (a) Desired step response; (b) circuit design.

Examining the waveform, in dc steady state we need a voltage gain of -8, which can be achieved by an inverting amplifier introduced in Fig. 3.15 with resistance ratio $R_F/R_A = 8$. To achieve a single exponential transition to steady state as required, we need a first-order circuit, so we must add a storage element. Choosing a capacitor, it cannot be placed in series with either resistor since the dc-blocking property of a capacitor makes it an effective infinite resistance and the dc steady-state gain will certainly be affected. Trying C in parallel with R_F, we have the circuit of Fig. 6.45(b). Noting that by the virtual short principle the noninverting input is at ground and applying KCL at the noninverting input node, we obtain

$$\frac{v_1}{R_A} + \frac{v_2}{R_F} + C\frac{dv_2}{dt} = 0$$

or

$$\frac{dv_2}{dt} + \frac{1}{R_F C}v_2 = \frac{-1}{R_A C}v_1$$

Setting $v_1(t) = u(t)$ V to find the step response, the last is a first-order differential equation with dc source as discussed in Section 6.4. The forced response will be a constant A satisfying

$$\frac{dA}{dt} + \frac{A}{R_F C} = \frac{-1}{R_A C}$$

or $A = -R_F/R_A$. The natural response will be an exponential with $\tau = R_F C$, so the total response for $t > 0$ must be

$$v_2(t) = Ke^{-t/R_F C} - \frac{R_F}{R_A}$$

The step response is always measured using zero initial conditions, so

$$v_2(0) = K - \frac{R_F}{R_A} = 0$$

or

$$v_2(t) = \frac{-R_F}{R_A}(1 - e^{-t/R_F C})u(t)$$

Returning to Fig. 6.45(a), we need only pick the resistance ratio $R_F/R_A = 8$ and the time constant $R_F C$ of the feedback network equal to $\frac{1}{2}$ s to complete the design. For instance, we may select $R_F = 80$ kΩ, $R_A = 10$ kΩ, and $C = 6.25$ μF.

Design Example 6.3

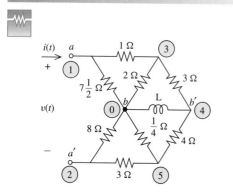

FIGURE 6.46 Circuit for Design Example 6.3.

A power company is supplying electrical power to a load region modeled by the resistive network shown in Fig. 6.46. If an L-henry inductance is introduced between the pair of nodes b and b' shown, find a formula for the time constant of the resulting RL load circuit as seen from the source terminals a–a'.

This design problem is a bit different from the previous ones, in that we must design many circuits at the same time, those with all possible time constants τ. Also, we have rigid structural constraints on our design; the form of the circuit layout is given. There is no single format for design problems, they come in many forms and with a great variety of constraints.

Since there is only one inductor, this is a first-order circuit with time constant L/R_{eq}, where R_{eq} is the equivalent resistance. Rather than compute the time constant using exact analytic methods, which would be time consuming since the resistors are neither in series nor parallel, we will use SPICE and numerical approximation. Suppose that we set $L = 1$ H. If we determine $\tau = \tau_1$ for $L = 1$ H, we would know $R_{\text{eq}} = 1/\tau_1$ and hence the desired formula $\tau(L) = L/R_{\text{eq}}$. A SPICE input file for this design strategy is shown below.

```
DESIGN EXAMPLE 6.3
*
I        2      1      DC              1
R1       1      0      7.5
R2       2      0      8
R3       1      3      1
R4       2      5      3
R5       3      0      2
R6       5      0      0.25
R7       3      4      3
R8       4      5      4
L        4      0      1       IC=0
.TRAN    0.2    5      UIC
.PLOT TRAN V(1,2)
.END
```

```
**** 03/04/95 15:47:55 ********* Evaluation PSpice (January 1991) ************

 DESIGN EXAMPLE 6.3

 ****      TRANSIENT ANALYSIS                 TEMPERATURE =   27.000 DEG C

******************************************************************************


  TIME         V(1,2)
(*)----------    4.0000E+00   4.0500E+00   4.1000E+00   4.1500E+00   4.2000E+00
                      - - - - - - - - - - - - - - - - - - - - - - - - - - - - - -
 0.000E+00  4.163E+00 .         .         .         . *       .
 2.000E-01  4.107E+00 .         .         .*        .         .
 4.000E-01  4.070E+00 .         .   *     .         .         .
 6.000E-01  4.047E+00 .        *.         .         .         .
 8.000E-01  4.032E+00 .      *  .         .         .         .
 1.000E+00  4.022E+00 .    *    .         .         .         .
 1.200E+00  4.016E+00 .   *     .         .         .         .
 1.400E+00  4.012E+00 .  *      .         .         .         .
 1.600E+00  4.009E+00 . *       .         .         .         .
 1.800E+00  4.008E+00 . *       .         .         .         .
 2.000E+00  4.007E+00 . *       .         .         .         .
 2.200E+00  4.006E+00 . *       .         .         .         .
 2.400E+00  4.005E+00 .*        .         .         .         .
 2.600E+00  4.005E+00 .*        .         .         .         .
 2.800E+00  4.005E+00 .*        .         .         .         .
 3.000E+00  4.005E+00 .*        .         .         .         .
 3.200E+00  4.005E+00 .*        .         .         .         .
 3.400E+00  4.005E+00 .*        .         .         .         .
 3.600E+00  4.005E+00 .*        .         .         .         .
 3.800E+00  4.005E+00 .*        .         .         .         .
 4.000E+00  4.005E+00 .*        .         .         .         .
 4.200E+00  4.005E+00 .*        .         .         .         .
 4.400E+00  4.005E+00 .*        .         .         .         .
 4.600E+00  4.005E+00 .*        .         .         .         .
 4.800E+00  4.005E+00 .*        .         .         .         .
 5.000E+00  4.005E+00 .*        .         .         .         .
                      - - - - - - - - - - - - - - - - - - - - - - - - - - - - - -


          JOB CONCLUDED

          TOTAL JOB TIME              .33
```

FIGURE 6.47(a) SPICE output.

The resulting plot, shown as Fig. 6.47(a), shows that the steady-state voltage response is 4.005 V. Since the initial voltage was 4.163 V, the size of the transient response is 4.163 − 4.005 = 0.158 V. The time constant τ is the time at which the transient is reduced to e^{-1} of its initial value, or

$$v_{12}(\tau) = 4.005 + e^{-1}(0.158) = 4.063$$

Figure 6.47(a) has too coarse time sampling to make out the time at which $v(1,2)$ takes on this value with acceptable accuracy. Expanding the scale using the edited transient mode control statement

```
.TRAN    0.01    0.5    UIC
```

results in the plot shown in Fig. 6.47(b), from which we see that

```
**** 03/04/95 16:02:44 ********* Evaluation PSpice (January 1991) ************
DESIGN EXAMPLE 6.3


  ****      TRANSIENT ANALYSIS              TEMPERATURE =   27.000 DEG C


***********************************************************************


    TIME         V(1,2)
  (*)----------   4.0500E+00    4.1000E+00    4.1500E+00    4.2000E+00    4.2500E+00
                -  -  -  -  -  -  -  -  -  -  -  -  -  -  -  -  -  -  -  -  -  -  -  -
    2.000E-01   4.106E+00  .                 .   *            .             .             .
    2.100E-01   4.104E+00  .                 .  *             .             .             .
    2.200E-01   4.102E+00  .               *                 .             .             .
    2.300E-01   4.100E+00  .               *                 .             .             .
    2.400E-01   4.098E+00  .             *  .                .             .             .
    2.500E-01   4.096E+00  .             *  .                .             .             .
    2.600E-01   4.094E+00  .           *    .                .             .             .
    2.700E-01   4.092E+00  .           *    .                .             .             .
    2.800E-01   4.090E+00  .         *      .                .             .             .
    2.900E-01   4.088E+00  .         *      .                .             .             .
    3.000E-01   4.086E+00  .       *        .                .             .             .
    3.100E-01   4.084E+00  .       *        .                .             .             .
    3.200E-01   4.083E+00  .      *         .                .             .             .
    3.300E-01   4.081E+00  .      *         .                .             .             .
    3.400E-01   4.079E+00  .      *         .                .             .             .
    3.500E-01   4.078E+00  .     *          .                .             .             .
    3.600E-01   4.076E+00  .     *          .                .             .             .
    3.700E-01   4.074E+00  .    *           .                .             .             .
    3.800E-01   4.073E+00  .    *           .                .             .             .
    3.900E-01   4.071E+00  .    *           .                .             .             .
    4.000E-01   4.070E+00  .   *            .                .             .             .
    4.100E-01   4.069E+00  .   *            .                .             .             .
    4.200E-01   4.067E+00  .  *             .                .             .             .
    4.300E-01   4.066E+00  .  *             .                .             .             .
    4.400E-01   4.064E+00  .  *             .                .             .             .
    4.500E-01   4.063E+00  . *              .                .             .             .
    4.600E-01   4.062E+00  . *              .                .             .             .
    4.700E-01   4.061E+00  . *              .                .             .             .
    4.800E-01   4.059E+00  . *              .                .             .             .
    4.900E-01   4.058E+00  . *              .                .             .             .
    5.000E-01   4.057E+00  . *              .                .             .             .
                -  -  -  -  -  -  -  -  -  -  -  -  -  -  -  -  -  -  -  -  -  -  -  -


     JOB CONCLUDED

     TOTAL JOB TIME            .35
```

FIGURE 6.47(b) SPICE output with different sampling interval and limits.

$\tau = 0.45$ s. Then $1/R_{eq} = 0.45$ and our final design result is the expression

$$\tau(L) = \frac{L}{R_{eq}} = 0.45L$$

Design Example 6.3 was the first in which SPICE was used to save labor. We proceeded by evaluating a candidate circuit, editing the input file, and evaluating again in a cyclic manner until a design goal was reached. Sometimes it is the circuit element values or locations that are changed; other times, output parameters such as the transient sampling interval and duration in this example. In either case the tedium of repetitive, compute-intensive hand circuit evaluations are bypassed. Most of the remaining designs will use SPICE in this fashion.

DESIGN EXERCISES

6.9.1. Design a circuit with time constant $\tau = 1$ s whose response to the voltage pulse $v_s(t) = 2u(t) - 2u(t-1)$ just reaches, but never exceeds, 1 V.
One of many possible solutions

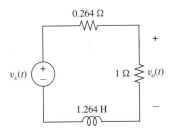

EXERCISE 6.9.1

6.9.2. A defibrillator is a bioelectric device for delivering a shock to stabilize the heart. If the defibrillator voltage is $v_C(t)$ in the figure, what is the minimum dc voltage source V_s that will charge the defibrillator from zero initial stored energy up to its working voltage of 400 V in 1 s?
Answer 1359.5 V

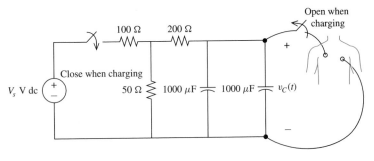

EXERCISE 6.9.2

SUMMARY

First-order circuits are those characterized by a single first-order differential equation. They can be identified by the presence of a single equivalent storage element (inductor or capacitor). The behavior of first-order circuits may be studied by writing and solving their differential equations in the time domain, which is the focus of this chapter.

- An unforced first-order differential equation can be solved by separation of variables or by the characteristic equation method. The solution is a real exponential time function characterized by a single time constant.

- The total solution to a forced first-order differential equation is the sum of a forced solution and an unforced (natural) solution. The natural solution can be found as above, with forcing terms removed from the differential equation. The forced solution can be found by use of a trial forced solution.

- The multiplier of the natural solution is determined by requiring that the total solution match the initial condition.

- If more than one independent source is present, superposition may be used to determine the overall forced response. For each source, all other sources are killed, and the resulting forced responses superposed.

- The unit step function $u(t)$ is defined as zero for $t < 0$, one for $t > 0$. It models the function of a switch which turns on at $t = 0$. The unit step response of a circuit is a common way to characterize its behavior.

- Inductors and capacitors with arbitrary initial conditions are supported by SPICE. The .TRAN control statement is used to generate a transient analysis.

Unlike resistive circuits, first-order circuits contain responses that do not simply mimic the waveshapes of the sources exciting them. These are the simplest circuits that can create something new, something not already inherent in the circuit inputs, namely real exponential time functions. Circuits with more than one storage element can produce a much broader range of new responses and are the subject of the next chapter.

PROBLEMS

6.1. Find $v(t)$ and $i(t)$ for $t > 0$ if $v(0) = 2$ V.

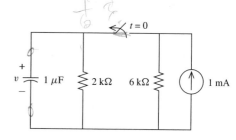

FIGURE P6.2

6.3. Determine $v_1(t)$ for $t > 0$ if $v_1(0-) = -12$ V.

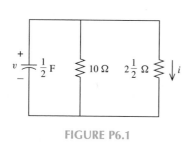

FIGURE P6.1

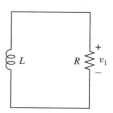

FIGURE P6.3

6.2. Find $v(t)$ for $t > 0$. Assume that the circuit is in dc steady state at $t = 0-$.

6.4. If the initial stored energy in the capacitor at time $t = 0-$ is 0.18 μJ, how much energy will remain stored at time $t = 20$ ms? At time $t = 200$ ms?

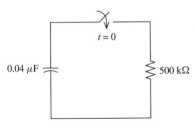

FIGURE P6.4

6.5. The energy $w = \frac{1}{2}Li^2$ stored in the inductor satisfies an enforced first-order differential equation

$$\frac{dw}{dt} + \alpha w = 0$$

Find α.

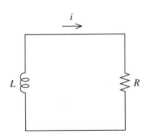

FIGURE P6.5

6.6. Design a sourceless RL circuit with a time constant of 10 ms using only 2-mH inductors and 1-kΩ resistors. Use the minimum number of circuit elements possible.

6.7. What is the time constant τ? Express $v(t)$ and $i(t)$ in terms of τ if $v(0) = +100$ V.

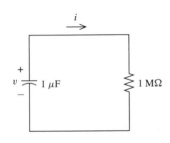

FIGURE P6.7

6.8. In terms of τ, how much time does it take a response in an unforced RC or RL circuit to decay by a factor of 2? By a factor of 10? By a factor of 1000?

6.9. Sketch $w(t)$ versus t for the circuit of Problem 6.7. What is the time constant of this waveform? Why does it differ from $\tau = RC$?

6.10. From Fig. 6.6, the initial slope of a waveform with time constant linearly extrapolated hits $v(t) = 0$ at $t = \tau$. Where does a similar construction from the point $v(\tau)$ hit the horizontal axis?

6.11. Design a source-free circuit with $\tau = 1$ μs. If we have $w(0) = 1$ J in this circuit, how much energy will it dissipate by time $t = 1$ μs? $t = 5$ μs?

6.12. A sourceless RL circuit has 4-mA current at time $t = 2$ ms and 1 mA at time $t = 4$ ms. What was the initial current at time $t = 0$?

6.13. Show that this circuit is characterized by a second-order differential equation, so it is not a first-order circuit.

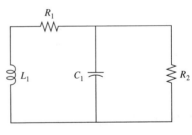

FIGURE P6.13

6.14. Repeat Problem 6.13 for the case in which C_1 is replaced by a second inductor L_2.

6.15. Find the time constant τ.

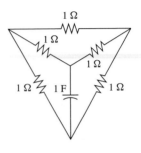

FIGURE P6.15

6.16. Find $v_2(t)$ for $t > 0$ if $v_c(0-) = +6$ V.

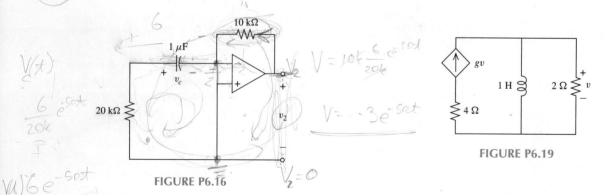

FIGURE P6.16

6.17. Find $i(t)$, $t > 0$, if the circuit is in dc steady state at time $t = 0-$.

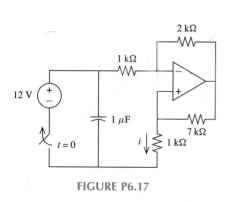

FIGURE P6.17

6.18. Find $i(t)$ for $t > 10$ s if $v(10-) = 2$ V. Repeat for $v(10-) = 0$ V.

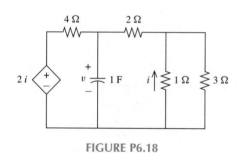

FIGURE P6.18

6.19. Find a value for g so that the time constant in this circuit is $\tau = 4$ s.

FIGURE P6.19

6.20. Find the time constant for this circuit.

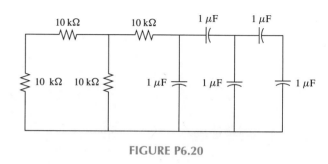

FIGURE P6.20

6.21. Find all currents and voltages for $t > 0$ in this circuit. Assume dc steady state at $t = 0-$.

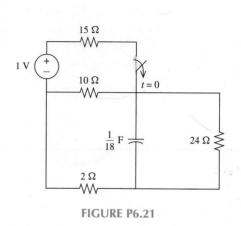

FIGURE P6.21

6.22. Find v and i for $t > 0$ if $i(0) = 1$ A.

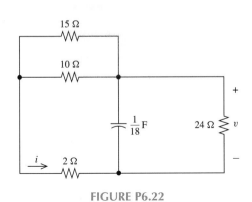

FIGURE P6.22

6.23. If the circuit is in steady state at $t = 0-$, find v for $t > 0$.

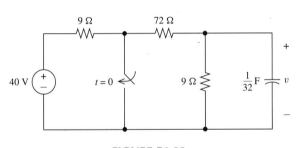

FIGURE P6.23

6.24. Find v for $t > 0$ if the circuit is in steady state at $t = 0-$.

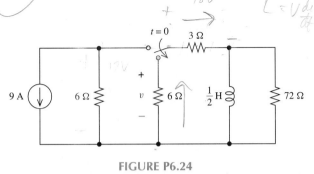

FIGURE P6.24

6.25. Find i for $t > 0$ if the circuit is in steady state at $t = 0-$.

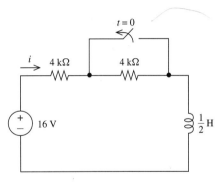

FIGURE P6.25

6.26. Solve the differential equation

$$\frac{di}{dt} + 6i = 24, \qquad i(0) = 1$$

6.27. Solve the equation

$$\int_0^t (i(\tau) + 2) \, d\tau + 3i(t) = 6$$

for $t > 0$ by converting it to a differential equation.

6.28. Find v for $t > 0$. Assume dc steady state at $t = 0$.

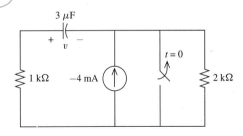

FIGURE P6.28

6.29. Find $v(t)$ for $t > 0$. Assume dc steady state at time $t = 0$.

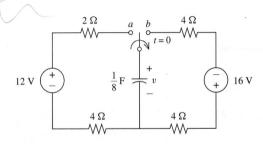

FIGURE P6.29

6.30. Find $i(0-)$ and $i(t)$ for $t > 0$. Assume dc steady state at $t = 0$.

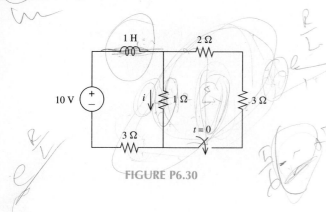

FIGURE P6.30

6.31. Find v for $t > 10$. Assume dc steady state at $t = 10-$.

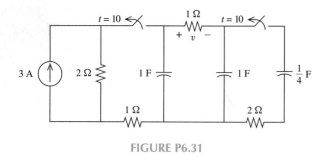

FIGURE P6.31

6.32. Find i for $t > 0$. Assume dc steady state at $t = 0-$. The controlled source has transresistance of $3\ \Omega$.

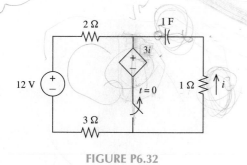

FIGURE P6.32

6.33. Repeat Problem 6.32 for the case in which the switch closes, rather than opens, at time $t = 0$.

6.34. Find i for $t > 0$. Assume dc steady state at $t = 0$.

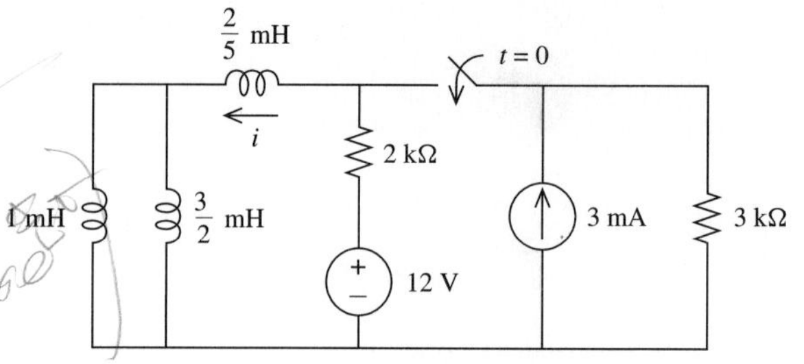

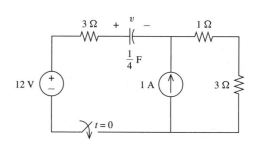

FIGURE P6.34

6.35. Find v for $t > 0$ using superposition. Assume the circuit is in dc steady state at $t = 0-$.

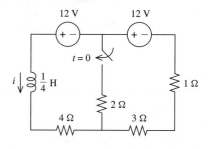

FIGURE P6.35

6.36. Repeat Problem 6.35 using the method of Thevenin transformation of all but the storage element.

6.37. Find i for $t > 0$. Assume dc steady state at $t = 0-$.

FIGURE P6.37

6.38. Find $v_{\text{out}}(t)$ for $t > 0$. Assume that all resistors are 10 kΩ and the capacitor is 20 μF and has no voltage across it at time $t = 0-$.

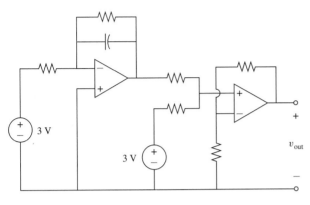

FIGURE P6.38

6.39. Express $v(t)$ in terms of step functions.

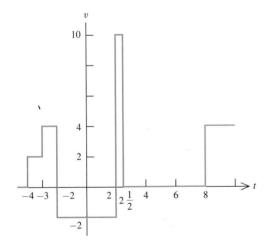

FIGURE P6.39

6.40. Sketch the voltage $v(t) = -3u(-t) + 3u(t+2) + 2u(t) - 2u(t-4)$.

6.41. Express $\int_{-\infty}^{t} v(\tau)d\tau$, $v(t)$ from 6.39, in terms of ramp functions.

6.42. Find the step response $v(t)$.

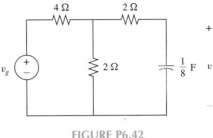

FIGURE P6.42

6.43. Repeat Problem 6.42 for the pulse response if $v_g(t) = u(t) - u(t-1)$.

6.44. Find v for $t > 0$ if $v_g = 3u(t)$ V.

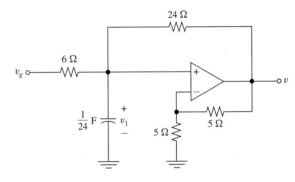

FIGURE P6.44

6.45. Find for $t > 0$ the current downward in the capacitor if $v(0) = 0$, and (a) $v_g = 4$ V, (b) $v_g = 2e^{-2t}$ V, and (c) $v_g = 2\cos 2t$ V.

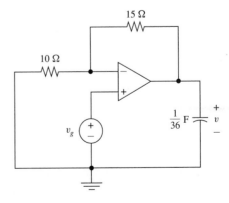

FIGURE P6.45

6.46. Find v for all time if $v_g = 2u(t)$ V.

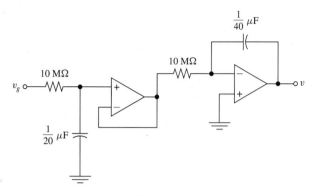

FIGURE P6.46

6.47. Find v for all t if (a) $v_g = 2u(t)$ V, and (b) $v_g = 2[u(t) - u(t-1)]$ V.

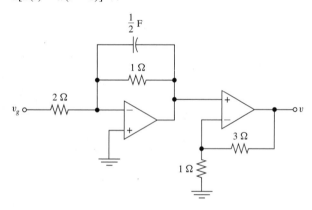

FIGURE P6.47

SPICE Problems

6.48. Find v and v_1 for $t > 0$. Assume dc steady state at $t = 0-$. Check using SPICE. The controlled source has transconductance $g = \frac{3}{8}$ S.

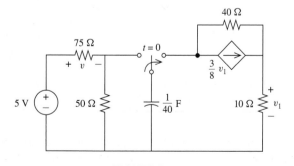

FIGURE P6.48

6.49. Find v_2 if the capacitive voltage at time $t = 0-$ is 0. Check using SPICE.

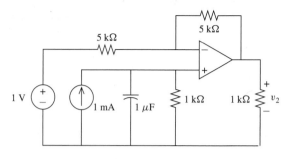

FIGURE P6.49

6.50. Find v_2 and v for $t > t_0$ if $v(t_0) = 4$ V. Check using SPICE.

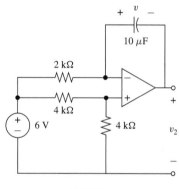

FIGURE P6.50

6.51. Solve Problem 6.18 using SPICE.

6.52. Use SPICE to determine the output of the circuit to a unit step input. Plot the response from $t = 0$ to $t = 20$ ms. Use subcircuits to describe the op amps and the integrators.

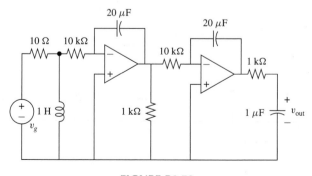

FIGURE P6.52

Design Problems

6.53. Interference from a car's ignition system induces 4000 identical equally spaced pulse voltage waveforms per second across the antenna wire when it is open circuited. The rise and fall times are $TR = TF = 1$ μs (see Fig. 6.36) and full-amplitude width is PW $= 8$ μs. The pulse has baseline V1 $= 0$ and amplitude V2 $= 1$ mV. If the radio is a 100-kΩ resistive load as seen by the antenna, determine an ideal inductor that can be placed in series with the radio input which reduces the peak-to-peak interference across the radio input to 0.1 mV $\pm$ 5%. This is called an *RF suppressor*. Model the antenna as having a 100-kΩ Thevenin equivalent resistance.

6.54. A certain loudspeaker is modeled as an 8-Ω resistor in series with a 2-mH inductor. Design a circuit to connect the loudspeaker to an independent voltage source V_g so that the time constant of the overall circuit is 1 ms and the power dissipated by the loudspeaker when $V_g = 10$ V dc is 1 W.

6.55. Design an *RC* circuit whose zero initial condition response to the current input $I_s(t) = u(t) - 2u(t-1)$ has a zero crossing of the capacitive voltage at time $t = 2$ s.

More Challenging Problems

6.56. Find the time constant τ.

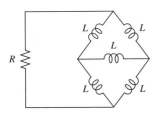

FIGURE P6.56

6.57. Find v_1, v_2, and v_3, for $t > 0$. Assume the circuit is in dc steady state at time $t = 0-$.

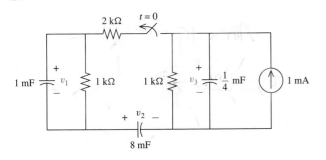

FIGURE P6.57

6.58. Find all currents and voltages in this circuit. Assume $i(0-) = 0$.

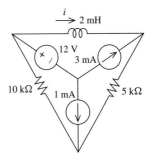

FIGURE P6.58

6.59. Find $v(t)$, $t > 0$ if the circuit is in dc steady state at $t = 0$.

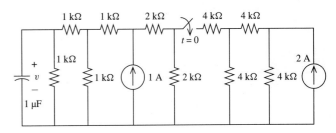

FIGURE P6.59

6.60. Find the voltage transfer equation for $v_2(t)$ in terms of $v_1(t)$.

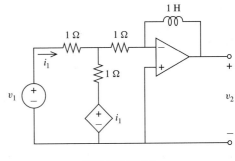

FIGURE P6.60

6.61. Repeat Problem 6.29 for the case in which the switch moves $a \rightarrow b$ at $t = 0$ and then moves back to a at time $t = 3$ s.

6.62. Find the unit step response $v_2(t)$ to the input $v_1(t)$.

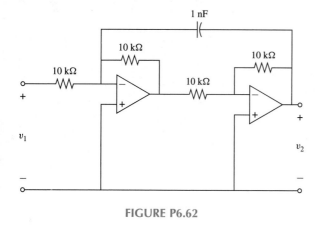

FIGURE P6.62

Samuel F. B. Morse
1791–1872

What hath God wrought!
[The famous message tapped
out on the first telegraph]

Samuel F. B. Morse

7

Second-Order Circuits

The first practical application of electricity is said by many to be the telegraph, developed by Samuel F. B. Morse, an American portrait painter and inventor. Morse built on the ideas of the famous American physicist Joseph Henry, using the opening and closing of relays to produce the dots and dashes (or Morse code) that represent letters and numbers.

Morse was born in Charlestown, Massachusetts, the son of a minister and author. He studied to be an artist at Yale and the Royal Academy of Arts in London, and by 1815 he was considered to be moderately successful. In 1826 he helped found and became the first president of the National Academy of Design. But the previous year his wife had died, in 1826 his father died, and in 1828 his mother died. The following year the distressed Morse went to Europe to recover and study further. In 1832, while returning home on board the passenger ship *Sully*, he met an eccentric inventor and became intrigued with developing a telegraph, the principle of which had already been considered by Henry. By 1836 Morse had a working model, and in 1837 he acquired a partner, Alfred Vail, who financed the project. Their efforts were rewarded with a patent and the financing by Congress of a telegraph in 1844, over which Morse—on May 24, 1844—sent his now-famous message, "What hath God wrought!"

Chapter Contents

The circuit analysis equations for linear circuits with energy storage elements may be expressed as linear differential equations, because the element laws are such that each term in the mesh or nodal equations is a derivative, integral, or multiple of the unknowns and the source variables. Evidently, a single differentiation will remove any integrals that it may contain, so in general the mesh or nodal analysis equations for a given circuit may be considered to be differential equations. The *describing equation*, a single equation whose only unknown is a selected output current or voltage, may be obtained by manipulation of these analysis equations.

The circuits containing storage elements that we have considered so far were first-order circuits. That is, they were described by first-order differential equations. This is always the case when there is only one storage element in the circuit, only one remains after series–parallel simplification, or when a switching action divides the circuit into two or more independent subcircuits each having at most one storage element.

In this chapter we consider second-order circuits, which, as we shall see, contain two storage elements and have describing equations that are second-order differential equations. We show that the total response is the sum of a natural and forced response, as with first-order circuits. Examination of the characteristic equation shows the natural response of a *RLC* circuit is not limited to real exponential decays, as with *RC* or *RL* circuits, but may also include sinusoids, damped sinusoids, and t-multiplied forms. The forced response is determined from a trial form as with first-order circuits, and the total response is computed by requiring that two initial conditions, derived from the initial energies stored in the two storage elements, be satisfied.

In general, nth-order circuits, containing n storage elements, are described by nth-order differential equations. The results for first- and second-order circuits may readily be extended to the general case using the differential-equations-based methods of Chapters 5 and 6, but we shall not do so here. There is a more efficient method for analyzing these higher-order circuits based on Laplace transform analysis of the circuit, which is introduced in Chapter 12.

The circuits considered in Chapter 6 gave rise to first-order describing equations. If the circuit has two storage elements, one inductor and one capacitor, or two of the same type but not series or parallel equivalent to a single element, a second-order differential equation will describe the circuit.

Example 7.1

Consider the circuit of Fig. 7.1, in which we seek the mesh current i_2. The circuit contains two inductors. The mesh equations are

$$2\frac{di_1}{dt} + 12i_1 - 4i_2 = v_g \tag{7.1a}$$

$$-4i_1 + \frac{di_2}{dt} + 4i_2 = 0 \tag{7.1b}$$

From the second of these we have

$$i_1 = \frac{1}{4}\left(\frac{di_2}{dt} + 4i_2\right) \tag{7.2}$$

which differentiated results in

$$\frac{di_1}{dt} = \frac{1}{4}\left(\frac{d^2i_2}{dt^2} + 4\frac{di_2}{dt}\right) \tag{7.3}$$

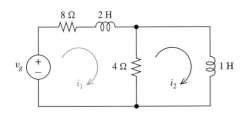

Substituting (7.2) and (7.3) into (7.1a) to eliminate i_1, we have, after multiplying through by 2, the describing equation for i_2:

$$\frac{d^2i_2}{dt^2} + 10\frac{di_2}{dt} + 16i_2 = 2v_g \tag{7.4}$$

FIGURE 7.1 Second-order circuit with two inductors.

This is a second-order differential equation (one in which the highest order of differentiation of the unknown is 2). For this reason we refer to Fig. 7.1 as a *second-order circuit* and note that, typically, second-order circuits contain two storage elements.

The two storage elements in a second-order circuit may be of the same type, as in Fig. 7.1, or one inductor and one capacitor, as in Fig. 7.2. This is the important *series RLC circuit*, which we return to throughout the chapter. By KVL,

$$Ri + \frac{1}{C}\int_{t_0}^{t} i(\tau)\,d\tau + v_c(t_0) + L\frac{di}{dt} = v_g$$

Differentiating and dividing by L yields

$$\frac{d^2i}{dt^2} + \frac{R}{L}\frac{di}{dt} + \frac{1}{LC}i = \frac{dv_g}{dt}$$

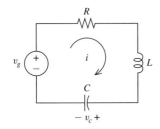

FIGURE 7.2 Series *RLC* circuit.

which is a second-order differential equation; hence the series *RLC* circuit is also a second-order circuit.

There are exceptions to the rule that two-storage-element circuits have second-order describing equations. If two or more storage elements of the same kind (inductors or capacitors) can be replaced by a single equivalent, they count as a single storage element in determining the describing equation order. In other cases, the storage elements may not interact. For example, consider the circuit of

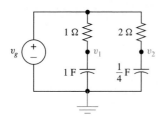

FIGURE 7.3 First-order circuit with two storage elements.

Fig. 7.3, which has two capacitors. Nodal analysis equations for this circuit are

$$\frac{dv_1}{dt} + v_1 = v_g$$

$$\frac{dv_2}{dt} + 2v_2 = 2v_g$$

These are *uncoupled* first-order differential equations; that is, neither unknown appears in both equations. Each may be solved using the methods of Chapter 6. Since the describing equations are first order, this is not a second-order circuit even though two storage elements irreducible to a single equivalent are in the circuit. Replacement of the ideal voltage source by a practical one with nonzero Thevenin equivalent resistance in this circuit, however, results in a second-order circuit, as shown in Exercise 7.1.2.

EXERCISES

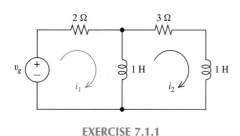

EXERCISE 7.1.1

7.1.1. Find the equation satisfied by the mesh current i_2.
Answer $d^2i_2/dt^2 + 7(di_2/dt) + 6i_2 = dv_g/dt$

7.1.2. Place a 1-Ω resistor in series with the ideal voltage source v_g in Fig. 7.3, and find the describing equations for v_1 and v_2.
Answer $\dfrac{d^2v_1}{dt^2} + \dfrac{11}{5}\dfrac{dv_1}{dt} + \dfrac{4}{5}v_1 = \dfrac{2}{5}\dfrac{dv_g}{dt} + \dfrac{4}{5}v_g;$

$\dfrac{d^2v_2}{dt^2} + \dfrac{11}{5}\dfrac{dv_2}{dt} + \dfrac{4}{5}v_2 = \dfrac{4}{5}\dfrac{dv_2}{dt} + \dfrac{4}{5}v_g$

7.1.3. Find v_1 and v_2 in Fig. 7.3 for $t > 0$ if v_g is a +6-V ideal dc source, $v_1(0-) = 1$ V, and $v_2(0-) = 4$ V.
Answer $6 - 5e^{-t}$ V; $6 - 2e^{-2t}$ V

7.2 SECOND-ORDER EQUATIONS

In Chapter 6 we considered first-order circuits and saw that their describing equations were first-order differential equations of the general form

$$\frac{dx}{dt} + ax = f(t) \tag{7.5}$$

In Section 7.1 we defined second-order circuits as those with describing equations that were second-order differential equations, given generally by

$$\frac{d^2x}{dt^2} + a_1\frac{dx}{dt} + a_0x = f(t) \tag{7.6}$$

In (7.5) and (7.6) the a's are real constants. x may be either a voltage or a current, and $f(t)$, the forcing function, is a known time function determined by the independent sources. As an example, for the circuit of Fig. 7.1, the describing equation is (7.4). Comparing this equation with (7.6), we see that $a_1 = 10$, $a_0 = 16$, $f(t) = 2v_g$, and $x = i_2$.

From Chapter 6 we know that the total response satisfying (7.5) is

$$x(t) = x_n(t) + x_f(t)$$

where $x_n(t)$ is the natural response, obtained by setting $f(t) = 0$, and $x_f(t)$ is the forced response, which satisfies the forced differential equation (7.5).

Let us see if this same procedure can be applied to the second-order equation (7.6). By a solution to (7.6), we shall mean any function x that satisfies (7.6) identically. That is, when x is substituted into (7.6), the left member becomes identically equal to $f(t)$ for all t within a specified solution region, usually $t > t_0$ for some initial time t_0.

If x_n is the natural response, it must satisfy the unforced (or *homogeneous*) differential equation

$$\frac{d^2 x_n}{dt^2} + a_1 \frac{dx_n}{dt} + a_0 x_n = 0 \qquad (7.7)$$

And if x_f is the forced solution, it satisfies the original forced equation

$$\frac{d^2 x_f}{dt_2} + a_1 \frac{dx_f}{dt} + a_0 x_f = f(t) \qquad (7.8)$$

Adding (7.7) and (7.8) and rearranging the terms, we may write

$$\frac{d^2}{dt^2}(x_n + x_f) + a_1 \frac{d}{dt}(x_n + x_f) + a_0(x_n + x_f) = f(t) \qquad (7.9)$$

Comparing (7.6) and (7.9), we see that the sum of x_n and x_f is indeed a solution, as it was in the first-order case. That is, x satisfying (7.6) is made up of two components, a natural response x_n satisfying the homogeneous equation (7.7) and a forced response x_f satisfying the original equation (7.8) or (7.6). As we shall see, the natural response will contain two arbitrary constants and, as in the first-order case, we will choose a trial form forced solution so that it will end up having no arbitrary constants. We consider methods of finding the natural and forced responses in the next two sections.

If the independent sources are such that $f(t) = 0$ in (7.6), the forced response is zero and the solution of the differential equation is simply the natural response. A reader who has had a course in differential equations will note that the natural response and the forced response may also be called, respectively, the *complementary solution* and the *particular solution*. The natural or complementary solution contains the arbitrary constants needed to match initial conditions, as will be discussed in Section 7.5, and the particular solution, as its name implies, contains no free constants.

EXERCISES

7.2.1. Show that $x_1 = K_1 e^{-2t}$ and $x_2 = K_2 e^{-3t}$ are each solutions of

$$\frac{d^2 x}{dt^2} + 5\frac{dx}{dt} + 6x = 0$$

regardless of the values of the constants K_1 and K_2.

7.2.2. Show that $x_1 + x_2 = K_1 e^{-2t} + K_2 e^{-3t}$ also satisifies the equation of Exercise 7.2.1, regardless of the values of K_1 and K_2.

7.2.3. Does the result of (7.7) to (7.9) that $x_n + x_f$ is a solution as long as x_n is a natural response and x_f a forced response hold for third-order circuits? For higher order?

Answer Yes; yes

7.2.4. For the *nonlinear* second-order differential equation

$$\frac{d^2x}{dt^2} = [1 + f(t)]x$$

show that $x_n = K_1 e^{+t} + K_2 e^{-t}$ is the natural response and $x_f = x_n = e^{+2t} - e^{-2t}$ is a forced response to the forcing term $f(t) = 3$; then see if $x_n + x_f$ satisfies the equation. Does the result of (7.7) to (7.9) that $x_n + x_f$ is a solution as long as x_n is a natural response and x_f a forced response hold for *nonlinear* circuits?

Answer No; no

7.3   NATURAL RESPONSE

The natural response x_n, a comonent of the total solution $x = x_n + x_f$, must satisfy the unforced equation, which we repeat from (7.7):

$$\frac{d^2x}{dt^2} + a_1 \frac{dx}{dt} + a_0 x = 0 \tag{7.10}$$

The trial form for the natural solution

$$x_n(t) = K e^{st} \tag{7.11}$$

worked well previously for the first-order case. Let us see if it will help with the second-order case. Any natural solution must satisfy (7.10) so substituting x_n from (7.11) gives

$$K s^2 e^{st} + K s a_1 e^{st} + K a_0 e^{st} = 0$$

or

$$K e^{st}(s^2 + a_1 s + a_0) = 0 \tag{7.12}$$

If the trial form works, this equation must be true for each t or else (7.10) would fail to hold. Equation (7.12) will hold if either $K = 0$ or the factor in parentheses is zero (e^{st} cannot be zero for any time t). If $K = 0$ we have the null solution, $x_n(t) = 0$ for each t. This corresponds to the special case of no stored energy. More generally, we will not have $x_n = 0$ at the initial time, so we cannot have $K = 0$. This leaves only the possibility that the other factor in (7.12) is zero, or

$$s^2 + a_1 s + a_0 s = 0 \tag{7.13}$$

This is the characteristic equation introduced in Chapter 6 for first-order circuits. Note that it can easily be derived from (7.10) simply by replacing derivatives by corresponding powers of s. That is, the second derivative of x is replaced by s^2, the first derivative by s, and the zeroth derivative of x, x itself, is replaced by $s^0 = 1$.

Since (7.13) is a quadratic equation, we have not one solution, as in the first-order case, but in general two solutions, say s_1 and s_2, given by the quadratic formula as

$$s_{1,2} = \frac{-a_1 \pm \sqrt{a_1^2 - 4a_0}}{2} \tag{7.14}$$

Unlike the first-order case, we have two natural solutions of the form Ke^{st}, which we denote by

$$x_{n1} = K_1 e^{s_1 t} \tag{7.15a}$$

$$x_{n2} = K_2 e^{s_2 t} \tag{7.15b}$$

Either of these two solutions (7.15) will satisfy the unforced equation for any value of K_1 or K_2, because substituting either one into (7.10) reduces it to (7.12). And we know that since s_1 and s_2 each satisfy the characteristic equation (7.13), (7.12) will indeed be true with either $s = s_1$ or $s = s_2$.

As a matter of fact, because (7.10) is a linear equation, *any linear combination* of the two distinct natural solutions (7.14) is also a solution. That is,

$$x_n = x_{n1} + x_{n2} = K_1 e^{s_1 t} + K_2 e^{s_2 t} \tag{7.16}$$

is a solution of (7.10) for any pair of values K_1 and K_2 as long as s_1 and s_2 satisfy the characteristic equation. To verify this, we need only substitute this expression for x into (7.10). This results in

$$\frac{d^2}{dt^2}(x_{n1} + x_{n2}) + a_1 \frac{d}{dt}(x_{n1} + x_{n2}) + a_0(x_{n1} + x_{n2})$$

$$= \left(\frac{d^2 x_{n1}}{dt^2} + a_1 \frac{dx_{n1}}{dt} + a_0 x_{n1} \right) + \left(\frac{d^2 x_{n2}}{dt^2} + a_1 \frac{dx_{n2}}{dt} + a_0 x_{n2} \right)$$

$$= 0 + 0 = 0$$

Since (7.16) includes the individual solutions (7.15a) and (7.15b) as special cases, it is called the *general solution of the homogeneous equation* if s_1 and s_2 are distinct (i.e., not equal) roots of the characteristic equation (7.12). Note that if we put $K_2 = 0$ in (7.16) we have x_{n1}, and $K_1 = 0$ results in x_{n2}.

Example 7.2

As an example, the homogeneous equation corresponding to (7.4) in Example 7.1 is repeated here,

$$\frac{d^2 i_2}{dt^2} + 10\frac{di_2}{dt} + 16i_2 = 0 \tag{7.17}$$

and the characteristic equation is

$$s^2 + 10s + 16 = 0$$

The roots are $s = -2$ and $s = -8$, so the general solution of the homogeneous equation is given by

$$i_2 = K_1 e^{s_1 t} + K_2 e^{s_2 t} = K_1 e^{-2t} + K_2 e^{-8t} \tag{7.18}$$

The reader may verify by direct substitution that (7.18) satisfies (7.17) regardless of the value of the unspecified constants K_1 and K_2.

The numbers s_1 and s_2 are called the *characteristic exponents* of the circuit. In Example 7.2 they are the negative reciprocals of the time constants to be considered in Chapter 8. There are two time constants in (7.18), compared to one in the first-order case.

For second-order systems the characteristic equation is a quadratic equation. Unlike first-order systems, whose characteristic equation always has one real solution, its solutions s_1 and s_2 may be real or complex numbers. The nature of the roots is determined by the discriminant in the characteristic equation, which may be positive (corresponding to real distinct roots), negative (complex roots), or zero (real repeated roots). To simplify discussion of these cases, we recast the characteristic equation (7.13) as

$$s^2 + 2\zeta\omega_0 s + \omega_0^2 = 0 \tag{7.19}$$

ω_0 is called the *undamped natural frequency* and ζ the *damping ratio*. The justification of these names will soon be apparent. Comparing (7.13) to (7.19), $\omega_0^2 = a_0$ and $2\zeta\omega_0 = a_1$. Substituting these into (7.14), s_1 and s_2 may be expressed in terms of the undamped natural frequency ω_0 and damping ratio ζ.

A circuit with characteristic equation $s^2 + a_1 s + a_0 = 0$ has:

Undamped natural frequency: $\omega_0 = +\sqrt{a_0}$

Damping ratio: $\zeta = \dfrac{a_1}{2\sqrt{a_0}}$ (7.20)

Characteristic exponents: $s_{1,2} = \left[-\zeta \pm \sqrt{(\zeta^2 - 1)} \right]\omega_0$

For instance, a circuit with characteristic equation $s^2 + 2s + 16$ has undamped natural frequency $\omega_0 = +\sqrt{16} = 4$ rad/s and damping ratio $\zeta = 2/(2)(4) = \frac{1}{4}$. The characteristic exponents for this circuit are

$$s_{1,2} = \left(-\tfrac{1}{4} \pm \sqrt{-\tfrac{15}{16}} \right)(4) = -1 \pm j\sqrt{15}$$

Here $j = \sqrt{-1}$ is the notation we use for the complex number of unit length directed along the imaginary axis in the complex plane (see Appendix B for a review of complex numbers).

Example 7.3

The circuit of Fig. 7.4 is called a *parallel RLC circuit*. The nodal equation for this circuit is

$$\frac{v}{R} + \frac{1}{L} \int_{t_0}^{t} v(\tau)\, d\tau + i_L(t_0) + C\frac{dv}{dt} = i_g$$

Differentiating and dividing both sides by C, the describing equation for v is

$$\frac{d^2 v}{dt^2} + \frac{1}{RC}\frac{dv}{dt} + \frac{1}{LC}v = \frac{1}{C}\frac{di_g}{dt}$$

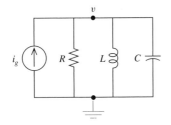

FIGURE 7.4 Parallel *RLC* circuit.

The characteristic equation is then

$$s^2 + \frac{1}{RC}s + \frac{1}{LC} = 0$$

By (7.20), the undamped natural frequency of this circuit is

$$\omega_0 = \frac{1}{\sqrt{LC}}$$

and the damping ratio of this parallel *RLC* circuit is

$$\zeta_p = \frac{1}{2R}\sqrt{\frac{L}{C}}$$

If the damping ratio $\zeta < 1$, the discriminant $\zeta^2 - 1 < 0$, and by (7.20), the characteristic exponents are complex. If $\zeta > 1$, by (7.20) there will be two real characteristic exponents. Because it forms the boundary between these distinct behaviors, $\zeta = 1$ is called *critical damping*.

Setting $\zeta = 1$ and solving for R, we may define the *critical parallel resistance* R_{cp}.

$$R_{cp} = \frac{1}{2}\sqrt{\frac{L}{C}}$$

For $R > R_{cp}$ the damping ratio $\zeta < 1$, which is described below as the underdamped case, and for $R < R_{cp}$ we have $\zeta > 1$, the overdamped case.

The three possible cases, that the characteristic exponents are real and distinct ($\zeta > 1$), real and repeated ($\zeta = 1$), or complex conjugates ($\zeta < 1$), lead to sharply different behavior of the natural response in a second-order circuit. We now consider these three cases, denoting the real part of the general characteristic exponent s as σ and the imaginary part ω so that

$$s = \sigma + j\omega$$

Overdamped Case ($\zeta > 1$)

With $\zeta > 1$ the characteristic exponents s_1 and s_2 in (7.20) are real and distinct, since the discriminant $\zeta^2 - 1 > 0$. In this case each s has only a real part,

$$s_1 = \sigma_1, \qquad s_2 = \sigma_2$$

and the natural response is given by a sum of two real exponentials,

$$x_n(t) = K_1 e^{\sigma_1 t} + K_2 e^{\sigma_2 t} \tag{7.21}$$

Examples include the circuit of Fig. 7.1 analyzed in Example 7.1 and the parallel *RLC* circuit discussed in Example 7.3 when R is less than the critical parallel resistance R_{cp}.

The two real characteristic exponents σ_1 and σ_2 may be positive, negative, or zero. For circuits consisting solely of *passive* elements driven by *independent* sources, however, it is not possible that either characteristic exponent be positive, for a positive σ_1 or σ_2 implies that $x_n(t)$ must increase in magnitude, eventually exceeding all bounds as t gets larger. This cannot happen, for in a passive circuit with independent sources killed (we are interested in the natural response), the total energy stored in the circuit can never increase beyond the initial stored energy.

Damping is the gradual loss of initial stored energy, that is, decay of the natural response toward zero. Passive circuits with distinct real roots have natural responses with two terms each exponentially decaying to zero. Thus $\zeta > 1$ is referred to as the *overdamped* case. In the overdamped case there is sufficient damping so that each term in the natural response steadily loses amplitude. In more lightly damped cases, the damping may be accompanied by oscillations, as we see next.

Underdamped Case ($\zeta < 1$)

If $\zeta < 1$, by (7.20) the characteristic exponents are complex:

$$s_1 = \sigma + j\omega, \qquad s_2 = \sigma - j\omega, \qquad \omega \neq 0$$

They must be complex *conjugates* (real parts equal, imaginary parts equal but opposite in sign). This follows immediately from application of the quadratic formula to the characteristic equation (7.14). Then

$$x_n(t) = e^{\sigma t}(K_1 e^{+j\omega t} + K_2 e^{-j\omega t}) \tag{7.22}$$

We need to use some properties of complex quantities to clarify this form and recognize its structure. For a real circuit, one with purely real element laws and sources, x_n will be real as well. This requires K_1 and K_2 to be complex conjugates, since otherwise the imaginary part of x_n in (7.22) would not be zero. Since the sum of a complex number and its conjugate is twice its real part,

$$x_n = \left(K_1 e^{+j\omega t} + K_1^* e^{-j\omega t}\right) e^{\sigma t} = 2\,\mathrm{Re}\left(K_1 e^{+j\omega t}\right) e^{\sigma t} \tag{7.23}$$

where $K_2 = K_1^*$ is the complex conjugate of K_1. The real part of a product of complex numbers is the product of their real parts minus the product of imaginary parts. Using the Euler identity,

$$e^{j\omega t} = \cos \omega t + j \sin \omega t \tag{7.24}$$

we may expand the real part of the product in (7.23) as

$$\mathrm{Re}\left(K_1 e^{+j\omega t}\right) = \tfrac{1}{2} B_1 \cos \omega t - \tfrac{1}{2} B_2 \sin \omega t$$

where for convenience we have defined

$$\mathrm{Re}(K_1) = \tfrac{1}{2} B_1, \qquad \mathrm{Im}(K_1) = \tfrac{1}{2} B_2$$

or $$B_1 = 2 \text{ Re}(K_1), \qquad B_2 = 2 \text{ Im}(K_1)$$

Equation (7.22) may then be rewritten in terms of real quantities as

$$x_n(t) = B_1 e^{\sigma t} \cos \omega t + B_2 e^{\sigma t} \sin \omega t \qquad (7.25)$$

While the overdamped case consists of two steadily decaying terms, in the present case (7.25) the decay is modulated by (multiplied by) sinusoidal oscillations of frequency ω rad/s. Since the damping is oscillatory rather than steady, this is called the *underdamped* case, which occurs when the damping ratio is less than its critical value of unity.

Note that the amplitude of these oscillations decays as $e^{\sigma t}$, so the damping rate varies inversely with σ. The extreme case is $\sigma = 0$, leading to no decay at all. This is *undamped* case, which occurs when the damping ratio $\zeta = 0$. The undamped natural response (7.25) is a constant-amplitude sinusoid with frequency, by (7.20), $\omega = \omega_0$. This explains why ω_0 is called the undamped natural frequency.

Critically Damped Case ($\zeta = 1$)

The last type we may have consists of characteristic exponents that are real and equal, a case that arises when the damping ratio ζ is unity in (7.20). This is the dividing line between the overdamped and underdamped cases and is referred to as the *critically damped* case

$$s_1 = s_2 = \sigma$$

In this case the general form for the natural solution given in (7.16) does not contain two free constants, since both terms may be collapsed into one. If we are to be able to meet initial conditions imposed by two storage elements, we must further generalize the trial form for the natural solution. Consider the trial form

$$x_n = h(t)e^{st} \qquad (7.26)$$

Thus far we have used $h(t) = 1$. To see if another $h(t)$ may also work, we substitute (7.25) into the unforced differential equation (7.10). Using the product rule to compute derivatives, after gathering like terms, we obtain

$$e^{st}\left[h(s^2 + a_1 s + a_0) + \frac{dh}{dt}(2s + a_1) + \frac{d^2h}{dt^2}\right] = 0 \qquad (7.27)$$

Evaluating this equation for the present repeated root case $s = \sigma$, the first term in brackets is zero since σ satisfies the characteristic equation, and the second term is zero since

$$s + a_1 s + a_0 = (s - \sigma)^2 = s^2 - 2\sigma s + \sigma^2$$

shows that $a_1 = -2\sigma$, so $2s + a_1 = 2\sigma - 2\sigma = 0$. Thus (7.26) is satisfied if and only if

$$\frac{d^2h}{dt^2} = 0$$

This equation in $h(t)$ has solutions

$$h(t) = K_1 + K_2 t$$

Our amended trial form for this case is, from (7.26) and the above,

$$x_n(t) = K_1 e^{\sigma t} + K_2 t e^{\sigma t} \qquad (7.28)$$

While the first term is the familiar exponential trial form, the second differs by an additional factor of t and is referred to as a *t-multiplied form*. The natural solution in the critically damped case consists of a sum of an exponential term and its associated *t*-multiplied form.

Example 7.4

Consider the series *RLC* circuit of Fig. 7.2. The describing equation was found to be

$$\frac{d^2 i}{dt^2} + \frac{R}{L}\frac{di}{dt} + \frac{1}{LC} i = \frac{dv_g}{dt}$$

The characteristic equation is

$$s^2 + \frac{R}{L} s + \frac{1}{LC} = 0$$

By (7.20), the undamped natural frequency of this circuit is

$$\omega_0 = \frac{1}{\sqrt{LC}}$$

and the damping ratio for the series *RLC* case is

$$\zeta_s = \frac{R}{2\sqrt{C/L}}$$

The *critical series resistance* R_{cs} is found by solving for R with $\zeta = 1$:

$$R_{cs} = 2\sqrt{\frac{L}{C}} \qquad (7.29)$$

As R increases the damping ratio ζ_s increases, and for values of R greater than R_{cs} the circuit is overdamped (two real characteristic exponents). The opposite relation holds for the parallel *RLC* circuit discussed in Example 7.3, in which the damping ratio ζ_p decreases with increased parallel resistance, and R greater than R_{cp} leads to underdamped behavior.

To illustrate the series RLC case, set $L = 1$ H and $C = \frac{1}{4}$ F in Fig. 7.2 and consider how the natural solution depends on R. The unforced equation in this case is

$$\frac{d^2i}{dt^2} + R\frac{di}{dt} + 4i = 0$$

with characteristic equation

$$s^2 + Rs + 4 = 0$$

By (7.20), $\omega_0 = 2$ rad/s and $\zeta = R/4$. The critically damped case is $R = 4\ \Omega$. In this case $s = \sigma = -2$ is the repeated characteristic exponent, and

$$i_n = K_1 e^{-2t} + K_2 t e^{-2t} \qquad \text{(critically damped)} \qquad (7.30)$$

The overdamped case requires that $\zeta = R/4 > 1$. Selecting $R = 5\ \Omega$ for illustration, the roots are $s_1 = -4$ and $s_2 = -1$, and

$$i_n = K_1 e^{-t} + K_2 e^{-4t} \qquad \text{(overdamped)} \qquad (7.31)$$

The underdamped case $\zeta < 1$ requires that $R < 4$. For instance, with $R = 2$, we have $s_1 = \sigma + j\omega = -1 + j\sqrt{3}$ and $s_2 = \sigma - j\omega = -1 - j\sqrt{3}$. The natural response in this case is

$$i_n = K_1 e^{-t} \cos \sqrt{3}t + K_2 e^{-t} \sin \sqrt{3}t \qquad \text{(underdamped)} \qquad (7.32)$$

For the undamped case $\zeta = 0$ we must have no series resistance $R = 0$. Then $s_1 = \sigma + j\omega = +j2$ and $s_2 = \sigma - j\omega = -j2$, and by (7.25),

$$i_n = B_1 \cos 2t + B_2 \sin 2t \qquad \text{(undamped)} \qquad (7.33)$$

Graphs for these four distinct responses are shown in Fig. 7.5.

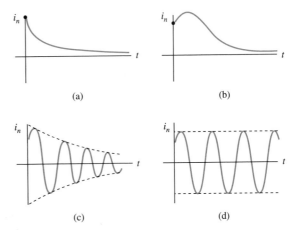

(a)   (b)

(c)   (d)

FIGURE 7.5 Natural responses for second-order circuit: (a) overdamped; (b) critically damped; (c) underdamped; (d) undamped.

Example 7.5

In the parallel *RLC* circuit of Fig. 7.4 and Example 7.3, let $R = \frac{5}{3}$ Ω, $L = 2$ H, and $C = \frac{1}{10}$ F with no source, $i_g = 0$. If $v(0-) = 4$ V and there is no current through the inductor at time $t = 0-$, find $v(t)$ for $t > 0$.

With no independent source, there will only be a natural response. The describing equation, from Example 7.3, is

$$\frac{d^2v}{dt^2} + \frac{1}{RC}\frac{dv}{dt} + \frac{1}{LC}v = \frac{1}{C}\frac{di_g}{dt}$$

or

$$\frac{d^2v}{dt^2} + 6\frac{dv}{dt} + 5v = 0$$

The characteristic equation is

$$s^2 + 6s + 5 = 0$$

or $s_1 = -1$ and $s_2 = -5$. This is an overdamped natural response,

$$v = K_1 e^{-t} + K_2 e^{-5t} \qquad (7.34)$$

The capacitive voltage is continuous, so $v(0+) = v(0-) = 4$ V. By KCL,

$$\frac{3}{5}v + i_L + \frac{1}{10}\frac{dv}{dt} = 0$$

where i_L is defined with its reference arrow pointing downward in Fig. 7.4. Evaluating this equation at $t = 0+$ yields

$$\frac{3}{5}(4) + 0 + \frac{1}{10}\frac{dv}{dt}\bigg|_{0+} = 0$$

where we have used continuity of inductive current to evaluate $i_L(0+)$. Then at $t = 0+$, $v = 4$ V and $dv/dt = -24$ V/s. Applying these conditions to (7.34), we have

$$v(0+) = 4 = K_1 + K_2$$

$$\frac{dv}{dt}\bigg|_{0+} = -24 = -K_1 - 5K_2$$

Adding gives

$$-20 = -4K_2$$

or $K_2 = 5$. Then back-substituting, $K_1 = -1$ and the solution is

$$v = 5e^{-5t} - e^{-t} \text{ V}$$

EXERCISES

7.3.1. What unforced second-order differential equation yields characteristic exponents $s_1 = -10$ and $s_2 = -3$? What are ω_0 and ζ? Repeat for $s_1 = -2 + j5$ and $s_2 = -2 - j5$.

Answer $d^2x/dt^2 + 13(dx/dt) + 30x = 0$; $\omega_0 = 5.48$ rad/s, $\zeta = 1.19$; $d^2x/dt^2 + 4(dx/dt) + 29x = 0$; $\omega_0 = 5.39$ rad/s; $\zeta = 0.37$

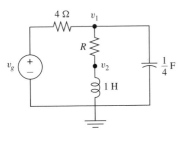

EXERCISE 7.3.3

7.3.2. In the series RLC circuit of Fig. 7.2, let $v_g = 0$, $R = \frac{20}{9}$ Ω, $L = \frac{1}{9}$ H, and $C = \frac{1}{4}$ F. Find the natural solution i_n.

Answer $i_n = K_1 e^{-2t} + K_2 e^{-18t}$

7.3.3. Write nodal equations for the circuit shown. Solve the v_1 nodal equation for v_2, and use this to eliminate v_2 from the differentiated v_2 nodal equation, yielding the describing equation for v_1. For what value of $R > 0$ is the circuit critically damped?

Answer $d^2 v_1/dt^2 + (R+1)(dv_1/dt) + (R+4)v_1 = dv_g/dt - Rv_g$; $R = 5$ Ω

7.3.4. Find the natural response i_{1n} for the circuit shown.

Answer $i_{1n} = K_1 e^{-(2+\sqrt{2})t} + K_2 e^{-(2-\sqrt{2})t}$

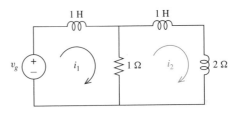

EXERCISE 7.3.4

7.4 FORCED RESPONSE

The forced response x_f of the general second-order circuit must satisfy the original, or forced, differential equation (7.8), repeated here:

$$\frac{d^2 x_f}{dt^2} + a_1 \frac{dx_f}{dt} + a_0 x_f = f(t) \qquad (7.35)$$

There are a number of methods for finding x_f, but for our purposes we use the procedure of guessing a trial form solution that has worked well for us thus far. Examining (7.35), we note that for this equation to be true, a linear combination of x_f and its derivatives must equal $f(t)$. Consider a trial solution in which x_f is guessed to be a linear combination of $f(t)$ and its derivatives. Since a linear combination of linear combinations is still a linear combination, this trial form might well work. In other words, the differentiations of the linear combination of $f(t)$ and its derivatives required on the left side of (7.35) may result in the cancellations required to leave only $f(t)$ itself, which is what is needed to satisfy the equation. The adequacy of any trial form will, in the final analysis, be judged by its ability to satisfy the differential equation.

Example 7.6

As an example, let us consider the dc source $v_g = 16$ V in the circuit of Fig. 7.1. Then, by (7.4),

$$\frac{d^2 i_2}{dt^2} + 10 \frac{di_2}{dt} + 16 i_2 = 32 \qquad (7.36)$$

The characteristic equation is

$$s^2 + 10s + 16 = (s+2)(s+8) = 0$$

with natural response

$$i_{2n} = K_1 e^{-2t} + K_2 e^{-8t} \qquad (7.37)$$

Since the forcing term $f(t)$ is a constant $f(t) = 32$, its first, second, and in fact all its derivatives, are zero. Then the most general linear combination of $f(t)$ and all its derivatives is simply a constant:

$$i_{2f} = A$$

Substituting this trial forced solution into (7.36) yields

$$0 + 0 + 16A = 32$$

The trial form works if we set $A = 2$. Then the total solution is

$$i_2 = i_{2n} + i_{2f} = K_1 e^{-2t} + K_2 e^{-8t} + 2$$

Knowledge of the initial energy stored in the inductors (or the initial inductive current) can be used to evaluate K_1 and K_2, as discussed in Section 7.5.

In the case of constant forcing functions, as in this example, we may obtain the forced solution directly from the circuit diagram. In dc steady state all currents and voltages will be constant, including the unknown in the describing equation. Thus the constant forced solution must be identical to the dc steady-state value. Recall that this may be found by replacing inductors by short circuits and capacitors by open circuits. This may easily be verified for the circuit of Fig. 7.1 used in Example 7.6 by replacing the inductors by short circuits, so by KVL around the outer loop

$$i_{2f} = \frac{v_g}{8} = \frac{16}{8} = 2 \text{ A}$$

Example 7.7

As another example, suppose that in Fig. 7.1 we have $v_g = 20 \cos 4t$ V. Then, by (7.4),

$$\frac{d^2 i_2}{dt^2} + 10 \frac{di_2}{dt} + 16 i_2 = 40 \cos 4t \qquad (7.38)$$

The natural response is given by (7.37), as in Example 7.6. To find the forced response, we need the general linear combination of $f(t)$ and all its derivatives. In the case under consideration, $f(t) = 40 \cos 4t$ and all its derivatives are of the form $C \cos 4t$ or $D \sin 4t$. Summing all these terms together, we arrive at a trial forced solution of the form

$$i_{2f} = A \cos 4t + B \sin 4t \qquad (7.39)$$

That is, every possible linear combination of $40 \cos 4t$ and all its derivatives must indeed be of this form.

We now determine whether this trial form works and, if so, what the values of the constants A and B must be. Substituting (7.39) into the forced differential equation (7.38), we have

$$40B \cos 4t - 40A \sin 4t = 40 \cos 4t$$

Since this must be true for all t, the coefficients of like terms must be the same on both sides of the equation. In the case of the $\cos 4t$ terms we have

$$40B = 40$$

and for the $\sin 4t$ terms we have

$$-40A = 0$$

Thus $A = 0$ and $B = 1$, so that

$$i_{2f} = 0 \cos 4t + 1 \sin 4t = \sin 4t$$

Superposing the natural and forced solutions, the general solution of (7.38) is given by

$$i_2 = i_{2n} + i_{2f} = K_1 e^{-2t} + K_2 e^{-8t} + \sin 4t$$

This may readily be verified by direct substitution.

Some of the more common forcing functions $f(t)$ are listed in the first column of Table 7.1. The general form of the corresponding forced response is given in the second column. This is the trial forced solution. In each case the trial form is the general linear combination of $f(t)$ and its derivatives. If the forcing term is a sum of entries in this table, superposition may be used to find the component of the forced response due to each separately. The forced response is then a sum of these components.

Table 7.1   Trial Forced Solutions for Common Forcing Functions

Forcing Term $f(t)$	Trial Forced Solution $x_f(t)$
k	A
t	$At + B$
t^2	$At^2 + Bt + C$
t^n	$At^n + Bt^{n-1} + \cdots + Ft + G$
$e^{\sigma t}$	$Ae^{\sigma t}$
e^{st}	Ae^{st}
$\sin \omega t, \cos \omega t$	$A \sin \omega t + B \cos \omega t$
$e^{\sigma t} \sin \omega t, e^{\sigma t} \cos \omega t$	$e^{\sigma t}(A \sin \omega t + B \cos \omega t)$
$te^{\sigma t} \sin \omega t, te^{\sigma t} \cos \omega t$	$te^{\sigma t}(A \sin \omega t + B \cos \omega t)$ $+ e^{\sigma t}(C \sin \omega t + D \cos \omega t)$

Example 7.8

Consider a circuit whose describing equation is

$$\frac{d^2i}{dt^2} + 3\frac{di}{dt} + 2i = 4 + 60e^{-7t}$$

Let i_{f1} and i_{f2} be the two components of the forced response satisfying

$$\frac{d^2i_{f1}}{dt^2} + 3\frac{di_{f1}}{dt} + 2i_{f1} = 4$$

$$\frac{d^2i_{f2}}{dt^2} + 3\frac{di_{f2}}{dt} + 2i_{f2} = 60e^{-7t}$$

The sum $i_f = i_{f1} + i_{f2}$ clearly satisfies the forced differential equation, as can be seen by adding the last two equations. Using the trial forms from Table 7.1, for i_{f1} we have a constant $i_{f1} = A$, from which

$$2A = 4$$

or $i_{f1} = 2$. For i_{f2} we have the trial form $i_{f2} = Ae^{-7t}$, or

$$Ae^{-7t}[+49 - 3(7) + 2] = 60e^{-7t}$$

or $A = 2$. The forced solution is then

$$i_f = i_{f1} + i_{f2} = 2 + 2e^{-7t}$$

which can be verified by direct substitution into the forced differential equation.

In the circuits and describing equations studied so far, we have used trial forms for the natural and forced solutions that have been chosen completely independently of one another. The form for the natural solution depended only on the left side of the describing equation (through the coefficients of the characteristic equation), while the form for the forced solution depended only on the right side [$f(t)$ itself]. There is one condition under which the natural and forced trial solution forms cannot, however, be determined independently.

Consider the following equation, for which *the trial forced solution happens to match a term in the natural solution.*

$$\frac{d^2v}{dt^2} + 3\frac{dv}{dt} + 2v = 6e^{-t} \tag{7.40}$$

The characteristic equation is

$$s^2 + 3s + 2 = (s+1)(s+2) = 0$$

and natural solution

$$v_n = K_1e^{-t} + K_2e^{-2t} \tag{7.41}$$

From Table 7.1, for $f(t) = 6e^{-t}$ the trial forced solution [consisting of the general linear combination of $f(t)$ and all its derivatives] is

$$v_f = Ae^{-t} \tag{7.42}$$

But by (7.41) we know that the function Ae^{-t} is in fact a special case of the *natural* solution ($K_1 = A$, $K_2 = 0$) and as such must satisfy the *unforced* version of (7.40), not the forced equation. The usual trial forced solution does not work in this circumstance. We must somehow modify (7.42) if we are to have a forced solution.

A similar problem arose with repeated roots of the characteristic equation, and the solution there proved to be to t-multiplication of the trial form. Guided by that experience, we will try the t-multiplied version of (7.42) as a new trial form,

$$v_f = Ate^{-t}$$

Computing the derivatives gives

$$\frac{dv_f}{dt} = Ae^{-t} - Ate^{-t}$$

$$\frac{d^2v_f}{dt^2} = -Ae^{-t} + Ate^{-t} - Ae^{-t}$$

Substituting these expressions for v_f and its derivatives into the forced equation (7.40), the t-multiplied forms cancel, leaving

$$Ae^{-t}(-2 + 3) = 6e^{-t}$$

The new trial form thus works if we set $A = 6$, or

$$v_f = 6te^{-t} \tag{7.43}$$

The existence of an A for which the new, t-multiplied trial forced solution works (satisfies the forced equation) is sufficient to justify our guess. Equation (7.43) is the forced solution, and the total solution is

$$v = v_n + v_f = K_1e^{-t} + K_2e^{-2t} + 6te^{-t}$$

Generalizing, *when the trial forced solution matches a term in the natural solution, it is necessary to t-multiply the trial forced solution.* If there is still a match, t-multiply again, and so on, until there is no match.

Finally, if the trial forced solution matches a *t-multiplied form* in the natural solution, it may be necessary to t-multiply the trial solution further, as demonstrated in Example 7.9 and in Exercise 7.7.3.

Example 7.9

$$\frac{d^2v}{dt^2} = f(t)$$

The characteristic equation is $s^2 = 0$, or $s_1 = s_2 = 0$. This is a repeated root case, and by (7.28) the natural solution is

$$v_n = K_1e^{st} + K_2te^{st} = K_1 + K_2t$$

First consider the case where the forcing term is $f(t) = 2$. The trial forced solution is $v_f = A$. This matches one of the terms (K_1) of the natural solution, so we t-multiply v_f, yielding $v_f = At$. This still matches a term in the natural solution (K_2t), so, recognizing that At would still satisfy the natural, not the forced, differential equation, we once again t-multiply, arriving at

$$v_f = At^2$$

This does not match a term in v_n. Substituting into the forced differential equation gives

$$2A = 2$$

or $A = 1$. The total solution is

$$v = K_1 + K_2 t + t^2$$

Differentiating twice, it is clear that v satisfies the original forced differential equation for any values of the free constants K_1 and K_2.

Next try the forcing term $f(t) = t$. The trial forced solution from Table 7.1 is $At + B$. Since this matches terms in the natural solution, we t-multiply to get the new trial form, $v_f = At^2$ (only the highest-order term in the t-multiplied form need be retained since the lower ones will satisfy the unforced, not the forced, differential equation). This trial form no longer matches a term in the natural solution, but upon substitution into the differential equation,

$$\frac{d^2 v_f}{dt^2} = \frac{d^2}{dt^2}(At^2) = 2A$$

which does not satisfy the forced equation for any A. The problem is that the original trial solution matched a t-multiplied term of the natural solution. Thus one final t-multiplication is needed, yielding the new trial form $v_f = At^3$. Upon substitution once more,

$$\frac{d^2}{dt^2}(At^3) = 6At$$

which satisfies the forced differential equation in this case [$f(t) = t$] with $A = \frac{1}{6}$. The forced solution is then $v_f = \frac{1}{6}t^3$ and the total solution is

$$v = K_1 + K_2 t + \frac{1}{6}t^3$$

Once again we may easily check by differentiating twice, noting that the result does indeed satisfy the original differential equation for any K_1 and K_2. Determining these values so that the solution not only satisfies the describing equation, but also agrees with the initial stored energy in the storage elements, will be taken up in the next section.

EXERCISES

7.4.1.   Find the forced response if

$$\frac{d^2 x}{dt^2} + 4\frac{dx}{dt} + 3x = f(t)$$

where $f(t)$ is given by (a) 6, (b) $2e^{-3t} + 6e^{-4t}$, and (c) $4e^{-t} + 2e^{-3t}$.
Answer (a) 2; (b) $2e^{-4t} - te^{-3t}$; (c) $t(2e^{-t} - e^{-3t})$

7.4.2. Find the forced response if

$$\frac{d^2x}{dt^2} + 4\frac{dx}{dt} + 4x = f(t)$$

where $f(t)$ is given by (a) $6e^{-2t}$ and (b) $6te^{-2t}$. [*Suggestion:* In (b), try $x_f = At^3e^{-2t}$.]

Answer (a) $3t^2e^{-2t}$; (b) t^3e^{-2t}

7.4.3. Find the total response if

$$\frac{d^2x}{dt^2} + 9x = 18\sin 3t$$

and $x(0) = dx(0)/dt = 0$.

Answer $\sin 3t - 3t\cos 3t$

7.4.4. Find the forced response for i_2 in the circuit of Fig. 7.1 if the source function is $v_g = 16t^2 + \frac{11}{2}$.

Answer $2t^2 - \frac{5}{2}t + 2$

7.5 TOTAL RESPONSE

The total response of a circuit is the sum of its natural and forced response, and since the natural response contains undetermined constants, so must the total response. In the first-order case studied in Chapter 6, there was a single undetermined constant in the natural response, and its value was specified by the need to have the total solution agree with the initial energy in the storage element (i.e., initial current in the inductor or initial voltage across the capacitor).

The same principle applies to second-order circuits, with two storage elements each supplying an initial value needed to define the two undetermined constants contained in the total response. These two initial inductive currents and/or capacitive voltages may be used to determine the two initial conditions required to solve the second-order describing equation.

Consider, for instance, the series *RLC* circuit of Fig. 7.6. Suppose that we are given the initial voltage $v_c(0) = -6$ V and inductive current $i(0) = 1$ A. By KVL,

$$5i + 4\int_0^t i(\tau)\,d\tau + v_c(0) + \frac{di}{dt} = 2e^{-2t}$$

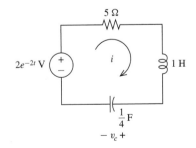

FIGURE 7.6 Forced series *RLC* circuit.

Differentiating, the describing equation for $i(t)$ is

$$\frac{d^2i}{dt^2} + 5\frac{di}{dt} + 4i = -4e^{-2t} \tag{7.44}$$

We will find the total solution by summing natural and forced solutions. The characteristic equation is

$$s^2 + 5s + 4 = (s+1)(s+4) = 0$$

with natural response for this overdamped (distinct real roots) series RLC circuit

$$i_n = K_1e^{-t} + K_2e^{-4t}$$

From Table 7.1, the trial forced solution is $i_f = Ae^{-2t}$, and substituting into (7.44) yields

$$(4 - 10 + 4)Ae^{-2t} = -4e^{-2t}$$

or $A = 2$. The total solution is then

$$i = i_n + i_f = K_1e^{-t} + K_2e^{-4t} + 2e^{-2t} \tag{7.45}$$

Evaluating the total solution (7.45) at $t = 0$ gives

$$i(0) = K_1 + K_2 + 2 \tag{7.46a}$$

Differentiating the total solution and evaluating at $t = 0$, the initial value of the derivative of the unknown must be

$$\left.\frac{di}{dt}\right|_0 = -K_1 - 4K_2 - 4 \tag{7.46b}$$

These two equations may be used to determine K_1 and K_2 if we can relate the initial conditions $i(0)$ and $di/dt|_0$ to the given initial circuit variables, the inductive current and capacitive voltage. The former is easy, since $i(0) = 1$ A is just the inductive current. To relate $di/dt|_0$ to the given values, we write KVL around the loop and evaluate at $t = 0$:

$$5i(0) + \left.\frac{di}{dt}\right|_0 + v_c(0) = 2$$

or

$$\left.\frac{di}{dt}\right|_0 = 2 + 6 - 5(1) = 3$$

Using these values, (7.46) may be rewritten as

$$K_1 + K_2 = -1 \tag{7.47a}$$

$$-K_1 - 4K_2 = 7 \tag{7.47b}$$

Adding (7.47a) and (7.47b) yields

$$-3K_2 = 6$$

or $K_2 = -2$. Substituting back into (7.47a), we have $K_1 = 1$. The total solution (7.45) is now fully specified as

$$i = e^{-t} - 2e^{-4t} + 2e^{-2t} \text{ A}$$

The initial capacitive voltages and/or inductive currents to determine the required initial conditions may be given, as in the previous illustration, or it may be necessary to determine them from the circuit diagram. As introduced in Chapter 6, initial conditions are often set by the action of switches in the circuit or of sources that suddenly step on or off. We next present two examples in which the initial values are computed rather than given, the first in a circuit containing a switch and the second containing a step source.

Example 7.10

We wish to determine the capacitive voltages v_1 and v_2 for $t > 0$ in the circuit of Fig. 7.7, given that it is in dc steady state at $t = 0-$, just before the switch acts. We begin by determining the capacitive voltages at $t = 0-$. For $t < 0$ the switch is closed, and in dc steady state, by KVL around the outer loop,

$$v_1(0-) - v_2(0-) = 12(1) = 12 \text{ V}$$

But since the switch has shorted out node 2 for $t < 0$, $v_2(0-) = 0 \text{ V}$, and hence $v_1(0-) = 12 \text{ V}$. By continuity of capacitive voltages, $v_1(0+) = v_1(0-) = 12 \text{ V}$ and $v_2(0+) = v_2(0-) = 0 \text{ V}$. These values permit us to compute the initial conditions at $t = 0+$ needed to solve the describing equation for $t > 0$.

At $t = 0$ the switch opens, and the two node equations for the circuit as it is configured in the desired solution interval $t > 0$ are

$$\frac{dv_1}{dt} + (v_1 - v_2) = 12 \tag{7.48a}$$

$$2\frac{dv_2}{dt} + 3v_2 - v_1 = 0 \tag{7.48b}$$

To combine these into a single describing equation, we solve (7.48a) for v_2,

$$v_2 = \frac{dv_1}{dt} + v_1 - 12 \tag{7.49}$$

and substitute this expression for v_2 into (7.48b), yielding

$$2\frac{d}{dt}\left(\frac{dv_1}{dt} + v_1 - 12\right) + 3\left(\frac{dv_1}{dt} + v_1 - 12\right) - v_1 = 0$$

or, after division by 2,

$$\frac{d^2v_1}{dt^2} + \frac{5}{2}\frac{dv_1}{dt} + v_1 = 18 \tag{7.50}$$

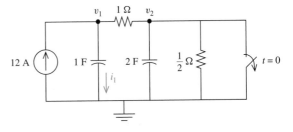

FIGURE 7.7   Circuit for Example 7.10.

This is the describing equation for v_1. Once the circuit is broken by solving this for v_1, v_2 will follow immediately from (7.49).

The characteristic equation for (7.50) is

$$s^2 + \tfrac{5}{2}s + 1 = (s+2)\left(s + \tfrac{1}{2}\right) = 0$$

and the natural solution is

$$v_{1n} = K_1 e^{-2t} + K_2 e^{-t/2}$$

The trial forced solution is $v_{1f} = A$, which, after substitution into (7.50), gives $A = 18$. The total solution is then the sum

$$v_1 = K_1 e^{-2t} + K_2 e^{-t/2} + 18 \qquad (7.51)$$

Evaluating the unknown v_1 and its derivative at $t = 0+$,

$$v_1(0+) = K_1 + K_2 + 18 \qquad (7.52a)$$

$$\left.\frac{dv_1}{dt}\right|_{0+} = -2K_1 - \frac{1}{2}K_2 \qquad (7.52b)$$

To help determine these initial conditions, note that the element law for a capacitor may be written

$$\frac{dv}{dt} = \frac{i}{C}$$

Then since v_1 is the voltage across the 1-F capacitor,

$$\left.\frac{dv_1}{dt}\right|_{0+} = \frac{i_1(0+)}{1} \qquad (7.53)$$

By KCL at node 1,

$$i_1(0+) + (v_1(0+) - v_2(0+)) = 12$$

or

$$i_1(0+) = 12 - 12 = 0$$

and by (7.53) $dv_1/dt|_{0+} = 0$. We are now ready to solve (7.52) for K_1 and K_2:

$$K_1 + K_2 = -6$$

$$-2K_1 - \frac{1}{2}K_2 = 0$$

Doubling the second equation and adding gives

$$-3K_1 = -6$$

or $K_1 = 2$, and from the first equation, $K_2 = -8$. The total solution is

$$v_1 = 2e^{-2t} - 8e^{-t/2} + 6 \text{ V}$$

for the unknown node voltage v_1. The other node voltage v_2 is given in (7.49) as

$$v_2 = \frac{d}{dt}(2e^{-2t} - 8e^{-t/2} + 6) + (2e^{-2t} - 8e^{-t/2} + 6) - 12$$

$$= -2e^{-2t} - 4e^{-t/2} - 6 \text{ V}$$

which completes the analysis.

A second-order describing equation will, in general, require initial conditions consisting of the values of the unknown and its first derivative. The steps required to determine the value of the first derivative at time $t = 0+$ are not always obvious. In Example 7.9 we used the element law for a capacitor in the form

$$\left.\frac{dv_c}{dt}\right|_{0+} = \left.\frac{1}{C}i_c\right|_{0+} \qquad (7.54a)$$

to determine the initial derivative of the unknown v_1. Similarly, for an inductor, the element law in the form

$$\left.\frac{di_L}{dt}\right|_{0+} = \left.\frac{1}{L}v_L\right|_{0+} \qquad (7.54b)$$

is often useful in determining the initial condition $di_L/dt|_{0+}$. To use these equations, it is necessary to know the circuit currents and voltages at $t = 0+$ so that their right-hand sides may be evaluated. In problems with switches or unit steps, we usually know only the capacitive voltages and inductive currents at time $t = 0+$, these by continuity from the moment $t = 0-$ (just before the switches act or the steps have their jump time). To determine the other circuit currents and voltages at time $t = 0+$, we may draw a circuit diagram for the time instant $t = 0+$ in which inductors are shown as current sources (their currents at $t = 0+$ are known) and capacitors as voltage sources. This $t = 0+$ circuit diagram will contain only resistors and sources, and it will be straightforward to determine all remaining currents and voltages at $t = 0+$, including those needed to evaluate (7.54). This procedure for determining initial conditions is illustrated in the next example.

Example 7.11

The circuit of Fig. 7.8 is in dc steady state at time $t = 0-$, and we require $i_1(t)$ for $t > 0$. Noting that the source is switched on for $t < 0$, $v_c(0-) = 12$ V and the inductive current is $i_1(0-) = 12$ A. For $t > 0$ the circuit is source-free, and the mesh equations are

$$\frac{1}{2}\frac{di_1}{dt} + (i_1 - i_2) = 0 \qquad (7.55a)$$

$$i_2 + 2\int_{0+}^{t} i_2(\tau)\, d\tau + v_c(0+) + (i_2 - i_1) = 0 \qquad (7.55b)$$

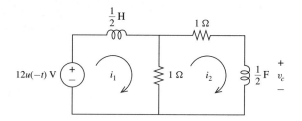

FIGURE 7.8   Circuit for Example 7.11.

Differentiating the second mesh equation, we obtain

$$2\frac{di_2}{dt} - \frac{di_1}{dt} + 2i_2 = 0 \tag{7.56}$$

To get the describing equation for i_1, solve (7.55a) for i_2 and substitute into (7.56):

$$2\frac{d}{dt}\left(\frac{1}{2}\frac{di_1}{dt} + i_1\right) - \frac{di_1}{dt} + 2\left(\frac{1}{2}\frac{di_1}{dt} + i_1\right) = 0$$

or

$$\frac{d^2i_1}{dt^2} + 2\frac{di_1}{dt} + 2i_1 = 0$$

The characteristic equation is

$$s^2 + 2s + 2 = 0$$

with complex conjugate roots $s_1 = -1 + j$ and $s_2 = -1 - j$. This is an underdamped system, and by (7.25) the natural solution is

$$i_{1n} = B_1 e^{-t}\cos t + B_2 e^{-t}\sin t \tag{7.57}$$

which is also the total solution since the forcing term in the describing equation is zero. To evaluate the constants B_1 and B_2, we need the initial conditions $i_1(0+)$ and $di_1/dt|_{0+}$. While $i_1(0+) = 12$ A is known, we will compute the initial derivative with the aid of the circuit diagram for $t = 0+$ shown in Fig. 7.9. Note that the inductor is shown as a current source of value $i_1(0-) = i_1(0+) = 12$ A, and the capacitor is shown as a voltage source of value $v_c(0-) = v_c(0+) = 12$ V. This is a resistive circuit, and all currents and voltages can be found easily. In particular, to use (7.54b) we need the voltage $v_L(0+)$. The right mesh equation in Fig. 7.9 is

$$i_2(0+) + 12 + (i_2(0+) - 12) = 0$$

or $i_2(0+) = 0$. Then by KVL around the outer loop,

$$v_L(0+) + 1(0) + 12 = 0$$

or $v_L(0+) = -12$ V. The required initial condition is found from the above and (7.54b) to be

$$\frac{di_1}{dt}\bigg|_{0+} = 2v_L(0+) = -24$$

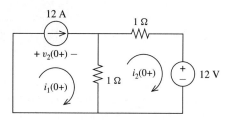

FIGURE 7.9   Circuit for $t = 0+$.

With the initial value of i_1 and its derivative now known, we can determine the constants in the total solution. Evaluating (7.57) at the initial time yields

$$i_1(0+) = B_1 = 12 \text{ A}$$

Differentiating (7.57) and evaluating at the initial time, we have

$$\left.\frac{di_1}{dt}\right|_{0+} = -B_1 + B_2 = -24$$

Using $B_1 = 12$, this implies that $B_2 = -12$, and the solution (7.57) is

$$i_1 = 12e^{-t}(\cos t - \sin t) \text{ A}$$

EXERCISES

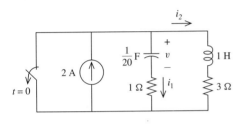

EXERCISE 7.5.2

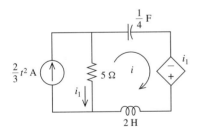

EXERCISE 7.5.4

7.5.1. Find $x(t)$ for $t > 0$, where

$$\frac{dx}{dt} + 2x + \int_0^t x \, dt = f(t)$$

$$x(0) = -1$$

and (a) $f(t) = 1$ and (b) $f(t) = t^2$.

Answer (a) $(2t - 1)e^{-t}$; (b) $3(1 + t)e^{-t} + 2t - 4$

7.5.2. Find i_1, i_2, and v for $t = 0-$ and $t = 0+$. Do any of these values change discontinuously at $t = 0$?

Answer At $t = 0-$, $i_1 = i_2 = v = 0$. At $t = 0+$, $i_1 = 2$ A, $i_2 = 0$, $v = 0$. i_1 changed discontinuously (it is neither a capacitive voltage nor inductive current and thus may do so).

7.5.3. Find a_1, a_0, and $f(t)$ so that the total response $x(t)$, $t > 0$ for

$$\frac{d^2x}{dt^2} + a_1\frac{dx}{dt} + a_0x = f(t)$$

is $x = e^{-t} - e^{-2t} - 3\cos t + \sin t$.

Answer 3; 2; $10 \sin t$

7.5.4. Find the total response i if the initial conditions are $i(0+) = 1$ and $di/dt|_{0+} = 0$. The controlled source has transresistance $r = 1 \, \Omega$.

Answer $6e^{-t} - 2e^{-2t} + 2t - 3$

7.6 UNIT STEP RESPONSE

The unit step response of second-order circuits exhibits much more variety than that of first-order circuits, shown in Chapter 6 to be characterized by a single exponential transient. The striking difference among the forms of the natural responses of over-damped, critically damped, and underdamped second-order circuits suggests that their step responses will show equal variety, since the natural response is a component of the step response as well.

Example 7.12

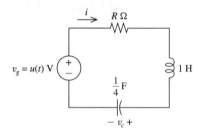

FIGURE 7.10   Circuit for Example 7.12.

Consider the series RLC circuit of Fig. 7.10 driven by a source $v_g = u(t)$. We will determine the response $v_c(t)$ to the unit step source. For $t < 0$ there is no source to stimulate a response, and $i(0-) = v_c(0-) = 0$. For $t > 0$

$$1\frac{di}{dt} + Ri + 4\int_{0+}^{t} i(\tau)\,d\tau + v_c(0+) = 1 \qquad (7.58)$$

or, after differentiating,

$$\frac{d^2i}{dt^2} + R\frac{di}{dt} + 4i = 0$$

This is an unforced second-order describing equation for the series current i, which was taken up in Example 7.4 for several different values of R. The critical resistance for this series RLC circuit is, by (7.29), $R_{cs} = 2\sqrt{4/1} = 4\ \Omega$. For this value of R, by (7.30),

$$i = K_1 e^{-2t} + K_2 t e^{-2t} \qquad \text{(critically damped)} \qquad (7.59)$$

The initial conditions $i(0+)$ and $di/dt|_{0+}$ are needed to determine the constants K_1 and K_2. By continuity of inductive current,

$$i(0+) = 0 \qquad (7.60a)$$

Evaluating (7.58) at $t = 0+$, the integral term is zero since its upper and lower limits are equal, $v_c(0+) = v_c(0-) = 0$ by continuity of capacitive voltage, and

$$\left.\frac{di}{dt}\right|_{0+} = 1 - Ri(0+) = 1 \qquad (7.60b)$$

Evaluation of (7.59) at $t = 0+$ with $i(0+) = 0$ yields $K_1 = 0$, and upon differentiation of (7.59), we have

$$\left.\frac{di}{dt}\right|_{0+} = 1 = K_2$$

so the mesh current in this critically damped case is

$$i = te^{-2t}\ \text{A} \qquad (7.61)$$

Note that the mesh current is purely transient; there is no steady-state component. This may be predicted directly from the circuit diagram Fig. 7.10, recalling that the dc steady-state response and the step response coincide as t goes to infinity. In dc steady state the capacitor is an open circuit; thus no current can flow, and the steady-state mesh current in response to a unit step must be zero. Here we are principally interested in the response v_c. By KVL,

$$Ri + \frac{di}{dt} + v_c = u(t) \qquad (7.62)$$

or, for $t > 0$,

$$v_c = 1 - Ri - \frac{di}{dt} \qquad (7.63)$$

For the critically damped case with $R = 4 \; \Omega$,

$$v_c = 1 - 4(te^{-2t}) - \frac{d}{dt}(te^{-2t})$$

or
$$v_c(t) = 1 - e^{-2t} - 2te^{-2t} \; \text{V} \qquad (7.64)$$

Note that the step response consists of a transient natural response of the form (7.59), plus a dc steady-state response of 1 V. In dc steady state, the capacitor is an open circuit and 1 V is the value of the unit step source voltage across the open circuit. This critically damped step response is shown in Fig. 7.11.

We next determine an overdamped unit step response v_c with $R = 5 \; \Omega$. From (7.31),

$$i = K_1 e^{-t} + K_2 e^{-4t} \qquad \text{(overdamped)}$$

and using the initial conditions we computed in (7.60) yields

$$i(0+) = 0 = K_1 + K_2$$

$$\left. \frac{di}{dt} \right|_{0+} = 1 = -K_1 - 4K_2$$

Adding these last two equations yields $K_2 = -\frac{1}{3}$, and substituting into the first, we have $K_1 = \frac{1}{3}$. The mesh current i in this overdamped case is

$$i = \tfrac{1}{3}(e^{-t} - e^{-4t}) \; \text{A}$$

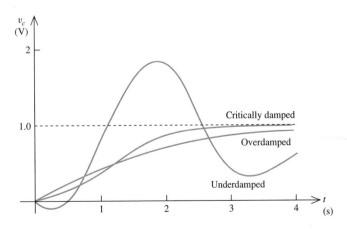

FIGURE 7.11 Three step responses compared.

Using this in (7.63), we have

$$v_c = 1 - \frac{5}{3}(e^{-t} - e^{-4t}) - \frac{1}{3}\frac{d}{dt}(e^{-t} - e^{-4t})$$

(7.65)

$$= 1 - \frac{4}{3}e^{-t} + \frac{1}{3}e^{-4t} \text{ V}$$

Finally, we determine the unit step response for an underdamped case with $R = \frac{1}{4}$ Ω. The characteristic equation is

$$s^2 + \frac{1}{4}s + 1 = 0$$

with complex conjugate characteristic exponents $s_{1,2} = -\frac{1}{8} \pm j\sqrt{255/8}$, or

$$i = K_1 e^{-t/8} \cos\sqrt{\frac{255}{8}}\, t + K_2 e^{-t/8} \sin\sqrt{\frac{255}{8}}\, t \quad \text{(underdamped)}$$

Using the same initial conditions as in the previous cases gives

$$i(0+) = 0 = K_1$$

$$\left.\frac{di}{dt}\right|_{0+} = 1 = -\frac{1}{8}K_1 + \sqrt{\frac{255}{8}}K_2$$

which yields the mesh current

$$i = \frac{8}{\sqrt{255}}e^{-t/8}\sin\sqrt{\frac{255}{8}}\, t \text{ A}$$

and, by (7.63),

$$v_c = 1 - \frac{1}{4}\left(\frac{8}{\sqrt{255}}e^{-t/8}\sin\sqrt{\frac{255}{8}}t\right) - \frac{d}{dt}\left(\frac{8}{\sqrt{255}}e^{-t/8}\sin\sqrt{\frac{255}{8}}t\right)$$

$$= 1 - e^{-t/8}\left(\cos\sqrt{\frac{255}{8}}t + \frac{1}{\sqrt{255}}\sin\sqrt{\frac{255}{8}}t\right) \text{ A}$$

(7.66)

The overdamped and underdamped step responses are also shown in Fig. 7.11.

Comparing the three step responses, we observe that the critically damped circuit demonstrates the fastest convergence to the steady state of 1 V. The transient in the overdamped case contains two terms, one of which has a longer time constant ($\tau = 1$ s compared to $\frac{1}{2}$ s for the critically damped case) and thus decays to zero more slowly than the critically damped case. The transient in the underdamped response is oscillatory, with the amplitude of these oscillations decaying to zero with an even longer time constant, $\tau = 8$ s. Thus, although the underdamped response first reaches the value of 1 V much faster than the other responses, it "overshoots" this value, and its actual convergence to steady state is even slower than the overdamped case.

These results may be confirmed easily using SPICE. The input file for the circuit of Fig. 7.10 is

```
Series RLC step response
VG      1    0    DC    1
*RS = 0.25 OHMS (UNDERDAMPED), 4 OHMS
*(CRIT. DAMPED), 5 OHMS (OVERDAMPED).
RS      1    2       0.25
LS      2    3       1        IC=0
CS      3    0       0.25     IC=0
.TRAN   1   10    UIC
.PROBE
.END
```

The results of three runs with the three different values for R_s are shown in Fig. 7.12. These graphs were generated by the PSpice graphics postprocessor program PROBE. They closely match the calculated expressions (7.64) to (7.66).

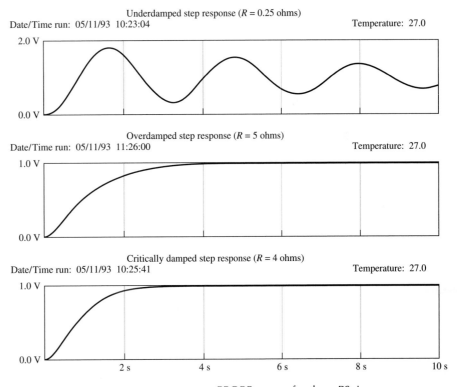

FIGURE 7.12   PROBE output for three PSpice runs.

While the methods introduced in this chapter and Chapter 6 may be generalized for studying higher-order circuits (those with three or more storage elements and characteristic exponents), there are more efficient methods for breaking such circuits. These techniques will be introduced in the context of Laplace transform analysis in Chapter 12. The time-domain methods of Chapters 6 and 7, based on the direct solution of differential equations, are effective tools for analyzing first- and second-order circuits. They are less frequently the method of choice for analyzing higher-order circuits.

EXERCISES

7.6.1. Find $v(0+)$, $dv/dt|_{0+}$, and the unit step response v of the parallel RLC circuit of Fig. 7.4 if $i_g = u(t)$, $R = 1\ \Omega$, $L = \frac{9}{14}$ H, $C = \frac{1}{9}$ F.
Answer $v(0+) = 0$; $dv/dt|_{0+} = 9$; $v = \frac{9}{5}(e^{-2t} - e^{-7t})$

7.6.2. Find $v(0+)$, $dv/dt|_{0+}$, and the unit step response v of the circuit shown.
Answer $v(0+) = 0$; $dv/dt|_{0+} = 0$; $v = 1 - \frac{1}{5}(e^{4t} + 4e^{-t})$

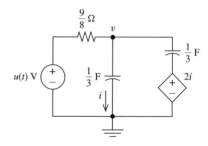

EXERCISE 7.6.2

7.6.3. Find the describing equation for v_2 and the unit step response.
Answer $d^2v_2/dt^2 + 2(dv_2/dt) + v_2 = 0$; $v_2 = -te^{-t}$

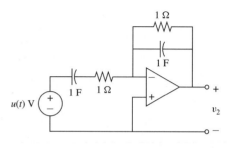

EXERCISE 7.6.3

7.7 DESIGN OF SECOND-ORDER CIRCUITS

Second-order circuits are characterized by second-order differential equations and may have undamped sinusoids, damped sinusoids, or pairs of real exponentials as their natural responses. In contrast to the single real exponential natural response found in first-order circuits, second-order circuits are capable of much richer behavior. In particular, the ability to generate a sinusoidal response where none was input to the circuit may be used to generate tones, as is demonstrated in the first design example.

Design Example 7.1

Touchtone telephone dialing uses combinations of two tones to identify each pushbutton on the keypad. One is from the low group, consisting of 697 Hz, 770 Hz, 852 Hz, and 941 Hz, and the other is from the high group, 1209 Hz, 1336 Hz, and 1477 Hz. For instance, when key one is depressed, a combined 697- and 1209-Hz tone is sent. Our goal here is to design a circuit whose natural (unforced) response is a voltage that is the sum of a 697-Hz sinusoid and a 1209-Hz sinusoid.

We begin this design by noting as in Fig. 7.5 that pure sinusoids are the natural solutions to undamped second-order differential equations. The equation

$$\frac{d^2}{dt^2}x(t) + \omega_0^2 x(t) = 0$$

has natural solution as in (7.25) with $\sigma = 0$,

$$x_n(t) = B_1 \cos \omega_0 t + B_2 \sin \omega_0 t$$

The parallel LC circuit in Fig. 7.4 with $\omega_0 = 1/\sqrt{LC}$ is one such circuit. Thus if we add together two such natural responses, one with

$$\omega_{01} = 2\pi(697) = 4379 \text{ rad/s} = \frac{1}{\sqrt{L_1 C_1}}$$

and the other

$$\omega_{02} = 2\pi(1209) = 7596 \text{ rad/s} = \frac{1}{\sqrt{L_2 C_2}}$$

we will produce the desired sum of tones. Setting $L_1 = L_2 = 1$ H for convenience, we get $C_1 = 52.1$ nF and $C_2 = 173$ nF. A noninverting summer that adds these tones together is shown in Fig. 7.13.

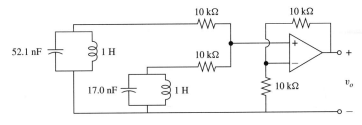

FIGURE 7.13 Touchtone pushbutton one-tone generator.

The reader should note that this is a purely *schematic* design, not intended as a practical physical one. A physical inductor will contain a significant amount of series resistance, as discussed in Chapter 5, and this will cause damping to attenuate the op amp inputs. Other circuit designs which are more practical as sinusoidal signal generators will be developed in Chapter 18.

Design a circuit, called an *analog computer* circuit, which solves the differential equation

$$\frac{d^2}{dt^2}y(t) + \rho\frac{d}{dt}y(t) + y(t) = 5\cos 10t \qquad (7.67)$$

with zero initial conditions. Use only passive elements and a single independent source. $y(t)$ should be a current somewhere in the circuit, and ρ a parameter of an element in the circuit. Show the SPICE plots of the solution for the interval $[0, 2 \text{ s}]$ for $\rho = 0.1$. Use the SIN independent source transient specification described in Appendix D.3.

Consider the series circuit of Fig. 7.14. In terms of general circuit parameters, the mesh equation is

$$Ri + L\frac{di}{dt} + \frac{1}{C}\int i(t)\,dt = v_s(t)$$

Differentiating and dividing by L yields

$$\frac{d^2i}{dt^2} + \frac{R}{L}\frac{di}{dt} + \frac{1}{LC}i = \frac{1}{L}\frac{d}{dt}v_s(t) \qquad (7.68)$$

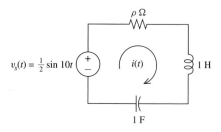

FIGURE 7.14 Circuit to compute a differential equation.

This equation is of the desired form. Setting $L = C = 1$ and $R = \rho$ and identifying the mesh current $i(t)$ as the unknown $y(t)$ in the differential equation (7.67) that we are solving by the simple analog computer circuit of Fig. 7.14, we need only match forcing terms. Since the right-hand sides of (7.67) and (7.68) match for $L = 1$ with

$$v_s(t) = \tfrac{1}{2}\sin 10t$$

this is the desired voltage source in the circuit. The SPICE input file is

```
Design Example 7.2
*
VS       1     0    SIN(0 0.5  1.59)
* (0     0.5    1.59) = (xoff xampl freq)
R        1     2    .1
L        2     3    1    IC=0
C        3     0    1    IC=0
.TRAN         .05    2    UIC
.PLOT         TRAN   I(VS)
.END
```

```
****      TRANSIENT ANALYSIS                 TEMPERATURE =    27.000 DEG C
*****************************************************************************


     TIME        I(VS)
    (*)----------  -1.0000E-01  -5.0000E-02   0.0000E+00   5.0000E-02   1.0000E-01
                   - - - - - - - - - - - - - - - - - - - - - - - - - - - - - - -
   0.000E+00  4.475E-08  .            .            *            .            .
   5.000E-02 -6.243E-03  .            .          *  .            .            .
   1.000E-01 -2.282E-02  .            .       *     .            .            .
   1.500E-01 -4.523E-02  .            .    *.       .            .            .
   2.000E-01 -6.797E-02  .         *   .            .            .            .
   2.500E-01 -8.546E-02  .     *       .            .            .            .
   3.000E-01 -9.361E-02  .  *          .            .            .            .
   3.500E-01 -8.936E-02  .   *         .            .            .            .
   4.000E-01 -7.349E-02  .        *    .            .            .            .
   4.500E-01 -5.030E-02  .           * .            .            .            .
   5.000E-01 -2.461E-02  .            .      *      .            .            .
   5.500E-01 -2.415E-03  .            .        *.   .            .            .
   6.000E-01  1.082E-02  .            .           * .            .            .
   6.500E-01  1.299E-02  .            .           * .            .            .
   7.000E-01  3.983E-03  .            .         * . .            .            .
   7.500E-01 -1.430E-02  .            .      *      .            .            .
   8.000E-01 -3.654E-02  .            .  *          .            .            .
   8.500E-01 -5.672E-02  .          *  .            .            .            .
   9.000E-01 -6.997E-02  .       *     .            .            .            .
   9.500E-01 -7.185E-02  .      *      .            .            .            .
   1.000E+00 -6.132E-02  .         *   .            .            .            .
   1.050E+00 -4.128E-02  .            * .           .            .            .
   1.100E+00 -1.570E-02  .            .      *      .            .            .
   1.150E+00  9.871E-03  .            .          *  .            .            .
   1.200E+00  2.922E-02  .            .            .      *      .            .
   1.250E+00  3.871E-02  .            .            .        *    .            .
   1.300E+00  3.673E-02  .            .            .        *    .            .
   1.350E+00  2.347E-02  .            .            .   *         .            .
   1.400E+00  2.987E-03  .            .            *.           .            .
   1.450E+00 -1.881E-02  .            .     *       .            .            .
   1.500E+00 -3.664E-02  .            . *           .            .            .
   1.550E+00 -4.511E-02  .           .*            .            .            .
   1.600E+00 -4.140E-02  .           . *           .            .            .
   1.650E+00 -2.678E-02  .            .    *        .            .            .
   1.700E+00 -4.017E-03  .            .        *.   .            .            .
   1.750E+00  2.220E-02  .            .            .   *         .            .
   1.800E+00  4.545E-02  .            .            .         * . .            .
   1.850E+00  6.089E-02  .            .            .            .  *          .
   1.900E+00  6.552E-02  .            .            .            .    *        .
   1.950E+00  5.780E-02  .            .            .            . *           .
   2.000E+00  4.019E-02  .            .            .          *  .            .
                   - - - - - - - - - - - - - - - - - - - - - - - - - - - - - - -
```

FIGURE 7.15   SPICE plot for Design Example 7.2.

The output is shown in Fig. 7.15. Note that the forced response, which is a sinusoid at $\omega = 10$ rad/s (1.59 Hz), is superposed with the natural response, an underdamped sinusoid with much lower frequency.

Design Example 7.3

Design an *active* analog computer circuit that solves the equation

$$\frac{d^2}{dt^2}y(t) + \frac{d}{dt}y(t) = x(t)$$

where $x(t)$ is an arbitrary source voltage and $y(t)$ another voltage somewhere in the circuit.

As with Design Example 7.2, we may get a second-order equation by differentiating a mesh or node equation containing both a derivative and an integral. In the present case, however, the integral must involve the input, not the response variable. A simple passive circuit cannot have just this source dependency. The active circuit of Fig. 7.16 does, however. Using the virtual short and open principles and summing currents into the inverting input node to zero yields

$$C\frac{dy}{dt} + \frac{y}{R} + \frac{1}{L}\int [-x(t)]\,dt = 0$$

which for $R = L = C = 1$ yields the desired equation. A unity-gain inverting amplifier has been used to get the proper sign for the input signal $x(t)$.

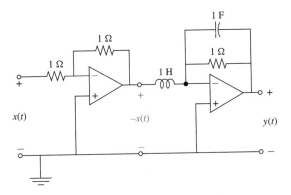

FIGURE 7.16 Circuit to compute another differential equation.

There are many other uses to which second-order circuits may be put. We did not consider filters, perhaps the most important use of all, as this topic is considered in detail in Chapter 14, and general methods for second-order filter design are discussed at length in Chapter 18. The design examples and exercises here, though, begin our exploration of second-order circuit design.

DESIGN EXERCISES

7.7.1. Design a circuit whose step response includes real exponentials with time constants $\tau = 10$ ms and $\tau = 1$ ms.

One of many possible solutions

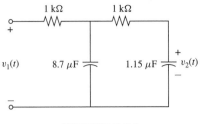

7.7.2. Using a single op amp, design a double-differentiator circuit, one that satisfies

$$\frac{d^2}{dt^2} v_2(t) = -50 v_1(t)$$

One of many possible solutions

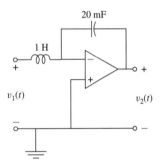

SUMMARY

Second-order circuits are those characterized by a single second-order differential equation. They can be identified by the presence of two equivalent storage elements (inductors and/or capacitors). Their behavior may be studied by writing and solving their describing equations in the time domain using the techniques introduced in Chapter 6 for first-order circuits. The natural behavior of second-order circuits is, however, much more diverse and interesting, ranging from pairs of real exponentials and t-multiplied exponentials to sinusoids both damped and undamped.

- The natural response of a second-order circuit is $K_1 e^{p_1 t} + K_2 e^{p_2 t}$, where p_1 and p_2 are the distinct roots of the characteristic equation, or if p_1 is a repeated root, the second term is replaced by the t-multiplied form $K_2 t e^{p_1 t}$.

- The natural response can also be expressed in terms of the undamped natural frequency ω_o and damping factor ζ. ω_0 is the sinusoidal frequency in the corresponding undamped case ($\zeta = 0$), and ζ indicates the rate of exponential growth or decay.

- An overdamped system ($\zeta > 1$) has two real exponentials as its natural response, an underdamped system ($\zeta < 1$) has a damped sinusoid, and a critically damped system ($\zeta = 1$) has a real exponential and its t-multiplied form. An undamped system, with $\zeta = 0$, has a pure sinusoid as its natural response.

- The forced response is found by substituting a trial forced solution, which depends on the forcing function, into the differential equation and matching coefficients.

- The total response is found by adding the natural and forced responses, using the initial conditions to determine the coefficients in the natural response terms.

- The unit step response settles to its steady-state value most rapidly in the critically damped case. Overdamped circuits have extended periods of undershoot or overshoot, while underdamped circuits exhibit damped oscillations (ringing) around the steady-state value.

PROBLEMS

7.1. Write describing equations for the indicated variables. Are these second-order circuits?

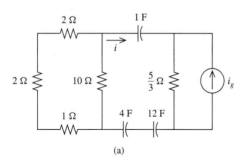

(a)

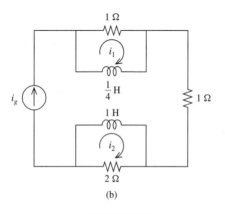

(b)

FIGURE P7.1

7.2. Write a second-order differential equation satisfied by v_1.

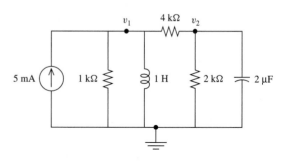

FIGURE P7.2

7.3. Write a second-order differential equation satisfied by v_2 in the diagram for Problem 7.2.

7.4. Repeat Problem 7.3 if the capacitor is replaced by a $\frac{1}{4}$-H inductor.

7.5. Insert a 1-Ω resistor in series with v_g in Fig. 7.3 thereby making the source a practical rather than an ideal one. Show that in this case v_2 satisfies the second-order equation,

$$5\frac{d^2v_2}{dt^2} + 11\frac{dv_g}{dt} + 4v_2 = 4\frac{dv_2}{dt} + 4v_2$$

7.6. Find the describing equation for i.

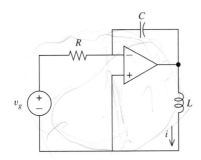

FIGURE P7.6

7.7. Find the describing equation for v_{out}. Is this a second-order circuit? A first-order circuit?

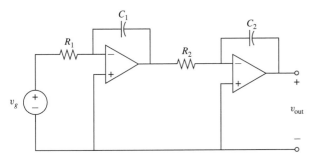

FIGURE P7.7

7.8. Consider the series RLC circuit shown in which the resistance is time varying. (a) Write the describing equation for i. (b) If i_n and i_f are natural and forced solutions, is their sum a solution?

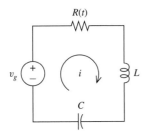

FIGURE P7.8

7.9. For a second-order differential equation

$$\frac{d^2x}{dt^2} + a_1\frac{dx}{dt} + a_0x = f(t)$$

let x_{n1} and x_{n2} be natural solutions and x_{f1}, x_{f2} be two distinct forced solutions. Show that *any* linear combination

of x_{n1} and x_{n2} is still a natural solution. What *specific* linear combinations of x_{f1} and x_{f2} will be forced solutions?

7.10. Suppose the general second-order describing equation of Problem 7.9 is forced with an $f(t)$ that is a linear combination of m forcing functions

$$f(t) = b_1 f_1(t) + b_2 f_2(t) + \cdots + b_m f_m(t)$$

Show that the forced solution is the same linear combination of forced solutions to each $f_i(t)$ separately.

7.11.

$$\frac{d^2x}{dt^2} + a_1\frac{dx}{dt} + a_0x = f(t)$$

Find the natural response $[f(t) = 0]$ if $a_1 = 14$, $a_0 = 49$, $x(0) = 0$, and $dx/dt|_0 = 2$.

7.12. Write a second-order unforced differential equation with natural response $k_1e^{-t}\sin 2t + k_2e^{-t}\cos 2t$. Find k_1 and k_2 if the initial conditions are $x(0) = 0$ and $dx/dt|_0 = 4$.

7.13. Find i for $t > 0$ if $i_1(0) = 3$ mA and $i(0) = 1$ mA.

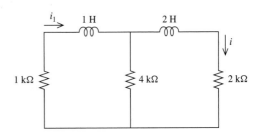

FIGURE P7.13

7.14. Find the natural response $i(t)$, $t > 0$ of the circuit of Problem 7.1(a) if $i(0) = 3$ A ($i_g = 0$).

7.15. Find i for $t > 0$ if $i_1(0) = 9$ A and $i(0) = 3$ A.

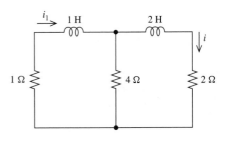

FIGURE P7.15

309

7.16. Find i for $t > 0$ if the circuit is in steady state at $t = 0-$.

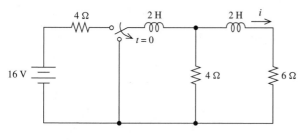

FIGURE P7.16

7.17. Find i for $t > 0$ if $i(0) = 4$ A and $v(0) = 8$ V.

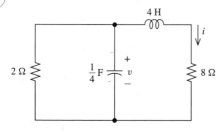

FIGURE P7.17

7.18. Find v in Problem 7.17 if $i(0)$ is changed to 1 A.

7.19. Find i for $t > 0$ if $v_1(0) = v_2(0) = 12$ V.

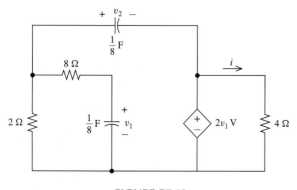

FIGURE P7.19

7.20. Find the forced solution $x_f(t)$ if $f(t) = 16e^{-2t} + 4$.

$$\frac{d^2x}{dt^2} + 6\frac{dx}{dt} + 8x = f(t)$$

7.21. Find the forced solution to the equation of Problem 7.20 if $f(t) = 4e^{-4t}$.

7.22. Find the forced solution to the equation of Problem 7.20 if $f(t) = 4te^{-4t} + 4e^{-t}$.

7.23. Sketch a circuit whose describing equation is $\frac{d^2v}{dt} + 2\frac{dv}{dt} + v = 2\sin 2t$.

7.24. Find the forced solution $i_f(t)$.

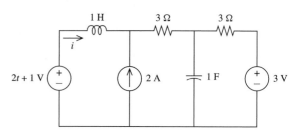

FIGURE P7.24

7.25. Find the forced solution $v_f(t)$.

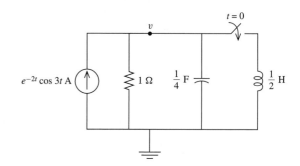

FIGURE P7.25

7.26. Find the forced solution $i_f(t)$.

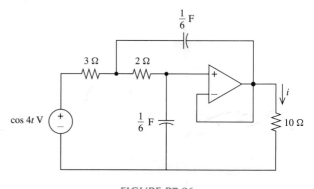

FIGURE P7.26

7.27. Find i_2 for $t > 0$ if $i_1(0) = 3$ A, $i_2(0) = -1$ A, and (a) $v_g = 15$ V, (b) $v_g = 10e^{-2t}$ V, and (c) $v_g = 5e^{-t}$ V.

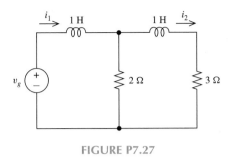

FIGURE P7.27

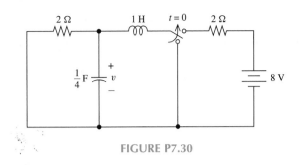

FIGURE P7.30

7.28. Find i for $t > 0$ if the circuit is in steady state at $t = 0-$.

7.31. Find i, $t > 0$, if there is no initial stored energy and (a) $R = 2\ \Omega$, $\mu = 2$; (b) $R = 2\ \Omega$, $\mu = 1$; and (c) $R = 1\ \Omega$, $\mu = 2$.

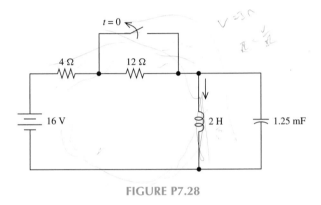

FIGURE P7.28

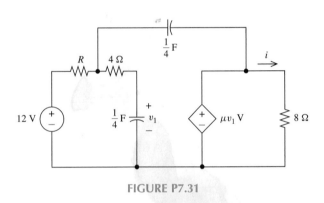

FIGURE P7.31

7.29. Find i for $t > 0$ if the circuit is in steady state when the switch is opened at $t = 0$.

7.32. Find i for $t > 0$ if $i(0) = 2$ A and (a) $v(0) = 6$ V.

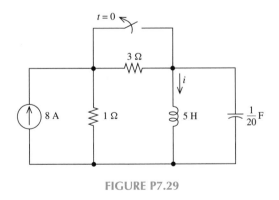

FIGURE P7.29

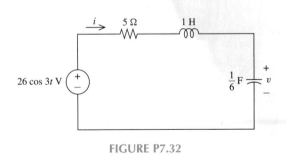

FIGURE P7.32

7.30. Find v for $t > 0$ if the circuit is in steady state at $t = 0-$.

For Problems 7.33 to 7.40, assume dc steady state at $t = 0-$ and find the total solution for $t > 0$ for the indicated variable.

7.33.

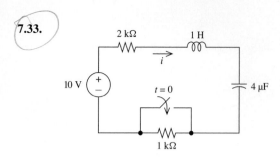

FIGURE P7.33

7.34. Solve Problem 7.33 if the switch closes at $t = 10$ s instead of $t = 0$ s.

7.35.

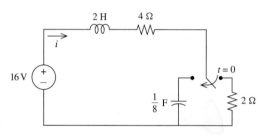

FIGURE P7.35

7.36.

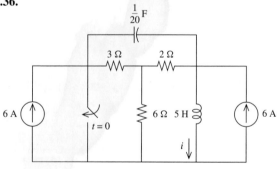

FIGURE P7.36

7.37.

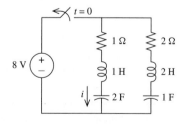

FIGURE P7.37

7.38.

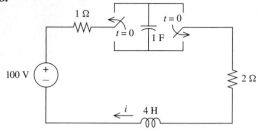

FIGURE P7.38

7.39.

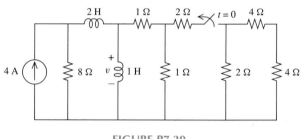

FIGURE P7.39

7.40. Find the total solution $v(t)$ if R is set to the critical resistance (critical damping).

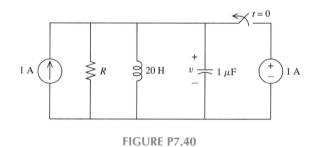

FIGURE P7.40

7.41. Find R so that the circuit is critically damped.

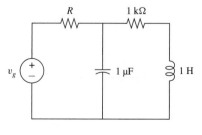

FIGURE P7.41

SPICE Problems

7.42. Find the unit step responses of the circuits of Problem 7.1. Check using SPICE.

7.43. Find the unit step response. Check using SPICE.

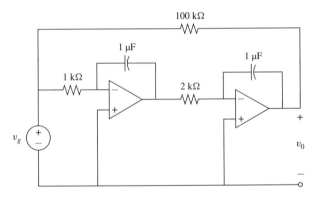

FIGURE P7.43

7.44. Find the unit step response. Check using SPICE.

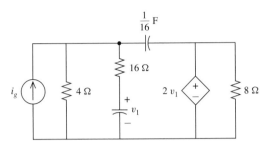

FIGURE P7.44

7.45. Find the unit step response. Check using SPICE.

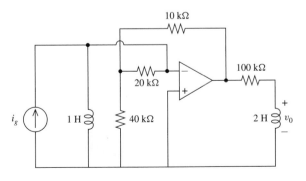

FIGURE P7.45

More Challenging Problems

7.46. Write a second-order differential equation satisfied by i_2.

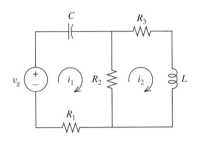

FIGURE P7.46

7.47. Find the mesh currents for $t > 0$ assuming all initial conditions are zero at $t = 0$.

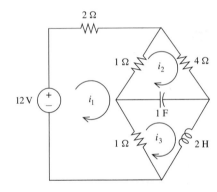

FIGURE P7.47

7.48. Find the forced response $v_2(t)$ if $v_1(t) = 4\cos 2t$.

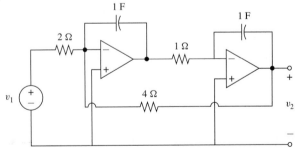

FIGURE P7.48

7.49. Write two different second-order differential equations that both have the same forced solution $x_{1f} = x_{2f} = 2t + 1$ but that have different forms for their natural solutions. Solve both equations assuming zero initial conditions.

7.50. Find the natural response $v_1(t)$, $t > 0$ (a) if $v_1(0) = v_2(0) = 0$; (b) if $v_1(0) = v_2(0) = 1$ V. [*Suggestion:* Solve node 1 equation for v_3, and use this to eliminate v_3 from the node 2 and 3 equations.]

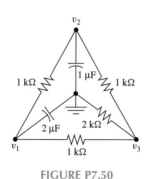

FIGURE P7.50

Design Problems

7.51. Find a value for R such that the natural responses of this circuit all consist of damped sinusoids $Ke^{\sigma t}\cos(\omega t + \theta)$ with $\omega = 3000$ rad/s. Use SPICE to help search for the solution.

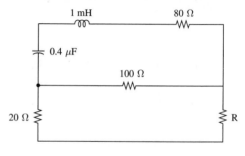

FIGURE P7.51

7.52. Design a circuit which solves the differential equation

$$\frac{d^2 y}{dt^2} - 2\frac{dy}{dt} + 2y = 0$$

Specify what variable in your circuit $y(t)$ is, and how the initial conditions $y(0)$, $dy/dt|_{t=0}$ are determined.

7.53. An electric motor is modelled as a 22-Ω resistor in parallel with a 1/2-H inductor. Design a passive circuit to connect the motor to an ideal voltage source so that the resulting circuit is critically damped at its natural frequency of 60 Hz.

8

Sinusoidal Sources and Phasors

Charles Proteus Steinmetz
1865–1923

I have found the equation that will enable us to transmit electricity through alternating current over thousands of miles. I have reduced it to a simple problem in algebra.

Charles Proteus Steinmetz

The use of complex numbers to solve ac circuit problems—the phasor method considered in this chapter—was first done by the German-Austrian mathematician and electrical engineer Charles Proteus Steinmetz in a paper presented in 1893. He is noted also for the laws of hysteresis and for his work in manufactured lightning.

Steinmetz was born in Breslau, Germany, the son of a government railway worker. He was deformed from birth and lost his mother when he was 1 year old, but this did not keep him from becoming a scientific genius. Just as his work on hysteresis later attracted the attention of the scientific community, his political activities while he was at the University of Breslau attracted the police. He was forced to flee the country just as he had finished the work for his doctorate, which he never received. He did electrical research in the United States, primarily with the General Electric Company. His paper on complex numbers revolutionized the analysis of ac circuits, although it was said at the time that no one but Steinmetz understood the method. In 1897 he also published the first book to reduce ac calculations to a science.

Chapter Contents

In Chapters 6 and 7 we analyzed circuits containing storage elements and have seen that the complete response is the sum of a natural and a forced response. The natural response is obtained after killing all independent sources and therefore is not a function of these sources, which are also known as *excitations* (unlike dependent sources, they excite responses without requiring coupling with other sources). The forced response, on the other hand, depends directly on the functional form of excitation applied to the circuit. In the case of a dc source, the forced response is a dc (constant) response, an exponential input evokes an exponential forced response, and so on.

Perhaps the single most important type of excitation is the sinusoid. Sinusoids are found everywhere in nature: for example, in the motion of a pendulum, the propagation of light or sound waves through space, and the vibration of strings or steel beams. As we have seen, any undamped second-order circuit produces a sinusoidal natural response, and any underdamped second-order circuit will have a decaying sinusoid as its natural response.

In electrical engineering and technology, sinusoidal time functions are found at the core of many, perhaps most, important applications. The carrier signals generated for communication systems are sinusoids, and the sinusoid is dominant in the electric power industry. Indeed, as we shall see later in the study of Fourier series, almost every useful signal in electrical engineering can be represented as a sum of sinusoidal components.

Because of their importance, circuits with sinusoidal excitation, or *ac circuits,* are considered in detail in this and several subsequent chapters. Since for all linear circuits the natural response is independent of the excitation and can be found by the methods of earlier chapters, we concentrate on finding the forced response to sinusoidal excitation. This response is important in itself since for stable circuits it is the ac steady-state response that remains after the time required for the transitory natural response has passed.

Being interested only in the forced response, we shall not limit ourselves, as we did in Chapters 6 and 7, to first- and second-order circuits. As we shall see, higher-order *RLC* circuits may be handled, insofar as the forced response is concerned, in much the same way as resistive circuits were handled in Chapter 2. The discovery of this technique for analyzing ac circuits with no more difficulty than dc circuits, called phasor analysis, is one of the greatest intellectual achievements in the field of engineering analysis. Modern engineering, electrical and otherwise, would be unthinkable without the power and versatility of phasor analysis, the principal subject of this chapter.

8.1 PROPERTIES OF SINUSOIDS

We devote this section to a review of some of the properties of sinusoidal functions. Let us begin with the sine wave,

$$v(t) = V_m \sin \omega t \tag{8.1}$$

which is sketched in Fig. 8.1. The *amplitude* of the sinusoid is V_m, which is the maximum value that the function attains. The *radian frequency,* or *angular frequency,* is ω, measured in radians per second (rad/s).

The sinusoid is a periodic function, defined generally by the property that there is a smallest number T such that for all t,

$$v(t + T) = v(t) \tag{8.2}$$

where T is the *period*. That is, the function goes through a complete cycle, or period, which is then repeated every T seconds. In the case of the sinusoid, the period is

$$T = \frac{2\pi}{\omega} \tag{8.3}$$

as may be seen from (8.1) and (8.2). Thus in 1 s the function goes through $1/T$ cycles, or periods. Its *frequency f* is then

$$f = \frac{1}{T} = \frac{\omega}{2\pi} \tag{8.4}$$

cycles per second, or *hertz* (abbreviated Hz). The latter term, named for the German

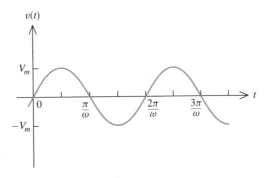

FIGURE 8.1   Sinusoidal function.

physicist Heinrich R. Hertz (1857–1894), is now the standard unit for frequency. The relation between frequency and radian frequency is seen by (8.4) to be

$$\omega = 2\pi f \tag{8.5}$$

A more general sinusoidal expression is given by

$$v(t) = V_m \sin(\omega t + \phi) \tag{8.6}$$

where ϕ is the *phase angle*, or simply the *phase*. To be consistent, since ωt is in radians, ϕ should be expressed in radians. However, degrees are a very familiar measure for angle. Therefore, we may write

$$v = V_m \sin\left(2t + \frac{\pi}{4}\right)$$

or

$$v = V_m \sin(2t + 45°)$$

interchangeably, even though the latter expression contains a formal mathematical inconsistency. In the absence of the small circle indicating degrees, the unit for angle will be taken to be radians.

A sketch of (8.6) is shown in Fig. 8.2 by the solid line, along with a sketch of (8.1), shown dashed. The solid curve is simply the dashed curve displaced ϕ/ω seconds, or ϕ radians to the left. Therefore, points on the solid curve, such as its peaks, occur ϕ rad, or ϕ/ω s, earlier than corresponding points on the dashed curve. Accordingly, we shall say that $V_m \sin(\omega t + \phi)$ *leads* $V_m \sin \omega t$ by ϕ rad (or degrees). Note that a positive leading phase ($\phi > 0$) implies a left shift of the graph of the function. In general, the sinusoid

$$v_1 = V_{m1} \sin(\omega t + \alpha)$$

leads the sinusoid

$$v_2 = V_{m2} \sin(\omega t + \beta)$$

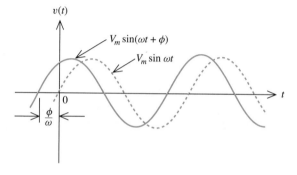

FIGURE 8.2 Two sinusoids with different phase.

by $\alpha - \beta$. An equivalent expression is that v_2 *lags* v_1 by $\alpha - \beta$. In Fig. 8.2, $V_m \sin \omega t$ lags $V_m \sin(\omega t + \phi)$ by ϕ radians. Positive lagging phase implies a right shift in the graph of the function.

As an example, consider

$$v_1 = 4 \sin(2t + 30°)$$

$$v_2 = 6 \sin(2t - 12°)$$

Then v_1 leads v_2 (or v_2 lags v_1) by $30 - (-12) = 42°$. The graph of v_1 is left shifted by $42°$ relative to v_2.

Thus far we have considered sine functions rather than cosine functions in defining sinusoids. It does not matter which form we use since

$$\cos\left(\omega t - \frac{\pi}{2}\right) = \sin \omega t \qquad (8.7a)$$

$$\sin\left(\omega t + \frac{\pi}{2}\right) = \cos \omega t \qquad (8.7b)$$

The only difference between sines and cosines is thus the phase angle. For example, we may write (8.6) as

$$v(t) = V_m \cos\left(\omega t + \phi - \frac{\pi}{2}\right)$$

Equations (8.7) are trigonometric identities for quarter-period shifts. Half-period shifts change sign,

$$\sin(\omega t \pm \pi) = -\sin \omega t \qquad (8.8a)$$

$$\cos(\omega t \pm \pi) = -\cos \omega t \qquad (8.8b)$$

while full-period shifts have no effect:

$$\sin(\omega t \pm 2\pi) = \sin \omega t \qquad (8.8c)$$

$$\cos(\omega t \pm 2\pi) = \cos \omega t \qquad (8.8d)$$

Example 8.1

To determine how much one sinusoid leads or lags another of the same frequency, we must first express both as sine waves or as cosine waves with positive amplitudes. For example, let

$$v_1 = 4 \cos(2t + 30°)$$

$$v_2 = -2 \sin(2t + 18°)$$

Then, by (8.8a),

$$-\sin \omega t = \sin(\omega t + 180°)$$

we have

$$v_2 = 2 \sin(2t + 18° + 180°)$$

$$= 2 \cos(2t + 18° + 180° - 90°)$$

$$= 2 \cos(2t + 108°)$$

Comparing this last expression with v_1, we see that v_1 leads v_2 by $30° - 108° = -78°$, which is the same as saying that v_1 lags v_2 by $78°$.

The sum of a sine wave and a cosine wave of the same frequency is another sinusoid of that frequency. To show this, consider

$$A \cos \omega t + B \sin \omega t = \sqrt{A^2 + B^2} \left(\frac{A}{\sqrt{A^2 + B^2}} \cos \omega t + \frac{B}{\sqrt{A^2 + B^2}} \sin \omega t \right)$$

which by Fig. 8.3 may be written

$$A \cos \omega t + B \sin \omega t = \sqrt{A^2 + B^2} \, (\cos \omega t \cos \theta + \sin \omega t \sin \theta)$$

By a formula from trigonometry, this is

$$A \cos \omega t + B \sin \omega t = \sqrt{A^2 + B^2} \cos(\omega t - \theta) \qquad (8.9\text{a})$$

where, by Fig. 8.3,

$$\theta = \tan^{-1} \frac{B}{A} \qquad (8.9\text{b})$$

A similar result may be established if the sine and cosine terms have phase angles other than zero, indicating that, in general, the sum of two sinusoids of a given frequency is another sinusoid of the same frequency.

The conversion of the sum of a sine and cosine may be reversed. By (8.9),

$$M \cos(\omega t - \theta) = A \cos \omega t + B \sin \omega t \qquad (8.10\text{a})$$

where

$$A = M \cos \theta \qquad (8.10\text{b})$$

$$B = M \sin \theta \qquad (8.10\text{c})$$

The decomposition of a sinusoid into sine and cosine components is called the *quadrature representation* of the sinusoid.

Note that we must be clear on what is meant by (8.9b), since some mathematics books take this expression as the principal value of the arctangent and place θ in a specific quadrant. We mean that the terminal side of the angle θ is in the quadrant where the point (A, B) is located.

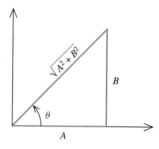

FIGURE 8.3   Triangle useful in adding two sinusoids.

Example 8.2

As an example, we have

$$-5\cos 3t + 12\sin 3t = \sqrt{5^2 + 12^2}\ \cos\left(3t - \tan^{-1}\frac{12}{-5}\right)$$

$$= 13\cos(3t - 112.6°)$$

since $\tan^{-1}(12/-5)$ is in the second quadrant, because $A = -5 < 0$ and $B = 12 > 0$.

EXERCISES

8.1.1. Find the period of the following sinusoids:
(a) $4\cos(5t + 33°)$
(b) $\cos(2t + \pi/4) + 3\sin(2t - \pi/6)$
(c) $6\cos 2\pi t$
Answer (a) $2\pi/5$; (b) π; (c) 1

8.1.2. Find the amplitude and phase of the following sinusoids:
(a) $3\cos 2t + 4\sin 2t$
(b) $(4\sqrt{3} - 3)\cos(2t + 30°) + (3\sqrt{3} - 4)\cos(2t + 60°)$ [*Suggestion:* In (b), expand both functions and use (8.9).]
Answer (a) $5, -53.1°$; (b) $5, 36.9°$

8.1.3. Find the frequency of the following sinusoids:
(a) $3\cos(6\pi t - 10°)$
(b) $4\sin 377t$
Answer (a) 3; (b) 60 Hz

8.2 *RLC* CIRCUIT EXAMPLE

As an example of a circuit with sinusoidal excitation, let us find the forced mesh current i in the series RLC circuit shown in Fig. 8.4. By KVL,

$$2\frac{di}{dt} + 2i + 10\int_{0+}^{t} i(\tau)\,d\tau + v_c(0+) = 15\cos 2t \qquad (8.11a)$$

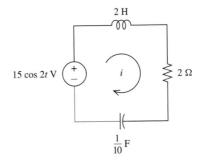

FIGURE 8.4 Series *RLC* circuit with sinusoidal excitation.

Differentiating and dividing by 2,

$$\frac{d^2i}{dt^2} + \frac{di}{dt} + 5i = -15 \sin 2t \tag{8.11b}$$

From Table 7.1, the trial forced solution is

$$i = A \cos 2t + B \sin 2t$$

Substituting this into (8.11) gives

$$(-4A \cos 2t - 4B \sin 2t) + (-2A \sin 2t + 2B \cos 2t) + 5(A \cos 2t + B \sin 2t) = -15 \sin 2t$$

and matching coefficients of the cos and sin terms,

$$A + 2B = 0 \tag{8.12a}$$

$$-2A + B = -15 \tag{8.12b}$$

Doubling the first of these equations and adding to the second yields $5B = -15$ or $B = -3$, and substituting into the first, $A = +6$. The forced response is

$$i = 6 \cos 2t - 3 \sin 2t \; A \tag{8.13}$$

These two quadrature terms may be combined into one using the conversion formulas (8.10):

$$M = \sqrt{6^2 + 3^2} = 3\sqrt{5}$$

$$\theta = \tan^{-1}\left(-\frac{3}{6}\right) = -26.6°$$

or

$$i = M \cos(\omega t - \theta) = 3\sqrt{5} \cos(2t + 26.6°) \; A \tag{8.14}$$

The method used to arrive at our solution (8.14) is straightforward but, the reader may agree, somewhat laborious for such a simple single-loop circuit. In particular, the necessity of solving a set of simultaneous equations in the unknown quadrature coefficients A and B is noted. For high-order circuits it may be anticipated that use of this procedure would be even more cumbersome. In the remainder of this chapter we develop an approach offering considerable computational advantage, which bypasses the need to solve two simultaneous equations for the two quadrature coefficients A and B and which allows us to treat RLC circuits in the same way that we treated purely resistive circuits in earlier chapters. The method used above, based on the time-domain solution of differential equations, will always be available, but for complex problems it will seldom be preferred to the method of phasors to be introduced next.

EXERCISES

8.2.1. Find the forced response v_f.
Answer $2 \cos(4t + 45°)$ V

322   Chapter 8   Sinusoidal Sources and Phasors

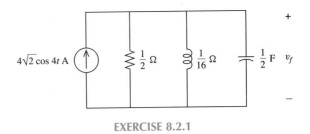

8.2.2. Repeat Exercise 8.2.1 if the inductor is removed from the circuit.
Answer $2\cos(4t - 45°)$ V

8.3 COMPLEX SOURCES

An alternative method for treating circuits with sinusoidal sources, which will be our focus of interest for this and the next several chapters, involves replacing the given sources by *complex sources*, those whose source functions have real and imaginary parts. Currents and voltages in circuits excited by complex sources will themselves be complex valued.

Since we rely heavily on complex numbers and their manipulation, readers unfamiliar with complex numbers or who seek to refresh their understanding should consult Appendix B. For convenience, we review several key definitions and properties of complex numbers before going on.

Each complex number is a point in the *complex plane*. The complex number A is written in *rectangular form* as

$$A = a + jb \tag{8.15}$$

where $j = \sqrt{-1}$ is the complex number of unit length along the imaginary axis in the complex plane. The real numbers a and b are the *real part*, denoted $a = \operatorname{Re} A$, and the *imaginary part*, $b = \operatorname{Im} A$, of the complex number A.

The same complex number A may be represented in *polar form* as

$$A = |A|\underline{/\alpha} \tag{8.16}$$

where $|A|$, the *magnitude* of A, and $\underline{/\alpha}$, the *angle* of A, are given by

$$|A| = \sqrt{a^2 + b^2} \tag{8.17a}$$

$$\alpha = \tan^{-1}\frac{b}{a} \tag{8.17b}$$

The relation between rectangular and polar forms is shown in Fig. 8.5.

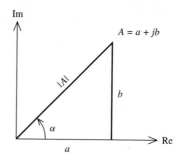

FIGURE 8.5   Rectangular and polar representations.

The important *Euler identity* is given by

$$e^{j\theta} = \cos\theta + j\sin\theta \qquad (8.18a)$$

or, using (8.18a) and (8.17), in its polar form,

$$e^{j\theta} = 1\underline{/\theta} \qquad (8.18b)$$

The latter form of the Euler identity may be used to generate a useful alternative way to write complex numbers in polar form. Since

$$A = |A|\underline{/\alpha} = |A|(1\underline{/\alpha})$$

then by (8.18b) we have the *exponential polar form* for a complex number:

$$A = |A|e^{j\alpha} \qquad (8.19)$$

Rectangular, polar, and exponential polar forms of the same complex number A will each prove convenient in different contexts, and it is essential to be able to move easily among them.

Example 8.3

Consider the complex number A given in polar form as $A = 4 + j3$. Then $|A| = \sqrt{4^2 + 3^2} = 5$ and $\alpha = \tan^{-1}\frac{3}{4} = 36.9°$. The polar form is thus

$$A = 5\underline{/36.9°}$$

and the exponential polar form is

$$A = 5e^{j36.9°}$$

The three forms each describe the same complex number A, the same point A in the complex plane.

The behavior of the *complex exponential function* $e^{j\omega t}$ is central to our studies. By the polar form of the Euler identity (8.18b),

$$e^{j\omega t} = 1\underline{/\omega t}$$

Examining this expression, the magnitude of the complex exponential is always unity, while its angle increases uniformly at the rate of ω radians per second. Thus *the complex exponential $e^{j\omega t}$ traces out unit circles in the complex plane, beginning on the positive real axis at time $t = 0$ and moving counterclockwise, completing one full circle (or period) every $T = 2\pi/\omega$ seconds.* The projection of this point onto the horizontal axis is its real part ($\cos\omega t$) and that onto the vertical axis is its imaginary part ($\sin\omega t$).

The general scaled and phase-shifted complex exponential $V_m e^{(j\omega t+\phi)}$ shown in Fig. 8.6(b) is similar, except that at $t = 0$ its initial phase is ϕ radians and it traces out circles of radius V_m. By horizontal projection,

$$\text{Re}\left[V_m e^{j(\omega t+\phi)}\right] = V_m \cos(\omega t + \phi)$$

and by vertical projection,

$$\text{Im}\left[V_m e^{j(\omega t+\phi)}\right] = V_m \sin(\omega t + \phi)$$

which can be verified by inspection of the rectangular form of the Euler identity.

We next turn to the subject of main interest, that is, the application of complex numbers to electric circuits. Consider the linear circuit C of Fig. 8.7, which has independent source v_g. We wish to determine the forced response to the excitation v_g. Assume that all other independent sources in C have been killed, along with all initial conditions. Now let $f_1(t)$ and $f_2(t)$ be any two real functions of time. Suppose that when we use the source $v_g = f_1(t)$ we measure the forced response $i_1(t)$, and when $v_g = f_2(t)$ is used, the forced response is $i_2(t)$, as shown in Fig. 8.7(a) and (b). These responses to the real inputs f_1 and f_2 are real. By the proportionality principle, scaling the excitation f_2 by the constant j will scale the response i_2 by the same factor. Then, by superposition, the response to the sum $f_1 + jf_2$ will be the corresponding sum of separate responses

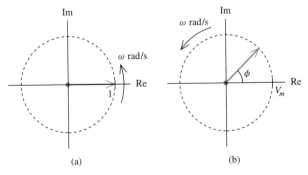

FIGURE 8.6 (a) Complex exponential $e^{j\omega t}$; (b) general complex exponential $V_m e^{(j\omega t+\phi)}$.

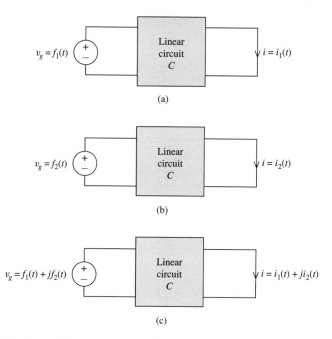

FIGURE 8.7 (a) Response to f_1; (b) response to f_2; (c) response to the linear combination $f_1 + jf_2$.

$i_1 + ji_2$, as shown in Fig. 8.7(c). Here we have used the fact that proportionality and superposition work equally well for complex quantities as for real ones.

Figure 8.7 reveals an important property of linear systems. $v_g = f_1 + jf_2$ is an arbitrary complex source, and $i = i_1 + ji_2$ its forced response. Since i_1 is also the separate response to f_1, which is the real part of the excitation v_g, *the response to the real part of a complex source is the real part of the response.* Similarly, noting that i_2 is the response to f_2, the imaginary part of v_g, *the response to the imaginary part of a complex source is the imaginary part of the response.*

This association is particularly significant when the complex source takes the form of a complex exponential $v_g = V_m e^{j(\omega t + \phi)}$. By Table 7.1 we know that the forced response to this source will be a complex exponential of the same frequency ω, which we denote as $i = I_m e^{j(\omega t + \theta)}$. *Once the response to the complex exponential source $V_m e^{j(\omega t + \phi)}$ has been found, we may immediately write down the response to the sinusoidal sources $V_m \cos(\omega t + \phi)$ as its real part and the response to $V_m \sin(\omega t + \phi)$ as its imaginary part.*

Example 8.4

Let us revisit the RLC circuit that was the subject of Section 8.2. Replacing the sinusoidal source $15 \cos 2t$ V by the complex source $15e^{j2t}$ V, (8.11a) is

$$2\frac{di}{dt} + 2i + 10 \int_{0+}^{t} i(\tau)\,d\tau + v_c(0+) = 15e^{j2t}$$

Differentiating and dividing by 2 as in (8.11b) gives

$$\frac{d^2 i}{dt^2} + \frac{di}{dt} + 5i = 15je^{j2t}$$

Substituting the trial form $i = Ae^{j2t}$, we have

$$(-4 + j2 + 5)Ae^{j2t} = 15je^{j2t}$$

Solving for A yields

$$A = \frac{15j}{1 + j2} = 3\sqrt{5}\underline{/26.6°}$$

The response i to this complex source is Ae^{j2t}, or

$$i = (3\sqrt{5}\underline{/26.6°})e^{j2t} = 3\sqrt{5}e^{j(2t+26.6°)} \text{ A}$$

where the exponential polar form (8.19) has been used in the last expression.

Since the real part of $15e^{j2t}$ is $15\cos 2t$, the response to the source $v_g = 15\cos 2t$ V must be

$$\text{Re}[3\sqrt{5}e^{j(2t+26.6°)}] = 3\sqrt{5}\cos(2t + 26.6°) \text{ A}$$

which agrees with our previous calculation. Moreover, since the imaginary part of $15e^{j2t}$ is $15\sin 2t$, had we instead applied the sinusoidal source $v_g = 15\sin 2t$, the response would have been

$$\text{Im}[3\sqrt{5}e^{j(2t+26.6°)}] = 3\sqrt{5}\sin(2t + 26.6°) \text{ A}$$

Example 8.5

As another example, let us find the forced response i_f of

$$\frac{d^2i}{dt^2} + 2\frac{di}{dt} + 8i = 12\sqrt{2}\cos(2t + 15°)$$

First we replace the real excitation by the complex excitation,

$$v_1 = 12\sqrt{2}e^{j(2t+15°)}$$

where for convenience the phase is written in degrees. The complex response i_1 satisfies

$$\frac{d^2i_1}{dt^2} + 2\frac{di_1}{dt} + 8i_1 = 12\sqrt{2}\ e^{j(2t+15°)}$$

and it must have the general form

$$i_1 = Ae^{j2t}$$

Therefore, we must have

$$(-4 + j4 + 8)Ae^{j2t} = 12\sqrt{2}e^{j2t}e^{j15°}$$

or

$$A = \frac{12\sqrt{2}\ e^{j15°}}{4 + j4} = \frac{12\sqrt{2}\ \underline{/15°}}{4\sqrt{2}\ \underline{/45°}} = 3\ \underline{/-30°}$$

which gives

$$i_1 = (3\underline{/-30°})e^{j2t}$$

The response to the original excitation is the real part of the response to this complex excitation:

$$i_f = \text{Re}\ (i_1) = 3\cos(2t - 30°)$$

In summary, given a circuit excited by a sinusoid, the forced response may be found by replacing the sinusoid by a complex exponential whose real (or imaginary) part is the given sinusoid. The describing equation will be easier to solve, since the trial forced solution will only have one undermined constant, rather than two, requiring the solution of simultaneous equations for the coefficients. Once solved, we may easily relate the complex excitation response to our desired sinusoidal response by taking the real part of the response if the original sinusoidal source is the real part of the complex exponential source, or the imaginary part if it is the imaginary part of the complex source.

EXERCISES

8.3.1. (a) From the time-domain equations, find the forced response v if $v_g = 10(e^{j8t})$ V. (b) Using the result in part (a), find the forced response v if $v_g = 10 \cos 8t$ V.

Answer (a) $2e^{j(8t-53.1°)}$ V; (b) $2\cos(8t - 53.1°)$ V

EXERCISE 8.3.1

8.3.2. Find the forced response v in Exercise 8.3.1 if $v_g = 10 \sin 8t$ V. [*Suggestion:* $\sin 8t = \text{Im}(e^{j8t})$.]

Answer $2\sin(8t - 53.1°)$ V

8.3.3. Using the method of complex excitation, find the forced response i if $v_g = 20 \cos 2t$ V.

Answer $2\cos(2t + 36.9°)$ A

8.3.4. Repeat Exercise 8.3.3 if $v_g = 16 \cos 4t$ V.

Answer $2 \cos 4t$ A

EXERCISE 8.3.3

8.4 PHASORS

The results obtained in the preceding section may be put in much more compact form by the use of quantities called *phasors*, which we introduce in this section. The phasor method of analyzing circuits is credited generally to Charles Proteus Steinmetz (1865–1923), a famous electrical engineer with the General Electric Company in the early part of the twentieth century.

We are concerned with the forced response of a circuit to sinusoidal excitation at frequency ω. Each sinusoidal source may be expressed as a cosine

$$v_g(t) = V_m \cos(\omega t + \phi)$$

Suppose that we replace each such source by a complex exponential source given by

$$v_{g1}(t) = V_m e^{j(\omega t + \phi)}$$

Comparing these two expressions, we see that *the complex source has the same frequency ω, and that the original source is the real part of the complex source we have chosen to replace it.*

Consider the forced response in the new circuit. Since it is excited by complex exponential sources of frequency ω, all currents and voltages will also be complex exponentials of frequency ω. This follows from the form of the trial forced solution $Ae^{j\omega t}$. Then each current will be of the form

$$i_1(t) = \mathbf{I}e^{j\omega t} \qquad\qquad\qquad (8.20a)$$

and each voltage of the form

$$v_1(t) = \mathbf{V}e^{j\omega t} \qquad\qquad\qquad (8.20b)$$

where $\mathbf{I}$ and $\mathbf{V}$ are complex numbers. We define $\mathbf{I}$ and $\mathbf{V}$ as *phasors*, that is, *the complex numbers that multiply $e^{j\omega t}$ in the expressions for currents and voltages.* To distinguish them from other quantities, phasors are printed in boldface. The units for phasors are the same as the currents and voltages they are associated with; thus in (8.20) $\mathbf{I}$ inherits the units of i_1, usually amperes, and $\mathbf{V}$ has the same units as the voltage v_1, usually volts.

Phasors are thus defined in terms of the response to complex excitations, but the importance of phasors lies in their direct link to sinusoidal responses. In the sinusoidal circuit with which we began this discussion, all forced responses are sinusoids at frequency ω. Let one such response be $v(t)$. After substituting complex exponential sources, the same response variable will be $v_1(t) = \mathbf{V}e^{j\omega t}$, where $\mathbf{V} = |\mathbf{V}|\underline{/\theta}$ is its voltage phasor. But we recall from the last section that the response to the real part of a source is just the real part of the response. Since the sinusoidal source $V_m \cos(\omega t + \phi)$ is the real part of the complex source

$$V_m \cos(\omega t + \phi) = \text{Re}[V_m e^{j(\omega t + \phi)}] \qquad\qquad\qquad (8.21)$$

it follows that the sinusoidal response $v(t)$ is just the real part of the complex response $v_1(t)$, or

$$v(t) = \text{Re}(\mathbf{V}e^{j\omega t})$$

Using the polar exponential form for $\mathbf{V}$ and taking the real part yields

$$v(t) = \text{Re}(|\mathbf{V}|e^{j\theta}e^{j\omega t}) = |\mathbf{V}| \cos(\omega t + \theta) \qquad\qquad\qquad (8.22)$$

Note the direct relationship between the the sinusoid $v(t)$ and its phasor **V**. By (8.22), $|\mathbf{V}|$ is the amplitude of $v(t)$ and θ its phase angle. *The amplitude of the sinusoid is the magnitude of its phasor, and the phase angle of the sinusoid is the angle of its phasor.* Thus we can immediately write down the sinusoidal current or voltage once its phasor has been computed.

Example 8.6

Suppose in Fig. 8.8 that $v_g = 6\cos 2t$ V. Since $v_g = \operatorname{Re}(6e^{j2t})$, we use $v_{g1} = 6e^{j2t}$ V as our complex exponential source. The describing equation is

$$\frac{di}{dt} + 4i = 6e^{j2t}$$

The trial forced solution is $i_1 = Ae^{j2t}$, and applying this to the equation above gives

$$(j2 + 4)Ae^{j2t} = 6e^{j2t}$$

or

$$A = \frac{6}{4 + j2} = 1.34\underline{/-26.6°}$$

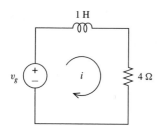

FIGURE 8.8   Circuit for Example 8.6.

The forced solution is $i_1 = Ae^{j2t}$, or

$$i_1 = (1.34\underline{/-26.6°})e^{j2t} \text{ A}$$

The phasor representation of i_1 is $\mathbf{I} = 1.34\underline{/-26.6°}$ and the sinusoidal response is

$$i = \operatorname{Re}(i_1) = 1.34\cos(2t - 26.6°) \text{ A}$$

Starting with the cosine form $V_m \cos(\omega t + \phi)$ for each source, we conclude that each response is also a cosine (8.22) of the same frequency, whose amplitude and phase are just the magnitude and angle of the associated phasor. To preserve this association, if the sinusoidal source is given in the sine form $V_m \sin(\omega t + \phi)$, we first convert to the cosine form using (8.7b), repeated here as

$$\cos(\omega t + \phi - 90°) = \sin(\omega t + \phi)$$

Table 8.1 shows the relation between these two sources and their phasors.

Table 8.1   Complex Sources and Source Phasors for Cosine and Sine Sources

Source	Complex Source	Source Phasor
$A\cos(\omega t + \phi)$	$Ae^{j(\omega t + \phi)}$	$A\underline{/\phi}$
$A\sin(\omega t + \phi)$	$Ae^{j(\omega t + \phi - 90°)}$	$A\underline{/\phi - 90°}$

Example 8.7

To find the forced response $i(t)$ in Fig. 8.9(a), we replace the source $36\cos(2t + 30°)$ V by the complex source $36e^{j(2t+30°)}$ V and the source $2\sin(2t - 15°)$ A, following Table 8.1, by $2e^{j(2t-105°)}$ A. Analyzing Fig. 8.9(b), the single mesh equation is

$$4i_1 + 3\frac{d}{dt}\left[i_1 - 2e^{j(2t-105°)}\right] = 36e^{j(2t+30°)}$$

or

$$\frac{di_1}{dt} + \tfrac{4}{3}i_1 = 12e^{j(2t+30°)} - j4e^{j(2t-105°)}$$

Combining terms on the right-hand side gives

$$(12e^{j30°} - j4e^{-j105°})e^{j2t} = (12\underline{/30°} - 4\underline{/-15°})e^{j2t} = (9.6\underline{/47.1°})e^{j2t}$$

Using the trial form $i_1 = Ae^{j2t}$, we obtain

$$(j2A + \tfrac{4}{3}A)e^{j2t} = (9.6\underline{/47.1°})e^{j2t}$$

and solving for A yields

$$A = \frac{9.6\underline{/47.1°}}{4/3 + j2} = 4.0\underline{/-9.2°} \tag{8.23}$$

Thus we have $i_1 = (4.0\underline{/-9.2°})e^{j2t}$ A. The phasor associated with this current is $4.0\underline{/-9.2°}$ A, and it follows that the corresponding sinusoidal current $i(t)$ has amplitude 4.0 A and phase angle $-9.2°$,

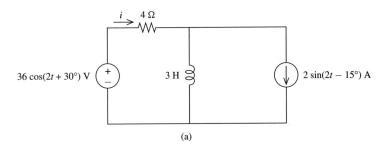

(a)

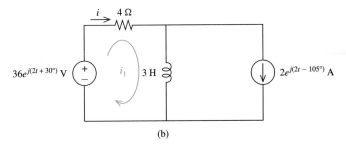

(b)

FIGURE 8.9   (a) Original circuit; (b) after source replacement.

or $\qquad i(t) = 4.0\cos(2t - 9.2°)$ A

One point of considerable practical significance is illustrated by Example 8.7. Our work with phasors will save us considerable time and effort compared to using real sinusoids, but only if we are equipped with the right tools. To use the phasor method, we need to be able to perform calculations such as (8.23),

$$\frac{9.6\underline{/47.1°}}{4/3 + j2} = 4.0\underline{/-9.2°}$$

rapidly and efficiently. This requires adequate computational support for the arithmetic operations required and repeated so frequently in the course of phasor circuit calculations: addition, subtraction, multiplication, and division of complex numbers, and also rectangular-to-polar and polar-to-rectangular conversion. *It is highly desirable to have access to an electronic calculator that supports complex data types, four-function complex arithmetic, and single-keystroke rectangular-to-polar and polar-to-rectangular conversions.* Although any calculator having sine, cosine, arctangent, and square root keys will permit us to do all required calculations, those explicitly set up to support complex arithmetic have a decided edge in convenience. Many fewer keystrokes will be required to complete a given problem, and keystrokes translate into time and diversion of mental focus from the concepts in the circuit problem at hand. Each major calculator manufacturer offers a line of suitable machines, beginning at relatively inexpensive prices.

Finally, in our work with phasors we rely on the principle that the response to the real part of a complex source (i.e., the response to a cosine source) can be computed as the real part of the response (i.e., the cosine part of a complex exponential). It is also possible to use another result from Section 8.3, that the response to the imaginary part of a complex source is the imaginary part of the response, to develop a second type of phasor analysis based on sines and imaginary parts rather than cosines and real parts. Since either type of phasor analysis is sufficient, the alternative form of phasor analysis is not considered in this book, except in one of the problems at the end of the chapter. We stick with phasors based on cosines and real parts, as in (8.22) and in the examples.

EXERCISES

8.4.1. Find the phasor representation of (a) $6\cos(2t + 45°)$, (b) $4\cos 2t + 3\sin 2t$, and (c) $-6\sin(5t - 65°)$.
$\qquad$ *Answer* (a) $6\underline{/45°}$; (b) $5\underline{/-36.9°}$; (c) $6\underline{/25°}$

8.4.2. Find the time-domain function represented by the phasors (a) $10\underline{/-17°}$, (b) $6 + j8$, and (c) $-j6$. In all cases, $\omega = 3$.
$\qquad$ *Answer* (a) $10\cos(3t - 17°)$; (b) $10\cos(3t + 53.1°)$; (c) $6\cos(3t - 90°)$

8.5 CURRENT–VOLTAGE LAWS FOR PHASORS

In this section we show that relationships between phasor voltage and phasor current for resistors, inductors, and capacitors are very similar to Ohm's law for resistors. In fact, the phasor voltage is proportional to the phasor current, as in Ohm's law. Consider a circuit in which all currents and voltages are of the form $Ae^{j\omega t}$. This will be the case when sinusoidal sources have been replaced by complex exponentials and we are interested in the forced response only. For the resistor of Fig. 8.10,

$$v = \mathbf{V}e^{j\omega t} \tag{8.24a}$$

$$i = \mathbf{I}e^{j\omega t} \tag{8.24b}$$

where $\mathbf{V}$ and $\mathbf{I}$ are phasors. By Ohm's law applied to (8.24),

$$\mathbf{V}e^{j\omega t} = R\mathbf{I}e^{j\omega t} \tag{8.25}$$

or, canceling the $e^{j\omega t}$ factors,

$$\mathbf{V} = R\mathbf{I} \tag{8.26}$$

Thus the phasor or frequency-domain relation for the resistor is exactly like the time-domain relation. The voltage–current relations for the resistor are illustrated in Fig. 8.11. With $\mathbf{V} = V_m e^{j\phi_v}$ and $\mathbf{I} = I_m e^{j\phi_I}$, (8.26) becomes

$$V_m e^{j\phi_v} = (RI_m)e^{j\phi_I}$$

that is, the magnitude of the voltage phasor equals the magnitude of the current phasor scaled by R, and the angles are the same. Recalling that the magnitude of a phasor is the amplitude of its sinusoid, and its angle is the phase angle of the sinusoid, we have Fig. 8.11. Note that since $\phi_V = \phi_I$, the current and voltage are *in phase*.

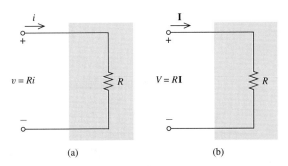

FIGURE 8.10 Voltage–current relations for a resistor R in the (a) time and (b) frequency domains.

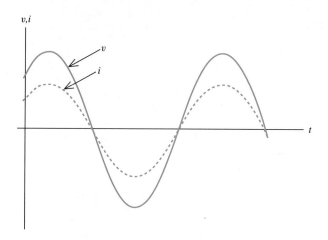

FIGURE 8.11   Voltage and current waveforms for a resistor.

Example 8.8

As an illustration, suppose that the voltage

$$v = 10\cos(100t + 30°) \text{ V}$$

is applied across a 5-Ω resistor, with the polarity indicated in Fig. 8.10(a). Then the phasor voltage is

$$\mathbf{V} = 10\underline{/30°} \text{ V}$$

and the phasor current is

$$\mathbf{I} = \frac{\mathbf{V}}{R} = \frac{10\underline{/30°}}{5} = 2\underline{/30°} \text{ A}$$

Therefore, in the time domain we have

$$i = 2\cos(100t + 30°) \text{ A}$$

This is simply the result we would have obtained using Ohm's law.

In the case of the inductor, substituting the complex current and voltage into the time-domain relation,

$$v = L\frac{di}{dt}$$

gives

$$V_m e^{j(\omega t + \phi_V)} = L\frac{d}{dt}\left[I_m e^{j(\omega t + \phi_I)}\right]$$

$$= j\omega L I_m e^{j(\omega t + \phi_I)}$$

Again, dividing out the factor $e^{j\omega t}$ and identifying the phasors, we obtain the phasor relation

$$\mathbf{V} = j\omega L\mathbf{I} \qquad\qquad (8.27)$$

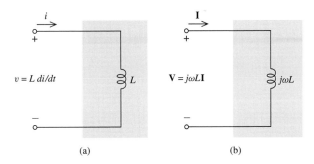

(a)   (b)

FIGURE 8.12 Voltage–current relations for an inductor L in the (a) time and (b) frequency domains.

Thus the phasor voltage $\mathbf{V}$, as in Ohm's law, is proportional to the phasor current $\mathbf{I}$, with the proportionality factor $j\omega L$. The voltage–current relations for the inductor are shown in Fig. 8.12.

If the current in the inductor is given by $i = I_m \cos(\omega t + \phi_I)$, then by (8.27) the phasor voltage is

$$\mathbf{V} = (j\omega L)(I_m \underline{/\phi_I})$$
$$= \omega L I_m \underline{/\phi_I + 90^\circ}$$

since $j = 1\underline{/90^\circ}$. Therefore, in the time domain we have

$$v = \omega L I_m \cos(\omega t + \phi_I + 90^\circ)$$

We see that in the case of an inductor the current *lags* the voltage by 90°. Another expression that is used is that the current and voltage are 90° out of phase. This is shown graphically in Fig. 8.13.

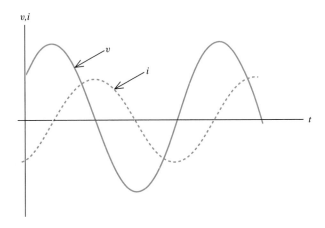

FIGURE 8.13 Voltage and current waveforms for an inductor.

Finally, let us consider the capacitor. Substituting the complex current and voltage into the time-domain relation,

$$i = C\frac{dv}{dt}$$

gives the complex relation

$$I_m e^{j(\omega t + \phi_I)} = C\frac{d}{dt}\left[V_m e^{j(\omega t + \phi_V)}\right]$$
$$= j\omega C V_m e^{j(\omega t + \phi_V)}$$

Again dividing by $e^{j\omega t}$ and identifying the phasors, we obtain the phasor relation

$$\mathbf{I} = j\omega C \mathbf{V} \qquad (8.28)$$

or

$$\mathbf{V} = \frac{\mathbf{I}}{j\omega C} \qquad (8.29)$$

Thus the phasor voltage $\mathbf{V}$ is proportional to the phasor current $\mathbf{I}$, with the proportionality factor given by $1/j\omega C$. The voltage–current relations for a capacitor in the time and frequency domains are shown in Fig. 8.14.

In the general case, if the capacitor voltage is given by $v = V_m \cos(\omega t + \phi_V)$, then by (8.28) the phasor current is

$$\mathbf{I} = (j\omega C)(V_m\underline{/\phi_V})$$
$$= \omega C V_m\underline{/\phi_V + 90°}$$

Therefore, in the time domain we have

$$i = \omega C V_m \cos(\omega t + \phi_V + 90°)$$

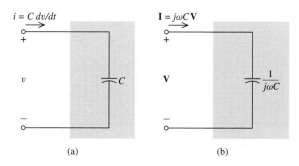

(a) (b)

FIGURE 8.14 Voltage–current relations for a capacitor in the (a) time and (b) frequency domains.

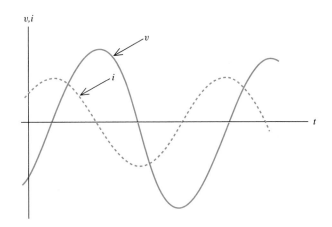

FIGURE 8.15   Voltage and current waveforms for a capacitor.

which indicates that in the case of a capacitor the current and voltage are out of phase, with the current *leading* the voltage by 90°. This is shown graphically in Fig. 8.15.

Example 8.9

As an example, if $v = 10\cos(100t + 30°)$ is applied across a 1-μF capacitor, then by (8.28) the phasor current is

$$\mathbf{I} = j(100)(10^{-6})(10\underline{/30°}) \text{ A}$$

$$= 1\underline{/120°} \text{ mA}$$

The time-domain current is then

$$i = \cos(100t + 120°) \text{ mA}$$

and therefore the current leads the voltage by 90°, as it must in all capacitors.

EXERCISES

8.5.1.   Using phasors, find the ac steady-state current i if $v = 12\cos(1000t + 30°)$ V in (a) Fig. 8.10(a) for $R = 4$ kΩ, (b) Fig. 8.12(a) for $L = 15$ mH, and (c) Fig. 8.14(a) for $C = \frac{1}{2} \mu$F.
  Answer (a) $3\cos(1000t + 30°)$ mA; (b) $0.8\cos(1000t - 60°)$ A; (c) $6\cos(1000t + 120°)$ mA

8.5.2.   In Exercise 8.5.1, find i in each case at $t = 1$ ms.
  Answer (a) 0.142 mA; (b) 0.799 A; (c) −5.993 mA

8.5.3.   For what L in Fig. 8.12 will a sinusoidal voltage of amplitude 20 V produce a 4-mA amplitude current at $\omega = 500$ rad/s?
  Answer 10 H

8.6   IMPEDANCE AND ADMITTANCE

Let us now consider a general circuit with two accessible terminals, as shown in Fig. 8.16. If the time-domain voltage and current at the terminals are given by

$$v = V_m \cos(\omega t + \phi_V) \tag{8.30a}$$

$$i = I_m \cos(\omega t + \phi_I) \tag{8.30b}$$

the phasor quantities at the terminals are

$$\mathbf{V} = V_m \underline{/\phi_V}$$
$$\mathbf{I} = I_m \underline{/\phi_I} \tag{8.31}$$

We define the ratio of the phasor voltage to the phasor current as the *impedance* of the circuit, which we denote by Z. That is,

$$\mathbf{V} = Z\mathbf{I} \tag{8.32}$$

and, by (8.31),

$$Z = |Z|\underline{/\phi_Z} = \frac{V_m}{I_m}\underline{/\phi_V - \phi_I} \tag{8.33}$$

where $|Z|$ is the magnitude and ϕ_Z the angle of Z. Evidently,

$$|Z| = \frac{V_m}{I_m}, \qquad \phi_Z = \phi_V - \phi_I$$

The magnitude of the impedance is the ratio of magnitudes of voltage and current phasors; the angle is the difference of the voltage and current phasor angles.

Impedance, as is seen from (8.32), plays the role, in a general circuit, played by resistance in resistive circuits. Indeed, (8.32) looks very much like Ohm's law; also like resistance, impedance is measured in ohms, being a ratio of volts to amperes.

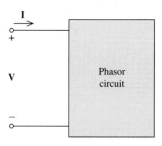

FIGURE 8.16   General phasor circuit.

It is important to stress that impedance is a complex number, being the ratio of two complex numbers, but it is *not* a phasor. That is, it has no corresponding sinusoidal time-domain function as current and voltage phasors do. Impedance is a complex constant that scales one phasor to produce another.

The impedance Z is written in polar form in (8.33); in rectangular form it is generally denoted by

$$Z = R + jX \qquad (8.34)$$

where $R = \text{Re } Z$ is the *resistive component*, or simply *resistance*, and $X = \text{Im } Z$ is the *reactive component*, or *reactance*. In general, $Z = Z(j\omega)$ is a complex function of $j\omega$, but $R = R(\omega)$ and $X = X(\omega)$ are real functions of ω. Like Z, both R and X are measured in ohms. Evidently, comparing (8.33) and (8.34), we may write

$$|Z| = \sqrt{R^2 + X^2}$$

$$\phi_Z = \tan^{-1} \frac{X}{R}$$

$$R = |Z| \cos \phi_Z$$

$$X = |Z| \sin \phi_Z$$

These relations are shown graphically in Fig. 8.17.

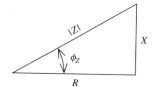

FIGURE 8.17 Graphical representation of impedance.

Example 8.10

As an example, suppose in Fig. 8.16 that $\mathbf{V} = 10\underline{/56.9°}$ V and $\mathbf{I} = 2\underline{/20°}$ A. Then we have

$$Z = \frac{10\underline{/56.9°}}{2\underline{/20°}} = 5\underline{/36.9°} \ \Omega$$

In rectangular form this is

$$Z = 5(\cos 36.9° + j \sin 36.9°)$$
$$= 4 + j3 \ \Omega$$

The impedances of resistors, inductors, and capacitors are readily found from their V–I relations of (8.26), (8.27), and (8.29). Distinguishing their impedances with subscripts R, L, and C, respectively, we have, from these equations and (8.32),

$$Z_R = R$$

$$Z_L = j\omega L = \omega L\underline{/90°} \qquad (8.35)$$

$$Z_C = \frac{1}{j\omega C} = -j\frac{1}{\omega C} = \frac{1}{\omega C}\underline{/-90°}$$

In the case of a resistor, the impedance is purely resistive, its reactance being zero. Impedances of inductors and capacitors are purely reactive, having zero resistive components. The *inductive reactance* is denoted by

$$X_L = \omega L \qquad (8.36)$$

so that

$$Z_L = j X_L$$

and the *capacitive reactance* is denoted by

$$X_C = -\frac{1}{\omega C} \qquad (8.37)$$

and thus

$$Z_C = j X_C$$

Since ω, L, and C are positive, we see that *inductive reactance is positive and capacitive reactance is negative*. In the general case of (8.34), we may have $X = 0$, in which case the circuit appears to be resistive; $X > 0$, in which case its reactance appears to be inductive; and $X < 0$, in which case its reactance appears to be capacitive. These cases are possible when resistance, inductance, and capacitance are all present in the circuit, as we shall see. As an example, the circuit with impedance given by $Z = 4 + j3$, which we have just considered, has reactance $X = 3$, which is of the inductive type. In all cases of passive circuits, the resistance R is nonnegative.

The reciprocal of impedance, denoted by

$$Y = \frac{1}{Z} \qquad (8.38)$$

is called *admittance* and is analogous to conductance (the reciprocal of resistance) in resistive circuits. Evidently, since Z is a complex number, then so is Y. The standard representation is

$$Y = G + j B \qquad (8.39)$$

The quantities $G = \text{Re}(Y)$ and $B = \text{Im}(Y)$ are called *conductance* and *susceptance*, respectively, and are related to the impedance components by

$$Y = G + jB = \frac{1}{Z} = \frac{1}{R + jX}$$

The units of Y, G, and B are all siemens, since in general Y is the ratio of a current to a voltage phasor.

To obtain the relation between components of Y and Z, we may rationalize the last member of the previous equation, which results in

$$G + jB = \frac{1}{R + jX}\frac{R - jX}{R - jX}$$

$$= \frac{R - jX}{R^2 + X^2}$$

Equating real and imaginary parts results in

$$G = \frac{R}{R^2 + X^2}$$

$$B = -\frac{X}{R^2 + X^2}$$

Therefore, we note that R and G are *not* reciprocals except in the purely resistive case ($X = 0$). Similarly, X and B are never reciprocals, but in the purely reactive case ($R = 0$) they are negative reciprocals.

Example 8.11

As an example, if we have

$$Z = 4 + j3$$

then

$$Y = \frac{1}{4 + j3} = \frac{4 - j3}{4^2 + 3^2} = \frac{4}{25} - j\frac{3}{25}$$

Therefore, $G = \frac{4}{25}$ and $B = -\frac{3}{25}$.
Further examples are

$$Y_R = G$$

$$Y_L = \frac{1}{j\omega L}$$

$$Y_C = j\omega C$$

which are the admittances of a resistor, with $R = 1/G$, an inductor, and a capacitor.

EXERCISES

8.6.1. Find the impedance seen at the terminals of a series RL subcircuit in both rectangular and polar form.

Answer $R + j\omega L$; $\sqrt{R^2 + \omega^2 L^2}\big/\tan^{-1}\omega L/R$

8.6.2. Find the admittance seen at the terminals of a series RL subcircuit in both rectangular and polar form.

Answer $\dfrac{R}{R^2 + \omega^2 L^2} - j\dfrac{\omega L}{R^2 + \omega^2 L^2}$;

$\dfrac{1}{\sqrt{R^2 + \omega^2 L^2}}\big/\!-\tan^{-1}(\omega L/R)$

8.6.3. Find the conductance and susceptance if Z is (a) $3 + j4$, (b) $0.4 + j0.3$, and (c) $(\sqrt{2}/2)/\underline{45°}$.

Answer (a) 0.12, -0.16; (b) 1.6, -1.2; (c) 1, -1

8.7 KIRCHHOFF'S LAWS AND IMPEDANCE EQUIVALENTS

If a complex excitation, say $V_m e^{j(\omega t + \theta)}$, is applied to a circuit, then complex voltages, such as $V_1 e^{j(\omega t + \theta_1)}$, $V_2 e^{j(\omega t + \theta_2)}$, and so on, appear across the elements in the circuit. KVL applied around a typical loop results in an equation such as

$$V_1 e^{j(\omega t + \theta_1)} + V_2 e^{j(\omega t + \theta_2)} + \cdots + V_N e^{j(\omega t + \theta_N)} = 0$$

Dividing out the common factor $e^{j\omega t}$, we have

$$\boxed{\mathbf{V}_1 + \mathbf{V}_2 + \cdots + \mathbf{V}_N = 0}$$

where

$$\mathbf{V}_n = V_n\big/\underline{\theta_n}, \qquad n = 1, 2, \ldots, N$$

are the phasor voltages around the loop. Thus *KVL holds for phasors*. A similar development will also establish KCL. At any node with N connected branches,

$$\boxed{\mathbf{I}_1 + \mathbf{I}_2 + \cdots + \mathbf{I}_N = 0}$$

where

$$\mathbf{I}_n = I_n\big/\underline{\theta_n}, \qquad n = 1, 2, \ldots, N$$

Thus *KCL holds for phasors*.

In circuits having sinusoidal excitations with a common frequency ω, if we are interested only in the forced, or ac steady-state, response, we may find the phasor voltages or currents of every element and use Kirchhoff's laws to complete the analysis. The analysis is therefore identical to the resistive circuit analysis of Chapters 2, 4, and 5, with impedances replacing resistances and phasors replacing time-domain currents and voltages. Once we have found the phasors, we can convert immediately to the time-domain sinusoidal answers in the usual fashion.

Example 8.12

As an example, consider the circuit of Fig. 8.18, which consists of N impedances connected in series. By KCL for phasors, the single phasor current $\mathbf{I}$ flows in each element. Therefore, the voltages shown across each element are

$$\mathbf{V}_1 = Z_1\mathbf{I}$$

$$\mathbf{V}_2 = Z_2\mathbf{I}$$

$$\vdots$$

$$\mathbf{V}_N = Z_N\mathbf{I}$$

and by KVL around the circuit,

$$\mathbf{V} = \mathbf{V}_1 + \mathbf{V}_2 + \cdots + \mathbf{V}_N$$

$$= (Z_1 + Z_2 + \cdots + Z_N)\mathbf{I}$$

Since we must also have, from Fig. 8.18,

$$\mathbf{V} = Z_{\text{eq}}\mathbf{I}$$

where Z_{eq} is the *equivalent* impedance seen at the terminals, it follows that

$$Z_{\text{eq}} = Z_1 + Z_2 + \cdots + Z_N \qquad (8.40)$$

as in the case of series resistors.

Similarly, as was the case for parallel conductances in Chapter 2, the equivalent admittance Y_{eq} of N parallel admittances is

$$Y_{\text{eq}} = Y_1 + Y_2 + \cdots + Y_N \qquad (8.41)$$

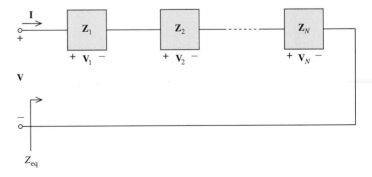

FIGURE 8.18   Impedances connected in series.

In the case of two parallel elements ($N = 2$), we have

$$Z_{eq} = \frac{1}{Y_{eq}} = \frac{1}{Y_1 + Y_2} = \frac{Z_1 Z_2}{Z_1 + Z_2} \qquad (8.42)$$

In like manner, voltage and current division rules hold for phasor circuits, with impedances and frequency-domain quantities, in exactly the same way that they held for resistive circuits, with resistances and time-domain quantities. The reader is asked to establish these rules in Exercise 8.7.2.

Example 8.13

For example, let us return to the RLC circuit considered in Section 8.2. The circuit and phasor circuit are shown in Fig. 8.19(a) and (b). By KVL in the phasor circuit,

$$Z_C\mathbf{I} + Z_L\mathbf{I} + R\mathbf{I} = 15\underline{/0}$$

or

$$(2 - j)\mathbf{I} = 15\underline{/0}$$

from which the phasor current is

$$\mathbf{I} = \frac{15}{2 - j} = 3\sqrt{5}\underline{/26.6°} \text{ A}$$

Therefore, in the time domain we have the same result as before, although now computed with considerably less effort:

$$i = 3\sqrt{5}\cos(2t + \underline{/26.6°}) \text{ A}$$

An alternative method of solution is to observe that the impedance Z seen at the source terminals is the impedance of the inductor, $Z_L = j\omega L = j4$, the resistor $Z_R = R = 2$, and the capacitor, $Z_C = 1/j\omega C = -j5$, all connected in series. Therefore,

$$Z = j4 + 2 + (-j5) = 2 - j \ \Omega$$

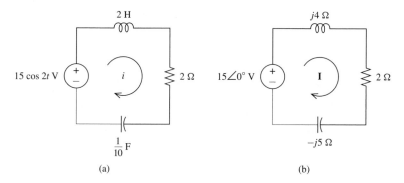

FIGURE 8.19   (a) Time-domain circuit; (b) phasor circuit.

and $\quad \mathbf{I} = \dfrac{\mathbf{V}}{\mathbf{Z}} = \dfrac{15\underline{/0}}{2 - j} = 3\sqrt{5}\underline{/26.6^\circ}$ A

as obtained earlier.

This ability to treat inductors, capacitors, and resistors alike as elements of the generic *impedance* without distinction among them when writing circuit equations is one of the main strengths of the phasor method. *RLC* circuits may be simplified using series–parallel equivalents, current–voltage dividers, and Thevenin–Norton transformations just as pure resistive circuits were earlier. Series impedances add, currents through parallel impedances divide in proportion to their admittances, and so on, regardless of the specific identities (*R*, *L*, or C) of the impedances involved.

Example 8.14

We wish to find $i(t)$ in steady state in Fig. 8.20. Using series–parallel impedance equivalents, we have

$$Z_1 = \dfrac{3(2 - j2)}{3 + (2 - j2)} = 1.45 - j0.621 \ \Omega$$

$$Z_2 = \dfrac{(-j1)(Z_1 + j1)}{-j1 + (Z_1 + j1)} = 0.583 - j0.75 \ \Omega$$

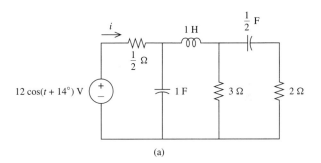

(a)

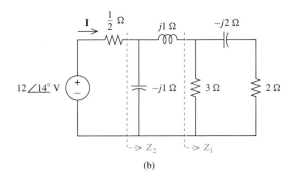

(b)

FIGURE 8.20 (a) Time-domain circuit; (b) phasor circuit.

so the equivalent impedance is

$$Z_{eq} = \tfrac{1}{2} + Z_2 = 1.083 - j0.75 \ \Omega$$

and

$$I = \frac{12\underline{/14^\circ}}{Z_{eq}} = \frac{12\underline{/14^\circ}}{1.083 - j0.75} = 9.11\underline{/48.7^\circ} \ A$$

Recalling that $\omega = 1$ rad/s yields

$$i(t) = 9.11\cos(t + 48.7^\circ) \ A$$

This is the desired ac steady-state current.

EXERCISES

8.7.1. Derive (8.41).

8.7.2. Show for circuit (a) that the voltage-division rule,

$$V = \frac{Z_2}{Z_1 + Z_2}V_g$$

and for circuit (b) that the current-division rule,

$$I = \frac{Y_2}{Y_1 + Y_2}I_g = \frac{Z_1}{Z_1 + Z_2}I_g$$

are valid, where $Z_1 = 1/Y_1$ and $Z_2 = 1/Y_2$.

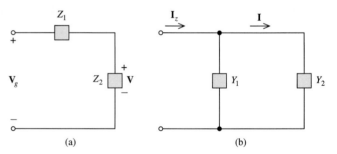

(a) (b)

EXERCISE 8.7.2

8.7.3. Find the steady-state current i using phasors.
Answer $2\cos(8t - 36.9^\circ)$ A

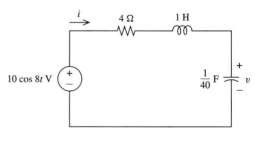

EXERCISE 8.7.3

8.7.4. Find the steady-state voltage v in Exercise 8.7.3 using phasors and voltage division.

Answer $10\cos(8t - 126.9°)$ V

8.8 PHASOR CIRCUITS

As the discussion in Section 8.7 suggests, we may omit the steps of finding the describing equation in the time domain, replacing the excitations and responses by their complex forcing functions and then dividing the equation through by $e^{j\omega t}$ to obtain the phasor equation. We may simply start with the *phasor circuit*, which we define as *the time-domain circuit with the voltages and currents replaced by their phasors and the elements identified by their impedances*. The describing equation obtained from this circuit is then the *phasor equation*. Solving this equation yields the phasor of the answer, which then may be converted to the time-domain answer.

The procedure from starting with the phasor circuit to obtaining the phasor answer is identical to that used earlier in resistive circuits. The only difference is that impedances replace resistances.

Example 8.15

As an example, let us find the steady-state current i in Fig. 8.21(a). The phasor circuit, shown in Fig. 8.21(b), is obtained by replacing the voltage source and the currents by their phasors and labeling the

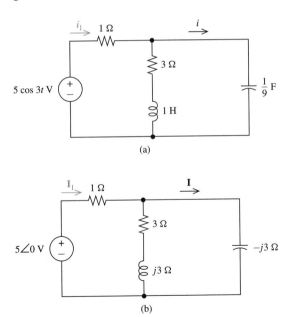

FIGURE 8.21 *RLC* time-domain and phasor circuits.

elements with their impedances. In the phasor circuit the impedance seen from the source terminals is

$$Z = 1 + \frac{(3+j3)(-j3)}{3+j3-j3}$$

$$= 4 - j3 \; \Omega$$

Therefore, we have

$$\mathbf{I}_1 = \frac{5\underline{/0}}{4-j3} = \frac{5\underline{/0}}{5\underline{/-36.9°}} = 1\underline{/36.9°}$$

and, by current division,

$$\mathbf{I} = \frac{3+j3}{3+j3-j3}\mathbf{I}_1 = \sqrt{2}\underline{/81.9°} \; \text{A}$$

In the time domain, the answer is

$$i = \sqrt{2}\cos(3t + 81.9°) \; \text{A}$$

In the case of a dependent source, such as a source kv_x volts controlled by a voltage v_x, it will appear in the phasor circuit as a source $k\mathbf{V}_x$, where $\mathbf{V}_x$ is the phasor representation of v_x, because $v_x = V_m\cos(\omega t + \phi)$ in the time domain will become $V_m e^{j(\omega t + \phi)}$ when a complex excitation is applied. Then dividing $e^{j\omega t}$ out of the equations leaves v_x represented by its phasor $V_m e^{j\phi}$. In the same way, $kv_x = kV_m\cos(\omega t + \phi)$ is represented by its phasor $kV_m e^{j\theta}$, which is k times the phasor of v_x.

Example 8.16

As an example of a circuit containing a dependent source, let us consider Fig. 8.22(a), in which it is required to find the steady-state value of i. The corresponding phasor circuit is shown in Fig. 8.22(b). Since phasor circuits are analyzed exactly like resistive circuits, we may apply KCL at node a in Fig. 8.22(b), resulting in

$$\mathbf{I} + \frac{\mathbf{V}_1 - \frac{1}{2}\mathbf{V}_1}{-j2} = 3\underline{/0} \tag{8.43}$$

By Ohm's law we have $\mathbf{V}_1 = 4\mathbf{I}$, which substituted into (8.43) yields

$$-j2\mathbf{I} + \tfrac{1}{2}(4\mathbf{I}) = -j6$$

or

$$\mathbf{I} = \frac{-j6}{2-j2} = \frac{6\underline{/-90°}}{2\sqrt{2}\underline{/-45°}} = \frac{3}{\sqrt{2}}\underline{/-45°} \; \text{A}$$

Therefore, we have

$$i = \frac{3}{\sqrt{2}}\cos(4t - 45°) \; \text{A}$$

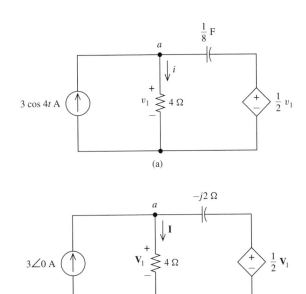

FIGURE 8.22 (a) Circuit containing a dependent source; (b) corresponding phasor circuit.

In the case of an op amp, the phasor circuit is the same as the time-domain circuit. That is, an ideal op amp in the time-domain circuit appears as an ideal op amp in the phasor circuit, because the time-domain equations

$$i = 0, \qquad v = 0$$

which characterize the current into and the voltage across the input terminals retain the identical form,

$$\mathbf{I} = 0, \qquad \mathbf{V} = 0$$

in the phasor equations.

Phasor analysis is indeed a tool of great power and versatility. This may already be apparent from our work in this chapter and will be further underscored by our reliance on phasor analysis to explore important issues such as ac steady-state power in the next several chapters. Like any tool, however, its range of applicability is limited, and it is useful to remind ourselves of these limits.

The first caution is that phasors are useful for finding the *forced* response only. If the natural response is desired, another tool must be used. Also, if the total response is required, phasor analysis is useful in determining the forced component only. The second caution concerns the use of phasors to determine steady-state response. For many circuits it is the case that natural responses all decay to zero as $t \to \infty$. In such circuits the forced response and the steady-state response are synonymous, and phasors (designed for determining the forced response) may be used to find the steady-state response. Such circuits are called *stable*. For the remaining or *unstable* circuits, the natural response does not in general decay to zero, and thus the steady-state response cannot be identified with the forced response. *Phasor analysis may be used to find the steady-state response only with stable circuits.* Examples of stable circuits include all circuits that are made up exclusively of passive elements and independent sources and that happen to have no undamped natural responses (no uncoupled LC subcircuits lacking resistance). Examples of unstable circuits are those with undamped natural response, and circuits possessing a natural response that grows, rather than decays, with time. Growing natural responses are possible only in linear circuits containing controlled sources. Stability is discussed in Chapter 14.

In general, unstable circuits do not possess steady-state responses at all (their natural responses either grow or oscillate periodically, so the circuit never goes to steady state no matter how long we wait). Thus it is meaningless to try to define a steady-state response for an unstable circuit, and for such circuits phasors may be used to compute the forced, but not the steady-state, response.

Example 8.17

Let us determine the describing equation for i in Fig. 8.23. The controlled source has transconductance $g = \frac{3}{2}$ S. By KVL around the left loop,

$$i + \int_{0+}^{t} \left(i - \frac{3v}{2} \right) d\tau + v_c(0+) + \frac{d}{dt} \left(i - \frac{3v}{2} \right) = +4 \cos 3t$$

Substituting $v = i$ and differentiating yields

$$\frac{d^2 i}{dt^2} - 2\frac{di}{dt} + i = -24 \sin 3t$$

The natural response is found from the characteristic equation

$$s^2 - 2s + 1 = (s - 1)^2 = 0$$

with repeated root at $s = +1$, so the natural response is of the form

$$i_n = K_1 e^t + K_2 t e^t$$

Note that except for the special case $K_1 = K_2 = 0$, all natural responses will grow rather than decay as $t \to \infty$, so the circuit cannot be expected to possess a steady state. This is an example of an unstable circuit. We may still use phasor analysis to determine the forced response, but since the circuit is unstable we will not equate the forced with the (nonexistent) steady-state response. To determine the forced response, consider the phasor circuit of Fig. 8.23(b). By KVL,

$$\mathbf{I} + \frac{j8}{3} \left(\mathbf{I} - \frac{3}{2}\mathbf{V} \right) = 4\underline{/0} \text{ V}$$

where we have summed the series impedances. Substituting $\mathbf{I} = \mathbf{V}$ and solving for $\mathbf{I}$ gives

$$\mathbf{I} = \frac{4\underline{/0}}{1 - j4/3} = 2.4\underline{/53.1°} \text{ A}$$

Thus the time-domain forced current is

$$i_f(t) = 2.4 \cos(3t + 53.1°) \text{ A}$$

For almost all initial conditions the total current $i = i_n + i_f$ will become larger and larger with time, never attaining steady state.

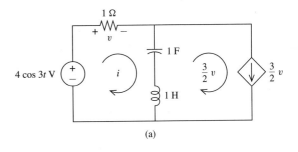

(a)

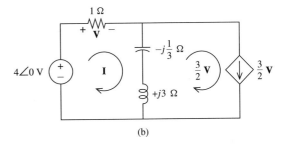

(b)

FIGURE 8.23   (a) Unstable circuit; (b) phasor circuit for computing forced (but not steady-state) response.

EXERCISES

8.8.1. Solve Exercise 8.2.1 by means of the phasor circuit.

8.8.2. Find the steady-state voltage v using the phasor circuit.
Answer $4\cos(8t - 53.1°)$ V

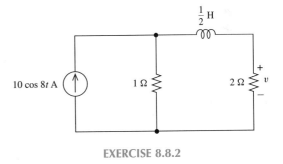

EXERCISE 8.8.2

8.8.3. Find the steady-state voltage v in Exercise 8.3.1 using the phasor circuit.
Answer $2\cos(8t - 53.1°)$

8.8.4. Find the steady-state voltage v using the phasor circuit given that $v_g = 4\cos 10t$ V.
Answer $\sqrt{2}\cos(10t + 135°)$ V

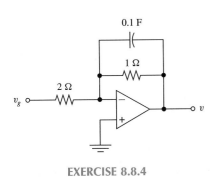

EXERCISE 8.8.4

The method of phasors permits efficient determination of the forced response in sinusoidal circuits while bypassing the difficulties of solving coupled differential equations. For stable circuits, the forced response is also the ac steady-state response. Ac steady-state reveals the long-term behavior of circuits excited by sinusoids, and as the excitation frequency varies, defines the circuit's frequency response.

- If a sinusoidal source is replaced by a complex exponential source whose real part is identical to original sinusoidal, the real part of the response will be the response to the original sinusoidal source.

- A phasor is the (generally complex) number multiplying $e^{j\omega t}$ in the forced response of a circuit excited by complex exponentials.

- The amplitude of a sinusoid is the magnitude of the phasor, and the phase angle of a sinusoid is the angle of its phasor.

- The impedance Z of an RLC element is the ratio of its voltage phasor $\mathbf{V}$ to its current phasor $\mathbf{I}$. $Z_R = R$, $Z_L = j\omega L$, and $Z_C = 1/j\omega C$.

- Admittance Y is the inverse of impedance, $Y = 1/Z$.

- A phasor circuit is identical to the original circuit, except that RLC elements are labeled by their impedances, and each sinusoidal source is replaced by the phasor of its corresponding complex exponential source.

- The phasor circuit may be analyzed using KVL, KCL, and $\mathbf{V} = Z\mathbf{I}$ exactly as if it were a dc circuit that happened to have constant but complex resistances and sources. There are no time-varying quantities in the phasor domain.

- The sinusoidal response is just the real part of the phasor response after multiplication by $e^{j\omega t}$.

- A stable circuit is one whose unforced responses decay to zero for all values of its initial conditions.

PROBLEMS

8.1. Consider two currents $i_1 = 4\cos(2t + 17°)$ and $i_2 = -3\sin(2t + 0.71)$. Find their amplitudes, frequencies, phase angles in both degrees and radians, and periods. Sketch both on the same plot, determining which leads and by how many degrees.

8.2. Find the quadrature representations of the two currents of Problem 8.1.

8.3. Using (8.8), find simple expressions for $\sin(\omega t + n\pi)$ and $\cos(\omega t + n\pi)$ where n is an integer. Show how your result follows from (8.8).

8.4. Suppose all currents and voltages have the same period. Find v_1 if $v_2 = 4\cos(3t - 70°)$, v_3 has the same amplitude as v_2 but leads v_2 by 28°, and v_4 has cosine quadrature component with amplitude 2 and sine component 0.

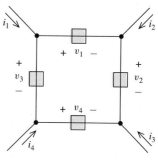

FIGURE P8.4

8.5. In the figure for Problem 8.4, is it true that the cosine quadrature components of the four currents sum to zero? The sine components? Justify.

8.6. Find the forced response i if $L = 4$ mH, $R = 6$ kΩ, $V_m = 5$ V, and $\omega = 2 \times 10^6$ rad/s.

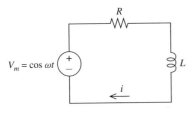

FIGURE P8.6

8.7. Write a differential equation for $i(t)$. Find the forced solution.

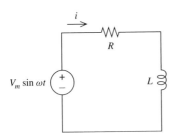

FIGURE P8.7

8.8. Write a differential equation for $v(t)$. Find the forced solution.

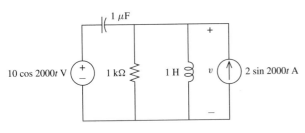

FIGURE P8.8

8.9. Write a single third-order differential equation for v_1 (*Hint*: Solve node equation 1 for v_2 and substitute into the other two node equations. Solve equation 3 for v_3.) Find the forced solution.

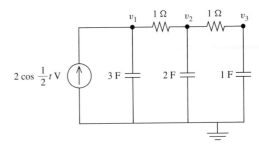

FIGURE P8.9

8.10. Sketch, in the complex plane, the locations of the complex exponential $(3/\underline{45°})e^{j4\pi t}$ at $t = 0$, $\frac{1}{8}$, $\frac{1}{4}$, $\frac{3}{8}$ and $\frac{1}{2}$ s.

8.11. If the complex exponential $Ae^{j\omega t}$ has value $2/\underline{80°}$ at $t = 0$ and first reaches $2/\underline{-100°}$ at $t = 2s$, what is its period? Find A and ω.

8.12. Express each complex number in three ways: rectangular form, polar form, and exponential polar form.
(a) $12/\underline{19°}$
(b) $-40 + j18$
(c) $\frac{1}{4}/\underline{-180°}$
(d) $4e^{j63°}$
(e) $16/\underline{90°}$
(f) $14 + j2$

8.13. Evaluate the following expressions. Give your answer in polar form.
(a) $\dfrac{(4+j2)(2e^{-j18°})}{(6-j)^2}$
(b) $\dfrac{3/\underline{20°}}{(2+j)+(3/\underline{120°})}$
(c) $(3 - j6)(3 - j6)^* + 4/\underline{18°}$
(d) $[(2 - j2) + (7e^{j45°})][6/\underline{-25°}]$

8.14. For each of the sinusoidal source functions specified, determine the complex exponential source that has given source as its *real* part.
(a) $4\cos 2t$
(b) $16\cos(100t - 26°)$
(c) $-3\sin(7t + 144°)$
(d) $2\cos(6t + 10°) + 3\sin(6t - 26°)$

8.15. If the forced response to an excitation $v_g = f_1(t) + jf_2(t)$ in some circuit is $i(t) = i_1(t) + ji_2(t)$, where f_1, f_2, i_1, and i_2 are all real functions, what would the response to $v_{g_1} = -f_1(t)$ be? To $v_{g_2} = 3f_2(t) + j[f_1(t) - f_2(t)]$?

8.16. Using complex excitations and superposition, show that if the sinusoidal source $v_g = A\cos(\omega t + \phi_1)$ is delayed by θ radians, the response $i(t) = \beta \cos(\omega t + \phi_2)$ to v_g is also delayed by θ radians.

8.17.

$$\frac{dx}{dt} + 2x = 4\cos 7t$$

Find the forced solution by replacing the sinusoidal forcing function by a complex exponential. Check by using the trial forced solution $A\cos 7t + B\sin 7t$.

8.18. Find i_4, using only the properties of sinusoids, if (a) $i_1 = 6\cos 3t$ A, $i_2 = 4\cos(3t - 30°)$ A, and $i_3 = -4\sqrt{3}\cos(3t + 60°)$ A, (b) $i_1 = 5\cos(3t + 30°)$ A, $i_2 = 5\sin 3t$ A, and $i_3 = 5\cos(3t + 150°)$ A, and (c) $i_1 = 25\cos(3t - 53.1°)$ A, $i_2 = 2\sin 3t$ A, and $i_3 = 13\cos(3t - 22.6°)$ A. (*Hint:* $\cos 22.6° = \frac{12}{13}$.)

FIGURE P8.18

8.19. If the phasor **V** associated with a certain voltage $v(t)$ in some circuit is $\mathbf{V} = V_m\underline{/\theta_V}$, what is the phasor if all independent sources are doubled? If all independent sources are lagged by 45°?

8.20.

$$\frac{d^2x}{dt^2} + \frac{dx}{dt} - 6x = 4\sin t - 2\cos(t - 27.1°)$$

Find the forced solutions by replacing the sinusoidal forcing function by a complex exponential. Check by using the trial forced solution $A\cos t + B\sin t$.

8.21. Write the describing equation as a second-order differential equation. Rewrite in terms of the phasor **I**. Solve for **I**, and then for the forced response $i(t)$.

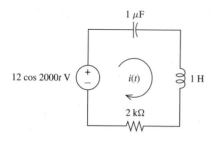

FIGURE P8.21

8.22. If $v(t) = 16\cos(4000t - 27°) + 3\sin(4000t - 114°)$ find the phasor associated with $v(t)$.

8.23. If $\omega = 20$ rad/s, find the time-domain functions represented by the phasors (a) $-5 + j5$, (b) $-4 - j3$, (c) $5 - j12$, (d) 10, and (e) $-j5$.

8.24. Find ω so that $i(t)$ lags $v(t)$ by 45°.

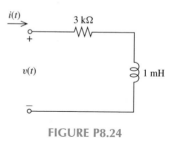

FIGURE P8.24

8.25. For what C will $i(t)$ lead $i_R(t)$ by 20° if the circuit is in steady state at frequency 2 kHz?

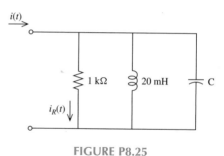

FIGURE P8.25

8.26. For phasor current $\mathbf{I} = 1\underline{/0}$ mA and $\omega = 1000$ rad/s, sketch $\mathbf{V}_R$, $\mathbf{V}_L$ and $\mathbf{V}_C$ in the complex plane.

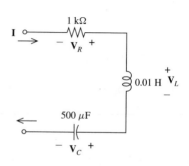

FIGURE P8.26

8.27. If $\omega = 1$ k rad, $\mathbf{V} = 20\underline{/16°}$ V and $\mathbf{I} = 6\underline{/-74°}$ A, what single element is in E (type and value)?

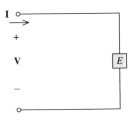

FIGURE P8.27

8.28. At what frequency ω are the magnitudes of the impedances of a 1-H inductor and 1-F capacitor equal?

8.29. Show that if Re $\mathbf{Z} = R$ is positive, then Re $\mathbf{Y} =$ Re$[1/\mathbf{Z}] = G$ is also positive.

8.30. For the phasor circuit shown, find Z_{eq} and use the result to find the phasor current $\mathbf{I}$. If $\omega = 7$ rad/s, find the forced response i corresponding to $\mathbf{I}$.

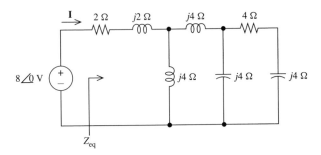

FIGURE P8.30

8.31. Find the reactance X so that the impedance seen by the source is real. For this case, find the steady-state current $i(t)$ corresponding to $\mathbf{I}$ if $\omega = 10$ rad/s.

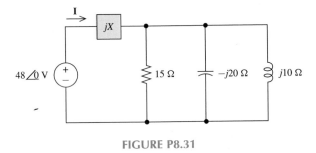

FIGURE P8.31

8.32. Find the impedance Z between a and b as a function of ω.

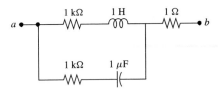

FIGURE P8.32

8.33. At what frequency ω is the susceptance B at terminals ab equal to $B = 2$ S?

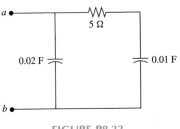

FIGURE P8.33

8.34. Find the steady-state value of i if (a) $\omega = 1$ rad/s and (b) $\omega = 2$ rad/s. Note that in the latter case the impedance seen at the terminals of the source is purely resistive.

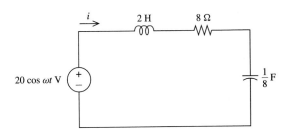

FIGURE P8.34

8.35. Convert all voltages to phasors. Use the fact that KVL holds for phasors to find $\mathbf{V}$, then $v(t)$.

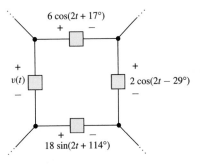

FIGURE P8.35

8.36. Repeat Problem 8.35 without using phasors.

8.37. Convert all currents to phasors. Use the fact that KCL holds for phasors to find **I**, then $i(t)$.

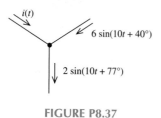

FIGURE P8.37

8.38. Find the steady-state value of v.

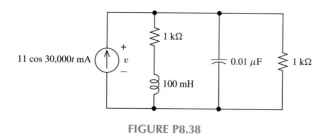

FIGURE P8.38

8.39. Find the steady-state values of v and v_1.

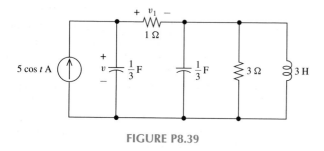

FIGURE P8.39

8.40. Find the steady-state values of i and v.

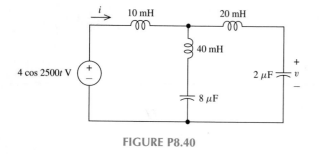

FIGURE P8.40

8.41. Find the steady-state current i.

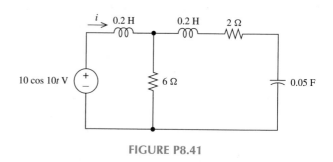

FIGURE P8.41

8.42. Find the steady-state voltage v.

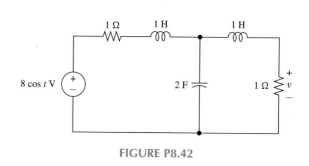

FIGURE P8.42

8.43. Find the steady-state value of i when (a) $\omega = 1$ rad/s, (b) $\omega = 2$ rad/s, and (c) $\omega = 4$ rad/s. [Note that (b) is the resonant case.]

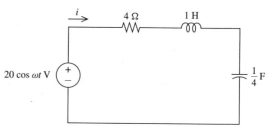

FIGURE P8.43

For Problems 8.44 through 8.48, sketch the phasor circuit and find the ac steady-state values of all indicated variables.

8.44.

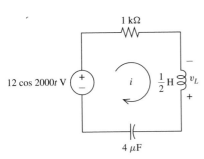

FIGURE P8.44

8.45.

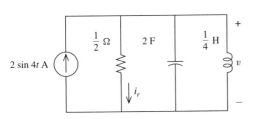

FIGURE P8.45

8.46.

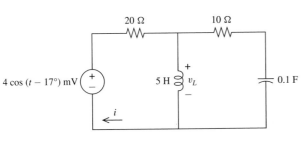

FIGURE P8.46

8.47.

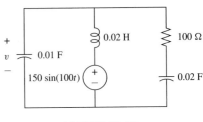

FIGURE P8.47

8.48.

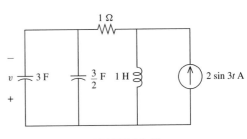

FIGURE P8.48

8.49. (a) Given a source $A \sin(\omega t + \phi)$ in an ac steady-state problem, suppose we use $A/\underline{\phi}$ as the corresponding phasor source (rather than $A/\underline{\phi - 90°}$ as prescribed in Table 8.1). It is no longer true that the ac steady-state response will be the real part of the complex response. How must this statement be modified, and why?
(b) Use this idea, sometimes called the "sine phasor" approach, to solve Problem 8.48 without using the 90° phase shift in the formula for the source phasor.

8.50. Find the *complete* response i if $i(0) = 2$ A and $v(0) = 6$ V. (*Suggestion:* Use phasors to get i_f and the differential equation to get i_n.)

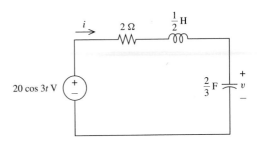

FIGURE P8.50

More Challenging Problems

8.51. Find i.

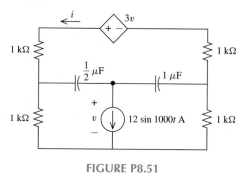

FIGURE P8.51

8.52. Find the ac steady-state value of $i(t)$ in terms of A and ϕ.

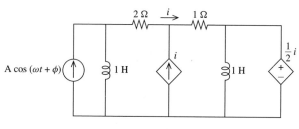

FIGURE P8.52

8.53. Suppose there is an element we call a *whizzer* which satisfies the element law $v = w\frac{d^2i}{dt^2}$. The constant w is called the whizzistance of the whizzer (units: scotts). What is the impedance of a w-scott whizzer? Derive laws for the equivalent whizzistance of whizzers in series and in parallel. The box is a 2-scott whizzer. Find the ac steady-state value of $v(t)$.

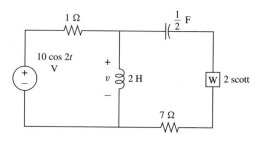

FIGURE P8.53

8.54. Find the ac steady-state value of $i_L(t)$.

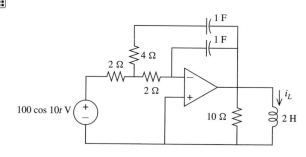

FIGURE P8.54

9

AC Steady-State Analysis

Thomas Alva Edison
1847–1931

Genius is one per cent inspiration and ninety-nine per cent perspiration.

Thomas A. Edison

The greatest American inventor and perhaps the greatest inventor in history was Thomas Alva Edison, who changed the lives of people everywhere with such inventions as the electric light and the phonograph. He patented over 1100 inventions of his own and improved many other persons' inventions, such as the telephone, the typewriter, the electric generator, and the motion picture. Perhaps most important of all, he was one of the first to organize research, at one time employing some 3000 helpers.

Edison was born in Milan, Ohio, the youngest of seven children. He had only three months of formal education because his mother took him out of school and taught him herself. He asked too many questions to get along with the schoolmaster. He was exempt from military service because of deafness, and during the Civil War he roamed from city to city as a telegraph operator. During this time he patented improvements on the stock ticker and sold the patents for the then astounding price of $40,000. In 1876 he moved to Menlo Park, New Jersey, and from there his steady stream of inventions made him world famous. The electric light was his greatest invention, but to supply it to the world he also designed the first electric power station. His discovery of the Edison effect, the movement of electrons in the vacuum of his light bulb, also marked the beginning of the age of electronics.

Chapter Contents

In Chapter 8 we saw that the steady-state response of circuits excited by sinusoidal sources (ac circuits) could be efficiently computed by converting the time-domain circuit to its corresponding phasor circuit. The circuits studied were relatively simple and were analyzed directly from Kirchhoff's laws and use of the basic notion of impedance. In the present chapter we move beyond these relatively simple circuits. We shall see, in fact, that all the tools of circuit analysis introduced thus far apply equally well to phasor circuits as to their time-domain counterparts. First we consider circuit simplification methods: series–parallel source and impedance equivalents, Thevenin–Norton equivalents, and current–voltage division may all be applied directly to phasor circuits. In the subsequent two sections the general methods of nodal and mesh analysis are shown to carry over to phasor circuits without change. In Section 9.4 the use of superposition to study circuits excited by multiple sources of different frequencies is developed. Finally, phasor diagrams are introduced as effective graphical aids to understanding phase relationships in ac steady state, and the use of SPICE in ac steady state is discussed.

9.1 CIRCUIT SIMPLIFICATIONS

In Chapter 2 several useful methods for simplifying resistive circuits were introduced. These include series–parallel equivalents for resistors and for sources, current–voltage division, and Thevenin–Norton equivalents. Each of these methods was derived from the same foundation: Ohm's law describing the I–V behavior of individual elements and Kirchhoff's current and voltage laws governing their interconnection.

In ac steady state we may replace the original circuit by its phasor circuit counterpart, in which the sinusoidal sources are represented by their corresponding phasor sources and RLC elements by their impedances Z. In the phasor circuit the same foundational laws are obeyed. Kirchhoff's laws apply equally well to phasors as to time-domain circuit variables, and the essence of Ohm's law, that current and voltage are linearly related through a simple constant of proportionality, holds equally well for impedances

$$\mathbf{V} = Z\mathbf{I} \tag{9.1}$$

where $\mathbf{V}$ and $\mathbf{I}$ are phasors. Since the laws upon which they are founded are equally valid for phasor circuits, we should anticipate that the same circuit simplifications derived for resistive circuits will hold for phasor circuits. Thus the full power of these methods may be brought to bear in ac steady-state circuit analysis, the only change being substitution of impedance Z for resistance R and the subsequent use of complex rather than real arithmetic.

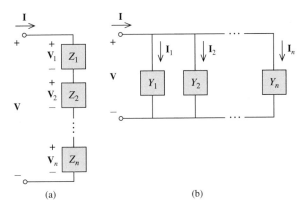

FIGURE 9.1 (a) Voltage division in phasor circuit; (b) current division.

We have already seen an important case of this principle in Chapter 8. Impedances in series were seen to be equivalent to a single impedance equal to the sum of the impedances, or more simply, series impedances add. Similarly, parallel admittances, in correspondence with parallel conductances in purely resistive circuits, also add. These equivalents may be used to derive the phasor form of current and voltage division. Consider the phasor circuit shown in Fig. 9.1(a). Since the n impedances are in series, $\mathbf{V} = Z\mathbf{I}$, where $Z = \mathbf{Z}_1 + \mathbf{Z}_2 + \cdots + \mathbf{Z}_n$. Then for each impedance, its voltage $\mathbf{V}_i = Z_i\mathbf{I}$, or

$$\mathbf{V}_i = \frac{Z_i}{Z_1 + Z_2 + \cdots + Z_n} \mathbf{V} \tag{9.2}$$

Comparing the voltage drops across two impedances gives Z_i and Z_j gives

$$\frac{\mathbf{V}_i}{\mathbf{V}_j} = \frac{Z_i}{Z_j} \tag{9.3}$$

The voltage across series impedances divides in direct proportion to their impedances. Similarly, consider Fig. 9.1(b), in which the elements are labeled by their admittances Y_i. Since parallel admittances add, from $\mathbf{I} = Y\mathbf{V}$ with $Y = Y_1 + Y_2 + \cdots + Y_n$, we have $\mathbf{I}_i = Y_i\mathbf{V}$, or

$$\mathbf{I}_i = \frac{Y_i}{Y_1 + Y_2 + \cdots + Y_n}\mathbf{I} \qquad (9.4)$$

and comparing two of these parallel currents,

$$\frac{\mathbf{I}_i}{\mathbf{I}_j} = \frac{Y_i}{Y_j} \qquad (9.5)$$

The current through parallel impedances divides in direct proportion to their admittances.

Example 9.1

We seek $\mathbf{V}_1$ and $\mathbf{I}_2$ in the phasor circuit of Fig. 9.2. The impedances for each element are shown in the circuit diagram. To get $\mathbf{V}_1$, we use voltage division. The admittance of the parallel elements is

$$Y = -j\tfrac{1}{10} + j\tfrac{1}{5} + \tfrac{1}{4} = \tfrac{1}{4} + j\tfrac{1}{10} \text{ S}$$

Thus the equivalent impedance of these elements is

$$Z = \frac{1}{Y} = 3.45 - j1.38 \ \Omega$$

Then, by (9.2),

$$\mathbf{V}_1 = \frac{2}{2 - j3 + (3.45 - j1.38)}(3\underline{/45^\circ})$$

$$= (0.286\underline{/38.8^\circ})(3\underline{/45^\circ})$$

$$= 0.858\underline{/83.8^\circ} \text{ V}$$

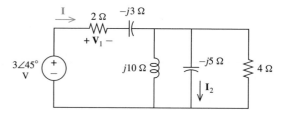

FIGURE 9.2   Circuit for Example 9.1.

We next use current division to get I_2. The current into the parallel combination is $\mathbf{I} = \mathbf{V}_1/2$ or $\mathbf{I} = 0.429\underline{/83.8°}$ A. Then, by (9.4),

$$\mathbf{I}_2 = \frac{j\frac{1}{5}}{\frac{1}{4} + j\frac{1}{10}}\,\mathbf{I}$$

$$= (0.743\underline{/68.2°})(0.429\underline{/83.8°})$$

$$= 0.319\underline{/152°}\ \text{A}$$

Series and parallel equivalents may be found for sources as well as impedances in phasor circuits. Application of KVL in Fig. 9.3(a) shows that voltage sources in series are equivalent to a single source whose source function is the sum of the individual source functions, or *series voltage sources add.* Using KCL in Fig. 9.3(b), we have the corresponding result: *parallel current sources add.*

Thevenin and Norton equivalents in phasor circuits are found exactly as described in Chapter 2 for resistive circuits, with only the substitution of impedance Z for resistance R and subsequent use of complex arithmetic. Following the development of Section 2.6 with only this change, we have, from (2.20), that the Thevenin and Norton forms shown in Fig. 9.4 are equivalent if the relations

$$\text{(a)} \quad Z_T = Z_N \tag{9.6a}$$

$$\text{(b)} \quad \mathbf{V}_T = Z_N \mathbf{I}_N \tag{9.6b}$$

hold between the circuits. To find the Thevenin or Norton equivalent of any two-terminal subcircuit of a phasor circuit, we follow the prescription of (2.24), again only replacing R by Z. The two-terminal circuit A with open-circuit phasor voltage $\mathbf{V}_{oc}$ and short-circuit phasor current $\mathbf{I}_{sc}$ shown in Fig. 9.5 is equivalent to the Thevenin and Norton forms

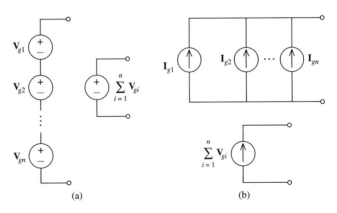

(a) (b)

FIGURE 9.3   (a) Series voltage sources and equivalent; (b) parallel current sources and equivalent.

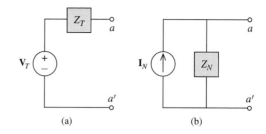

FIGURE 9.4 (a) Thevenin form; (b) Norton form.

shown in Fig. 9.4 if

$$(a)\ \mathbf{V}_T = \mathbf{V}_{oc}$$

$$(b)\ \mathbf{I}_N = \mathbf{I}_{sc}$$

$$(c)\ Z_T = Z_N = \frac{\mathbf{V}_{oc}}{\mathbf{I}_{sc}}$$

Z_T and Z_N can also be found as the impedance looking into terminals a–a' with all independent sources in the phasor circuit A killed.

Thevenin and Norton equivalents are used in phasor circuits to reduce complicated multielements circuits to simple two-element circuits, just as they were used in the resistive circuits studied previously. These equivalent circuits are even more generally useful in phasor circuits, since the restriction that the passive elements all be of the same type does not apply. Any mixture of RLC elements, that is, impedances, will do.

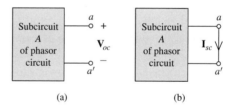

FIGURE 9.5 $\mathbf{V}_{oc}$ and $\mathbf{I}_{sc}$ for computing Thevenin and Norton equivalents of the circuit A.

Example 9.2

We wish to determine the value of R in Fig. 9.6(a) that will cause a sinusoidal current of amplitude 1 A to flow through this resistor. Our strategy will be to reduce this to the simple one-loop circuit shown in Fig. 9.6(b) by finding the Thevenin equivalent of everything except the resistor. The open-circuit voltage $\mathbf{V}_{oc} = \mathbf{V}_{12}$ with R removed is found by voltage division to be

$$\mathbf{V}_{oc} = \frac{j4(10\underline{/17^\circ})}{j4 - j2} = 20\underline{/17^\circ}\ \text{V}$$

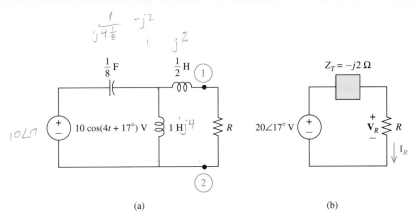

FIGURE 9.6 (a) Circuit for Example 9.2; (b) after Thevenin equivalent of circuit to left of terminals 1–2.

where $j4\ \Omega$ is the inductive impedance, $-j2\ \Omega$ is the capacitive, and the phasor source is $20/\underline{17^\circ}$ V. $\mathbf{V}_{oc}$ is the Thevenin source shown in Fig. 9.6(b). The short-circuit current is found by shorting terminals 1 and 2 together. The two inductors in this case are in parallel, and are equivalent to an inductance of $\frac{1}{3}$ H, with impedance

$$Z_L = j\omega L = j\tfrac{4}{3}\ \Omega$$

The common voltage across these inductors is, by voltage division,

$$\mathbf{V}_L = \frac{j4/3}{(j\frac{4}{3} - j2)(10/\underline{17^\circ})} = 20/\underline{-163^\circ}\ \text{V}$$

The current through the $\frac{1}{2}$-H inductor is the short-circuit current

$$\mathbf{I}_{sc} = \frac{\mathbf{V}_L}{j(4)(\frac{1}{2})} = \frac{20/\underline{-163^\circ}}{(2/\underline{90^\circ})} = 10/\underline{107^\circ}\ \text{A}$$

Thus Z_T in Fig. 9.6(b) is found to be

$$Z_T = \frac{\mathbf{V}_{oc}}{\mathbf{I}_{sc}} = \frac{20/\underline{17^\circ}}{10/\underline{107^\circ}} = 2/\underline{-90^\circ}\ \Omega$$

Having reduced the problem to one with a simple circuit, we turn to Fig. 9.6(b). The current through the resistor is

$$\mathbf{I}_R = \frac{20/\underline{17^\circ}}{R - j2}$$

The magnitude of the phasor $\mathbf{I}_R$ is given by the ratio of magnitudes of its numerator and denominator,

$$|\mathbf{I}_R| = \frac{20}{\sqrt{R^2 + 4}}$$

If $|\mathbf{I}_R| = 1$ A, the amplitude of the corresponding sinusoidal current i_R will be 1 A as well. Thus we require that

$$\frac{20}{\sqrt{R^2 + 4}} = 1$$

or $R = 6\sqrt{11}\ \Omega$.

EXERCISES

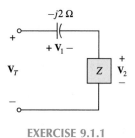

EXERCISE 9.1.1

9.1.1. (a) Find $\mathbf{V}_2$ for $Z = 1 + j2$ Ω and $\mathbf{V}_T = 1\underline{/0}$ V. (b) If $\mathbf{V}_1 = 3 + j$ Ω, for what Z will $\mathbf{V}_2 = 7 - j3$ V?
   Answer (a) $2.24\underline{/63.4°}$ V; (b) $4.82\underline{/-132°}$ V

9.1.2. (a) Find $\mathbf{I}_1$ for $Y_1 = 2 - j$ S and $\mathbf{I}_T = 1\underline{/0}$ A. (b) If $\mathbf{I}_3 = 3\underline{/18°}$ A, for what Y_1 will $\mathbf{I}_1 = 4 - j$ A?
   Answer (a) $0.35\underline{/24.8°}$ A; (b) $2.75\underline{/-32.0°}$ S

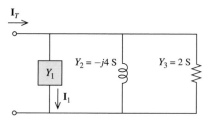

EXERCISE 9.1.2

9.1.3. Find the Norton equivalent of the circuit shown.
   Answer $Z_N = 3 + j4$ Ω; $\mathbf{I}_N = 2\underline{/10°}$ A

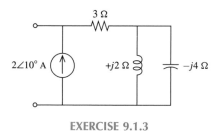

EXERCISE 9.1.3

9.1.4. Find the Thevenin equivalent of the circuit of Exercise 9.1.3.
   Answer $Z_T = 3 + j4$ Ω; $\mathbf{V}_T = 10\underline{/63.1°}$ V

9.2 NODAL ANALYSIS

As we have seen, the voltage–current relation (9.1)

$$\mathbf{V} = Z\mathbf{I}$$

for passive elements is identical in form to Ohm's law, and KVL and KCL hold in phasor circuits exactly as they did in resistive circuits. Therefore, the only difference in analyzing phasor circuits and resistive circuits is that the excitations and responses are complex quantities in the former case and real quantities in the latter case. Thus we may analyze phasor circuits in exactly the same manner in which we analyzed resistive circuits. Specifically, nodal and mesh analysis methods apply. We shall illustrate nodal analysis in this section and mesh analysis in the following section.

Example 9.3

To illustrate the nodal method, let us find the ac steady-state voltages v_1 and v_2 of Fig. 9.7. First we obtain the phasor circuit by replacing the element values by their impedances for $\omega = 2$ rad/s and the sources and node voltages by their phasors. This results in the circuit of Fig. 9.8(a). Since we are interested in finding $\mathbf{V}_1$ and $\mathbf{V}_2$, the node voltage phasors, we may replace the two sets of parallel impedances by their equivalent impedances, resulting in the simpler equivalent circuit of Fig. 9.8(b).

The nodal equations, from Fig. 9.8(b), are

$$2(\mathbf{V}_1 - 5\underline{/0}) + \frac{\mathbf{V}_1}{j1} + \frac{\mathbf{V}_1 - \mathbf{V}_2}{-j1} = 0$$

$$\frac{\mathbf{V}_2 - \mathbf{V}_1}{-j1} + \frac{\mathbf{V}_2}{(1+j2)/5} = 5\underline{/0}$$

which in simplified form are

$$(2+j2)\mathbf{V}_1 - j1\mathbf{V}_2 = 10$$

$$-j1\mathbf{V}_1 + (1-j1)\mathbf{V}_2 = 5$$

Solving these equations by Cramer's rule, we have

$$\mathbf{V}_1 = \frac{\begin{vmatrix} 10 & -j1 \\ 5 & 1-j1 \end{vmatrix}}{\begin{vmatrix} 2+j2 & -j1 \\ -j1 & 1-j1 \end{vmatrix}} = \frac{10-j5}{5} = 2 - j1 \text{ V}$$

$$\mathbf{V}_2 = \frac{\begin{vmatrix} 2+j2 & 10 \\ -j1 & 5 \end{vmatrix}}{5} = \frac{10+j20}{5} = 2 + j4 \text{ V}$$

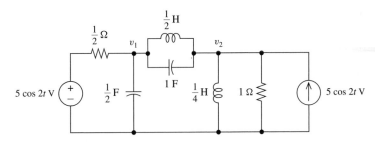

FIGURE 9.7 Circuit to be analyzed by the phasor method.

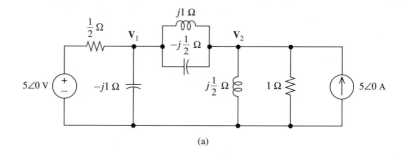

(a)

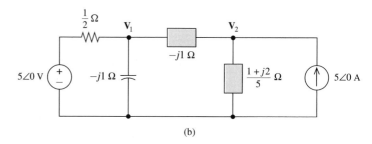

(b)

FIGURE 9.8 Two versions of the phasor circuit corresponding to Fig. 9.7.

In polar form these quantities are

$$\mathbf{V}_1 = \sqrt{5}\underline{/-26.6°} \text{ V}$$

$$\mathbf{V}_2 = 2\sqrt{5}\underline{/63.4°} \text{ V}$$

Therefore, the time-domain solutions are

$$v_1 = \sqrt{5}\cos(2t - 26.6°) \text{ V}$$

$$v_2 = 2\sqrt{5}\cos(2t + 63.4°) \text{ V}$$

Example 9.3 illustrates an important feature of phasor analysis. By reducing element laws for storage elements that involve integrals and derivatives to the simpler $\mathbf{V} = \mathbf{ZI}$, we replace *calculus* by *algebra*. The subsequent nodal or mesh equations may always be formulated as a single vector-matrix equation and solved by the general methods of linear algebra rather than the more complicated methods involving differential equations. Of course, for simple circuits we need not rewrite as vector-matrix equations if we choose, but this general method will always be available for all phasor analysis problems.

Example 9.4

As an example involving a dependent source, let us consider Fig. 9.9, in which it is required to find the forced response i. Taking the ground node as shown, we have the two unknown node voltages v and $v + 3000i$, as indicated. The phasor circuit in its simplest form is shown in Fig. 9.10, from which we may observe that only one nodal equation is needed. Writing KCL at the supernode, shown

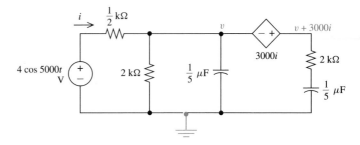

FIGURE 9.9 Circuit containing a dependent source.

dashed, we have

$$\frac{\mathbf{V} - 4}{\frac{1}{2}(10^3)} + \frac{\mathbf{V}}{\frac{2}{5}(1 - j2)(10^3)} + \frac{\mathbf{V} + 3000\mathbf{I}}{(2 - j1)(10^3)} = 0$$

Note that the gain of the dependent source is 3000 V/A. Also, from the phasor circuit we have

$$\mathbf{I} = \frac{4 - \mathbf{V}}{\frac{1}{2}(10^3)}$$

Eliminating $\mathbf{V}$ between these two equations and solving for $\mathbf{I}$, we have

$$\mathbf{I} = 24 \times 10^{-3} \underline{/53.1^\circ} \ \text{A}$$

$$= 24 \underline{/53.1^\circ} \ \text{mA}$$

Therefore, in the time domain, we have

$$i = 24\cos(5000t + 53.1^\circ) \ \text{mA}$$

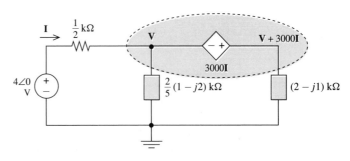

FIGURE 9.10 Phasor circuit of Fig. 9.9.

Example 9.5

As a final example illustrating nodal analysis, let us find the forced response v in Fig. 9.11 if

$$v_g = V_m \cos \omega t \ \text{V}$$

We note first that the op amp and the two 2-kΩ resistors constitute a noninverting amplifier with gain $1 + \frac{2000}{2000} = 2$ (see Section 3.5).

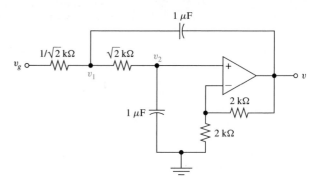

FIGURE 9.11 Circuit containing an op amp.

Therefore, $v = 2v_2$, or $v_2 = v/2$, as indicated by the phasor $\mathbf{V}/2$ in the phasor circuit of Fig. 9.12.

Writing nodal equations at the nodes labeled $\mathbf{V}_1$ and $\mathbf{V}/2$, we have

$$\frac{\mathbf{V}_1 - V_m\underline{/0}}{(1/\sqrt{2})(10^3)} + \frac{\mathbf{V}_1 - (\mathbf{V}/2)}{\sqrt{2}(10^3)} + \frac{\mathbf{V}_1 - \mathbf{V}}{-j\,10^6/\omega} = 0$$

$$\frac{(\mathbf{V}/2) - \mathbf{V}_1}{\sqrt{2}(10^3)} + \frac{\mathbf{V}/2}{-j\,10^6/\omega} = 0$$

Eliminating $\mathbf{V}_1$, by solving the last for $\mathbf{V}_1$ and substituting into the previous equation, and solving for $\mathbf{V}$ results in

$$\mathbf{V} = \frac{2V_m}{[1 - (\omega^2/10^6)] + j\,(\sqrt{2}\omega/10^3)}$$

In polar form this is

$$\mathbf{V} = \frac{2V_m\underline{/\theta}}{\sqrt{1 + (\omega/1000)^4}} \tag{9.7}$$

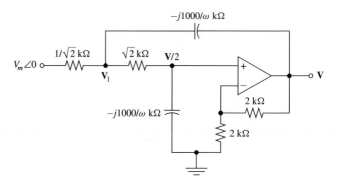

FIGURE 9.12 Phasor circuit of Fig. 9.11.

where

$$\theta = -\tan^{-1} \frac{\sqrt{2}\omega/1000}{1 - (\omega/1000)^2} \tag{9.8}$$

In the time domain we have

$$v = \frac{2V_m}{\sqrt{1 + (\omega/1000)^4}} \cos(\omega t + \theta) \tag{9.9}$$

We might note in this example that for low frequencies, say $0 < \omega < 1000$, the amplitude of the output voltage v is relatively large, and for higher frequencies, its amplitude is relatively small. Thus the circuit of Fig. 9.11 *filters* out higher frequencies and allows lower frequencies to "pass." Such a circuit is called a *filter* and is considered in more detail in Chapter 14.

EXERCISES

9.2.1. Find the forced response v using nodal analysis.
Answer $10 \sin 3t$ V

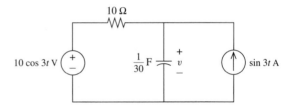

EXERCISE 9.2.1

9.2.2. Find the steady-state value of v using nodal analysis.
Answer $25\sqrt{2}\cos(2t - 81.9°)$ V

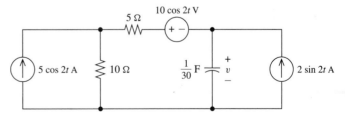

EXERCISE 9.2.2

9.2.3. Find the amplitude of v in Example 9.5 if $V_m = 10$ V for (a) $\omega = 0$, (b) $\omega = 1000$ rad/s, (c) $\omega = 10,000$ rad/s, and (d) $\omega = 100,000$ rad/s.
Answer (a) 20; (b) 14.14; (c) 0.2; (d) 0.002 V

Find the steady-state value of v using nodal analysis.

Answer $3\sqrt{2}\cos(2t - 135°)$ V

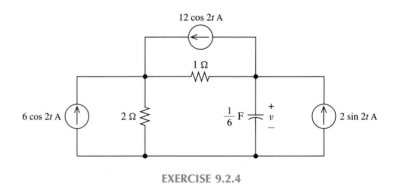

12 cos 2t A

1 Ω

6 cos 2t A   2 Ω $\frac{1}{6}$ F $+\ v\ -$ 2 sin 2t A

EXERCISE 9.2.4

9.3 MESH ANALYSIS

As suggested in Section 9.2, the general methods of circuit analysis apply to phasor circuits as if they were resistive circuits, with resistance R replaced by the more general impedance Z. In this section we illustrate the application of mesh analysis to phasor circuits. Once again we note that the generalized form of Ohm's law using impedance $\mathbf{V} = \mathbf{ZI}$ permits replacement of calculus (differentiation and integration) by the simpler operations of algebra (multiplication and division).

Example 9.6

To illustrate mesh analysis of an ac steady-state circuit, let us find v_1 in Fig. 9.7, which was obtained, using nodal analysis, in Section 9.2. We shall use the phasor circuit of Fig. 9.8(b), which is redrawn in Fig. 9.13, with mesh currents $\mathbf{I}_1$ and $\mathbf{I}_2$, as indicated. Once the circuit is broken, the phasor voltage $\mathbf{V}_1$ may be obtained as

$$\mathbf{V}_1 = 5 - \frac{\mathbf{I}_1}{2} \qquad (9.10)$$

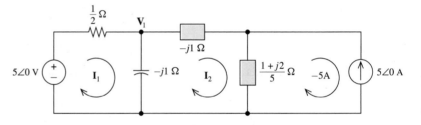

$\frac{1}{2}$ Ω $\mathbf{V}_1$ $-j1$ Ω

$5\angle0$ V $\mathbf{I}_1$ $-j1$ Ω $\mathbf{I}_2$ $\frac{1+j2}{5}$ Ω $-5A$ $5\angle0$ A

FIGURE 9.13 Circuit of Fig. 9.8 redrawn for mesh analysis.

The two mesh equations are

$$\frac{1}{2}\mathbf{I}_1 - j1(\mathbf{I}_1 - \mathbf{I}_2) = 5$$

$$-j1(\mathbf{I}_2 - \mathbf{I}_1) - j1\mathbf{I}_2 + \left(\frac{1+j2}{5}\right)(\mathbf{I}_2 + 5) = 0 \qquad (9.11)$$

Solving these equations for $\mathbf{I}_1$, we have

$$\mathbf{I}_1 = 6 + j2 \text{ A}$$

which substituted into (9.10) yields

$$\mathbf{V}_1 = 2 - j1 \text{ V}$$

This is the same result that was obtained in Section 9.2 and may be used to obtain the time-domain voltage v_1.

The same shortcut procedures for writing nodal and mesh equations discussed in Sections 4.3 and 4.5 for resistive circuits apply to phasor circuits. For example, in Fig. 9.13, if $\mathbf{I}_3 = -5$ is the mesh current in the right mesh in the clockwise direction, the two mesh equations are written by inspection as

$$\left(\frac{1}{2} - j1\right)\mathbf{I}_1 - (-j1)\mathbf{I}_2 = 5$$

$$-(-j1)\mathbf{I}_1 + \left(-j1 - j1 + \frac{1+j2}{5}\right)\mathbf{I}_2 - \left(\frac{1+j2}{5}\right)\mathbf{I}_3 = 0$$

These are equivalent to (9.11) and are formed exactly as in the resistive circuit case. That is, in the first equation the coefficient of the first variable is the sum of the impedances around the first mesh. The other coefficients are the negatives of the impedances common to the first mesh and the meshes whose numbers correspond to the currents. The right member is the sum of the voltage sources in the mesh with polarities consistent with the direction of the mesh current. Replacing "first" by "second" applies to the next equation, and so on. The dual development, as described in Section 4.3, holds for nodal equations.

Example 9.7

This example illustrates that series–parallel impedance conversions, together with Thevenin–Norton transformations, may be of great benefit in simplifying a circuit before writing the general analysis equations (nodal or mesh). We shall apply mesh analysis to the circuit of Fig. 9.14(a), in which the desired response is the voltage v across the 2-H inductor. Since this circuit contains six meshes and one current source, simultaneous solution of five equations in five unknowns would be required if the mesh equations were written directly. By series–parallel conversions, the circuit may be redrawn

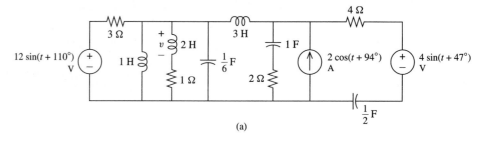

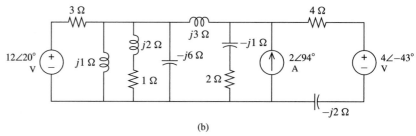

FIGURE 9.14 (a) Circuit for Example 9.7; (b) corresponding phasor circuit.

as in Fig. 9.15(a), where

$$Z_1 = j1||(1 + j2)||(-j6) = 0.128 + j0.79 \ \Omega$$

$$Z_2 = 2 - j1 \ \Omega$$

$$Z_3 = 4 - j2 \ \Omega$$

We have also converted the series combination of Z_3 and the $4/{-43°}$ V voltage source in Fig. 9.14(b) to the Norton form shown

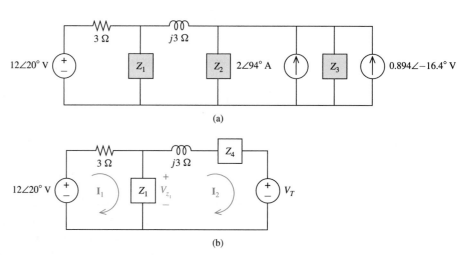

FIGURE 9.15 (a) Simplified phasor circuit; (b) further simplified.

in Fig. 9.15(a), where

$$\mathbf{I}_N = \frac{4\underline{/-43^\circ}}{Z_3} = 0.894\underline{/-16.4^\circ} \text{ A}$$

As a final simplification, we combine the parallel impedances Z_2 and Z_3 and the parallel current sources, and convert the result back to the Thevenin form, leading to the circuit shown in Fig. 9.15(b).

$$Z_4 = Z_2 \| Z_3 = \frac{(2-j1)(4-j2)}{6-j3} = \frac{4}{3} - j\frac{2}{3} \text{ } \Omega$$

$$\mathbf{V}_T = Z_4(2\underline{/94^\circ} + 0.894\underline{/-16.4^\circ}) = 2.81\underline{/41^\circ} \text{ V}$$

The mesh equations, in vector-matrix form, are

$$\begin{bmatrix} 3.13 + j0.79 & -0.128 - j0.79 \\ -0.128 - j0.79 & 1.45 + j3.12 \end{bmatrix} \begin{bmatrix} \mathbf{I}_1 \\ \mathbf{I}_2 \end{bmatrix} = \begin{bmatrix} 12\underline{/20^\circ} \\ 2.81\underline{/-139^\circ} \end{bmatrix}$$

The determinant of the matrix is

$$\Delta = (3.13 + j0.79)(1.45 + j3.12) - (-0.128 - j0.79)^2 = 11.1\underline{/75.8^\circ}$$

Thus

$$\begin{bmatrix} \mathbf{I}_1 \\ \mathbf{I}_2 \end{bmatrix}$$

$$= \frac{1}{11.1\underline{/75.8^\circ}} \begin{bmatrix} 1.45 + j3.12 & 0.128 + j0.79 \\ 0.128 - j0.79 & 3.13 + j0.79 \end{bmatrix} \begin{bmatrix} 12\underline{/20^\circ} \\ 2.81\underline{/-139^\circ} \end{bmatrix}$$

$$= \begin{bmatrix} 3.58\underline{/7.2^\circ} \\ 0.657\underline{/88.3^\circ} \end{bmatrix}$$

Having broken the circuit by finding the mesh currents, the desired unknowns are next found in terms of the mesh currents. The voltage across Z_1 in Fig. 9.15(b) is

$$\mathbf{V}_{Z1} = Z_1(\mathbf{I}_1 - \mathbf{I}_2) = (0.128 + j0.79)(3.58\underline{/7.2^\circ} - 0.657\underline{/88.3^\circ})$$

$$= 2.83\underline{/77.4^\circ} \text{ V}$$

From Fig. 9.14 we see that this voltage is across an impedance of $j2 \text{ } \Omega$ (the desired voltage, $\mathbf{V}$) in series with another impedance of $1 \text{ } \Omega$. By voltage division,

$$\mathbf{V} = \frac{j2}{1+j2} 2.83\underline{/77.4^\circ} = 2.53\underline{/104^\circ} \text{ V}$$

which is the desired unknown phasor. The final time-domain result is

$$v(t) = 2.53\cos(t + 104^\circ) \text{ V}$$

While some effort was required to arrive at this result, it should be remembered that the original circuit had a total of 10 passive elements and 3 independent sources distributed in 6 meshes. Had we tried to solve the phasor circuit directly with no simplifications, or worse had we not used phasors at all, much more computational effort would have been required.

Example 9.8

As a final example, let us consider the circuit of Fig. 9.16(a), where the response is the steady-state value of v_1. The controlled source has transconductance $g = 2$ S. The phasor circuit is shown in Fig. 9.16(b), with the mesh currents as indicated.

Applying KVL around the supermesh labeled $\mathbf{I}$, we have

$$-\mathbf{V}_1 - j1(-j1 + \mathbf{I}) + (1 + j2)(\mathbf{I} + 2\mathbf{V}_1) = 0$$

Also, from the figure we see that

$$\mathbf{V}_1 = j1(4 - \mathbf{I})$$

Eliminating $\mathbf{I}$ from these equations and solving for $\mathbf{V}_1$, we have

$$\mathbf{V}_1 = \frac{-4 + j3}{5} = 1\underline{/143.1°} \text{ V}$$

Therefore, in the time domain, the voltage is

$$v_1 = \cos(2t + 143.1°) \text{ V}$$

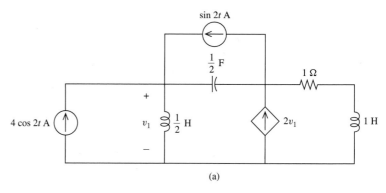

(a)

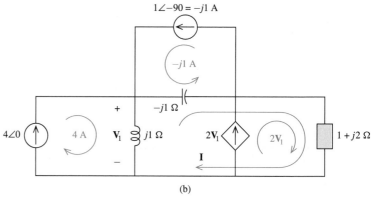

(b)

FIGURE 9.16 (a) Time-domain circuit; (b) phasor counterpart.

EXERCISES

9.3.1. Find the forced response i in Fig. 9.9 using mesh analysis.

9.3.2. Solve Exercise 9.2.4 using mesh analysis.

9.3.3. Find the steady-state current i using mesh analysis.
Answer $\sqrt{2}\cos(2t - 45°)$ A

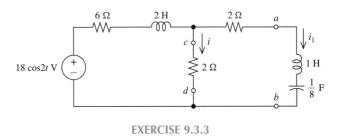

EXERCISE 9.3.3

9.4 SOURCES WITH DIFFERENT FREQUENCIES

An ac circuit is, we recall, a circuit whose independent sources are all sinusoids. If all sources in an ac circuit are of the same frequency ω, the corresponding phasor circuit may be used to determine the forced response in the manner described in Sections 9.1 to 9.3. It may be computed by a single unified nodal or mesh analysis or by use of superposition (summing the component responses to each individual source or group of sources with all other independent sources killed).

Superposition is a general principle that may always be used to find the response of a linear circuit containing more than one source. Even if an ac circuit contains sources with different frequencies, superposition may still be used to find the forced response. For purposes of superposition, the sources are grouped so that each component problem contains only sources of a single frequency. Then, for each resulting component problem, a phasor circuit may be used to determine the phasor response, which is then converted to a sinusoidal response and added with the other component responses as the principle of superposition requires.

Example 9.9

To find v in Fig. 9.17(a), we will use superposition. The component phasor circuits are shown in Fig. 9.17(b) and (c). Note that the impedances in the two component problems are different for all elements but the resistors, since for the other (storage) elements impedance is frequency dependent. By voltage division,

$$\mathbf{V}_1 = -\left(\frac{-j1}{-j1 + Z_a}\right)(12\underline{/0}) \qquad (9.12)$$

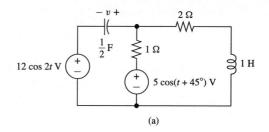

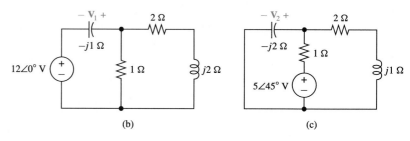

FIGURE 9.17 (a) Ac circuit; (b) phasor circuit for the $\omega = 2$ component; (c) phasor circuit for the $\omega = 1$ component.

where Z_a is the equivalent of the rightmost three impedances in Fig. 9.17(b):

$$Z_a = \frac{(1)(2+j2)}{3+j2} = 0.769 + j0.154 \ \Omega$$

Using this in (9.12), $\mathbf{V}_1 = 10.5\underline{/138°}$ V. Thus

$$v_1 = 10.5\cos(2t + 138°) \ \text{V}$$

Turning to Fig. 9.17(c), again by voltage division,

$$\mathbf{V}_2 = \frac{Z_b}{Z_b + 1} \ (5\underline{/45°})$$

where Z_b is the parallel equivalent of the impedances $-j2$ and $2 + j1$, or

$$Z_b = \frac{(-j2)(2+j1)}{2-j1} = 1.6 - j1.2 \ \Omega$$

which, upon substitution, yields

$$\mathbf{V}_2 = \frac{1.6 - j1.2}{2.6 - j1.2}(5\underline{/45°}) = 3.5\underline{/32.9°} \ \text{V}$$

The corresponding sinusoidal component is

$$v_2 = 3.5\cos(t + 32.9°) \ \text{V}$$

and by superposition, the forced response v is the sum of its components:

$$v = v_1 + v_2 = 10.5\cos(2t + 138°) + 3.5\cos(t + 32.9°) \ \text{V}$$

Note in Example 9.9 that $\mathbf{V}_1$ and $\mathbf{V}_2$ were converted to sinusoids before being added together. *Phasors corresponding to different frequencies cannot be superposed; only their corresponding sinusoids can be superposed.* Recall that a given phasor corresponds to amplitude and phase information on a sinusoid of a specific frequency ω. If the frequencies of two phasors differ, it makes no sense to add them together. The magnitude of their phasor sum does not correspond to any sinusoidal amplitude, nor does the angle of their sum to any sinusoidal phase angle. When frequencies differ, the principle of superposition applies to the summing of time-domain components, not phasors. Within a component problem corresponding to a single frequency, however, phasors may also be superposed. This is illustrated in the next example.

Example 9.10

We seek the current i through the voltage source in the ac circuit shown in Fig. 9.18(a). We will use superposition, grouping the two sources at $\omega = 10$ rad/s together and calling this component of the response i_1; the remaining component due to the source at $\omega = 5$ rad/s is i_2. Figure 9.18(b) shows the phasor circuit for computing i_1 and Fig. 9.18(c) that for i_2. In each case the other sources have been killed, and the value of ω corresponding to the active sources used to compute the impedances. Using superposition, each component problem may be set up and solved independently.

The right mesh equation in Fig. 9.18(b) is

$$\mathbf{I}_1(-j2 + 1 + j4) + j2(\mathbf{I}_1 - 4\underline{/0}) = 5\underline{/30°}\ \text{A}$$

or

$$\mathbf{I}_1 = \frac{5\underline{/30°} + j8}{1 + j4} = 2.76\underline{/-8.4°}\ \text{A}$$

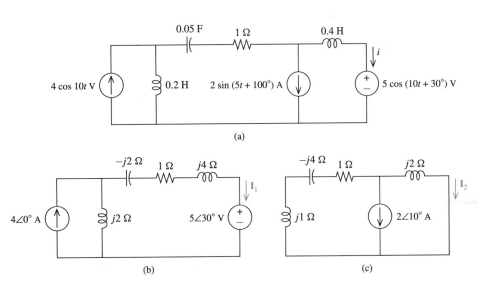

(a)

(b)   (c)

FIGURE 9.18   (a) Ac circuit; (b) phasor circuit for the $\omega = 10$ component; (c) phasor circuit for the $\omega = 5$ component.

The corresponding ac component is

$$i_1 = 2.76 \cos(10t - 8.4°) \text{ A}$$

Turning to the other component, use of current division in Fig. 9.18(c) yields

$$\mathbf{I}_2 = \frac{1 - j3}{1 - j1}(-2\underline{/10°}) = 4.47\underline{/163°} \text{ A}$$

The second ac component is then

$$i_2 = 4.47 \cos(5t + 163°) \text{ A}$$

The overall forced response, or ac steady-state response, is the sum of these components:

$$i = i_1 + i_2 = 2.76 \cos(10t - 8.4°) + 4.47 \cos(5t + 163°) \text{ A}$$

In Example 9.10 we computed one component for each source frequency and summed these components. Note that since there are two sources in the phasor circuit Fig. 9.18(b) corresponding to the frequency $\omega = 10$ rad/s, we may choose to solve this component problem by once again invoking the principle of superposition. We could compute the subcomponents of $\mathbf{I}_1$ due to each of the two sources with the other killed and then sum these two phasors to get $\mathbf{I}_1$. Superposition may be freely applied within a single-phasor circuit, that is, one in which there is a single frequency ω.

In this section we have seen that superposition may be used to decompose ac steady-state problems involving independent sources at more than one frequency into component problems each containing sources at the same frequency. The component problems may then be solved with the help of phasors. In other instances thus far where superposition was discussed, it was also possible to choose to bypass superposition, solving in a single unified manner with all sources in place. This same choice is also available for ac circuits with multiple frequencies; but if we are to use phasors, we cannot choose to solve the problem all at once. The difficulty is that a single unified phasor circuit involving sources of different frequencies cannot be defined.

Phasors are defined as those complex numbers $\mathbf{I}$ and $\mathbf{V}$ used to specify currents and voltages when they are of the specific form $i = \mathbf{I}e^{j\omega t}$ and $v = \mathbf{V}e^{j\omega t}$. In circuits with distinct frequencies ω_1 and ω_2, the currents and voltages will not be of the required form. They will in fact be the sums of complex exponentials at each of the different frequencies. Thus, in ac circuits containing sources with different frequencies, we cannot hope to define a unified overall phasor associated with a given current or voltage. If we were to try to do so by coverting the original circuit to a phasor circuit containing sources at both ω_1 and ω_2, the dilemma would be apparent when we tried to assign values to the impedances. What would we use as the value of ω in computing impedance values for those elements whose impedance depends on ω? Clearly, neither ω_1 nor ω_2 by itself would do. A single phasor circuit with sources at different frequencies is meaningless. *For sources at different frequencies, superposition is not just a choice; it must be used to determine the ac steady-state response.*

EXERCISES

9.4.1. Find the steady-state current i.

Answer $2\cos(2t - 36.9°) + 3\cos(t + 73.8°)$ A

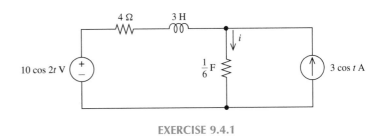

EXERCISE 9.4.1

9.4.2. For the phasor circuit corresponding to Exercise 9.3.3, replace the part to the left of terminals a–b by its Thevenin equivalent and find the steady-state current i_1.

Answer $\mathbf{V}_{oc} = \frac{9}{5}(2 - j1)$ V, $Z_{th} = \frac{1}{5}(18 + j1)$ Ω, $i_1 = \cos 2t$ A

9.4.3. Find $\mathbf{V}_1$, $\mathbf{I}_1$, and $\mathbf{I}_2$. Use the ladder method, assuming $\mathbf{V} = 1\underline{/0}$.

Answer 3 V; $3 - j3$ A; 3 A

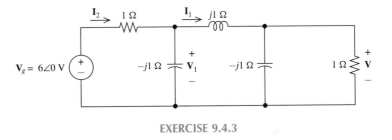

EXERCISE 9.4.3

9.5 PHASOR DIAGRAMS

Since phasors are complex numbers, they may be represented by vectors in a plane, where operations such as addition of phasors may be carried out geometrically. Such a sketch is called a *phasor diagram* and may be helpful in analyzing ac steady-state circuits.

Example 9.11

To illustrate, let us consider the phasor circuit of Fig. 9.19, for which we shall draw all the voltages and currents on a phasor diagram. To begin, let us observe that the current $\mathbf{I}$ is common to all elements and take it as our *reference* phasor, denoting it by

$$\mathbf{I} = |\mathbf{I}|\underline{/0}$$

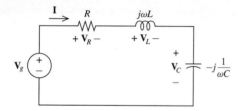

FIGURE 9.19   *RLC* series phasor circuit.

We have taken the angle of **I** arbitrarily to be zero, since we want **I** to be our reference. We may always adjust this assumed value to the true value by the proportionality principle, which permits the addition of a constant phase shift θ to every current and voltage (scaling of every phasor by $1\underline{/\theta}$).

The voltage phasors of the circuit are

$$\mathbf{V}_R = R\mathbf{I} = R|\mathbf{I}|$$

$$\mathbf{V}_L = j\omega L\mathbf{I} = \omega L|\mathbf{I}|\underline{/90°}$$

$$\mathbf{V}_C = -j\frac{1}{\omega C}\mathbf{I} = \frac{1}{\omega C}|\mathbf{I}|\underline{/-90°}$$

$$\mathbf{V}_g = \mathbf{V}_R + \mathbf{V}_L + \mathbf{V}_C$$

These are shown in the phasor diagram of Fig. 9.20(a), where it is assumed that $|\mathbf{V}_L| > |\mathbf{V}_C|$. The cases $|\mathbf{V}_L| < |\mathbf{V}_C|$ and $|\mathbf{V}_L| = |\mathbf{V}_C|$ are shown in Fig. 9.20(b) and (c), respectively. In all cases the lengths representing the units of current and voltage are not necessarily the same, so for clarity **I** is shown longer than $\mathbf{V}_R$.

Recall that the phasor diagram is a snapshot at time $t = 0$ of circular motion in the complex plane. All vectors rotate together in the counterclockwise direction; thus a more counterclockwise phasor leads a less counterclockwise one in this motion.

In case (a) the net reactance is inductive, and the current lags the source voltage by the angle θ that can be measured. In (b) the circuit has a net capacitive reactance, and the current leads the voltage. Finally, in (c) the current and voltage are in phase, since the inductive and capacitive reactance components exactly cancel each

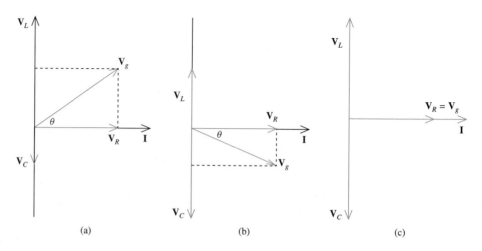

FIGURE 9.20   Phasor diagrams for Fig. 9.19.

other. These conclusions follow also from the equation

$$\mathbf{I} = \frac{\mathbf{V}_g}{Z} = \frac{\mathbf{V}_g}{R + j[\omega L - (1/\omega C)]} \tag{9.13}$$

Case (c) is characterized by

$$\omega L - \frac{1}{\omega C} = 0$$

or

$$\omega = \frac{1}{\sqrt{LC}} \tag{9.14}$$

If the current in Fig. 9.19 is fixed, the real component of the voltage $\mathbf{V}_g$ is fixed, since it is $R|\mathbf{I}|$. In this case the *locus* of the phasor $\mathbf{V}_g$ (its possible location on the phasor diagram) is the dashed line of Fig. 9.21. The voltage phasor varies up and down this line as ω varies between zero and infinity. The minimum amplitude of the voltage occurs when $\omega = 1/\sqrt{LC}$, as seen from the figure. For any other frequency, a larger voltage is required to produce the same current.

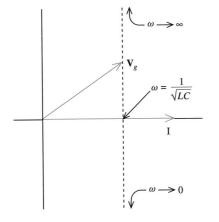

FIGURE 9.21   Locus of the voltage phasor for a fixed current response.

Example 9.12

We wish to determine $\mathbf{V}_C$ in Fig. 9.22 and to visualize the phase relationships among all the currents and voltages. We will use $\mathbf{V}_C$ as our reference phasor, and since its magnitude is also unknown, we will temporarily set $\mathbf{V}_C = 1/\underline{0}$ as shown in Fig. 9.22(b). Since the current through a capacitor leads its voltage by 90°, $\mathbf{I}_3$ must lead $\mathbf{V}_C$ by 90° as shown in the phasor diagram. Using the reference value for $\mathbf{V}_C$,

$$\mathbf{I}_3 = \frac{\mathbf{V}_C}{-j2} = \frac{-1}{j2} = \frac{1}{2}\underline{/90°}$$

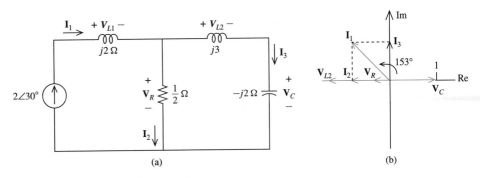

FIGURE 9.22   (a) Phasor circuit; (b) phasor diagram using $\mathbf{V}_C$ as the reference phasor.

$\mathbf{V}_{L2}$, the voltage across an inductor, leads its current $\mathbf{I}_3$ by 90°, so it falls on the negative real axis in the phasor diagram as shown.

$$\mathbf{V}_{L2} = (j3)\left(\tfrac{1}{2}\underline{/90°}\right) = \tfrac{3}{2}\underline{/180°}$$

$\mathbf{V}_R$ is, by KVL, the vector sum of $\mathbf{V}_C$ and $\mathbf{V}_{L2}$. Since they both lie on the real axis the sum must as well:

$$\mathbf{V}_R = 1\underline{/0} + \tfrac{3}{2}\underline{/180°} = \tfrac{1}{2}\underline{/180°}$$

Current and voltage in a resistor are in phase, so $\mathbf{I}_2$ has the same angle in the phasor diagram as $\mathbf{V}_R$:

$$\mathbf{I}_2 = 2\mathbf{V}_R = 1\underline{/180°}$$

Then since $\mathbf{I}_1$ is the vector sum of $\mathbf{I}_2$ and $\mathbf{I}_3$ as shown in the phasor diagram,

$$\mathbf{I}_1 = 1\underline{/180°} + \tfrac{1}{2}\underline{/90°} = 1.12\underline{/153°}$$

Thus we see that $\mathbf{V}_C = 1\underline{/0}$, together with all the other phasors we computed based on this reference phasor, would have been produced if the current source were of value $1.12\underline{/153°}$. By the proportionality principle, scaling the source will scale all phasors by the same factor. Choosing the scale factor α to match the actual source value $2\underline{/30°}$,

$$\alpha = \frac{2\underline{/30°}}{1.12\underline{/153°}} = 1.79\underline{/-123°}$$

we see that all the phasors in the phasor diagram must be enlarged by a factor of 1.79 and rotated 123° clockwise to complete the solution. In particular,

$$\mathbf{V}_C = \alpha(1\underline{/0}) = 1.79\underline{/-123°} \text{ V}$$

Note that rotating all the phasors in the vector diagram the same amount does not change their phase relationships: $\mathbf{I}_2$ still leads $\mathbf{I}_3$ by 90° and is still 180° out of phase with $\mathbf{V}_C$, and so on. The arbitrarily selected angle of the reference phasor is adequate to determine all phase relationships without the need for corrections. Corrections are needed only to determine the phase angle of a response rather than phase shift between two response variables.

Example 9.13

As a final example illustrating the use of phasor diagrams, let us find the locus of $\mathbf{I}$ as R varies in Fig. 9.23. The current is given by

$$\mathbf{I} = \frac{V_m}{R + j\omega L} = \frac{V_m(R - j\omega L)}{R^2 + \omega^2 L^2}$$

Therefore, if

$$\mathbf{I} = x + jy \qquad (9.15)$$

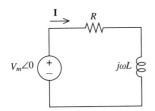

FIGURE 9.23   *RL* phasor circuit.

we have

$$x = \text{Re}\,(\mathbf{I}) = \frac{RV_m}{R^2 + \omega^2 L^2} \qquad (9.16)$$

$$y = \text{Im}\,(\mathbf{I}) = \frac{-\omega L V_m}{R^2 + \omega^2 L^2} \qquad (9.17)$$

The equation of the locus is the equation satisfied by x and y as R varies; thus we need to eliminate R between (9.16) and (9.17).

If we divide the first of these two equations by the second, we have

$$\frac{x}{y} = -\frac{R}{\omega L}$$

from which

$$R = -\frac{\omega L x}{y}$$

Substituting this value of R into (9.16), we have, after some simplification,

$$x^2 + y^2 = -\frac{V_m y}{\omega L}$$

This result may be rewritten as

$$x^2 + \left(y + \frac{V_m}{2\omega L}\right)^2 = \left(\frac{V_m}{2\omega L}\right)^2 \qquad (9.18)$$

which is the equation of a circle with center at $[0, -(V_m/2\omega L)]$ and radius $V_m/2\omega L$.

The circle (9.18) appears to be the locus, as R varies, of the phasor $\mathbf{I} = x + jy$. However, by (9.16), $x \geq 0$; thus the locus is actually the semicircle shown dashed on the phasor diagram of Fig. 9.24. The voltage $V_m\underline{/0}$, taken as reference, is also shown, along with the phasor $\mathbf{I}$. If $R = 0$, we have, from (9.16) and (9.17), $x = 0$ and $y = -V_m/\omega L$. If $R \to \infty$, then $x \to 0$ and $y \to 0$. Thus as R varies from 0 to ∞, the current phasor moves counterclockwise along the circle.

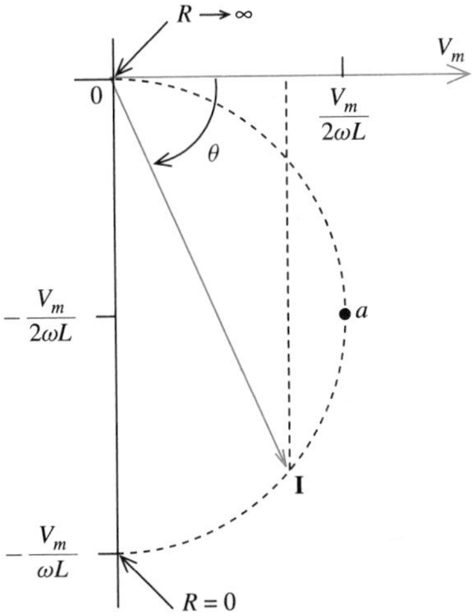

If $\mathbf{I}$ is as shown in Fig. 9.24, the current phasor may be resolved into two components, one having amplitude $I_m \cos\theta$ in phase with the voltage and one with amplitude $I_m \sin\theta$, which is $90°$ out of phase with the voltage. This construction is indicated by the dashed vertical line. As we shall see in Chapter 10, the in-phase component of the current is important in calculating the average power delivered by the source. Thus the phasor diagram gives us a method of seeing at a glance the maximum in-phase component of current. Evidently, this occurs at point a, which corresponds to $\theta = 45°$. This is the case $x = -y$, or $R = \omega L$.

FIGURE 9.24   Locus of the phasor $\mathbf{I}$.

9.5.1. Eliminate ωL in (9.16) and (9.17) and show that as ωL varies, the locus of the phasor $\mathbf{I} = x + jy$ is a semicircle.

$$\textit{Answer} \quad \left(x - \frac{V_m}{2R}\right)^2 + y^2 = \left(\frac{V_m}{2R}\right)^2, \qquad y \le 0$$

9.5.2. Find ωL in Exercise 9.5.1 so that Im $\mathbf{I}$ has its largest negative value. Also find $\mathbf{I}$ for this case.

$$\textit{Answer} \quad R; \ (V_m/\sqrt{2}\,R)\underline{/-45^\circ}$$

9.6   SPICE AND AC STEADY STATE

The use of phasors substantially reduces the computational burden of ac steady-state circuit analysis compared to formulating and solving the describing equation as a high-order differential equation. Still, for *RLC* circuits with several nodes and loops, substantial computing is required. The complex arithmetic used to develop the solution becomes time consuming and tedious, even with the support of a calculator well suited to the purpose.

Fortunately, SPICE is equipped with an ac control statement, which, if included in the SPICE input file, invokes ac steady-state analysis. In this mode the SPICE program determines the solution to a phasor circuit that you have specified. The format for this control statement is

```
.AC    FVAR    NP    FLOW    FHIGH
```

In the ac analysis mode invoked by this statement, SPICE is set up to perform multiple ac steady-state analyses at user-specified sets of frequencies. We will have occasion to make full use of this capability when studying frequency response in subsequent chapters, but for present purposes, determination of the ac steady state at a single source frequency is sufficient. Since the control statement is set up to accommodate the more general purpose, however, we must be aware of the general format. `FVAR` must be replaced by one of the keywords `DEC`, `OCT`, or `LIN`. This keyword describes the manner in which the frequency variation of the set of frequencies to be used for repeated ac analysis is to be done: by decade, by octave, or linearly. `NP` is a number specifying the number of frequencies per decade, per octave, or in the `LIN` case, the total number of frequencies in the set. `FLOW` is the lowest frequency, and `FHIGH` the highest frequency to be analyzed, with units of hertz (Hz). For instance,

```
.AC    DEC    5    10    1000
```

specifies that an ac analysis is to be performed for each of five evenly spaced frequencies per decade, from 10 Hz to 1 kHz. The logarithmic units of decade and octave are not needed for the present purpose of ac steady-state analysis; they will be taken up again in Chapter 14. The control statement

```
.AC   LIN   1   30K   30K
```

specifies that a single ac analysis be done at a frequency of 30 kHz. When NP is set to 1 as above, there is only one frequency in the analysis set. The resulting statement may seem like a roundabout way simply to specify an ac analysis at the single frequency of 30 kHz, but these are the current rules of SPICE. Perhaps in some future version improvements in this statement format will be made.

The phasors for all independent sources in the ac analysis must be specified on their element statements. The format is

```
VXXXX   N1   N2   AC   MAG   ANG
IXXXX   N1   N2   AC   MAG   ANG
```

V stands for independent voltage source, XXXX its name, and N1 and N2 the positive and negative nodes. Inclusion of the keyword AC indicates that this source is to be used in an ac analysis. MAG is the magnitude and ANG the angle of the phasor voltage source. The units for all angular quantities, such as ANG, throughout SPICE are degrees. The independent current source statement is similar to the voltage source, with the current source function reference arrow pointing out of node N1, through the source, and into node N2.

The output of a SPICE ac analysis is a set of phasor response variables. These may be printed as a table using the statement

```
.PRINT   AC   CVLIST
```

where CVLIST is a list of circuit variables. This list is formatted exactly as described in Chapter 4 for dc analysis, except that we specify the magnitude or phase of a variable by including M or P after the V (for voltage) or I (for current). For instance,

```
.PRINT   AC   VM(2)   VP(2)   IM(VDUMMY)
```

results in the printing of the magnitude and phase of the node voltage phasor V_2 and the magnitude of the current phasor through the voltage source VDUMMY. If rectangular components of the output are preferred, substitution of R or I for M or P will result in the printing of the real or imaginary parts.

Recall that in some versions of SPICE, only currents through voltage sources may be output, and it is necessary to install a zero volt "look-in" or "dummy" voltage source in series with any other element whose current is required and that does not happen to have a voltage source already in series. If the rectangular rather than polar representation is desired, inclusion of R or I after the leading V (voltage) or I (current) will result in printing of the real part or the imaginary part, respectively. SPICE solves phasor circuits only; it is the responsibility of the user to convert the original sinusoidal sources to their phasor representations and then convert the response phasors back to sinusoids.

To illustrate the application of SPICE, consider the circuit of Fig. 9.7. Let us find the voltage of node 1 in polar form and the current of the 1-Ω resistor in rectangular form. A SPICE deck for calculating these values is

```
AC STEADY-STATE SOLUTION FOR CIRCUIT OF FIG. 9.7
*DATA STATEMENTS
V1 100 0 AC 5 0
R1 100 1 .5
C1 1 0 .5
C2 1 2 1
L1 1 2 .5
L2 2 0 .25
R2 2 0 1
I1 0 2 AC 5 0
*SOLUTION CONTROL STATEMENT FOR AC ANALYSIS [f = 2/(2*PI) Hz]
.AC LIN 1 .3183 .3183
*OUTPUT CONTROL STATEMENT FOR V(1) & I(R2)
.PRINT AC VM(1) VP(1) IR(R2) II(R2)
.END
```

The .PRINT statement is formatted for versions of SPICE that accept current references such as I(R2). If your version of SPICE only outputs currents through voltage sources I(VXXXXX), a dummy source should be inserted in series with R1. The solution printed in this case is

FREQ	VM(1)	VP(1)	IR(R2)	II(R2)
3.183E-01	2.263E+00	-2.657E+01	2.000E+00	4.000E+00

The corresponding sinusoids are $v = 2.26\cos(2t - 26.6°)$, and since $\mathbf{I}_R = 2 + j4 = 4.47\underline{/63.4°}$, $i_R = 4.47\cos(2t + 63.4°)$.

As a second example, consider finding the phasor current $\mathbf{I}$ in the circuit of Fig. 9.9, which contains a controlled source of transresistance $r = 3000\ \Omega$.

```
AC STEADY-STATE SOLUTION FOR FIG. 9.9.
*DATA STATEMENTS
V 1 0 AC 4 0
R1 1 2 0.5K
R2 2 0 2K
C1 2 0 0.2UF
H 3 2 V -3000
R3 3 4 2K
C2 4 0 0.2UF
```

```
*SOLUTION CONTROL STATEMENT FOR f = 5000/(2*3.1416)
.AC LIN 1 795.77 795.77
.PRINT AC IM(R1) IP(R1)
.END
```

The solution is

```
       FREQ          IM(R1)        IP(R1)
    7.958E+02      2.400E-02     5.313E+01
```

Thus the ac steady-state current is $0.024 \cos(5000t + 53.1°)$.

In Chapter 4 the use of subcircuits in SPICE was introduced. The subcircuit definition, a set of statements enclosed by .SUBCKT and .ENDS control lines, may be included in the SPICE input file for the circuit containing the subcircuit, or it may be stored as a separate library file. The latter is particularly useful if the subcircuit is to be used in several different circuits. Inclusion of the control statement

```
.LIB    FILENM
```

will cause the contents of the text file FILENM to be linked to the main SPICE input file. FILENM must contain only subcircuits and, if we wish to link other library files to FILENM, .LIB control statements. For instance, an op amp model introduced as a subcircuit in Example 4.23 and repeated here is

```
.SUBCKT &    OPAMP   1   2   3
*NODE 1 is the + in, 2 the - in, and 3 the output
RIN   1   2   1MEG
E1    4   0   1   2   100K
RO    4   3   30
.ENDS
```

Since we will have occasion to use this model frequently, assume we have stored these five lines as a separate file named OPAMP.CKT. Example 9.16 illustrates how this library file can be used.

SPICE Example 9.16

Let us find the phasor output voltage of the op amp circuit of Fig. 9.11 if the input voltage is $v_g = 10 \cos(1000t + 30°)$ V. A circuit file for the nodes of the op amp inverting input, op amp output, and input source, assigned as 3, 4, and 10, respectively, with nodes 1 and 2 as shown, is

```
AC STEADY-STATE SOLUTION OF FIG. 9.11
*DATA STATEMENTS USING OPAMP.CKT OF CHAPTER 4
.LIB OPAMP.CKT
VG 10 0 AC 10 30
R1 10 1 0.707K
R2 1 2 1.414K
C1 1 4 1UF
C2 2 0 1UF
R3 3 0 2K
R4 3 4 2K
XOPAMP 3 2 4 OPAMP
*SOLUTION CONTROL STATEMENT [f = 1000/(2*3.1416) Hz]
.AC LIN 1 159.15 159.15
.PRINT AC VM(4) VP(4)
.END
```

This circuit file gives a solution

```
        FREQ         VM(4)          VP(4)
    1.592E+02     1.414E+01     -5.999E+01
```

For ac circuits containing sources at different frequencies, recall that a separate phasor analysis is required for the subset of sources at each distinct frequency, with all other sources killed. After conversion of the desired response phasors to sinusoids, the component results are superposed. Each phasor analysis requires, in general, a separate SPICE run with some editing of the SPICE input file in between. The phasor-to-sinusoid conversions and final addition must be done by hand. Based as it is on phasor analysis, SPICE has no facility to handle ac analysis of circuits with different frequencies more directly.

In Chapter 8 a fundamental caution in the use of phasors to compute ac steady state was pointed out. Phasors directly compute the forced response in ac circuits, which can be equated with ac steady-state response only if the circuit is stable. Stable circuits are those whose natural responses all decay to zero over time. If a circuit is not stable, its natural response will not all die away, and thus it will not in general possess a steady state regardless of how long we wait. If the circuit is not stable, we will get a numerical result using phasor analysis, but we should interpret this as the forced response only, not the ac steady state (which does not exist for unstable circuits).

Ac analysis in SPICE, which performs phasor analysis, inherits the same limitation. The result of an ac analysis in SPICE will always equate to the forced response, but this is also the ac steady-state response only if the circuit happens to be stable (which SPICE ac analysis does not tell us). Given the output of SPICE ac analysis, it is up to the user to verify independently that the circuit is stable before calling it the ac steady state. This requirement is illustrated in Example 9.17.

SPICE Example 9.17

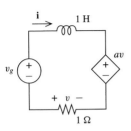

FIGURE 9.25 Circuit for Example 9.17.

Consider the single-loop circuit of Fig. 9.25, with describing equation for the mesh current i as follows:

$$\frac{di}{dt} + (1-a)i = v_g$$

The characteristic equation is

$$s + (1-a) = 0$$

leading to the natural solution

$$i_n = Ke^{-(1-a)t}$$

If the controlled source voltage gain $a < 1$, then i_n is a damped exponential, which goes to zero, and the system is stable. If $a > 1$, then i_n goes to (plus or minus) infinity, the circuit is unstable, and there is no steady state.

Let us perform ac analysis on this circuit using SPICE, with $v_g = \cos t$. Setting $a = 0.5$, our SPICE input file is

```
Circuit of Fig. 9.25
*
V       1       0       AC      1       0
L       1       2       1
H       2       3       V       0.5
R       3       0       1
.AC     LIN     1       .159    .159
.PRINT AC IM(V)         IP(V)
.END
```

which when run yields the output

```
FREQ            IM(V)           IP(V)
1.590E-01       8.951E-01       1.166E+02
```

The indicated forced response, $\mathbf{I} = 0.895\underline{/117°}$ A or $i = 0.895 \cos(t + 117°)$ A, is the ac steady-state response of this stable circuit. Changing to the value $a = 2$ in the SPICE input file and rerunning, the resulting output is

```
FREQ            IM(V)           IP(V)
1.590E-01       7.075E-01       4.497E+01
```

Since this circuit is unstable, the forced response $\mathbf{I} = 0.707\underline{/45°}$ A or $i = 0.707 \cos(t + 45°)$ A computed by SPICE can no longer also be identified as the ac steady-state response. There is no ac steady state in this unstable circuit.

The determination of whether a network is stable will be taken up more systematically in Chapter 14. For now, recall that *RLC* circuits with no dependent sources and no undamped natural responses (some resistance in every loop) are all stable. Also, op amp circuits with negative feedback, such as the building block circuits of Chapter 3, are stable. For other circuits, we should interpret the results of any phasor analysis, including ac analysis by SPICE, with the proper caution. If an ac steady state exists, SPICE will find it, but we will not be warned if it does not.

EXERCISES

9.6.1. Use SPICE to find the phasor representation of v in Exercise 9.2.4 for $\omega = 5$ rad/s.

Answer $2.224 \underline{/-158.2°}$ V

9.6.2. Use SPICE to find the phasor current of the 1-H inductor.

Answer $0.331\underline{/47.8°}$ A

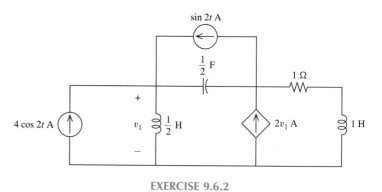

EXERCISE 9.6.2

SUMMARY

In this chapter phasor analysis is systematically applied to ac steady-state circuits. All of our familiar tools, nodal analysis, mesh analysis, current and voltage division, series and parallel equivalents, and so on, pass over to the phasor domain unchanged. Their use in the phasor domain is greatly simplified by the fact that all currents and voltages are constants, and all nonsource elements satisfy the same simple equation $\mathbf{V} = \mathbf{ZI}$.

- Series impedances add, parallel admittances add.
- Thevenin–Norton transformations and current or voltage division are unchanged in the phasor domain.
- Nodal and mesh analysis are unchanged in the phasor domain.
- For circuits with sinusoidal sources at two or more frequencies, superposition must be invoked to define a separate phasor circuit for each frequency. Each is solved independently. The overall ac steady-state response is the sum of the ac steady-state (sinusoidal) responses of all these phasor circuits.

- A phasor diagram is the plot of one or more phasors in the complex plane. The length of each phasor is the amplitude of its corresponding sinusoid, and the angle of a phasor is the phase angle of its corresponding sinusoid. All phasors rotate together in a counterclockwise direction at ω rad/s.

- SPICE may be used to analyze phasor circuits. The frequency, magnitude, and phase of each sinusoidal source is entered in its element statement, and a $\cdot AC$ control statement entered to invoke phasor analysis.

PROBLEMS

9.1. Find $\mathbf{V}_2$ using the voltage divider principle (9.2). Check by computing $\mathbf{V}_2 = Z_2\mathbf{I}$.

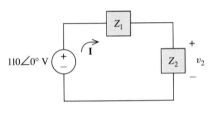

FIGURE P9.1

9.2. Find v_1 using voltage division.

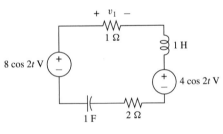

FIGURE P9.2

9.3. Specify three impedances Z_1, Z_2, Z_3 so that in this circuit $\mathbf{V}_1 = 6/\underline{60°}$, $\mathbf{V}_2 = 6/\underline{0}$, and $\mathbf{V}_3 = 6/\underline{-60°}$ V.

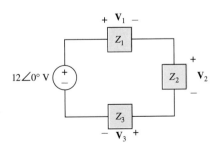

FIGURE P9.3

9.4. Find $\mathbf{I}_2$ using the current-divider principle (9.4). Check by computing $\mathbf{I}_2 = Y_2\mathbf{V}$.

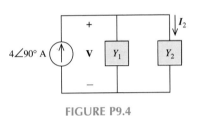

FIGURE P9.4

9.5. Find i_1 using current division.

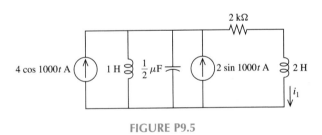

FIGURE P9.5

9.6. Specify three admittances Y_1, Y_2, Y_3 so that in this circuit $\mathbf{I}_1 = 1/\underline{0}$ A, $\mathbf{I}_2 = 1/\underline{60°}$ A, and $\mathbf{I}_3 = 1/\underline{-60°}$ A.

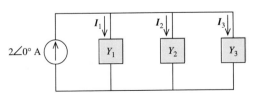

FIGURE P9.6

9.7. Find the Thevenin and the Norton equivalents of this subcircuit.

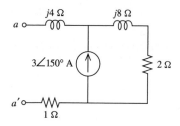

FIGURE P9.7

9.8. Find the Thevenin equivalent to the right of the dashed line; then use the voltage-divider principle to find $\mathbf{V}_1$.

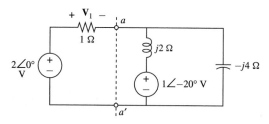

FIGURE P9.8

9.9. For what Z_1 is $\mathbf{V}_1 = 1\underline{/90°}$ V?

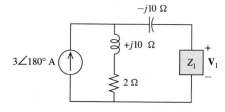

FIGURE P9.9

9.10. Find the steady-state current i using nodal analysis.

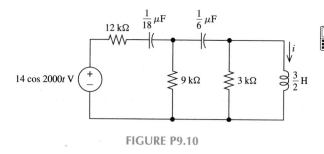

FIGURE P9.10

9.11. Find the steady-state voltage v using nodal analysis.

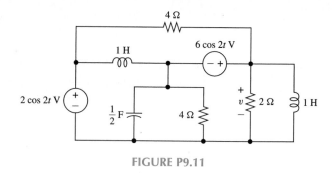

FIGURE P9.11

9.12. Find the steady-state voltage v using nodal analysis.

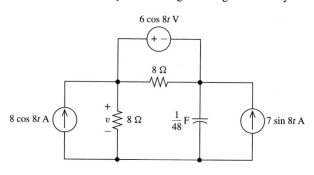

FIGURE P9.12

9.13. Find the steady-state current i_1 using nodal analysis.

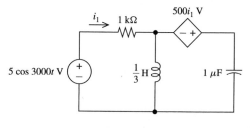

FIGURE P9.13

For Problems 9.14 through 9.18, solve for the indicated variables using nodal analysis.

9.14.

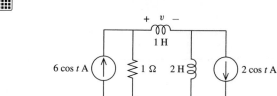

FIGURE P9.14

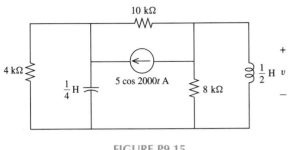

FIGURE P9.15

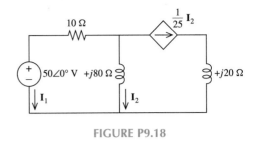

FIGURE P9.18

9.19. Find the steady-state voltage v if $v_g = 2 \cos 2000t$ V.

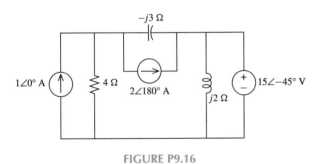

FIGURE P9.16

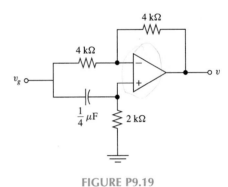

FIGURE P9.19

9.17.

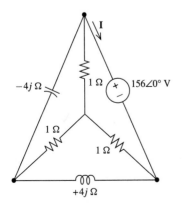

FIGURE P9.17

9.20. Find the steady-state voltage v if $v_g = 2 \cos 5000t$ V.

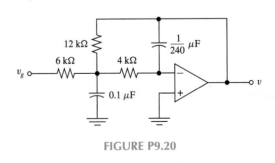

FIGURE P9.20

9.21. Solve Exercise 9.2.1 using mesh analysis.

9.22. Solve Exercise 9.2.2 using mesh analysis.

9.23. Find the steady-state voltage v.

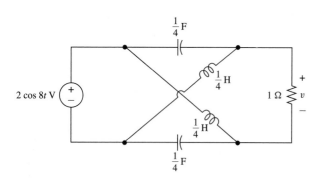

FIGURE P9.23

9.24. Find the steady-state current i.

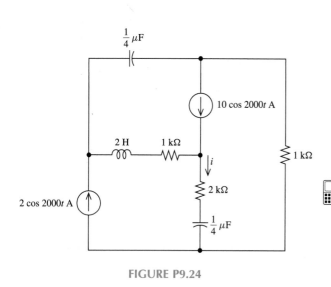

FIGURE P9.24

9.25. Show that if $\mathbf{Z}_1 \mathbf{Z}_4 = \mathbf{Z}_2 \mathbf{Z}_3$ in the "bridge" circuit shown, then $\mathbf{I} = \mathbf{V} = 0$ and therefore all the other currents and voltages remain unchanged for any value of $\mathbf{Z}_5$. Thus it may be replaced by an open circuit, a short circuit, etc. In this case, the circuit is said to be a *balanced bridge*.

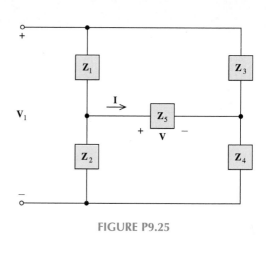

FIGURE P9.25

For Problems 9.26 through 9.30, solve for the indicated variables using mesh analysis.

9.26.

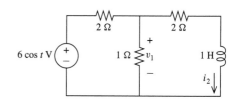

FIGURE P9.26

9.27.

FIGURE P9.27

9.28.

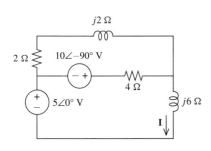

FIGURE P9.28

9.29. The controlled source has transconductance $g = (1/2)$ S.

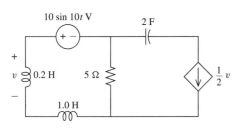

FIGURE P9.29

9.30.

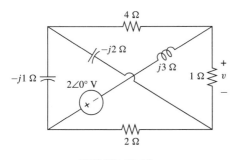

FIGURE P9.30

9.31. Solve Problem 9.18 using mesh analysis.

9.32. Mesh analysis is restricted to planar circuits. Sketch a circuit that cannot be analyzed by the mesh method and that contains the minimum number of nodes for a nonplanar circuit.

9.33. Find the steady-state voltage v.

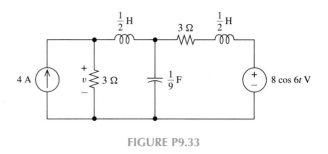

FIGURE P9.33

9.34. Find the steady-state value of v.

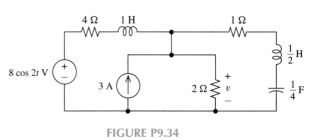

FIGURE P9.34

9.35. Find v.

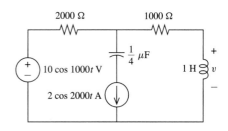

FIGURE P9.35

9.36. The sinusoidal sources v_g and i_g operate at $\omega = 1000$ and $\omega = 2000$ rad/A, respectively. If $v = 4\cos 1000t - 2\sin 2000t$, find v_g and i_g.

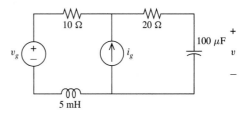

FIGURE P9.36

9.37. Sketch the three given voltages as a phasor diagram, and find **V**, the phasor corresponding to v, from the diagram. Find v.

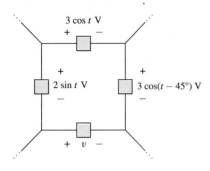

FIGURE P9.37

9.38. Sketch the impedances of the three elements in a phasor diagram. What must L be if the magnitude of the overall impedance is $|Z| = 14\ \Omega$? Assume $\omega = 10$ rad/s.

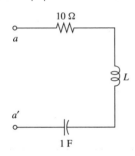

FIGURE P9.38

9.39. Sketch the locus of the total series impedance shown in Problem 9.38 in the complex plane as L varies from 0 to 1 H.

9.40. Find the locus of $\mathbf{I}$, the phasor associated with i, as ω varies from 0 to ∞. For which values of ω is $|\mathbf{I}|$ largest? Smallest?

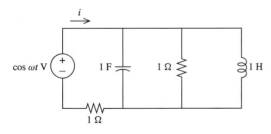

FIGURE P9.40

SPICE Problems

9.41. Use SPICE to solve Problem 9.15.

9.42. Use SPICE to solve Problem 9.27.

9.43. Find the phasor current $\mathbf{I}$ in this nonplanar circuit using SPICE. $R = 10\ k\Omega$, $C = 1\ \mu F$, and $i_s = 4\cos 377t$.

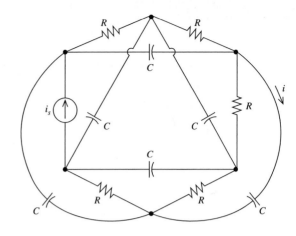

FIGURE P9.43

9.44. Find $v(t)$ for $\omega = 1, 2, 3, 4, 5$ rad/s. Use only one SPICE run, and use the ideal voltage-controlled voltage source op amp model of Fig. 3.7 with $A = 100,000$.

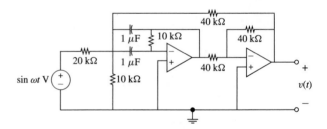

FIGURE P9.44

More Challenging Problems

9.45. Solve Problem 9.44 as specified, but use the improved op amp model of Fig. 3.9 with $A = 100,000$, $R_i = 1\ M\Omega$, and $R_0 = 30\ \Omega$. Enter the op amp model in the SPICE input file as a $\cdot$ SUBCKT.

9.46. For what L will $i(t) = 0$?

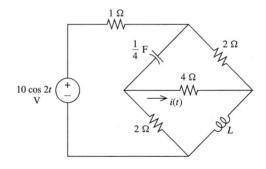

FIGURE P9.46

9.47. Find $v(t)$ in ac steady state.

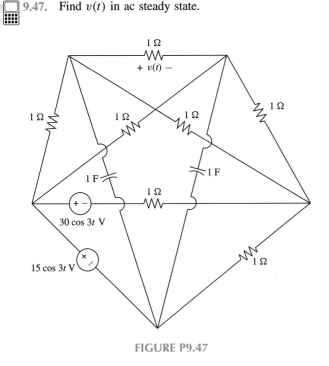

FIGURE P9.47

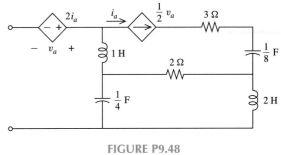

FIGURE P9.48

9.49. Find v_2 in ac steady state if $v_1 = \cos \omega t$ V.

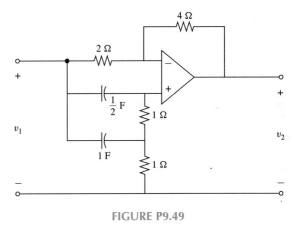

FIGURE P9.49

9.48. Find the ac steady-state Thevenin equivalent, specifying $\mathbf{V}_T$ and Z_T.

10

AC Steady-State Power

James Prescott Joule
1818–1889

The heating of a conductor depends upon its resistance and the square of the current passing through it.

James P. Joule

The man to whom we are indebted for the familiar expression i^2R for the power dissipated in a resistor is the English physicist James Prescott Joule, who published the result as Joule's law in 1841. He also shared in the famous discovery of the conservation of energy.

Joule was born in Salford, England, the second of five children of a wealthy brewer. He taught himself electricity and magnetism at home as a young boy and obtained his formal education at nearby Manchester University. His experiments on heat were conducted in his home laboratory, and to maintain the accuracy of his measurements he was forced to develop his own system of units. His chief claim to fame is that he did more than any other person to establish the idea that heat is a form of energy. Throughout most of his life Joule was an isolated amateur scientist, but toward the end of his years his work was recognized by honorary doctorates from Dublin and Oxford. In his honor the unit of energy was named the *joule*.

Chapter Contents

In this chapter we consider power relationships for networks that are excited by periodic currents and voltages. We concern ourselves primarily with sinusoidal currents and voltages since nearly all electrical power is generated in this form. Instantaneous power, as we now well know, is the rate at which energy is absorbed by an element, and it varies as a function of time. The instantaneous power is an important quantity in engineering applications because its maximum value must be limited for all physical devices. For this reason, the maximum instantaneous power, or *peak power*, is a commonly used specification for characterizing electrical devices. In an electronic amplifier, for instance, if the specified peak power at the input is exceeded, the output signal will be distorted. Greatly exceeding this input rating may even damage the amplifier permanently.

A more important measure of power, particularly for periodic currents and voltages, is that of *average power*. The average power is equal to the average rate at which energy is absorbed by an element, and it is independent of time. This power, for example, is what is monitored by the electric company in determining monthly electricity bills. Average powers may range from a few picowatts, in applications such as satellite communications, to megawatts, in applications such as supplying the electrical needs of a large city.

Our discussion will begin with a study of average power. After introducing the convenient root-mean-square (rms) metric, complex power is defined and its uses are discussed. Superposition is related to power, and the notions of maximum power transfer and conservation of power explored. The chapter ends with power factor, an idea of great practical interest to companies supplying electrical power, and a comment on the use of SPICE for ac power calculations.

10.1   AVERAGE POWER

In linear networks that have inputs that are periodic functions of time, the steady-state currents and voltages produced are periodic, each having identical periods. Consider an instantaneous power

$$p = vi \tag{10.1}$$

where v and i are periodic of period T. That is, $v(t + T) = v(t)$, and $i(t + T) = i(t)$. In this case

$$
\begin{aligned}
p(t + T) &= v(t + T)i(t + T) \\
&= v(t)i(t) \tag{10.2} \\
&= p(t)
\end{aligned}
$$

Therefore, the instantaneous power is also periodic and p repeats itself every T seconds.

The period T_p of p (the *minimum* time in which p repeats itself) is not necessarily equal to T, however, but T must contain an integral number of periods T_p. In other words,

$$
T = nT_p \tag{10.3}
$$

where n is a positive integer.

Example 10.1

As an example, suppose that a resistor R carries a current $i = I_m \cos \omega t$ with period $T = 2\pi/\omega$. Then

$$
\begin{aligned}
p &= Ri^2 \\
&= RI_m^2 \cos^2 \omega t \\
&= \frac{RI_m^2}{2}(1 + \cos 2\omega t)
\end{aligned}
$$

Evidently, $T_p = \pi/\omega$, and therefore $T = 2T_p$. Thus, for this case, $n = 2$ in (10.3). This is illustrated by the graph of p and i shown in Fig. 10.1.

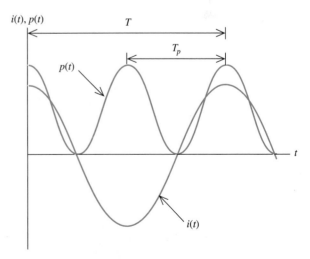

FIGURE 10.1 Instantaneous power and current waveforms (sinusoidal case).

If we now take $i = I_m(1 + \cos \omega t)$, then i is still periodic with period T.

$$p = R I_m^2 (1 + \cos \omega t)^2$$

In this case, $T_p = 2\pi/\omega$, and $n = 1$ in (10.3).

Mathematically, the average value of a periodic function is defined as the time integral of the function over a complete period, divided by the period. Therefore, the average power P for a periodic instantaneous power p is given by

$$P = \frac{1}{T_p} \int_{t_1}^{t_1 + T_p} p \, dt \tag{10.4}$$

where t_1 is arbitrary (Fig. 10.2). If we integrate over an integral number of periods, say mT_p (where m is a positive integer), the total area is simply m times that of the integral in (10.4). Thus we may write

$$P = \frac{1}{mT_p} \int_{t_1}^{t_1 + mT_p} p \, dt \tag{10.5}$$

If we select m such that $T = mT_p$ (the period of v or i), then

$$P = \frac{1}{T} \int_{t_1}^{t_1 + T} p \, dt \tag{10.6}$$

Therefore, we may obtain the average power by integrating over the period of p, as in (10.4), or over the period of v or i.

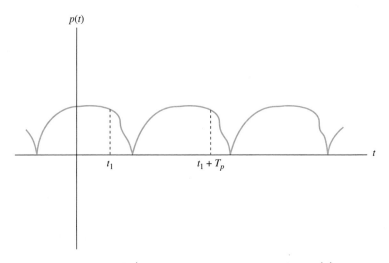

FIGURE 10.2 Periodic instantaneous power (nonsinusoidal case).

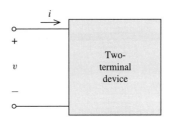

FIGURE 10.3   General two-terminal device.

Consider the general two-terminal device of Fig. 10.3, which is assumed to be in ac steady state. If

$$v = V_m \cos(\omega t + \phi_v) \qquad (10.7a)$$

we have

$$i = I_m \cos(\omega t + \phi_i) \qquad (10.7b)$$

for some I_m and ϕ_i, since in ac steady state all currents and voltages have the same frequency.

The average power delivered to the device, taking $t_1 = 0$ for convenience in (10.6), is

$$P = \frac{\omega V_m I_m}{2\pi} \int_0^{2\pi/\omega} \cos(\omega t + \phi_v) \cos(\omega t + \phi_i) \, dt$$

Using the trigonometric identity

$$\cos \alpha \cos \beta = \tfrac{1}{2} \cos(\alpha + \beta) + \tfrac{1}{2} \cos(\alpha - \beta) \qquad (10.8)$$

this may be rewritten

$$P = \frac{\omega V_m I_m}{2\pi} \int_0^{2\pi/\omega} \left[\frac{1}{2} \cos(2\omega t + \phi_v + \phi_i) + \frac{1}{2} \cos(\phi_v - \phi_i) \right] dt \qquad (10.9)$$

Since the integral runs over two periods of the first term and the average value of any sinusoid (with $\omega \neq 0$) is zero, the integral of the first term is zero. The second term is constant, and we have

$$P = \frac{V_m I_m}{2} \cos(\phi_v - \phi_i) \qquad (10.10)$$

Thus the average power absorbed by a two-terminal device is determined by the amplitudes V_m and I_m and the angle θ by which the voltage v leads the current i.

In terms of the phasors of v and i,

$$\mathbf{V} = V_m \underline{/\phi_v} = |\mathbf{V}| \underline{/\phi_v}$$
$$\mathbf{I} = I_m \underline{/\phi_i} = |\mathbf{I}| \underline{/\phi_i}$$

and we have, from (10.10),

$$P = \tfrac{1}{2}|\mathbf{V}|\,|\mathbf{I}|\cos(\phi_v - \phi_i) \qquad (10.11)$$

The average power absorbed by a two-terminal subcircuit is one-half the product of the magnitude of their current and voltage phasors times the cosine of the angle between them.

If the two-terminal device is a resistor R, then $\phi_v - \phi_i = 0$, and $V_m = RI_m$, so (10.10) becomes

$$P_R = \tfrac{1}{2}RI_m^2$$

It is worth noting at this point that if $i = I_{dc}$, a constant (dc) current, then $\omega = \phi_v = \phi_i = 0$, and $I_m = I_{dc}$ in (10.7b). The instantaneous power $vi = (RI_{dc})(I_{dc})$ is constant in this special case, so it equals its average, or

$$P_R = RI_{dc}^2$$

Example 10.2

Let us determine the average power absorbed by various parts of the ac steady-state circuit shown in Fig. 10.4(a), whose phasor circuit is drawn in Fig. 10.4(b). Combining impedances yields

$$\mathbf{Z} = 2 + (j2)(1 - j2) = 6 + j2 \ \Omega$$

so

$$\mathbf{I} = \frac{4\underline{/0}}{6 + j2} = 0.632\underline{/-18.4^\circ} \ \text{A}$$

By current division,

$$\mathbf{I}_1 = \frac{j2}{1}\mathbf{I} = 1.26\underline{/71.6^\circ} \ \text{A}$$

so $\mathbf{V}_R = 1.26\underline{/71.6^\circ}$ V and

$$\mathbf{V}_C = (-j2)\mathbf{I}_1 = 2.52\underline{/-18.4^\circ} \ \text{V}$$

The phasor diagram for these currents and voltages is shown in Fig. 10.5.

We first determine the average power absorbed by the 1-Ω resistor. The magnitude of its voltage is $|V_R| = 1.26$ V, its current is $|I_1| = 1.26$ A, and the angle between these phasors is 0; they are

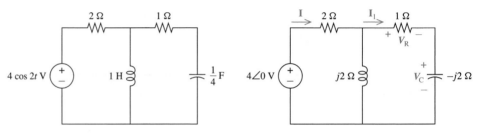

FIGURE 10.4 (a) Circuit for Example 10.2; (b) phasor circuit.

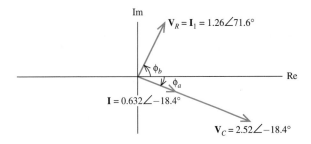

FIGURE 10.5 Phasor diagram for selected currents and voltages from the circuit of Fig. 10.4.

in phase. Hence by (10.11),

$$P_R = \tfrac{1}{2}|\mathbf{V}_R||\mathbf{I}_1|\cos(\underline{/\mathbf{V}_R} - \underline{/\mathbf{I}_1}) = \tfrac{1}{2}(1.26)^2 = 0.80 \text{ W}$$

For the capacitor we have

$$P_C = \tfrac{1}{2}|\mathbf{V}_C||\mathbf{I}_1|\cos\theta_b = 0$$

because the angle between voltage and current is $\theta_b = \underline{/\mathbf{V}_C} - \underline{/\mathbf{I}_1} = -90°$. The net average power absorbed by all the impedances is determined by the voltage $4\underline{/0}$ across the impedance subcircuit and current I into it to be

$$P_Z = \tfrac{1}{2}(4)(0.632)\cos\theta_a = 1.26\cos(-18.4°) = 1.20 \text{ W}$$

Finally, let us compute the power delivered by the source. Note from Fig. 10.3 that the terminal current and voltage used to compute power absorbed by a subcircuit using (10.11) must satisfy the passive sign convention. If they do not, (10.11) measures the *negative* of the power absorbed, that is, the power *delivered by* the subcircuit to the rest of the circuit. Considering the voltage source as a subcircuit, the terminal voltage $4\underline{/0}$ V and current $\mathbf{I}$ violate the passive sign convention since the current reference direction arrow points out of the positive end of the voltage reference direction. Thus the 1.20 W absorbed by the impedances above is also the power delivered by the source. Indeed, the total power delivered to all the elements absorbing power in a circuit is balanced by the power supplied, an intuitively reasonable conservation result to be discussed in more detail in Section 10.6.

Example 10.3

Let us find the power supplied by the two sources in Fig. 10.6. Combining the two parallel impedances gives

$$\mathbf{Z} = \frac{(1)(-j1)}{1 - j1} = \frac{1}{2} - j\frac{1}{2} \ \Omega$$

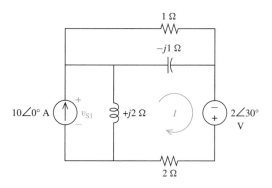

FIGURE 10.6 Circuit for Example 10.3.

The resulting circuit has two meshes and one current source; thus only one mesh equation is needed. Writing the mesh equation for **I**, we have

$$\left(\tfrac{1}{2} - j\tfrac{1}{2} + 2 + j\right)\mathbf{I} - j2(10\underline{/0}) = 2\underline{/30^\circ}$$

or

$$\mathbf{I} = 7.23\underline{/54.3^\circ}\ \text{A}$$

The voltage $\mathbf{V}_{S1}$ is identical to the voltage across the inductor:

$$\mathbf{V}_{S1} = j2(10\underline{/0} - \mathbf{I}) = 16.5\underline{/44.6^\circ}\ \text{V}$$

The voltage $\mathbf{V}_{S1}$ and the current source function $10\underline{/0}$ together violate the passive sign convention relative to the current source. Thus the power delivered by this source, rather than absorbed by it, is given by (10.11):

$$P_{S1} = \tfrac{1}{2}|\mathbf{V}_{S1}||10\underline{/0}|\cos 44.6^\circ = \tfrac{1}{2}(16.5)(10)(0.712) = 58.7\ \text{W}$$

Since **I** and the voltage source function $2\underline{/30^\circ}$ also violate the passive sign convention, the power delivered by the voltage source is

$$P_{S2} = \tfrac{1}{2}|\mathbf{I}||2\underline{/30^\circ}|\cos(30^\circ - 54.3^\circ) = \tfrac{1}{2}(7.23)(2)(0.911) = 6.59\ \text{W}$$

Both sources supply net power to the rest of the circuit. The power is in turn dissipated by the pair of resistors.

Since the current and voltage phasors are always in phase for resistors, by (10.11) resistors always absorb net power. For capacitors and inductors, the plus or minus 90° phase angle of their impedances drives their current and voltage 90° out of phase, so the storage elements neither absorb nor supply steady-state average power in any circuit. It is no surprise that none of the *RLC* elements supplies average power, since we know them to be passive elements. Because inductors and capacitors do not absorb net power either, they are referred to as *lossless* elements. Any element or subcircuit that absorbs average power is referred to as *lossy*. Clearly, any *LC* subcircuit (one containing only inductors and capacitors) is lossless, and any *RLC* subcircuit in which there is current flow into at least one resistor is lossy.

As we saw in Example 10.3, there is no such uniform rule concerning delivery or absorption of power that we may apply to sources. Sources can do either, depending on the circuit contexts in which they are placed. For instance, the same automobile battery will supply power to the starter motor subcircuit or absorb power from the alternator subcircuit depending on the switch positions in the rest of the circuit.

EXERCISES

10.1.1. Determine the period T_p of the instantaneous power $p(t)$ and the average power P.

Answer 6 S, −6 W

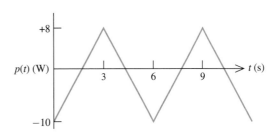

EXERCISE 10.1.1

10.1.2. For a capacitor of C farads carrying a current $i = I_m \cos \omega t$, verify that the average power is zero from (10.6). Repeat this for an inductor of L henrys.

10.1.3. Find the average power delivered to a 10-Ω resistor carrying a current of:

(a) $i = 5|\sin 10t|$ mA

(b) $i = 10 \sin 10t$ mA, $0 \le t < \pi/10$ s
$\quad = 0, \pi/10 \le t < \pi/5$ s; $T = \pi/5$ s

(c) $i = 5$ mA, $0 \le t < 10$ ms
$\quad = -5$ mA, $10 \le t < 20$ ms; $T = 20$ ms

(d) $i = 2t, 0 \le t < 2$ s; $T = 2$ s

Answer (a) 125 μW; (b) 0.25 mW; (c) 0.25 mW; (d) $\frac{160}{3}$ W

10.1.4. Find the average power absorbed by the capacitor, the two resistors, and the source.

Answer 0; $\frac{10}{3}$, $\frac{2}{3}$; −4 W

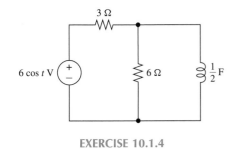

EXERCISE 10.1.4

10.1.5. If $f_1(t)$ is periodic of period T_1 and $f_2(t)$ is periodic of period T_2, show that $f_1(t) + f_2(t)$ is periodic of period T if relatively prime positive integers m and n exist such that

$$T = mT_1 = nT_2$$

Extend this result to the function $(1 + \cos \omega t)^2$, considered in this section, to find its period, $T = 2\pi/\omega$.

10.2 RMS VALUES

We have seen in Section 10.1 that periodic currents and voltages deliver an average power to resistive loads. The amount of power that is delivered depends on the characteristics of the particular waveform. A method of comparing the power delivered by different waveforms is therefore very useful. One such method is the use of *rms* or *effective* values for periodic currents or voltages.

The *rms* value of a periodic current (voltage) is a constant that is equal to the dc current (voltage) that would deliver the same average power to a resistance R. Thus, if I_{rms} is the *rms* value of i, we may write

$$P = RI_{rms}^2 = \frac{1}{T} \int_0^T Ri^2 \, dt$$

from which the rms current is

$$I_{rms} = \sqrt{\frac{1}{T} \int_0^T i^2 \, dt} \qquad (10.12a)$$

In a similar manner, the rms voltage is

$$V_{rms} = \sqrt{\frac{1}{T} \int_0^T v^2 \, dt} \qquad (10.12b)$$

The term rms is an abbreviation for *root mean square*. Inspecting (10.12a), we see that we are indeed taking the square *root* of the average, or *mean*, value of the *square* of the current. From our definition, the rms value of a constant (dc) is simply the constant itself. The dc case is a special case ($\omega = 0$) of the sinusoidal current or voltage.

Suppose that we now consider a sinusoidal current $i = I_m \cos(\omega t + \phi)$. Then, from (10.12a) and (10.8), we find

$$I_{rms} = \sqrt{\frac{\omega I_m^2}{2\pi} \int_0^{2\pi/\omega} \cos^2(\omega t + \phi) \, dt} = \frac{I_m}{\sqrt{2}} \qquad (10.13a)$$

Thus a sinusoidal current having an amplitude I_m delivers the same average power to a resistance R as does a dc current that is equal to $I_m/\sqrt{2}$. We also see that the rms current is independent of the frequency ω or the phase ϕ of the current i. Similarly, in the case of a sinusoidal voltage, we find that

$$V_{rms} = \frac{V_m}{\sqrt{2}} \qquad (10.13b)$$

Substituting these values into (10.10), we have for any two-terminal subcircuit,

$$P = V_{rms} I_{rms} \cos(\phi_v - \phi_i) \qquad (10.14)$$

where P is the average power absorbed if the passive sign convention is satisfied and the power delivered by the subcircuit if it is not.

Rms values are often used in the fields of power generation and distribution. For instance, the nominal 115-V ac power which is commonly used for household appliances is an rms value. Thus the power supplied to most of our homes is provided by a 60-Hz voltage having a maximum value of $115\sqrt{2} \approx 163$ V. On the other hand, maximum values are more commonly used in electronics and communications.

Thus far we have defined the rms value of time-domain currents and voltages. Next let $\mathbf{V}$ and $\mathbf{I}$ be phasors, and define their associated *rms phasors* by

$$\mathbf{V}_{\text{rms}} = \frac{1}{\sqrt{2}}\mathbf{V} \tag{10.15a}$$

$$\mathbf{I}_{\text{rms}} = \frac{1}{\sqrt{2}}\mathbf{I} \tag{10.15b}$$

RMS phasors thus have the same angle as our usual phasors, but their magnitudes are reduced by the factor $1/\sqrt{2}$. Thus while the magnitude of a standard or amplitude phasor equals the amplitude of the associated sinusoid, the magnitude of an rms phasor equals the rms value of the associated sinusoid [by (10.12a) and (10.13a)]. An immediate use of the rms phasor may be seen by substituting (10.13) into (10.10), revealing that

$$P = |\mathbf{V}_{\text{rms}}|\,|\mathbf{I}_{\text{rms}}|\cos(\phi_v - \phi_i) \tag{10.16}$$

Average power may be computed as the product of rms phasor magnitudes times the cosine of the angle between current and voltage phasors without the factor of $\frac{1}{2}$ necessary with amplitude phasors as in (10.10).

Note that by dividing all phasors by $\sqrt{2}$, it is easy to show that rms phasors satisfy Kirchhoff's current and voltage laws, and also that

$$\mathbf{V}_{\text{rms}} = Z\mathbf{I}_{\text{rms}} \tag{10.17}$$

for any impedance. Thus, by designating all independent sources by their rms phasors and elements by their usual impedances, we have an *rms phasor circuit* in which rms currents and voltage phasors may be computed exactly like amplitude current and voltage phasors in amplitude phasor circuits.

Example 10.4

Let us find the power delivered by the 60-Hz ac generator to the load impedance in the circuit of Fig. 10.7(a). The rms source phasor is $(325/\sqrt{2})\underline{/0} = 230\underline{/0}$ V rms, and the rms phasor current is shown in Fig. 10.7(b). The equivalent load impedance is computed in the usual way:

$$Z_L = \frac{1}{j\omega c} \,\|\, 30 = \frac{(-j53)(30)}{30 - j53} = 22.7 - j12.9 \ \Omega$$

In an rms phasor diagram, all phasors are rms, and the mesh current is

$$\mathbf{I}_{\text{rms}} = 230\underline{/0}/(24.7 - j11.9) = 8.39\underline{/25.7^\circ} \ \text{A rms}$$

so

$$\mathbf{V}_{\text{rms}} = Z_L\mathbf{I}_{\text{rms}} = (22.7 - j12.9)(8.39\underline{/25.7^\circ}) = 219\underline{/-3.91^\circ} \ \text{V rms}$$

Then, by (10.16), the average power *absorbed* by the load impedance Z_L is

$$P = |\mathbf{V}_{\text{rms}}|\,|\mathbf{I}_{\text{rms}}|\cos(\phi_v - \phi_i) = (219)(8.39)\cos(-3.91^\circ - 25.7^\circ)$$

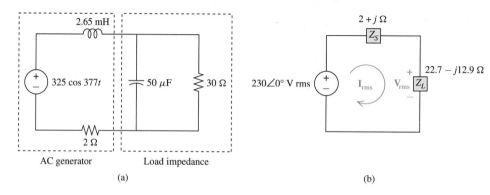

(a) AC generator   Load impedance

(a)

(b)

FIGURE 10.7 (a) Circuit for Example 10.4; (b) phasor circuit.

or $P = 1598$ W. This is also the average power delivered by the ac generator.

EXERCISES

10.2.1. Find the rms value of a periodic current for which one period is defined by:
(a) $i = I,$ $0 \le t < 2$ s
    $= -I,$ $2 \le t < 4$ s
(b) $i = 2t,$ $0 \le t < T$
(c) $i = I_m \sin \omega t,$ $0 \le t \le \pi/\omega$
    $= 0,$ $\pi/\omega \le t \le 2\pi/\omega$ $(T = 2\pi/\omega).$
    Answer (a) I; (b) $2T/\sqrt{3}$; (c) $I_m/2$

10.2.2. Find the rms values of (a) $i = 10 \sin \omega t + 20 \cos(\omega t + 30°)$, (b) $i = 8 \sin \omega t + 6 \cos(2\omega t + 10°)$, and (c) $i = I(1 + \cos 377t)$.
    Answer (a) 12.25; (b) 7.07; (c) $I\sqrt{\frac{3}{2}}$

10.2.3. Find V_{rms}.
    Answer 4 V

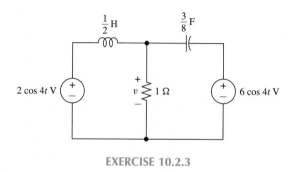

EXERCISE 10.2.3

10.3 COMPLEX POWER

We have seen that the use of phasors reaps great benefits in the study of ac steady-state circuits. By using complex numbers to represent real time-domain currents and voltages, we have dispensed with the frequent invocation of those identities from trigonometry needed to combine sinusoids and, in addition, replaced the calculus operations of differentiation and integration by the far easier arithmetic operations of multiplication and division in the complex plane.

In our present efforts to extend ac steady-state analysis to include power calculations, the complications of trigonometry and calculus have reappeared, for instance in (10.8) and (10.9). To extend the full benefits of phasor analysis to the study of power in ac steady-state circuits, we will once again define a new complex quantity, which we will call complex power. Similar to the concise way in which phasors represent sinusoids, complex power will encapsulate two essential power-related quantities in one compact, easily manipulated representation.

Given a two-terminal subcircuit with rms voltage and current phasors $\mathbf{V}_{rms}$ and $\mathbf{I}_{rms}$, where

$$\mathbf{V}_{rms} = |\mathbf{V}_{rms}| \underline{/\phi_v}$$

$$\mathbf{I}_{rms} = |\mathbf{I}_{rms}| \underline{/\phi_i}$$

we define the *complex power* into the subcircuit as

$$\mathbf{S} = \mathbf{V}_{rms}\mathbf{I}_{rms}^* \qquad (10.18)$$

where $\mathbf{I}_{rms}^*$ is the complex conjugate of the rms current phasor. Exploring this definition, the magnitude of the complex number $\mathbf{S}$ is given by

$$|\mathbf{S}| = |\mathbf{V}_{rms}||\mathbf{I}_{rms}^*| = |\mathbf{V}_{rms}||\mathbf{I}_{rms}| \qquad (10.19a)$$

and its angle is

$$\underline{/\mathbf{S}} = \underline{/\mathbf{V}_{rms}} + \underline{/\mathbf{I}_{rms}^*} \qquad (10.19b)$$

Since the angle of complex conjugate of any complex number is the negative of the angle of the number itself,

$$\underline{/\mathbf{S}} = \underline{/\mathbf{V}_{rms}} - \underline{/\mathbf{I}_{rms}} = \phi_v - \phi_i$$

Turning to the rectangular form for $\mathbf{S}$, its real part is $\text{Re}(\mathbf{S}) = |\mathbf{S}| \cos \underline{/\mathbf{S}}$, which, by (10.19), is

$$\text{Re}(\mathbf{S}) = |\mathbf{V}_{rms}||\mathbf{I}_{rms}| \cos(\phi_v - \phi_i) \qquad (10.20)$$

Comparing the last with (10.6), we have an important connection with our earlier results. *The real part of the complex power is the average power,*

$$\text{Re}(\mathbf{S}) = P \qquad (10.21)$$

Thus we have another way to compute the average power: compute the complex power $\mathbf{S}$; then take the real part to get P.

To avoid sign errors, the rms current and voltage phasors used to define the complex power $\mathbf{S}$ into a subcircuit will be assumed to satisfy the passive sign convention. In this case the real part of **S** is interpreted as the power *absorbed* by the subcircuit. Thus if Re(**S**) is positive, the subcircuit is absorbing average power, and, if negative, it is delivering average power to the rest of the circuit. Occasionally, where the meaning is clear, we will use in (10.18) current and voltage phasors that together violate the passive sign convention, in which case the real part must be interpreted as the power delivered by the subcircuit to the rest of the circuit.

Example 10.5

In Example 10.3 we computed the average power for the two sources in the standard (non-rms) phasor circuit shown in Fig. 10.6. The rms phasor circuit corresponding to this problem is shown in Fig. 10.8. Note that the only difference is that the source phasors have been divided by a factor of $\sqrt{2}$. By proportionality, we may get the desired phasors in the rms circuit by scaling the results of Example 10.4 by the same factor, so from the Example 10.4 results we have

$$\mathbf{I}_{rms} = \frac{7.23}{\sqrt{2}}\underline{/54.3°} = 5.11\underline{/54.3°} \ \text{A rms}$$

$$\mathbf{V}_{S1 \ rms} = \frac{16.5}{\sqrt{2}}\underline{/44.6°} = 11.7\underline{/44.6°} \ \text{V rms}$$

The complex power into the voltage source is

$$\mathbf{S}_{S2} = (\sqrt{2}\underline{/30°})(5.11\underline{/54.3°})^* = 7.23\underline{/-24.3°}$$

and the complex power into the current source is, after reversing the sign of the current in order to comply with the passive sign convention,

$$\mathbf{S}_{S1} = \mathbf{V}_{S1 \ rms}(5\sqrt{2}\underline{/180°})^* = (11.7\underline{/44.6°})(7.07\underline{/180°})^*$$

$$= (11.7\underline{/44.6°})(7.07\underline{/-180°})$$

$$= 82.7\underline{/-135°}$$

Note that reversing the sign of a complex number adds or subtracts 180° to its angle in the complex plane, while taking complex con-

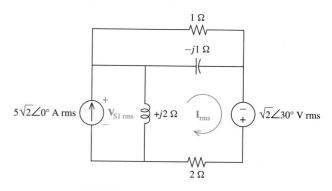

FIGURE 10.8 rms phasor circuit

jugates reverses the sign of its angle. In the present case both these operations were performed.

The average power of the sources is then

$$P_{S2} = \text{Re}(\mathbf{S}_{S2}) = 7.23\cos(-24.3°) = 6.59 \text{ W}$$

$$P_{S1} = \text{Re}(\mathbf{S}_{S1}) = 82.7\cos(-135°) = -58.7 \text{ W}$$

Observing that the current and voltage used for P_{S2} violate the passive sign convention, the voltage source supplies 6.59 W of average power while the current source supplies 58.7 W to the rest of the circuit. This agrees with results of Example 10.4.

The definition of complex power permits useful expressions for the power absorbed by impedances to be easily derived. Given a single impedance or two-terminal subcircuit containing only impedances, using terminal rms phasors $\mathbf{V}_{\text{rms}}$ and $\mathbf{I}_{\text{rms}}$, which satisfy the passive sign convention,

$$\mathbf{V}_{\text{rms}} = Z\mathbf{I}_{\text{rms}} \tag{10.22}$$

and the power into the subcircuit is

$$\mathbf{S} = \mathbf{V}_{\text{rms}}\mathbf{I}_{\text{rms}}^* = Z\mathbf{I}_{\text{rms}}\mathbf{I}_{\text{rms}}^*$$

Recalling that the product of a complex number and its complex conjugate is the real number equal to its magnitude squared, we have

$$\mathbf{S} = Z|\mathbf{I}_{\text{rms}}|^2 \tag{10.23}$$

One implication of this useful formula is that the angle of $\mathbf{S}$ for an impedance is equal to angle of the impedance itself. Since $\text{Re}(\mathbf{S}) = P$, another is that, taking real parts,

$$P = \text{Re}(Z)|\mathbf{I}_{\text{rms}}|^2 \tag{10.24a}$$

or, substituting the generalized Ohm's law (10.22A),

$$P = \frac{\text{Re}(Z)}{|Z|^2}|V_{\text{rms}}|^2 \tag{10.24b}$$

Note that if the impedance has an angle of plus or minus 90°, $P = 0$. This is the lossless case discussed earlier. If the real part of Z is positive, the impedance absorbs net power and is lossy. Combining these cases, impedances with angles from minus to plus 90° correspond to passive elements, while those with angles outside this range must be active elements. Any interconnection of passive elements is passive; thus we expect the angles of impedances constructed from passive elements to be 90° or less in magnitude.

In terms of the admittance, if the subcircuit satisfies

$$\mathbf{I}_{\text{rms}} = Y\mathbf{V}_{\text{rms}}$$

then, by the definition of complex power,

$$\mathbf{S} = \mathbf{V}_{\text{rms}}\mathbf{I}_{\text{rms}}^* = Y^*|\mathbf{V}_{\text{rms}}|^2 \tag{10.25}$$

and

$$P = \text{Re}(Y^*)|\mathbf{V}_{\text{rms}}|^2 = \text{Re}(Y)|\mathbf{V}_{\text{rms}}|^2 \tag{10.26a}$$

with (10.26a) following from the fact that taking complex conjugates does not change the real part of a complex number. Again using the generalized Ohm's law, (10.26a) may be expressed as

$$P = \frac{\text{Re}(Y)}{|Y|^2}|I_{\text{rms}}|^2 \qquad (10.26\text{b})$$

Example 10.6

To determine the power into the load in Fig. 10.9(a), we use the rms phasor circuit shown, where

$$Z_1 = \frac{(3)(-j2)}{3 - j2} = 0.923 - j1.38 \ \Omega$$

$$Z_L = 2 + j1 \ \Omega$$

By voltage division,

$$\mathbf{V}_{L \ \text{rms}} = Z_L\frac{50/\sqrt{2}}{Z_L + Z_1} = \frac{35.4(2 + j1)}{2.92 - j0.38}$$

$$= 26.9\underline{/34.1^\circ} \ \text{V rms}$$

Then by (10.26a) with

$$Y_L = \frac{1}{Z_L} = \frac{1}{2 + j1} = 0.4 - j0.2 \ \text{S}$$

we have

$$P_L = \text{Re}(Y)|\mathbf{V}_{L \ \text{rms}}|^2 = 0.4(26.9^2) = 289 \ \text{W}$$

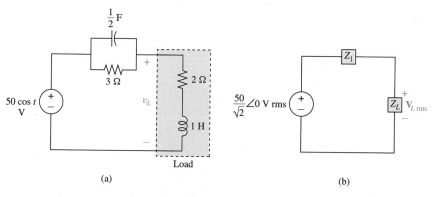

(a)   (b)

FIGURE 10.9   (a) Circuit for Example 10.6; (b) rms phasor circuit.

Example 10.7

We wish to find a value for the conductance G in Fig. 10.10 so that the source will deliver 1 W to the rest of the circuit. Since the power absorbed by the rest of the circuit must equal this value, by (10.26b), with Y the total admittance of all the RLC elements,

$$P = \frac{4\,\text{Re}(\mathbf{Y})}{|\mathbf{Y}|^2} = 1.0 \qquad (10.27)$$

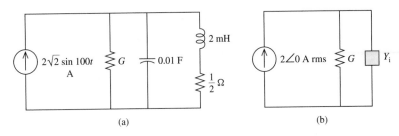

FIGURE 10.10   (a) Circuit for Example 10.7; (b) rms phasor circuit.

Combining the other admittances gives

$$Y_1 = j1 + \frac{(-j5)(2)}{2 - j5} = 1.72 + j0.31 \text{ S}$$

Since G and Y_1 are in parallel, the total admittance is

$$Y = G + Y_1 = (G + 1.72) + j0.31 \text{ S}$$

Substituting this into (10.27) yields

$$\frac{4(G + 1.72)}{(G + 1.72)^2 + 0.31^2} = 1$$

This yields a quadratic equation for G:

$$G^2 - 0.56G - 3.83 = 0$$

The roots of this quadratic are $G = 2.25$ and $G = -3.39$. Since $G > 0$ for passive conductors, the unique solution is $G = 2.25$ S, which is a $1/2.25 = 0.444$-Ω resistor.

Having identified the real part of the complex power $\mathbf{S}$ with the average power P, we next turn to the complex part, which we define as the *reactive power* Q. Thus

$$\mathbf{S} = \mathbf{V}_{\text{rms}}\mathbf{I}^*_{\text{rms}} = P + jQ \qquad (10.28)$$

where P is the average power in watts and Q the reactive power,

$$P = |\mathbf{V}_{\text{rms}}||\mathbf{I}_{\text{rms}}| \cos(\phi_v - \phi_i) \qquad (10.29a)$$

$$Q = |\mathbf{V}_{\text{rms}}||\mathbf{I}_{\text{rms}}| \sin(\phi_v - \phi_i) \qquad (10.29b)$$

The unit of reactive power is the *var*, or volt–ampere reactive.

To explore the meaning of reactive power, consider the phasor diagram shown as Fig. 10.11, in which the rms current and voltage phasors for the subcircuit A are shown. For convenience we take $\mathbf{I}_{\text{rms}}$ as the reference phasor and designate the angle between these phasors as $\theta = \phi_v - \phi_i$. Suppose that we resolve $\mathbf{V}_{\text{rms}}$ into two orthogonal components, one in phase with the current, designated $\mathbf{V}_i$, and the other 90° out of phase (or in *quadrature* with the current), $\mathbf{V}_q$. The magnitudes of these two components are

$$|\mathbf{V}_i| = |\mathbf{V}_{\text{rms}}| \cos \theta$$

$$|\mathbf{V}_q| = |\mathbf{V}_{\text{rms}}| \sin \theta$$

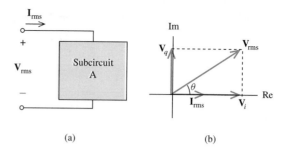

FIGURE 10.11   (a) Subcircuit; (b) rms phasor diagram.

Now the complex power absorbed by the subcircuit is

$$\mathbf{S} = \mathbf{V}_{rms}\mathbf{I}_{rms}^* = \mathbf{V}_i\mathbf{I}_{rms}^* + \mathbf{V}_q\mathbf{I}_{rms}^*$$

The first term, or in-phase component, is purely real since the current and voltage are in phase, that is, have equal angles. The second term, or quadrature component, is purely imaginary. Thus the average power P is determined solely by the in-phase component and the reactive power Q solely by the quadrature component. Since we are interested here in the reactive power, we will focus on the quadrature component. In terms of the corresponding time functions,

$$i(t) = \sqrt{2}|\mathbf{I}_{rms}|\cos\omega t$$

$$v_q(t) = \sqrt{2}|\mathbf{V}_q|\cos(\omega t \pm 90°)$$

$$= \sqrt{2}|\mathbf{V}_{rms}|\sin\theta\cos(\omega t \pm 90°)$$

The phase angle is written as $\pm 90°$ since the sign of this term depends on whether $\mathbf{V}_{rms}$ happens to lead $\mathbf{I}_{rms}$, the case shown in Fig. 10.11, or lag. The instantaneous power in the quadrature component is then the product

$$p_q(t) = v_q(t)i(t) = [2|\mathbf{V}_{rms}||\mathbf{I}_{rms}|\sin\theta]\cos\omega t\cos(\omega t \pm 90°)$$

which may be simplified using the trigonometric identity introduced in Section 10.1 to be

$$p_q(t) = |\mathbf{V}_{rms}||\mathbf{I}_{rms}|\sin\theta\cos(2\omega t \pm 90°)$$

From this expression we see that the instantaneous power in the quadrature component $p_q(t)$ is a sinusoid with amplitude equal to $|\mathbf{V}_{rms}||\mathbf{I}_{rms}||\sin\theta|$. By (10.29), this equals the magnitude of Q. Thus we arrive at a physical interpretation for the quantity Q: *the magnitude of the reactive power Q is the amplitude of the instantaneous power in the quadrature i–v component.* Since the instantaneous power in the quadrature component is a sinusoid centered at zero, it indicates energy being transferred periodically to the subcircuit and then back from the subcircuit to the rest of the network in equal amount. From earlier chapters we know this behavior to be characteristic of energy storage elements, neither supplying nor dissipating power on average, but rather exchanging it back and forth with the rest of the circuit. *Reactive power Q is thus a measure of the amount of periodic energy exchange taking place between a given subcircuit and the rest of the circuit.* By (10.29b), the sign of Q is positive for inductive reactances and negative for capacitive ones.

Example 10.8

Consider the series connection of a 1-Ω resistor, a $\frac{1}{2}$-F capacitor, a 1-H inductor, and an independent voltage source of $12\sqrt{2}\cos t$ V. The rms source phasor is

$$\mathbf{V}_{s \text{ rms}} = 12\underline{/0}$$

and the mesh current is

$$\mathbf{I}_{\text{rms}} = \frac{12\underline{/0}}{1 - j2 + j1} = 8.5\underline{/45°}$$

The complex power absorbed by each impedance is

$$\mathbf{S} = \mathbf{V}_{\text{rms}}\mathbf{I}_{\text{rms}}^* = \mathbf{Z}|\mathbf{I}_{\text{rms}}|^2$$

For the resistor, this is purely real with

$$\mathbf{S}_R = P_R + jQ_R = (1)(8.5^2) = 72 + j0$$

while for the capacitor

$$\mathbf{S}_C = P_C + jQ_C = (-j2)(8.5^2) = 0 - j144$$

and for the inductor

$$\mathbf{S}_L = P_L + jQ_L = (j1)(8.5^2) = 0 + j72$$

Reversing the reference direction of $\mathbf{I}_{\text{rms}}$ to make it satisfy the passive sign convention together with $\mathbf{V}_{s \text{ rm}}$ for calculation of the power absorbed by the source,

$$\mathbf{S}_V = P_V + jQ_V = (12\underline{/0})(8.5\underline{/-135°})^* = -72 + j72$$

The reactive power of the resistor is zero, as anticipated since current and voltage in resistors are in phase. The storage elements are pure reactances and have 90° phase shift between their currents and voltages, and thus the complex power is purely reactive (zero average power). Note that the magnitude of the reactive power Q_C of the capacitor is twice that of the inductor Q_L. Since the impedance is larger by a factor of 2, the voltage across the capacitor is twice that across the inductor, and since both these voltages are in quadrature with the mesh current, the amplitude of the instantaneous power in the quadrature component is twice as large for the capacitor as for the inductor. The capacitor exchanges more energy per period, back and forth with the rest of the circuit, by a factor of 2. Finally, note that since the reactive power summed over the passive elements does not equal zero, the source must supply vars (as well as watts). That is, the source must develop an out-of-phase as well as an in-phase component. How the quadrature component may be adjusted to reduce the total current–voltage requirements at the source is discussed in Section 10.7.

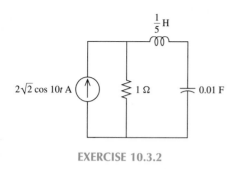

EXERCISE 10.3.2

10.3.1. The voltage across an impedance $\mathbf{Z} = 14\underline{/37°}\ \Omega$ has rms phasor $110\underline{/30°}$ V rms. Find the rms current phasor into the impedance and the complex power delivered to the impedance.

Answer $\mathbf{I}_{rms} = 7.86\underline{/-7°}$ A rms; $\mathbf{S} = 690.5 + j520.3$

10.3.2. Find the complex power delivered by each of the elements in this circuit.

Answer $\mathbf{S}_{source} = -3.93 + j0.489$; $\mathbf{S}_R = 3.93 + j0$; $\mathbf{S}_L = j0.123$; $\mathbf{S}_C = -j0.612$

10.3.3. A parallel combination of 1-Ω resistor and L-H inductor is found to absorb equal watts and vars when $\omega = 10$ rad/s. What must L be?

Answer 100 mH

10.4 SUPERPOSITION AND POWER

In this section we consider power in networks containing two or more independent sources. Let i and v be the terminal current and voltage into a two-terminal subcircuit in a circuit containing two independent sources. The principle of superposition always applies to currents and voltages in any linear circuit, so

$$i = i_1 + i_2$$

$$v = v_1 + v_2$$

where the subscripts refer to the responses due to each source separately with the other killed. The instantaneous power absorbed by the subnetwork is, assuming i and v have reference directions that satisfy the passive sign convention,

$$p = iv = (i_1 + i_2)(v_1 + v_2)$$

or

$$p = i_1 v_1 + i_2 v_2 + (i_1 v_2 + i_2 v_1) \tag{10.30a}$$

$$p = p_1 + p_2 + (i_1 v_2 + v_1 i_2) \tag{10.30b}$$

where p_1 and p_2 are the instantaneous powers due to each source separately. If superposition were to apply to power, we would have p as the sum of p_1 and p_2, which from the last equation is not in general the case. *Superposition does not in general hold for instantaneous power.* That is, the instantaneous power into a subnetwork is not in general simply the sum of the instantaneous powers due to each excitation separately. Failure of superposition to work in general for power is a consequence of the nonlinear nature of power, its definition as the product of two circuit variables.

Next consider the same circuit in ac steady state. Suppose that both sources operate at the same frequency ω. If $\mathbf{I}_1$ and $\mathbf{I}_2$ are the rms current phasors and $\mathbf{V}_1$ and $\mathbf{V}_2$ the corresponding voltage phasors produced by the two sources, applying superposition to current and voltage in the rms phasor circuit,

$$\mathbf{I} = \mathbf{I}_1 + \mathbf{I}_2$$

$$\mathbf{V} = \mathbf{V}_1 + \mathbf{V}_2$$

Thus the complex power into the subcircuit is

$$\mathbf{S} = \mathbf{VI}^* = (\mathbf{V}_1 + \mathbf{V}_2)(\mathbf{I}_1 + \mathbf{I}_2)^* \tag{10.31}$$

or
$$\mathbf{S} = \mathbf{V}_1\mathbf{I}_1^* + \mathbf{V}_2\mathbf{I}_2^* + (\mathbf{V}_1\mathbf{I}_2^* + \mathbf{V}_2\mathbf{I}_1^*) \tag{10.32a}$$

$$\mathbf{S} = \mathbf{S}_1 + \mathbf{S}_2 + (\mathbf{V}_1\mathbf{I}_2^* + \mathbf{V}_2\mathbf{I}_1^*) \tag{10.32b}$$

where $\mathbf{S}_1$ and $\mathbf{S}_2$ and the separate complex powers due to each source. Since the term in parentheses in (10.32b) is not in general zero, *superposition does not hold in general for complex power.* Examining the real parts of both sides of this expression,

$$P = P_1 + P_2 + \text{Re}(\mathbf{V}_1\mathbf{I}_2^* + \mathbf{V}_2\mathbf{I}_1^*)$$

Since the last term in general is not zero, *superposition does not in general hold for average power P.* That is, the watts absorbed will not be the simple sum of those due to each source separately. Similarly, equating the imaginary parts of (10.32b) yields

$$Q = Q_1 + Q_2 + \text{Im}(\mathbf{V}_1\mathbf{I}_2^* + \mathbf{V}_2\mathbf{I}_1^*)$$

Superposition does not in general hold for reactive power Q.

The failure of superposition shown above clearly extends to the case of more than two sources. Thus, in ac steady-state circuits containing multiple sources at the same frequency, power cannot be computed by superposing powers due to individual sources. This applies equally to all forms of power: instantaneous, average, reactive, or complex. Superposition may be used to find currents and voltages, but for sources at the same frequency, these components must be superposed *before* power is computed.

Example 10.9

To determine the net power supplied by the voltage source in Fig. 10.12(a), we note that since both sources in the circuit are at the same frequency $\omega = 2$ rad/s, we may determine the ac steady-state response using a single phasor circuit. The rms phasor circuit for $\omega = 2$ rad/s is shown in Fig. 10.12(b). Let $\mathbf{I}_1$ be the component of $\mathbf{I}$

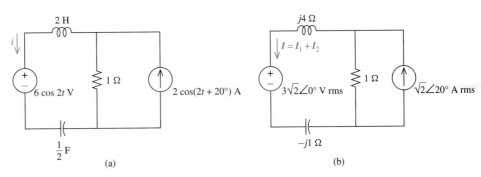

FIGURE 10.12   (a) Circuit diagram; (b) rms phasor diagram.

due to the voltage source and $\mathbf{I}_2$ that due to the current source. Then

$$\mathbf{I}_1 = -\frac{3\sqrt{2}\underline{/0})}{1 + j3} = -0.424 + j1.27 \text{ A rms}$$

and by current division

$$\mathbf{I}_2 = \frac{\sqrt{2}\underline{/20°}}{1 + j3} = 0.278 - j0.35 \text{ A rms}$$

Thus

$$\mathbf{I} = \mathbf{I}_1 + \mathbf{I}_2 = -0.146 + j0.92 \text{ A rms}$$

and the complex power absorbed by the voltage source is

$$\mathbf{S} = \mathbf{VI}^* = (3\sqrt{2}\underline{/0})(-0.146 - j0.92) = -0.62 - j3.9$$

We conclude that the voltage source delivers 620 mW of average power and 3.9 vars of reactive power to the rest of the circuit.

The failure of superposition for ac steady-state power must be reconsidered when the sources are of different frequencies. Consider a subcircuit with terminal variables i and v in a circuit containing two sinusoidal sources at frequencies $\omega_1 \neq \omega_2$. By (10.30b), the instantaneous power is

$$p = p_1 + p_2 + (i_1 v_2 + v_1 i_2) \qquad (10.33)$$

where p_1 and p_2 are the separate instantaneous powers due to each source with the other killed. Since the last term is a nonzero time function, superposition does not apply to instantaneous power in this case, any more than the previous case considered in which the frequencies were equal, $\omega_1 = \omega_2$.

When the frequencies are unequal, however, we cannot use (10.31) any longer since current or voltage phasors such as $\mathbf{V}_1$ and $\mathbf{V}_2$, which must be defined in different phasor circuits (corresponding to different frequencies ω_i) cannot be superposed. In this case, the average power is computed from (10.33) as

$$P = \frac{1}{T} \int_0^T p \, dt = \frac{1}{T} \int_0^T (p_1 + p_2 + i_1 v_2 + i_2 v_1) \, dt$$

or
$$P = P_1 + P_2 + \frac{1}{T} \int_0^T (i_1 v_2 + i_2 v_1) \, dt \qquad (10.34)$$

Evaluating the first term in the integral, let

$$i_1 = I_{m1} \cos(\omega_1 t + \phi_{i1})$$

$$v_2 = V_{m2} \cos(\omega_2 t + \phi_{v2})$$

where we are interested in the case $\omega_1 \neq \omega_2$. Then

$$\frac{1}{T} \int_0^T (i_1 v_2) \, dt = \frac{I_{m1} V_{m2}}{T} \int_0^T \cos(\omega_1 t + \phi_{i1}) \cos(\omega_2 t + \phi_{v2}) \, dt \qquad (10.35)$$

Using the identity (10.8), for $\omega_1 \neq \omega_2$, both resulting terms in the integrand are sinusoids with nonzero frequency, both have zero average values, and thus (10.35) evaluates to zero. By identical reasoning, the second term in the integral of (10.34) also evaluates to zero, and (10.35) simplifies to

$$P = P_1 + P_2$$

In other words, *superposition holds for average power P when the sources are of different frequencies.* Thus to compute the average ac steady-state power in a circuit excited by sources at two or more frequencies, we compute the average power for each frequency separately and then add average powers. By a similar argument, it can be shown that *superposition also holds for reactive power Q when the sources are of different frequencies.* Interestingly, when phasors can be superposed ($\omega_1 = \omega_2$), neither P nor Q can be, and when phasors cannot be superposed ($\omega_1 \neq \omega_2$), superposition works for both P and Q.

Finally, since both P and Q obey the superposition principle for distinct frequencies, so must $\mathbf{S}$, the complex power. Two examples illustrating the use of superposition of power in ac steady-state circuits excited by sources of different frequencies follow.

Example 10.10

We wish to determine the average power absorbed by the resistor in Fig. 10.13(a). This is the same circuit used in Example 10.9 [see Fig. 10.12(a)], except the frequency of one of the sources has been changed. Since the frequencies of the sources in the present case

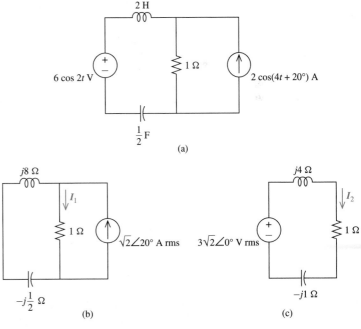

FIGURE 10.13 (a) Circuit diagram; (b) rms phasor diagram ($\omega = 4$ rad/s); (c) rms phasor diagram ($\omega = 2$ rad/s).

differ, we must compute their individual responses from different phasor circuits. Then we may apply superposition to the average powers produced at each frequency.

By current division

$$\mathbf{I}_1 = (\sqrt{2}\underline{/20^\circ})\frac{j15/2}{1 + j15/2} = 1.40\underline{/27.6^\circ} \text{ A rms}$$

and the power into the resistor at $\omega = 4$ is

$$P_1 = \text{Re}(\mathbf{Z})|\mathbf{I}_1|^2 = R|\mathbf{I}_1|^2$$
$$= (1)(1.40)^2 = 1.96 \text{ W}$$

For $\omega = 2$ the mesh current is

$$\mathbf{I}_2 = \frac{3\sqrt{2}\underline{/0}}{1 + j3} = 1.34\underline{/-71.6^\circ} \text{ A rms}$$

and the power into the resistor at $\omega = 2$ is

$$P_2 = R|\mathbf{I}_2|^2$$
$$= (1)(1.34)^2 = 1.80 \text{ W}$$

Thus the average power absorbed by the resistor is $1.96 + 1.80 = 3.76$ W.

Example 10.11

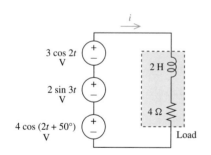

FIGURE 10.14 Circuit for Example 10.11.

We wish to find the average and reactive power absorbed by the load in Fig. 10.14. To determine the power at $\omega = 2$, we kill the source at $\omega = 3$, and since the two remaining voltage sources are in series, the rms phasor current at $\omega = 2$ is

$$\mathbf{I}_1 = \frac{1.5\sqrt{2}\underline{/0} + 2\sqrt{2}\underline{/50^\circ}}{4 + j4} = 0.795\underline{/-16.2^\circ} \text{ A rms}$$

The complex power absorbed by the load at $\omega = 2$ is then

$$\mathbf{S}_1 = \mathbf{Z}|\mathbf{I}_1|^2 = 0.795^2(4 + j4) = 2.53 + j2.53$$

At $\omega = 3$, the rms source phasor is $\sqrt{2}\underline{/0}$ (selecting sine phasors for convenience) and impedance $\mathbf{Z} = 4 + j\frac{3}{2}$. Then

$$\mathbf{I}_2 = \frac{\sqrt{2}\underline{/0}}{4 + j6} = 0.196\underline{/-56.3^\circ} \text{ A rms}$$

$$\mathbf{S}_2 = \mathbf{Z}|\mathbf{I}_2|^2 = 0.196^2(4 + j6) = 0.784 + j0.118$$

Since complex power at different frequencies superposes,

$$\mathbf{S} = \mathbf{S}_1 + \mathbf{S}_2 = 3.31 + j3.71$$

so the load absorbs 3.31 W of average power and 3.71 vars of reactive power.

In summary, for each frequency at which an ac steady-state circuit is excited, superposition may be applied at the level of currents and voltages, but not power. Then the average, reactive, or complex power may be found by superposing the power computed at each distinct frequency.

EXERCISES

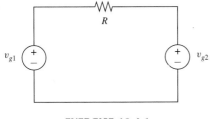

EXERCISE 10.4.1

10.4.1. Find the average power delivered to the resistor if $R = 10 \ \Omega$ and:
(a) $v_{g1} = 10 \cos 100t$ and $v_{g2} = 20 \cos(100t + 60°)$ V
(b) $v_{g1} = 100 \cos(t + 60°)$ and $v_{g2} = 50 \sin(2t - 30°)$ V
(c) $v_{g1} = 50 \cos(t + 30°)$ and $v_{g2} = 100 \sin(t + 30°)$ V
(d) $v_{g1} = 20 \cos(t + 25°)$ and $v_{g2} = 30 \sin(5t - 35°)$ V
Answer (a) 15 W; (b) 625 W; (c) 625 W; (d) 65 W

10.4.2. Find the average power absorbed by each resistor and each source.
Answer 8 W; 24 W; −8 W; −24 W

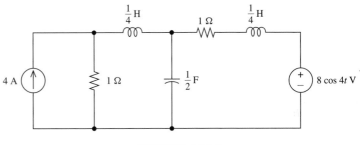

EXERCISE 10.4.2

10.4.3. Find the average power absorbed by the resistor and each source.
Answer 8 W; −4 W; −4 W

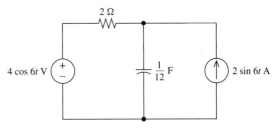

EXERCISE 10.4.3

10.5 MAXIMUM POWER TRANSFER

When designing a circuit, it is frequently desirable to arrange for the maximum possible power transfer to the load from the rest of the circuit. Consider the circuit of Fig. 10.15, consisting of a given two-terminal subcircuit connected to a load of impedance Z_L. We wish to specify Z_L so that the average power absorbed by this impedance is a maximum. Such problems arise, for instance, when specifying the impedance of an audio speaker system that will draw maximum power from the audio amplifier to

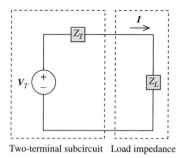

Two-terminal subcircuit   Load impedance

FIGURE 10.15   Load impedance connected to a subcircuit.

which it is attached. The process of specifying the load for maximum power from a given subcircuit is sometimes called impedance matching, for reasons that will soon be evident.

To simplify the analysis, we assume that the given two-terminal subcircuit is specified by its Thevenin equivalent, as shown in Fig. 10.15. The current into the load is given in terms of rms phasors by

$$\mathbf{I} = \frac{\mathbf{V}_T}{Z_L + Z_T}$$

and the average power absorbed by the load is, using (10.24b),

$$P_L = \text{Re}(Z_L)|\mathbf{I}|^2 \tag{10.36a}$$

$$= \frac{\text{Re}(Z_L)|\mathbf{V}_T|^2}{|Z_T + Z_L|^2} \tag{10.36b}$$

Let $Z_L = R_L + jX_L$ and $Z_T = R_T + jX_T$. In terms of these variables, (10.36b) may be written

$$P_L = \frac{R_L}{(R_L + R_T)^2 + (X_L + X_T)^2}|V_T|^2$$

We wish to know the values of R_L and X_L that maximize P for fixed values of the remaining quantities. The maximizing choice for X_L is clearly $X_L = -X_T$, since this makes the denominator as small as possible. Setting $X_L = -X_T$ gives

$$\left.\frac{\partial}{\partial R_L}P_L\right|_{X_L = X_T} = \frac{\partial}{\partial R_L}\frac{R_L}{(R_L + R_T)^2}$$

the maximizing choice for R_L is found by setting this partial derivative to zero. It is easily checked that the partial above is zero for $R_L = R_T$ (and that the second partial derivative is negative, indicating a maximum). We conclude that the maximum power is transferred to a load impedance Z_L when the real parts of the source impedance $R_T = \text{Re}(Z_T)$ and load impedance $R_L = \text{Re}(Z_L)$ are equal, and their imaginary parts $X_T = \text{Im}(Z_T)$ and $X_L = \text{Im}(Z_L)$ are equal and opposite in sign. We may state both requirements together by concluding that *the maximum power is transferred to a load Z_L from a given source with Thevenin equivalent impedance Z_T if the load and Thevenin equivalent impedance* *are complex conjugates:*

$$Z_L = Z_T^* \qquad (10.37)$$

Thus maximum power is transferred to a load if the magnitude of its impedance is matched to that of the source impedance Z_T and the angle of the load impedance matched to the negative of the angle of the source impedance. Load impedances whose magnitudes are large will sharply limit the current **I**, resulting in less power delivery, while those with small impedance will have only a small fraction of the voltage $\mathbf{V}_T$ dropped across it, again reducing the power draw.

To determine the maximum power draw, using $Z_L = Z_T^*$ in (10.36b),

$$P_{\text{max}} = \frac{\text{Re}(Z_T)|\mathbf{V}_T|^2}{|2\,\text{Re}(Z_T)|^2}$$

$$= \frac{|\mathbf{V}_T|^2}{4\,\text{Re}(Z_T)}$$

or, in terms of $Y_T = 1/Z_T$,

$$P_{\text{max}} = \frac{\text{Re}(Y_T)|\mathbf{V}_T|^2}{4}$$

$$= \frac{|\mathbf{I}_T|^2}{4\,\text{Re}(Y_T)}$$

It should be remembered that rms phasors are used in these expressions.

Example 10.12

Given the rms phasor circuit of Fig. 10.16(a), we first convert all but the load impedance to its Thevenin equivalent, as shown in Fig. 10.16(b). Z_T is the series equivalent of the impedances 3 and $j4$, or $3 + j4$. The open-circuit voltage with Z_L removed is

$$\mathbf{V}_T = (3)(4\underline{/0}) = 12\underline{/0} \text{ V rms}$$

Thus, for maximum power transfer, we select $Z_L = Z_T^* = 3 - j4\ \Omega$. The maximum power draw is

$$P_{\text{max}} = \frac{12^2}{(4)(3)} = 12 \text{ W}$$

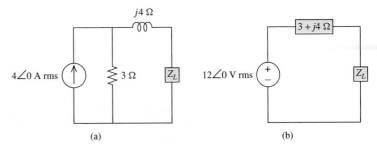

FIGURE 10.16   (a) Circuit for Example 10.12; (b) Thevenin equivalent.

EXERCISES

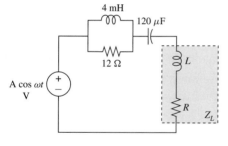

EXERCISE 10.5.1

10.5.1. Find values for load resistance R and and load inductance L so that the source with amplitude $A = 2\sqrt{2}$ V and frequency $\omega = 1000$ rad/s delivers maximum power to the load. Find the value of the power delivered to the load at these maximizing values for R and L.

Answer 1.2 Ω; 4.73 mH; 833 mW

10.5.2. Repeat Exercise 10.5.1 for $A = 4\sqrt{2}$ V and $\omega = 1000$ rad/s. Note that the optimal values for the load impedance do not depend on the source strength A, but the maximum power delivered does.

Answer 1.2 Ω; 4.73 mH; 3.33 W

10.5.3. For the given rms phasor circuit, determine the load impedance Z_L that would absorb maximum power.

Answer 4.0$\underline{/41.6°}$ Ω

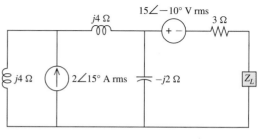

EXERCISE 10.5.3

10.6 CONSERVATION OF POWER

Conservation laws play key roles throughout the physical sciences. In circuit theory, no principle is more fundamental than the conservation of electrical charge, an immediate consequence of which is Kirchhoff's current law. In this section we see how the general principles of conservation of power and energy apply to electrical circuits in ac steady state.

Consider an arbitrary $n + 1$ node circuit with nodes numbered $0, 1, \ldots, n$. Let $\mathbf{V}_j$ be the node voltage rms phasor at node $j = 1, \ldots, n$ and node 0 be the reference node. The current flow from node j to node k will be designated $\mathbf{I}_{jk}$. By KCL at node j,

$$\sum_{k=0}^{n} \mathbf{I}_{jk} = 0$$

where $\mathbf{I}_{j0}$ is the current flow into the reference node and $\mathbf{I}_{jj} = 0$. Taking the complex conjugate of this KCL equation and multiplying by the node voltage $\mathbf{V}_j$ yields

$$\sum_{k=1}^{n} \mathbf{V}_j \mathbf{I}_{jk}^* + \mathbf{V}_j \mathbf{I}_{j0}^* = 0$$

where the term containing the reference node current is shown separately. There is one such equation for each node $j = 1, \ldots, n$. Adding all n equations together gives

$$\sum_{j=1}^{n} \left[\sum_{k=1}^{n} \mathbf{V}_j \mathbf{I}_{jk}^* + \mathbf{V}_j \mathbf{I}_{j0}^* \right] = 0$$

The current between every pair of nonreference nodes appears twice in the first double sum, as $\mathbf{I}_{jk}$ and its negative, $\mathbf{I}_{kj}$. Replacing $\mathbf{I}_{kj}$ by $-\mathbf{I}_{jk}$ for $k < j$, these two terms may be combined, producing the equation

$$\sum_{j=1}^{n} \left[\sum_{k=j+1}^{n} (\mathbf{V}_j - \mathbf{V}_k)\mathbf{I}_{jk}^* + \mathbf{V}_j \mathbf{I}_{j0}^* \right] = 0$$

Each term in this equation is recognized to be the complex power $\mathbf{S}_{jk}$ absorbed by the subcircuit connecting a particular node pair (j, k), and each node pair in the circuit appears exactly once. Thus this equation may be simplified to

$$\sum_{\text{node pairs } (j,k)} \mathbf{S}_{jk} = 0 \tag{10.38}$$

Finally, we note that since we have already counted all nodes in the circuit, the subcircuit connecting nodes j and k can only consist of elements in parallel. Suppose that there are m such, for $m > 0$. These individual elements are shown in Fig. 10.17. The complex power into this parallel subcircuit is

$$\mathbf{S}_{jk} = (\mathbf{V}_j - \mathbf{V}_k)\mathbf{I}_{jk}^*$$

which, by KCL applied to the upper node, is equal to

$$\mathbf{S}_{jk} = \sum_{l=1}^{m} (\mathbf{V}_j - \mathbf{V}_k)\mathbf{I}_{jkl}^*$$

The terms inside the summation are the complex power absorbed by each of the m elements separately. Combining this with (10.38), we have the principle of *conservation of complex power: the sum of the complex powers absorbed by all the elements in a circuit equals zero.*

Conservation of complex power immediately implies conservation of average power P and reactive power Q, since these are the real and imaginary parts of $\mathbf{S}$. By putting

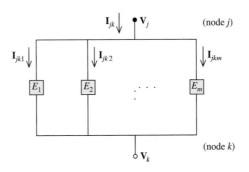

FIGURE 10.17 Elements connecting two nodes.

positive and negative terms on opposite sides of the equation, this conservation principle may be restated: *the net power absorbed by all elements in a circuit equals the net power delivered by all sources.* This principle, which holds equally for complex, average, and reactive power, is intuitively satisfying, since power absorbed must have come from somewhere and power delivered is delivered to somewhere. But it is important to have satisfied ourselves that such intuitively reasonable results follow from our basic definitions rather than simply to trust that they are probably true.

Example 10.13

We wish to determine the "power budgets," that is, the power absorbed by or delivered by each element, in the rms phasor circuit shown in Fig. 10.18. The equivalent impedance is

$$\mathbf{Z} = j2 + \frac{(2)(-j1)}{2 - j1} = 0.4 + j1.2 \ \Omega$$

and the current $\mathbf{I}$ is

$$\mathbf{I} = \frac{1/0}{0.4 + j1.2} = 0.791/{-71.6°} \ \text{A rms}$$

At the source,

$$\mathbf{S}_V = \mathbf{VI}^* = (1/0)(0.791/{+71.6°}) = 0.25 + j0.75$$

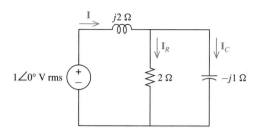

FIGURE 10.18 Circuit for Example 10.13.

This complex power must be interpreted as the power delivered by the source, since $\mathbf{V}$ and $\mathbf{I}$ violate the passive sign convention. The power absorbed by the inductor is

$$\mathbf{S}_L = \mathbf{Z}_L|\mathbf{I}|^2 = (j2)(0.791)^2 = 0 + j1.25$$

We know that the complex power absorbed by a resistor is purely real and by a capacitor it is purely imaginary. By conservation of average power, the 0.25 W produced by the source must then be absorbed by the resistor, or

$$\mathbf{S}_R = 0.25 + j0$$

The source absorbs -0.750 var (produces $+0.75$ var) and the inductor absorbs an additional 1.25 var, so by conservation of reactive power the capacitor must produce 0.50 var or, in terms of power absorbed,

$$\mathbf{S}_C = 0 - j0.50$$

The last two results may be verified by direct calculations, since by current division

$$\mathbf{I}_R = \frac{(0.791/{-71.6°})(-j1)}{2 - j1} = 0.354/{-135°} \ \text{A rms}$$

$$\mathbf{S}_R = Z_R|\mathbf{I}_R|^2 = (2)(0.354)^2 = 0.25 + j0$$

$$\mathbf{I}_C = \mathbf{I} - \mathbf{I}_R = 0.707\underline{/-45^\circ} \text{ A rms}$$

$$\mathbf{S}_C = Z_C|\mathbf{I}_C|^2 = (-j1)(0.707)^2 = -j0.50$$

Finally, we note that since energy is the time integral of power and power is conserved, energy is conserved as well. The net energy produced by the sources in a circuit over a given period of time is absorbed by the other elements. Each source producing net energy may be studied further to determine the primary cause responsible for this energy production. In a storage battery, for instance, the primary cause is a chemical reaction converting electrochemical potential energy to electrical energy delivered by the battery to the rest of the electrical circuit. In a hydro generator it is the work done by the gravitational potential energy of falling water on a spinning shaft. Our study of electrical circuits, limited as it is to the electrical variables of current and voltage, will determine the amount of electrical energy produced by a source, but it is mute concerning its origins. All we can say from our circuit analysis is that if an element is a net producer of electrical energy, this energy must have been derived from some other, nonelectrical prime source.

EXERCISES

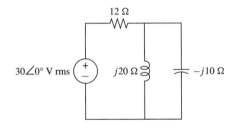

EXERCISE 10.6.1   Rms phasor circuit.

10.6.1. Determine the complex power absorbed by the resistor, inductor, and capacitor and show that their sum equals the complex power delivered by the source.
Answer Power absorbed by each of the elements: $\mathbf{S}_R = 19.9$; $\mathbf{S}_L = +j33.1$; $\mathbf{S}_C = -j66.2$; $\mathbf{S}_V = -19.9 + j33.1$

10.6.2. A circuit consists of a 110-V rms source, a resistor, and an inductor. If the resistor absorbs 25 W and the inductor absorbs 40 var, what is the rms current produced by the source?
Answer 429 mA rms

10.6.3. Use conservation of power to find the power delivered to the resistor by first finding the real part of the complex power supplied by the source.
Answer 36 W

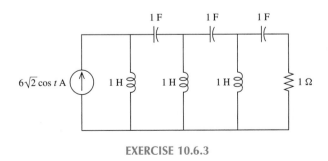

EXERCISE 10.6.3

10.7 REACTIVE POWER AND POWER FACTOR

The average power delivered to a load in the ac steady state, repeating (10.14), is

$$P = V_{rms} I_{rms} \cos\theta$$

where $\theta = \phi_v - \phi_i$. The power is thus equal to the product of the rms voltage, the rms current, and the cosine of the angle between the voltage and current phasors. In practice, the rms current and voltage are easily measured and their product, $V_{rms} I_{rms}$, is called the *apparent power*. The apparent power is usually referred to in terms of its units, volt-amperes (VA) or kilovolt-amperes (kVA), to avoid confusing it with the unit of average power, the watt. It is clear that the average power can never be greater than the apparent power.

The ratio of the average power to the apparent power is defined as the *power factor*. Thus if we denote the power factor by pf, then in the sinusoidal case

$$pf = \frac{P}{V_{rms} I_{rms}} = \cos\theta \qquad (10.39)$$

which is dimensionless. The angle θ, in this case, is often referred to as the *pf angle*. We also recognize it as the angle of the impedance Z of the load.

In the case of purely resistive loads, the voltage and current are in phase. Therefore, $\theta = 0$, $pf = 1$, and the average and apparent powers are equal. A unity power factor ($pf = 1$) can also exist for loads that contain inductors and capacitors if the reactances of these elements are such that they cancel one another. Adjusting the reactance of loads so as to approximate this condition is very important in electrical power systems, as we shall see shortly.

In a purely reactive load, $\theta = \pm90°$, $pf = 0$, and the average power is zero. In this case the equivalent load is an inductance ($\theta = 90°$) or a capacitance ($\theta = -90°$), and the current and voltage differ in phase by $90°$.

A load for which $-90° < \theta < 0$ is equivalent to an *RC* combination, whereas one having $0 < \theta < 90°$ is an equivalent *RL* combination. Since $\cos\theta = \cos(-\theta)$, it is evident that the pf for an *RC* load having $\theta = -\theta_1$, where $0 < \theta_1 < 90°$, is equal to that of an *RL* load with $\theta = \theta_1$. To avoid this difficulty in identifying such loads, the pf is characterized as *leading* or *lagging* by the *phase of the current with respect to that of the voltage*. Therefore, an *RC* load has a leading pf and an *RL* load has a lagging pf. For example, the series connection of a 100-Ω resistor and a 0.1-H inductor at 60 Hz has $Z = 100 + j37.7 + 106.9\underline{/20.66°}\ \Omega$ and a pf of $\cos 20.66° = 0.936$ lagging.

Example 10.14

In practice, the power factor of a load is very important. In industrial applications, for instance, loads may require thousands of watts to operate, and the power factor greatly affects the electric bill. Suppose, for example, that a mill consumes 100 kW from a 220-V rms line. At a pf of 0.85 lagging, we see that the rms current into the mill is

$$I_{\text{rms}} = \frac{P}{V_{\text{rms}}pf} = \frac{10^5}{(220)(0.85)} = 534.8 \text{ A}$$

which means that the apparent power supplied is

$$V_{\text{rms}}I_{\text{rms}} = (220)(534.8) \text{ VA} = 117.66 \text{ kVA}$$

Now suppose that the pf by some means is increased to 0.95 lagging. Then

$$I_{\text{rms}} = \frac{10^5}{(220)(0.95)} = 478.5 \text{ A}$$

and the apparent power is reduced to

$$V_{\text{rms}}I_{\text{rms}} = 105.3 \text{ kVA}$$

Comparing the latter case with the former, we see that I_{rms} was reduced by 56.3 A (10.5%). Therefore, the generating station must generate a larger current in the case of the lower pf. Since the transmission lines supplying the power have resistance, the generator must produce a larger average power to supply the 100 kW to the load. If the resistance is 0.1 Ω, for instance, then the power generated by the source must be

$$P_g = 10^5 + 0.1I_{\text{rms}}^2$$

Therefore, we find

$$P_g = \begin{cases} 128.6 \text{ kW}, & pf = 0.85 \\ 122.9 \text{ kW}, & pf = 0.95 \end{cases}$$

which requires that the power station produce 5.7 kW (4.64%) more power to supply the lower pf load. It is for this reason that power companies encourage a pf exceeding, say, 0.9 and impose a penalty on large industrial users who do not comply.

Let us now consider a method of correcting the power factor of a load having an impedance Z or admittance $Y = 1/Z$. We may alter the power factor by connecting an element in parallel with the given load, as in Fig. 10.19. By placing the new element in parallel, the voltage delivered to the load will not be divided between the elements, as it would if the new element were placed in series.

Designate the new element by its admittance Y_1. The total admittance of the parallel combination is the sum of admittances

$$Y_T = Y + Y_1$$

We wish the new element to absorb no average power itself, so it must be a pure reactance, $\text{Re}(Y_1) = 0$ shown as a vertical vector in the phasor diagram. If our goal is to adjust the power factor pf to some desired value $pf = PF$, then, as shown in the diagram with $\cos \theta_d = PF$,

$$Y_1 = j \left[\text{Im}(Y) - \text{Im}(Y_T) \right]$$

where

$$\text{Im}(Y_T) = \text{Re}(Y) \tan \theta_d$$

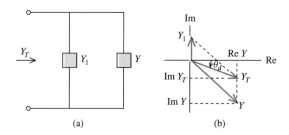

FIGURE 10.19 (a) Circuit for power factor correction; (b) phasor diagram.

In other words, Y_1 must reduce the imaginary part of Y so that with fixed real part the angle of Y_T will have the desired power factor PF.

Example 10.15

As an example, let us change the power factor for the circuit of Example 10.14 (shown in Fig. 10.18) to 0.98 leading. The impedance of this circuit was found in Example 10.13 to be $Z = 0.4 + j1.2$ Ω, so its admittance is

$$Y = \frac{1}{Z} = 0.25 - j0.75 = 0.791\underline{/-71.6°} \text{ S}$$

Thus the circuit as given has a power factor of $\cos(71.6°) = 0.32$, which is a lagging power factor (the angle of an admittance in this case $-71.6°$, is the angle by which the current leads the voltage). The desired Y_T must have an angle of $\theta_d = \cos^{-1} 0.98 = +11.5°$ (Y_T of phase angle $-11.5°$ would be 0.98 *lagging* rather than leading). The unique complex number that has real part 0.25 and angle $+11.5°$ has as its imaginary part

$$\text{Im}(Y_T) = (0.25)(\tan 11.5°) = 0.051 \text{ S}$$

Thus we must change the admittance from $Y = 0.25 - j0.75$ S to $Y_T = 0.25 + j0.051$ S. This requires that

$$Y_1 = Y_T - Y = j(0.051 + 0.75) = j0.801 \text{ S}$$

Thus, to correct the power factor to the desired value, we need to install an admittance of $j0.801$ S or, equivalently, an impedance of $1/j0.801 = -j1.25$ Ω, in parallel with the given impedance. For instance, if the working frequency for the circuit is $\omega = 1$ rad/s, then the power factor correction would be achieved by using a capacitor with impedance

$$\frac{1}{j(1)C} = -j1.25 \text{ Ω}$$

or $C = 0.80$ F.

EXERCISES

10.7.1. Find the apparent power for (a) a load that requires 20 A rms from a 115-V rms line and (b) a load consisting of a 100-Ω resistor in parallel with a 25-μF capacitor connected to a 120-V rms 60-Hz source.

Answer (a) 2.3 kVA; (b) 197.9 VA

10.7.2. Find the power factor for (a) a load consisting of a series connection of a 10-Ω resistor and a 10-mH inductor operating at 60 Hz, (b) a capacitive load requiring 25 A rms and 5 kW at 230 V rms, and (c) a load that is a parallel connection of a 5-kW load with a power factor of 0.9 leading and a 10-kW load with a 0.95 lagging power factor.

Answer (a) 0.936 lagging; (b) 0.87 leading; (c) 0.998 lagging

10.7.3. A load consisting of a series combination of a 100-Ω resistor and 1-H inductor is connected to a source delivering power at $\omega = 100$ rad/s. What capacitor C across this load will correct the power factor to (a) 0.9 leading; (b) 0.9 lagging?

Answer (a) $C = 74.22 \ \mu$F; (b) $C = 25.78 \ \mu$F

10.8 SPICE AND AC STEADY-STATE POWER

When an ac analysis is requested of SPICE by inclusion of the .AC control statement, some additional information labeled "small-signal bias solution" accompanies the output. This includes a "total power dissipated" value. For instance, Example 9.17 used the source file

```
Circuit of Fig. 9.25
*
V     1    0   AC    1    0
L     1    2   1
H     2    3   V     0.5
R     3    0   1
.AC  LIN   1   .159  .159
.PRINT  AC  IM(V)   IP(V)
.END
```

The following data were found in the SPICE printout above the printout that we requested (the magnitude and phase angle of the current though the voltage source labeled V).

```
*** SMALL SIGNAL BIAS SOLUTION TEMP = 27.0 DEG C ***
NODE VOLTAGE          NODE VOLTAGE          NODE VOLTAGE
( 1) 0.0000           ( 2) 0.0000           ( 3) 0.0000

VOLTAGE SOURCE CURRENTS
NAME      CURRENT
V         0.000E+00

TOTAL POWER DISSIPATION 0.00E+00 WATTS
```

The power dissipation reported in this portion of the printout is *not* related to the ac power dissipated by the elements of the circuit. Whenever an ac analysis is requested, SPICE automatically first determines the dc operating point of the circuit, that is, the dc or constant values of all currents and voltages when the ac sources are killed. This operating point information is of great value in the case that nonlinear elements, such as transistors and diodes, are included in the circuit, and dc supplies are used to power them. This is the "small-signal bias solution" reported. The "total power dissipation" reported refers to steady-state dc power required to support this operating point. In a circuit with no dc sources all these values will be zero. Thus the power dissipation automatically reported by SPICE is of no help in determining ac steady-state power dissipations at nonzero frequencies.

Sadly, SPICE offers no specific control card or keyword for printing out ac steady-state power at any other frequency than $\omega = 0$ (dc) as described above. To determine the ac power absorbed by a circuit element using SPICE, we must request output of the current and voltage phasors; then use the formulas studied in this chapter to get complex, average, or reactive power (as required) from knowledge of the element current and voltage.

SPICE Example 10.16

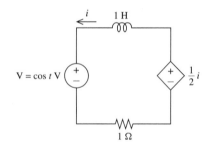

FIGURE 10.20 Circuit for Example 10.16.

We wish to determine the complex power supplied by the two sources in the circuit of Fig. 10.20. The SPICE input file for this circuit was listed at the beginning of this section. The output from this SPICE run is

```
FREQ          IM(V)         IP(V)
1.590E-01     8.951E-01     1.166E+02
```

The current phasor associated with the counterclockwise mesh current i in Fig. 10.20 is thus $\mathbf{I} = 0.895\underline{/116.6°}$ A. Since this current phasor satisfies the passive sign convention relative to the independent source phasor voltage $\mathbf{V} = 1\underline{/0}$, the complex power absorbed by the independent source is given by

$$\tfrac{1}{2}\mathbf{VI}^* = \tfrac{1}{2}(1\underline{/0})(0.895\underline{/-116.6°}) = -0.2 - j0.4 \qquad (10.40a)$$

and the complex power absorbed by the dependent source is given by

$$\tfrac{1}{2}\left(\tfrac{1}{2}\mathbf{I}\right)(-\mathbf{I})^* = -0.2 + j0 \qquad (10.40b)$$

Note that we used $-\mathbf{I}$, not $\mathbf{I}$, in the last calculation, since $-\mathbf{I}$ satisfies the passive sign convention for the dependent source (together with its voltage $\tfrac{1}{2}\mathbf{I}$). Also note that the phasors reported by SPICE are amplitude, not rms, phasors since we used amplitude phasors to describe the source. Thus we must include the factor $\tfrac{1}{2}$ in the power formulas (10.40) consistent with the use of non-rms standard or amplitude phasors. Examining our result, we see that the independent and dependent sources each deliver 200 mW to the rest of the circuit. By conservation of average power, the resistor must be absorbing this 400 mW. In addition, the independent source supplies 0.4 var to the inductor.

Finally, note that rms values may be used to describe the magnitude of sources in SPICE input files. In this case all calculated current and voltage phasor magnitudes will be in rms as well, and the factor of $\frac{1}{2}$ should be omitted from calculations of complex power such as (10.40a).

EXERCISES

10.8.1. Find the power delivered to the 80-Ω resistor.
Answer 13.0 mW

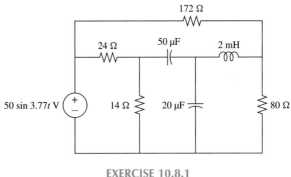

EXERCISE 10.8.1

10.8.2. Check the results of Example 10.11 (Fig. 10.14) using SPICE.
Answer 3.02 W, 2.74 var. Note that two SPICE runs are required, one at $\omega = 2$ rad/s and one at $\omega = 3$ rad/s (0.318 and 0.477 Hz).

SUMMARY

The power calculations are important not just to the electrical power generation and distribution industries, but also in the design of efficient circuits of all kinds and of miniaturized circuits which must dissipate heat over very small areas. Just as was the case with currents and voltages in Chapters 8 and 9, the calculation of power in ac steady-state circuits is greatly simplified by the use of phasors.

- For terminal variables $v(t)$ and $i(t)$ that satisfy the passive sign convention, instantaneous power $p(t) = v(t)i(t)$ greater than zero implies the circuit between the terminals is absorbing power at t, less than zero that it is supplying power to the rest of the circuit at t.

- Average power equals the product of current amplitude, voltage amplitude, and the cosine of the difference of their phase angles. If current and voltage satisfy the passive sign convention, the average power greater than zero implies that power is on average dissipated, less than zero, and that it is being supplied to the rest of the circuit.

- The rms value of a sinusoid equals its amplitude divided by $\sqrt{2}$. An rms phasor is a phasor whose magnitude equals the rms value of the corresponding sinusoid.

- The real part of the complex power $S = \mathbf{V}_{rms}\mathbf{I}^*_{rms}$ is the average power (SI unit watt); the imaginary part is the reactive power (SI unit var).

- In a circuit containing sources at different frequencies, the total power for each component is the sum of the power delivery or dissipation at each frequency.

- Maximum power is delivered to a load impedance if it is the complex conjugate of the Thevenin equivalent impedance of the rest of the circuit.

- All forms of power are conserved: instantaneous, average, reactive, and complex.

- The power factor is the cosine of the phase angle between voltage and current. Unity power factor requires the least rms current to deliver a fixed amount of average power to a load at a given voltage level.

PROBLEMS

10.1. $i(t)$ flows through a 1-kΩ resistor. Find T, the period of $i(t)$, T_p, the period of the instantaneous power, and P, the average power.

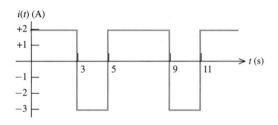

FIGURE P10.1

10.2. Let $v(t) = 2\cos \pi t$ V for $3n \le t < 3n + 1$, $n = 0$, $\pm 1, \pm 2, \ldots$, and $v(t) = 0$ elsewhere. If $v(t)$ is across a 1-Ω resistor, sketch the instantaneous power $p(t)$ and find the average power P.

10.3. One cycle of a periodic current is given by

$$i = 10 \text{ A}, \quad 0 \le t < 1 \text{ ms}$$

$$= 0, \qquad 1 \le t < 4 \text{ ms}$$

If the current flows in a 20-Ω resistor, find the average power.

10.4. One period of a periodic current is as shown. If current flows in a 20-Ω resistor, find the average power.

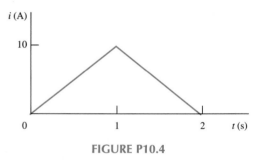

FIGURE P10.4

10.5. Find the average power dissipated in a resistor of resistance R if the current is $i = I_m(1 + \cos \omega t)$ A.

10.6. Find the average power absorbed by the resistors, the inductor, and the source.

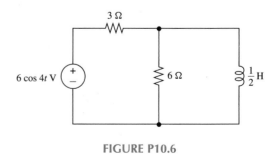

FIGURE P10.6

10.7. Find the average power delivered to the resistor when $\phi = 0$. How does the average power depend on the phase angle ϕ?

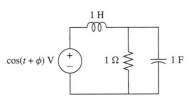

FIGURE P10.7

10.8. Find the average power delivered to each *RLC* element and the average power delivered by the source.

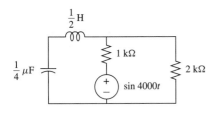

FIGURE P10.8

10.9. For what value of ϕ is the average power delivered to the 1-Ω resistor smallest?

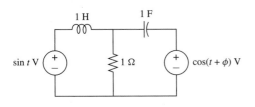

FIGURE P10.9

10.10. Find the average power absorbed by the 3-kΩ resistor and the dependent source.

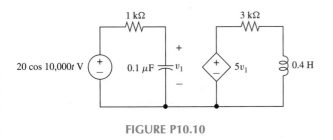

FIGURE P10.10

10.11. Sketch the rms phasor circuit and find the average power supplied by the source.

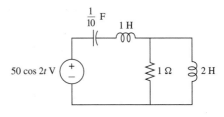

FIGURE P10.11

10.12. Find the rms value of the voltage
(a) $v = 5 + 12\sqrt{2}\cos t$ V
(b) $v = 12\cos(2t - 30°) + 4\cos 5t + \sqrt{2}\cos(8t + 60°)$ V
(c) $v = 4\cos t - 6\sqrt{2}\cos(2t + 45°) + \sqrt{10}\cos(5t - 10°)$ V

10.13. Find the rms value of a periodic current for which one period is given by
(a) $i = 4$ A, $\quad 0 < t < 1$ s
  $= 0$, $\quad\quad 0 < t < 2$ s
(b) $i = 3t$ A, $\quad\quad\quad 0 < t < 2$ s
  $= -3(t - 4)$ A, $\quad 0 < t < 4$ s
(c) $i = I_m \sin \dfrac{2\pi t}{T}$ A, $\quad 0 < t < \dfrac{T}{2}$ s
  $= 0$, $\quad\quad\quad\quad\quad \dfrac{T}{2} < t < T$ s
(d) $i = I_m \sin \dfrac{\pi t}{T}$ A, $\quad 0 < t < T$ s

10.14. Find an expression for P, the average power absorbed by an impedance Z, in terms of $|V_{\text{rms}}|^2$, where V_{rms} is the rms voltage phasor across the impedance.

10.15. Sketch the rms phasor circuit and find the average power delivered to each element.

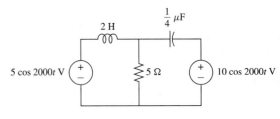

FIGURE P10.15

10.16. Determine the rms current; $i(t)$ is periodic with period 1 ms, and equals $\sin 2\pi t$ for $0 \le t < 0.5$, and zero for $0.5 \le t < 1$ ms.

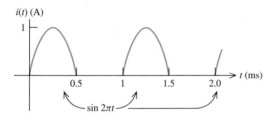

FIGURE P10.16

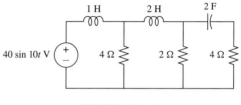

FIGURE P10.22

10.17. Show that the average power absorbed by an impedance Z may be expressed as $P = |\mathbf{I}_{rms}|^2 \operatorname{Re} Z$ where $\mathbf{I}_{rms}$ is the rms current phasor.

10.18. An impedance $Z = 2 - 3j$ Ω carries a current $i(t) = 3\sin(10t - 14°)$ A. Find the complex power absorbed by Z.

10.19. A series combination of a resistor and capacitor absorbs 100 W and −42 vars at 60 Hz. If $R = 100$ Ω, what must C be?

10.20. $\mathbf{V}_s$ is the rms source phasor. What must $|\mathbf{V}_s|$ be in order that the average power absorbed by the load be 25 W?

10.23. Solve Problem 10.8 by finding the complex power for each element and taking the real part.

10.24. Find the reactive power absorbed by the inductor if (a) $\omega = 1$ rad/s; (b) $\omega = 0$.

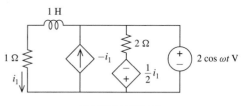

FIGURE P10.24

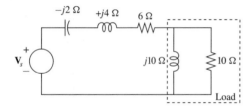

FIGURE P10.20

10.21. Find the complex power delivered to each inductor and resistor if $v_g = 0.10\sin(2t + 30°)$ V.

10.25. Find the rms value of the steady-state voltage v.

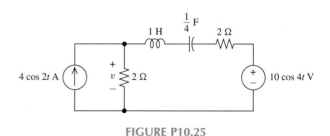

FIGURE P10.25

10.26. Find the rms value of the steady-state current in R.

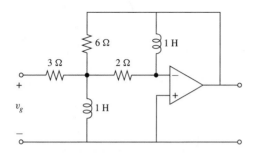

FIGURE P10.21

10.22. Determine the complex power supplied by the source. Use the ladder method to find the equivalent impedance.

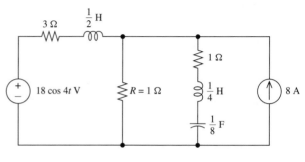

FIGURE P10.26

10.27. Find the complex power delivered by each source.

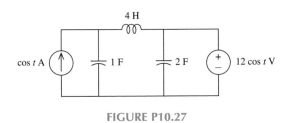

FIGURE P10.27

10.28. For the rms phasor circuit shown, find the complex power supplied by the controlled source.

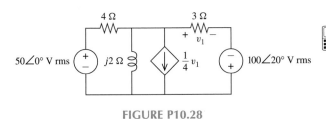

FIGURE P10.28

10.29. Find P_a, the average power absorbed by the 4-Ω resistor due to $v_a(t)$ with $v_b(t)$ killed, and P_b, the power absorbed by the same elements due to $v_b(t)$ with $v_a(t)$ killed. Show that the power P absorbed by this element in this circuit is *not* equal to the sum $P_a + P_b$.

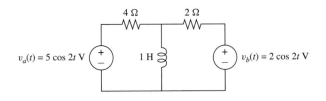

FIGURE P10.29

10.30. Compute the watts supplied to the load Z_L.

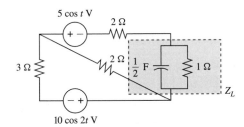

FIGURE P10.30

10.31. Determine the average power P delivered by each of the three sources.

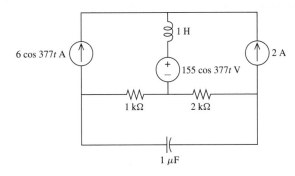

FIGURE P10.31

10.32. Find the power delivered to the 100-Ω load resistor.

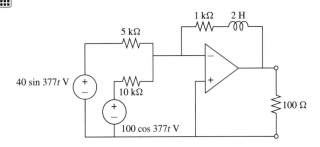

FIGURE P10.32

10.33. For a Thevenin equivalent circuit consisting of a voltage source $\mathbf{V}_g$ and an impedance $\mathbf{Z}_g = R_g + jX_g$, (a) show that the circuit delivers maximum average power to a load $\mathbf{Z}_L = R_L + jX_L$ when $R_L = R_g$ and $X_L = -X_g$, and (b) show that the maximum average power is delivered to a load R_L when $R_L = |\mathbf{Z}_g|$. This is the *maximum power transfer theorem* for ac circuits. (In both cases, $\mathbf{V}_g$ and $\mathbf{Z}_g$ are fixed, and the load is variable.)

10.34. What must the amplitude A of the source be to supply 100 W to the purely resistive load $Z_L = 2\ \Omega$? What would it be reduced to if we could adjust Z_L to absorb 100 W using the lowest possible value source amplitude A?

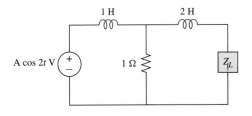

FIGURE P10.34

10.35. Find the value of transconductance g for which the real average power delivered by the independent source is maximized. Repeat for apparent power.

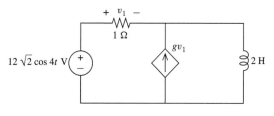

FIGURE P10.35

10.36. Find the load impedance Z_L that will absorb maximum real average power. Find the value of the maximum real average power into the load.

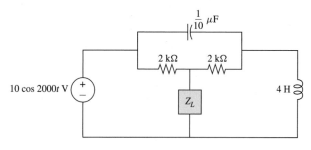

FIGURE P10.36

10.37. For maximum power transfer to the load Z_L, should Z_L be constructed as an RL circuit or an RC circuit?

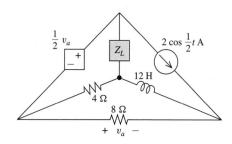

FIGURE P10.37

10.38. Find the reactive power absorbed by the inductor. (*Hint*: Find the complex power supplied by the source.)

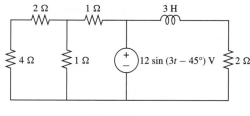

FIGURE P10.38

10.39. Verify conservation of complex power for this rms phasor circuit.

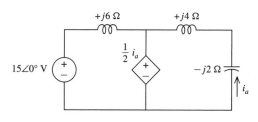

FIGURE P10.39

10.40. An ideal 220-V rms voltage source is put across a load impedance of $Z = 11 + j11$ Ω. How many vars does the source supply? If $\omega = 377$ rad/s, what capacitor C connected across the load will reduce the vars supplied by the source to zero?

10.41. Show that for all $R > 0$, $C > 0$, and $\omega > 0$, the power factor of a series RC circuit and of a parallel RC circuit are leading.

10.42. Show that for all $R > 0$, $L > 0$, and $\omega > 0$, the power factor of a series RL circuit and of a parallel RL circuit are lagging.

10.43. Find the rms value of the steady-state current i and the power factor seen from the source terminals. What element connected in parallel with the source will correct the power factor to 0.8 lagging?

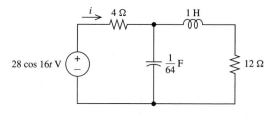

FIGURE P10.43

10.44. Find the power factor seen from the terminals of the source and the reactance necessary to connect in parallel with the source to change the power factor to unity.

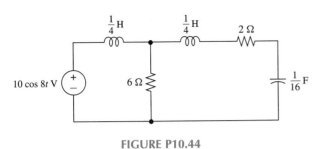

FIGURE P10.44

10.45. Find the power factor seen from the terminals of the independent source and the reactance which must be connected in parallel with the independent source to change the power factor to 0.8 lagging.

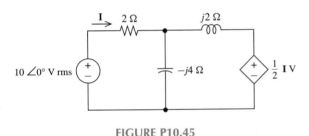

FIGURE P10.45

10.46. Three parallel passive loads, Z_1, Z_2, and Z_3, are receiving complex power values of $6 - j5$, $8 + j10$, and $2 + j7$ VA, respectively. If a voltage source of $50\underline{/0}$ V rms is connected across these loads, find the rms value of the current that flows from the source and the power factor seen by the source.

10.47. Two loads in parallel draw a total of 3 kW at a 0.9 lagging power factor from a 115-V rms 60-Hz line. One load is known to absorb 1000 W at a 0.8 lagging power factor. Find (a) the power factor of the second load and (b) the parallel reactive element necessary to correct the power factor to 0.95 lagging for the combined load.

SPICE Problems

10.48. A load impedance $Z_L = 1 + j0.5$ is connected to a 10-mA rms ideal current source. What is the power factor of the load? How much could the voltage rating of the current source be reduced if a capacitor was put in series with the load, creating an overall power factor of 0.95 lagging? Use SPICE to determine the capacitance which corrects the pf to 0.95 lagging.

10.49. Use SPICE to help determine the value of R for which the source supplies 1.00 mW to the rest of the circuit.

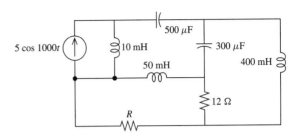

FIGURE P10.49

10.50. Repeat Problem 10.21 using SPICE. Use the ideal voltage amplifier op amp model of Fig. 3.7 with open-loop gain $A = 10^6$.

More Challenging Problems

10.51. At what frequency ω is the rms value of v_2 maximized?

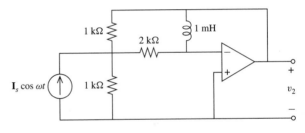

FIGURE P10.51

10.52. (a) Determine the complex power delivered by the independent source. (b) Does the answer to (a) depend upon the source phase angle ($-17°$)? Explain.

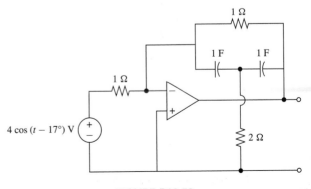

FIGURE P10.52

10.53. For what range ω is the current supplying more real average power than the voltage source?

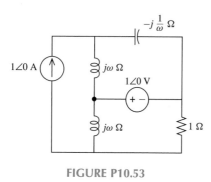

FIGURE P10.53

10.54. Find the complex power delivered by each source. $i_1 = 2\cos 2t$ A, $i_2 = 6\sin 2t$ A, $v_1 = 12\sin(3t - 45°)$ V, $v_2 = 24\sin(3t - 45°)$ V.

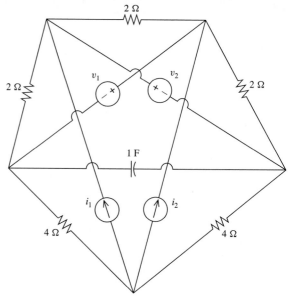

FIGURE P10.54

11

Three-Phase Circuits

Nikola Tesla
1856–1943

He [Tesla] was the greatest inventor in the realm of electrical engineering.

W. H. Eccles

If Thomas A. Edison has a rival for the title of the world's greatest inventor, it is certainly the Croatian-American engineer, Nikola Tesla. When the tall, lanky Tesla arrived in the United States in 1884, the country was in the middle of the "battle of the currents" between Thomas A. Edison, promoting dc, and George Westinghouse, leading the ac forces. Tesla quickly settled the argument in favor of ac with his marvelous inventions, such as the polyphase ac power system, the induction motor, the Tesla coil, and fluorescent lights.

Tesla was born in Smiljan, Austria-Hungary (now Croatia), the son of a clergyman of the Greek Orthodox Church. As a boy Tesla had a talent for mathematics and an incredible memory, with the ability to recite by heart entire books and poems. He spent two years at the Polytechnic Institute of Graz, Austria, where he conceived the idea of the rotating magnetic field that was the later basis for his induction motor. At this point in Tesla's life his father died, and he decided to leave school, taking a job in Paris with the Continental Edison Company. Two years later he came to America, where he remained until his death. During his remarkable lifetime he held over 700 patents, settled the ac versus dc dispute, and was primarily responsible for the selection of 60 Hz as the standard ac frequency in the United States and throughout much of the world. After his death he was honored by the choice of *tesla* as the unit of magnetic flux density.

Chapter Contents

As we have already noted, one very important use of ac steady-state analysis is its application to power systems, most of which are alternating current systems. One principal reason for this is that it is economically feasible to transmit power over long distances only if the voltages involved are very high, and it is easier to raise and lower voltages in ac systems than in dc systems. Alternating voltage can be stepped up for transmission and stepped down for distribution with transformers, as we shall see in Chapter 15. Transformers have no moving parts and are relatively simple to construct. In a dc system, on the other hand, the waveforms must first be converted to ac before transformers may be used to change voltage levels and then converted back to dc. This process requires expensive high-power switches and dissipates some of the energy it is converting.

Also, for reasons of economics and performance, almost all electric power is produced by *polyphase* sources (those generating voltages with more than one phase). In a single-phase circuit, the instantaneous power delivered to a load is pulsating, even if the current and voltage are in phase. A polyphase system, on the other hand, is somewhat like a multicylinder automobile engine in that the power delivered is much steadier. An economic advantage is that the weight of the conductors and associated components required in a polyphase system is appreciably less than that required in a single-phase system that delivers the same power. Virtually all the power produced in the world is polyphase power at 50 or 60 Hz. In the United States, 60 Hz is the standard frequency.

In this chapter we begin with single-phase three-wire systems, but we concentrate on three-phase circuits, which are by far the most common of the polyphase systems. In the latter case the sources are three-phase generators that produce a *balanced* set of voltages, by which we mean three sinusoidal voltages having the same amplitude and frequency but displaced in phase by 120°. Thus the three-phase source is equivalent to three interconnected single-phase sources, each generating a voltage with a different phase. If the three currents drawn from the sources also constitute a balanced set, the system is said to be a *balanced* three-phase system. A balanced system is required to reap the full benefits of polyphase power, and this is the case on which we concentrate our attention.

11.1 SINGLE-PHASE THREE-WIRE SYSTEMS

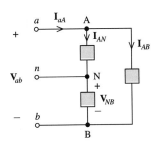

FIGURE 11.1 Double-subscript notation.

Before proceeding to the three-phase case, let us digress in this section to establish our notation and consider an example of a single-phase system that is in common household use. The circuits we have studied thus far are all single phase and are a good frame of reference for understanding the significance of multiphase.

In this chapter we shall find extremely useful the double-subscript notation introduced in Chapter 1 for voltages. In the case of phasors, the notation is $\mathbf{V}_{ab}$ for the voltage of point a with respect to point b. We shall also use a double-subscript notation for current, taking, for example, $\mathbf{I}_{aA}$ as the current flowing in the *direct* path from node a to node A. These quantities are illustrated in Fig. 11.1, where the direct path (no intermediate nodes) from A to B is distinguished from an indirect path from A to B through N.

Because of the simpler expressions for average power that result, we shall use rms values of voltage and current, and rms phasors, throughout this chapter. That is, if

$$\mathbf{V} = |\mathbf{V}|\underline{/0} \text{ V rms}$$

$$\mathbf{I} = |\mathbf{I}|\underline{/-\theta} \text{ A rms} \tag{11.1}$$

are the rms phasors associated with an element having impedance,

$$Z = |Z|\underline{/\theta} \text{ } \Omega \tag{11.2}$$

the average power delivered to the element is

$$P = |\mathbf{V}| \cdot |\mathbf{I}| \cos\theta$$

$$= |\mathbf{I}|^2 \text{ Re}(Z) \text{ W} \tag{11.3}$$

In the time domain the voltage and current are

$$v = \sqrt{2}|\mathbf{V}| \cos\omega t \text{ V}$$

$$i = \sqrt{2}|\mathbf{I}| \cos(\omega t - \theta) \text{ A}$$

All phasors in this chapter will be understood to be rms phasors, whose magnitude corresponds to the rms value of the associated current or voltage, as introduced in Section 10.2.

Example 11.1

The use of double subscripts makes it easier to handle phasors both analytically and geometrically. For example, in Fig. 11.2(a), the voltage $\mathbf{V}_{ab}$ is

$$\mathbf{V}_{ab} = \mathbf{V}_{an} + \mathbf{V}_{nb}$$

This is evident without referring to a circuit since by KVL the voltage between two points a and b is the same regardless of the path, which in this case is the path a–n–b. Also, since $\mathbf{V}_{nb} = -\mathbf{V}_{bn}$, we have

$$\mathbf{V}_{ab} = \mathbf{V}_{an} - \mathbf{V}_{bn}$$

$$= 100 - 100\underline{/-120°}$$

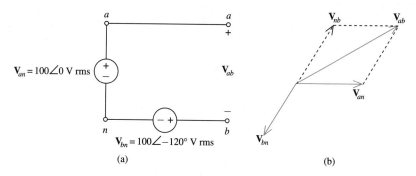

(a)   (b)

FIGURE 11.2   (a) Phasor circuit; (b) corresponding phasor diagram.

which, after simplification, is

$$\mathbf{V}_{ab} = 100\sqrt{3}\underline{/30°} \text{ V rms}$$

These steps are shown graphically in Fig. 11.2(b).

A single-phase three-wire source, as shown in Fig. 11.3, is one having three output terminals a, b, and a *neutral* terminal n that center-taps the voltage $\mathbf{V}_{ab}$:

$$\mathbf{V}_{an} = \mathbf{V}_{nb} = \mathbf{V}_1 \tag{11.4}$$

This is a common arrangement in a North American residence supplied with both 115 V and 230 V rms, since if $|\mathbf{V}_{an}| = |\mathbf{V}_1| = 115$ V rms, then $|\mathbf{V}_{ab}| = |2\,\mathbf{V}_1| = 230$ V rms.

Let us now consider the source of Fig. 11.3(a) loaded with two identical loads, both having an impedance Z_1, as shown in Fig. 11.3(b). The currents in the lines aA

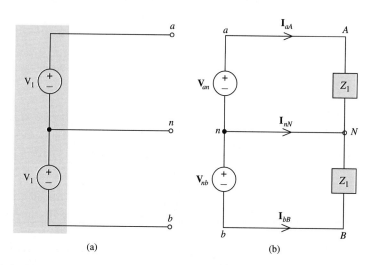

(a)   (b)

FIGURE 11.3   (a) Single-phase three-wire circuit; (b) single-phase three-wire circuit with two identical loads.

and bB are

$$\mathbf{I}_{aA} = \frac{\mathbf{V}_{an}}{Z_1} = \frac{\mathbf{V}_1}{Z_1}$$

$$\mathbf{I}_{bB} = \frac{\mathbf{V}_{bn}}{Z_1} = -\frac{\mathbf{V}_1}{Z_1} = -\mathbf{I}_{aA}$$

Therefore, the current in the neutral wire, nN, by KCL is

$$\mathbf{I}_{nN} = -(\mathbf{I}_{aA} + \mathbf{I}_{bB}) = 0$$

Now it is always the case that a wire carrying no current may be freely removed in any circuit without affecting any current or voltage, since its removal changes no KVL, KCL, or element law equation for that circuit. Thus in Fig. 11.3(b), the neutral could be removed without changing any current or voltage in the system. We use this procedure to simplify circuits throughout the chapter.

If the lines aA and bB are not perfect conductors but have equal impedances Z_2, then $\mathbf{I}_{nN}$ is still zero because we may simply add the series impedances Z_1 and Z_2 and have essentially the same situation as in Fig. 11.3(b). Indeed, in the more general case shown in Fig. 11.4, the neutral current $\mathbf{I}_{nN}$ is still zero. This may be seen by writing the two mesh equations

$$(Z_1 + Z_2 + Z_3)\mathbf{I}_{aA} + Z_3\mathbf{I}_{bB} - Z_1\mathbf{I}_3 = \mathbf{V}_1$$

$$Z_3\mathbf{I}_{aA} + (Z_1 + Z_2 + Z_3)\mathbf{I}_{bB} + Z_1\mathbf{I}_3 = -\mathbf{V}_1$$

and adding the result, which yields

$$(Z_1 + Z_2 + Z_3)(\mathbf{I}_{aA} + \mathbf{I}_{bB}) + Z_3(\mathbf{I}_{aA} + \mathbf{I}_{bB}) = 0$$

or
$$\mathbf{I}_{aA} + \mathbf{I}_{bB} = 0 \qquad\qquad (11.5)$$

Since by KCL the left side of (11.5) equals $-\mathbf{I}_{nN}$, the neutral current is zero. This is a consequence of the symmetry of Fig. 11.4.

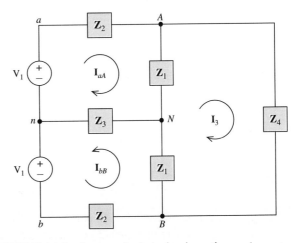

FIGURE 11.4 Symmetrical single-phase three-wire system.

If the symmetry of Fig. 11.4 is broken by having unequal loads at terminals A–N and N–B or unequal line impedances in lines aA and bB, there will be a neutral current.

Example 11.2

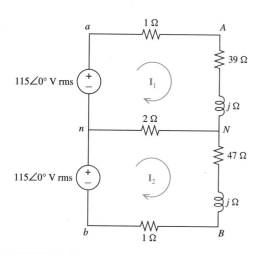

$115\angle 0°$ V rms

$115\angle 0°$ V rms

FIGURE 11.5 Unsymmetrical single-phase three-wire system.

For example, let us consider the circuit of Fig. 11.5. Note that the loads across terminals AN and NB are not equal, and the symmetry of Fig. 11.4 is broken. The mesh equations, in vector-matrix form, are

$$\begin{bmatrix} 42+j & -2 \\ -2 & 50+j \end{bmatrix}\begin{bmatrix} \mathbf{I}_1 \\ \mathbf{I}_2 \end{bmatrix}\begin{bmatrix} 115\underline{/0} \\ 115\underline{/0} \end{bmatrix}$$

The determinant is

$$\Delta = (42+j)(50+j) - 4 = 2097\underline{/2.51°}$$

and solving by matrix inversion yields

$$\begin{bmatrix} \mathbf{I}_1 \\ \mathbf{I}_2 \end{bmatrix} = \frac{115\underline{/0}}{\Delta}\begin{bmatrix} 50+j & +2 \\ +2 & 42+j \end{bmatrix}\begin{bmatrix} 1 \\ 1 \end{bmatrix} = \begin{bmatrix} 2.85\underline{/-1.41°} \\ 2.41\underline{/-1.21°} \end{bmatrix}$$

so the neutral current $\mathbf{I}_{nN} = \mathbf{I}_2 - \mathbf{I}_1 = 0.44\underline{/177°}$ A rms and is not zero.

EXERCISES

11.1.1. Derive (11.5) by superposition applied to Fig. 11.4.

11.1.2. Find the complex power $\mathbf{S}_{AN}$ and $\mathbf{S}_{NB}$ delivered to the loads $Z_{AN} = 39 + j$ Ω, $Z_{NB} = 47 + j$ Ω, respectively, of Fig. 11.5.
Answer $317 + j8.12$; $273 + j5.81$

11.1.3. Find the real average power P_{aA}, P_{bB}, and P_{nN} lost in the lines in Fig. 11.5.
Answer 8.12 W; 5.81 W; 0.39 W

11.1.4. Find the real average power P_{an} and P_{nb} delivered by the two sources in Fig. 11.5. Check the results in Exercises 11.1.2 and 11.1.3 for conservation of power.
Answer 328 W; 277 W

11.2 THREE-PHASE WYE–WYE SYSTEMS

Let us consider the three-phase source of Fig. 11.6(a), which has *line* terminals a, b, and c and a *neutral* terminal n. In this case the source is said to be *wye (Y) connected*. An equivalent representation is that of Fig. 11.6(b), which is somewhat easier to draw.

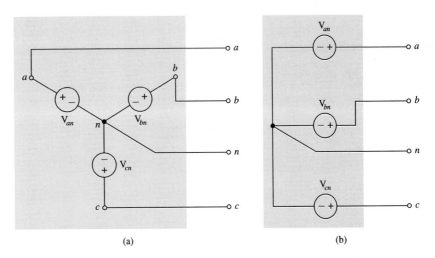

FIGURE 11.6   Two representations of a wye-connected source.

The voltages $\mathbf{V}_{an}$, $\mathbf{V}_{bn}$, and $\mathbf{V}_{cn}$ between the line terminals and the neutral terminal are called *phase voltages* and in most cases we shall consider are given by

$$
\begin{aligned}
\mathbf{V}_{an} &= V_p\underline{/0} \\
\mathbf{V}_{bn} &= V_p\underline{/-120°} \\
\mathbf{V}_{cn} &= V_p\underline{/120°}
\end{aligned}
\tag{11.6}
$$

or

$$
\begin{aligned}
\mathbf{V}_{an} &= V_p\underline{/0} \\
\mathbf{V}_{bn} &= V_p\underline{/120°} \\
\mathbf{V}_{cn} &= V_p\underline{/-120°}
\end{aligned}
\tag{11.7}
$$

In both cases, each phase voltage has the same rms magnitude V_p, and the phases are displaced 120°, with $\mathbf{V}_{an}$ arbitrarily selected as the reference phasor. Such a set of voltages is called a *balanced set* and is characterized by

$$
\mathbf{V}_{an} + \mathbf{V}_{bn} + \mathbf{V}_{cn} = 0
\tag{11.8}
$$

as may be seen from (11.6) or (11.7).

The sequence of voltages in (11.6) is called the *positive*, or *abc, sequence*, while that of (11.7) is called the *negative*, or *acb, sequence*. Phasor diagrams of the two sequences are shown in Fig. 11.7, where we may see by inspection that (11.8) holds. Evidently, the only difference between positive and negative sequences is the arbitrary

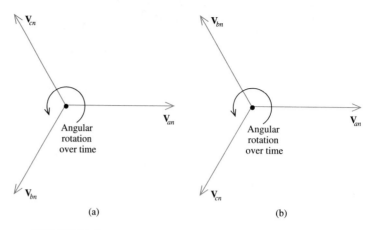

FIGURE 11.7   (a) Positive and (b) negative phase sequences.

choice of the terminal labels, a, b, and c. Thus without loss in generality we consider only the positive sequence.

By Fig. 11.7(a), the voltages in the abc sequence may each be related to $\mathbf{V}_{an}$. The relationships, which will be useful later, are

$$\mathbf{V}_{bn} = \mathbf{V}_{an}\underline{/-120^\circ}$$

$$\mathbf{V}_{cn} = \mathbf{V}_{an}\underline{/120^\circ} \tag{11.9}$$

The *line-to-line* voltages, or simply *line* voltages, in Fig. 11.6 are $\mathbf{V}_{ab}$, $\mathbf{V}_{bc}$, and $\mathbf{V}_{ca}$, which may be found from the phase voltages. For example,

$$\mathbf{V}_{ab} = \mathbf{V}_{an} + \mathbf{V}_{nb}$$

$$= V_p\underline{/0} + V_p\underline{/+60^\circ}$$

$$= (V_p) + (V_p)\left(\frac{1}{2} + j\frac{\sqrt{3}}{2}\right)$$

$$= \sqrt{3}\, V_p\left(\frac{\sqrt{3}}{2} + j\frac{1}{2}\right)$$

$$= \sqrt{3}\, V_p\underline{/30^\circ}$$

In like manner,

$$\mathbf{V}_{bc} = \sqrt{3}\, V_p\underline{/-90^\circ}$$

$$\mathbf{V}_{ca} = \sqrt{3}\, V_p\underline{/-210^\circ}$$

If we denote the magnitude of the line voltages by V_L, we have

$$V_L = \sqrt{3}\, V_p \tag{11.10a}$$

and thus

$$\mathbf{V}_{ab} = V_L\underline{/30^\circ}, \qquad \mathbf{V}_{bc} = V_L\underline{/-90^\circ}, \qquad \mathbf{V}_{ca} = V_L\underline{/-210^\circ} \tag{11.10b}$$

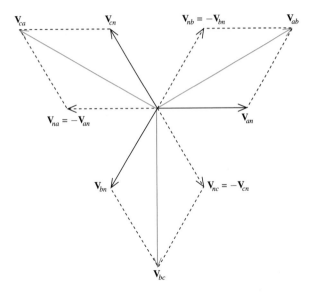

FIGURE 11.8 Phasor diagram showing phase and line voltages.

Examining Fig. 11.8, the line voltages are also a balanced set, whose magnitudes exceed the phase voltages by a factor of $\sqrt{3}$ and which are 30° out of phase with them (line voltage $\mathbf{V}_{ab}$ leads phase voltage $\mathbf{V}_{an}$ by 30°, $\mathbf{V}_{ca}$ leads $\mathbf{V}_{cn}$ by 30°, and $\mathbf{V}_{bc}$ leads $\mathbf{V}_{bn}$ by 30°). These results also may be obtained graphically from the phasor diagram shown in Fig. 11.8.

Let us now consider the system of Fig. 11.9, which is a *balanced* wye–wye three-phase four-wire system if the source voltages are given by (11.6). The term *wye–wye* applies since both the source and the load are wye-connected. The system is said to be balanced since the source voltages constitute a balanced set and the load is balanced (each *phase impedance* is equal, in this case, to the common value Z_p). The fourth wire

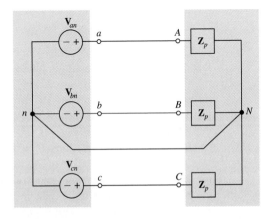

FIGURE 11.9 Balanced wye–wye system.

is the neutral line n–N, which since it carries no current in this symmetric circuit, may be omitted to form a three-phase three-wire system.

Applying KVL around loops containing the neutral wire, the line currents of Fig. 11.9 are evidently

$$\mathbf{I}_{aA} = \frac{\mathbf{V}_{an}}{Z_p}$$

$$\mathbf{I}_{bB} = \frac{\mathbf{V}_{bn}}{Z_p} = \frac{\mathbf{V}_{an}\underline{/-120°}}{Z_p} = \mathbf{I}_{aA}\underline{/-120°}$$ (11.11)

$$\mathbf{I}_{cC} = \frac{\mathbf{V}_{cn}}{Z_p} = \frac{\mathbf{V}_{an}\underline{/120°}}{Z_p} = \mathbf{I}_{aA}\underline{/120°}$$

The last two results are a consequence of (11.9) and show that the line currents also form a balanced set. Therefore, their sum is

$$-\mathbf{I}_{nN} = \mathbf{I}_{aA} + \mathbf{I}_{bB} + \mathbf{I}_{cC} = 0$$

confirming that the neutral carries no current in a balanced wye–wye four-wire system.

In the case of wye-connected loads, the currents in the lines aA, bB, and cC are also the *phase currents* (the currents carried by the phase impedances). If the magnitudes of the phase and line currents are I_p and I_L, respectively, then $I_L = I_p$, and (11.11) becomes

$$\mathbf{I}_{aA} = I_L\underline{/-\theta} = I_p\underline{/-\theta}$$

$$\mathbf{I}_{bB} = I_L\underline{/-\theta - 120°} = I_p\underline{/-\theta - 120°}$$ (11.12)

$$\mathbf{I}_{cC} = I_L\underline{/-\theta + 120°} = I_p\underline{/-\theta + 120°}$$

where θ is the angle of Z_p.

The average power P_p delivered to each phase of Fig. 11.9 is

$$P_p = V_p I_p \cos\theta$$
$$= I_p^2 \operatorname{Re}(Z_p)$$ (11.13)

and the total power delivered to the load is

$$P = 3P_p$$

The angle θ of the phase impedance is thus the power factor angle of the three-phase load as well as that of a single phase.

Suppose now that instead of assuming that they are perfect conductors, a line impedance Z_L is used to model each of the lines aA, bB, and cC and that a neutral line impedance Z_N, not necessarily equal to Z_L, is inserted in line nN in series with the phase impedances. The two sets of impedances may be combined to form perfect conducting lines aA, bB, and cC with a load impedance $Z_p + Z_L$ in each phase.

To determine the effect of Z_N, consider the full set of KCL, KVL, and element laws for this circuit. The currents and voltages solving these equations for the case $Z_N = 0$ include $I_{nN} = 0$, since the circuit in this case is balanced. But these same currents and voltages must solve the circuit equations for *any* Z_N, not just $Z_N = 0$. This follows from the fact that substituting the solutions for the case $Z_N = 0$ into the equations containing arbitrary Z_N, no term in any equation is changed, since every term containing Z_n is multiplied by zero (I_{nN} or V_{nN}). By this argument we see that *in any circuit containing an impedance Z that draws no current, Z may be changed arbitrarily without affecting any current or voltage.* In particular, we may replace Z by a short circuit ($Z = 0$) or an open circuit ($Z = \infty$).

Example 11.3

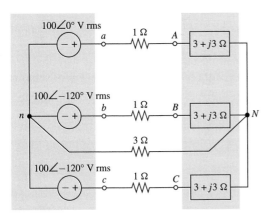

FIGURE 11.10 Balanced system with line impedances.

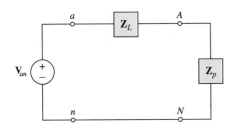

FIGURE 11.11 Single phase for a per-phase analysis.

As an example, let us find the line currents in Fig. 11.10. We may combine the 1-Ω line impedance and $(3 + j3)$-Ω phase impedance to obtain

$$Z_p = 4 + j3 = 5\underline{/36.9°}\ \Omega$$

as the effective phase load. Since by the foregoing discussion there is no neutral current, we have

$$\mathbf{I}_{aA} = \frac{100\underline{/0}}{5\underline{/36.9°}} = 20\underline{/-36.9°}\ \text{A rms}$$

The currents form a balanced, positive sequence set, so we also have

$$\mathbf{I}_{bB} = 20\underline{/-156.9°}\ \text{A rms}, \qquad \mathbf{I}_{cC} = 20\underline{/-276.9°}\ \text{A rms}$$

This example was solved on a "per-phase" basis. Since the impedance in the neutral is immaterial in a balanced wye–wye system, we may imagine the neutral line to be a short circuit. We may do this if it contains an impedance or even if the neutral wire is not present (a three-wire system). We may then look at only one phase, say phase A, consisting of the source $\mathbf{V}_{an}$ in series with Z_L and Z_p, as shown in Fig. 11.11. (The line nN is replaced by a short circuit.) The line current $\mathbf{I}_{aA}$, the phase voltage $\mathbf{I}_{aA}Z_p$, and the voltage drop in the line $\mathbf{I}_{aA}Z_L$ may all be found from this single-phase analysis. The other voltages and currents in the system may be found similarly, or from the previous results, since the system is balanced.

Example 11.4

As another example, suppose that we have a balanced wye-connected source, having line voltage $V_L = 200$ V rms that is supplying a balanced wye-connected load with $P = 900$ W at a power factor of 0.9 lagging. Let us find the line current I_L and the phase impedance Z_p. Since the power supplied to the load is 900 W, the power supplied to each phase is $P_p = \frac{900}{3} = 300$ W, and from

$$P_p = V_p I_p \cos\theta$$

we have

$$300 = \frac{200}{\sqrt{3}} I_p (0.9)$$

Therefore, since for a wye-connected load the phase current is also the line current, we have

$$I_L = I_p = \frac{3\sqrt{3}}{2(0.9)} = 2.89 \text{ A rms}$$

The magnitude of Z_p is given by

$$|Z_p| = \frac{V_p}{I_p} = \frac{200/\sqrt{3}}{3\sqrt{3}/(2)(0.9)} = 40 \ \Omega$$

and since $\theta = \cos^{-1} 0.9 = 25.84°$ is the angle of Z_p, we have

$$Z_p = 40/\underline{25.84°} \ \Omega$$

If the load is unbalanced but there is a neutral wire that is a perfect conductor, we may still use the per-phase method of solution for each phase. However, if this is not the case, this shortcut method does not apply.

EXERCISES

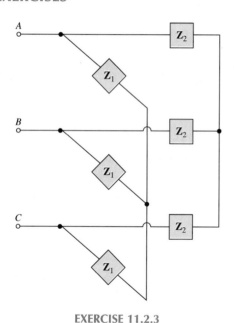

EXERCISE 11.2.3

11.2.1. $\mathbf{V}_{ab} = 100/\underline{0}$ V rms is a line voltage of a balanced wye-connected three-phase source. (a) If the phase sequence is abc, find the phase voltages. (b) Repeat for the acb phase sequence.
Answer (a) $57.7/\underline{-30°}$, $57.7/\underline{-150°}$, $57.7/\underline{+90°}$ V rms; (b) $57.7°/\underline{-30°}$, $57.7°/\underline{+90°}$, $57.7/\underline{-150°}$ V rms

11.2.2. In Fig. 11.9 the source voltages are determined by Exercise 11.2.1(a) and the load in each phase is a series combination of a 30-Ω resistor, a 500-μF capacitor, and a 0.25-H inductor. The frequency is $\omega = 200$ rad/s. Find the line currents and the power delivered to the load.
Answer $1.15/\underline{-83.1°}$; $1.15/\underline{-203.1°}$; $1.15/\underline{36.9°}$ A rms; 120 W

11.2.3. Show that if a balanced three-phase three-wire system has two balanced three-phase loads connected in parallel, as shown, the load is equivalent to that of Fig. 11.9 with

$$Z_p = \frac{Z_1 Z_2}{Z_1 + Z_2}$$

11.2.4. If, in Exercise 11.2.3, $Z_1 = 4 + j3 \ \Omega$, $Z_2 = 4 - j3 \ \Omega$, and the line voltage is $V_L = 200\sqrt{3}$ V rms, find the current I_L in each line.
Answer 64 A rms

11.3 SINGLE-PHASE VERSUS THREE-PHASE POWER DELIVERY

It is interesting to compare the relative merits of the single-phase three-wire circuit of Section 11.1 and the three-phase circuit of Section 11.2 for delivering power to a load. Figure 11.12 shows the two circuits to be considered. Examining this figure, one

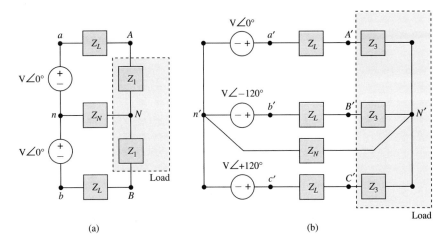

FIGURE 11.12 (a) Single- and (b) three-phase circuits delivering power to a load.

immediate observation is that the three-phase circuit has a higher element count: more wires, more phase sources, and more phase impedances. Simplicity favors the single-phase circuit. But we will look beyond element count in comparing the relative virtues of single- and three-phase systems.

Consider the line losses incurred by each system while producing the same amount of power at the same voltage level and power factor. In the single-phase case, $I_{nN} = 0$, and replacing Z_N by a short circuit in Fig. 11.12(a) gives

$$\mathbf{I}_{aN} = \frac{V\underline{/0}}{Z_1 + Z_L}$$

$$P_{aN} = |\mathbf{V}_{aN}| \, |\mathbf{I}_{aN}| \cos \theta = \frac{V^2 \cos \theta}{|Z_1 + Z_L|}$$

where $\cos \theta$ is the power factor of the series line/load impedance. The total power delivered is

$$P_{aN} + P_{bN} = 2P_{aN} = \frac{2V^2 \cos \theta}{|Z_1 + Z_L|} \tag{11.14}$$

In the three-phase case of Fig. 11.12(b) again, the neutral current $I_{n'N'} = 0$ and the per-phase calculation is as above. The total power delivered by the three-phase source is

$$P_{a'N'} + P_{b'N'} + P_{c'N'} = 3P_{a'N'} = \frac{3V^2 \cos \theta}{|Z_3 + Z_L|} \tag{11.15}$$

The powers delivered by the single- and three-phase sources are equal if (11.14) and (11.15) are equal, or

$$\frac{|Z_3 + Z_L|}{|Z_1 + Z_L|} = \frac{3}{2} \tag{11.16}$$

For equal power delivery, the series phase impedance in the three-phase circuit, $Z_3 + Z_L$, is larger than the single-phase impedance by half. Since the phase voltages are the same, the rms current in the single-phase lines must be larger by the same ratio:

$$\frac{|\mathbf{I}_{aN}|}{|\mathbf{I}_{a'N'}|} = \frac{V/|Z_1 + Z_L|}{V/|Z_3 + Z_L|} = \frac{3}{2} \tag{11.17}$$

The line losses are $|\mathbf{I}_{aN}|^2 \, \text{Re}(Z_L)$ for each nonneutral line, or in the single-phase case the total line losses are

$$P_L = 2|\mathbf{I}_{aN}|^2 \, \text{Re}(Z_L) \tag{11.18}$$

while in the three-phase circuit we have

$$P_{L'} = 3|\mathbf{I}_{a'N'}|^2 \, \text{Re}(Z_L)$$

Substituting (11.17), this is

$$P_{L'} = 3 \cdot \left(\tfrac{2}{3}\right)^2 |\mathbf{I}_{aN}|^2 \, \text{Re}(Z_L) = \tfrac{4}{3}|\mathbf{I}_{aN}|^2 \, \text{Re}(Z_L) \tag{11.19}$$

Comparing (11.18) and (11.19), we conclude that $P_L = \tfrac{3}{2}P_{L'}$; in other words, with other factors equal (power produced, voltage level, power factor), *the single-phase circuit delivers power at the cost of half again as many watts dissipated in the lines. Three-phase power delivery is more efficient.*

Example 11.5

In Figure 11.12, let $V = 4$ kV rms, $Z_L = 2 \, \Omega$, and the single-phase load impedance be $30 + j8 \, \Omega$. Determine the phase impedance Z_3 for equal power from the single- and three-phase sources at equal power factors and the power delivered to the lines and loads. In the single-phase circuit, the rms line current I_{aN} is

$$\mathbf{I}_{aN} = \frac{4000\underline{/0}}{32 + j8} = 121\underline{/-14.0°} \tag{11.20}$$

The power delivered to the load is $2|\mathbf{I}_{aN}|^2 \, \text{Re}(Z_1) = 2(121)^2(30) = 878$ kW and that dissipated by the lines $2|\mathbf{I}_{aN}|^2 \, \text{Re}(Z_L) = 2(121)^2 (2) = 58.6$ kW. In the three-phase case, the power dissipated by the lines at the same power factor is scaled by a factor of $\tfrac{2}{3}$; that is, $(\tfrac{2}{3})(58.6) = 39.0$ kW. Since powers delivered by the sources are equal, the power to the three-phase load must be increased by the line losses saved, or three-phase load power of $878 + \tfrac{1}{3}(58.6) = 897$ kW. The impedance Z_3 required must, by (11.16), satisfy

$$\frac{|Z_3 + 2|}{|32 + j8|} = \frac{3}{2}$$

or

$$|Z_3 + 2| = 49.5$$

For equal power factors, the angle of the impedance $Z_3 + 2$ must agree with that of $Z_1 + 2 = 32 + j8$, which is $\theta = 14.0°$ by (11.20).

$$Z_3 + 2 = |Z_3 + 2|\underline{/\theta} = 49.5\underline{/14.0°}$$

$$Z_3 = 49.5\underline{/14.0°} - 2 = 46.0 + j12.0 \, \Omega$$

Examination of the instantaneous power waveforms associated with these two circuits reveals another difference relevant to many practical applications. From Fig. 11.12(a), in the single-phase case

$$\mathbf{I}_{aN} = \mathbf{I}_{nB} = \frac{V\underline{/0}}{Z_1 + Z_L} \tag{11.21a}$$

and

$$i_{aN}(t) = i_{nB}(t) = \frac{\sqrt{2}V}{|Z_1 + Z_L|} \cos(\omega t - \theta) \tag{11.21b}$$

where θ is, as before, the angle of the series impedance, $Z_L + Z_1$, and the factor of $\sqrt{2}$ is required since V is the magnitude of the rms phasor. The instantaneous power delivered to each of the two parts of the load is $p(t) = i_{aN}^2 \, \mathrm{Re}(Z_1)$ and is shown in Fig. 11.13(a). Note the instantaneous power delivered to the two load impedances is in phase, and the total power delivered to the load varies between its maximum and zero

$$p_{1T}(t) = 2 \left(\frac{\sqrt{2}\,V}{|Z_1 + Z_L|} \right)^2 \cos^2(\omega t - \theta) \cdot \mathrm{Re}(Z_1)$$

$$= \frac{2V^2 \, \mathrm{Re}(Z_1)}{|Z_1 + Z_L|^2} [1 + \cos 2(\omega t - \theta)]$$

periodically.

From Fig. 11.12(b), in the three-phase case,

$$\mathbf{I}_{a'N'} = \frac{V\underline{/0}}{Z_3 + Z_L}$$

$$i_{aN}(t) = \frac{\sqrt{2}\,V}{|Z_3 + Z_L|} \cos(\omega t - \theta)$$

The three line currents form a balanced set, as shown in the phasor diagram in Fig. 11.14. These currents and hence the instantaneous powers to the three-phase impedances are $120°$ out of phase, as shown in Fig. 11.13(b). The total instantaneous power is the sum over

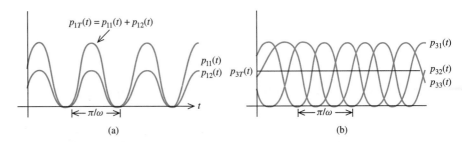

(a)

(b)

FIGURE 11.13 Instantaneous power delivery to single-phase three-wire and three-phase loads. (a) $p_{11}(t)$ and $p_{12}(t)$ are power to the single-phase load impedances and $p_{1T}(t)$ is total power. (b) $p_{31}(t)$, $p_{32}(t)$, and $p_{33}(t)$ are power to the three-phase impedances and $p_{3T}(t)$ is the total power.

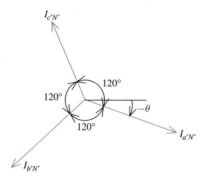

FIGURE 11.14 Line currents associated with the three-phasor circuit of Fig. 11.12(b). θ is the angle of the phase impedance.

the three individual phases:

$$
\begin{aligned}
p_{3T}(t) &= \left(\frac{\sqrt{2}\,V}{|Z_3 + Z_L|} \right)^2 \\
&\quad \times \operatorname{Re}(Z_3)\left[\cos^2(\omega t + \theta) + \cos^2(\omega t + \theta + 120°) + \cos^2(\omega t + \theta - 120°) \right] \\
&= \frac{2V^2 \operatorname{Re}(Z_3)}{|Z_3 + Z_L|^2} \\
&\quad \times \left[\tfrac{3}{2} + \cos 2(\omega t + \theta) + \cos 2(\omega t + \theta + 120°) + \cos 2(\omega t + \theta - 120°) \right]
\end{aligned}
$$

(11.22)

The cosine terms in (11.22) are evidently time functions representing a balanced set of phasors. Since a balanced set of phasors sums to zero, so must these three terms. The remaining term is constant, and thus $p_{3T}(t)$, the total instantaneous three-phase power delivered to the load, is constant, as shown in Fig. 11.13(b).

We conclude that single-phase power is delivered unevenly, varying periodically between zero and its maximum, while three-phase power is constant over time. Suppose that we were delivering power to an electric motor. Since the torque produced varies with the power delivered to it, a single-phase motor shakes as the power delivery fluctuates periodically, while a three-phase motor "pulls" as hard each instant of time as every other, with no induced vibrations. Or consider powering fluorescent lights, whose brightness varies with instantaneous power delivery. A pair of single-phase three-wire fluorescent fixtures would flicker periodically, while three three-phase fixtures mounted together would illuminate without flicker.

Thus not only does three-phase power delivery have an efficiency advantage over single phase, but a smooth power delivery advantage as well. It is no surprise that three-phase is favored the world over for delivery of bulk power. Only where small amounts of power and small fractional horsepower motors are sufficient is single-phase power the method of choice. This is the case for most residential wiring, for which the simplicity and low element weight of single-phase circuits are decisive. Dc is sometimes used to transport power over long distances, but multiphase ac is dominant in almost all power delivery systems.

11.3.1. A three-phase source and load 100 km apart are connected by lines with impedance 0.01 Ω/km, as shown in the figure. What is the *efficiency* of this system, defined as power delivered to the load divided by total power produced by the source? At what distance is the efficiency 90%?

Answer 0.95; 211 km

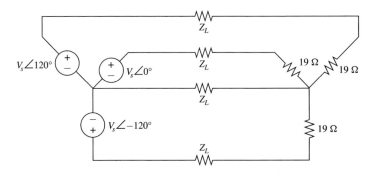

EXERCISE 11.3.1

11.3.2. The figure shows a single-phase system of the same source voltage $|V_s|$ as in Exercise 11.3.1. For 100-km separation, for which R_L would the same power be delivered to the load as delivered by the three-phase system of Exercise 11.3.1 to its three-phase load? What is the efficiency of this system? (See Exercise 11.3.1.)

Answer 11.95 Ω or 0.0837 Ω; 0.923 for $R_L = 11.95$ Ω; 0.0772 for $R_L = 0.0837$ Ω

11.3.3. Consider a motor powered by a single-phase 50-Hz source. What is the period of the vibrational motion the motor will experience? Repeat for a balanced three-phase source and an unbalanced three-phase source both at 50 Hz.

Answer 10 ms; no vibrational motion; 10 ms

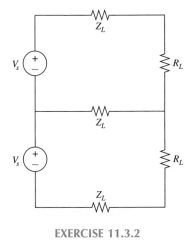

EXERCISE 11.3.2

11.4 DELTA CONNECTION

Another method of connecting a three-phase load to a line is the *delta*, or Δ, *connection*. A balanced Δ-connected load (with equal phase impedance Z_P) is shown in Fig. 11.15(a), in a way that resembles a Δ, and in an equivalent way in Fig. 11.15(b). If the source is Y- or Δ-connected, the system is a Y–Δ or Δ–Δ system.

An advantage of a Δ-connected load over a Y-connected load is that loads may be added or removed more readily on a single phase of a Δ, since the loads are connected directly across the lines. This may not be possible in the Y connection, since the neutral may not be accessible. Also, for a given power delivered to the load, the phase currents in a Δ are smaller than those in a Y. On the other hand, the Δ phase voltages are

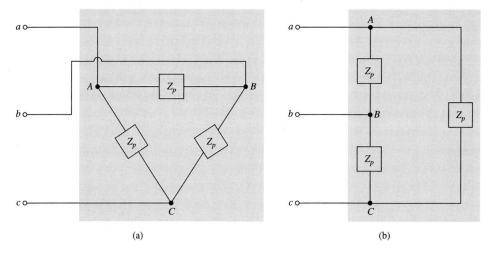

FIGURE 11.15 Two diagrams of a delta-connected load.

higher than those of the Y connection. Sources are rarely Δ-connected, because if the voltages are not perfectly balanced, there will be a net source voltage, and consequently a circulating current, around the delta. This causes undesirable heating effects in the generating machinery. Also, the phase voltages are lower in the Y-connected generator, and thus less insulation is required. Obviously, systems with Δ-connected loads are three-wire systems, since there is no neutral connection.

From Fig. 11.16 we see that in the case of a Δ-connected load the line voltages are the same as the phase voltages. Therefore, if the line voltages are given by (11.10b), as before, the phase voltages are

$$\mathbf{V}_{AB} = V_L\underline{/30°}, \qquad \mathbf{V}_{BC} = V_L\underline{/-90°}, \qquad \mathbf{V}_{CA} = V_L\underline{/150°} \qquad (11.23)$$

where

$$V_L = V_p \qquad\qquad (11.24)$$

If $Z_p = |Z_p|\underline{/\theta}$, the phase currents are

$$\mathbf{I}_{AB} = \frac{\mathbf{V}_{AB}}{Z_p} = I_p\underline{/30° - \theta}$$

$$\mathbf{I}_{BC} = \frac{\mathbf{V}_{BC}}{Z_p} = I_p\underline{/-90° - \theta} \qquad (11.25)$$

$$\mathbf{I}_{CA} = \frac{\mathbf{V}_{CA}}{Z_p} = I_p\underline{/150° - \theta}$$

where the rms phase current magnitude is

$$I_p = \frac{V_L}{|Z_p|} \qquad\qquad (11.26)$$

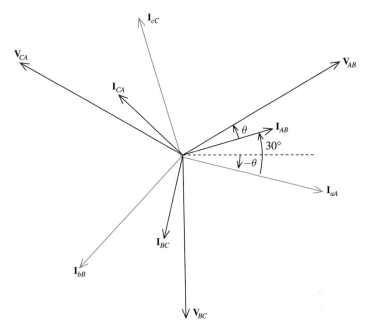

FIGURE 11.16 Phasor diagram for a delta-connected load.

The current in line aA is

$$\mathbf{I}_{aA} = \mathbf{I}_{AB} - \mathbf{I}_{CA}$$

which after some simplification is

$$\mathbf{I}_{aA} = \sqrt{3}\ I_p\underline{/-\theta}$$

The other line currents, obtained similarly, are

$$\mathbf{I}_{bB} = \sqrt{3}\ I_p\underline{/-120° - \theta}$$

$$\mathbf{I}_{cC} = \sqrt{3}\ I_p\underline{/-240° - \theta}$$

Evidently, the relation between the line and phase current magnitudes in the Δ case is

$$I_L = \sqrt{3}\ I_p \qquad\qquad (11.27)$$

and the line currents are thus

$$\mathbf{I}_{aA} = I_L\underline{/-\theta}, \qquad \mathbf{I}_{bB} = I_L\underline{/-120° - \theta}, \qquad \mathbf{I}_{cC} = I_L\underline{/-240° - \theta} \qquad (11.28)$$

Thus the currents and voltages are balanced sets, as expected. The relations between line and phase currents for the Δ-connected load are summed up in the phasor diagram of Fig. 11.16.

Example 11.6

As an example of a three-phase circuit with a delta-connected load, let us find the line current I_L in Fig. 11.15 if the line voltage is 250 V rms and the load draws 1.5 kW at a lagging power factor of 0.8. For one phase, $P_p = \frac{1500}{3} = 500$ W, and thus

$$500 = 250 I_p(0.8)$$

or
$$I_p = 2.5 \text{ A rms}$$

Therefore, we have

$$I_L = \sqrt{3} \, I_p = 4.33 \text{ A rms}$$

Finally, in this section, let us derive a formula for the power delivered to a balanced three-phase load with a power factor angle θ. Whether the load is Y-connected or Δ-connected, we have

$$P = 3P_p = 3V_p I_p \cos\theta$$

In the Y-connected case, $V_p = V_L/\sqrt{3}$ and $I_p = I_L$, and in the Δ-connected case, $V_p = V_L$ and $I_p = I_L/\sqrt{3}$. In either case, then,

$$P = 3\frac{V_L I_L}{\sqrt{3}} \cos\theta$$

or

$$P = \sqrt{3} \, V_L I_L \cos\theta \qquad (11.29)$$

As a check on Example 11.6, (11.29) yields

$$1500 = \sqrt{3}(250) I_L(0.8)$$

or, as before,

$$I_L = 4.33 \text{ A rms}$$

EXERCISES

11.4.1. Solve Exercise 11.2.2 if the source and load are unchanged except that the load is Δ-connected. [*Suggestion:* Note that in (11.23), (11.25), and (11.28), 30° must be subtracted from every angle.]
Answer $2\sqrt{3}\underline{/-83.1°}$; $2\sqrt{3}\underline{/156.9°}$; $2\sqrt{3}\underline{/36.9°}$ A rms, 360 W

11.4.2. A balanced Δ-connected load has $Z_p = 4 + j3$ Ω, and the line voltage is $V_L = 200$ V rms at the load terminals. Find the total power delivered to the load.
Answer 19.2 kW

11.4.3. A balanced delta-connected load has a line voltage of $V_L = 100$ V rms at the load terminals and absorbs a total power of 4.8 kW. If the power factor of the load is 0.8 leading, find the phase impedance.
Answer $4 - j3$ Ω

In many power systems applications it is important to be able to convert from a wye-connected load to an equivalent delta-connected load, and vice versa. For example, suppose that we have a Y-connected load in parallel with a Δ-connected load, as shown in Fig. 11.17, and wish to replace the combination by an equivalent three-phase load. If both loads were Δ-connected, this would be relatively easy, since corresponding phase impedances would be in parallel. Also, as we saw in Exercise 11.2.3, if both loads are Y-connected and balanced, the phase impedances may also be combined as parallel impedances.

To obtain Y-to-Δ or Δ-to-Y conversion formulas, let us consider the Y and Δ connections of Fig. 11.18. To effect a Y–Δ transformation, we need expressions for Y_{ab}, Y_{bc}, and Y_{ca} of the Δ in terms of Y_a, Y_b, and Y_c of the Y so that the Δ connection is equivalent to the Y connection at the terminals A, B, and C. That is, if the Y is replaced by the Δ, the same node voltages $\mathbf{V}_A$, $\mathbf{V}_B$, and $\mathbf{V}_C$ will appear if the same currents $\mathbf{I}_1$ and $\mathbf{I}_2$ are caused to flow. Conversely, a Δ–Y transformation is an expression of the Y parameters in terms of the Δ parameters.

Let us begin by writing nodal equations for both circuits. If node C is taken as reference, in the case of the Y network we have

$$Y_a\mathbf{V}_A - Y_a\mathbf{V}_D = \mathbf{I}_1$$

$$Y_b\mathbf{V}_B - Y_b\mathbf{V}_D = \mathbf{I}_2$$

$$-Y_a\mathbf{V}_A - Y_b\mathbf{V}_B + (Y_a + Y_b + Y_c)\mathbf{V}_D = 0$$

Solving for $\mathbf{V}_D$ in the third equation and substituting its value into the first two equations, we have, after simplification,

$$\frac{Y_aY_b + Y_aY_c}{Y_a + Y_b + Y_c}\mathbf{V}_A - \frac{Y_aY_b}{Y_a + Y_b + Y_c}\mathbf{V}_B = \mathbf{I}_1$$

$$-\frac{Y_aY_b}{Y_a + Y_b + Y_c}\mathbf{V}_A + \frac{Y_aY_b + Y_bY_c}{Y_a + Y_b + Y_c}\mathbf{V}_B = \mathbf{I}_2$$

$$(11.30)$$

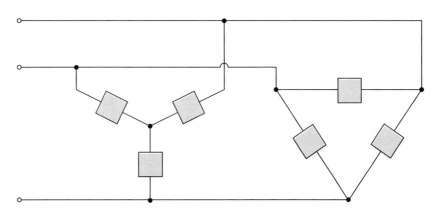

FIGURE 11.17 Wye- and delta-connected loads in parallel.

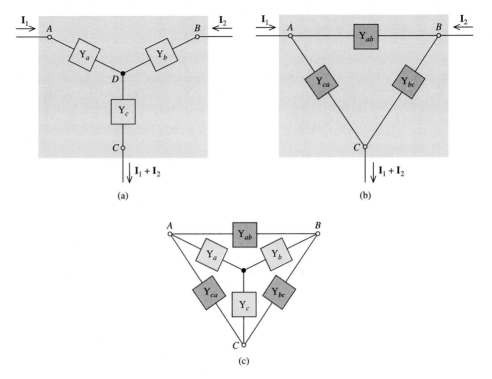

FIGURE 11.18 (a) Wye connection; (b) delta connection; (c) the two superimposed.

The nodal equations for the Δ circuit are

$$(Y_{ab} + Y_{ca})\mathbf{V}_A - Y_{ab}\mathbf{V}_B = \mathbf{I}_1$$

$$-Y_{ab}\mathbf{V}_A + (Y_{ab} + Y_{bc})\mathbf{V}_B = \mathbf{I}_2$$

Equating coefficients of like terms in these equations and (11.30), we have the Y–Δ transformation:

$$Y_{ab} = \frac{Y_a Y_b}{Y_a + Y_b + Y_c}$$

$$Y_{bc} = \frac{Y_b Y_c}{Y_a + Y_b + Y_c} \qquad (11.31)$$

$$Y_{ca} = \frac{Y_c Y_a}{Y_a + Y_b + Y_c}$$

If we imagine the Y and Δ circuits superimposed on a single diagram as in Fig. 11.18(c), then Y_a and Y_b are *adjacent* to Y_{ab}, Y_b and Y_c are *adjacent* to Y_{bc}, and so on. Thus we may state (11.31) in words, as follows: *The admittance of an arm of the delta is equal to the product of the admittances of the adjacent arms of the wye divided by the sum of the wye admittances.*

To obtain the Δ–Y transformation, we may solve (11.31) for the Y admittances, a difficult task, or we may write two sets of loop equations for the Y and Δ circuits. In the latter case we shall have the dual of the procedure that led to (11.31). In either case, as the reader is asked to show in Problem 11.51, the Δ–Y transformation is

$$Z_a = \frac{Z_{ab}Z_{ca}}{Z_{ab} + Z_{bc} + Z_{ca}}$$

$$Z_b = \frac{Z_{bc}Z_{ab}}{Z_{ab} + Z_{bc} + Z_{ca}} \qquad (11.32)$$

$$Z_c = \frac{Z_{ca}Z_{bc}}{Z_{ab} + Z_{bc} + Z_{ca}}$$

where the Z's are the reciprocals of the Y's of Fig. 11.18. The rule is as follows: *The impedance of an arm of the wye is equal to the product of the impedances of the adjacent arms of the delta divided by the sum of the delta impedances.* (By *adjacent* here, we mean "on each side of and terminating on the same node as." For example, in the superimposed drawing of the Y and Δ, Fig. 11.18(c), Z_a lies between Z_{ab} and Z_{ca} and all three have a common terminal A. Thus Z_{ab} and Z_{ca} are *adjacent* arms of Z_a.)

The Y–Δ conversion formulas (11.31) and (11.32) do not require balanced loads on even three-phase circuits. They are often of considerable use in simplifying general circuits containing impedances that are neither in series nor parallel, as illustrated in the next example.

Example 11.7

We seek the source phasor current **I**. Note in Fig. 11.19 that using fundamental methods (nodal or loop analysis) three simultaneous equations would have to be solved. Comparing Fig. 11.18, we have

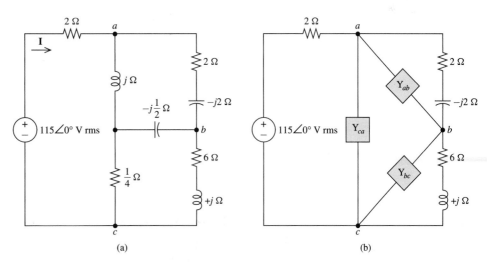

(a)   (b)

FIGURE 11.19   Using Y–Δ impedance conversion to simplify a circuit.

$Y_a = -j1$, $Y_b = j2$, and $Y_c = 4$ S. Then $Y_a + Y_b + Y_c = 4 + j$ and

$$Y_{ab} = \frac{(-j1)(j2)}{4+j} = \frac{2}{4+j}$$

$$Y_{bc} = \frac{j8}{4+j}$$

$$Y_{ca} = \frac{-4j}{4+j}$$

By Fig. 11.19(b) we see that after this Y–Δ transformation, Y_{ab} is in parallel with the admittance $1/(2 - j2)$, leading to a parallel equivalent admittance of

$$\frac{2}{4+j} + \frac{1}{2-j2} = \frac{8-3j}{10-6j}$$

Similarly, the equivalent admittance to Y_{bc} in parallel with the admittance $1/(6 + j)$ is

$$\frac{j8}{4+j} + \frac{1}{6+j} = \frac{-4+49j}{23+10j}$$

This yields the equivalent circuit with impedances shown in Fig. 11.20. Using series–parallel combinations, we have

$$Z_1 = \frac{10-6j}{8-3j} + \frac{23+10j}{-4+49j}$$

$$= 1.51 - j0.73$$

$$Z = 2 + \frac{[(-4-j)/4j]Z_1}{[(-4-j)/4j] + Z_1}$$

$$= 2 + \frac{(-0.25+j1)(1.51-j0.73)}{1.26+j0.27}$$

$$= 2.54 + j1.23$$

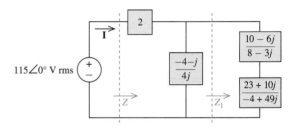

FIGURE 11.20 Equivalent of Fig. 11.19(b) with impedances shown.

Thus

$$\mathbf{I} = \frac{115\underline{/0}}{Z} = \frac{115\underline{/0}}{2.54 + j1.23} = 40.7\underline{/-25.8^\circ} \text{ A rms}$$

The use of Y–Δ conversion in this unbalanced single-phase circuit permitted us to bypass writing and solving three complex equations in three unknowns.

If the load is balanced, the Y–Δ conversion rules simplify greatly. Setting $Y_a = Y_b = Y_c = Y_y$ in (11.31), where Y_y is the common phase admittance of the Y-connected balanced load, we see from (11.31) that $Y_{ab} = Y_{bc} = Y_{ca} = Y_\Delta$, the common admittance of the Δ-connected load, or

$$Y_\Delta = \frac{(Y_Y)^2}{3Y_Y} = \frac{1}{3}Y_Y \tag{11.33}$$

or, inverting, we have

$$Z_\Delta = 3Z_Y \tag{11.34}$$

The balanced Δ-connected phase impedance is three times the equivalent balanced Y-connected phase impedance.

Example 11.8

Find the magnitude of the balanced line voltages $\mathbf{V}_{AB}$, $\mathbf{V}_{BC}$, and $\mathbf{V}_{CA}$ (Fig. 11.21). This circuit is Y–Δ connected. We will convert to a Y–Y connected circuit so that simple per-phase calculations may be done. The load has Y equivalent impedances by (11.34):

$$Z_Y = \tfrac{1}{3}Z_\Delta = \tfrac{1}{3}(6 + j3) = 2 + j$$

Converting to this Y-connected load, the voltage across the equivalent phase a load impedance is, by voltage division,

$$\mathbf{V}_{AN} = \frac{2 + j}{3 + j}(12\underline{/0}) = 8.49\underline{/8.13^\circ}$$

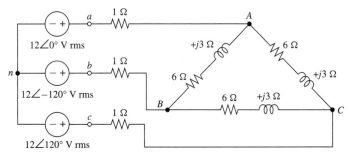

FIGURE 11.21 Circuit for Example 11.8.

The line voltages are, by (11.10a), $\sqrt{3}$ larger, so

$$|\mathbf{V}_{AB}| = \sqrt{3}\,|\mathbf{V}_{AN}| = \sqrt{3}\,(8.49) = 14.7 \text{ V rms}$$

which is the common magnitude of the balanced line voltages $\mathbf{V}_{AB}$, $\mathbf{V}_{BC}$, and $\mathbf{V}_{CA}$.

As demonstrated in this example, it is usually helpful to convert Y–Δ-connected systems to Y–Y so that currents, voltage, and power may be computed on the per-phase basis discussed in Section 11.2.

EXERCISES

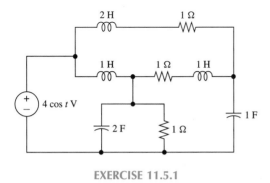

EXERCISE 11.5.1

11.5.1. Find the input impedance seen by the source using a Y–Δ or Δ–Y transformation to simplify the circuit. From this result find the average power delivered by the source.
Answer $(1 + j2)/5$ Ω; 8 W

11.5.2. A balanced three-phase source with $V_L = 100$ V rms is delivering power to a balanced Y-connected load with phase impedance $Z_1 = 8 + j6$ Ω in parallel with a balanced Δ-connected load with phase impedance $Z_2 = 12 + j9$ Ω. Find the power delivered by the source.
Answer 2.4 kW

11.5.3. Show that the Y–Δ transformation of (11.31) is equivalent to

$$Z_{ab} = \frac{Z_a Z_b + Z_b Z_c + Z_c Z_a}{Z_c}$$

$$Z_{bc} = \frac{Z_a Z_b + Z_b Z_c + Z_c Z_a}{Z_a}$$

$$Z_{ca} = \frac{Z_a Z_b + Z_b Z_c + Z_c Z_a}{Z_b}$$

or in words: The impedance of an arm of the delta is equal to the sum of the products of the impedances of the wye, taken two at a time, divided by the impedance of the *opposite* arm of the wye.

11.5.4. If the lines in Exercise 11.4.2 each have a resistance of 0.1 Ω, find the power lost in the lines.
Answer 1.44 kW

11.6 SPICE AND THREE-PHASE CIRCUITS

The analysis of three-phase networks presented previously has been restricted to balanced systems whose solutions can be expressed in terms of a single phase. SPICE is easily used for both balanced or unbalanced systems when applied to the entire network. SPICE makes no distinction between single- and three-phase circuits.

As a first example, consider finding the line voltages and phase currents at the load for the balanced Y–Y system of Fig. 11.22. The transmission line for interconnecting the generator and load has losses that are represented by 2-Ω resistors. A circuit file for this network is

```
3-PHASE Y-Y SYSTEM WITH TRANSMISSION LINE LOSSES
*DATA STATEMENTS VOLTAGES EXPRESSED IN RMS
VAN 1 0 AC 120 0
VBN 2 0 AC 120 -120
VCN 3 0 AC 120 120
RLOSSA 1 4 2
RLOSSB 2 5 2
RLOSSC 3 6 2
RLOSSN 10 0 2
RA 4 7 10
LA 7 10 0.1
RB 5 8 10
LB 8 10 0.1
RC 6 9 10
LC 9 10 0.1
.AC LIN 1 60 60
.PRINT AC VM(4,5) VP(4,5) IM(VAN) IP(VAN)
.PRINT AC VM(5,6) VP(5,6) IM(VBN) IP(VBN)
.PRINT AC VM(6,4) VP(6,4) IM(VCN) IP(VCN)
.PRINT AC IM(RLOSSN) IP(RLOSSN)
.END
```

The resulting output lists VM(4, 5) = VM(5, 6) = VM(6, 4) = 204.9 V rms, and VP(4, 5) = 32.8°, VP(5, 6) = −87.2°, VP(6, 4) = 152.8°, a balanced, positive sequence set as expected.

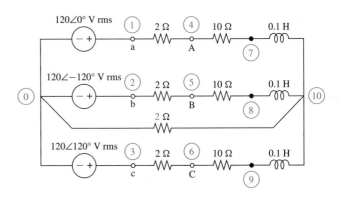

FIGURE 11.22   Balanced Y–Y system for SPICE analysis.

Also, $\text{IM(VAN)} = \text{IM(VBN)} = \text{IM(VCN)} = 3.033$ A rms; $\text{IP(VAN)} = 107.7°$, $\text{IP(VBN)} = -12.3°$, $\text{IP(VCN)} = -132.3°$ forms a balanced set of phase currents. The magnitude of the neutral current is computed to be $\text{IM(RLOSSN)} = 2.08 \times 10^{-15}$ A rms, very close to its exact value of zero.

SPICE Example 11.10

In this example we investigate the effects of unbalancing the circuit of Example 11.9. SPICE does not check for balanced conditions or perform "per-phase" calculations; hence it can deal with unbalanced circuits as easily as balanced ones.

Unbalanced conditions may arise from either an unbalanced source or load. If the source is unbalanced by reducing the phase *a* source magnitude by 10% to 108 V rms, replacing the VAN element line in the previous circuit file by

```
VAN 1 0 AC 108 0
```

the resulting currents and voltages are unbalanced as well. The output shows line voltages $\mathbf{V}_{AB} = 194.7\underline{/34.5°}$, $\mathbf{V}_{BC} = 204.9\underline{/-87.2°}$, and $\mathbf{V}_{CA} = 194.7\underline{/151.1}$ [$\mathbf{V}_{AB} = \text{V}(4, 5)$, and so on; see Fig. 11.22]. Note that an imbalance of the phase source magnitudes leads to unbalanced phase as well as magnitude of the outputs. The line currents are also unbalanced in both magnitude and phase $\mathbf{I}_{Aa} = 2.74\underline{/107.4°}$, $\mathbf{I}_{Bb} = 3.04\underline{/-12.1°}$, $\mathbf{I}_{Cc} = 3.02\underline{/-132.3°}$. The neutral current is no longer zero, with the output listing $\mathbf{I}_{Nn} = 0.29\underline{/115.5°}$. This implies that if the neutral wire were removed, conditions would change. This is unlike the balanced case in which $\mathbf{I}_{Nn} = 0$, and this wire can be removed without altering any output. Removing the neutral wire requires only removing (or "commenting out" by adding an asterisk in the first column) the RLOSSN element line. With the neutral wire removed, rerunning SPICE, the currents have changed to $\mathbf{I}_{Aa} = 2.83\underline{/107.7°}$, $\mathbf{I}_{Bb} = 2.99\underline{/-10.7°}$, and $\mathbf{I}_{Cc} = 2.99\underline{/-134°}$. Note that the degree of imbalance caused by this modest (10%) source imbalance is itself relatively small. Assuming balanced circuit conditions will not lead to large errors in the presence of minor source, line, or load imbalances.

EXERCISES

11.6.1. Find the line voltage and phase current for the load of phase *A* of the system of Fig. 11.22 if the load of phase *C* is short-circuited (called a phase fault).

Answer $204.9\underline{/32.8°}$ V; $4.07\underline{/90.54°}$ A

11.6.2. Repeat Exercise 11.6.1 if the 2-Ω neutral line between nodes 0 and 10 is removed.

Answer 204.9$\underline{/32.8°}$ V; 5.33$\underline{/82.41°}$ A

SUMMARY

Three-phase power is by far the most common form of power distribution. Its advantages when compared to single-phase power delivery include smoother and more efficient power delivery. These benefits are maximized if the circuit is balanced; that is, corresponding circuit variables in the three individual phases are equal in magnitude and 120° out of phase. This chapter presents an introduction to balanced three-phase circuits.

- A Y-connected source is balanced if its phase voltages are equal in magnitude and 120° out of phase.

- The line voltages in a balanced wye-connected three-phase source are $\sqrt{3}$ times bigger than its phase voltages.

- Wye- or Δ-connected loads are said to be balanced when their three component impedances are identical.

- In a balanced Y–Y circuit, the neutral wire may be removed or inserted without affecting any current or voltage.

- In a balanced four-wire Y–Y circuit, calculations may all be done on a per-phase basis, that is, with no interaction between the circuit variables of different phases.

- The equivalent of a balanced Y-connected load with phase impedances Z_p is a Δ-connected load with impedances $Z_1 = 3Z_p$.

- A Y–Δ system is best analyzed by first performing a Δ–Y load transformation, then using per-phase calculations.

- SPICE has no special routines for three-phase circuits. It does not recognize balanced circuits nor do per-phase calculations.

PROBLEMS

11.1. If in Fig. 11.3 $\mathbf{V}_{an} = \mathbf{V}_{nb} = 100\underline{/0}$ V rms, the impedance between terminals A–N is $10\underline{/60°}$ Ω, and that between terminals N–B is $10\underline{/-60°}$ Ω, find the neutral current $\mathbf{I}_{nN}$.

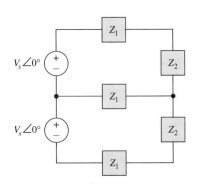

11.2. For the symmetrical single-phase power delivery circuit shown, derive an expression for the loss fraction l, defined as the ratio of the real average power dissipated by the line impedances Z_1, to the total real average power produced by the sources, in terms of Z_1 and Z_2.

FIGURE P11.2

11.3. Find the currents $\mathbf{I}_{aA}$, $\mathbf{I}_{bB}$, and $\mathbf{I}_{nN}$ in this symmetric single-phase three-wire system. Sketch in a phasor diagram.

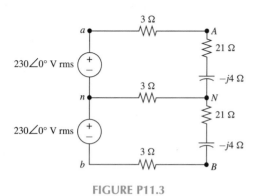

FIGURE P11.3

11.4. Find the real average power supplied by each of two voltage sources in the circuit of Problem 11.3.

11.5. Determine the currents $\mathbf{I}_{aA}$, $\mathbf{I}_{bB}$, and $\mathbf{I}_{nN}$, and sketch in a phasor diagram. The source frequency is 60 Hz.

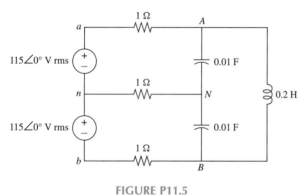

FIGURE P11.5

11.6. Determine the Thevenin equivalent of the circuit of Problem 11.5 at the terminals AB when the inductor is removed. What value of inductance connected between A and B causes the complex power delivered by the sources to be purely real?

11.7. In Fig. 11.9 a balanced, positive-sequence source has $\mathbf{V}_{ab} = 120\underline{/0}$ V rms and $\mathbf{I}_{aA} = 10\underline{/-60°}$ A rms. Find $\mathbf{Z}_p$ and the power delivered to the three-phase load.

11.8. A balanced three-phase Y-connected load draws 1.2 kW at a power factor of 0.6 leading. If the line voltages are a balanced 200-V rms set, find the line current I_L.

11.9. A balanced Y–Y system with $\mathbf{Z}_p = 3\sqrt{3}\underline{/30°}$ Ω delivers 9.6 kW to the load. Find the line voltage V_L and the line current I_L.

11.10. In Fig. 11.9, the source is balanced, with positive phase sequence, and $\mathbf{V}_{an} = 100\underline{/0}$ V rms. Find $\mathbf{Z}_p$ if the source delivers 3.6 kW at a power factor of 0.6 leading.

11.11. For the balanced wye–wye system shown, with $\mathbf{V}_{an} = 115\underline{/0}$ V rms, $\mathbf{Z}_l = 0$ Ω, $\mathbf{Z}_p = 2 + j3$ Ω and positive phase sequence, compute the line voltages $\mathbf{V}_{ab}$, $\mathbf{V}_{bc}$, $\mathbf{V}_{ca}$ and line currents $\mathbf{I}_{aA}$, $\mathbf{I}_{bB}$, $\mathbf{I}_{cC}$, and sketch them in phasor diagrams.

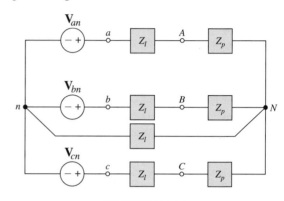

FIGURE P11.11

11.12. Repeat Problem 11.11 for negative phase sequence.

11.13. If in the balanced three-phase circuit shown in Problem 11.11, we have phase sources of magnitude 4400 V rms, $\mathbf{Z}_l = 0.01$ Ω/km, and $\mathbf{Z}_p = 20 + j4$, find an equation relating line length in kilometers to rms line current, and sketch as a graph.

11.14. A balanced wye–wye system with 60-Hz phase sources of 1200 V rms and no line losses delivers 3.00 kW to a load consisting of a series combination of a 10-Ω resistor and L-henry inductor. Find L.

11.15. What is the magnitude of the source phase voltage needed to deliver 100 W at 60 Hz to a small three-phase motor whose $\mathbf{Z}_p$ is equivalent to a series combination of a 12-Ω resistor and 10-mH inductor?

11.16. A balanced wye–wye three-phase system with phase voltage of 115 V rms supplies power at a power factor of 0.85 lagging. Find the complex power supplied to the three-phase circuit if the line current is 2.3 A rms.

11.17. A balanced Y–Y three-wire, positive-sequence system has $\mathbf{V}_{an} = 200\underline{/0}$ V rms and $\mathbf{Z}_p = 3 + j4$ Ω. The lines each have a resistance of 1 Ω. Find the line current I_L, the power delivered to the load, and the power dissipated in the lines.

11.18. A balanced Y-connected source, $\mathbf{V}_{an} = 200\underline{/0}$ V rms, positive sequence, is connected by four perfect conductors (having zero impedance) to an unbalanced Y-connected load, $\mathbf{Z}_{AN} = 10\ \Omega$, $\mathbf{Z}_{BN} = 10\underline{/-30°}\ \Omega$, and $\mathbf{Z}_{CN} = 20\sqrt{2}\underline{/75°}\ \Omega$. Find the neutral current and the power absorbed by the load.

11.19. In Fig. 11.12, let $V = 115$ V rms, $Z_L = Z_N = 0$ and $Z_3 = 20\underline{/+15°}$. For what $Z_1 = |Z_1|\underline{/+15°}$ will the line currents in the single- and three-phase circuits be equal? Compute the complex power delivered to the load for the single-phase current with this Z_1 and the complex power delivered to the three-phase load.

11.20. Consider Fig. 11.12 with V, Z_L, Z_N, and Z_3 as in Problem 11.19. For what $Z_1 = |Z_1|\underline{/15°}$ will the real average power delivered to the single-phase load and three-phase loads be equal? Compute the line current magnitudes in both circuits if this Z_1 is used in the single-phase case.

11.21. Why is the frequency of mechanical vibration of a single-phase motor twice that of its electrical supply?

11.22. Consider Fig. 11.12 with $V = 4.4$ kV rms, each source operating at a power factor of 1.0, and each circuit supplying 660 kW to its load. Sketch phasor diagrams for the line currents $\mathbf{I}_{aA}$, $\mathbf{I}_{bB}$; $\mathbf{I}_{a'A'}$, $\mathbf{I}_{b'B'}$, $\mathbf{I}_{c'C'}$ and the total instantaneous power absorbed by the single-phase and three-phase loads as a function of time.

11.23. In Fig. 11.15 the source is balanced with positive phase sequence and $\mathbf{V}_{an} = 100\underline{/0}$ V rms. If the phase impedance is $3\sqrt{3}\underline{/30°}\ \Omega$, find the line current and the power delivered to the load.

11.24. In Fig. 11.15 the positive sequence system has $\mathbf{V}_{an} = 200\underline{/0}$ V rms. Find $\mathbf{Z}_p$ if the source delivers 2.4 kW at a power factor of 0.8 lagging.

11.25. In the Y-Δ system shown, the source is positive sequence with $\mathbf{V}_{an} = 100\underline{/0}$ V rms and the phase impedance is $\mathbf{Z}_p = 3 - j4\ \Omega$. Find the line voltage V_L, the line current I_L, and the power delivered to the load.

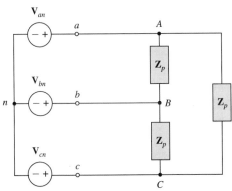

FIGURE P11.25

11.26. For the Y-Δ system of Problem 11.25, $\mathbf{Z}_p = 4 + j3\ \Omega$ and the power delivered to the load is 19.2 kW. Find the line current I_L and the source phase voltages if $\mathbf{V}_{an}$ is the reference.

11.27. What advantages and disadvantages might six-phase power delivery systems have compared to three-phase?

11.28. For the delta-connected load shown in Fig. 11.15, find the line currents and real average power to the load if $\mathbf{Z}_p = 3 - j$ and the line voltages are $\mathbf{V}_{ab} = 25\underline{/0}$, $\mathbf{V}_{bc} = 25\underline{/-120°}$, $\mathbf{V}_{ca} = 25\underline{/+120°}$.

11.29. For the balanced, positive phase sequence wye–delta circuit shown, assume $\mathbf{V}_{an} = 1200\underline{/0}$ V rms, $Z_l = 0.5\ \Omega$, and $Z_p = 8 + j2\ \Omega$. Find the line currents $\mathbf{I}_{aA}$, $\mathbf{I}_{bB}$, and $\mathbf{I}_{cC}$ and the complex power delivered to the load. Do not use wye–delta transformation.

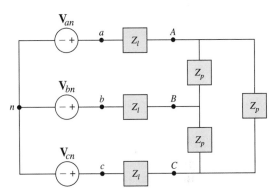

FIGURE P11.29

11.30. A balanced wye-connected source with line voltage 115 V rms supplies power to a balanced delta-connected load with phase loads Z_p each consisting of a parallel combination of a 75-Ω resistor, 2.5-mH inductor, and 10-μF capacitor. Find the complex power delivered to the load and the power factor at 60 Hz and at 400 Hz.

11.31. For the circuit of Problem 11.29, suppose $\mathbf{V}_{an} = 1200\underline{/0}$, $Z_l = 0\ \Omega$, and the three-phase source delivers +2 kvars of reactive power to the load. Find Z_p and the real average power in watts delivered if the power factor is 0.92 lagging.

11.32. In the **Y**–Δ system of Problem 11.25 the source voltage $\mathbf{V}_{an} = 100\underline{/0}$ V rms and $\mathbf{Z}_p = 10\underline{/60°}$ Ω. Find the line voltage, the line current, the load current magnitudes, and the power delivered to the load.

11.33. Find the line current I_L.

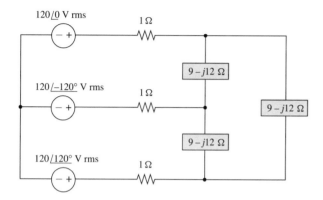

120$\underline{/0}$ V rms   1 Ω

9 − j12 Ω

120$\underline{/-120°}$ V rms   1 Ω

9 − j12 Ω

9 − j12 Ω

120$\underline{/120°}$ V rms   1 Ω

FIGURE P11.33

11.34. Find the power delivered to the load in Problem 11.33 if the magnitude of the source voltages is 200 V rms and $\mathbf{Z}_p = 9 - j9$ Ω.

11.35. A balanced three-phase, positive-sequence source with $\mathbf{V}_{ab} = 200\underline{/0}$ V rms is supplying a Δ-connected load, $\mathbf{Z}_{AB} = 50$ Ω, $\mathbf{Z}_{BC} = 20 + j20$ Ω, and $\mathbf{Z}_{CA} = 30 - j40$ Ω. Find the line currents and the power absorbed by the load.

11.36. A balanced three-phase positive-sequence source with $\mathbf{V}_{ab} = 240\underline{/0}$ V rms is supplying a parallel combination of a **Y**-connected load and a Δ-connected load. If the **Y** and Δ loads are balanced with phase impedances of $8 - j8$ Ω and $24 + j24$ Ω, respectively, find the line current I_L and the power supplied by the source, assuming perfectly conducting lines.

11.37. A balanced delta-connected load has $\mathrm{Re}(\mathbf{Z}_p) = 14$ Ω. If a balanced wye-connected source delivers 1 kW at a power factor of 0.5 leading, find $\mathbf{Z}_p$ and the reactive power delivered to the load.

11.38. Draw an *RLC* circuit diagram for the wye-connected load that is equivalent at 60 Hz to a delta-connected load with each phase load consisting of a 300-Ω resistor in series with a 2-mH inductor. Repeat at 600 Hz.

11.39. Find the equivalent delta load if (a) the load is balanced $\mathbf{Z}_p = 2 + j$; (b) $\mathbf{Z}_{p1} = \mathbf{Z}_{p2} = 2 + j$, $\mathbf{Z}_{p3} = 2 - j$.

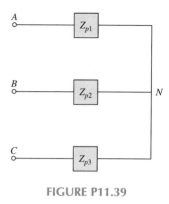

FIGURE P11.39

11.40. Find the equivalent delta- and wye-connected loads if $Z_1 = -j30$ Ω and $Z_2 = 60 + j15$ Ω.

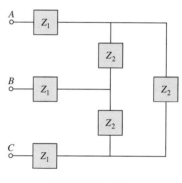

FIGURE P11.40

11.41. Solve Problem 11.5 by use of delta–wye load transformation.

11.42. A balanced three-phase wye-connected source with line currents 2 A rms is supplying 300 W to a balanced delta-connected load with phase impedance angle $\angle Z_p = +20°$. Find Z_p and the line voltage.

SPICE Problems

11.43. Each $Z = 1$ Ω. Find the delta-connected three-phase equivalent load. Check using SPICE.

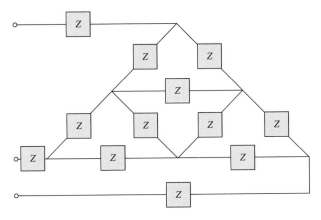

FIGURE P11.43

11.44. Find the total power delivered to the three-phase load by the unbalanced source. All resistors are 10 Ω. Check using SPICE.

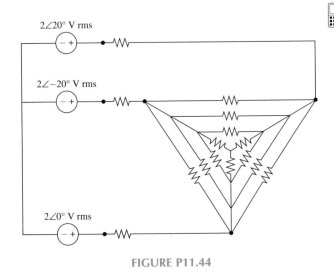

2∠20° V rms

2∠−20° V rms

2∠0° V rms

FIGURE P11.44

11.45. Solve Problem 11.5 above using SPICE.

11.46. The circuit shown is operated at 60 Hz. Find the phase load currents I_{AB}, I_{BC}, and I_{CA} using SPICE.

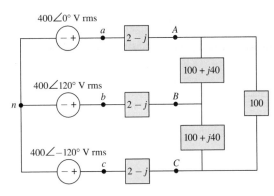

400∠0° V rms

400∠120° V rms

400∠−120° V rms

FIGURE P11.46

11.47. For the circuit of Problem 11.46, suppose the lines are protected by circuit breakers set to trip when a line current I_{aA}, I_{bB}, or I_{cC} reaches 100 A. If the 100-Ω phase impedance between A and C were replaced by an R-Ω resistor, for what approximate value of R will a circuit breaker trip? Solve using SPICE. Guess at a trial value for R, reduce it if the limit is not exceeded and increase if it is. Find R to two significant digits.

More Challenging Problems

11.48. Write three loop equations for this circuit using line current I_1, I_2, and I_3. Include the neutral line in each loop. Rewrite for the case $\rho = 0$, calling the resulting balanced current variables I_{1b}, I_{2b}, and I_{3b}. Now expressing the unbalanced currents I_1, I_2, and I_3 in terms of a balanced and unbalanced component, $I_1 = I_{1b} + i_1$, $I_2 = I_{2b} + i_2$, $I_3 = I_{3b} + i_3$, subtract the second (balanced) set of equations from the first. The resulting equations may be easily solved for the unbalanced components. Find I_1, I_2, and I_3 for $V = 115$ V rms, $R = 1$ Ω, $Z = 1 + j$ Ω and $\rho = -\frac{1}{2}$ Ω.

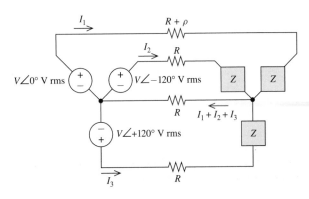

FIGURE P11.48

11.49. A balanced three-phase three-wire source is supplying power to a balanced delta-connected three-phase load with pure resistive phase impedances $R_p = 100\ \Omega$. If one of the three phase loads is reduced to $R'_p < 100\ \Omega$, find a value for R'_p so that the line current imbalance, defined as the ratio of the minimum to maximum rms line currents, is 0.50 to two significant digits. Repeat for 0.10 to two significant digits. Use SPICE.

11.50. What is the equivalent of the three-phase wye–delta transformation result $Z_\Delta = 3Zy$ in the six-phase case? Sketch the two equivalents and derive their relationship.

11.51. Derive Equation 11.32.

11.52. Find the impedance Z looking into this two-terminal subnetwork. Use a wye–delta transform. $Z_1 = 12 + j6$, $Z_2 = +j12$, $Z_3 = 4 + j4$, and $Z_4 = 16$ (all in ohms).

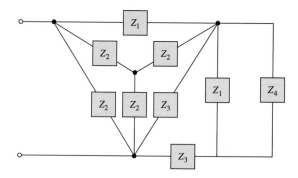

FIGURE P11.52

11.53. A three-phase load consists of a parallel combination of a balanced wye-connected resistive load with phase impedances $R\ \Omega$, and a balanced delta-connected inductive load with phase impedances of 2 H. If the load has pf 0.8 lagging at 60 Hz, find R.

The Laplace Transform

Pierre Simon Laplace
1749–1827

There was no need for God in my hypothesis [to Napoleon, who had to ask].

Pierre Simon Laplace

Pierre Simon, Marquis de Laplace, the famous French astronomer and mathematician, is credited with the transform that bears his name and allows us to further generalize the generalized phasor method to analyze circuits with non-sinusoidal inputs. Laplace was better known, however, for *Celestial Mechanics*, his master work, which summarized the achievements in astronomy from the time of Newton.

Laplace was born in Beaumont-en-Auge, Normandy, France. Little is known of his early life other than that his father was a farmer, because the snobbish Laplace, after he became famous, did not like to speak of his humble origins. Rich neighbors, it is said, recognized his talent and helped finance his education, first at Caen and later at the military school in Beaumont. Through the efforts of the famous physicist d'Alembert, who was impressed by his abilities and his effrontery, Laplace became a professor of mathematics in Paris at age 20. He was an opportunist, shifting his political allegiance as required so that his career successfully spanned three regimes in revolutionary France—the republic, the empire of Napoleon, and the Bourbon restoration. Napoleon made him a count and Louis XVIII made him a marquis. His mathematical abilities, however, were genuine, inspiring the great mathematician Simeon Poisson to label him the Isaac Newton of France.

Chapter Contents

In Chapters 8 to 11 our goal was to determine the behavior of ac steady-state circuits. We began with basic circuit analysis equations containing time integrals and derivatives, or *integrodifferential* equations. While solving these sets of coupled integrodifferential equations directly proved somewhat tedious, we discovered an indirect strategy that paid great dividends. By first transforming the integrodifferential equations into the phasor domain, we managed to replace the derivative and integral operations of calculus with the much simpler multiplication and division operations of ordinary algebra. After solving the circuit equations in this more convenient *algebraic* phasor form, the results were then transformed back into the original sinusoidal time-domain format without difficulty. The strategy of transformation of circuit equations to phasor form proved effective in reducing computations and brought us a conceptual bonus as well. It permitted extension of the idea of resistance R into impedance $Z(j\omega)$, which forms the basis for a unified treatment of all RLC elements in ac steady state.

Our goal in this chapter is to introduce a tool that will permit us to generalize this successful strategy beyond the narrow confines of ac steady state. That tool is *Laplace transformation*. Clearly, such a tool is needed. The same sets of coupled integrodifferential equations that proved inconvenient in ac steady-state problems also govern the behavior of circuits containing nonsinusoidal sources and circuits in which transients are to be determined. We cannot transform these circuit problems using phasors, which apply in the ac steady-state case. The Laplace transform supplies what is needed to convert from integrodifferential to algebraic equations and to generalize the important notion of impedance to all linear circuit problems, in all cases, not just ac steady state.

We first define the Laplace transform and the *s*-domain in which it takes value. The singularity functions are next introduced, a set of functions useful in describing transient behavior. In succeeding sections, the Laplace transforms of common time functions are calculated and properties of the Laplace transform explored. Return back from the *s*-domain to the time domain (inverse transformation) by partial fractions is next considered, and the chapter concludes with the solution of coupled integrodifferential equations by this method.

THE s-DOMAIN

Laplace transformation L is defined as a rule $L[f(t)] = \mathbf{F}(s)$ that associates with the time function $f(t)$ a function $\mathbf{F}(s)$, of a complex variable s, which is given by

$$\mathbf{F}(s) = \int_{0-}^{\infty} f(t)e^{-st}\,dt \qquad (12.1)$$

$\mathbf{F}(s)$ is called the *Laplace transform* of $f(t)$, and $f(t)$ the *inverse Laplace transform* of $\mathbf{F}(s)$. Together, $f(t)$ and $\mathbf{F}(s)$ are called a *Laplace transform pair*. The complex plane over which s ranges and $\mathbf{F}(s)$ takes value is called the *s-domain*.

Several comments are in order with respect to the definition (12.1). Note first that the lower limit is specified as 0−. The quantities 0− and 0+ were introduced in Chapter 5 as times just before and after $t = 0$. The domain of integration in (12.1) begins just to the left of $t = 0$ and continues to $t = \infty$. For ordinary functions $f(t)$, the lower limit 0− can freely be replaced by either 0 (or even 0+) without changing the value of the integral, since there can be no nonzero area under any finite integrand in the infinitesimal interval $t = 0-$ to $t = 0+$. However, there is a class of decidedly unordinary functions called the *singularity functions* to be introduced in the next section. In a singularity function, area may "pile up" right at a single time instant such as $t = 0$, and we choose to include this area in the definition of the Laplace transform by setting the lower limit to 0−.

Since values of $f(t)$ for $t < 0-$ do not enter into (12.1), *the Laplace transform of a time function $f(t)$ does not depend on values of $f(t)$ prior to 0−*. In circuits with transients, there is good justification for making transforms depend only on values of currents and voltages after a specific time. In transient circuits there is some initial time, which we usually call $t = 0$, before which the detailed time histories of the currents and voltages are not of interest. For example, a circuit may have no sources acting until $t = 0$, at which time a voltage source is switched in and a transient response begins. Before $t = 0$, nothing of interest is happening. Our choice of lower limit makes (12.1) well suited to study such circuits, making the Laplace transform properly indifferent to behavior prior to the initial time of interest. The earlier behavior will be summarized effectively by specifying initial conditions at time 0−. Equation (12.1) is sometimes referred to as the "one-sided" Laplace transform to distinguish it from the "two-sided" Laplace transform defined with lower limit equal to $-\infty$. We will have no need for the two-sided Laplace transform in this book and will take *Laplace transform* to be synonymous with *one-sided Laplace transform* as defined in (12.1).

By *existence* of the Laplace transform $\mathbf{F}(s)$ of a given $f(t)$, we mean that the defining integral (12.1) exists for at least some values of its argument s. That is, there exists a region of the s-domain, called the *region of convergence*, for which the integral in (12.1) is well defined. A function $f(t)$ whose Laplace transform $\mathbf{F}(s)$ exists is said to be *Laplace transformable*. Sufficient conditions for Laplace transformability of $f(t)$ are:

1. $f(t)$ is a piecewise continuous function; that is, $f(t)$ is continuous except perhaps at a set of isolated finite discontinuities [points $\{t_i\}$, where $f(t_i+)$ and $f(t_i-)$ are distinct finite numbers].

2. $f(t)$ is of exponential order; that is, there is some real exponential function $Me^{\sigma t}$ that for all t sufficiently large is *greater* in magnitude than $f(t)$:

$$|f(t)| \le Me^{\sigma t} \qquad (12.2)$$

Fortunately, almost all functions of practical interest satisfy these conditions and are therefore Laplace transformable.* Those few that are not contain infinite discontinuities, such as $f(t) = 1/t^2$, and thus are not piecewise continuous or grow too fast to be of exponential order, such as $f(t) = t^t$. While they are of interest mathematically, we are not likely to encounter such exotic functions in practical circuits, and we will not consider non-Laplace-transformable functions further.

As a first Laplace transform calculation, consider the real function of time $f(t) = e^{-at}u(t)$ shown in Fig. 12.1(a). $u(t)$ is the unit step function introduced in Section 6.6. Recall that multiplication by the unit step function has the effect of forcing the product to zero for $t < 0$. Using (12.1), $L[f(t)] = F(s)$ is

$$\mathbf{F}(s) = \int_{0-}^{\infty} e^{-(s+a)t} \, dt \qquad (12.3a)$$

Evaluating this integral yields

$$\mathbf{F}(s) = -\frac{1}{s+a} e^{-(s+a)t} \bigg|_{0-}^{\infty} = \frac{1}{s+a} \left[e^{-(s+a)0-} - e^{-(s+a)\infty} \right] \qquad (12.3b)$$

Since $e^{-(s+a)t}$ is a continuous function, its value at $0-$ is the same as its value at 0, which is unity. To evaluate at $t = \infty$, write s in rectangular form as $s = \sigma + j\omega$.

$$e^{-(s+a)t} = e^{-(\sigma+a)t} e^{-j\omega t}$$

The magnitude of this complex number is $e^{-(\sigma+a)t}$. For $\sigma + a > 0$, this value goes to zero as t goes to infinity. Thus the last term in (12.3b) is zero for $\sigma = \mathrm{Re}(s) > -a$ and

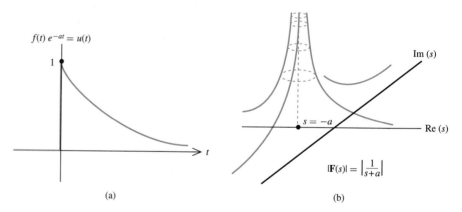

FIGURE 12.1 $f(t)$ and the magnitude of its Laplace transform $\mathbf{F}(s)$.

---

*Even some functions failing these two rather weak sufficient conditions are still Laplace transformable. $f(t)$ can be *almost* piecewise continuous; that is, there may be isolated times at which $f(t)$ has infinite discontinuities as long as $f(t)$ is absolutely integrable on such intervals. Similarly, there are a few functions that are not of exponential order which are still Laplace transformable.

we have determined the Laplace transform of $f(t) = e^{-at}u(t)$ to be

$$F(s) = \frac{1}{s+a} \qquad (12.4)$$

A sketch of the magnitude of $F(s)$ over the s-domain is shown in Fig. 12.1(b). The region of convergence for $F(s)$ is the portion of the s-domain $\text{Re}(s) > -a$. Note that the function $f(t) = e^{-at}$ has the same $F(s)$ as that for $e^{-at}u(t)$ calculated above, since these two functions disagree only in negative time. Behavior in negative time is never accounted for by the one-sided Laplace transform.

With $a = 0$, this result (12.4) reveals a particularly useful transform pair: the Laplace transform of the unit step function $f(t) = u(t)$ is $F(s) = 1/s$, or

$$L[u(t)] = \frac{1}{s} \qquad (12.5)$$

For convenience in graphing Fig. 12.1, we have assumed that a is a real number, making $f(t) = e^{-at}u(t)$ a real time function. More generally, if a is complex, the steps leading to the transform (12.4) are unchanged [except that the region of convergence changes from $\text{Re}(s) > -a$ to $\text{Re}(s) > -\text{Re}(a)$]. So the transform pair identified above is equally valid for the complex a case as the real a case.

Points in the s-domain at which $F(s)$ fails to be finite, such as the point $s = -a$ in Fig. 12.1(b), or $s = 0$ in the transform of the unit step above, are called the *poles* of $F(s)$. The name derives from the appearance of $F(s)$, as if a pole were supporting the function at $s = -a$ in Fig. 12.1(b). Also important are the *zeros* of $F(s)$, those points in the s-domain at which $F(s) = 0$. $F(s) = 1/(s + a)$ has one pole (at $s = -a$) and no zeros. The following example has both poles and zeros.

Example 12.1

We seek the Laplace transform of $f(t) = (e^{-t} + e^{-2t})u(t)$. By (12.1),

$$F(s) = \int_{0-}^{\infty} \left(e^{-t} + e^{-2t} \right) e^{-st}\, dt \qquad (12.6a)$$

$$= \int_{0-}^{\infty} e^{-t} e^{-st}\, dt + \int_{0-}^{\infty} e^{-2t} e^{-st}\, dt \qquad (12.6b)$$

The first integral is (12.4) with $a = 1$, and the second (12.4) with $a = 2$. Thus

$$F(s) = \frac{1}{s+1} + \frac{1}{s+2} = \frac{2s+3}{(s+1)(s+2)} \qquad (12.7)$$

which is the desired Laplace transform. $F(s)$ has poles at $s = -1$ and $s = -2$ and a single zero at $s = -\frac{3}{2}$. As we shall see later in this chapter, knowledge of the poles of $F(s)$ is useful in finding the inverse transform $f(t)$.

Example 12.1 suggests a general property of Laplace transforms that will greatly aid in their calculation. Let $f(t) = c_1 f_1(t) + c_2 f_2(t)$. Then, by the defining integral (12.1),

$$\mathbf{F}(s) = \int_{0-}^{\infty} \left[c_1 f_1(t) + c_2 f_2(t) \right] e^{-st} \, dt$$

$$= c_1 \int_{0-}^{\infty} f_1(t) e^{-st} \, dt + c_2 \int_{0-}^{\infty} f_2(t) e^{-st} \, dt$$

Recognizing the two integrals as $\mathbf{F}_1(s)$ and $\mathbf{F}_2(s)$, we have the *linearity property*.

LINEARITY

$$L[c_1 f_1(t) + c_2 f_2(t)] = c_1 \mathbf{F}_1(s) + c_2 \mathbf{F}_2(s) \qquad (12.8)$$

That is, *the Laplace transform of any linear combination of time functions is the same linear combination of their individual Laplace transforms.* This property derives directly from the linearity of any integral such as (12.1) in its integrand. The linearity property allows us to use known Laplace transforms to create new ones.

Example 12.2

Let us find the Laplace transforms of $\sin \omega t$ and $\cos \omega t$. Since the Laplace transforms of $e^{+j\omega t}$ and $e^{-j\omega t}$ are $1/(s - j\omega)$ and $1/(s + j\omega)$, respectively, then by linearity

$$L(\sin \omega t) = L \left[\frac{1}{2j} (e^{+j\omega t} - e^{-j\omega t}) \right] = \frac{1}{2j} \left(\frac{1}{s - j\omega} - \frac{1}{s + j\omega} \right)$$

or

$$L[\sin \omega t] = \frac{\omega}{s^2 + \omega^2} \qquad (12.9a)$$

Similarly,

$$L(\cos \omega t) = L \left[\frac{1}{2} (e^{+j\omega t} + e^{-j\omega t}) \right] = \frac{1}{2} \left(\frac{1}{s - j\omega} + \frac{1}{s + j\omega} \right)$$

or

$$L(\cos \omega t) = \frac{s}{s^2 + \omega^2} \qquad (12.9b)$$

The same technique will identify the Laplace transforms of $\cosh at$ and $\sinh at$, the hyperbolic cosine and hyperbolic sine functions defined for a real as

$$\cosh at = \tfrac{1}{2} (e^{+at} + e^{-at})$$

$$\sinh at = \tfrac{1}{2} (e^{+at} - e^{-at})$$

Then, by linearity,

$$L(\cosh at) = \frac{1}{2} \left(\frac{1}{s - a} + \frac{1}{s + a} \right) = \frac{s}{s^2 - a^2}$$

$$L(\sinh at) = \frac{1}{2} \left(\frac{1}{s - a} + \frac{1}{s + a} \right) = \frac{a}{s^2 - a^2}$$

In subsequent sections of this chapter we introduce other properties that will be helpful in finding Laplace transforms and their inverse transforms and solving circuit equations. These properties, beginning with the linearity property above, are summarized in Table 12.1. Table 12.2 gathers together all the important transform pairs we derive throughout this chapter. These tables are located in Section 12.3, and expanded versions of these tables are found on the inside of the back cover and its facing page.

Definition (12.1) of the Laplace transform permits us to move from the time-domain description $f(t)$ of a time function to its s-domain description $F(s)$. The relation that explicitly defines passage back from the s-domain to the time domain is the *Laplace inversion integral*

$$f(t) = \frac{1}{2\pi} \int_{-\infty}^{+\infty} F(s)e^{st}\, d\omega$$

where $s = \sigma + j\omega$. σ is any value within the region of convergence of $F(s)$ and the principal value of the integral is taken. The Laplace inversion integral is seldom easy to apply, and there is a much more convenient method for recovering $f(t)$ from $F(s)$ based on partial fraction expansion, the subject of Section 12.4. For our purposes the significance of the inversion integral is that, together with (12.1), it shows that *there is a 1:1 relation between Laplace transformable functions and their transforms*. If $f(t)$'s and their $F(s)$'s were not bonded together uniquely in this way, we could not find much use for the Laplace transform. For instance, suppose that we wish to recover $f(t)$ from its transform $F(s) = 1/(s + a)$. By (12.4) we know that $e^{-at}u(t)$ has this transform. But if there were other time functions with this same transform, we could not recover $f(t)$ uniquely. The inversion integral shows that there is only one $f(t)$ for a given $F(s)$, so the unique inverse transform must be $e^{-at}u(t)$.

EXERCISES

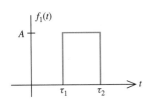

EXERCISE 12.1.1

12.1.1. Use the defining integral (12.1) to determine $L[f_1(t)]$ for the $f_1(t)$ shown.

Answer $A/s(e^{-s\tau_1} - e^{-s\tau_2})$

12.1.2. Use the results of Exercise 12.1.1 and linearity to find $L[f_2(t)]$ for the $f_2(t)$ shown.

Answer $1/s(1 + 3e^{-s} - e^{-2s} - 2e^{-4s} - e^{-7s})$

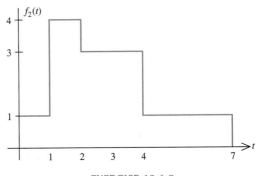

EXERCISE 12.1.2

12.1.3. Is $f(t) = t^2 e^{+t}$ Laplace transformable? If so, find suitable M and σ in (12.2).

Answer Yes; for instance, $M = 1$ and any $\sigma > +1$. $Me^{\sigma t} > |f(t)|$ for all

$$\frac{t}{\ln t} > \frac{2}{\sigma - 1}$$

12.2   SINGULARITY FUNCTIONS

Repeated integration and differentiation of the unit step function produce a family of functions called the *singularity functions*. In this section we introduce this family, develop general rules for finding the Laplace transform of derivatives and integrals, and use these rules to find the Laplace transforms of the singularity functions. In the application of Laplace transforms to circuit analysis, singularity functions arise naturally and play a central role. Indeed, this family of functions is prominent in the study of all manner of physical systems that are modeled by integrodifferential equations.

The only singularity function we have encountered so far is the unit step $u(t)$. We next consider singularity functions produced by repeated integration of $u(t)$. The first integral of the unit step is referred to as the *unit ramp* function $r(t)$. Since $u(t) = 0$ for $t < 0$, its integral $r(t) = 0$ for $t < 0$. For $t \geq 0$,

$$r(t) = \int_{0-}^{t} u(\tau)\, d\tau = \int_{0-}^{t} 1\, d\tau = t, \qquad t \geq 0 \qquad (12.10)$$

Examining Fig. 12.2(b), $r(t)$ is zero until $t = 0$, then "ramps" upward at a constant slope of unity. Integrating once again, we have the *unit parabola* function $p(t)$. Since $r(t)$ is zero for $t < 0$, its integral $p(t)$ also will be zero for $t < 0$, and for $t \geq 0$:

$$p(t) = \int_{0-}^{t} r(\tau)\, d\tau = \int_{0-}^{t} \tau\, d\tau = \frac{t^2}{2}, \qquad t \geq 0 \qquad (12.11)$$

The unit parabola is sketched in Fig 12.2(c).

Note in Fig. 12.2 that the unit step, ramp, and parabola are each zero for $t < 0$. Functions produced by further integrations will share this property, as will those produced by differentiation, however many times repeated. *The singularity functions are all zero for $t < 0$.* Thus $r(t) = tu(t)$ and $p(t) = \frac{1}{2}t^2 u(t)$. Note also that each is infinitely smooth, that is, possesses derivatives of all orders, everywhere except at the point $t = 0$. This is the *singular point* of these functions from which their family name derives. Continuing,

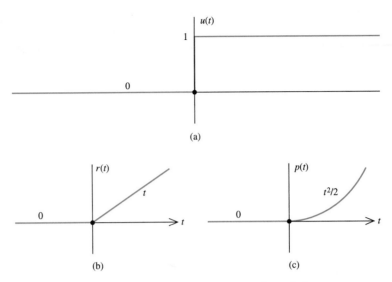

(a)

(b)   (c)

FIGURE 12.2   Unit step, ramp, and parabola.

further integration of the unit parabola yields the unit cubic function, equal to $\frac{t^3}{6}u(t)$, then the unit quartic, $\frac{t^4}{24}u(t)$, and so on. Each is a scaled power of t in positive time.

One practical application of this branch of the family of singularity functions is in representing *piecewise polynomial functions*, those functions that are fixed polynomials over subintervals. This use is shown in the next example and the exercises at the end of this section.

Example 12.3

Consider the function $f(t)$ of Fig. 12.3, given by

$$f(t) = \begin{cases} 0, & -\infty < t < 1 \\ t^2 - 2t + 1, & 1 \le t < 3 \\ -2t + 10, & 3 \le t < 6 \\ 2, & 6 \le t < \infty \end{cases} \qquad (12.12)$$

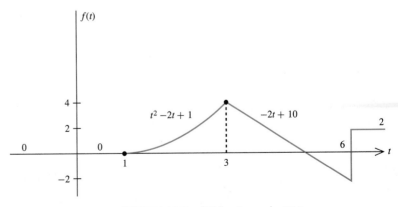

FIGURE 12.3 $f(t)$ for Example 12.3.

We wish to express $f(t)$ as a linear combination of steps, ramps, and parabolas. Starting from the left, note that in the subinterval $1 \leq t \leq 3$,

$$f(t) = 2p(t-1) = 2\left[\tfrac{1}{2}(t-1)^2\right], \qquad 1 \leq t \leq 3$$

In the subinterval beginning at $t = 3$, the t^2 term must be canceled, requiring that we add the term $-2p(t-3)$. But

$$2p(t-1) - 2p(t-3) = (t-1)^2 - (t-3)2 = 4t - 8, \qquad t \geq 3$$

To produce the desired slope -2 requires that we add $-6r(t-3)$. Then up to the end of this subinterval, $t = 6$,

$$f(t) = 2p(t-1) - 2p(t-3) - 6r(t-3), \qquad t \leq 6$$

Finally, adding $+2r(t-6)$ corrects the slope to 0 in the final subinterval, and the constant term is corrected from $+10 + 2(-6) = -2$ to $+2$ by adding $+4u(t-6)$. This is the desired description, valid for all t, of $f(t)$ as the weighted sum:

$$f(t) = 2p(t-1) - 2p(t-3) - 6r(t-3) + 2r(t-6) + 4u(t-6)$$

We next consider the remaining singularity functions, those formed by repeated differentiation of the unit step. The first derivative of the unit step function is called the *unit impulse* function $\delta(t)$:

$$\delta(t) = \frac{d}{dt}u(t) \tag{12.13}$$

Consider the graph of the unit step shown in Fig. 12.2(a). Since $u(t)$ is constant for all t negative, its derivative $\delta(t)$ must equal zero there, and the same is true for all t positive. So surely the unit impulse function $\delta(t)$ must be equal to zero for all t, except perhaps at $t = 0$. What is its behavior at the singular point $t = 0$? The unit step $u(t)$ has a jump at $t = 0$, and we recall that a function is not formally differentiable at a point of discontinuity. Thus the unit impulse, being its derivative, cannot be understood to be an ordinary function, since its value at $t = 0$ is not well defined in the usual mathematical sense.

Let us examine the situation at $t = 0$ more closely. Consider the unit steplike function $v(t)$ shown in Fig. 12.4(a),

$$v(t) = \frac{1}{2\Delta}[r(t-\Delta) - r(t+\Delta)] \tag{12.14a}$$

where $r(t)$ is the unit ramp. Let $d(t) = (d/dt)v(t)$ be its derivative. Since $u(t) = (d/dt)r(t)$, it follows from (12.14a) that

$$d(t) = \frac{d}{dt}v(t) = \frac{1}{2\Delta}[u(t-\Delta) - u(t+\Delta)] \tag{12.14b}$$

As Δ gets smaller and smaller, $v(t)$ approaches the unit step, and its derivative $d(t)$ thus approaches the unit impulse $\delta(t)$. The unit impulselike $d(t)$ is a pulse with height $1/2\Delta$

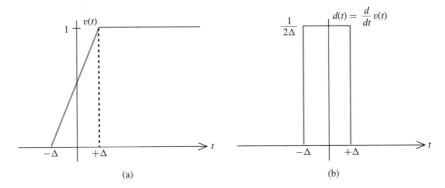

FIGURE 12.4 Functions approaching the unit step and impulse.

and base 2Δ (and thus area always equal to 1) centered at $t = 0$. These functions are sketched in Fig. 12.4. Note two characteristics of $\delta(t)$ that emerge by noting that $\delta(t)$ is the limit of $d(t)$ as Δ goes to zero:

$$\delta(t) = 0 \qquad \text{for all } t \neq 0 \tag{12.15a}$$

$$\int_{-\infty}^{+\infty} \delta(t)\, dt = 1 \tag{12.15.b}$$

Examining Fig. 12.4(b) as $\Delta \to 0$, *the unit impulse $\delta(t)$ is an infinitely tall, infinitely narrow pulse of unit area.* The graphic symbol we will use for it is an upward-pointing arrow as shown in Fig. 12.5. The number written alongside is not its height, which is infinite, but its *area*, which is finite. From Figure 12.5 it is clear why the unit impulse $\delta(t)$ is not an ordinary function well defined at $t = 0$. Its value is infinite there, and infinity is not a real number.

While $\delta(t)$ is not an ordinary function, it can still be usefully defined as a *generalized function*. Generalized functions are careful extensions to the class of ordinary functions that if used in the proper settings, yield accurate and rigorous results.

Generalized objects in mathematics are not new to us. We are familiar with the idea that infinity (∞) is not an ordinary real number (a point on the real line). There is no real number that x approaches when we say "let x approach infinity." Yet we can work with infinity quite effectively and rigorously, treating it as an extension to the set of real numbers, a *generalized* real number. We can do arithmetic with it; for instance,

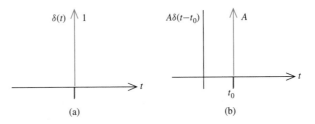

FIGURE 12.5 (a) Unit impulse function $\delta(t)$; (b) shifted impulse of area A, $A\delta(t - t_0)$.

the equation $1/\infty = 0$ is correct and obvious. However, we must be careful when using the generalized number ∞. For instance, although the equation $x/x = 1$ is valid for all real numbers $x \neq 0$, it is not valid for the generalized number ∞. Even though ∞ is not zero, ∞/∞ is not in general equal to 1.

$\delta(t)$ is to ordinary functions as ∞ is to ordinary numbers, an extension that if used in the right equations will yield accurate results. Which are the right equations for $\delta(t)$? To answer this question thoroughly and rigorously would require an extended digression into an area of mathematical analysis called the theory of generalized functions. Indeed, the status of the unit impulse was the subject of very lively debate among great thinkers for a considerable period of time. Two of the most prominent analysts of the last hundred years, Oliver Heaviside (1850–1925), a self-educated English engineer, and Paul Dirac (1902–), English physicist and Nobel Prize winner, played key roles in developing the theory and reconciling the controversies. A thumbnail biography of Heaviside introduces Chapter 13.

Here we limit ourselves to identifying some specific conditions under which the unit impulse may be used with accuracy and rigor. The generalized function $\delta(t)$ may be integrated, as in (12.15b), and evaluated at all times except its singular time, as in (12.15a). Identifying the unit impulse as the derivative of the unit step (12.13) is correct, with the understanding that since $\delta(t)$ is a generalized function, we mean the *generalized derivative* of the unit step (its ordinary derivative does not exist). Often, for brevity, we will omit the word *generalized*, referring to $\delta(t)$ as simply the derivative of the unit step.

An immediate consequence of the fact that the unit impulse is the derivative of the unit step is that the *unit step is the integral of the unit impulse*:

$$u(t) = \int_{-\infty}^{t} \delta(\tau)\, dt \tag{12.16}$$

For any upper limit $t < 0$, there is no area under the impulse within the integral's limits, and as t passes through zero, the unit area "piled up" at $t = 0$ is added, so the value of the integral jumps to 1. The function that is zero for $t < 0$ and 1 for $t > 0$ is, of course, the unit step.

A consequence of the fact that $\delta(t) = 0$ for all t except $t = 0$ is that if we have a product $f(t)\delta(t)$, where $f(t)$ is an ordinary function continuous at $t = 0$, then

$$f(t)\delta(t) = f(0)\delta(t)$$

This follows by comparing the two sides for t not 0 (they are both 0) and for $t = 0$ (the two sides are identical). If the impulse is time shifted so that its singular point is $t = t_0$, then, by reasoning identical to the above, the continuous function $f(t)$ may again be replaced by its value at the singular point of the impulse, t_0:

$$f(t)\delta(t - t_0) = f(t_0)\delta(t - t_0) \tag{12.17}$$

This rule, the *impulse product rule*, will be helpful in simplifying expressions containing $\delta(t)$.

Example 12.4

Evaluate the generalized derivative $(d/dt)[\cos tu(t)]$ and show that it obeys the fundamental theorem of calculus; that is,

$$\int_{-\infty}^{t} \frac{d}{d\tau}[\cos(\tau)u(\tau)]\,d\tau = \cos(t)u(t) \qquad (12.18)$$

We first note that $\cos(t)u(t)$ is not differentiable in the ordinary sense, since it has a jump discontinuity at $t = 0$. One approach is to express this function as the sum of a continuous function whose derivative is an ordinary derivative and a unit step whose generalized derivative is known to be the unit impulse $\delta(t)$.

$$\cos(t)u(t) = f(t) + u(t)$$

where $f(t)$ is the continuous function:

$$f(t) = \begin{cases} 0, & t < 0 \\ \cos(t) - 1, & t \geq 0 \end{cases}$$

$f(t)$ has no jumps and is thus differentiable in the ordinary sense, with derivative equal to zero in negative time and $-\sin t$ in positive time. Thus

$$\frac{d}{dt}\cos(t)u(t) = \frac{d}{dt}f(t) + \frac{d}{dt}u(t) \qquad (12.19)$$
$$= -\sin(t)u(t) + \delta(t)$$

As shown in Fig. 12.6, the derivative has a unit impulse located where the function being differentiated has a jump discontinuity.

Finally, integrating this derivative we have calculated yields

$$\int_{-\infty}^{t} \frac{d}{d\tau}[\cos(\tau)u(\tau)]d(\tau) = -\int_{-\infty}^{t} \sin(\tau)u(\tau)d(\tau) + \int_{-\infty}^{t} \delta(\tau)\,d\tau$$

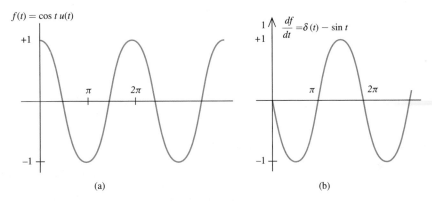

FIGURE 12.6 Functions for Example 12.4.

The first integral evaluates to 0 in negative time and $\cos(t) - 1$ in positive time. The second is the integral of $\delta(t)$, which is $u(t)$.

$$\int_{-\infty}^{t} \frac{d}{d\tau}[\cos(\tau)u(\tau)]\,d\tau = [\cos(t) - 1]u(t) + u(t) = \cos(t)u(t)$$

and (12.18) is verified. This completes the example.

The previous calculations could have been speeded by use of Leibniz's rule. *Leibniz's rule (the product rule for differentiation) may be freely applied to products involving singularity functions and ordinary functions.*

$$\frac{d}{dt}\cos(t)u(t) = \left[\frac{d}{dt}\cos(t)\right]u(t) + \cos(t)\left[\frac{d}{dt}u(t)\right]$$

$$= -\sin(t)u(t) + \cos(t)\delta(t)$$

The second term can be simplified to $\cos(0)\delta(t) = \delta(t)$ by the impulse product rule (12.17), thus arriving at (12.19) somewhat more efficiently.

A final property we will have need for is the *sifting property*. Consider the integral

$$\int_{-\infty}^{+\infty} f(t)\delta(t - t_0)\,dt$$

where $f(t)$ is assumed continuous at $t = t_0$. Applying the impulse product rule, we obtain

$$\int_{-\infty}^{+\infty} f(t)\delta(t - t_0)\,dt = \int_{-\infty}^{+\infty} f(t_0)\delta(t - t_0)\,dt$$

Taking the constant $f(t_0)$ out of the integral and noting that the integral is just the total area under this (shifted) unit impulse, which is 1, we arrive at

SIFTING PROPERTY

$$\int_{-\infty}^{+\infty} f(t)\delta(t - t_0)\,dt = f(t_0) \tag{12.20}$$

The sifting property (12.20) shows that *the effect of the impulse in the integral is to sift out, or select, the value of the rest of the integrand at its singular point as the value of the integral.* This results holds for all ordinary functions $f(t)$ continuous at t_0.

Example 12.5

Evaluate the following integral (log means log base 10):

$$\int_{-9}^{+\infty} 5(\log t)\delta(t - 10)\,dt$$

The lower limit of the integral may be extended to $-\infty$ since the impulse function is 0 in the added interval. Then, applying the sifting property (2.20) with singular point $t_0 = 10$ gives

$$\int_{-9}^{+\infty} 5 \log(t) \delta(t - 10) \, dt = 5 \log(10) = 5$$

Thus far we have discussed the singularity functions produced by integrating the unit step any number of times and by differentiating once. It remains to describe those produced by repeated differentiation. These, collectively called the *higher-order singularity functions*, while interesting from a mathematical point of view, are less often needed in circuit applications. Thus we will be content with mentioning only those few facts relevant to the present purpose.

The derivative of the unit impulse is called the *unit doublet function* $\delta'(t)$:

$$\delta'(t) = \frac{d}{dt} \delta(t)$$

$\delta(t)$ is a generalized function, the generalized derivative of $u(t)$, so all order derivatives of $\delta(t)$ must be generalized functions as well. Since $\delta(t) = 0$ everywhere except its singular point, the unit doublet $\delta'(t)$, its derivative, must inherit this property. $\delta'(t)$ has a sifting property,

$$\int_{-\infty}^{+\infty} f(t) \delta'(t - t_0) \, dt = -f'(t_0) \tag{12.21}$$

which can be verified by integration by parts and use of the impulse sifting property (2.20). *Just as the unit impulse sifts out the value of a function at its singular point, the unit doublet, its derivative, sifts out the derivative there (multiplied by −1).*

The remaining higher-order singularity functions $\delta^{(2)}(t)$, $\delta^{(3)}(t)$, ..., are each derivatives of the previous one. Each has a sifting property, selecting out the values of the higher-order derivatives at the singular point (see Problem 12.47).

Having introduced the family of singularity functions, we turn to computing their Laplace transforms. To do so, we first develop rules for writing down the Laplace transform of the derivative and of the integral of any function whose Laplace transform is known.

DIFFERENTIATION

$$L\left[\frac{d}{dt} f(t)\right] = s\mathbf{F}(s) - f(0-) \tag{12.22}$$

The Laplace transform of a derivative of a function is s times the Laplace transform of the function minus its initial value. Integrating by parts gives

$$\int_{0-}^{\infty} \frac{d}{dt} f(t) e^{-st} \, dt = f(t) e^{-st} \Big|_{0-}^{\infty} + s \int_{0-}^{\infty} f(t) e^{-st} \, dt$$

The last term is clearly $s\mathbf{F}(s)$, and evaluation of the first at infinity yields 0 for all s within the region of convergence of the transform. Equation (12.22) follows.

INTEGRATION

$$L\left[\int_{0-}^{t} f(\tau)\,dt\right] = \frac{\mathbf{F}(s)}{s} \qquad (12.23)$$

The Laplace transform of the integral of a function is the Laplace transform of the function divided by s. This may also be verified by an integration of the defining integral by parts as above. (The reader is asked to do so in Problem 12.19.)

Armed with these rules, we may easily evaluate the Laplace transforms of all the singularity functions. We begin with $1/s$, the transform of $u(t)$, derived in Section 12.1. Since the unit ramp is the integral of the unit step, by the integration rule the transform of $r(t)$ is $\mathbf{R}(s) = 1/s^2$. Similarly, the transform of the unit parabola is $\mathbf{P}(s) = 1/s^3$, and so on, for those singularity functions produced by further integrations. To find the transform of $\delta(t)$, apply the differentiation rule with $f(t) = u(t)$:

$$\mathbf{L}[\delta(t)] = s\frac{1}{s} - 0 = 1 \qquad (12.24)$$

The transform of the unit impulse is simply the constant 1. The unit doublet $\delta'(t)$, being the derivative of $\delta(t)$, has transform $s(1) - 0 = s$. Similarly, $\delta^{(2)}(t)$ has transform s^2, $\delta^{(3)}(t)$ has transform s^3, and so on. *The transforms of the singularity functions are all powers of s.*

Example 12.6

Let us determine the Laplace transforms of $\dfrac{d}{dt}[\cosh tu(t)]$ and $\dfrac{d^2}{dt^2}[\sinh tu(t)]$, where cosh and sinh are the hyperbolic sine and cosine functions discussed in Example 12.2. Noting the transforms given there and by the differentiation rule we have

$$L\left[\frac{d}{dt}\cosh tu(t)\right] = sL[\cosh tu(t)] - \cosh(0-)u(0-)$$

$$= \frac{s^2}{s^2 - 1^2}$$

since $u(0-) = 0$. Similarly,

$$L\left[\frac{d}{dt}\sinh tu(t)\right] = sL[\sinh tu(t)] - \sinh(0-)u(0-)$$

$$= \frac{s}{s^2 - 1^2}$$

and differentiating once again yields

$$L\left[\frac{d^2}{dt^2}\sinh tu(t)\right] = sL\left[\frac{d}{dt}\sinh tu(t)\right] - \left.\frac{d}{dt}\sinh(t)\right|_{t=0-} u(0-)$$

$$= \frac{s^2}{s^2 - 1^2}$$

We may verify these calculations by differentiating first,

$$\frac{d}{dt}\cosh atu(t) = \frac{d}{dt}\left[\frac{1}{2}(e^t + e^{-t})u(t)\right]$$

$$= \frac{1}{2}[(e^t - e^{-t})u(t) + (e^t + e^{-t})\delta(t)]$$

$$= \sinh tu(t) + \delta(t)$$

and then transforming.

$$L[\sinh tu(t) + \delta(t)] = \frac{1}{s^2 - 1} + 1 = \frac{s^2}{s^2 - 1}$$

which, of course, gives the same result. Note that the transforms of $(d/dt)\cosh tu(t)$ and $d^2/dt^2 \sinh tu(t)$ are the same, so the inverse transforms must be identical. Indeed, these two time functions are identical, as can be verified by direct calculation.

EXERCISES

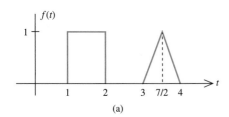

(a)

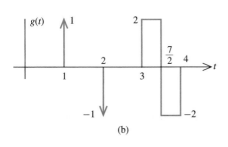

(b)

EXERCISE 12.2.2

12.2.1. Find the Laplace transform of $\delta(t - t_0)$ using the sifting theorem. Note that the lower limit on the integral in the sifting theorem is $-\infty$. Find the Laplace transform of $u(t-t_0)$ from the defining integral (12.1). Consider $t_0 < 0$ and $t_0 \geq 0$ as separate cases.
  Answer $L[\delta(t - t_0] = 0$ for $t_0 < 0$, e^{-st_0} for $t_0 \geq 0$; $L[u(t - t_0] = 1/s$ for $t_0 < 0$, e^{-st_0}/s for $t_0 \geq 0$

12.2.2. Sketch $g(t) = df/dt$ for the $f(t)$ shown. Express $f(t)$ and $g(t)$ as linear combinations of shifted singularity functions. Use Exercise 12.2.1 to find $G(s)$.
  Answer $f(t) = u(t-1) - u(t-2) + 2r(t-3) - 4r(t-\frac{7}{2}) + 2r(t-4);$ $g(t) = \delta(t - 1) - \delta(t - 2) + 2u(t - 3) - 4u(t - \frac{7}{2}) + 2u(t - 4);$ $G(s) = e^{-s} - e^{-2s} + 2e^{-3s}/s - 4e^{-7/2s}/s + 2e^{-4s}/s$

12.2.3. Express $e(t)$, the integral of $f(t)$ in Exercise 12.2.2,

$$e(t) = \int_{0-}^{t} f(\tau)\,d\tau$$

as a linear combination of shifted singularity functions. Find $E(s)$.
  Answer $e(t) = r(t-1) - r(t-2) + 2p(t-3) - 4p(t-\frac{7}{2}) + 2p(t-4);$ $E(s) = e^{-s}/s^2 - e^{-2s}/s^2 + 4e^{-3s}/s^3 - 8e^{7/2s}/s^3 + 4e^{-4s}/s^3$

12.2.4. Compute $d^3/dt^3[\sin^3 tu(t)]$. For what values $n > 0$ do the nth derivatives of $\sin^3 tu(t)$ exist as ordinary functions? For what values as generalized functions?

Answer $6\cos^3 tu(t) - 21\sin^2 t \cos tu(t); n = 1, 2, 3; n > 3$

12.3 OTHER TRANSFORM PROPERTIES AND PAIRS

We have seen the usefulness of linearity and the integration and differentiation properties in computing Laplace transforms. In this section we develop a few other properties that facilitate calculation of new transforms from known ones and apply them to functions of interest to us in the study of circuits.

TIME SHIFT

$$L[f(t - t_0)u(t - t_0)] = e^{-st_0}\mathbf{F}(s), \qquad t_0 > 0 \qquad (12.25)$$

The Laplace transform of a time function made zero for $t < 0$, then shifted to the right (delayed), is multiplied by a complex exponential. To verify, we evaluate the transform of $f(t - t_0)u(t - t_0)$:

$$\int_{0-}^{\infty} f(t - t_0)u(t - t_0)e^{-st}\,dt = \int_{t_0}^{\infty} f(t - t_0)e^{-st}\,dt$$

Let $\tau = t - t_0$ and change variables in this integral.

$$\int_0^{\infty} f(\tau)e^{-st_0}\,d\tau = e^{-st_0}\mathbf{F}(s)$$

$f(t)$ and its time-shifted version $f(t - t_0)u(t - t_0)$ are shown in Fig. 12.7.

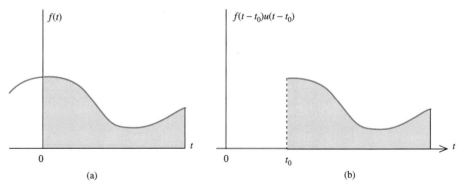

FIGURE 12.7 (a) Function $f(t)$; (b) $f(t)$ made zero for $t < 0$ and delayed by t_0.

$$L[e^{-s_0 t} f(t)] = \mathbf{F}(s + s_0) \qquad\qquad (12.26)$$

Frequency shift in the s-domain corresponds to multiplication in the time domain by a complex exponential. To verify (12.26), evaluate the transform of $e^{-s_0 t} f(t)$.

Example 12.7

The transform of the staircase pulse $g(t)$ sketched in Fig. 12.8(a) may be found by using the time-shift property. Write $g(t)$ as

$$g(t) = 2u(t - 1) + 2u(t - 3) - 2u(t - 5) - 2u(t - 7)$$

Then time shifting the unit step $f(t) = u(t)$ by delays of $t_0 = 1, 3, 5,$ and 7, we obtain

$$\mathbf{G}(s) = \tfrac{2}{s}(e^{-s} + e^{-3s} - e^{-5s} - e^{-7s})$$

To evaluate the transform of the product $\cos \omega t f(t)$, where $f(t)$ is any Laplace transformable function, note that

$$\cos \omega t f(t) = \tfrac{1}{2} e^{j\omega t} f(t) + \tfrac{1}{2} e^{-j\omega t} f(t)$$

Using the frequency shift property with $s_0 = \pm j\omega$ gives

$$L[\cos \omega t f(t)] = \tfrac{1}{2}[\mathbf{F}(s - j\omega) + \mathbf{F}(s + j\omega)] \qquad\qquad (12.27)$$

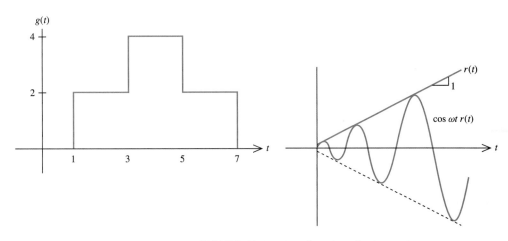

FIGURE 12.8 Time functions for Example 12.7.

For instance, the transform of $f(t)$, the unit ramp $f(t) = r(t) = tu(t)$, is $\mathbf{F}(s) = 1/s^2$, as shown in Section 12.2, and using (12.27), we obtain

$$L[t \cos \omega t u(t)] = \frac{1}{2}\left[\frac{1}{(s - j\omega)^2} + \frac{1}{(s + j\omega)^2}\right]$$

$$= \frac{s^2 - \omega^2}{(s^2 + \omega^2)^2}$$

(12.28)

which the desired Laplace transform for Fig. 12.8(b). Although intended to illustrate the use of the time and frequency shift properties, both these results required the use of the linearity property as well.

TIME–FREQUENCY SCALING

$$L[f(ct)] = \frac{1}{c}\mathbf{F}\left(\frac{s}{c}\right), \qquad c > 0$$

(12.29)

Compressing (expanding) the time scale expands (compresses) the frequency scale by an equal factor. To verify, compute the left side of (12.29) with $\tau = ct$:

$$L[f(ct)] = \int_{0-}^{\infty} f(ct)e^{-st}\,dt$$

$$= \int_{0-}^{\infty} f(\tau)e^{-s\tau/c}\frac{d\tau}{c}$$

Taking $1/c$ out of the integral, the result follows. Note that values of $c > 1$ correspond to time compression and $c < 1$ to time expansion. For instance, the graph of the function $g(t) = f(3t)$ is compressed by a factor of 3 relative to that of $f(t)$. The real positive constant c in (12.29) is sometimes called the *time-compression factor*.

Example 12.8

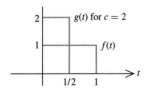

FIGURE 12.9 $f(t)$ and a compressed $f(t)$ for Example 12.8.

Consider the square pulse $f(t) = u(t) - u(t - 1)$ shown. If we compress it by a factor of $c > 1$ and at the same time amplitude-scale by the same factor c, we get $g(t) = cf(ct)$. $g(t)$ for $c = 2$ is shown in Fig. 12.9. The transform of $f(t)$, using the time-shift property (12.25), is

$$\mathbf{F}(s) = \frac{1}{s} - \frac{e^{-s}}{s}$$

Then, with $g(t) = cf(ct)$, by time–frequency scaling,

$$\mathbf{G}(s) = c\left[\frac{1}{c}\left(\frac{1}{s/c} - \frac{e^{-s/c}}{s/c}\right)\right] = \frac{c}{s}(1 - e^{-s/c})$$

Thus for a time-compression factor $c = 2$, $\mathbf{G}(s) = (2/s)(1 - e^{-s/2})$. It is interesting to note that as c gets larger and larger, $g(t)$ approaches the unit impulse function $\delta(t)$ [its area is always 1 for any c, and $g(t)$ goes to zero for any nonzero t]. To evaluate its transform $\mathbf{G}(s)$ as c gets very large, we may apply the well-known Taylor series expansion of the exponential function,

$$e^x = 1 + x + \frac{x^2}{2!} + \frac{x^3}{3!} + \cdots$$

with $x = -s/c$ to get

$$\mathbf{G}(s) = \frac{c}{s}\left[1 - \left(1 - \frac{s}{c} + \frac{s^2}{2!c^2} - \frac{s^3}{3!c^3} + \cdots\right)\right]$$

$$= 1 - \frac{s}{2!c} + \frac{s^2}{3!c^2} - \cdots$$

As c gets very large, $\mathbf{G}(s)$ converges to 1. This is consistent with the transform of the function that $g(t)$ is approaching, since we know that $\mathbf{L}[\delta(t)] = 1$.

T-MULTIPLICATION

$$\mathbf{L}[tf(t)] = \frac{-d}{ds}\mathbf{F}(s) \qquad (12.30)$$

t-multiplication corresponds to negative differentiation in the s-domain. This property follows by computing the negative derivative of both sides of (12.1) with respect to s:

$$-\frac{d}{ds}\mathbf{F}(s) = -\frac{d}{ds}\left[\int_{0-}^{\infty} f(t)e^{-st}\,dt\right]$$

$$= \int_{0-}^{\infty} tf(t)e^{-st}\,dt$$

Example 12.9

Let us find the Laplace transforms of $t^n e^{-at}$ for $n = 1, 2, 3, \ldots$. In Section 12.1 we determined that $f(t) = e^{-at}$ and $\mathbf{F}(s) = 1/(s + a)$ were a transform pair. Then, by the t-multiplication property,

$$\mathbf{L}[te^{-at}] = -\frac{d}{ds}\frac{1}{s+a} = \frac{1}{(s+a)^2}$$

t-multiplying once again gives

$$\mathbf{L}[t^2 e^{-at}] = \mathbf{L}[t(te^{-at})]$$

$$= -\frac{d}{ds}\frac{1}{(s+a)^2} = \frac{2}{(s+a)^3}$$

and one final time,

$$L[t^3 e^{-at}] = -\frac{d}{ds}\frac{2}{(s+a)^3} = \frac{(3)(2)}{(s+a)^4}$$

The pattern, which can be verified formally using finite induction, is

$$L[t^m e^{-at}] = \frac{m!}{(s+a)^{m+1}} \tag{12.31}$$

The transform properties discussed to this point are collected together and summarized in Table 12.1. A few entries in the table, such as n-fold differentiation and n-fold t-multiplication, have not been derived explicitly but are closely related to other properties that have been derived. Their derivations have been assigned as exercises and end-of-chapter problems.

A collection of the most useful of the specific transform pairs we have determined so far is included as Table 12.2. Once again we have left derivation of a few of these entries to the exercises and problems. Slightly expanded versions of these tables may be found on the inside of the back cover and its facing page. The added entries there, those not found in Table 12.1 or Table 12.2, correspond to results that will be introduced in later chapters.

Note that a few entries in Table 12.2 are listed in two places, as both singularity functions and ordinary functions. This is true of the unit step and the ordinary function $f(t) = 1$, both of which have transform $1/s$. This is because any functions that differ only in negative time will have the same Laplace transform. For the same reason, *any of the f(t)'s in Table 12.2 may be multiplied by u(t) without changing its transform F(s).*

These tables can be used in conjunction to produce many more transform pairs. Given a function $f(t)$ whose transform $\mathbf{F}(s)$ is to be found, the table of Laplace transform

Table 12.1 Laplace Transform Properties

	$f(t)$	$\mathbf{F}(s)$
1. Linearity	$c_1 f_1(t) + c_2 f_2(t)$	$c_1 \mathbf{F}_1(s) + c_2 \mathbf{F}_2(s)$
2. Differentiation	$\dfrac{d}{dt} f(t)$	$s\mathbf{F}(s) - f(0-)$
3. n-Fold differentiation	$\dfrac{d^n}{dt^n} f(t)$	$s^n \mathbf{F}(s) - s^{n-1} f(0-) - s^{n-2} f'(0-)$
		$- \cdots - s f^{(n-2)}(0-) - f^{(n-1)}(0-)$
4. Integration	$\displaystyle\int_{0-}^{t} f(\tau)\, d\tau$	$\dfrac{\mathbf{F}(s)}{s}$
5. Time shift	$f(t - t_0)u(t - t_0), t_0 > 0$	$e^{-st_0} \mathbf{F}(s)$
6. Frequency shift	$e^{-s_0 t} f(t)$	$\mathbf{F}(s + s_0)$
7. Time–frequency scaling	$f(ct), c > 0$	$\dfrac{1}{c}\mathbf{F}\left(\dfrac{s}{c}\right)$
8. t-Multiplication	$t f(t)$	$\dfrac{-d}{ds}\mathbf{F}(s)$
9. n-Fold t-multiplication	$t^n f(t)$	$(-1)^n \dfrac{d^n}{ds^n}\mathbf{F}(s)$

Table 12.2   Laplace Transform Pairs

	$f(t)$	$\mathbf{F}(s)$
	Singularity Functions	
1. Unit impulse	$\delta(t)$	1
2. Unit step	$u(t)$	$\dfrac{1}{s}$
3. Unit ramp	$r(t) = tu(t)$	$\dfrac{1}{s^2}$
4. Unit parabola	$p(t) = \frac{1}{2}t^2 u(t)$	$\dfrac{1}{s^3}$
5. nth integral of impulse	$\delta^{(-n)}(t)$	$\dfrac{1}{s^n}$
6. Unit doublet	$\delta'(t)$	s
7. nth derivative of impulse	$\delta^{(n)}(t)$	s^n
	Ordinary Functions	
8. Constant	1	$\dfrac{1}{s}$
9. t	t	$\dfrac{1}{s^2}$
10. Power of t	$\dfrac{t^{n-1}}{(n-1)!}$	$\dfrac{1}{s^n}$
11. Exponential	e^{-at}	$\dfrac{1}{s+a}$
12. t-Multiplication exponential	te^{-at}	$\dfrac{1}{(s+a)^2}$
13. Repeated t-multiplication exponential	$\dfrac{1}{(n-1)!}t^{n-1}e^{-at}$	$\dfrac{1}{(s+a)^n}$
14. Sine	$\sin \omega t$	$\dfrac{\omega}{s^2 + \omega^2}$
15. Cosine	$\cos \omega t$	$\dfrac{s}{s^2 + \omega^2}$
16. Sinusoid	$\sqrt{c^2 + d^2}\cos\left(\omega t - \tan^{-1}\dfrac{d}{c}\right)$	$\dfrac{cs + d\omega}{s^2 + \omega^2}$
17. Damped sine	$e^{-at}\sin \omega t$	$\dfrac{\omega}{(s+a)^2 + \omega^2}$
18. Damped cosine	$e^{-at}\cos \omega t$	$\dfrac{s+a}{(s+a)^2 + \omega^2}$
19. Damped sinusoid	$\sqrt{c^2 + a^2}e^{-at}\cos\left(\omega t - \tan^{-1}\dfrac{d}{c}\right)$	$\dfrac{c(s+a) + d\omega}{(s+a)^2 + \omega^2}$
20. t-Multiplicated sine	$t\sin \omega t$	$\dfrac{2\omega s}{(s^2 + \omega^2)^2}$
21. t-Multiplicated cosine	$t\cos \omega t$	$\dfrac{s^2 - \omega^2}{(s^2 + \omega^2)^2}$

pairs may be scanned to find a similar table entry, say $g(t)$. Properties that may be used to transform $g(t)$ to $f(t)$ are then sought in consultation with the table of Laplace transform properties. Once a sequence of transformations yielding $f(t)$ from $g(t)$ has been discovered, the properties table specifies the corresponding changes to $G(s)$ that yield $F(s)$. The same strategy may be used given a transform $F(s)$ whose inverse transform $f(t)$ is sought. We conclude this section with examples of the use of these tables.

Example 12.10

Find $F(s)$ for the time function $f(t) = (t-3)^2 e^{-2t} u(t-3)$. The closest entry in Table 12.2 appears to be the repeated t-multiplied exponential. With $n = 3$ and $a = 2$,

$$L\left[\frac{1}{2}t^2 e^{-2t}\right] = \frac{1}{(s+2)^3}$$

Consulting Table 12.1, a time shift by $t_0 = 3$ yields multiplication of the transform by $e^{-st_0} = e^{-3s}$, or

$$L\left[\frac{1}{2}(t-3)^2 e^{-2(t-3)} u(t-3)\right] = \frac{e^{-3s}}{(s+2)^3}$$

Comparing our current time function with $f(t)$, we need only scale by 2 and change the factor $e^{-2(t-3)}$ to e^{-2t}, which is easily done by multiplying by e^{-6}. Then by the linearity property with $c_1 = 2e^{-6}$ and $c_2 = 0$,

$$L[(t-3)^2 e^{-2t} u(t-3)] = \frac{2e^{-3(s+2)}}{(s+2)^3}$$

Example 12.11

Find the inverse transform $f(t)$ for $F(s) = (s^2+3)/(s^2+4)$. Noting that the nearest entries in Table 12.2 are for sine and cosine, rewrite $F(s)$ as the sum

$$F(s) = \frac{s^2}{s^2+4} + \frac{3}{s^2+4}$$

The second term is recognizable as the transform of $\frac{3}{2}\sin 2t$. The first term is close to the cos entry but must be s-multiplied. Noting the differentiation property, $g(t) = (d/dt)\cos 2t$ will have transform

$$G(s) = s\frac{s}{s^2+4} - \cos 0- = \frac{s^2}{s^2+4} - 1$$

Summing the two terms yields

$$L^{-1}\left[-1 + \frac{s^2}{s^2+4} + \frac{3}{s^2+4}\right] = \frac{d}{dt}\cos 2t + \frac{3}{2}\sin 2t$$

But we know that adding 1 to the transform corresponds to adding $\delta(t)$ to the time function, or

$$L^{-1}\left[1 - 1 + \frac{s^2}{s^2+4} + \frac{3}{s^2+4}\right] = \delta(t) + \frac{d}{dt}\cos 2t + \frac{3}{2}\sin 2t$$

or

$$L^{-1}\left[\frac{s^2+3}{s^2+4}\right] = \delta(t) - \frac{1}{2}\sin 2t$$

This result can also be gained by long division of $\mathbf{F}(s)$, as in Problem 12.28.

12.3.1. Use Table 12.2 entries 14 and 15 for pure sine and cosine, together with linearity, to write formulas for $L[\sin(\omega t + \theta)]$ and $L[\cos(\omega t + \theta)]$.

Answer $(s \sin\theta + \omega \cos\theta)/(s^2 + \omega^2)$; $(s \cos\theta - \omega \sin\theta)/(s^2 + \omega^2)$

12.3.2. Use time-shift to compute $\mathbf{L}\left[\frac{1}{2}(t-1)^2 e^{-(t-1)} u(t-1)\right]$.

Answer $e^{-s}/(s+1)^2$

12.3.3. Use the t-multiplication property to find $\mathbf{F}(s)$ for $f(t) = te^{-2t}\cos 4t$.

Answer $([s+2]^2 - 16)/([s+2]^2 + 16)^2$

12.4 PARTIAL FRACTION EXPANSION

We have seen how a table of basic transform pairs may be used together with a table of Laplace transform properties to generate additional transform pairs. Given $f(t)$, this strategy may be used to find its transform $\mathbf{F}(s)$, or given $\mathbf{F}(s)$, to find the inverse transform $f(t)$.

In the case that $\mathbf{F}(s)$ is a *rational function*, or ratio of polynomials in s, there is a convenient procedure for finding the inverse transform based on *partial fraction expansion*. After developing the method in this section, in Section 12.5 we explore the main application of this procedure. Rational functions arise whenever Laplace transforms are used to solve differential equations, which are the basic equations of circuit analysis.

A rational function $\mathbf{F}(s)$,

$$\mathbf{F}(s) = \frac{a_m s^m + a_{m-1}s^{m-1} + \cdots + a_1 s + a_0}{s^n + b_{n-1}s^{n-1} + \cdots + b_1 s + b_0} \tag{12.32}$$

is said to be *proper* if the order of its denominator polynomial exceeds that of the numerator, $n > m$. An improper $\mathbf{F}(s)$ must first be prepared for partial fraction expansion by performing the operation of long division, as illustrated in the following example.

Example 12.12

Perform long division on the improper $\mathbf{F}(s) = (3s^3 + 2s + 1)/(s^2 + s + 2)$.

$$
\begin{array}{r}
3s - 3 \\
s^2 + s + 2 \overline{\smash{)}\, 3s^3 + 2s + 1} \\
-\underline{(3s^3 + 3s^2 + 6s)} \\
-3s^2 - 4s + 1 \\
-\underline{(-3s^2 - 3s - 6)} \\
- s + 7
\end{array}
$$

The result of the long division is $3s - 3$ with remainder $-s + 7$. Thus

$$\mathbf{F}(s) = 3s - 3 + \frac{-s + 7}{s_2 + s + 2}$$

Long division converts an improper rational function into the sum of a polynomial ($3s - 3$ in the example) and a proper rational function. Finding the inverse Laplace transform of the polynomial part is easy, since all powers of s have singularity functions as their inverse transforms (entry 7 in Table 12.2). For instance, the polynomial $\mathbf{F}_1(s) = 3s - 3$ has inverse transform $f_1(t) = 3\delta'(t) - 3\delta(t)$. Thus the problem reduces to finding the inverse transform of the other part, the proper rational function, for which there is no obvious table entry. For the remainder of this section we assume that $\mathbf{F}(s)$ is a proper rational function.

The basic idea of partial fraction expansion is to write $\mathbf{F}(s)$ as a sum of simple rational functions, or "partial fractions," each of which may be then easily inverse transformed. As an illustration, let us invert the transform

$$\mathbf{F}(s) = \frac{2(s + 10)}{(s + 1)(s + 4)}$$

Obtaining a partial fraction expansion is the opposite operation of getting a common denominator. That is, we ask ourselves what simple fractions add together to yield $\mathbf{F}(s)$. Since the transform is proper, the partial fractions will be proper and must therefore be of the form

$$\mathbf{F}(s) = \frac{2(s + 10)}{(s + 1)(s + 4)} = \frac{A}{s + 1} + \frac{B}{s + 4}$$

The constants A and B are determined so as to make this an identity in s.

The simplest means of determining A and B is to note that

$$(s + 1)\mathbf{F}(s) = \frac{2(s + 10)}{s + 4} = A + \frac{B(s + 1)}{s + 4}$$

$$(s + 4)\mathbf{F}(s) = \frac{2(s + 10)}{s + 1} = \frac{A(s + 4)}{s + 1} + B$$

Since these must be identities for all s, let us evaluate the first at $s = -1$, which eliminates B, and the second at $s = -4$, which eliminates A. The results are

$$A = (s + 1)\mathbf{F}(s)\Big|_{s=-1} = \frac{2(9)}{3} = 6$$

$$B = (s + 4)\mathbf{F}(s)\Big|_{s=-4} = \frac{2(6)}{-3} = -4$$

Therefore, we have

$$\mathbf{F}(s) = \frac{6}{s + 1} - \frac{4}{s + 4}$$

and by Table 12.2,

$$f(t) = 6e^{-t} - 4e^{-4t}$$

The poles $p = -1$ and $p = -4$ in this case are *simple* poles or poles of *order* 1, and in general for each such simple pole the partial fraction expansion contains a term $A/(s - p)$. In the case of a denominator factor $(s - p)^n$, where $n = 2, 3, 4, \ldots$, the pole $s = p$ is a *multiple* pole, or a pole of order n. The factors comprising $(s - p)^n$ are thus not distinct, and the partial fraction expansion must be modified. We consider multiple poles later.

SIMPLE REAL POLE

For each simple real pole p in $\mathbf{F}(s)$, there will be a term $A/(s-p)$ in its partial fraction expansion, where $A = (s-p)\mathbf{F}(s)|_{s=p}$, and a term Ae^{+pt} in its inverse transform.

Complex poles occur in conjugate pairs and their corresponding coefficients in the expansion are complex conjugates. For example, if $\mathbf{F}(s)$ has simple poles $s = \alpha \pm j\beta$ and an expansion

$$\mathbf{F}(s) = \frac{A}{s - \alpha - j\beta} + \frac{B}{s - \alpha + j\beta}$$

then

$$A = (s - \alpha - j\beta)\mathbf{F}(s)\Big|_{s=\alpha+j\beta}$$

$$B = (s - \alpha + j\beta)\mathbf{F}(s)\Big|_{s=\alpha-j\beta}$$

From this we see that $B = A^*$, because $\mathbf{F}(s)$ is a ratio of polynomials in s with real coefficients. The inverse transform is

$$f(t) = Ae^{(\alpha+j\beta)t} + A^*e^{(\alpha-j\beta)t}$$

which is the sum of a complex number and its conjugate. Therefore,

$$f(t) = 2\,\mathrm{Re}[Ae^{(\alpha+j\beta)t}]$$

If $A = |A|e^{j\theta}$, we have

$$f(t) = 2\,\mathrm{Re}[|A|e^{\alpha t}e^{j(\beta t+\theta)}]$$

$$= 2|A|e^{\alpha t}\cos(\beta t + \theta)$$

SIMPLE PAIR OF COMPLEX CONJUGATE POLES

For each simple (unrepeated) pair of complex conjugate poles $p = \alpha + j\beta$, $p^* = \alpha - j\beta$ in $\mathbf{F}(s)$, there will be a pair of terms $A/(s - p) + A^*/(s - p^*)$ in its partial fraction expansion, where $A = (s - p)\mathbf{F}(s)|_{s=p}$. There will be a single term $2|A|e^{\alpha t}\cos(\beta t + \underline{/A})$ in its inverse transform due to the pair of poles.

Example 12.13

As an example, the transform

$$\mathbf{F}(s) = \frac{s}{(s + 1)(s^2 + 2s + 2)}$$

has the partial fraction expansion

$$\mathbf{F}(s) = \frac{s}{(s + 1)(s + 1 - j1)(s + 1 + j1)}$$

$$= \frac{A}{s + 1} + \frac{B}{s + 1 - j1} + \frac{B^*}{s + 1 + j1}$$

where

$$A = \frac{s}{s^2 + 2s + 2}\Big|_{s=-1} = -1$$

$$B = \frac{s}{(s + 1)(s + 1 + j1)}\Big|_{s=-1+j1} = \frac{1 - j1}{2} = \frac{1}{\sqrt{2}}\underline{/-45°}$$

Thus we have

$$f(t) = Ae^{-t} + 2\operatorname{Re}\left[Be^{(-1+j1)t}\right]$$

$$= -e^{-t} + 2\operatorname{Re}\left(\frac{1}{\sqrt{2}}e^{-t}\underline{/1t - 45°}\right)$$

$$= -e^{-t} + \sqrt{2}e^{-t}\cos(t - 45°)$$

An alternative form is

$$f(t) = -e^{-t} + \sqrt{2}e^{-t}(\cos t \cos 45° + \sin t \sin 45°)$$

$$= -e^{-t} + e^{-t}(\cos t + \sin t)$$

Associated with a repeated pole p of order r, there will be r terms in the partial fraction expansion:

$$\mathbf{F}(s) = \frac{\mathbf{N}(s)}{\mathbf{D}_1(s)(s-p)^r} = \frac{A_r}{(s-p)^r} + \frac{A_{r-1}}{(s-p)^{r-1}} + \cdots + \frac{A_1}{s-p} + \mathbf{F}_1(s) \qquad (12.33)$$

Multiplying both sides of (12.33) by $(s-p)^r$ and evaluating at $s = p$, each term on the right is zero except the first, which is A_r. To get the remaining A_{r-1}, A_{r-2}, and so on, either of two procedures may be used. The term just found involving A_r may be subtracted from both sides, leaving a new equation similar to the original (12.33) except that the pole at $s = p$ is now repeated only $r - 1$ times. A_{r-1} may then be found as described above, by multiplying by $(s-p)^{r-1}$ and evaluating at $s = p$. The process descends through the A_k's until A_1 is found. We refer to this as *repeated subtraction*.

The second procedure for finding A_{r-1}, A_{r-2}, involves *repeated differentiation*. If both sides of (12.33) are multiplied by $(s-p)^r$ and the result is differentiated with respect to s, we have

$$\frac{d}{ds}(s-p)^r\mathbf{F}(s) = \frac{d}{ds}(A_r + (s-p)A_{r-1} + (s-p)^2A_{r-2} + \cdots)$$

$$= (A_{r-1} + 2(s-p)A_{r-2} + \cdots)$$

Evaluating at $s = p$, each of the terms but the first is zero, and thus we have found A_{r-1}. In a similar manner, A_{r-2} is found by differentiating twice, and so on for each A-coefficient.

REPEATED POLES

For each pole p in $\mathbf{F}(s)$ of order r, there will be r terms in its partial fraction expansion, of the form $A_k/(s-p)^k$, $k = 1, \ldots, r$. The numerators may be found from $A_k = [1/(r-k)!]d^{r-k}/ds^{r-k}(s-p)^r\mathbf{F}(s)|_{s=p}$. There will be r corresponding terms in its inverse transform, each of the form $[A_k/(k-1)!]t^{(k-1)}e^{pt}$, $k = 1, \ldots, r$.

Note that for brevity we have mentioned only the differentiation method in the last summary statement. Both methods are illustrated in the next example.

Example 12.14

We wish to find the inverse transform of $F(s) = 3/(s + 1)^2(s + 2)$. There is a simple pole at -2 and a repeated pole of order 2 at -1. Thus the partial fraction expansion is of the form

$$F(s) = \frac{A}{(s + 1)^2} + \frac{B}{s + 1} + \frac{C}{s + 2} \qquad (12.34)$$

Both methods for computing the terms associated with the repeated pole begin by evaluating A as

$$A = (s + 1)^2 F(s) \Big|_{s=-1} = \frac{3}{s + 2} \Big|_{s=-1} = 3 \qquad (12.35)$$

The repeated subtraction method continues by subtracting $3/(s + 1)^2$ from both sides of (12.34), leaving

$$\frac{3}{(s + 1)^2(s + 2)} - \frac{3}{(s + 1)^2} = \frac{B}{s + 1} + \frac{C}{s + 2}$$

Combining terms on the left yields

$$\frac{-3}{(s + 1)(s + 2)} = \frac{B}{s + 1} + \frac{C}{s + 2} \qquad (12.36)$$

But (12.36) is just the partial fraction expansion of $F_1(s) = -3/(s + 1)(s + 2)$, so

$$B = (s + 1)F_1(s) \Big|_{s=-1} = \frac{-3}{s + 2} \Big|_{s=-1} = -3 \qquad (12.37)$$

$$C = (s + 2)F_1(s) \Big|_{s=-2} = \frac{-3}{s + 1} \Big|_{s=-2} = +3 \qquad (12.38)$$

The repeated differentiation method finds B by computing the derivative,

$$\frac{d}{ds}(s + 1)^2 F(s) = \frac{d}{ds}\frac{3}{s + 2} = \frac{-3}{(s + 2)^2}$$

and evaluating at the repeated pole:

$$B = \frac{-3}{(s + 2)^2} \Big|_{s=-1} = -3 \qquad (12.39)$$

The remaining coefficient C is then determined by the simple pole formula

$$C = (s + 2)F(s) \Big|_{s=-2} = \frac{3}{(s + 1)^2} \Big|_{s=-2} = +3 \qquad (12.40)$$

Comparing (12.37)–(12.38) with (12.39)–(12.40), both methods agree that the partial fraction expansion is

$$F(s) = \frac{3}{(s + 1)^2} - \frac{3}{s + 1} + \frac{3}{s + 2}$$

whence

$$f(t) = 3te^{-t} - 3e^{-t} + 3e^{-2t} = 3[(t - 1)e^{-t} + e^{-2t}]$$

Example 12.15

Next we shall find the inverse transform of $\mathbf{F}(s) = 1/(s+1)(s^2 + 2s+2)^2$. The roots of the quadratic are $-1 \pm j$, so there are repeated complex conjugate poles. Let $p = -1+j$. Then

$$\mathbf{F}(s) = \frac{A}{(s-p)^2} + \frac{B}{s-p} + \frac{A^*}{(s-p^*)^2} + \frac{B^*}{s-p^*} + \frac{C}{s+1} \quad (12.41)$$

First compute A:

$$A = \left.\frac{1}{(s+1)(s-p^*)^2}\right|_{s=p} = \frac{1}{(p+1)(p-p^*)^2}$$

$$= \frac{1}{(+j)(2j)^2} = j\frac{1}{4}$$

Continuing by the repeated differentiation method yields

$$\frac{d}{ds}\frac{1}{(s+1)(s-p^*)^2} = \frac{-[(s-p^*)^2 + 2(s+1)(s-p^*)]}{(s+1)^2(s-p^*)^4}$$

Evaluating at $s = p$, we obtain

$$B = \frac{-[(p-p^*)^2 + 2(p+1)(p-p^*)]}{(p+1)^2(p-p^*)^4}$$

Using the fact that $p - p^* = j2\,\mathrm{Im}(p) = 2j$ and $p+1 = j$ gives

$$B = \frac{-[-4 + 2(j)(2j)]}{(-1)(16)} = \frac{1}{2}$$

We also need C:

$$C = \left.\frac{1}{s^2 + 2s + 2}\right|_{s=-1} = 1$$

We now have the full partial fraction expansion (12.41). It remains to find the inverse transform. From Table 12.2,

$$f(t) = Ate^{pt} + Be^{pt} + A^*te^{p^*t} + B^*e^{p^*t} + Ce^{-t}$$

We recognize that the terms involving A and A^* are conjugate terms, as are those involving B and B^*, so

$$f(t) = 2\,\mathrm{Re}(Ate^{pt} + Be^{pt}) + Ce^{-t}$$

To simplify the final substitution, converting to polar form, we have $A = \frac{1}{4}\underline{/90^\circ}$, $B = \frac{1}{2}\underline{/0}$. Then

$$f(t) = 2\,\mathrm{Re}\left[t\left(\tfrac{1}{4}\underline{/90^\circ}\right)(e^{-t}e^{+jt}) + \left(\tfrac{1}{2}\underline{/0}\right)(e^{-t}e^{+jt})\right] + e^{-t}$$

$$= 1/2te^{-t}\,\mathrm{Re}(e^{+jt+90^\circ}) + e^{-t}\,\mathrm{Re}(e^{+jt}) + e^{-t}$$

$$= te^{-t}\cos(t+90^\circ) + e^{-t}\cos(t) + e^{-t}$$

The first term in this solution may be further simplified to $-te^{-t}\sin t$.

As illustrated in Example 12.15, repeated poles may be treated the same whether real or complex. If they are complex, they will come in conjugate pairs and this will facilitate their calculation, since half the coefficients define the other half. Terms in the inverse transform due to these poles will also always come in complex conjugate pairs, as in the example, and can conveniently be combined into a single phase-shifted sinusoidal form.

EXERCISES

12.4.1. Find the PFE for $F(s) = (s^3 - 1)/[s(s + 2)]$.

Answer $s - 2 + \dfrac{\frac{1}{2}}{s} + \dfrac{\frac{7}{2}}{s + 2}$

12.4.2. Find the inverse transform of $F(s) = 5(s + 2)/[(s - 3)(s^2 + 2s + 10)]$.

Answer $e^{3t} + 1.05e^{-t}\cos(3t - 161.6°)$

12.4.3. Find the inverse transform of $F(s) = (2s^2 + s + 2)/(s^5 + 2s^3 + s)$.

Answer $2 - 2\cos t + \frac{1}{2}(\sin t - t\cos t)$

12.5 SOLVING INTEGRODIFFERENTIAL EQUATIONS

In this chapter we have introduced the basic tools needed to work in the s-domain: definitions, transform properties and pairs, and partial fractions. Thus armed, we close the chapter by discussing a most important application of these ideas.

Laplace transforms may be used to solve differential and integrodifferential equations, such as the equations of basic electric circuit analysis. The equation or equations are first Laplace transformed and then solved by straightforward algebraic means. The inverse transform of the solution is identified by partial fraction expansion.

Example 12.16

As a first example, we seek the solution of the second-order differential equation

$$\frac{d^2x}{dt^2} + 4\frac{dx}{dt} + 3x = e^{-2t} \qquad (12.42)$$

for $t > 0$, given the initial conditions $x(0-) = 1, x'(0-) = 2$. First, note that (12.42) is an equality between time functions; thus their transforms must be equal. *Both sides of a time-domain equation may be Laplace transformed, or an s-domain equation inverse transformed, while retaining equality.* Transforming (12.42) to the s-domain and noting that the linearity property allows transforming a weighted sum as a weighted sum of transforms,

$$L\left[\frac{d^2x}{dt^2}\right] + 4L\left[\frac{dx}{dt}\right] + 3L[x] = L[e^{-2t}] = \frac{1}{s + 2}$$

Using the n-fold differentiation property in Table 12.1, we have

$$[s^2X(s) - s(1) - 2] + 4[sX(s) - 1] + 3X(s) = \frac{1}{s + 2}$$

Solving for the unknown transform $\mathbf{X}(s)$ yields

$$\mathbf{X}(s) = \frac{s^2 + 8s + 13}{(s + 1)(s + 2)(s + 3)}$$

The partial fraction expansion is, using the method of Section 12.4,

$$\mathbf{X}(s) = \frac{3}{s + 1} - \frac{1}{s + 2} - \frac{1}{s + 3}$$

which has inverse transform $x(t)$ solving the original differential equation

$$x(t) = 3e^{-t} - e^{-2t} - e^{-3t}$$

This value for $x(t)$ is not necessarily valid for negative time, since the differential equation we solved is required to be valid only for $t \geq 0$. We may verify this solution by direct substitution into (12.42).

$$\frac{d^2}{dt^2}(3e^{-t} - e^{-2t} - e^{-3t}) + 4\frac{d}{dt}(3e^{-t} - e^{-2t} - e^{-3t})$$

$$+ 3(3e^{-t} - e^{-2t} - e^{-3t}) = (3e^{-t} - 4e^{-2t} - 9e^{-3t})$$

$$+ 4(-3e^{-t} + 2e^{-2t} + 3e^{-3t}) + 3(3e^{-t} - e^{-2t} - e^{-3t}) = e^{-2t}$$

as (12.42) requires. Checking the initial conditions, we note that $x(t)$ and its derivative are continuous at $t = 0$, so $x(0-) = x(0) = 1$ and $x'(0-) = x'(0) = 2$, as required.

In Example 12.17, the initial conditions were specified at time $t = 0-$ and the differential equation was required to be satisfied for $t > 0$. This is a useful way to specify the initial conditions for a circuit problem, since we may compute the initial values at $t = 0-$ using steady-state calculations based on the $t < 0$ circuit (the circuit as it appears before the switches and unit steps act at $t = 0$), as described in Chapters 5 and 6. This procedure is shown in the next two examples.

Example 12.17

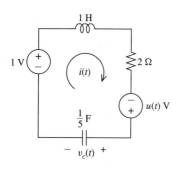

FIGURE 12.10   Circuit for Example 12.17.

Consider the single-mesh circuit of Fig. 12.10. Its mesh equation for $t > 0$ is

$$\frac{di}{dt} + 2i + \left[5\int_{0-}^{t} i(\tau)\, d\tau + 1\right] = 1 + u(t) \qquad (12.43)$$

Just before the unit step acts, the $t < 0$ circuit is in dc steady state with $i(0-) = 0$ A and $v_c(0-) = 1$ V. The net source voltage for $t > 0$ is the sum of the sources $1 + u(t)$. Transforming (12.43) gives

$$[s\mathbf{I}(s) - 0] + 2\mathbf{I}(s) + \frac{5\mathbf{I}(s)}{s} + \frac{1}{s} = \frac{1}{s} + \frac{1}{s}$$

or

$$\mathbf{I}(s) = \frac{1}{s^2 + 2s + 5}$$

This transform has complex conjugate poles at $-1 \pm j2$ in the s-domain. Then

$$I(s) = \frac{A}{s + 1 - j2} + \frac{A^*}{s + 1 + j2}$$

where

$$A = \frac{1}{s + 1 + j2}\Big|_{s=p=-1+j2} = -\frac{j}{4}$$

Then $i(t) = Me^{\alpha t} \cos(\beta t + \theta)$, where $p = \alpha + j\beta = -1 + j2$, $M = 2|A|$, $\theta = \underline{/A}$, and

$$i(t) = \tfrac{1}{2}e^{-t} \cos(2t - 90°) = \tfrac{1}{2}e^{-t} \sin 2t$$

Example 12.18

As our final example, we will find the node voltages v_1 and v_2 in the circuit shown in Fig. 12.11. The nodal analysis equations are

$$\frac{1}{2}(v_1 - 3) + \frac{d}{dt}v_1 + 2\frac{d}{dt}(v_1 - v_2) = 0$$

$$2\frac{d}{dt}(v_2 - v_1) + \frac{1}{3}v_2 = r(t)$$

Since neither source switches on until $t = 0$, the values of all currents and voltages at $t = 0-$ are zero. Transforming the equations with these initial conditions gives

$$\frac{1}{2}\left[V_1(s) - \frac{3}{s}\right] + sV_1(s) + 2s[V_1(s) - V_2(s)] = 0$$

$$2s[V_2(s) - V_1(s)] + \frac{1}{3}V_2(s) = \frac{1}{s^2}$$

This is a pair of coupled algebraic equations in the unknowns $V_1(s)$ and $V_2(s)$, which may be solved by several methods. Choosing matrix inversion, we rewrite the node equations as

$$\begin{bmatrix} 3s + \frac{1}{2} & -2s \\ -2s & 2s + \frac{1}{3} \end{bmatrix} \begin{bmatrix} V_1(s) \\ V_2(s) \end{bmatrix} = \begin{bmatrix} \dfrac{3}{2s} \\ \dfrac{1}{s^2} \end{bmatrix}$$

Performing the matrix inversion, the determinant is $2s^2 + 2s + \frac{1}{6}$ and

$$\begin{bmatrix} V_1(s) \\ V_2(s) \end{bmatrix} = \frac{1}{2s^2 + 2s + \frac{1}{6}} \begin{bmatrix} 2s + \frac{1}{3} & 2s \\ 2s & 3s + \frac{1}{2} \end{bmatrix} \begin{bmatrix} \dfrac{3}{2s} \\ \dfrac{1}{s^2} \end{bmatrix}$$

$$= \begin{bmatrix} \dfrac{\frac{3}{2}s + \frac{5}{4}}{s(s^2 + s + \frac{1}{12})} \\ \dfrac{\frac{3}{2}s^2 + \frac{3}{2}s + \frac{1}{4}}{s^2(s^2 + s + \frac{1}{12})} \end{bmatrix}$$

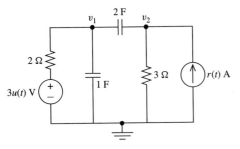

FIGURE 12.11   Circuit for Example 12.18.

The partial fraction expansion for $\mathbf{V}_1(s)$, with real poles at 0, $-\frac{1}{2} \pm \sqrt{\frac{1}{6}}$, is

$$\mathbf{V}_1(s) = \frac{A}{s} + \frac{B}{s + \frac{1}{2} - \sqrt{\frac{1}{6}}} + \frac{C}{s + \frac{1}{2} + \sqrt{\frac{1}{6}}}$$

where

$$A = \frac{5}{4} \div \frac{1}{12} = 15$$

and resorting to finite precision on our calculator,

$$B = \frac{\left(\frac{3}{2}\right)(-0.0918) + \frac{5}{4}}{(-0.0918)(0.816)} = -14.848$$

$$C = \frac{\left(\frac{3}{2}\right)(-0.908) + \frac{5}{4}}{(-0.908)(-0.816)} = -0.152$$

or $\qquad v_1(t) = 15 - 14.848e^{-0.0918t} - 0.152e^{-0.908t}$

The partial fraction expansion for $\mathbf{V}_2(s)$ is

$$\mathbf{V}_2(s) = \frac{A_2}{s^2} + \frac{A_1}{s} + \frac{B}{s + \frac{1}{2} - \sqrt{\frac{1}{6}}} + \frac{C}{s + \frac{1}{2} + \sqrt{\frac{1}{6}}}$$

where

$$A_2 = \frac{1}{4} \div \frac{1}{12} = 3$$

Subtracting yields

$$\frac{\frac{3}{2}s^2 + \frac{3}{2}s + \frac{1}{4}}{s^2\left(s^2 + s + \frac{1}{12}\right)} - \frac{3}{s^2} = \frac{-\frac{3}{2}s - \frac{3}{2}}{s\left(s^2 + s + \frac{1}{12}\right)}$$

The remaining values A_1, B, and C may now be found using the simple real pole formula on the rational function written on the right side of the equation above, and the inverse transform is computed to be

$$v_2(t) = 3t - 18 + 18.19e^{-0.0918t} - 0.190e^{-0.908t}$$

The s-domain approach to the solution of differential or integrodifferential equations has some clear advantages over classical differential equation techniques worked in the time domain. First and foremost, derivatives and integrals of unknown time functions transform to powers of s times unknown transforms, effectively converting a problem in *calculus* to one in *algebra*. Initial conditions at $t = 0-$ are taken into account naturally by the method and do not have to be "fit" after a general form for the solution is found. The price paid for these advantages is that the equations must first be transformed and then, at the end, their s-domain solutions returned back to the time domain (inverse transformed). This price is seldom very high, since the s-domain version of the original equations can always be written by inspection, and partial fractions can easily be computed, at least for low-order rational functions. High-order rational functions imply high-order integrodifferential equations or complicated source functions, either of which would add greatly to the complexity of time-domain solutions as well. The s-domain approach to solving these equations is almost always more efficient than the conventional time-domain approach.

12.5.1. Find $x(t)$, $t > 0$: $d^2x/dt^2 + 3(dx/dt) + 2x = 2$, $x(0-) = 0$, $dx/dt|_{0-} = 0$.

Answer $1 - 2e^{-2t} + e^{-t}$

12.5.2. Find $x(t)$, $t > 0$: $d^2x/dt^2 = 2t + 1$, $x(0-) = 0$, $dx/dt|_{0-} = 1$.

Answer $\frac{1}{3}t^3 + \frac{1}{2}t^2 + t$

12.5.3. Solve the following pair of differential equations for $t > 0$:

$$\frac{d^2x}{dt^2} - x - 5y = 0, \ x(0-) = 0, \ \frac{dx}{dt}\bigg|_{0-} = 0$$

$$\frac{d^2y}{dt^2} - 3x + y = 0, \ y(0-) = 0, \ \frac{dy}{dt}\bigg|_{0-} = 64$$

Answer $x(t) = 10e^{2t} - 10e^{-2t} - 20\sin 2t$

$$y(t) = 6e^{2t} - 6e^{-2t} + 20\sin 2t$$

SUMMARY

Laplace transformation is a way of converting functions of time into functions of a complex variable s. The calculus operations of differentiation and integration in the time domain go over to simpler algebraic operations of multiplication and division in the s domain. Thus a key use of Laplace transformation is to convert sets of integrodifferential equations, such as circuit equations, to the s-domain, solve using simple algebra, and convert the solution back to its desired time-domain format. Conversion back, or inverse transformation, is accomplished by a straightforward manipulation called partial fraction expansion.

■ The Laplace transform of $f(t)$ is defined by the integral transform

$$F(s) = \int_{0-}^{\infty} f(t)e^{-st}\, dt$$

■ The Laplace transform of a sum is the sum of Laplace transforms; scaling a time function scales its transform equally. Laplace transformation is a linear operation.

■ The unit step $u(t)$ and its integrals and derivatives are called singularity functions. Their Laplace transforms are s^n, where the integer n is 1 for the unit step, 2 for the unit ramp, 3 for the unit parabola, and 0 for the unit impulse.

■ The unit impulse $\delta(t)$ is the generalized derivative of the unit step. $\delta(t) = 0$ for all $t \neq 0$, and has a total area of unity.

■ The Laplace transform of a derivative df/dt is s times the Laplace transform of the function $f(t)$ minus its initial value $f(0-)$.

- Unknown transforms may often be found by applying the Laplace transform properties of Table 12.1 to known transforms listed in Table 12.2.

- Partial fraction expansion (PFE) is the process of rewriting a rational function as a sum of simple terms each one of which has a known inverse transform.

- The PFE term due to a simple pole p is $A/(s - p)$.

- The PFE terms due to an r-fold repeated pole p are of the form $A_k/(s - p)^k$ for $k = 1, \ldots, r$.

- Integrodifferential equations may be solved by converting the equations to the s-domain term by term, solving for the unknown transforms, and using partial fraction expansion to recover the inverse transforms.

PROBLEMS

12.1. Determine $\mathbf{F}(s)$ for the following $f(t)$'s:
(a) $5e^{-9t} - 2e^{-t} + e^{+t}$
(b) $(5e^{-9t} - 2e^{-t} + e^{+t})u(t + 1)$
(c) $e^{j2t} - e^{-j2t}$

12.2. The lower limit on the integral defining the Laplace transform (12.1) is sometimes taken to be 0+. How would the $\mathbf{F}(s)$'s calculated in Problem 12.1(a)–(c) change if we adopt that definition?

12.3. Is e^{+t^2} Laplace transformable? Explain.

12.4. Show that there is no M and σ so that, for all $t > 0, |t^t| \leq Me^{\sigma t}$. This proves that $f(t) = t^t$ is not of exponential order.

12.5. Explain why $f(t) = 1/(t + 1)$ is Laplace transformable but $1/(t - 1)$ is not.

12.6. Specify a periodic function $f(t)$ that is not Laplace transformable. Explain why not.

12.7. Which function converges to zero faster: $t^{10}e^{-10t}$ or e^{-9t}? Find a t^* such that the one you select is always less than the other for $t > t^*$. This defines faster convergence.

12.8. Considering the defining integral, what is the value of $\mathbf{F}(s)$ if $f(t) = 0$ everywhere except $f(t_0) = 1, t_0 \geq 0$? What does this imply about the difference $f_1(t) - f_2(t)$ of two time functions with the same Laplace transform $\mathbf{F}_1(s) = \mathbf{F}_2(s)$?

12.9. Use the defining integral (12.1) to find the Laplace transform of $5e^{-2t}\cos(3t + 12°)$. *Hint:* Use the Euler identity and linearity.

12.10. Determine $\mathbf{F}(s)$ for the following $f(t)$'s:

(a)

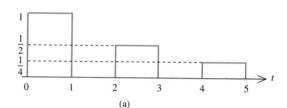

(a)

(b)

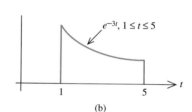

(b)

(c)

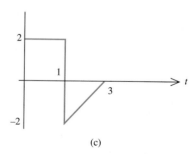

(c)

FIGURE P12.10

12.11. Determine $\mathbf{F}(s)$ and the region of convergence for the following $f(t)$'s:

(a)

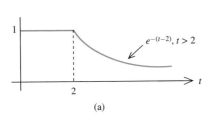

(a)

(b)

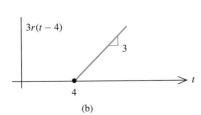

(b)

(c)

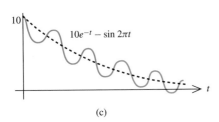

(c)

FIGURE P12.11

12.12. Express the following $f(t)$'s as linear combinations of time-shifted unit steps $u(t)$, ramps $r(t)$, and parabolas $p(t)$. All are zero for $t < 0$.

(a)

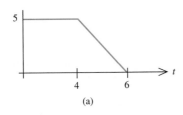

(a)

(b)

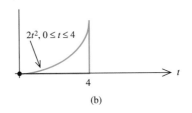

(b)

(c)

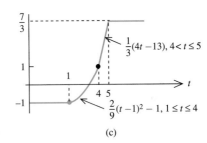

(c)

FIGURE P12.12

12.13. Sketch the following functions, labeling their values at $t = 1$ and $t = 4$.
(a) $2[u(t) - u(t - 15)]$
(b) $6r(t) - u(t - 2)$
(c) $4[u(t - \frac{1}{2}) - u(t - \frac{3}{2})] + p(t)$
(d) $r(t) - r(t - 1) - r(t - 2) + r(t - 3)$

12.14. Evaluate the following and determine if they are ordinary derivatives or generalized derivatives.
(a) $\dfrac{d}{dt}[\sin^2 tu(t)]$
(b) $\dfrac{d}{dt}[u(t) - te^{-t}u(t)]$
(c) $\dfrac{d^2}{dt^2}\cos 2t\, u(t)$

12.15. Evaluate the first and second derivatives.
(a) $f(t)$ of Problem 12.1(a)
(b) $f(t)$ of Problem 12.1(b)
(c) $f(t)$ of Problem 12.1(c)

12.16. Evaluate the first three derivatives df/dt, d^2f/dt^2, and d^2f/dt^3.
(a) $f(t)$ of Problem 12.12(a)
(b) $f(t)$ of Problem 12.12(b)
(c) $f(t)$ of Problem 12.12(c)

12.17. Determine the transforms of the following time functions.
(a) df/dt for $f(t)$ of Problem 12.12(a)
(b) $f(t)$ of Problem 12.12(b)
(c) df/dt for $f(t)$ of Problem 12.12(b)

12.18. Show the unit doublet sifting theorem (integrate the impulse sifting theorem by parts).

$$\int_{-\infty}^{+\infty} \delta'(t - t_0) f(t) \, dt = -f'(t_0)$$

12.19. Evaluate $\int_{-\infty}^{+\infty} \delta^{(3)}(t - 1) f(t - 2) \, dt$ if $f(t) = t^2 e^{-2t} u(t)$.

12.20. Find the Laplace transform of
(a) $u(t) - u(t - 2)$
(b) $e^{-2t}[u(t) - u(t - 1)]$
(c) $(t + 1)u(t - 1)$
(d) $e^{-2t} u(t - 2)$

12.21. Find the Laplace transform.
(a) $\dfrac{d}{dt}[e^{-t} u(t)]$
(b) $\dfrac{d}{dt}(e^{-t})$
(c) $\dfrac{d}{dt}[e^{-t} \sin 2t \, u(t)]$
(d) $\dfrac{d}{dt}[e^{-t} \sin 2t]$

12.22. Find the Laplace transform.
(a) $(2 - t)$
(b) $(2 - t)e^{+3t}$
(c) $te^{-3t} \cos(2t + 30°)$

12.23. Find the Laplace transform of the functions.
(a) $e^{-2t} (t + 2) \, u(t)$
(b) $e^{-4t} \sin 2t \, u(t)$
(c) $e^{-3t} \cosh 2t \, u(t)$
(d) $e^{-t} \cos 4t \, u(t)$

12.24. Find the inverse Laplace transform.
(a) $\dfrac{s^2 + 1}{s}$
(b) $\dfrac{s^2 + 3s + 2}{s^2}$
(c) $\dfrac{s^3 + 3s + 2}{s}$

12.25. Starting from $L\left[e^{-t}u(t)\right] = 1/(s + 1)$, verify the relation $L\left[e^{-at}u(t)\right] = 1/(s + a)$ by the
(a) Time-frequency scaling property
(b) Frequency shift property

12.26. Starting from $L\left[e^{-t}u(t)\right] = 1/(s + 1)$, verify the relation $L(t^{n-1}e^{-t}) = (n - 1)!/(s + 1)^n$ using the t-multiplication property.

12.27. Find the inverse Laplace transforms.
(a) $\dfrac{s + 2}{s + 1}$
(b) $\dfrac{s^2 + 2s}{s + 1}$
(c) $\dfrac{s^5}{s^2 - 1}$

12.28. Find the inverse Laplace transform of $F(s) = \dfrac{s^2 + 3}{s^2 + 4}$ as in Example 12.11, but by first using long division.

12.29. Solve for $y(t), t > 0$:
(a) $y(t) = \cos t + \int_0^t e^{-(t-\tau)} y(\tau) \, d\tau$
(b) $y(t) = \sin t + \int_0^t e^{-(t-\tau)} y(\tau) \, d\tau$

12.30. Find the inverse Laplace transform of
(a) $\dfrac{s}{(s + a)(s + b)}, b \neq a$
(b) $\dfrac{s + 4}{(s + 1)(s + 2)}$
(c) $\dfrac{s^2 + 9s + 6}{s^3 + 4s^2 + 3s}$
(d) $\dfrac{5s^3 - 3s^2 + 2s - 1}{s^4 + s^2}$

12.31. Find the inverse Laplace transform of
(a) $\dfrac{s^2 - 2s + 5}{(s + 1)(s^2 + 2s + 5)}$
(b) $\dfrac{s + 2}{(s^2 + 2s + 2)(s + 1)^2}$
(c) $\dfrac{4(s^3 - s^2 + 3s - 15)}{(s^2 + 9)(s^2 + 4s + 13)}$
(d) $\dfrac{100}{(s + 2)(s^2 + 2s + 5)^2}$

12.32. Find the inverse transform of
(a) $\dfrac{27}{(s + 1)^3(s + 4)}$
(b) $\dfrac{1}{s(s + 1)^4}$

12.33. Using Laplace transforms, solve the following for $t > 0$:
(a) $x'' + x = 0, x(0) = -1, x'(0) = 1$
(b) $x'' + 2x' + 2x = 0, x(0) = 0, x'(0) = 1$
(c) $x''' - 2x'' + 2x' = 0, x(0) = x'(0) = 1, x''(0) = 2$
(d) $x'' + 4x' + 3x = 4 \sin t + 8 \cos t, x(0) = 3, x'(0) = -1$
(e) $x'' + 4x' + 3x = 4e^{-3t}, x(0) = x'(0) = 0$
(f) $x'' + 4x' + 3x = 4e^{-t} + 8e^{-3t}, x(0) = x'(0) = 0$

12.34. Using Laplace transforms, solve the following for $t > 0$:

(a) $x' + 4x + 3 \int_0^t x(\tau)\, d\tau = 5, x(0) = 1$

(b) $x' + 4 \int_0^t x(\tau)\, d\tau = 3 \sin t, x(0) = 4$

12.35. Find the inverse transforms (these functions have real distinct poles).

(a) $\dfrac{2}{s^2 + 3s + 2}$

(b) $\dfrac{5s + 1}{s^3 - s}$

(c) $\dfrac{6s^4}{s^3 - s^2}$

12.36. Find the inverse Laplace transforms (these functions have distinct complex conjugate poles).

(a) $\dfrac{5}{2s^2 + 5}$

(b) $\dfrac{3s + 2}{(s + 2)^2 + 9}$

(c) $\dfrac{6s}{(s^2 + 1)(s^2 + 4)}$

12.37. Find the inverse Laplace transforms (these functions have real repeated poles).

(a) $\dfrac{6s}{(2s + 3)^2}$

(b) $\dfrac{s - 4}{s^2 + 6s + 9}$

(c) $\dfrac{7s^2}{s^2 + 3s^2 + 3s + 1}$

12.38. Solve for x for $t > 0$:

$$x' + x + y' + y = 1$$
$$-2x + y' - y = 0$$
$$x(0) = 0, \quad y(0) = 1$$

12.39. Find i for $t > 0$ if $v(0) = 6$ V by using Laplace transforms.

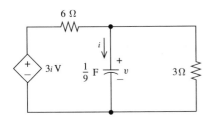

FIGURE P12.39

12.40. Find $i_1(t)$ and $i_2(t)$, $t > 0$, by transforming the two mesh equations. Take $v_g = 14e^{-2t}$ V, $i_1(0+) = 6$ A, and $i_2(0+) = 2$ A.

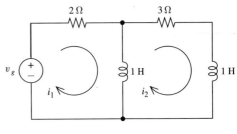

FIGURE P12.40

12.41. Solve for $x(t)$, $t > 0$, by s-domain methods. Verify your solution by substitution.

(a) $\dfrac{dx}{dt} + \int_{0-}^t x(\tau)\, d\tau = 0, x(0-) = 1$

(b) $\dfrac{dx}{dt} + \int_{0-}^t x(\tau)\, d\tau = 1, x(0-) = 0$

(c) $\dfrac{dx}{dt} + 3x + 2 \int_{0-}^t x(\tau)\, d\tau = e^{-t} + 1, x(0-) = 0$

12.42. Show that the integrodifferential equation, for $t > 0$,

$$\frac{dx}{dt} + ax + b \int_{0-}^t x(\tau)\, d\tau = y(t) \qquad x(0-) = x_0$$

is equivalent to the differential equation

$$\frac{d^2x}{dt^2} + a\frac{dx}{dt} + bx = z(t), \qquad x(0-) = x_0, \quad \dot{x}(0-) = \dot{x}_0$$

where $z(t) = dy/dt$ and $\dot{x}_0 = y(0-) - ax_0$.

12.43. Write and solve the nodal equation for this circuit, assuming $i_L(0-) = 0$.

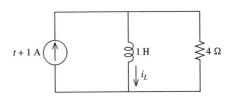

FIGURE P12.43

12.44. Solve the following differential equations for $x(\delta)$, $t > 0$, by s-domain methods. Check by time-domain methods.

(a) $\dfrac{dx}{dt} + 3x = 4e^{-t}, x(0-) = 1$

(b) $\dfrac{dx}{dt} - 7x = 2e^{-2t} + 6, x(0-) = 3$

(c) $\dfrac{dx}{dt} + 16x = te^{+6t}, x(0-) = 0$

12.45. Write and solve the coupled mesh equations for this circuit, assuming all initial conditions $= 0$ at $t = 0-$.

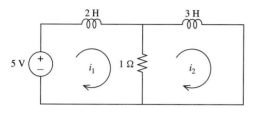

FIGURE P12.45

More Challenging Problems

12.46. Find the Laplace transform.
(a) $t^2 e^{-4t} \sin 7t$
(b) $t^2 e^{+6t} \cos(5t - 118°)$

12.47. Find:
(a) $\mathcal{L}[\sin 2t \sinh t\, u(t)]$
(b) $\mathcal{L}[\sin t \cosh 2t u(t)]$
(c) $\mathcal{L}[\sin t \cos 2t\, u(t)]$

12.48. Find the Laplace transform of the periodic function $f(t) = t, -1 \le t < +1$ which has period $T = 2$. Suggestion: write as a sum of shifted functions each one period long, use Taylor's series.

12.49. Find the inverse Laplace transform of $\mathbf{F}(s) = 2s/(s^8 - 1)$.

12.50. Solve for $x_1(t)$, $x_2(t)$, $x_3(t)$, $t > 0$, by s-domain methods. Assume all initial conditions are zero except $x_1(0-) = 1$.

$$\frac{d^2 x_1}{dt^2} + \frac{dx_3}{dt} = 1$$

$$2\frac{dx_1}{dt} + x_1 + x_2 - x_3 = t$$

$$\frac{dx_1}{dt} + x_2 + \frac{dx_3}{dt} = 0$$

Oliver Heaviside
1850–1925

[Electromagnetics] has been said to be too complicated. This probably came from a simple-minded man.

Oliver Heaviside

13

Circuit Analysis in the *s*-Domain

Oliver Heaviside is from the tradition of great scientific thinkers whose vision and determination ultimately overcame an almost total lack of formal training. While having never graduated beyond the elementary school level, Heaviside taught himself enough mathematics, physics, and electrical engineering to make many fundamental contributions in both theoretical and experimental domains. His lack of educational credentials made the intellectual establishment of his day slow to accept some of his more controversial ideas, but in the end Heaviside left a lasting imprint on mathematics, atmospheric physics, telegraphy, and many other areas.

Born in 1850 in London, England, Heaviside was the nephew of Charles Wheatstone, early telegrapher and inventor of the Wheatstone bridge. After retiring at age 24 from employment as a telegraph dispatcher due to deafness, Heaviside became fascinated with the electromagnetic theories of Maxwell and discovered novel vector formulations for Maxwell's equations. Turning to electrical circuits, he employed divergent series and daring generalizations of standard mathematical functions, creating "operational" methods of calculus. The lack of detailed rigor in support of these operational methods, combined no doubt with Heaviside's lack of advanced degrees, caused some influential figures in the scientific elite initially to dismiss the work. But its remarkable ability to predict the results of electrical experiments won them over. Heaviside generalized resistance to impedance and showed the importance of inductive reactance in the transmission of telegraph signals along long lines. He predicted the existence of the ionosphere and showed its properties as a reflector of electromagnetic waves. Appropriately enough, Oliver Heaviside was the first winner of the Faraday Medal, a prestigious award named in honor of another great self-taught scientist.

Chapter Contents

Laplace transformation was used in Chapter 12 to convert integrodifferential equations in the time domain into more easily solved algebraic equations in the s-domain. Since the basic equations of circuit analysis are of the integrodifferential type, this defines a clear path to improve our methods for studying electrical circuits. Indeed, this path is almost identical to one already followed when we employed phasors and phasor circuits containing impedances in ac steady state. As we shall discover, the s-domain permits generalization of this highly effective approach to non-ac-steady-state conditions where phasors cannot be applied. Circuit analysis in the s-domain, that is, after Laplace transformation of circuit variables and equations, is the topic of this chapter.

We begin by showing that all the essential circuits laws carry over from the time domain to the s-domain: Kirchhoff's laws, element laws, series–parallel equivalents, Thevenin–Norton equivalents, and so on. The forms of these equations in the s-domain prove very similar to those of phasor analysis, with the Laplace variable s replacing the phasor variable $j\omega$. It is therefore not surprising that the same general methods we found effective earlier in the context of phasors will apply in the s-domain as well, most notably nodal and mesh analysis using impedance to characterize the *RLC* elements.

The s-domain circuit diagram is next introduced. It is used in the same fashion in which the phasor circuit diagram was earlier, as an aid in writing the analysis equations. One difference with the phasor circuit diagram is the introduction of initial condition generators, whose inclusion underscores our ability to determine transient behavior in the s-domain. The transfer function, a basic tool of input–output analysis, is defined and the information it contains is next examined. Two useful s-domain laws, the initial and final value theorems, are developed. The chapter concludes with a discussion of the unit impulse response, its close relationship to the transfer function, and the time-domain operation called convolution, which permits us to determine a circuit's response to any input from knowledge of its response to a unit impulse input.

In this section we examine the forms taken by the most familiar circuit laws when transformed to the s-domain. Consider the mesh of Fig. 13.1(a) and the node of Fig. 13.1(b). Applying the Kirchhoff laws gives

$$v_1(t) + v_2(t) + v_3(t) + v_4(t) = 0 \tag{13.1a}$$

$$i_1(t) + i_2(t) + i_3(t) + i_4(t) = 0 \tag{13.1b}$$

Transforming these equations yields

$$\mathbf{V}_1(s) + \mathbf{V}_2(s) + \mathbf{V}_3(s) + \mathbf{V}_4(s) = 0 \tag{13.2a}$$

$$\mathbf{I}_1(s) + \mathbf{I}_2(s) + \mathbf{I}_3(s) + \mathbf{I}_4(s) = 0 \tag{13.2b}$$

Voltage drops around a closed loop sum to zero, and currents entering a node sum to zero, after Laplace transformation equally well as before. *Kirchhoff's voltage and current laws apply unchanged in the s-domain.*

Next consider the *RLC* element laws. Transforming both sides of Ohm's law,

$$v_R(t) = R i_R(t) \tag{13.3a}$$

yields

$$\mathbf{V}_R(s) = R \mathbf{I}_R(s) \tag{13.3b}$$

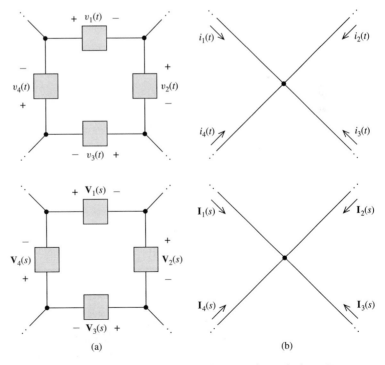

FIGURE 13.1 (a) Closed loop for KVL; (b) node for KCL.

The element law for an inductor is

$$v_L(t) = L\frac{d}{dt}i(t) \tag{13.4a}$$

Using the derivative rule, in the s-domain this equation becomes

$$\mathbf{V}_L(s) = sL\mathbf{I}_L(s) - L\, i_L(0-) \tag{13.4b}$$

The element law for a capacitor can be written

$$i_C(t) = C\frac{d}{dt}v_C(t) \tag{13.5a}$$

Transforming gives

$$\mathbf{I}_C(s) = sC\mathbf{V}_C(s) - Cv_C(0-) \tag{13.5b}$$

Solving the last for $\mathbf{V}_C(s)$, we obtain

$$\mathbf{V}_C(s) = \frac{1}{sC}\mathbf{I}_C(s) + \frac{1}{s}\, v_C(0-) \tag{13.5c}$$

It has been assumed in writing these element laws that the current and voltage variables satisfy the passive sign convention.

Equations (13.3b), (13.4b), and (13.5c) show that the voltage $\mathbf{V}(s)$ across the *RLC* elements is the sum of a term proportional to its current $\mathbf{I}(s)$ and a term that depends on its initial condition (this term is zero in the case of a resistor).

The factor of proportionality between voltage and current in the first term is the *impedance* $\mathbf{Z}(s)$. We define the impedance $\mathbf{Z}(s)$ of an *RLC* element as *the ratio of* $\mathbf{V}(s)$ *to* $\mathbf{I}(s)$ *when all initial conditions are zero.*

$$\mathbf{V}(s) = \mathbf{Z}(s)\mathbf{I}(s) \qquad \text{(all ICs} = 0) \tag{13.6}$$

Applying this to the equations cited above yields impedances

$$\mathbf{Z}_R(s) = R \tag{13.7a}$$

$$\mathbf{Z}_L(s) = sL \tag{13.7b}$$

$$\mathbf{Z}_C(s) = \frac{1}{sC} \tag{13.7c}$$

as shown in Fig. 13.2.

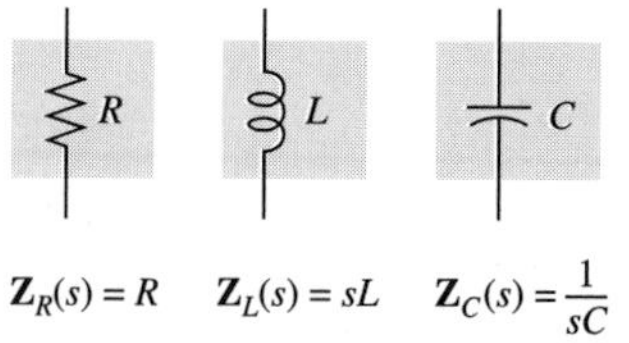

$$\mathbf{Z}_R(s) = R \qquad \mathbf{Z}_L(s) = sL \qquad \mathbf{Z}_C(s) = \frac{1}{sC}$$

FIGURE 13.2 Impedance $\mathbf{Z}(s)$.

Example 13.1

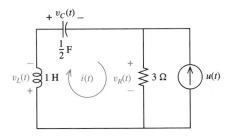

FIGURE 13.3 Circuit for Example 13.1.

We wish to find $v_C(t)$, $t > 0$ for the circuit shown in Fig. 13.3. By KVL in the s-domain, summing voltage drops around the loop, we obtain

$$\mathbf{V}_C(s) + \mathbf{V}_R(s) + \mathbf{V}_L(s) = 0 \qquad (13.8)$$

Since there are no sources in the circuit for $t < 0$, all initial conditions at 0− are zero. Each voltage above is therefore the product of impedance $\mathbf{Z}(s)$ and current $\mathbf{I}(s)$. The impedance of the capacitor is $\mathbf{Z}_C(s) = 1/sC = 2/s$, for the inductor it is $\mathbf{Z}_L(s) = sL = s$, and for the resistor, $\mathbf{Z}_R(s) = R = 3$. Then

$$\frac{2}{s}\mathbf{I}(s) + 3\left[\mathbf{I}(s) + \frac{1}{s}\right] + s\mathbf{I}(s) = 0$$

$$\qquad (13.9)$$

$$\mathbf{I}(s) = \frac{-3/s}{s + 3 + (2/s)} = \frac{-3}{s^2 + 3s + 2}$$

$\mathbf{V}_C(s)$ may be recovered by multiplication of $\mathbf{I}(s)$ by the impedance of the capacitor.

$$\mathbf{V}_C(s) = \mathbf{Z}_C(s)\mathbf{I}(s) = \frac{-6}{s(s^2 + 3s + 2)}$$

This completes the s-domain solution for the capacitive voltage $\mathbf{V}_C(s)$. To recover the time-domain solution $v_c(t)$, we require the inverse transform. By partial fractions we have that

$$\mathbf{V}_C(s) = \frac{-6}{s(s + 1)(s + 2)} = \frac{-3}{s} + \frac{6}{s + 1} - \frac{3}{s + 2}$$

so

$$v_C(t) = 6e^{-t} - 3e^{-2t} - 3 \qquad (t > 0) \qquad (13.10)$$

The inverse of the impedance $\mathbf{Z}(s)$ is called the *admittance* $\mathbf{Y}(s)$, where

$$\mathbf{I}(s) = \mathbf{Y}(s)\mathbf{V}(s) \qquad \text{(all ICs} = 0) \qquad (13.11)$$

Admittance $\mathbf{Y}(s)$ *is the ratio of current* $\mathbf{I}(s)$ *to voltage* $\mathbf{V}(s)$ *when all initial conditions are zero.* In light of the corresponding impedances in (13.7), the admittances of a resistor R, inductor L, and capacitor C are

$$\mathbf{Y}_R(s) = \frac{1}{R} = G \qquad (13.12a)$$

$$\mathbf{Y}_L(s) = \frac{1}{sL} \qquad (13.12b)$$

$$\mathbf{Y}_C(s) = sC \qquad (13.12c)$$

The Kirchhoff laws and the element laws are the basis for all other useful circuit theorems, such as the rules for series–parallel combination, current and voltage division, and Thevenin–Norton equivalents. Since the Kirchhoff laws are identical in the time domain and s-domain, and the element law $\mathbf{V}(s) = \mathbf{Z}(s)\mathbf{I}(s)$ is of the same form as Ohm's law in the time domain, we expect that these rules would carry over to the s-domain unchanged. This is indeed the case.

Example 13.2

Find the equivalent impedance $\mathbf{Z}(s)$ of the subnetwork shown in Fig. 13.4. Since the 1-H inductor and 2-Ω resistor are in series, their impedances $sL = s$ and $R = 2$ add:

$$\mathbf{Z}_1(s) = s + 2$$

The impedance $\mathbf{Z}_1(s)$ is in parallel with a $\frac{1}{3}$-F capacitor whose impedance is $\mathbf{Z}_C(s) = 1/sC = 3/s$. The equivalent is the inverse of the sum of inverses, or

$$\mathbf{Z}_2(s) = \frac{1}{[1/(s+2)] + (s/3)} = \frac{3(s+2)}{s^2 + 2s + 3}$$

$\mathbf{Z}_2(s)$ could also have been found by adding parallel admittances $\mathbf{Y}_1(s) = 1/\mathbf{Z}_1(s) = 1/(s+2)$ and $\mathbf{Y}_C(s) = s/3$ and then inverting the equivalent admittance $\mathbf{Y}_2(s)$. Or, since there are exactly two impedances $\mathbf{Z}_1(s)$ and $\mathbf{Z}_C(s)$ in parallel, the equivalent impedance $\mathbf{Z}_2(s)$ could have been computed as the product of these impedances divided by their sum.

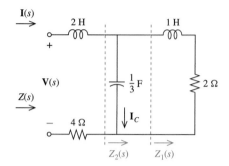

FIGURE 13.4 Circuit for Example 13.2.

Combining $\mathbf{Z}_2(s)$ with the impedances it is in series with gives

$$\mathbf{Z}(s) = \mathbf{Z}_2(s) + 2s + 4 \tag{13.13a}$$

$$= \frac{3(s+2) + (2s+4)(s^2 + 2s + 3)}{s^2 + 2s + 3}$$

$$= \frac{2s^3 + 8s^2 + 17s + 18}{s^2 + 2s + 3} \tag{13.13b}$$

Note that *all the equivalent forms for combining series–parallel elements are valid for impedances $\mathbf{Z}(s)$: the product-by-sum shortcut for two parallel impedances, summing admittances for parallel admittances, and so on.*

Example 13.3

Find $\mathbf{I}_C(s)$ in the circuit of Example 13.2 for an input voltage $\mathbf{V}(s) = 1/s^2$, assuming that all initial conditions are zero. Using (13.13b), we have

$$\mathbf{I}(s) = \mathbf{Y}(s)\mathbf{V}(s) = \frac{s^2 + 2s + 3}{2s^3 + 8s^2 + 17s + 18} \frac{1}{s^2}$$

By the current divider rule, current divides in proportion to admittance. The admittance of the capacitor is $sC = s/3$, and for the other arm of the current divider dividing $\mathbf{I}(s)$ it is $1/(sL + R) = 1/(s+2)$, or

$$\mathbf{I}_C(s) = \frac{s/3}{(s/3) + [1/(s+2)]}\mathbf{I}(s) = \frac{s+2}{s(2s^3 + 8s^2 + 17s + 18)}$$

In summary, the Kirchhoff laws are unchanged in s-domain analysis, and the notion of impedance, first defined for phasor circuits, carries over into the s-domain as

Chapter 13   Circuit Analysis in the s-Domain

$Z(s)$. Impedances may be combined in series and parallel by the usual rules. Other familiar workhorse theorems of circuit analysis, such as current and voltage division and Thevenin and Norton transformation, carry over into the s-domain as well. If the initial conditions are all zero, they carry over unchanged in form. If the initial conditions are not identically zero, an initial condition term must be added to the corresponding element law $V(s) = Z(s)I(s)$. In Section 13.2 we shall see how to take nonzero initial conditions into account without difficulty.

EXERCISES

13.1.1. Find $I(s)$ in Fig. 13.3 by s-domain current division. Assume zero initial conditions.
Answer

$$I(s) = \frac{s/(s^2 + 2)}{\left[s/(s^2 + 2)\right] + \frac{1}{3}}\left(-\frac{1}{s}\right) = -\frac{3}{s^2 + 3s + 2}$$

13.1.2. What is the equivalent admittance $Y(s)$ of a 10-Ω resistor, 1-F capacitor, and 5-H inductor all in series? Express as a rational function.
Answer $s/(5s^2 + 10s + 1)$

13.1.3. Find the Thevenin and Norton equivalents in the s-domain at the terminals of the inductor of Fig. 13.3. Assume zero initial conditions.
Answer

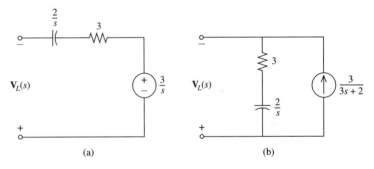

(a)   (b)

EXERCISE 13.1.3

13.2 THE s-DOMAIN CIRCUIT

Since the Kirchhoff laws remain valid in the s-domain, we can write circuit equations in the s-domain directly by inspection of the circuit diagram. Voltage drops around loops in the s-domain still sum to zero, as do s-domain currents into each node or closed region. We need only relabel the original time-domain circuit diagram to reflect the necessary changes: s replaces t in the unknown currents and voltages, and independent source functions are replaced by their s-domain transform pairs.

One additional change is necessary before the s-domain circuit diagram is complete. Let us return to the s-domain RLC element laws (13.3) to (13.5). Repeating the *v*-forms of these element laws,

$$\mathbf{V}_R(s) = R\mathbf{I_R}(s) \tag{13.14a}$$

$$\mathbf{V}_L(s) = sL\mathbf{I_L}(s) - Li_L(0-) \tag{13.14b}$$

$$\mathbf{V}_C(s) = \frac{1}{sC}\,\mathbf{I}_C(s) + \frac{1}{s}\,v_C(0-) \tag{13.14c}$$

The first of these equations, that for a resistor, is of the form $\mathbf{V}(s) = \mathbf{Z}(s)\mathbf{I}(s)$, where $\mathbf{Z}(s) = R$ is the impedance of an R-ohm resistor. The other two are of a form which suggests that the storage elements in the original time-domain circuit are each equivalent to two elements in series in the s-domain: a pure impedance and an *initial condition generator*. Circuit diagrams for these equivalents are shown in Fig. 13.5. Each labeled impedance, whether R or sL or $1/sC$, is considered a two-terminal device satisfying a pure impedance relationship $\mathbf{V}(s) = \mathbf{Z}(s)\mathbf{I}(s)$. The initial condition is accounted for separately as if a second element, the *initial condition generator*, was present.

This is the only other change between time-domain and s-domain. *The s-domain circuit is simply the original time-domain circuit with s-domain unknowns and source functions replacing their time-domain counterparts and combinations of impedances and initial condition generators replacing the RLC elements.* The equations of s-domain circuit analysis follow from inspection of this circuit diagram.

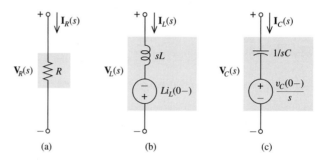

(a) (b) (c)

FIGURE 13.5 *s*-Domain equivalents of *RLC* elements.

Example 13.4

We seek the voltage across the capacitor in Fig. 13.6(a) for $t > 0$. Noting that the source jumps from 12 to 24 V at $t = 0$, the dc steady-state values at $0-$ are computed to be $i_L(0-) = 2$ A and $v_C(0-) = 12$ V. The s-domain circuit is shown in Fig. 13.6(b). Note that impedances are labeled here, together with initial condition generators $Li_L(0-) = 60$ and $v_C(0-)/s = 12/s$ for the storage elements. The node equation at the upper-right node in Fig. 13.6(b) is then

$$\frac{\mathbf{V}(s) - (12/s)}{5/s} + \frac{\mathbf{V}(s)}{6} + \frac{\mathbf{V}(s) - [60 + (24/s)]}{30s} = 0$$

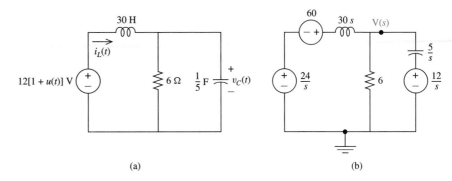

(a)   (b)

FIGURE 13.6 (a) Original circuit; (b) s-domain circuit.

$$\mathbf{V}(s) = \frac{12s^2 + 10s + 4}{s\left(s^2 + \frac{5}{6}s + \frac{1}{6}\right)}$$

Since $\mathbf{V}_c(s) = \mathbf{V}(s)$, this completes the solution in the s-domain. To recover the time-domain value, the inverse transform is computed by the usual method.

$$\mathbf{V}(s) = \frac{24}{s} + \frac{24}{s + \frac{1}{2}} - \frac{36}{s + \frac{1}{3}}$$

and for $t > 0$,

$$v_c(t) = 12\left(2 + 2e^{-t/2} - 3e^{-t/3}\right) \tag{13.15}$$

Example 13.5

Let us find the s-domain mesh currents $\mathbf{I}_1(s)$ and $\mathbf{I}_2(s)$. The original circuit is shown in Fig. 13.7(a) and the corresponding s-domain circuit in Fig. 13.7(b).

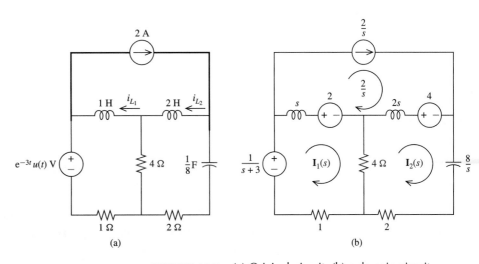

(a)   (b)

FIGURE 13.7 (a) Original circuit; (b) s-domain circuit.

To determine the equivalent circuit for each inductor, its initial current at $0-$ is needed. For $t < 0$, the voltage source in Fig. 13.7(a) is zero and the inductors act as short circuits, so $i_{L1}(0-) = i_{L2}(0-) = 2$ A. Thus the initial condition generator for L_1 is $L_1 i_{L1}(0-) = (1)(2) = 2$ and for L_2 is $L_2 i_{L2}(0-) = (2)(2) = 4$. As shown in Fig. 13.5, these initial condition generators are in series with their respective inductive impedances $\mathbf{Z}_1(s) = sL_1 = s$ and $\mathbf{Z}_2(s) = sL_2 = 2s$. The capacitive voltage at $0-$, again examining Fig. 13.7(a), is 0 V. No initial condition generator is needed for this element. Noting that the mesh current in the upper mesh is $2/s$, the mesh equations are

$$\begin{bmatrix} s+5 & -4 \\ -4 & 2s+6+\dfrac{8}{s} \end{bmatrix} \begin{bmatrix} \mathbf{I}_1(s) \\ \mathbf{I}_2(s) \end{bmatrix} = \begin{bmatrix} \dfrac{1}{s+3} \\ 0 \end{bmatrix}$$

Solving by matrix inversion gives

$$\begin{bmatrix} \mathbf{I}_1(s) \\ \mathbf{I}_2(s) \end{bmatrix} = \frac{1}{\Delta(s)} \begin{bmatrix} 2s+6+\dfrac{8}{s} & 4 \\ 4 & s+5 \end{bmatrix} \begin{bmatrix} \dfrac{1}{s+3} \\ 0 \end{bmatrix}$$

where $\Delta(s)$ is the determinant

$$\Delta(s) = (s+5)\left(2s+6+\frac{8}{s}\right) - (-4)(-4)$$

Carrying out the algebra, the vector of mesh currents is

$$\begin{bmatrix} \mathbf{I}_1(s) \\ \mathbf{I}_2(s) \end{bmatrix} = \frac{1}{(s+3)(s^3+8s^2+11s+20)} \begin{bmatrix} s^2+3s+4 \\ 2s \end{bmatrix} \quad (13.16)$$

which completes the required calculation. Note that each mesh current has the same four poles but different zeros, so they will contain the same four time functions but with different partial fraction coefficients. One pole is due to the independent source. The other three are due to the three storage elements in the circuit.

Thevenin–Norton transformation in the s-domain works exactly as described in the time domain in Chapter 2, with $\mathbf{Z}(s)$ replacing resistance R. Consider, for instance, the networks shown in Fig. 13.5. Since each is a series combination of impedance and voltage source, we will call these the *Thevenin forms* of the s-domain equivalents (the Thevenin equivalent voltage source in the resistive case is defined to be 0). If we perform Thevenin–Norton transformation on these circuits, the equivalent Norton forms have current sources equal to the ratio of the Thevenin equivalent voltage divided by the impedance and are in parallel with the same impedances. The resulting *Norton forms* of the s-domain equivalent circuits are shown in Fig. 13.8. As with all equivalents, either may be freely used. Note that each Thevenin-form s-domain equivalent circuit with a nonzero initial condition generator adds a node to the circuit, whereas each Norton form adds a mesh. The Thevenin form is often preferred when there are other impedances in series, and the Norton form when other impedances are in parallel.

Finally, it is always helpful to combine series impedances before writing nodal analysis equations in the s-domain and parallel impedances if mesh analysis is planned.

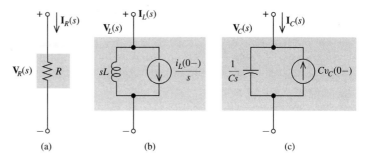

(a)　　　　　　(b)　　　　　　(c)

FIGURE 13.8 *s*-Domain equivalents of *RLC* elements: alternative forms.

The number of equations and unknowns is reduced and our ability to recover all currents and voltages once the circuit is broken is unimpaired.

EXERCISES

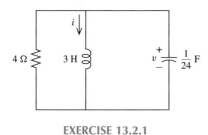

EXERCISE 13.2.1

13.2.1. Find $v(t)$, $t > 0$ for $i(0-) = 1$ A and $v(0-) = 4$ V.
Answer $20e^{-4t} - 16e^{-2t}$

13.2.2. Write the mesh equations for $\mathbf{I}(s)$ and solve for $i(t)$, $t > 0$.
Answer $i(t) = (1 - t)e^{-t}$

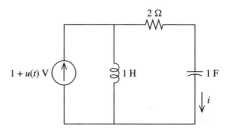

EXERCISE 13.2.2

13.2.3. Using series–parallel equivalents in the *s*-domain for both impedances and sources, reduce the given circuit to one containing only two unknown node voltages $\mathbf{V}_1(s)$ and $\mathbf{V}_2(s)$, and write the node equations in vector matrix form.
Answer

$$
\begin{bmatrix}
\dfrac{1}{2s+6} + s + \dfrac{4s}{4s+1} + \dfrac{1}{3} & -\left(s + \dfrac{1}{2s+6}\right) \\[3mm]
-\left(s + \dfrac{1}{2s+6}\right) & \dfrac{1}{2s+6} + s + \dfrac{1}{s} + \dfrac{1}{3s+2}
\end{bmatrix}
\begin{bmatrix} \mathbf{V}_1(s) \\ \mathbf{V}_2(s) \end{bmatrix}
$$

$$
=
\begin{bmatrix}
1 - \left(8 + \dfrac{5}{s}\right)\dfrac{1}{2s+6} + \left(\dfrac{2}{s}\right)\dfrac{4s}{4s+1} \\[3mm]
-1 + \left(8 + \dfrac{5}{s}\right)\dfrac{1}{2s+6} - \dfrac{1}{s} + \dfrac{3}{3s+2}
\end{bmatrix}
$$

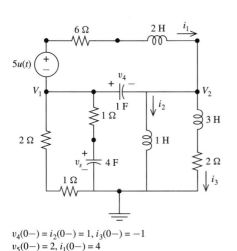

$v_4(0-) = i_2(0-) = 1$, $i_3(0-) = -1$
$v_5(0-) = 2$, $i_1(0-) = 4$

EXERCISE 13.2.3

13.3 TRANSFER FUNCTIONS

Often our goal in analyzing a circuit is to determine its complete behavior: the values of all its currents and voltages in response to all its sources and initial stored energy. Most of our work has been oriented toward this completely general goal. The two systematic circuit analysis approaches emphasized in this book, mesh and nodal analysis, each solve for a set of key variables (the mesh currents and node voltages, respectively), from which all remaining circuit variables can be easily derived. Even if in a given problem only a subset of all circuit variables is of interest, our strategy has been to first "break the circuit." Once the circuit is broken, in other words once the key variables have been found, every other circuit variable of interest becomes easy to compute.

Sometimes, however, our interest in studying a given circuit is much more narrowly defined than that of finding all responses to all sources. We may wish to understand the relationship between one particular source or other circuit variable we will call the *input* and another circuit variable, the *output* current or voltage that input produces. The goal of *input–output analysis* is to understand their relationship, in other words to be able to predict the output for each possible value of the input. If only a single input–output pair is of interest, it is unlikely that methods oriented to determining all currents and all voltages in response to all sources and initial conditions will be very efficient when applied to this narrower purpose.

Analysis in the s-domain has already shown its value in speeding calculation of the complete behavior of a circuit. It turns out that the s-domain approach is also highly suited to input–output analysis, as we consider next. For a circuit with input $\mathbf{V}_i(s)$ and output $\mathbf{V}_o(s)$, *define the transfer function $\mathbf{H}(s)$ as the s-domain ratio of the output to the input when all initial conditions are zero*:

$$\mathbf{H}(s) = \frac{\mathbf{V}_o(s)}{\mathbf{V}_i(s)} \qquad \text{(all ICs} = 0) \qquad\qquad (13.17)$$

Here $\mathbf{V}_i(s)$ and $\mathbf{V}_o(s)$ may be both currents, both voltages, or one of each. We require that all initial conditions be set to zero to isolate the selected input as the cause of the output we are computing.

Rewriting (13.17), we see the significance of this definition:

$$\mathbf{V}_o(s) = \mathbf{H}(s)\mathbf{V}_i(s) \qquad \text{(all ICs} = 0) \qquad\qquad (13.18)$$

The value of the s-domain output for any input is just the product of the transfer function and the s-domain input. Thus, once $\mathbf{H}(s)$ is known, we may predict the output $\mathbf{V}_o(s)$ for any input $\mathbf{V}_i(s)$ by simple multiplication of two known s-domain quantities. If the time-domain output $v_o(t)$ is required, the additional step of inverse transformation will be required.

To determine $\mathbf{H}(s)$, all initial conditions are set to zero and the s-domain circuit drawn. A consequence of the zero initial condition requirement is that there will never be initial-condition generators in this circuit. The value of the output selected is then computed. Division by the input, as in (13.17), yields the transfer function $\mathbf{H}(s)$.

Example 13.6

As a first example we compute the transfer function $\mathbf{H}(s)$ for the input $v_s(t)$ and output $v_1(t)$. The time-domain and s-domain circuits are shown in Fig. 13.9. Note the absence of initial condition generators in the s-domain circuit, a consequence of setting all initial conditions to zero.

We proceed by computing the specified output $\mathbf{V}_1(s)$. The equivalent impedance of the three rightmost elements is

$$\mathbf{Z}_1(s) = \frac{(s)(2s+4)}{s+(2s+4)} = \frac{2s^2+4s}{3s+4}$$

Then, by voltage division,

$$\mathbf{V}_1(s) = \frac{4/s}{(4/s)+\mathbf{Z}_1(s)}\mathbf{V}_s(s)$$

$$= \frac{6s+8}{s^3+2s^2+6s+8}\mathbf{V}_s(s)$$

(13.19)

Having solved for the output in terms of the input, $\mathbf{H}(s)$ is found by division of both sides by the input $\mathbf{V}_s(s)$.

$$\mathbf{H}(s) = \frac{\mathbf{V}_1(s)}{\mathbf{V}_s(s)} = \frac{6s+8}{s^3+2s^2+6s+8}$$

(13.20)

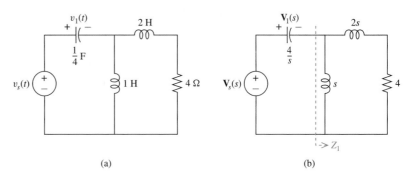

(a)   (b)

FIGURE 13.9   (a) Circuit for Example 13.6; (b) corresponding s-domain circuit.

A consequence of the linear relationship between input and output seen in (13.18) is that if the input $\mathbf{V}_i(s) = 0$, the output must be zero also. Since this would not be the case if there were other independent sources contributing to the output $\mathbf{V}_o(s)$, they must be killed before computing its transfer function $\mathbf{H}(s)$ relative to the specified input. *The transfer function measures the output due to a specified input, with all other sources and initial conditions set to zero.*

Example 13.7

Find the transfer function $\mathbf{H}_1(s) = \mathbf{V}_2(s)/\mathbf{I}_{s1}(s)$ for the circuit in Fig. 13.10. The nodal equations with $V_{s2}(s)$ killed are, in vector-matrix form,

$$\begin{bmatrix} s+2 & -2 \\ -2 & 2s+3 \end{bmatrix}\begin{bmatrix} \mathbf{V}_1(s) \\ \mathbf{V}_2(s) \end{bmatrix} = \begin{bmatrix} -\mathbf{I}_{s1}(s) \\ 0 \end{bmatrix}$$

Then, using Cramer's rule to recover $\mathbf{V}_2(s)$ gives

$$\mathbf{V}_2(s) = \frac{\begin{vmatrix} s+2 & -\mathbf{I}_{s1}(s) \\ -2 & 0 \end{vmatrix}}{\begin{vmatrix} s+2 & -2 \\ -2 & 2s+3 \end{vmatrix}} = -\frac{-2}{2s^2 + 7s + 2}\mathbf{I}_{s1}(s)$$

Note that the linearity of the last equation in $\mathbf{I}_{s1}(s)$ verifies that if the input $\mathbf{I}_{s1}(s)$ were zero, the output would be zero. We are computing the component of the output due to the specified source. Dividing the output by the input $\mathbf{I}_{s1}(s)$, we have the desired transfer function:

$$\mathbf{H}_1(s) = \frac{\mathbf{V}_2(s)}{\mathbf{I}_{s1}(s)} = \frac{-2}{2s^2 + 7s + 2} \tag{13.21}$$

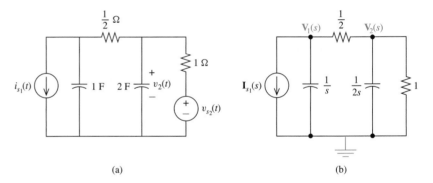

(a) (b)

FIGURE 13.10 (a) Circuit for Example 13.7; (b) s-domain circuit with $\mathbf{V}_{s2}(s)$ killed.

In determining a transfer function in a circuit containing more than one independent source, the other sources must be killed before the output is computed. Let $\mathbf{V}_{i1}(s)$, $\mathbf{V}_{i2}(s)$, $\ldots$, $\mathbf{V}_{in}(s)$ be the n separate independent sources in a circuit, and $\mathbf{V}_o(s)$ be the designated output variable. Then, with all sources $j \neq k$ killed,

$$\mathbf{V}_{ok}(s) = \mathbf{H}_k(s)\mathbf{V}_{ik}(s) \qquad \text{[all ICs} = 0, \text{ all other sources} \mathbf{V}_{ij}(s) = 0] \tag{13.22}$$

where $\mathbf{H}_k(s)$ is the transfer function between output $\mathbf{V}_o(s)$ and input $\mathbf{V}_{ik}(s)$, and $\mathbf{V}_{ok}(s)$ is the value of the output produced under these conditions.

Now suppose that we are computing the total output $\mathbf{V}_o(s)$ due to all sources $\mathbf{V}_{i1}(s)$, $\mathbf{V}_{i2}(s)$, $\ldots$, $\mathbf{V}_{in}(s)$, keeping the initial conditions at zero. Using superposition, the component due to source $\mathbf{V}_{ik}(s)$ would be $\mathbf{V}_{ok}(s)$, the value of the output computed under precisely the same conditions of (13.22). Thus the total output would be the superposition

$$\mathbf{V}_o(s) = \mathbf{V}_{o1}(s) + \mathbf{V}_{o2}(s) + \cdots + \mathbf{V}_{on}(s) \tag{13.23a}$$

or

$$\mathbf{V}_o(s) = \mathbf{H}_1(s)\mathbf{V}_1(s) + \mathbf{H}_2(s)\mathbf{V}_2(s) + \cdots + \mathbf{H}_n(s)\mathbf{V}_n(s) \qquad (13.23b)$$

In a circuit with multiple inputs, the forced s-domain output due to all sources is the superposition of the separate transfer functions for each input times the value of the corresponding s-domain inputs.

Example 13.8

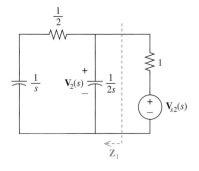

FIGURE 13.11 s-Domain circuit for computing $\mathbf{H}_2(s)$ in Example 13.8.

Continuing Example 13.7, we wish to determine the transfer function $\mathbf{H}_2(s) = \mathbf{V}_2(s)/\mathbf{V}_{s2}(s)$ due to the input $\mathbf{V}_{s2}(s)$, and the total output $\mathbf{V}_2(s)$ due to all inputs. To compute $\mathbf{H}_2(s)$ of Fig. 13.11, we kill the other source $\mathbf{I}_{s1}(s)$ and determine the component of the output due solely to the input $\mathbf{V}_{s2}(s)$.

The impedance of the leftmost three elements is found by series–parallel to be

$$\mathbf{Z}_1(s) = \frac{s+2}{2s^2 + 6s}$$

Using $\mathbf{Z}_1(s)$ and the 1-Ω resistor as the arms of a voltage divider yields

$$\mathbf{V}_2(s) = \frac{\mathbf{Z}_1(s)}{1 + \mathbf{Z}_1(s)} \mathbf{V}_{s2}(s)$$

or

$$\mathbf{H}_2(s) = \frac{\mathbf{V}_2(s)}{\mathbf{V}_{s2}(s)} = \frac{(s+2)/(2s^2+6s)}{1 + \left[(s+2)/(2s^2+6s)\right]}$$

$$= \frac{s+2}{2s^2 + 7s + 2} \qquad (13.24)$$

The output is then the superposition of the outputs due to the two separate inputs

$$\mathbf{V}_2(s) = \mathbf{H}_1(s)\mathbf{I}_{s1}(s) + \mathbf{H}_2(s)\mathbf{V}_{s2}(s) \qquad (13.25a)$$

$$\mathbf{V}_2(s) = \frac{-2}{2s^2 + 7s + 2}\mathbf{I}_{s1}(s) + \frac{s+2}{2s^2 + 7s + 2}\mathbf{V}_{s2}(s) \quad (13.25b)$$

From this input–output equation (13.25b), we may now deduce the output for any input conditions directly, without computing any unneeded circuit variables along the way. For instance, if the inputs are set as $i_{s1}(t) = u(t)$ and $v_{s2}(t) = e^{-t}u(t)$, then

$$\mathbf{V}_2(s) = \frac{-2}{2s^2 + 7s + 2}\left(\frac{1}{s}\right) + \frac{s+2}{2s^2 + 7s + 2}\left(\frac{1}{s+1}\right)$$

and $v_2(t)$ would be computed by partial fractions. (These calculations assume zero initial conditions.)

Once $\mathbf{H}(s)$ is known, it is easy to vary the input and determine the effect on the output. The transfer function $\mathbf{H}(s)$ will prove central to input–output analysis, that is, analysis focusing on the cause–effect relations between chosen inputs and outputs. We will exploit this capability, for instance, in studying frequency response in Chapter 14, the dependence of the output on the frequency of the input.

Although useful, transfer functions do not tell the entire story. Transfer functions cannot tell us about the full range of a circuit's potential behaviors, since they are defined with initial conditions zero. For nonzero initial conditions, the *natural response*, that is, the output when all independent sources are killed, must be superposed with the output due to the inputs or *forced response* computed using $\mathbf{H}(s)$ if the total output is desired. The natural response may be computed from the s-domain circuit with initial condition generators present and all other independent sources killed.

EXERCISES

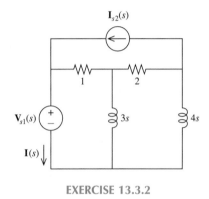

EXERCISE 13.3.2

13.3.1. Denote the source in Fig. 13.3 $i_s(t)$ and find $\mathbf{H}(s)$ for input $\mathbf{I}_s(s)$ and output $\mathbf{V}_L(s)$.
Answer $s^2/s^2 + 3s + 2$

13.3.2. Find the transfer functions between output $\mathbf{I}(s)$ and inputs $\mathbf{V}_{s1}(s)$, $\mathbf{I}_{s2}(s)$.
Answer $-(7s+2)/(12s^2+13s+2)$, $(13s+2)/(12s^2+13s+2)$

13.3.3. Find $\mathbf{H}(s)$ with input $\mathbf{V}_R(s)$ and output $\mathbf{V}_L(s)$. [*Hint*: Find both in terms of $\mathbf{V}_s(s)$.]
Answer $-Ls/R$

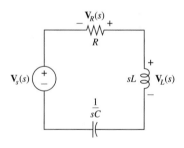

EXERCISE 13.3.3

13.4 POLES AND STABILITY

In Chapter 12 we saw that the poles of a transform $\mathbf{F}(s)$ determine, up to a multiplying constant, the terms in the partial fraction expansion and therefore the time functions that make up its inverse transform $f(t)$. A simple pole at $s = p$ gives rise to a term Ke^{pt} in $f(t)$, while repeated poles at $s = p$ create t-multiplied forms of the same exponential time function.

The dependence of the term Ke^{pt} in $f(t)$ on the location of the pole p in its transform $\mathbf{F}(s)$ is illustrated in Fig. 13.12. Let $p = \sigma + j\omega$, so $e^{pt} = e^{\sigma t}e^{j\omega t}$. For $\omega = 0$, we have a purely real pole located on the horizontal axis in the figure. The corresponding

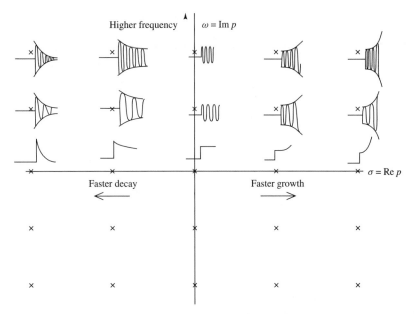

FIGURE 13.12 Dependence of the term e^{pt} in $f(t)$ on the location of the pole $p = \sigma + j\omega$ in the complex plane.

term $e^{\sigma t}$ decays to zero for $\sigma < 0$, is constant for $\sigma = 0$, and grows without limit for $\sigma > 0$. For $\omega \neq 0$, the pole is complex, and the corresponding exponential $e^{pt} = e^{\sigma t} e^{j\omega t}$ is complex. In this case, Fig. 13.12 shows the term e^{pt}, combined with the term e^{p^*t}, which is due to the presence of its complex conjugate pole $p^* = \sigma - j\omega$. Since

$$e^{pt} + e^{p^*t} = e^{\sigma t} e^{j\omega t} + e^{\sigma t} e^{-j\omega t} = 2e^{\sigma t} \cos \omega t \qquad (13.26)$$

we see that *the rate of growth or decay in each term of f(t) is governed by the real part of its pole. If Re (p)* $= \sigma < 0$, *the corresponding term in f(t) decays to zero.* This conclusion also holds for repeated poles, since t-multiplied forms of $e^{\sigma t}$ eventually decay to zero with time as long as $e^{\sigma t}$ itself does. Note in Fig. 13.12 that only the upper half of the complex plane responses are sketched, since each response sketched at the complex-pole location p actually combines terms due to a complex conjugate pair of poles p and p^*. The responses not sketched in the lower half of Fig. 13.12 are identical to those shown at the complex conjugate locations in the upper half of the complex plane.

We are particularly interested in the location of the poles of the currents and voltages in the natural response. For any linear circuit, we may kill all independent sources and write a complete set of network analysis equations, either mesh or nodal analysis, in the s-domain. To be specific, let us assume that mesh analysis has been selected. The equations will be of the form

$$\mathbf{M}(s)\mathbf{I}(s) = \mathbf{V}_{IC}(s) \qquad (13.27)$$

where $\mathbf{I}(s)$ is the vector of unknown mesh currents, $\mathbf{V}_{IC}(s)$ the vector of initial condition generator source terms, and $\mathbf{M}(s)$ the square connection matrix. *The values of s for which the determinant of M (s) equals zero are called the poles of the circuit.*

Example 13.9

Consider the circuit shown in Fig. 13.13(a) transformed to the s-domain in Fig. 13.13(b). The mesh equations with $\mathbf{V}_o(s) = 0$ are

$$\begin{bmatrix} 2s + 4 & -4 \\ -4 & 4 + \dfrac{8}{s} \end{bmatrix} \begin{bmatrix} I_1(s) \\ I_2(s) \end{bmatrix} = \begin{bmatrix} 2i_1(0-) \\ \dfrac{-v_c(0-)}{s} \end{bmatrix}$$

Then the determinant $\delta(s)$ is

$$\Delta(s) = (2s + 4)\left(4 + \dfrac{8}{s}\right) - 16 = \dfrac{8(s^2 + 2s + 4)}{s} \qquad (13.28)$$

The poles of the circuit are the zeros of $\Delta(s)$, which are the quadratic roots p_1, $p_2 = -1 \pm j\sqrt{3}$. Solving (13.28) yields

$$\begin{bmatrix} \mathbf{I}_1(s) \\ \mathbf{I}_2(s) \end{bmatrix} = \dfrac{1}{\Delta(s)} \begin{bmatrix} 4 + \dfrac{8}{s} & 4 \\ 4 & 2s + 4 \end{bmatrix} \begin{bmatrix} 2i_1(0-) \\ \dfrac{-v_c(0-)}{s} \end{bmatrix} \qquad (13.29)$$

$$= \dfrac{1}{s^2 + 2s + 4} \begin{bmatrix} i_1(0-)s + 2i_1(0-) - \frac{1}{2}v_c(0-) \\ (i_1(0-) - \frac{1}{4}v_c(0-))s - \frac{1}{2}v_c(0-) \end{bmatrix}$$

From (13.29) we see that the current through each element in the circuit has as its poles the poles of the circuit p_1 and p_2 determined above. The same is also true of each voltage in the circuit, as can easily be verified by substituting the mesh currents above into the element laws for each element in the circuit.

Performing nodal analysis on this same circuit, the only unknown node voltage is $\mathbf{V}(s)$, and summing currents at this node gives

$$\dfrac{s}{8}\left[\mathbf{V}(s) - \dfrac{v_c(0-)}{s}\right] + \dfrac{1}{4}\mathbf{V}(s) + \dfrac{1}{2}s[\mathbf{V}(s) - 2i_1(0-)] = 0$$

$$V(s) = \dfrac{sv_c(0-) + 8i_1(0-)}{s^2 + 2s + 4}$$

Mesh and nodal analysis arrive at the same set of poles for the circuit, as they must.

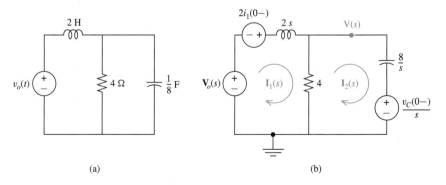

(a) (b)

FIGURE 13.13 (a) Circuit for Example 13.9; (b) in the s-domain.

The poles of a system dictate the fate of its natural response. As shown in Fig. 13.12, each pole whose real part is negative will lead to natural response terms that decay to zero. Those with zero real parts will neither grow nor decay, while poles with positive real parts generate terms that grow without bounds.

We define a circuit to be *stable* if *all natural responses in the circuit decay to zero with time*. Since only circuit poles with negative real parts produce natural responses that decay to zero, *the stability of a circuit is equivalent to* Re(p) < 0 *for every circuit pole p*. Stated another way, *stability requires that all the poles of the circuit lie in the left half of the complex plane* as shown in Fig. 13.12. The circuit of Example 13.9 is stable, since the complex conjugate system poles have real part −1. The next example illustrates a circuit that for different ranges of a circuit parameter may be either stable or unstable.

Example 13.10

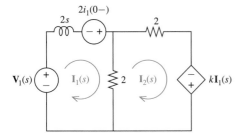

FIGURE 13.14 Circuit for Example 13.10.

Consider the s-domain circuit of Fig. 13.14, which contains a current-controlled voltage source of gain k ohms and a single inductor producing the initial condition generator shown. Determine if this circuit is stable.

The mesh equations are

$$\begin{bmatrix} 2s+2 & -2 \\ -2-k & 3 \end{bmatrix} \begin{bmatrix} \mathbf{I}_1(s) \\ \mathbf{I}_2(s) \end{bmatrix} = \begin{bmatrix} \mathbf{V}_1(s) + 2i_1(0-) \\ 0 \end{bmatrix} \qquad (13.30a)$$

The determinant is

$$\Delta(s) = (2s+2)(3) - 2(2+k) = 6s + 2(1-k) \qquad (13.30b)$$

So there is a single circuit pole located at $p = \frac{1}{3}(k-1)$. For $k < 1$, this pole has negative real part. We conclude that for $k < 1$ this circuit is stable; for $k \geq 1$, it is not stable.

Stability has several implications with respect to the behavior of a circuit. Here we point out two that are of immediate importance: the relationship between stability and steady-state and that between stability and transfer function analysis. We earlier defined steady state as the condition in which all transients had decayed to zero and the overall response of the circuit was equal to its forced response. But what if the transients in a circuit never tend to zero? The circuit will never attain steady state, and steady-state analysis, including phasor analysis of ac steady state, will not determine the long-term values of currents and voltages. *Steady-state analysis is applicable only to stable circuits.* The long-term values of currents and voltages in unstable circuits depend strongly on their initial conditions. There is no steady state for such circuits.

Example 13.11

Determine the ac steady-state value of $i_1(t)$ in the circuit shown in Fig. 13.14 for the source $v_1(t) = 40\cos 2t$ V. Applying Cramer's rule to solve (13.30a) gives

$$\mathbf{I}_1(s) = \frac{\begin{vmatrix} \mathbf{V}_1(s) + 2i_1(0-) & -2 \\ 0 & 3 \end{vmatrix}}{\begin{vmatrix} 2s+2 & -2 \\ -2-k & 3 \end{vmatrix}} = \frac{i_1(0-)}{s + \frac{1}{3}(1-k)} + \frac{\frac{1}{2}\mathbf{V}_1(s)}{s + \frac{1}{3}(1-k)}$$

$$(13.31)$$

The single pole of the circuit is at $p = \frac{1}{3}(k-1)$. If $k \geq 1$, the pole is not in the left half-plane and the circuit is unstable. For instance, with $k = 4$ the natural response [first term in (13.31)] is $i_1(0-)e^{+t}$, which certainly does not decay to zero as t gets large. For $k \geq 1$, the circuit never "settles down" to a steady-state value; steady state does not exist in this circuit for $k \geq 1$. For $k < 1$, the natural response does go to zero, regardless of the initial condition, and we will have steady state. Using the source $V_4(s) = 40s/(s_2 + 4)$, the partial fraction expansion for the forced response [second term in (13.31)] is

$$\frac{20s}{\left[s + \frac{1}{3}(1-k)\right](s^2 + 4)} = \frac{A}{s + \frac{1}{3}(1-k)} + \frac{B}{s - j2} + \frac{B^*}{s + j2}$$

For $k < 1$, the first term on the right will decay to zero with time, so since only steady state is required, we need not compute A.

$$B = \frac{20(j2)}{\left[j2 + \frac{1}{3}(1-k)\right](j4)} = \frac{10}{\frac{1}{3}(1-k) + j2}$$

and the required steady-state value for any $k < 1$ is

$$i_1(t) = 2|B|\cos(2t + \theta)$$

where

$$|B| = \frac{10}{\sqrt{\frac{1}{9}(1-k)^2 + 4}}, \qquad \theta = -\arctan\frac{6}{1-k}$$

The inapplicability of steady-state analysis for circuits that are not stable suggests that we ought to investigate stability before launching a steady-state calculation. This can be done by determining if all the poles of the circuit have negative real parts as described above. In practice, circuits lacking controlled sources are almost always stable. Currents and voltages due to initial conditions can build indefinitely only if there is an element supplying power to the rest of the circuit, which cannot be an RLC element (which we know never supplies average power). Since independent sources are killed in determining the natural response, the only remaining candidate is a controlled source. The only circuits lacking controlled sources that may not be stable are those containing isolated resistance-free LC loops, which correspond to circuit poles exactly on the $j\omega$ axis [$\text{Re}(p) = 0$].

Finally, consider the implications of stability in the use of transfer functions. The transfer function $H(s)$ permits calculation of an output $V_o(s)$ for a given input $V_i(s)$ by the relationship

$$V_o(s) = H(s)V_i(s) \qquad (\text{all ICs} = 0)$$

Suppose, however, that the initial conditions in the circuit are not exactly equal to zero.* To what extent is this relation still applicable? If the circuit is stable, the natural response will decay to zero. The forced response determined by the equation above will approach the actual total response to arbitrary accuracy as $t \to \infty$. On the other hand, if the circuit

---

*Which will always be the case for physical circuits, if due only to unavoidable thermal noise voltages and currents.

is unstable, the total response will differ from the forced response determined by transfer function analysis by a term (the natural response) that may well be going to infinity.

We conclude that the property of stability has important bearing on the usefulness of both steady-state and transfer function analysis. In circuits that are not stable, steady-state analysis does not apply, and transfer function analysis, being limited to determining the forced part of a response, may yield results quite different from the actual total response. Stability may be conveniently determined in the s-domain by examining the real parts of the circuit poles, which must all be negative for the circuit to be stable. The poles of the circuit are the zeros of the determinant of the nodal or mesh analysis matrix.

EXERCISES

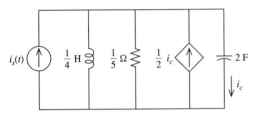

EXERCISE 13.4.2

13.4.1. Show that a simple two-element parallel LC circuit is not stable. *Answer* Circuit poles at $\pm j\sqrt{1/LC}$ do not have negative real parts.

13.4.2. Determine the poles of the circuit. Is this system stable? *Answer* $-1, -4$; yes

13.4.3. What is the largest value of k for which this circuit is not stable? *Answer* -4

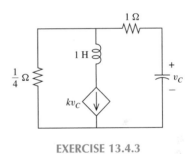

EXERCISE 13.4.3

13.5 INITIAL AND FINAL VALUE THEOREMS

We next develop two useful theorems that permit us to deduce the initial value of a time function, $f(0+)$, and its final value, $f(\infty)$, from its transform $\mathbf{F}(s)$. Consider any $f(t)$ that has no impulses or higher-order singularities at $t = 0$. Write $f(t)$ in the form $f(t) = g(t) + du(t)$, where $g(t)$ is continuous at $t = 0$, and $d = f(0+) - f(0-)$ is the size of the jump discontinuity in $f(t)$ at $t = 0$. Differentiating gives

$$\frac{df}{dt} = \frac{dg}{dt} + d\delta(t)$$

By the derivative rule applied to this equation, we obtain

$$s\mathbf{F}(s) - f(0-) = \int_{0-}^{\infty} \frac{df}{dt} e^{-st}\, dt = \int_{0-}^{\infty} \frac{dg}{dt} e^{-st}\, dt + \int_{0-}^{\infty} d\delta(t) e^{-st}\, dt \qquad (13.32)$$

Using the sifting theorem, the last integral on the right is just d, regardless of the value of s. As $s \to \infty$, the integral summing with it must go to zero, since the integrand is becoming arbitrarily small at each t. Thus, passing to the limit as $s \to \infty$, we obtain

$$\lim_{s \to \infty} s\mathbf{F}(s) - f(0-) = d = f(0+) - f(0-)$$

Canceling common terms on both sides, we have the following theorem.

INITIAL VALUE THEOREM

For any $f(t)$ with no impulses or higher-order singularities at $t = 0$,

$$\lim_{s \to \infty} s\mathbf{F}(s) = f(0+)$$

The condition of no impulses or higher-order singularities is satisfied, for instance, by any proper rational function $\mathbf{F}(s)$.

Next consider (13.32) as $s \to 0$. The last integral is still d, and the integral summing with it is

$$\lim_{s \to 0} \int_{0-}^{\infty} \frac{dg}{dt} e^{-st} \, dt = \int_{0-}^{\infty} \frac{dg}{dt} \, dt = g(\infty) - g(0-)$$

The last equality requires that $g(\infty)$ exist for the given function. Then if $g(\infty)$ exists,

$$\lim_{s \to 0} s\mathbf{F}(s) - f(0-) = g(\infty) - g(0-) + d \qquad (13.33)$$

Since $f(t) = g(t) + du(t)$, then $g(\infty) = f(\infty) - du(\infty) = f(\infty) - d$. Substituting into (13.33) and noting that $f(0-) = g(0-)$, we have the following theorem.

FINAL VALUE THEOREM

For any $f(t)$ such that the limit $f(\infty)$ exists,

$$\lim_{s \to 0} s\mathbf{F}(s) = f(\infty)$$

Example 13.12

$\mathbf{I}(s) = (7s^3 + 2)/(s^4 + 3s^3 + 5s^2 + 2s + 2)$. Find the initial value $i(0+)$ and the final value $i(\infty)$. As $s \to \infty$ the highest power terms dominate both numerator and denominator, while the constant terms dominate as $s \to 0$. Thus

$$\lim_{s \to \infty} s\mathbf{I}(s) = i(0+) = 7, \qquad \lim_{s \to 0} s\mathbf{I}(s) = i(\infty) = 0$$

A principal benefit of these two theorems is that we do not have to compute an inverse transform in order to discover its initial and final values. In the foregoing example, for instance, this would require factoring a quartic to discover the poles and then computing the partial fraction expansion numerators. These difficulties are bypassed if only initial and final values of the inverse transform are needed.

13.5.1. Compute $L^{-1}[(-40s+20)/(s^3+6s^2+5s)]$ and check by applying the initial and final value theorems.

Answer $4-15e^{-t}+11e^{-5t}$; IVT: $f(0+) = 0$; FVT: $f(\infty) = 4$

13.5.2. If $i_s(t) = u(t)$ and the initial conditions at 0− are zero in the circuit of Exercise 13.4.2, find $i_c(0+)$ and $i_c(\infty)$.

Answer 2 A; 0 A

13.5.3. A circuit has two independent sources and, for the output variable $v_o(t)$, transfer functions

$$\mathbf{H}_1(s) = \frac{5s-3}{s^2+2s}, \qquad \mathbf{H}_2(s) = \frac{s^2+1}{s^3+1}$$

relative to these sources. Find $v_o(0+)$ and $v_o(\infty)$ if $\delta(t)$ is applied at each input and all initial conditions at 0− are zero.

Answer $+6$; $\frac{-3}{2}$

13.6   IMPULSE RESPONSE AND CONVOLUTION

The transfer function $\mathbf{H}(s)$ permits calculation of an output $\mathbf{V}_o(s)$ for a given input $\mathbf{V}_i(s)$ by the relationship

$$\mathbf{V}_o(s) = \mathbf{H}(s)\mathbf{V}_i(s) \qquad \text{(all ICs = 0)} \tag{13.34}$$

If we select the input $v_i(t)$ to be the unit impulse function,

$$v_i(t) = \delta(t) \tag{13.35}$$

then since the transform of $\delta(t)$ is the constant 1, the response under these conditions will be simply

$$\mathbf{V}_o(s) = \mathbf{H}(s)(1) = \mathbf{H}(s) \tag{13.36a}$$

or in the time domain

$$v_o(t) = L^{-1}[\mathbf{H}(s)] = h(t) \tag{13.36b}$$

The time function $h(t)$, being the output caused by a unit impulse input when all initial conditions equal zero, is reasonably enough called the *unit impulse response* or sometimes just the *impulse response*. By (13.36b) we see that *the unit impulse response $h(t)$ and the transfer function $\mathbf{H}(s)$ are a Laplace transform pair.* Since all practical circuits are *causal*, that is, they cannot begin responding to an input until that input itself begins, we will require that the impulse response be zero in negative time; $h(t) = 0$ for $t < 0$.

Example 13.13

Find the transfer function $\mathbf{H}(s)$ and unit impulse response $h(t)$ for the input $v_{s_1}(t)$ and output $v_L(t)$ for the circuit of Fig. 13.15.

The s-domain circuit is shown, prepared for computing $\mathbf{H}(s)$ by killing the other independent source and setting initial condition generators to zero. The equations for nodal analysis in this circuit

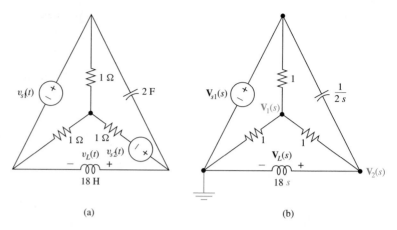

(a)   (b)

FIGURE 13.15 (a) Circuit for Example 13.13; (b) s-domain circuit with the other source killed.

are

$$\mathbf{V}_1(s) + [\mathbf{V}_1(s) - \mathbf{V}_{s1}(s)] + [\mathbf{V}_1(s) - \mathbf{V}_2(s)] = 0 \qquad (13.37a)$$

$$\tfrac{1}{18s}\mathbf{V}_2(s) + [\mathbf{V}_2(s) - \mathbf{V}_1(s)] + 2s[\mathbf{V}_2(s) - \mathbf{V}_{s1}(s)] = 0 \qquad (13.37b)$$

Multiplying (13.37b) by 3 and adding to (13.37a) eliminates $\mathbf{V}_1(s)$:

$$\left(\frac{1}{6s} + 3 + 6s - 1\right)\mathbf{V}_2(s) = (6s + 1)\mathbf{V}_{s1}(s)$$

$$\mathbf{H}(s) = \frac{\mathbf{V}_2(s)}{\mathbf{V}_{s1}(s)} = \frac{s}{s + \frac{1}{6}}$$

To determine the impulse response $h(t)$ we need the inverse transform of the transfer function $\mathbf{H}(s)$. After long division,

$$\mathbf{H}(s) = 1 - \frac{\frac{1}{6}}{s + \frac{1}{6}} \qquad (13.38)$$

$$h(t) = \delta(t) - \tfrac{1}{6}e^{-t/6}u(t)$$

Note that we have multiplied by $u(t)$ since we require $h(t) = 0$ for $t < 0$.

Example 13.14

When the input $v_i(t) = 3\delta(t - 5)$ is applied to a circuit with zero initial conditions and all other independent sources killed, the output is measured to be

$$v_o(t) = e^{-4t}\cos[7(t - 9)]u(t - 9)$$

Determine the transfer function $\mathbf{H}(s)$ and impulse response $h(t)$. Transforming the input $v_i(t)$ gives

$$\mathbf{V}_i(s) = 3e^{-5s}$$

Using the Table 12.2 entry for the damped cosine and time shifting by 9, we have

$$L\{e^{-4(t-9)}\cos[7(t-9)]u(t-9)\} = \frac{(s+4)e^{-9s}}{[s+4]^2+49}$$

and since $v_o(t) = e^{-36}\{e^{-4(t-9)}\cos[7(t-9)]u(t-9)\}$,

$$\mathbf{V}_o(s) = \frac{(s+4)e^{-9(s+4)}}{(s+4)^2+49}$$

The transfer function $H(s)$ is the ratio of output to input under these conditions:

$$\mathbf{H}(s) = \frac{\mathbf{V}_o(s)}{\mathbf{V}_i(s)} = \frac{\frac{1}{3}(s+4)e^{-4s}e^{-36}}{(s+4)^2+49}$$

To get $h(t)$, begin again from the table entry for damped cosine, time shifted by 4:

$$L\{e^{-4(t-4)}\cos[7(t-4)]u(t-4)\} = \frac{(s+4)e^{-4s}}{(s+4)^2+49}$$

Comparing this with $\mathbf{H}(s)$, we must amplitude scale by the factor $e^{-36}/3$:

$$h(t) = \tfrac{1}{3}e^{-4(t+5)}\cos[7(t-4)]u(t-4) \qquad (13.39)$$

We see that in this circuit the response $h(t)$ to the unit impulse $\delta(t)$ does not begin until time $t = 4$, even though $\delta(t)$ itself "fires" at $t = 0$.

Since the impulse response $h(t)$ and the transfer function $\mathbf{H}(s)$ are a transform pair, we may use knowledge of one to gain the other. As we saw in Example 13.14, an input other than the impulse may be used, the transfer function found, and from it the impulse response. In the case of physical circuits, it is usually more convenient to input a unit step than a unit impulse. Then from the resulting *unit step response s(t)*, both $\mathbf{H}(s)$ and $h(t)$ may be computed. *We define the unit step response s(t) to be the output when a unit step input u(t) is applied, assuming zero initial conditions and all other independent sources killed.*

The unit step response $s(t)$ and its transform $\mathbf{S}(s)$ are closely related to the unit impulse response $h(t)$ and its transform $\mathbf{H}(s)$. With unit step input $\mathbf{V}_i(s) = 1/s$ in (13.34), the output $\mathbf{V}_o(s) = \mathbf{S}(s)$, and we have from that equation

$$\mathbf{S}(s) = \frac{\mathbf{H}(s)}{s} \qquad (13.40)$$

Using the integration rule on (13.40) gives

$$s(t) = \int_{0-}^{t} h(\tau)\,d\tau \qquad (13.41a)$$

Since $h(t) = 0$ for $t < 0$, the lower limit may be extended to $-\infty$. From this it is clear that $s(t) = 0$ for $t < 0$ just as $h(t)$ does. The step response cannot begin before the step input begins. Finally, differentiating both sides, we have

$$h(t) = \frac{d}{dt}s(t) \qquad (13.41\text{b})$$

The impulse response is the derivative of the step response (the step response is the integral of the impulse response). This result could have been anticipated by noting that the unit step is the integral of the unit impulse, and by linearity, the response of a linear circuit to the integral of an input is the integral of its response to the original input.

Example 13.15

Determine the unit step response for the circuit of Example 13.13. Using (13.40) and the result of that example, we have

$$\mathbf{S}(s) = \frac{\mathbf{H}(s)}{s} = \frac{1}{s + \frac{1}{6}}$$

The step response has an additional pole at the origin, which cancels the zero in $\mathbf{H}(s)$ at that location. Then the step response is

$$s(t) = e^{-t/6}u(t)$$

Note that we have multiplied by the unit step, since $s(t)$ is required to be zero for $t < 0$.

The transfer function analysis equation (13.34) concisely links the input, output, and transfer function in the s-domain. What is the time-domain equivalent of this expression? Let us begin by defining an operation between two time functions $f(t)$ and $g(t)$ that produces a third time function $d(t)$ via the integral

$$d(t) = \int_{-\infty}^{\infty} f(\tau)g(t - \tau)\,d\tau \qquad (13.42)$$

We refer to $d(t)$ as the *convolution* of $f(t)$ and $g(t)$. Here we will restrict ourselves to convolutions between functions $f(t)$ and $g(t)$ that are both 0 for $t < 0$. Transforming yields

$$\mathbf{D}(s) = \int_{0-}^{\infty} \left[\int_{0-}^{\infty} f(\tau)g(t - \tau)\,d\tau\right] e^{-st}\,dt$$

where the inner lower limit has been replaced by $0-$, since $f(\tau) = 0$ for $\tau < 0$. Reversing the order of integration, we obtain

$$\mathbf{D}(s) = \int_{0-}^{\infty} f(\tau) \left[\int_{0-}^{\infty} g(t - \tau)e^{-st}\,dt\right] d\tau$$

The inner integral is the transform of $g(t - \tau)$, which, by the time shift property, yields

$$\mathbf{D}(s) = \int_{0-}^{\infty} f(\tau)[e^{-s\tau}G(s)]\,d\tau$$

Taking $\mathbf{G}(s)$ out of the integral, what remains is $\mathbf{F}(s)$, and we have the following property:

CONVOLUTION

$$\mathbf{L}\left[\int_{-\infty}^{\infty} f(\tau)g(t-\tau)\ dt\right] = \mathbf{F}(s)\mathbf{G}(s) \qquad (13.43)$$

The Laplace transform of the convolution of two time functions is the product of their Laplace transforms. Then, comparing (13.34) with (13.43), we have

$$\mathbf{L}\left[\int_{-\infty}^{\infty} h(\tau)v_i(t-\tau)\ dt\right] = \mathbf{H}(s)\mathbf{V}_i(s) = \mathbf{V}_o(s) \qquad \text{(all ICs} = 0) \qquad (13.44)$$

or, taking inverse transforms of both sides,

$$v_o(t) = \int_{-\infty}^{\infty} h(\tau)v_i(t-\tau)\ d\tau \qquad \text{(all ICs} = 0) \qquad (13.45)$$

The output equals the convolution of the impulse response and the input, assuming zero initial conditions. This is the desired time-domain form of the transfer function equation specifying that multiplication of impulse response $\mathbf{H}(s)$ and input $\mathbf{V}_i(s)$ in the s-domain goes over to convolution of $h(t)$ and $v_i(t)$ for calculation of the forced output in the time domain.

A useful observation is that since the order of the functions $\mathbf{F}(s)$ and $\mathbf{G}(s)$ on the right side of (13.43) may be reversed without changing its value, so may the order of $f(t)$ and $g(t)$ on the left. In other words, *in computing the convolution integral (13.45), either function in the integrand may be written with argument τ and the other with argument $t-\tau$.*

Example 13.16

Determine $v_2(t)$, $t > 0$, if $v_s(t)$ is the "drooping" pulse shown in Fig. 13.16, and $v_2(0-) = 0$. The transfer function is found by voltage division:

$$\mathbf{H}(s) = \frac{\mathbf{V}_2(s)}{\mathbf{V}_s(s)} = \frac{[3(15/s)]/[3+(15/s)]}{\{[3(15/s)]/[3+(15/s)]\}+5} = \frac{3}{s+8}$$

So the impulse response is

$$h(t) = 3e^{-8t}u(t)$$

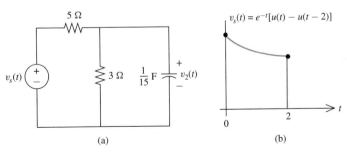

FIGURE 13.16   (a) Circuit for Example 13.16; (b) input $v_s(t)$.

Convolving $h(t)$ and $v_s(t)$, the desired output is

$$v_2(t) = \int_{-\infty}^{\infty} h(t-\tau)v_s(\tau)\,d\tau$$

The input $v_s(\tau)$ is nonzero only for $0 \le \tau \le 2$, so the limits of this integral may be adjusted accordingly. Within those limits, $v_s(\tau) = e^{-\tau}$, so

$$v_2(t) = \int_0^2 3e^{-8(t-\tau)}u(t-\tau)e^{-\tau}\,d\tau$$

The unit step in the integrand has different effects depending on the value of t. If $t < 0$, then $u(t-\tau) = 0$ within the specified limits of integration and $v_2(t) = 0$ for all $t < 0$. If $t > 2$, then $u(t-\tau) = 1$ within the specified limits of integration, and

$$v_2(t) = \int_0^2 3e^{-8(t-\tau)}e^{-t}\,d\tau = \frac{3}{7}(e^{-2}-e^{-16})e^{-8(t-2)} \qquad t > 2$$

For values of t between the limits of the integral, $0 \le \tau \le 2$, the unit step reduces the interval over which the integrand is nonzero to $0 \le \tau \le t$, since $u(t-\tau) = 0$ for $\tau > t$.

$$v_2(t) = \int_0^t 3e^{-8(t-\tau)}e^{-\tau}\,d\tau = \frac{3}{7}(e^{-t}-e^{-8t}) \qquad 0 \le t \le 2$$

A sketch of this function is shown as Fig. 13.17.

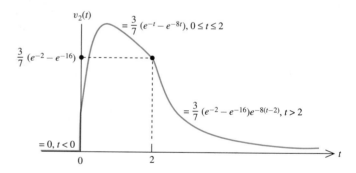

FIGURE 13.17 Output $v_2(t)$ for Example 13.16.

For each fixed instant of time t, the convolution (13.45) equals the integral over all τ of the product of the two τ-functions which make up its integrand. Since the value of any real integral is just the area under its integrand, we may interpret a convolution integral as *the area under its integrand product curve as a function of the parameter* t. It is often helpful to visualize the process of evaluating this area by fixing t and sketching (versus τ) a graph of each of the two functions in the integrand, $v_i(t-\tau)$ and $h(\tau)$. Only in τ-regions where their nonzero values overlap will the product, and thus the contribution to the convolution, or area under the product curve, be nonzero. This is sometimes referred to as the *graphical method* for evaluating a convolution and is illustrated in the final example in this section.

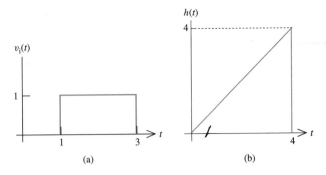

FIGURE 13.18   (a) $v_i(t)$; (b) $h(t)$.

Example 13.17

Let us compute the convolution (13.45) of the functions $v_i(t)$ and $h(t)$ shown in Fig. 3.18. We will graph $v_i(t - \tau)$ in two steps. First time-reverse $v_i(t)$ by flipping it about the vertical axis as shown in Fig. 13.19(a), relabeling the independent variable τ, to get the graph of $v_i(-\tau)$. Then shift this function by t units on the τ axis by sliding it t units to the right, producing $v_i(t - \tau)$ as in Fig. 13.19(b). In that same figure $h(\tau)$ is included, identical to $h(t)$ with t replaced by τ.

As t increases, the square pulse $v_i(t - \tau)$ moves from left to right along the τ axis. We may identify several distinct regions, depending on the type of overlap between $v_i(t - \tau)$ and $h(\tau)$. The first region corresponds to locations of the square pulse (i.e., values of t) for which the pulse is entirely to the left of the nonzero range of values of $h(\tau)$. There is no overlap for any τ, so the product of these two functions, hence the area under the product curve, is zero. The convolution $v_o(t) = 0$ for all t in this region, which is marked by the rightmost edge of the square pulse being to the left of the origin in Fig. 13.19(b), or

$$v_0(t) = 0, \quad t - 1 < 0 \quad (t < 1)$$

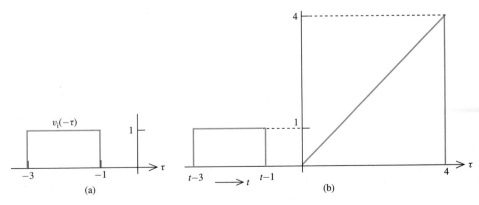

FIGURE 13.19   (a) $v_i(-\tau)$; (b) $v_i(t - \tau)$ and $h(\tau)$ vs. τ.

In the next region $t - 1 > 0$, so there is overlap, but $t - 3 < 0$, so the overlap is partial. In this case the product curve is shown in Fig. 13.20(a), and its triangular area is given by

$$v_o(t) = \tfrac{1}{2}(t - 1)^2, \quad 1 < t < 3$$

In the next region the overlap is complete, with $t - 1 < 4$ but $t - 3 > 0$. There, the product of the two functions is a trapezoid as in Fig. 13.20(b), and the area under this product curve is

$$\text{area} = [(t - 1) - (t - 3)](t - 3) + \tfrac{1}{2}[(t - 1) - (t - 3)]^2$$

or $\quad v_o(t) = 2(t - 2), \qquad 3 < t < 5$

As the right end of the square pulse moves past the triangle $h(\tau)$, the overlap is partial once again and the product is the trapezoid of

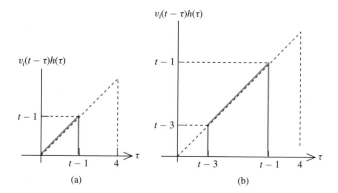

(a) (b)

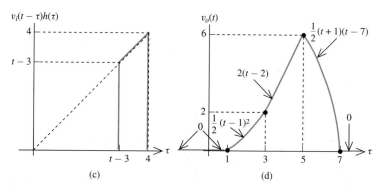

(c) (d)

FIGURE 13.20 (a) Integrand for the region $1 < t < 3$; (b) integrand for region $3 < t < 5$; (c) integrand for region $5 < t < 7$; (d) convolution $v_o(t)$ for all regions.

Fig. 13.20(c), so for $t - 1 > 4$ but $t - 3 < 4$,

$$\text{area} = [4 - (t - 3)](t - 3) + \tfrac{1}{2}[4 - (t - 3)]^2$$

or $\qquad v_o(t) = -\tfrac{1}{2}(t + 1)(t - 7), \qquad 5 < t < 7$

Finally, for $t - 3 > 4$, there is no overlap remaining, so

$$v_o(t) = 0, \qquad t > 7$$

The graph of the result $v_o(t)$ is shown in Fig. 13.20(d).

EXERCISES

13.6.1. A circuit has impulse response $h(t) = 2[u(t - 1) - u(t - 2)]$. Find the zero initial condition response $v_o(t)$ to the input $v_i(t) = (t - 1)u(t)$ by convolution.

Answer 0 for $t < 1$; $2t - 5$ for $1 \le t \le 2$; $t^2 - 4t + 3$ for $t > 2$

13.6.2. Suppose that two copies of the circuit of Exercise 13.6.1 are connected in cascade so that the output from one becomes the input to the other. Find the impulse response of the cascaded circuits.

Answer $4[r(t - 2) - 2r(t - 3) + r(t - 4)]$, where $r(t)$ is the unit ramp

13.6.3. Verify the result of Example 13.16 by finding $\mathbf{V}_2(s)$ from the transfer function $\mathbf{H}(s) = 3/(s + 8)$. What is the required source $\mathbf{V}_s(s)$?

Answer $\mathbf{V}_s(s) = \dfrac{1 - e^{-2(s+1)}}{s + 1}$

SUMMARY

In Chapter 8 we saw that the use of phasors greatly simplified electric circuit analysis where applicable but was restricted to finding the forced response in sinusoidal circuits. Here we see that the s-domain reaps the same benefits but is applicable to all circuit problems, forced or unforced, steady state or transient.

- KVL, KCL, and Ohm's law carry over from the time to the s-domain unchanged.

- The impedance $\mathbf{Z}(s)$ of an RLC element or subcircuit is the ratio of its s-domain voltage and current when all initial conditions are zero.

- The impedance of the RLC elements is R, sL, and $1/sC$, respectively.

- The s-domain circuit is the original circuit diagram with the RLC elements labeled by impedance and independent sources labeled by their transforms.

- The s-domain circuit may be analyzed as if it were a dc resistive circuit, with impedance treated as resistance, and s-domain source functions treated as constants.

549

- Transfer functions specify the ratio of two s-domain circuit variables when all initial conditions are zero.

- Stable circuits have all their poles in the left half-plane.

- Transfer functions may be found by solving the s-domain circuit for the output variable in terms of the input variable and dividing.

- The initial and final value theorems permit us to determine initial ($t = 0+$) and final ($t = \infty$) values of a time function from its transform without computing the inverse transform.

- The forced response of any circuit is the convolution of the impulse response and the input. The impulse response is the inverse transform of the transfer function.

PROBLEMS

13.1. Find the equivalent impedance $Z(s)$.

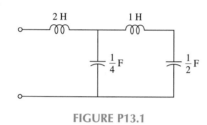

FIGURE P13.1

13.2. Find the equivalent impedance $Z(s)$.

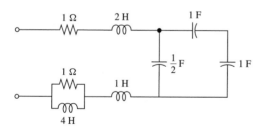

FIGURE P13.2

13.3. Find the equivalent admittance $Y(s)$.

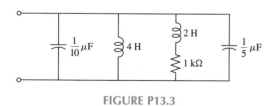

FIGURE P13.3

13.4. Find $\mathbf{V}_2(s)$ by voltage division. Assume all ICs $= 0$.

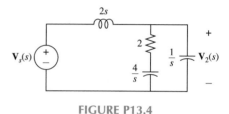

FIGURE P13.4

13.5. Find $\mathbf{I}_2(s)$ by current division. Assume all ICs $= 0$.

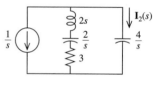

FIGURE P13.5

13.6. Find the s-domain Thevenin and Norton equivalents, assuming $v_1(0-) = i_2(0-) = 0$.

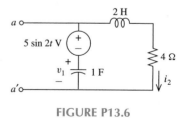

FIGURE P13.6

13.7. Find the s-domain Thevenin equivalent assuming all IC's $= 0$.

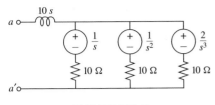

FIGURE P13.7

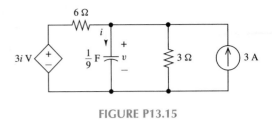

FIGURE P13.15

13.8. Find a two-terminal RLC subnetwork for which $Z(s) = (s^2 + 7s + 3)/s$.

13.9. Find a two-terminal RLC subnetwork for which $Z(s) = (s^2 + s + 2)/(s^2 + 2)$.

13.10. Find a two-terminal RLC subnetwork for which $Y(s) = 2 + (2/s) + (s/2) + [2/(s + 2)]$.

13.11. Determine the location of the poles of $Z(s)$ and $Y(s)$ for all two-element series–parallel combinations of the RLC elements.

13.12. Solve 13.4 by first taking the Thevenin equivalent of everything except the right-most capacitor.

13.13. Use Laplace transforms to find i for $t > 0$ if $v(0) = 6$ V, $i(0) = 2$ A, and (a) $v_g = 10$ V and (b) $v_g = 4 \cos t$ V.

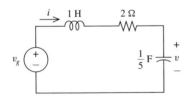

FIGURE P13.13

13.14. Use Laplace transforms to find v for $t > 0$ if (a) $i_g = 2u(t)$ A and (b) $i_g = 2e^{-1}u(t)$ A.

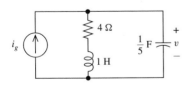

FIGURE P13.14

13.15. Use Laplace transforms to find v for $t > 0$ if $v(0) = 4$ V.

13.16. Repeat Problem 13.6 for the case $v_1(0-) = 2$ V and $i_2(0-) = -1$ A.

13.17. A certain two-terminal subnetwork has open-circuit voltage $\mathbf{V}_{oc}(s) = 3/(2s + 3)$ and short-circuit current $\mathbf{I}_{sc}(s) = 1/(2s + 2)$. Sketch the circuit in the time domain, specifying all element and source values. Is there more than one such circuit? Explain.

13.18. Find $v_1(t)$ assuming all ICs $= 0$ at $t = 0-$.

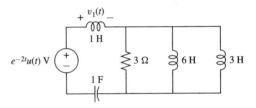

FIGURE P13.18

13.19. Draw the s-domain circuit. Find the node voltages $\mathbf{V}_1(s)$, $\mathbf{V}_2(s)$.

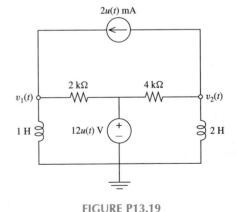

FIGURE P13.19

13.20. Repeat Problem 13.19 for the case that the voltage source function is replaced by 12 V.

13.21. Draw the s-domain circuit. Combining series elements, write and solve a single node equation for $\mathbf{V}(s)$.

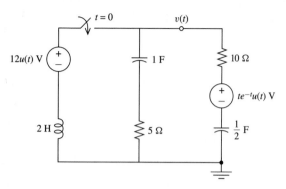

FIGURE P13.21

13.22. Draw the s-domain circuit. Combining parallel elements, write and solve a single mesh equation for $\mathbf{I}(s)$.

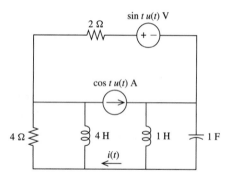

FIGURE P13.22

13.23. If a unit step of current $i(t) = u(t)$ is injected into this circuit, the voltage for $t > 0$ is measured to be $v(t) = 4t + 3 - 6e^{-2t}$. What are the initial conditions $v_C(0-), i_L(0-)$?

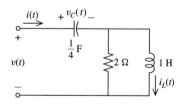

FIGURE P13.23

13.24. Find $i(t), t > 0$.

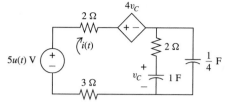

FIGURE P13.24

13.25. Find $v_C(t), t > 0$, if $v_C(0-) = 1$ and $i_L(0-) = 0$ A.

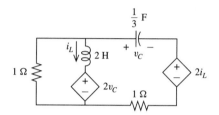

FIGURE P13.25

13.26. Find $v_{c_1}(t)$ and $v_{c_2}(t)$ for $t > 0$.

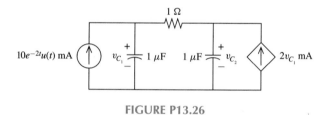

FIGURE P13.26

13.27. Find the transfer function $\mathbf{H}(s) = \mathbf{V}_1(s)/\mathbf{I}_1(s)$.

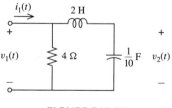

FIGURE P13.27

13.28. Repeat Problem 13.27 for the transfer function $\mathbf{V}_2(s)/\mathbf{V}_1(s)$.

13.29. Repeat Problem 13.27 for the transfer function $\mathbf{V}_2(s)/\mathbf{I}_1(s)$.

13.30. Find $\mathbf{H}(s) = \mathbf{V}_2(s)/\mathbf{V}_s(s)$ for the circuit of Problem 13.4.

552

13.31. Find $\mathbf{H}(s) = \mathbf{I}_1(s)/\mathbf{V}_s(s)$.

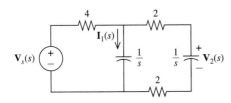

FIGURE P13.31

13.32. Find $\mathbf{H}(s) = \mathbf{V}_2(s)/\mathbf{V}_s(s)$; and find the unit impulse response $h(t)$ for the circuit of Problem 13.31.

13.33. Find $i_1(t)$ in Problem 13.31 if $v_s(t) = 15u(t)$ V.

13.34. Find the transfer functions between output $v_2(t)$ and inputs $v_{s1}(t)$ [$\mathbf{H}_1(s)$], $i_{s2}(t)$ [$\mathbf{H}_2(s)$].

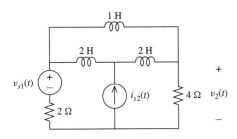

FIGURE P13.34

13.35. Find the transfer function $\mathbf{H}(s)$ between input $i_s(t)$ and output $v_2(t)$.

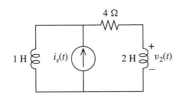

FIGURE P13.35

13.36. In a certain circuit the zero in initial condition output $v_0(t)$ for $t > 0$ is measured to be $1 + 3te^{-2t} - e^{-2t}$ when the input is $v_i(t) = 5u(t)$. What is $\mathbf{H}(s)$? What is the forced response $v_0(t)$, $t > 0$, for $v_i(t) = e^{-5t}u(t)$?

13.37. In the circuit described in Problem 13.36, there is a single storage element, a capacitor. When $v_c(0-) = 1$ V and the input $v_i(t) = 0$, the output is found to be $(t - 1)e^{-2t}$ V. Using the data given in Problem 13.36 also, find $v_0(t)$, $t > 0$, if $v_i(t) = 5\sin t$ and $v_c(0-) = 4$ V.

13.38. Find the transfer functions $\mathbf{H}_1(s)$ and $\mathbf{H}_2(s)$ between output $V_c(t)$ and inputs $V_{s1}(t)$ and $V_{s2}(t)$. Use these to help find the zero initial condition response $V_c(t)$ when $v_{s1}(t) = 10$ V and $v_{s2}(t) = e^{-3t}$ V.

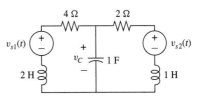

FIGURE P13.38

13.39. Repeat Problem 13.38 if $v_{s1}(t) = te^{-t}$ V, $v_{s2}(t) = 0$, the initial inductive currents are both zero, and $v_c(0-) = +12$ V.

13.40. Find all the circuit poles. Is this system stable?

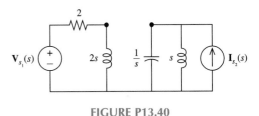

FIGURE P13.40

13.41. Find the range of values for k for which this circuit is stable.

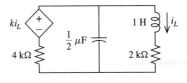

FIGURE P13.41

13.42. Find a value for R for which this circuit is stable and another for which it is unstable. Assume $R \geq 0$.

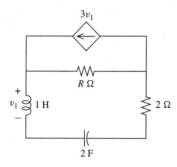

FIGURE P13.42

13.43. Find the poles of the circuit. Is this circuit stable?

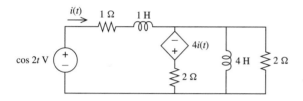

FIGURE P13.43

13.44. Find the poles of the circuit of Problem 13.26 and determine if the circuit is stable.

13.45. Find the ac steady-state responses $i_1(t)$, $v_2(t)$, of the circuit of Problem 13.31 if $\mathbf{V}_s(s) = 1/(s^2 + 1)$.

13.46. If a circuit is stable, its impulse response must decay to zero as $t \to \infty$. Show why this is true.

13.47. If a circuit is not stable, its impulse response may still decay to zero as $t \to \infty$. Give an example of such a circuit.

13.48. $\mathbf{F}(s) = (3s + 1)/(s^2 + 7s + 6)$. Use the initial and final value theorems to find $f(0+)$ and $f(\infty)$, and verify by finding $f(t)$ by partial fractions.

13.49. $\mathbf{F}(s) = -s^2/(s+1)^3$. Use the initial and final value theorems to find $f(0+)$ and $f(\infty)$, and verify by finding $f(t)$ by partial fractions.

13.50. Find $v(0+)$ in Problem 13.21 by use of the initial value theorem.

13.51. Assume $Z(s)$ is a rational function. Under what conditions $Z(s)$ will the current $i(t)$ be continuous at $t = 0$?

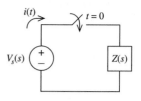

FIGURE P13.51

13.52. A circuit with impulse response $h(t) = (3 + e^{-t})u(t)$ has forced response $v_0(t) = te^{-t}u(t)$ to a certain input $v_i(t)$. Find $v_i(t)$.

13.53. Find the convolution of $f(t) = -2u(t)$ with $g(t) = e^{-t}u(t - 1)$.

13.54. Find the convolution of the time functions $f(t)$ and $g(t)$ shown.

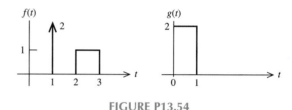

FIGURE P13.54

13.55. Show that the convolution of $df(t)/dt$ and $g(t)$ is the same as the derivative of the convolution of $f(t)$ and $g(t)$.

13.56. Determine the unit step response $s(t) = v_c(t)$ to the input $v(t) = u(t)$ in the circuit of Problem 13.23. Use convolution to compute $s(t)$.

13.57. Find the transfer function $\mathbf{H}(s)$ with $v_2(t)$ as output and $i_1(t)$ as input (*Hint:* Find both in terms of $\mathbf{V}_s(s)$; then divide). Determine the impulse response $h(t)$ and the forced response to the input $i_1(t) = 2(1 - e^{-t})$.

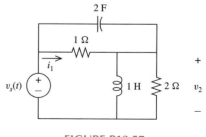

FIGURE P13.57

13.58. If $x(t)$ is the input and $y(t)$ is the output, find the step response $r(t)$ and the impulse response $h(t)$ for the following:

(a) $y'' + 6y' + 5y = 20x$.

(b) $y'' + 4y' + 13y = 13x$.

13.59. If a network has an impulse response

$$h(t) = te^{-2t}u(t)$$

find the forced response to an input

$$f(t) = e^{-2t}\cos tu(t)$$

More Challenging Problems

13.60. Find a two-terminal subnetwork with impedance $Z(s) = (s^3 + 3s^2 + 6s + 7)/(s^2 + 3s + 2)$. Suggestion: use partial fractions.

13.61. Find i, for $t > 0$, using the transformed circuit, if $v(0) = 4$ V and $i(0) = 2$ A.

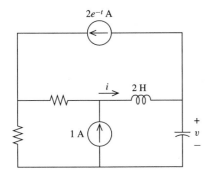

FIGURE P13.61

13.62. Replace the independent current source in the circuit of Problem 13.19 by a voltage controlled current source (same reference direction) of value $kv_1(t)$ A. For what values of the real parameter k is this circuit stable?

13.63. Find the impulse response $h(t)$ with input $v_s(t)$, output $v(t)$.

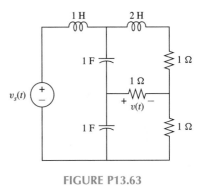

FIGURE P13.63

13.64. Sketch a circuit whose impulse response is $h(t) = \delta'(t) + e^{-t}u(t)$, specifying which circuit voltage is the output, and using an independent voltage source as the input. Repeat using an independent current source as the input.

14

Frequency Response

Heinrich Rudolf Hertz
1857–1894

That was no mean performance. [On his verification of Maxwell's theories]

Heinrich Hertz

The way for the development of radio, television, and radar was opened by the German physicist Heinrich Rudolf Hertz with his discovery in 1886–1888 of electromagnetic waves. His work confirmed the 1864 theory of the great English physicist James Clerk Maxwell that such waves existed.

Hertz was born in Hamburg, the oldest of five children in a prominent and prosperous family. After graduation from high school he spent a year with an engineering firm in Frankfurt, a year of volunteer military service in Berlin, and then a year at the University of Munich. Finally he entered the University of Berlin as a student of the great physicist Hermann von Helmholtz. Later Hertz received his doctorate and was a professor at Karlsruhe when he began his quest for electromagnetic waves. It was there that he met Elizabeth Doll, the daughter of one of his fellow professors, and after a three-month courtship they were married. Only a few years after his famous discovery, Hertz died on New Year's Day in 1894 of a bone malignancy at the young age of 37. His researches ushered in the modern communication age, and in his honor the unit of frequency (cycles per second) was named the hertz.

Chapter Contents

Since sinusoidal currents and voltages are easily produced, one convenient way to test a linear circuit is to inject a sinusoid as input and observe the sinusoidal steady-state response. Changing the phase or amplitude of the sinusoid will not test the circuit further, since the response will predictably scale with the amplitude and phase shift along with the input. At each possible input frequency, however, the response of the circuit, both amplitude and phase, will in general be different. These response variations with frequency form the *frequency response* of the circuit.

We may be acquainted with some aspects of frequency response from the manner in which audio equipment is often advertised. A speaker whose "3-dB bandwidth" is, say, 50 Hz to 20 kHz is better than one of bandwidth 200 Hz to 5 kHz, since it will reproduce bass and treble frequencies better. In this chapter we introduce the concept of frequency response, using it first for analysis and then as the key descriptor of what a linear circuit "does," the function that it is designed to perform and the quality with which it performs that function. Frequency response is the critical link between the analysis and design of linear circuits.

First we introduce the frequency response function $\mathbf{H}(j\omega)$ and show its close relationship with the transfer function $\mathbf{H}(s)$ of Chapter 13. The decibel scale, a convenient way to measure nonnegative numbers, is defined and used to produce Bode gain plots, which describe the changes in the input–output amplification ratio (gain) with frequency. Using these tools, the frequency response of resonant circuits and of op amps is next investigated, and the very important class of frequency-selective circuits called filters studied. The chapter concludes with a discussion of frequency response and the use of SPICE.

14.1 FREQUENCY RESPONSE FUNCTION

The transfer function $\mathbf{H}(s)$ permits us to compute the s-domain output for any input via the basic input–output equation (13.8), repeated here.

$$\mathbf{V}_o(s) = \mathbf{H}(s)\mathbf{V}_i(s) \qquad \text{(all ICs} = 0) \tag{14.1}$$

We apply this simple but general relationship to phasor analysis, which was first introduced in Chapter 8 in the context of ac steady state. Suppose that the time-domain independent sources in a circuit are each complex exponentials $\mathbf{V}_i e^{j\omega t}$ at the same frequency ω, as required for phasor analysis. $\mathbf{V}_i$ is the phasor associated with the source $\mathbf{V}_i e^{j\omega t}$. Since

$$L(\mathbf{V}_i e^{j\omega t}) = \frac{\mathbf{V}_i}{s - j\omega}$$

then

$$\mathbf{V}_o(s) = \frac{\mathbf{H}(s)\mathbf{V}_i}{s - j\omega} \tag{14.2}$$

Since the linear circuits we consider contain only fixed *RLC* elements and sources, $\mathbf{H}(s)$ will be a rational function

$$\mathbf{H}(s) = \frac{\mathbf{N}(s)}{\mathbf{D}(s)} = \frac{\mathbf{N}(s)}{(s - p_1)^{r_1} \cdots (s - p_n)^{r_n}} \tag{14.3}$$

where the p_i are poles of the circuit, as discussed in Section 13.4. Then

$$\mathbf{V}_o(s) = \frac{\mathbf{N}(s)\mathbf{V}_i}{(s - p_1)^{r_1} \cdots (s - p_n)^{r_n}(s - j\omega)} \tag{14.4}$$

Assuming that there are no poles on the $j\omega$ axis, $\text{Re}(p_i) \neq 0$ for each $i = 1, \ldots, n$, by partial fraction expansion

$$\mathbf{V}_o(s) = \left[\frac{A}{(s - p_1)^{r_1}} + \frac{B}{(s - p_1)^{r_1 - 1}} + \cdots + \frac{Q}{s - p_n} \right] + \frac{R}{s - j\omega} \tag{14.5}$$

where

$$R = (s - j\omega)\mathbf{V}_o(s)\Big|_{s=j\omega} \tag{14.6}$$

which, by (14.4), evaluates to

$$R = \mathbf{H}(j\omega)\mathbf{V}_i \tag{14.7}$$

Examining the (14.5), $v_o(t)$ will, in general, contain the term $Re^{j\omega t}$ together with terms due to the poles of circuit shown in brackets. Recall that phasor analysis is applicable only in circuits for which all currents and voltages are of the form $Re^{j\omega t}$. This is the case in (14.4) if the circuit is stable and if we are considering the steady-state response $v_o(t)$ only. Stability, signaled by negative real parts of all the poles p_i of the circuit, guarantees that the inverse transform of each term in parentheses in (14.4) decays to zero as t goes to infinity. This leaves, in steady state, only the final term, which by (14.5) and (14.7) is the forced response

$$v_o(t) = L^{-1}\frac{\mathbf{H}(j\omega)\mathbf{V}_i}{s - j\omega} = \mathbf{H}(j\omega)\mathbf{V}_i e^{j\omega t} \tag{14.8}$$

By definition, the phasor $\mathbf{V}_o$ associated with this output $v_o(t)$ is the quantity multiplying $e^{j\omega t}$, or, by (14.8),

$$\mathbf{V}_o = \mathbf{H}(j\omega)\mathbf{V}_i \qquad (14.9)$$

$\mathbf{H}(j\omega)$, the transfer function $\mathbf{H}(s)$ with s replaced by $j\omega$, scales the input phasor to yield the output phasor. We define $\mathbf{H}(j\omega)$ as the *frequency response function*. Then (14.9), which we refer to as the *frequency response equation*, asserts that *for stable circuits, the output phasor is the product of the input phasor and the frequency response function $H(j\omega)$*. Given $\mathbf{H}(s)$, or equivalently $\mathbf{H}(j\omega)$, we can determine the sinusoidal steady-state output of any stable linear circuit at any frequency. Our restriction in (14.5) that there are no $j\omega$-axis poles does not further limit this result, since such systems are not stable and thus have no steady state.

Example 14.1

Consider a circuit with transfer function

$$\mathbf{H}(s) = \frac{2s}{s^2 + 4s + 1}$$

Find the steady-state output $v_o(t)$ if the input $v_i(t) = 12\cos(2t - 30°)$. The input phasor is $\mathbf{V}_i = 12\underline{/-30°}$, and by the frequency response equation (14.9),

$$\mathbf{V}_o = \mathbf{H}(j2)\mathbf{V}_i = \frac{2(j2)}{(j2)^2 + 4(j2) + 1}(12\underline{/-30°})$$

Evaluating on a calculator, we obtain

$$\mathbf{V}_o = 5.62\underline{/-50.6°}$$

Thus

$$v_o(t) = 5.62\cos(2t - 50.6°) \qquad (14.10)$$

Example 14.2

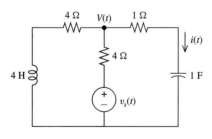

FIGURE 14.1 Circuit for Example 14.2.

Find the forced response $i(t)$ if $v_s(t) = 6\cos t$ V. We will determine $\mathbf{H}(s)$ and replace s by $j\omega$, $\omega = 1$ rad/s, to evaluate the frequency response function for this circuit operating at this frequency. The s-domain node equation at the node labeled $V(t)$ in Fig. 14.1 is

$$\frac{1}{4s + 4}V(s) + \frac{s}{s + 1}V(s) + \frac{1}{4}[V(s) - V_s(s)] = 0$$

and solving yields

$$V(s) = \frac{s + 1}{5s + 2}V_s(s) \qquad (14.11)$$

The required output $I(s)$ is related to $V(s)$ by the series admittance

$$I(s) = \frac{s}{s + 1}V(s) = \frac{s}{5s + 2}V_s(s)$$

Thus the transfer function $\mathbf{H}(s) = I(s)/V_s(s)$ is

$$\mathbf{H}(s) = \frac{s}{5s + 2}$$

The source frequency is $\omega = 1$ rad/s, the source phasor is $\mathbf{V}_s = 6\underline{/0}$ V, and, by the frequency response equation, the output phasor $\mathbf{I}$ equals the input phasor scaled by the frequency response function evaluated at $\omega = 1$, or

$$\mathbf{I} = \mathbf{H}(j1)\mathbf{V}_s = \frac{j1}{5(j1)+2}6\underline{/0} \text{ A}$$

Using a calculator once again, we have

$$\mathbf{I} = (0.186\underline{/21.8°})(6\underline{/0}) = 1.12\underline{/21.8°} \text{ A}$$

whence

$$i(t) = 1.12\cos(t + 21.8°) \text{ A}$$

which is the required steady-state current.

Taking magnitudes of both sides of the frequency response equation (14.9),

$$|\mathbf{V}_o| = |\mathbf{H}(j\omega)||\mathbf{V}_i| \qquad (14.12)$$

Since the magnitude of a phasor is the amplitude of the corresponding sinusoid, (14.12) suggests that *the ratio by which the amplitude of the output sinusoid exceeds the input sinusoid is given by the magnitude of the frequency response function.* Thus $|\mathbf{H}(j\omega)|$ is reasonably enough called the *gain* of the circuit. Similarly, taking angles of both sides of (14.9),

$$\underline{/\mathbf{V}_o} = \underline{/\mathbf{H}(j\omega)} + \underline{/\mathbf{V}_i} \qquad (14.13)$$

Since the angle of a phasor is identical to the phase angle of the corresponding sinusoid, *the amount by which the phase angle of the output exceeds the phase angle of the input is given by the angle of the frequency response function.* This input–output phase difference $\underline{/\mathbf{H}(j\omega)}$ is called the *phase shift.* Taken together, the gain and phase shift, both functions of frequency ω, tell us how the amplitude and phase angle of any input sinusoid is changed by passage through the circuit from input to output. Thus, *given the gain and phase shift, or equivalently given the frequency response function* $\mathbf{H}(j\omega)$, *we have a complete description of how the circuit responds to inputs at any frequency.* This is why $\mathbf{H}(j\omega)$ is referred to as the frequency response function.

Example 14.3

Plot the gain and phase shift of the circuit in Fig. 14.2. The output is $v(t)$. By voltage division,

$$\mathbf{V}(s) = \frac{1/sC}{R + (1/sC)}\mathbf{V}_s(s) = \frac{1/RC}{s + (1/RC)}\mathbf{V}_s(s)$$

The transfer function is

$$\mathbf{H}(s) = \frac{\mathbf{V}(s)}{\mathbf{V}_s(s)} = \frac{1/RC}{s + 1/RC}$$

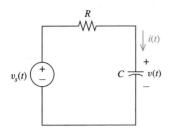

FIGURE 14.2 Circuit for Example 14.3.

The circuit has a single pole at $s = -1/RC$, so it is stable. We may replace s by $j\omega$ and determine the steady-state behavior of this circuit via its frequency response function $\mathbf{H}(j\omega)$. The gain as a function of frequency is given by

$$|\mathbf{H}(j\omega)| = \frac{|1/RC|}{|j\omega + (1/RC)|} = \frac{1/RC}{\sqrt{\omega^2 + (1/RC)^2}} \qquad (14.14)$$

which is shown in Fig. 14.3(a). Note that the dc gain is unity and declines with increasing frequency. The impedance of the capacitor is infinite at dc, so as the frequency increases the voltage divider will completely favor the capacitor at $\omega = 0$. Its impedance declines to zero with increasing frequency, so as frequency increases the voltage divider ratio shifts steadily toward the resistor and away from the C. These facts are consistent with the gain shown in Fig. 14.3(a). The phase shift, shown in Fig. 14.3(b), is given by

$$\underline{/\mathbf{H}(j\omega)} = -\tan^{-1}\omega RC \qquad (14.15)$$

The output lags the input by an increasing phase shift as frequency increases. Note that since at dc the input and output are equal (the capacitor is an open circuit), it is consistent that the dc phase shift be zero.

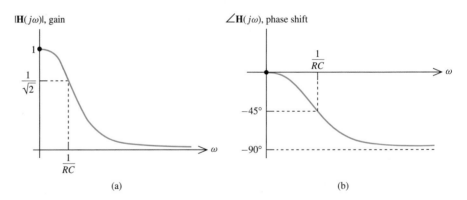

FIGURE 14.3 (a) Gain versus ω; (b) phase shift versus ω.

The frequency response equation (14.9) is a fundamental relationship linking input sinusoids and the resulting steady-state output sinusoids. It applies to stable circuits, those guaranteed to have a steady state. Unstable circuits never "settle down" to steady-state conditions; hence their steady-state gain or phase shift has no meaning. Even in the unstable case (14.9) still has a useful interpretation, however. Since (14.6) is a forced response of the system regardless of the values of $A, B, \ldots, Q$ we may set all these numerators to zero leaving only the term deduced in (14.9). Thus the frequency response equation is valid for the *forced* response in any system, whether stable or not, and *steady-state* response in stable systems. Since the forced response may be a negligible part of the total response for unstable systems, it is the stable steady-state case that we pursue.

14.1.1   For $\mathbf{H}(s) = (s+1)/(s^2+s+3)$, what are the gain and phase shift at $\omega = 2$ rad/s? At $f = 1$ Hz?

        Answer   1.00, $-53.1°$; 0.172, $-89.3°$

14.1.2   For $\mathbf{H}(s) = -2/(s+4)$, find all frequencies $\omega > 0$ at which the amplitude of the output is $\frac{1}{2}$ when the input amplitude is 2. Repeat for phase shift $+114°$.

        Answer   One ω for each: 6.93 rad/s, 8.98 rad/s

14.1.3   In the circuit of Fig. 14.2, if the output is defined to be the voltage across the resistor, find the gain.

$$Answer \quad \frac{\omega}{\sqrt{\omega^2 + (1/RC)^2}}$$

14.2   THE DECIBEL SCALE

The use of a linear scale to measure gain has its limitations. One is the small dynamic range it makes available for graphing. We cannot simultaneously make out the detailed behavior of the gain curve where it is very high and very low. In Fig. 14.3(a), for instance, we cannot choose a linear scale that permits us to determine the frequency at which the gain is, say, 10% greater than its value at $\omega = 1/RC$ and another ω whose gain is 10% greater than the much smaller value at $100/RC$. Detail in the baseline is lost if we scale for detail elsewhere.

A related problem with a direct linear scale is that it is very often *fractional* changes in a quantity such as gain that are of interest, not the change itself. For instance, experiments with human listeners show that we perceive one sound as significantly louder than another not when their strengths differ by some fixed value, but when the ratios of their strengths differ by a fixed value. We would like a scale that measures, for instance, 50% changes in gain between two frequencies as equal displacements, regardless of their absolute level.

The mapping of equal ratios into equal displacements is a logarithmic scale. If we use a log scale, say $Y = \log X$, to measure two numbers X_1 and X_2 whose ratio is $X_2/X_1 = R$, their difference on this scale is fixed:

$$Y_2 - Y_1 = \log X_2 - \log X_1 = \log \frac{X_2}{X_1} = \log R$$

regardless of their actual value. So numbers such as $X_1 = 10^{10}$ and $X_2 = 10^{11}$ will be displaced on this scale equally as much as numbers such as $X_1 = 10^{-4}$ and $X_2 = 10^{-3}$, and we can make out both these fluctuations simultaneously on the same graph.

The most common scale used to measure gain is a logarithmic scale called the *decibel scale*, which is defined for positive real numbers R by

$$R \text{ (dB)} = 20 \log_{10} R \qquad\qquad (14.16)$$

Equation (14.16) is to be read *the number R measured in decibels equals 20 times log base 10 of R*. Historically, the *bel* came first, originally defined by Alexander Graham Bell (1847–1922), the inventor of the telephone, as a measurement of the ratio of two powers P_1 and P_2.

$$\text{power ratio (bels)} = \log_{10} \frac{P_2}{P_1} \tag{14.17}$$

Thus a resistor dissipating 4 W of power is operating at a power level 3 bels above one dissipating only 4 mW. Bell later decided this scale was too coarse and scaled it by a factor of 10 to create the decibel scale for the measurement of power:

$$\text{power ratio (dB)} = 10 \log_{10} \frac{P_2}{P_1} \tag{14.18}$$

To use this scale to measure gain, note that the ratio of output amplitudes A_2 and A_1 produced by the same input at two frequencies whose gains are $|\mathbf{H}(j\omega_2)|$ and $|\mathbf{H}(j\omega_1)|$ is

$$\frac{A_2}{A_1} = \frac{|\mathbf{H}(j\omega_2)|}{|\mathbf{H}(j\omega_1)|} \tag{14.19}$$

so the ratio of their powers P_2 and P_1 must be as the square of the gains,

$$\frac{P_2}{P_1} = \frac{|\mathbf{H}(j\omega_2)|^2}{|\mathbf{H}(j\omega_1)|^2} \tag{14.20}$$

Then, applying (14.20) to (14.18), we have

$$\text{gain (dB)} = 10 \log_{10} \frac{|\mathbf{H}(j\omega_2)|^2}{|\mathbf{H}(j\omega_1)|^2} \tag{14.21}$$

$$\text{gain (dB)} = 20 \log_{10} \frac{|\mathbf{H}(j\omega_2)|}{|\mathbf{H}(j\omega_1)|} \tag{14.22}$$

which agrees with our definition (14.16) of the decibel scale.

Example 14.4

Graph the gain $|\mathbf{H}(j\omega)|$ of the circuit of Example 14.3 in decibels. By (14.14),

$$|\mathbf{H}(j\omega)| \, (\text{dB}) = 20 \log_{10} \frac{1}{RC} - 20 \log_{10} \sqrt{\omega^2 + \left(\frac{1}{RC}\right)^2} \tag{14.23a}$$

$$= 20 \log_{10} \frac{1}{RC} - 10 \log_{10} \left[\omega^2 + \left(\frac{1}{RC}\right)^2\right] \tag{14.23b}$$

The graph is shown in Fig. 14.4. Note that there is 3-dB difference between the gains at $\omega = 0$ and $\omega = RC$ and also 3-dB between $\omega = 100/RC$ and $\omega = 100\sqrt{2}/RC$. The latter gains are so small that they are very hard to make out on a linear scale [see Fig. 14.3(a)].

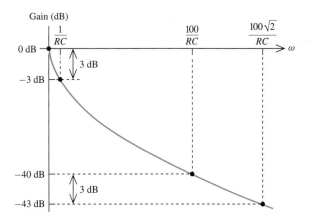

FIGURE 14.4   Gain in dB for Example 14.4.

There are a few numbers R whose decibel equivalents are worth remembering. The number 1 measured in decibels is clearly 0 dB. The ratio 10:1, or number $R = 10$, measured in decibels is

$$10 \text{ (dB)} = 20 \log_{10} 10 = +20 \text{ dB} \qquad (14.24)$$

Note that when dB is put in parentheses after a number R, it is to be read as "the number R, when converted to decibels, is," while dB next to a number without parentheses indicates that the value already has units of decibels, such as $+20$ dB in (14.23). The implication of (14.24) is that *multiplication by 10 always adds 20 dB*. This follows since *the product of two numbers when converted to decibels, equals the sum of their decibel equivalents.*

$$RS \text{ (dB)} = 20 \log_{10} (RS) = 20 \log_{10} R + 20 \log_{10} S$$

$$RS \text{ (dB)} = R \text{ (dB)} + S \text{ (dB)} \qquad (14.25)$$

If $R = 10$, then R (dB) $= +20$ dB, and by (14.24) we add 20 dB if we scale S by a factor of 10. Replacing S by $1/S$ in (14.24), we have

$$\frac{R}{S} \text{ (dB)} = 20 \log_{10} R - 20 \log_{10} S = R \text{ (dB)} - S \text{ (dB)} \qquad (14.26)$$

When converted to decibels, the ratio of two numbers equals the difference of their decibel equivalents. So, since multiplication by 10 adds 20 dB, *division by 10 subtracts 20 dB.*

In addition, the number 2 converted to decibels is almost exactly equal to 6 dB:

$$2 \text{ (dB)} = 20 \log_{10} 2 = 6.02 \text{ dB} \cong 6 \text{ dB} \qquad (14.27)$$

So *multiplication by 2 adds 6 dB and division by 2 subtracts 6 dB.* Finally, since raising to the $\frac{1}{2}$ power scales the log by $\frac{1}{2}$, *multiplication by $\sqrt{2}$ adds 3 dB, and division by $\sqrt{2}$ subtracts 3 dB.*

Example 14.5

The number 84.4 measured in decibels is $20 \log_{10} 84.4 = 38.5$ dB (to three significant digits). Then

$$844 \text{ (dB)} = 38.5 \text{ dB} + 20 \text{ dB} = 58.5 \text{ dB}$$

$$844{,}000 \text{ (dB)} = 38.5 \text{ dB} + 4(20 \text{ dB}) = 118.5 \text{ dB}$$

Also,

$$8.44 \text{ (dB)} = 38.5 \text{ dB} - 20 \text{ dB} = 18.5 \text{ dB}$$

$$8.44 \times 10^{-10} \text{ (dB)} = 38.5 \text{ dB} - 11(20 \text{ dB}) = -181.5 \text{ dB}$$

Also,

$$16.88 \text{ (dB)} = 38.5 \text{ dB} - 20 \text{ dB} + 6 \text{ dB} = 24.5 \text{ dB}$$

$$4.22(\text{dB}) = 38.5 \text{ dB} - 20 \text{ dB} - 6 \text{ dB} = 12.5 \text{ dB}$$

and so on.

EXERCISES

14.2.1 Without using a calculator, determine 0.001 (dB), 0.004 (dB), and $500\sqrt{2}$ (dB). Then verify with a calculator.
Answer -60 dB; -48 dB; $+57$ dB

14.2.2 If an amplifier section has gain of $+18$ dB, how many must be connected back to back (in cascade) to lift a 1-mV amplitude input to 250-V amplitude?
Answer 6

14.2.3 If the input and output phasors at a certain frequency ω are $320/\underline{30°}$ and $210/\underline{-41°}$, respectively, what is the gain in decibels at this frequency?
Answer -3.66 dB

14.3 BODE GAIN (AMPLITUDE) PLOTS

The frequency response function $\mathbf{H}(j\omega)$ reveals the steady-state behavior of a circuit at any single frequency and, by superposition, due to a sum of inputs at different frequencies. It is very useful to be able to visualize the dependence of $\mathbf{H}(j\omega)$ on frequency. Where in the frequency domain, for instance, is the gain strong and where is it weak? An input sum of two sinusoids may have its two components equal in amplitude coming into a circuit, but in the output the sinusoid whose gain is high will dominate one whose frequency corresponds to low gain. One is effectively blocked and the other passed. This type of behavior can be told at a glance from the gain curve.

An exact plot of gain versus frequency, whether gain is measured on a linear scale or in decibels, is somewhat tedious to produce, as can be seen from (14.14) and (14.23). Among his several important contributions to circuit theory and control theory, the German-born engineer Hendrick Bode (1905–1982), while working at Bell Labs in the

United States in the 1930s, devised a simple but accurate method for graphing gain and phase-shift plots. They bear his name, *Bode gain plot* and *Bode phase plot* (pronounced *boh'-duh*, not the frequently heard *boh-dee* or *bowd*).

Given a transfer function $\mathbf{H}(s)$, consider its factored form in which numerator and denominator of $\mathbf{H}(s)$ are factored separately:

$$\mathbf{H}(s) = \frac{K(s - z_1)^{q_1} \cdots (s - z_m)^{q_m}}{(s - p_1)^{r_1} \cdots (s - p_n)^{r_n}} \tag{14.28}$$

where p_i are the poles, of multiplicity r_i, and z_i the zeros, with multiplicity q_i. In general, some of the poles and zeros will be real and others complex. It is always the case that if p_i is a complex pole of order r_i, its complex conjugate p_i^* will also be an order r_i pole; that is, $p_j = p_i^*$ for some $j = 1, 2, \ldots, n$ and $r_j = r_i$. This is also true of complex zeros (the appearance of complex poles or zeros only in conjugate pairs was first noted in Section 12.4 in the discussion of partial fraction expansions). Recombining the conjugate pair factors in (14.28), let p be the value of a complex pole or zero:

$$(s - p)(s - p^*) = s^2 + as + b \tag{14.29a}$$

Each pair of complex conjugate poles or zeros corresponds to a quadratic factor in $\mathbf{H}(s)$. The coefficients are unity (for s^2), and we label the s-coefficient a and the constant b. For each quadratic factor, define two parameters

$$\omega_n = \sqrt{b} \tag{14.29b}$$

$$\zeta = \frac{a}{2\omega_n} \tag{14.29c}$$

or

$$s^2 + as + b = s^2 + (2\zeta\omega_n)s + \omega_n^2 \tag{14.30}$$

ω_n is called the *natural frequency* of the complex root pair factor (14.29a), and ζ is its *damping factor*. The resulting form for $\mathbf{H}(s)$, after recombination of complex root pair factors into quadratic factors (14.29a), will be called the *standard form*. We begin the process of Bode plotting by putting $\mathbf{H}(s)$ in standard form.

Example 14.6

Put the transfer function $H(s) = (2s + 6)/(s^3 + 2s^2 + 2s)$ into standard form. First factoring, we find one zero and three distinct poles each of order 1,

$$\mathbf{H}(s) = \frac{2(s + 3)}{s(s + 1 + j)(s + 1 - j)}$$

Recombining the conjugate terms yields

$$\mathbf{H}(s) = \frac{2(s+3)}{s(s^2+2s+2)}$$

We may identify a constant $K = 2$, a single zero at $s = -3$, a pole at $s = 0$, and a quadratic factor with natural frequency $\omega_n = \sqrt{2}$ and damping factor $\zeta = \sqrt{2}/2$.

In standard form, the transfer function is a ratio of products,

$$\mathbf{H}(s) = \frac{K\mathbf{N}_1(s)\cdots\mathbf{N}_\mu(s)}{\mathbf{D}_1(s)\cdots\mathbf{D}_\eta(s)} \qquad (14.31)$$

where, after recombination of complex factors, each $N_i(s)$ and $D_i(s)$ is either of the form $(s+a)^r$ with a real, a *real root* factor, or $(s^2+2\zeta\omega_n s+\omega_n^2)^r$, a *complex root pair* factor. Replacing s by $j\omega$ and taking magnitudes, the gain is

$$|\mathbf{H}(j\omega)| = \frac{|K|\,|\mathbf{N}_1(j\omega)|\cdots|\mathbf{N}_\mu(j\omega)|}{|\mathbf{D}_1(j\omega)|\cdots|\mathbf{D}_\eta(j\omega)|} \qquad (14.32)$$

By (14.25), the decibel value of a product is the product of decibel values, so

$$|\mathbf{H}(j\omega)|\,(\text{dB}) = |K|\,(\text{dB}) + \sum_{i=1}^{\mu}[|\mathbf{N}_i(j\omega)|\,(\text{dB})] - \sum_{i=1}^{\eta}[|\mathbf{D}_i(j\omega)|\,(\text{dB})] \qquad (14.33)$$

The strategy for plotting the gain $|\mathbf{H}(j\omega)|$ in decibels will be to plot each term on the right of (14.33) separately and then add these component plots graphically. The result will be the desired gain in decibels. There are only two kinds of terms, real root terms and complex root pair terms, to learn how to plot. More precisely, there is a third, the constant term $|K|$ (dB), but the graph of this term will clearly be a flat straight line at the level $20\log_{10}|K|$.

Turning first to the real root case, assume for a moment that this term is in the numerator. Then

$$|\mathbf{N}_i(j\omega)|\,(\text{dB}) = 20\log_{10}|\mathbf{N}_i(j\omega)|$$
$$= 20\log_{10}|j\omega+a|^r \qquad (14.34)$$

where r is the multiplicity of the real zero $-a$. This can be simplified slightly to

$$|\mathbf{N}_i(j\omega)|\,(\text{dB}) = 20r\log_{10}|j\omega+a| \qquad (14.35)$$

For $\omega \ll a$, $|j\omega+a|$ may, with vanishingly small error as $\omega/a \to 0$, be replaced by $|a|$. The gain of this term is then

$$|\mathbf{N}_i(j\omega)|\,(\text{dB}) \cong 20r\log_{10}|a|, \qquad \omega \ll a \qquad (14.36)$$

Thus, for $\omega \ll a$, the graph will be a flat straight line at the level $20r\log_{10}|a|$. For

$\omega \gg a$, $|j\omega + a|$ may, with vanishingly small error as $\omega/a \to \infty$, be replaced by $|j\omega|$, which is just ω itself. So

$$|\mathbf{N}_i(j\omega)| \text{ (dB)} \cong 20r \log_{10} |j\omega| = 20r \log_{10} \omega, \qquad \omega \gg a \qquad (14.37)$$

The graph for $\omega \gg a$ will take on a particularly simple form if we choose to plot gain not versus ω but versus $\log_{10} \omega$. In this case, (14.37) indicates that we have an equation of the form $y = mx$ with y the gain in decibels and x the log base 10 of ω, that is, a simple linear relationship between these variables. The graph for $\omega \gg a$ will be a straight line with slope m of $20r$ dB gain change per unit change in the log base 10 of ω. A change of one unit in $\log_{10} \omega$ corresponds to a change of a factor of 10 in ω itself, so we refer to this slope as $20r$ *dB per decade*.

The two straight-line segments for $w \ll a$ and $\omega \gg a$ meet at $\omega = |a|$, as can be seen by comparing (14.36) and (14.37). For this reason, $\omega = |a|$ is referred to as the *break frequency*. A plot of the piecewise linear approximation to the gain curve that we have determined, a pair of straight lines meeting at the break frequency, is called the *uncorrected Bode gain plot* and is shown in Fig. 14.5 along with the exact or *corrected Bode gain plot* computed by evaluating (14.35). Note that the maximum error occurs at the break frequency and is $3r$ dB. Far from the break frequency the uncorrected and corrected plots merge smoothly.

The plot of the same factor, $(s + a)^r$, when moved to the denominator of $H(s)$, simply flips around the horizontal axis. This follows, since the decibel value of $1/X$ is the negative of the decibel value of X. The initial value is then $-[|a| \text{ (dB)}]$, the negative of the decibel equivalent of $|a|$, and the slope after the break is $-20r$ dB/decade.

Note that there is a special case of the real root factor when the pole or zero $a = 0$. In this case the form given in (14.37) is not approximate for $\omega \gg a$ but is exact for all ω. This plot passes through the value 0 dB at $\omega = 1$. The plot for this factor is shown in Fig. 14.6. The technique for using these component graphs to generate the Bode gain plots, both uncorrected and corrected, is illustrated in Example 14.7.

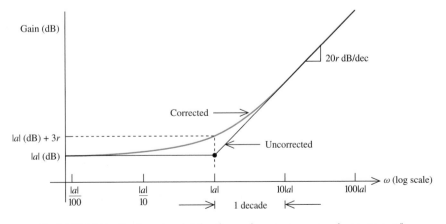

FIGURE 14.5   Bode gain plot for the real root numerator factor $(s + a)^r$.

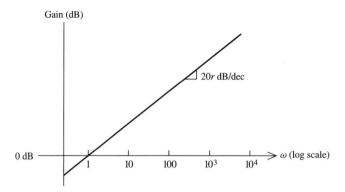

Gain (dB)

20r dB/dec

0 dB

ω (log scale)

1   10   100 10^3 10^4

FIGURE 14.6 Bode gain plot for the numerator factor s^r.

Example 14.7

Let us determine the uncorrected and corrected Bode gain plots for $H(s) = (-5s + 10)/(s^2 + 20s)$. In standard form,

$$\mathbf{H}(s) = \frac{(-5)(s + 2)}{s(s + 20)} \tag{14.38}$$

There is a zero at $s = -2$, with poles at $s = -20$ and $s = 0$. The real factor $(s + 2)$ has its break frequency at $\omega = 2$ and slope 20 dB/decade after the break $(r = 1)$. The initial level value of this factor is $|2|$ (dB) $= 20\log_{10}|2| = 6$ dB. The real denominator factor $(s + 20)$ has its break frequency at $\omega = 20$, slope -20 dB/decade after the break, and initial level value $-20\log_{10} 20 = -26$ dB. The factor of s in the denominator has a slope of -20 dB/decade everywhere and passes through 0 dB at $\omega = 1$. Finally, the constant factor $K = -5$ in (14.38) has a constant graph at the level $|K|$ (dB) $= |-5|$ (dB) $= 5$ (dB) $= 14$ dB. The component plots for each of these factors is shown in Fig. 14.7. Note that no corrections are required for the s^r factor or the constant factor component plots.

Next we add the uncorrected component plots graphically. First we consider the slopes in each region. Recall that the slope of a sum of functions equals the sum of the slopes. Working left to right, the initial segments all have slopes of zero except the $1/s$ factor, which has slope of -20 dB/decade. Then the slope of the sum will be -20 dB/decade. This will change, that is, the slope will *break*, at the leftmost (lowest) break frequency $\omega = 2$. Adding slopes at $\omega = 2+$ results in 0 dB/decade, since the net slope was -20 dB/decade at $\omega = 2-$, one slope increased (or "broke") $+20$ dB/decade at 2, and the others did not change. There is a downward break of $+20$ dB/decade at the second break frequency $\omega = 20$ and no further breaks in slope. This yields the black curve in Fig. 14.8.

We are ready to determine the decibel values. Since all slopes are known, we need only know the decibel value at a single point in order to determine all other decibel values. Examining the com-

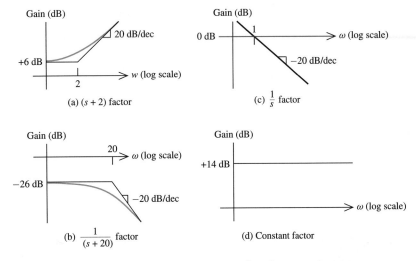

FIGURE 14.7 Component plots for Example 14.7.

ponent graphs, all values are known at $\omega = 1$. Adding, the gain in decibels at $\omega = 1$ is $+6 - 26 + 0 + 14 = -6$ dB. The gain at the break frequency $\omega = 2$ is then the value at $\omega = 1$ reduced by -20 dB/decade times the number of decades between frequencies 1 and 2. Since decades measure the factors of 10 separating two frequencies,

$$D = \log_{10} \frac{\omega_2}{\omega_1} \qquad (14.39)$$

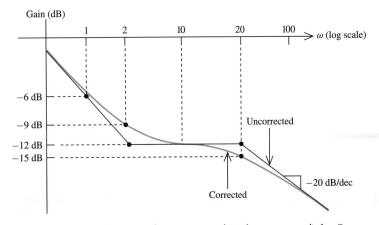

FIGURE 14.8 Bode gain plots (corrected and uncorrected) for Example 14.7.

where D is the number of decades between ω_2 and ω_1. Thus, in the example, there are

$$D = \log_{10} \frac{2}{1} = 0.3 \text{ decade}$$

separating the frequencies $\omega_2 = 2$ and $\omega_1 = 1$, and the gain in dB at $\omega = 2$ is

$$\begin{aligned}
|\mathbf{H}(j2)| \text{ (dB)} &= |\mathbf{H}(j1)| \text{ (dB)} + (-20 \text{ dB/decade })(0.3 \text{ decade}) \\
&= -6 - 20(0.3) \\
&= -12 \text{ dB}
\end{aligned}$$

Thus we have the final uncorrected Bode plot. The corrected plot is the sum of the corrected component plots, which will differ by 3 dB inside the knee at each break frequency. Both uncorrected and corrected Bode gain plots are shown in Fig. 14.8. This completes the example.

Having seen how the constant and real root terms are plotted, we turn last to the complex root pair term $(s^2 + 2\zeta\omega_n s + \omega_n^2)^r$. Taking the same path that we followed in understanding the real root term, we replace s by $j\omega$ and take magnitudes. For the case that the term is in the numerator, the gain in decibels is

$$|\mathbf{N}_i(j\omega)| \text{ (dB)} = 20r \log_{10} |(\omega_n^2 - \omega^2) + j(2\zeta\omega_n\omega)| \qquad (14.40)$$

The approximate values for $\omega \ll \omega_n$ and $\omega \gg \omega_n$ are

$$|\mathbf{N}_i(j\omega)| \text{ (dB)} \cong \begin{cases} 20r \log_{10} \omega_n^2, & \omega \ll \omega_n \\ 20r \log_{10} \omega^2, & \omega \gg \omega_n \end{cases}$$

Comparing these two, the straight-line segments comprising the uncorrected plot meet at $\omega = \omega_n$, so the break frequency for the complex root pair term is its natural frequency ω_n. The slope before the break is 0, and beyond the break is $40r$ dB/decade. The graph of the uncorrected Bode plot is shown in Fig. 14.9 in black.

While the uncorrected Bode plot for this term is independent of its damping factor ζ, the correction to be applied depends strongly on its damping factor. The gain curves for various values of ζ ranging from its minimum (for complex roots) absolute value of 1 upward is shown in Fig. 14.9. The maximum correction occurs near, but not precisely at, the break frequency ω_n. To determine the correction at the break frequency ω_n, evaluate the exact gain expression at $\omega = \omega_n$:

$$|\mathbf{N}_i(j\omega_n)| \text{ (dB)} = 20r \log_{10} |j2\zeta\omega_n^2| = 20r \log_{10} 2|\zeta|\omega_n^2$$

or $$|\mathbf{N}_i(j\omega_n)| \text{ (dB)} = 20r \log_{10} 2|\zeta| + 20r \log_{10} \omega_n^2 \qquad (14.41)$$

Since the second term in (14.41) is the uncorrected value at ω_n, the correction must be the first term, or $20r \log_{10} 2|\zeta|$. The reference direction for this correction is upward, or inside the knee of the uncorrected plot. If $|\zeta| < \frac{1}{2}$, this correction will be negative, and the graph will be corrected downward. Noting that the correction $20r \log_{10} 2|\zeta|$ is r times the number $2|\zeta|$ measured in decibels, a general rule for correcting the complex pair term that works for numerator and denominator terms alike is this. *The correction to any complex root pair graph at $\omega = \omega_n$ is r times the decibel equivalent of $2/\zeta$, with reference*

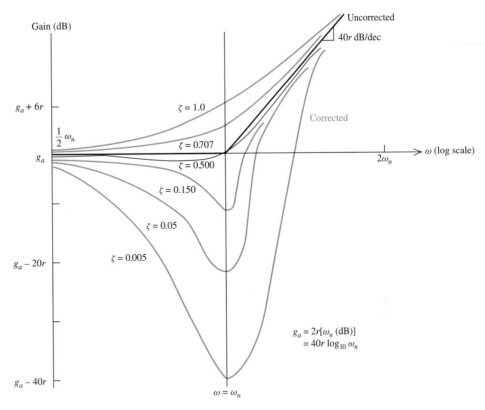

Gain (dB)

Uncorrected

40r dB/dec

$g_a + 6r$

$\frac{1}{2}\omega_n$

$\zeta = 1.0$

g_a

$\zeta = 0.707$

Corrected

$2\omega_n$

ω (log scale)

$\zeta = 0.500$

$\zeta = 0.150$

$\zeta = 0.05$

$g_a - 20r$

$\zeta = 0.005$

$g_a = 2r[\omega_n \text{ (dB)}]$
$= 40r \log_{10} \omega_n$

$g_a - 40r$

$\omega = \omega_n$

FIGURE 14.9   Bode gain plot for the complex root pair numerator factor
$(s^2 + 2\zeta\omega_n s + \omega_n^2)^r$.

direction inside the knee of the uncorrected plot. Examining Fig. 14.9, the correction at
$\omega = \omega_n$ is very close to the desired maximum correction for both lightly and heavily
damped systems, those with damping factors $\zeta < 0.3$ or $\zeta > 0.6$. For moderately damped
systems, $0.3 \leq \zeta \leq 0.6$, the calculated correction, while accurate at $\omega = \omega_n$, is not a
good measure of the maximum correction, which occurs far from $\omega = \omega_n$. Fortunately,
the maximum correction is not large for moderately damped systems.

Example 14.8

Determine the uncorrected and corrected Bode gain plots for

$$\mathbf{H}(s) = \frac{1}{\left(s^2 + \frac{3}{2}s + 900\right)^3}$$

Checking the radical of the denominator quadratic, we find it is
negative, the roots are complex, and this is a complex root pair term.
Its natural frequency and damping factor are, by (14.30), found from

$$\omega_n^2 = 900$$

$$2\zeta\omega_n = \frac{3}{2}$$

or $\omega_n = 30$, $\zeta = 3/[(2)(2)(30)] = \frac{1}{40}$. The break frequency is $\omega_n =$
30, and slope is $-40r = -120$ dB/decade after the break (this is a

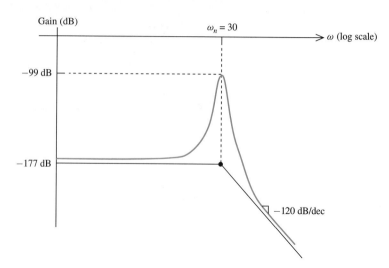

Gain (dB)

$\omega_n = 30$

ω (log scale)

−99 dB

−177 dB

−120 dB/dec

FIGURE 14.10 Bode gain plots for Example 14.8.

denominator factor). The initial value is, by (14.40), $-20r \log_{10} \omega_n^2$ (or $-2r$ times the decibel equivalent of ω_n) with minus sign due to the denominator location. This initial value is

$$-20r \log_{10} \omega_n^2 = (2)(-60) \log_{10} 30 = -177 \text{ dB}$$

This completes the uncorrected plot, shown in Fig. 14.10. The correction is $r = 3$ times the decibel equivalent of $2|\zeta| = \frac{1}{20}$, which is $3\left[\frac{1}{20} \text{ (dB)}\right] = 3(-26 \text{ dB}) = -78 \text{ dB}$. The reference direction for this correction is inside the knee (downward in this case), so since the correction is negative, it goes outside the knee (upward). The corrected plot is also shown in Fig. 14.10.

Any number of real root and complex root pair terms may appear in the same transfer function, as shown in the following example. Their component plots are simply added together, as are any other pair of terms in $\mathbf{H}(s)$.

Example 14.9

Let us find the Bode gain plots for a circuit with transfer function

$$\mathbf{H}(s) = \frac{200(s^2 + 2s + 16)}{(s + 50)(s + 2500)}$$

In the numerator there is a constant factor and a complex root pair factor with $\omega_n^2 = 16$ and $2\zeta\omega_n = 2$, or $\omega_n = 4$, $\zeta = \frac{1}{4}$. In the denominator we see real root factors with break frequencies of 50 and 2500. The component plots are shown in Fig. 14.11. Starting from the left, the uncorrected Bode plot will have a slope that is the sum of the slopes of the components; $0 + 0 + 0 + 0 = 0$ dB/decade. The plot breaks upward $+40$ dB/decade at $\omega = 4$, the break frequency (natural frequency) of the complex root pair factor in the numerator, down -20 dB/decade at $\omega = 50$, and again at $\omega = 2500$. The dB value at $\omega = 4$ in the uncorrected plot is the sum of the initial

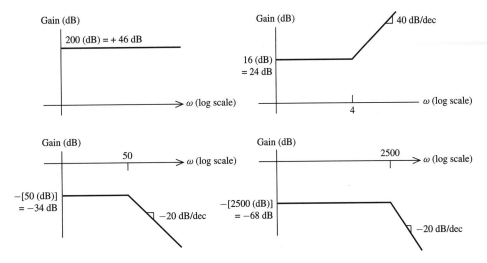

FIGURE 14.11   Component plots for Example 14.9.

values,

$$46 + 24 - 34 - 68 = -32 \text{ dB}$$

Since there are $\log_{10} 50/4 = 1.1$ decades between $\omega = 4$ and $\omega = 50$, the uncorrected gain at $\omega = 50$ is -32 dB $+(40$ dB/decade$)$ $(1.1$ decade$) = +12$ dB, and at $\omega = 2500$ it is similarly found to be $+12$ dB $+ (20$ dB/decade$)$ $(1.7$ decade$) = 46$ dB. The correction due to the complex root term at $\omega = 4$ is $2|\zeta|$ (dB) $= \frac{1}{2}$ (dB) $= -6$ dB or 6 dB outside the knee (downward). That at the other two break frequencies is 3 dB inside the knee (downward). The uncorrected and corrected Bode gain plots are shown in Fig. 14.12.

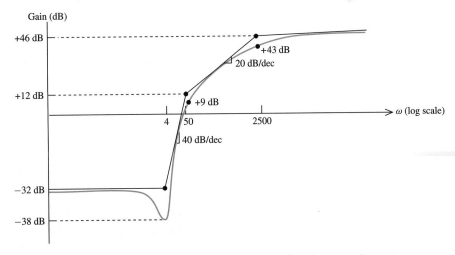

FIGURE 14.12   Bode gain plots for Example 14.9.

This completes our introduction to Bode's gain plotting method. There is a related method for plotting the phase shift by piecewise linear approximation, graphical addition, and correction, producing the *Bode phase plot*. There are several reasons why the Bode phase plotting technique is not as widely used as the Bode gain plot described above. The Bode phase plot method is more cumbersome, since there are two, not one, break frequencies for each real root term. The corrections spread over a much wider frequency band than in the gain plot and are more difficult to estimate accurately by freehand sketch. In addition, for many applications the variation of circuit gain with frequency is more important to its intended function than are phase variations. For these reasons, while it is not without its uses, we will not discuss the Bode phase plotting method further.

In summary, while the process is not entirely without effort, the Bode gain plot permits relatively quick and accurate freehand plots of gain versus frequency in a circuit, and gain is often a circuit's most important single characteristic. We can get considerable insight into what a circuit "does," the function it performs, by identifying the frequency bands of high and low gain. If a computer running SPICE is handy, an ac analysis will get us a more accurate gain plot with less work. But insight into how the Bode procedure works informs us of how the circuit gain curve will change if we add or delete poles or zeros or if we shift their locations. When confronted with a design problem, this understanding can be used to guide our selection of a transfer function and ultimately a circuit design. Neither SPICE, nor any other computer-based simulation tool, can substitute for a circuit designer's own understanding of how the frequency response of circuits can be manipulated. This is the real payoff for studying the Bode gain plot method.

EXERCISES

14.3.1 For each component factor in $\mathbf{H}(s) = 4000(s^2-1)/[(s+20)^2(s^2+3s+36)]$, specify the initial slope, final slope, initial value, and location of break frequency. Also specify ω_n and ζ for complex pole pair factors.

Answer

4000: 0 dB/decade, 0 dB/decade, +72 dB, no break frequency

$(s-1)$: 0 dB/decade, +20 dB/decade, 0 dB, 1 rad/s

$(s+1)$: 0 dB/decade, +20 dB/decade, 0 dB, 1 rad/s

$1/(s+20)^2$: 0 dB/decade, −40 dB/decade, −52 dB, 20 rad/s

$1/(s^2+3s+36)$: 0 dB/decade, −40 dB/decade, −31.1 dB, 6 rad/s,

$$\omega_n = 6, \quad \zeta = 1/4$$

14.3.2 Sketch the uncorrected and corrected Bode gain plots for the $\mathbf{H}(s)$ of Exercise 14.3.1, labeling all slopes and both coordinates at each breakpoint.

Answer

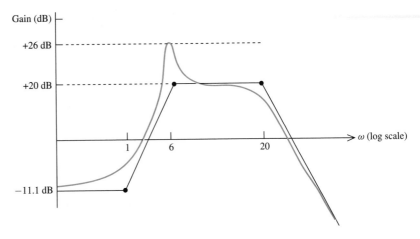

EXERCISE 14.3.2

14.3.3 Sketch the Bode gain plots for $\mathbf{H}(s) = (s^2 + s + 100)/(s^2 + 20s + 100)$.

Answer

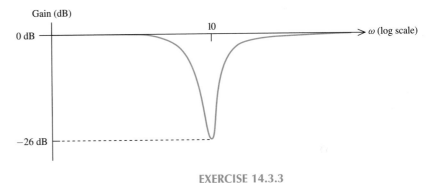

EXERCISE 14.3.3

14.4 RESONANCE

The frequency response function $\mathbf{H}(j\omega)$ reveals the frequency-dependent behavior of a circuit, and Bode gain plots offer a convenient means to access this information. We next examine specific circuits to see what sort of frequency-dependent behavior we may have and why we might want to have it.

Of considerable interest are circuits that treat a narrow range of frequencies very differently than all other frequencies. These are referred to as *resonant circuits*. The gain of a highly resonant circuit attains a sharp maximum or minimum at its *resonant frequency*, while its variations outside a small band of frequencies around the resonant frequency are much lower.

Resonance is an extremely useful property for many circuit applications. For instance, suppose that we wish to build a circuit to receive a sinusoidal input at frequency ω_0. The total input to our circuit will in general contain many other stray frequency components due to noise and other interfering signals being transmitted at the same time. If the gain of our receiver is very high at ω_0 and very low elsewhere, we will effectively block the unwanted components, which will pass through the circuit from input to output much reduced in amplitude (low gain at those frequencies), while strongly passing the desired component (high gain at its frequency ω_0). Or consider the laser, a device that amplifies those light waves passing through it that are of a specific frequency ω_0 very strongly, while sharply attenuating all others, thus producing very pure light output with almost all its energy at a single frequency. These are examples of resonant circuits.

Resonance is not a property of electrical circuits only. It is found throughout nature in many physical forms. When a pitch pipe is played near a guitar, one of the guitar strings will continue to "sing" long afterward if it is tuned to the pipe's pitch (frequency). Each of the instrument's strings is a resonant system with different resonant frequency, and only the one matched to the input frequency will respond strongly. Similarly, a quartz crystal will "sing" at its resonant frequency when stimulated electrically, and the resulting pure output is used to time computer circuits and digital watches. Some automobiles, while accelerating, will resonate strongly (and annoyingly) as the engine speed passes through the resonant frequency of the car's body or chassis. At other speeds the engine vibrations are just as strong but do not produce strong vibrations in the passenger compartment due to the low gain off resonance.

One circuit capable of exhibiting resonance is the *series RLC circuit*. Consider the input to this circuit to be its voltage $v(t)$ and its output to be its current $i(t)$ in Fig. 14.13(a). Define its transfer function as $\mathbf{H}(s) = \mathbf{I}(s)/\mathbf{V}(s)$, which is identical with its admittance $\mathbf{Y}(s)$.

$$\mathbf{H}(s) = \mathbf{Y}(s) = \frac{1}{sL + R + (1/sC)} \tag{14.42}$$

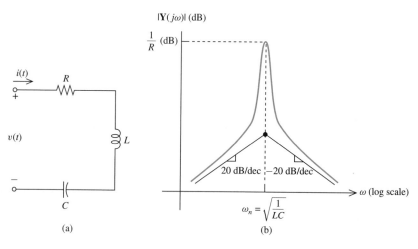

FIGURE 14.13 (a) Series *RLC* circuit; (b) Bode gain plot.

In standard form, $\mathbf{H}(s)$ may be written

$$\mathbf{H}(s) = \frac{(1/L)s}{s^2 + (R/L)s + (1/LC)} \tag{14.43}$$

The Bode gain plot will consist of a constant term due to the factor $1/L$, a real zero at the origin, and a quadratic term that may contain either complex or real roots. The roots will be complex if the radical of the quadratic is negative:

$$\left(\frac{R}{L}\right)^2 - \frac{4}{LC} < 0$$

which may be rewritten

$$\frac{R}{2}\sqrt{\frac{C}{L}} < 1 \tag{14.44}$$

We will assume this inequality is satisfied, so the roots are complex. Then, by (14.30),

$$H(s) = \frac{N(s)}{s^2 + 2\zeta\omega_n s + \omega_n^2}$$

where the denominator is an underdamped quadratic function. By (14.29), the natural frequency ω_n and damping factors are found from the coefficients of the quadratic factor via $\omega_n^2 = 1/LC, 2\zeta\omega_n = R/L$ to be

$$\omega_n = \sqrt{\frac{1}{LC}}$$

$$\zeta = \frac{R}{2}\sqrt{\frac{C}{L}} \tag{14.45}$$

The resulting Bode plot is shown in Fig. 14.13(b). There is resonant peak at ω_n, and the corrected gain at ω_n is $1/R$. Far from ω_n, the gain rolls off at 20 dB/decade.

Note from (14.44) and (14.45) that the requirement for complex poles is that the damping factor $\zeta < 1$. Define the *resonant frequency* ω_r as the frequency at which the gain $|\mathbf{H}(j\omega)|$ passes through its maximum or minimum (in this case maximum). For the series RLC circuit, the impedance

$$Z(j\omega) = R + j\omega L + \frac{1}{j\omega C} = R + j\left(\omega L - \frac{1}{\omega C}\right)$$

consists of a fixed real part and imaginary part that varies with frequency. At the frequency at which $\omega L - 1/(\omega C) = 0$ the imaginary part vanishes and the impedance is at a minimum magnitude. This is the frequency $\omega = \omega_n = \sqrt{1/LC}$. If the impedance is a minimum, the admittance is a maximum, and thus

$$\omega_r = \sqrt{\frac{1}{LC}} \tag{14.46}$$

This is the resonant frequency of the series RLC circuit and coincides with the natural frequency ω_n.

The width of the resonant peak is a useful measure of the frequency selectivity of the resonant circuit. Define the *bandwidth B* of a resonant circuit to be the difference

$$B = \omega_u - \omega_l \qquad (14.47)$$

where ω_u and ω_l are the frequencies at which the gain is 3 dB below its value at ω_r (or 3 dB above its value at ω_r if the resonance is a minimum of the gain). $\omega_u > \omega_r$ is called the *upper half-power frequency* and $\omega_l < \omega_r$ the *lower half-power frequency*. Note that if the amplitude of a sinusoid is reduced by 3 dB, or by a factor of $1/\sqrt{2}$, the power in that sinusoid is reduced by the amplitude ratio squared, or $\frac{1}{2}$. This explains the terminology for ω_u and ω_l. For the present case, their values can be found by solving the equation

$$|Y(j\omega)| = \frac{1}{\sqrt{2}} \frac{1}{R} \qquad (14.48)$$

for ω, since $|\mathbf{Y}(j\omega)|$ is the gain and its maximum value is $1/R$. Leaving the algebra to the exercises at the end of this section, after subtraction of the two solutions to (14.48), the bandwidth for the series RLC is simply

$$B = \frac{R}{L} \qquad (14.49)$$

While the bandwidth B measures the width of the frequency band within which the circuit is behaving in a near-resonant fashion, that is, with gains within 3 dB of the resonant gain, B is not always a good measure of the sharpness of the resonant peak. Consider two resonant circuits having the same bandwidth, say 1 rad/s, with resonant frequencies of 1 and 10^3 rad/s. Gain curves for these two circuits are shown in Fig. 14.14. Which resonance is more frequency selective? If an input to each circuit at their respective resonant frequencies drifts, say, 1% from its ω_r, its gain will be essentially indistinguishable

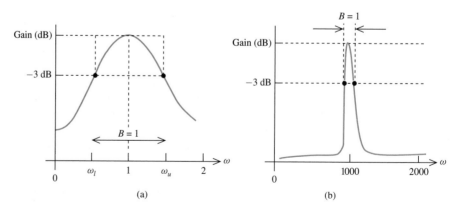

FIGURE 14.14 Two gain curves with same bandwidth B but different quality factor Q.

from the maximum gain in the first case, Fig. 14.14(a), but it will have dropped well out of the resonant peak of the second circuit, Fig. 14.14(b). Clearly, the second circuit is more frequency selective in the fractional sense. What matters in the sense of fractional frequency deviation is the bandwidth normalized by the resonant frequency itself.

Define the *quality factor* Q as the ratio of resonant frequency to bandwidth,

$$Q = \frac{\omega_r}{B} \qquad (14.50)$$

The larger the quality factor Q, the more frequency selective in the fractional sense is the circuit. For the series RLC, substituting (14.49) and (14.50), we have

$$Q = \frac{1}{R}\sqrt{\frac{L}{C}} \qquad (14.51)$$

Circuits with high quality factors are very frequency selective, and as we see from (14.52), this implies low values of resistance R. Since practical inductors include significant series resistance included in their own device models, it is difficult and expensive to design high-Q resonant circuits passively, that is, solely with RLC elements.

Example 14.10

We wish to design a series RLC resonant circuit with resonant frequency in hertz equal to 1 kHz and quality factor $Q = 100$. Since there are two constraints on three parameters RLC, we may take a trial value for one arbitrarily. Suppose that we fix a convenient value for the capacitor, $C = 1\ \mu F$. Then, by (14.46),

$$2\pi(10^3) = \sqrt{\frac{1}{L(10^{-6})}}$$

or squaring both sides and solving for L yields

$$L = \frac{1}{(2\pi)^2} = 25.3\ \text{mH}$$

Solving (14.51) for R gives

$$R = \frac{1}{Q}\sqrt{\frac{L}{C}} = 10^{-2}\sqrt{\frac{25.3 \times 10^{-3}}{10^{-6}}} = 1.59\ \Omega \qquad (14.52)$$

If the length of wire used to wind the given inductor contains more than 1.59 Ω of equivalent resistance, then, even in the absence of any external resistor, we could not meet these specifications. Note that by (14.51) we could increase the permissible series resistance at fixed quality factor Q and resonant frequency ω_r by increasing L and decreasing C so that their product remained the same. The increased inductance needed, however, requires more turns of wire on our inductor, which will inevitably carry with it increased series resistance.

For high-Q circuits, we will find it more convenient to use active circuits, those containing op amps. Their principles will be explored later in this chapter. Thus far we

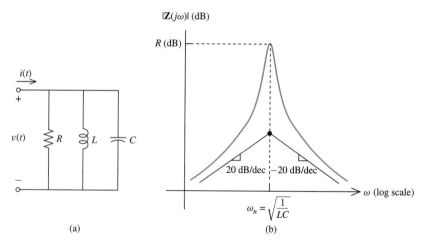

FIGURE 14.15 (a) Parallel RLC circuit; (b) Bode gain plot.

have considered only series RLC resonant circuits. Many other circuit configurations can be used to produce resonant behavior. In the case of the *parallel RLC circuit* (Fig. 14.15), taking $i(t)$ as the input and $v(t)$ as the output, the transfer function characterizing the gain is

$$\mathbf{H}(s) = \mathbf{Z}(s) = \frac{1}{sC + (1/R) + (1/sL)} \qquad (14.53)$$

We could study this resonant circuit by repeating the steps we took in the series RLC case, but there is an easier way. Comparing (14.53) with (14.42), the transfer functions are identical after substitution of C for L, L for C, and $1/R$ for R. Since all our results are computed from the transfer function, we may simply repeat the results noted for the series RLC case while making these substitutions. The results are summarized in Table 14.1.

A final observation concerning series and parallel RLC circuits is that at resonance the phase shift of their frequency response functions $\mathbf{H}(j\omega)|_{\omega=\omega r}$ is zero. In both cases, $|\mathbf{Z}(j\omega_r)|$ and $|\mathbf{Y}(j\omega_r)|$ at resonance are each purely real, and voltage and current are in phase. The imaginary parts of the impedances and the admittances cancel completely at the resonant frequency, explaining the location of a minimum or a maximum there.

Table 14.1 Properties of Series and Parallel RLC Circuits

Property	Series RLC	Parallel RLC
ω_r (resonant frequency)	$\dfrac{1}{\sqrt{LC}}$	$\dfrac{1}{\sqrt{LC}}$
B (bandwidth)	$\dfrac{R}{L}$	$\dfrac{1}{RC}$
Q (quality factor)	$\dfrac{1}{R}\sqrt{\dfrac{L}{C}}$	$R\sqrt{\dfrac{C}{L}}$

Many other circuits, both passive and active, exhibit resonance. In each case their resonant behavior can be determined from a Bode gain plot and the resonant frequency found by determining the location of the maximum (or minimum) gain. Unlike the series and parallel RLC cases, it is frequently difficult to determine the resonant frequency exactly in the general case. For lightly damped systems, those exhibiting significant resonance, the resonant frequency falls very near the point of maximum correction in the Bode plot, which is, in turn, very near the natural frequency ω_n of the complex root pair Bode plot term. The natural frequency ω_n of this term can be used as a convenient approximation to the resonant frequency ω_r in the lightly damped case.

Example 14.11

Consider the resonant circuit shown in Fig. 14.16. It is neither series nor parallel RLC, so the formulas in Table 14.1 do not apply. With the specified input and output, we have $\mathbf{H}(s) = \mathbf{Z}(s)$ and

$$Z(s) = \frac{(10/s)(10s + 1)}{(10/s) + 10s + 1} = \frac{10\left(s + \frac{1}{10}\right)}{s^2 + \frac{1}{10}s + 1}$$

There is a constant factor 10, a real zero at $s = -\frac{1}{10}$, and a complex root pair factor with $\omega_n = 1, 2\zeta\omega_n = \frac{1}{10}$; that is, $\zeta = \frac{1}{20}$. The corrected and uncorrected Bode gain plots are shown in Fig. 14.16(b). The corrected plot shows that the circuit is resonant near $\omega_n = 1$ rad/s. This is not exactly the location ω_r of maximum gain, but it is close. Another estimate of the resonant frequency is that frequency at which the phase of $Z(j\omega)$ is zero. In fact, some authors define this as the resonant frequency for general resonant circuits. In the present case, this results in the equation

$$\tan^{-1} 10\omega = \tan^{-1} \frac{\omega}{10(1 - \omega^2)}$$

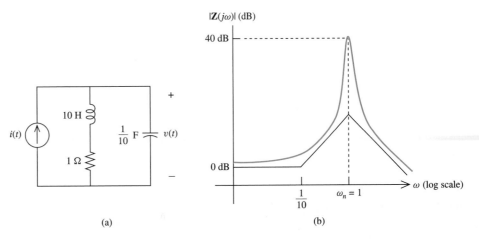

FIGURE 14.16   (a) Circuit for Example 14.11; (b) Bode gain plot.

Taking tangents of both sides gives

$$10\omega = \frac{\omega}{10(1-\omega^2)}$$

and solving, $\omega = (\sqrt{99})/10$. These two approximations to ω_r are very close. Neither is the exact frequency ω_r at maximum gain, which may be found (by the ambitious) as the exact solution of the equation

$$\frac{d}{d\omega}|Z(j\omega)| = 0 \qquad (14.54)$$

EXERCISES

14.4.1 Show that the solutions to (14.48) satisfy

$$\left(\frac{1}{LC} - \omega^2\right)^2 = \left(\frac{\omega R}{L}\right)^2$$

14.4.2 Show that the equation in Exercise 14.4.1 has four real solutions, two of which are positive and two negative, and that the difference of the positive solutions is the bandwidth $B = R/L$ given in (14.50).

14.4.3 Design a series or parallel RLC resonant circuit whose admittance goes through a minimum at $\omega_r = 100$ rad/s and that has a quality factor Q of 50. Use a 100-μF capacitor.
Answer Parallel, $R = 5$ kΩ, $L = 1$ H; series, $R = 2$ Ω, $L = 1$ H.

14.5   FREQUENCY RESPONSE OF OP AMPS

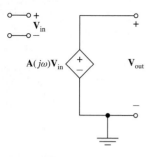

FIGURE 14.17 Phasor circuit for op-amp model with frequency-dependent open-loop gain $\mathbf{A}(j\omega)$.

The op amp was introduced in Chapter 3 as an active device containing many transistors and other elements that, under proper circumstances, behaves simply like a voltage-controlled voltage source with high gain. The models used thus far to study the op amp's behavior in various circuits and thereby determine the "proper circumstances" for it to behave as simply as described, have consisted entirely of frequency-independent elements: resistors and controlled sources with constant gain.

We are now prepared to consider another dimension of the behavior of practical op amps, their frequency dependence. Returning to the ideal voltage amplifier model of Fig. 3.7, repeated as Fig. 14.17, we will modify the open-loop op amp gain A. Fixed as a constant in the original model of Chapter 3, we modify it to reflect the actual rolloff of open-loop gain with frequency

$$\mathbf{A}(j\omega) = \frac{A_0\omega_0}{j\omega + \omega_0} \qquad (14.55)$$

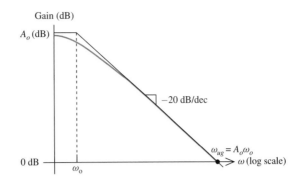

FIGURE 14.18 Bode plot of $|\mathbf{A}(j\omega)|$ for a typical op amp.

Fortunately, the dominant frequency-dependent characteristics of many practical op amps, such as the popular 741 family, are captured by this simple modification. The value of this gain at dc is A_0, the *dc open-loop gain* of the op amp, typically on the order of 10^5 to 10^6. The single pole at $s = -\omega_0$ causes the Bode plot of $|\mathbf{A}(j\omega)|$ shown in Fig. 14.18 to break at $\omega = \omega_0$, typically 5 to 50 rad/s, and roll off at -20 dB/decade thereafter. The frequency at which the gain is reduced to unity (0 dB) is called the *unity-gain bandwidth*, ω_{ug}. By (14.55), $\omega_{ug} = A_0\omega_0$ to an excellent approximation. The product $A_0\omega_o$ of the dc gain and the -3-dB bandwidth is referred to as the *gain–bandwidth product*, an important parameter of the op amp which we will see again shortly.

Considering Fig. 14.18, when used open-loop the op amp is capable of delivering very high gain, but only over an extremely narrow bandwidth of roughly dc to $\omega_o = 1$ Hz. For the op amp to perform in circuits such as the inverting and noninverting amplifiers, which are usually required to work over a large part or all of the audio range from dc to 40 kHz, its bandwidth must be much extended. We do not often work with signals whose frequencies fall conveniently below 1 Hz.

In Chapter 3 it was demonstrated that the use of negative feedback greatly reduces the effects of changes in the open-loop gain A on the input–output voltage transfer ratio for the circuit. Here we are introducing a new mechanism for variation in open-loop gain, the decline in gain with frequency. Perhaps negative feedback will protect the circuit from this new cause of variation in $\mathbf{A}$ as well as it protected it against the random piece-to-piece variations in $\mathbf{A}$ considered in Chapter 3. Fortunately, this is indeed the case, as we now demonstrate.

Consider a typical building-block circuit, the inverting amplifier shown in Fig. 14.19. By our previous analysis its circuit gain with input v_1 and output v_2 is $-R_F/R_A$. This result assumes the open-loop gain A to be fixed, large, and frequency independent, hardly the case with the gain function $\mathbf{A}(j\omega)$ that we are now considering. Repeating the analysis of this circuit in Chapter 3, we have the previous result (3.12), but with $\mathbf{A}(j\omega)$ and phasors in place of A and time-domain quantities. By that equation, the input–output equation for this circuit is

$$\mathbf{V}_2 = \mathbf{H}(j\omega)\mathbf{V}_1 \tag{14.56}$$

where

$$\mathbf{H}(j\omega) = \frac{-R_F}{R_A + [1/\mathbf{A}(j\omega)](R_A + R_F)}$$

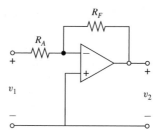

FIGURE 14.19 Inverting amplifier.

Factoring out $-R_F/R_A$ and substituting $\mathbf{A}(j\omega)$, we obtain

$$\mathbf{H}(j\omega) = \frac{-R_F}{R_A} \frac{1}{1 + [(j\omega + \omega_0)/A_0\omega_0](1 + R_F/R_A)}$$

Separating the real and imaginary parts in the denominator yields

$$= \frac{-R_F}{R_A} \frac{1}{j\omega\{[1 + (R_F/R_A)]/A_0\omega_0\} + \{1 + [1 + (R_F/R_A)]/A_0\}}$$

If we agree, as before, to limit our desired circuit gains $R_F/R_A \ll A_0$, not a very restrictive assumption since A_0 is very large, then the real term in the denominator is well approximated by 1, and we have

$$\mathbf{H}(j\omega) = \frac{-R_F}{R_A} \frac{1}{j\omega\{[1 + (R_F/R_A)]/A_0\omega_0\} + 1}$$

The Bode plot for this single-pole gain function is shown in Fig. 14.20. It shows us the relationship between the gain $|-R_F/R_A| = R_F/R_A$ and the bandwidth ω_c for the inverting amplifier. For any desired gain R_F/R_A, the circuit will have bandwidth ω_c much larger than the open-loop bandwidth ω_0. Indeed, the reduction in gain due to negative feedback, of ratio A_0 to R_F/R_A, has been returned almost completely in expansion of the bandwidth, ratio $\omega_c/\omega_0 = A_0\omega_0/[1 + (R_F/R_A)]$. Consider, for instance, an op amp with

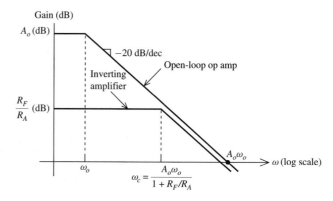

FIGURE 14.20 Bode plot for an inverting amplifier.

$A_0 = 10^5$ and $\omega_0 = 10$ rad/s used in an inverting amplifier with gain $R_F/R_A = 4$. We sustain a gain reduction from 10^5 to 4, or a factor of 25,000, in passing from open-loop op amp gain to circuit gain. The bandwidth, however, has increased from 10 rad/s to $\omega_c = (10^5)(10)/(1 + 4) = 200{,}000$, a factor of 20,000. Thus we have gotten back in bandwidth almost all we gave up in gain. This trade-off greatly favors our purpose, since we need a wider bandwidth and only want the indicated gain of $R_F/R_A = 4$. The new bandwidth, 200,000 rad/s, is almost 40 kHz, large enough to accommodate most audio circuits tasks.

Finally, note that the gain–bandwidth product of the inverting amp is

$$\text{GBW} = \frac{R_F}{R_A}\omega_c = A_0\omega_0\frac{R_F/R_A}{1 + (R_F/R_A)}$$

As the gain R_F/R_A ranges over moderate values, say 1 to 10, the gain–bandwidth product does not stray far from the open-loop gain–bandwidth product $\text{GBW} = A_0\omega_0$. This suggests that gain changes are compensated roughly equally by bandwidth changes in a manner limited only by the GBW of the op amp itself. If a larger combination of gain and bandwidth is needed, the gain must be split between two stages or a different, probably more expensive, op amp with higher gain–bandwidth product is used.

In this section we have seen that the op amp is highly frequency dependent, with too little bandwidth to be of much practical use open loop. When configured in a negative-feedback circuit, however, such as the building-block circuits introduced earlier (single- and multi-input inverting and noninverting amplifiers and voltage followers), the bandwidth will increase roughly in proportion to the decrease in gain. Thus if we restrict ourselves to modest circuit gains for single op amp stages, we can secure the bandwidth we need to build practical circuits despite the very small bandwidth the op amp itself presents us with open loop. Within their bandwidth, the op amps once again appear to be frequency independent, just as we have been assuming thus far.

EXERCISES

14.5.1 An op amp with 1-MHz gain–bandwidth product is used as an inverting amp. Find the bandwidth of this circuit in hertz if the gain R_F/R_A is set to 1, 10, 100, 1000, and 10^4.
   Answer 500 kHz; 90.9 kHz; 9.9 kHz; 999 Hz; 100 Hz

14.5.2 Suppose that two copies of the circuit of Exercise 14.5.1 are put in cascade, creating a circuit with gain $(R_F/R_A)^2$. What is the bandwidth of this new circuit for gains $(R_F/R_A)^2 = 100, 10^4$? Note from Fig. 14.5 that a corrected gain of $+1.5$ dB occurs at $\omega = 0.642\omega_c$.
   Answer 58.4 kHz; 6.36 kHz

14.5.3 Starting from (3.19b) and using the approximation $A_0 \gg [1 + (R_F/R_A)]$, show that the bandwidth ω_c of the noninverting amp is the same as that for the inverting amp shown in Fig. 14.20. What is its gain–bandwidth product?
   Answer $A_0\omega_0$

14.6 FILTERS

In ordinary usage, a filter is a device designed to block some things while letting others pass through. A car's oil filter stops abrasive metal particles and other contaminants from circulating through the engine, while permitting the oil to pass unimpeded. In circuit parlance, *a filter is a device that impedes the passage of signals whose frequencies fall within a band called the stopband, while permitting those in another band, the passband, to pass from input to output relatively unchanged.*

Filters are among the most common modules found in general circuit designs. Almost every practical electronic circuit of any complexity contains one or more filters. This includes radio, telephone, television and other receivers and transmitters, control systems for industrial machinery, computer circuits, power supplies, and so on. A common use of filters is to suppress noise. If we know that the signals in our circuit are limited to specific bands, we can put the other bands in the stopband of a filter and block noise from degrading the performance of our circuit. Another common use is to enhance an important range of frequencies. If we are tuned to a certain radio station, the band of frequencies containing that station's signal can be made the passband of a high-gain filter so that the desired signal becomes very prominent in the filter output.

The degree to which a signal of a particular frequency ω passes from input to output is measured by the frequency response function $\mathbf{H}(j\omega)$. An *ideal filter* is one for which $\mathbf{H}(j\omega) = 0$ in the stopband and $1/\underline{0}$ in the passband. *Ideal filters block stopband signals completely while passing signals in the passband without any change.* $\mathbf{H}(j\omega) = 1/\underline{0}$ in the passband means that the output is the same amplitude as the input and there is no phase shift between them.

Filters are named by the location of their passband. If it consists of all $\omega \leq \omega_c$ for some ω_c we have a *low-pass filter* with *cutoff frequency* ω_c. If, instead, the passband is the set of frequencies above a given cutoff frequency ω_c, $\omega \geq \omega_c$, we have a *high-pass filter*. In the case of an ideal filter, the value of the cutoff frequency ω_c is obvious; it is the frequency at which the gain jumps between 0 and 1. For nonideal filters we need to define the cutoff frequency ω_c in some reasonable way. We will usually define it as the half-power frequency; in other words, *the cutoff frequency ω_c is the frequency at which the gain is 3 dB below its maximum value.*

Example 14.12

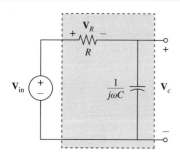

FIGURE 14.21   *RC* low-pass or high-pass filter circuit.

The simple *RC* voltage-divider circuit of Fig. 14.21 can be configured as either a low- or a high-pass filter. Defining the phasor $\mathbf{V}_c$ as output, we obtain

$$\mathbf{H}_{\text{LP}}(j\omega) = \frac{\mathbf{V}_c}{\mathbf{V}_{\text{in}}} = \frac{1/(j\omega C)}{R + [1/(j\omega C)]} = \frac{1}{j\omega RC + 1} \qquad (14.57)$$

The gain of this circuit is

$$|\mathbf{H}_{\text{LP}}(j\omega)| = \frac{1}{\sqrt{R^2 C^2 \omega^2 + 1}} \qquad (14.58)$$

and is plotted in Fig. 14.22(a). Since the high-gain region consists of all frequencies below a certain value, this is a low-pass filter. By (14.58), the maximum gain occurs at dc and equals unity. Three decibels below that is -3 dB, so the gain at ω_c must be $1/\sqrt{2}$. Also

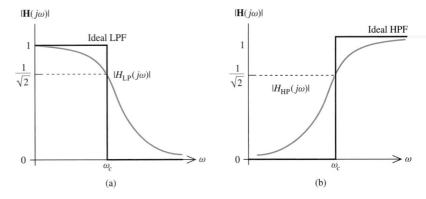

FIGURE 14.22 Low- and high-pass filter gain curves.

by (14.58), this occurs at

$$\omega_c = \frac{1}{RC}$$

If, instead, we take $\mathbf{V}_R$ as the output,

$$\mathbf{H}_{\mathrm{HP}}(j\omega) = \frac{\mathbf{V}_R}{\mathbf{V}_{\mathrm{in}}}$$

$$= \frac{R}{R + [1/(j\omega C)]} = \frac{j\omega RC}{j\omega RC + 1} \qquad (14.59)$$

$$|\mathbf{H}_{\mathrm{HP}}(j\omega)| = \frac{RC\omega}{\sqrt{R^2 C^2 \omega^2 + 1}}$$

This is a high-pass filter with maximum gain as $\omega \to \infty$ equal to unity. The -3-dB point is once again $\omega_c = 1/RC$. To illustrate the degree of nonideality in these RC filters, suppose that we move 1 decade into the passband; for instance, in the low-pass case take $\omega = \omega_c/10 = 1/(10RC)$. Situated comfortably in the passband, the frequency response $H_{\mathrm{LP}}(j\omega_c/10)$ is

$$\mathbf{H}_{\mathrm{LP}} \frac{j}{10RC} = 0.995\underline{/-5.7}^{\circ}$$

The ideal filter would have gain of unity and phase shift of 0 since we are within the passband. We are within $\frac{1}{2}\%$ of unity gain and within $6°$ of zero phase shift. For many, but not all, purposes this performance is sufficient.

There are other filters designed to pass not all frequencies lower or greater than a limit, but to pass some intermediate band. A *bandpass filter* has passband $\omega_l \leq \omega \leq \omega_u$, where ω_u *and* ω_l *are the upper and lower half-power frequencies.* A *bandreject* or *bandstop filter* has a split passband, all $\omega \leq \omega_l$ and all $\omega \geq \omega_u$, as shown in Fig. 14.23.

A bandpass filter gain characteristic can be produced by multiplying low- and high-pass filter gain curves whose passbands have been set to overlap. Suppose that we have

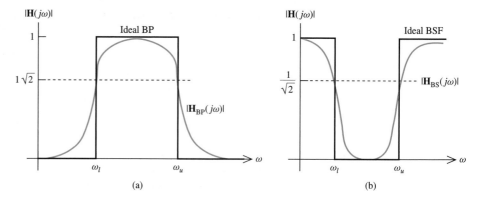

FIGURE 14.23 Bandpass and bandstop filter gain curves.

a circuit for which

$$\mathbf{H}_{BP}(j\omega) = \mathbf{H}_{LP}(j\omega)\mathbf{H}_{HP}(j\omega) \tag{14.60}$$

with the cutoff frequency of the high-pass filter ω_{cHP} set lower than that of the low pass, $\omega_{cHP} < \omega_{cLP}$. If these two filters are ideal, we will have the ideal bandpass frequency response, $\mathbf{H}(j\omega) = 1$ for all ω between $\omega_l = \omega_{cHP}$ and $\omega_u = \omega_{cLP}$, and $\mathbf{H}(j\omega) = 0$ elsewhere. For nonideal low- and high-pass filters, the bandpass filter gain (14.60) will be close to unity within its passband $\omega_l \le \omega \le \omega_u$ and have gain lower than $1/\sqrt{2}$ outside the passband.

How do we produce a circuit satisfying (14.60)? Suppose that two circuits with voltage transfer functions $\mathbf{H}_1(s) = \mathbf{V}_2(s)/\mathbf{V}_1(s)$ and $\mathbf{H}_2(s) = \mathbf{V}_4(s)/\mathbf{V}_3(s)$ are connected in cascade, the output of one to the input of the other, through a switch as shown in Fig. 14.24. The transfer function $\mathbf{H}_1(s)$ is defined for $\mathbf{I}_2(s) = 0$; so as long as the switch is open, $\mathbf{V}_2(s) = \mathbf{H}_2(s)\mathbf{V}_1(s)$.

Assume for a moment that $\mathbf{V}_2(s)$ is exactly the same whether the switch in Fig. 14.24 is open or closed. In that case, with the switch closed,

$$\mathbf{V}_3(s) = \mathbf{V}_2(s) = \mathbf{H}_1(s)\mathbf{V}_1(s) \tag{14.61}$$

and since $\mathbf{V}_3(s)$ is the input to a circuit with transfer function $\mathbf{H}_2(s)$,

$$\mathbf{V}_4(s) = \mathbf{H}_2(s)\mathbf{V}_3(s) = \mathbf{H}_1(s)\mathbf{H}_2(s)\mathbf{V}_1(s) \tag{14.62}$$

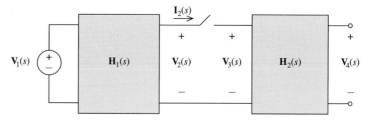

FIGURE 14.24 Circuits in cascade.

Defining the overall voltage transfer function of the cascade interconnection shown in Fig. 14.24 as $\mathbf{H}(s)$, by (14.62) we have, under these circumstances,

$$\mathbf{H}(s) = \mathbf{H_1}(s)\mathbf{H_2}(s) \tag{14.63}$$

The condition that the output voltage $\mathbf{V}_2(s)$ change significantly when the circuits are connected is called *loading* and was first raised in Chapter 3. We see from (14.63) that *in the absence of loading the overall transfer function of a cascade interconnection is the product of individual transfer functions.* If loading does in fact occur, the modified $\mathbf{V}_2(s)$ will not equal its unloaded value $\mathbf{H_1}(s)\mathbf{V}_1(s)$, (14.61) will not be true, and the product rule for cascaded circuits (14.63) no longer follows.

Replacing all to the left of the switch by its Thevenin equivalent and all to the right likewise, assuming no internal sources in $\mathbf{H_2}(s)$, we have Fig. 14.25. The impedance $\mathbf{Z}_1(s)$, the Thevenin impedance of the left circuit looking back into its output terminals, is called its *output impedance*, and $\mathbf{Z}_2(s)$ is the *input impedance* of the right circuit. With the switch open, $\mathbf{V}_2(s) = \mathbf{V}_T(s)$. With it closed,

$$\mathbf{V}_2(s) = \frac{\mathbf{Z}_2(s)}{\mathbf{Z}_1(s) + \mathbf{Z}_2(s)}\mathbf{V}_T(s) \tag{14.64}$$

$\mathbf{V}_2(s)$ will not change significantly when the circuits are connected together if the impedance ratio in (14.64) is close to unity. *Loading will not occur when two circuits are cascaded if the input impedance of the load circuit is much greater than the output impedance of the source circuit.* In this case we can use the product rule (14.63); otherwise, we cannot.

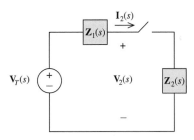

FIGURE 14.25 Thevenin equivalents of cascaded circuits.

Example 14.13

Suppose that we cascade an RC low-pass filter with $\omega_{cLP} = 10$ and an RC high-pass with $\omega_{cHP} = 1$ rad/s as shown in Fig. 14.26. If we assume no loading, we will have a bandpass filter with half-power frequencies very near to $\omega_l = 1$ and $\omega_u = 10$ rad/s. The actual overall transfer function, however, can be computed from the s-domain mesh equations:

$$\begin{bmatrix} 1 + \dfrac{10}{s} & \dfrac{-10}{s} \\ \dfrac{-10}{s} & 1 + \dfrac{11}{s} \end{bmatrix} \begin{bmatrix} \mathbf{I}_1(s) \\ \mathbf{I}_2(s) \end{bmatrix} = \begin{bmatrix} \mathbf{V}_1(s) \\ 0 \end{bmatrix}$$

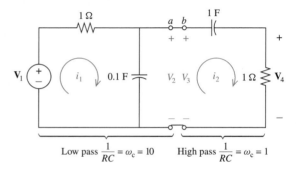

FIGURE 14.26   Cascade of passive low- and high-pass filters.

By Cramer's rule,

$$\mathbf{I}_2(s) = \frac{(10/s)\mathbf{V}_1(s)}{1 + (21/s) + (10/s^2)}$$

$$= \frac{10s\mathbf{V}_1(s)}{s^2 + 21s + 10}$$

Since $\mathbf{I}_2(s) = \mathbf{V}_4(s)$, the true overall transfer function is

$$\mathbf{H}(s) = \frac{10s}{s^2 + 21s + 10} \qquad (14.65)$$

Assuming no loading, the overall transfer function would be the product of the low- and high-pass transfer functions:

$$\mathbf{H}_{\mathrm{LP}}(s)\mathbf{H}_{\mathrm{HP}}(s) = \frac{1}{0.1s + 1}\left(\frac{s}{s + 1}\right)$$

$$= \frac{10s}{s^2 + 11s + 10} \qquad (14.66)$$

Clearly, (14.65) and (14.66) are quite different. For instance, when $s = j\omega = j$, the actual gain from (14.65) is $|\mathbf{H}(j)| = 0.438$, while the product rule in (14.66) predicts a gain of $|\mathbf{H}_{\mathrm{LP}}(j)||\mathbf{H}_{\mathrm{HP}}(j)| = 0.704$. *The effect of loading is to reduce the overall response.*

   If we insert a voltage follower, introduced in Chapter 3, into this circuit as a buffer amplifier between a and b to eliminate loading, then in this case we will have the input impedance of the circuit to the right of a essentially infinite (the virtual open principle for op amp analysis), so loading can no longer occur. The overall transfer function for Fig. 14.27 will be the product of transfer functions of the cascaded circuits as given by (14.66).

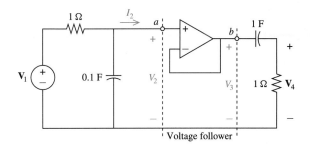

FIGURE 14.27   Inserting a voltage follower to eliminate loading.

Thus, if we take care to prevent loading, perhaps by separating stages with a buffer amplifier with very high input impedance such as the voltage follower, we can get cascaded circuits whose transfer functions are the products of the individual transfer functions in cascade. The circuit of Fig. 14.27 has the desired bandpass gain characteristic shown in Fig. 14.23.

Finally, a bandstop filter characteristic can be produced by subtracting a bandpass frequency response from unity:

$$\mathbf{H}_{BS}(j\omega) = 1 - \mathbf{H}_{BP}(j\omega)$$

So wherever the gain of the bandpass is near 1, that of the bandstop will be near 0, and vice versa, as shown in Fig. 14.23(b).

With what physical elements are filters designed? A filter built entirely from passive (RLC) elements is a *passive filter*. If op amps or other circuits with electronic devices requiring power supplies are used, we have an *active filter*. While having obvious disadvantages of component count and power consumption, active filters are very often the choice for practical filter design applications. There are serious limitations to the use of passive filters, as we now describe.

Passive filters corresponding to transfer functions containing complex root pair factors require both inductors and capacitors, and inductors tend to be undesirable elements in modern miniaturized circuit design. They are bulky and expensive and have sufficient series resistance to make them generally useful only in relatively low Q applications. In addition, inductors cannot be included in thin-film integrated circuits effectively using current technology. Passive filters are *dissipative*; in other words, they cannot have gains greater than unity at any frequency. Moreover, as we saw in Example 14.13, cascading passive filters leads to loading, which makes the design process more difficult, since overall transfer functions cannot be predicted accurately from the product of the individual filter transfer functions. Active filters do not require inductors, may have any desired gain, and can be easily designed to be highly resistant to loading, as we have seen.

Example 14.14

Design a third-order low-pass filter with transfer function

$$\mathbf{H}(s) = \frac{1}{(s+1)(s+5)(s+20)}$$

We can cascade together three low-pass filters with transfer functions $1/(s+1)$, $1/(s+5)$, and $1/(s+20)$ and get the overall transfer function above, taking care to use a voltage follower to buffer the

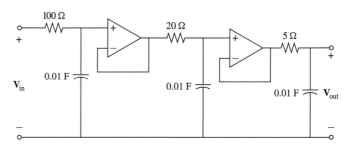

FIGURE 14.28 Circuit for Example 14.14.

filter sections. Using (14.59), we can select either R or C in a given section arbitrarily and then select the other so that the required pole is located at $s = -1/RC$. Picking the capacitors to each be 0.01 F, we have the design shown in Fig. 14.28. While this transfer function may also be realized using passive RC elements, it is less straightforward to design since successive stages will load one another.

In summary, filters are devices that selectively suppress certain ranges of frequencies while promoting others. They are frequently used to suppress random noise, enhance bands where important signals reside, and eliminate specific interfering signals such as 60-Hz "hum." Filters may be cascaded to achieve a product transfer function if there is no loading, and loading can be prevented by the use of buffer amplifiers such as the voltage follower. Simple RC filters with and without buffering have been analyzed. The four basic filters—low-pass, high-pass, bandpass, and bandstop—described here can be augmented by many more: all-pass filters that operate to manipulate phase rather than gain, comb filters with multiple equally spaced passbands, and so on.

EXERCISES

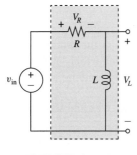

EXERCISE 14.6.1

14.6.1. What kind of filter is this passive RL circuit with $\mathbf{H}(s) = \mathbf{V}_L(s)/\mathbf{V}_{in}(s)$? What is its half-power frequency ω_c? Repeat for $\mathbf{H}(s) = \mathbf{V}_R(s)/\mathbf{V}_{in}(s)$.

Answer High pass, R/L; low pass, R/L

14.6.2. What kind of filter is $\mathbf{H}(s) = s^2/(s^2 + 2\sqrt{2}\omega_n s + \omega_n^2)$? What is its half-power frequency?

Answer High-pass; ω_n

14.6.3. Loading is frequency dependent. If two identical RC low-pass filters (see Fig. 14.21) with $R = C = 1$ are cascaded, at what frequency will there be no loading at all? What is the ratio of the actual gain at $\omega = 1$ rad/s to the product of the two individual filter gains? At $\omega = 100$ rad/s?

Answer dc; $\frac{2}{3}$; 0.9998

14.7 SCALING

Throughout this book, numerical values such as $R = 1\ \Omega$, $C = 2$ F, and $\omega = 1$ rad/s have been used in almost all examples, exercises, and problems. Although convenient for pedagogical purposes and easing the labor of hand calculation, the use of these values may concern the reader since they are not practical circuit parameters for realistic applications.

Fortunately, we can have the best of both worlds. We can frame a problem in a simple numerical setting, solve it, and then convert the solution to a more practical range of circuit operation. The process of conversion is called *scaling*. In this section we discuss two types of scaling, impedance scaling and frequency scaling.

For a given circuit, *impedance scaling* by α is the specification of a new set of circuit parameters for the circuit such that voltages are unchanged, but currents are all scaled by the constant $1/\alpha$. To see how a circuit may be impedance scaled, consider its mesh equations in vector-matrix form:

$$\mathbf{ZI} = \mathbf{V_s} \tag{14.67}$$

where $\mathbf{I}$ is the vector of mesh currents, $\mathbf{V_s}$ the independent source vector, and $\mathbf{Z}$ a square connection matrix. If we can find a new circuit whose connection matrix is scaled to $\alpha\mathbf{Z}$ but whose source vector $\mathbf{V_s}$ is unchanged, then since (14.67) implies that

$$(\alpha\mathbf{Z})\frac{\mathbf{I}}{\alpha} = \mathbf{V_s} \tag{14.68}$$

the currents will each be scaled by $1/\alpha$. Having studied mesh analysis, we know that the on-diagonal (i, i) element of $\mathbf{Z}$ contains the sum of all impedances around the ith mesh, and the off-diagonal elements, the negative sum of impedances on the boundary between the ith and jth meshes. If there are no controlled sources in the circuit, $\mathbf{Z}$ consists entirely of these impedances and will be scaled as in (14.68) if each impedance is scaled by α. That is, we must replace each RLC in the original circuit by

$$R_s = \alpha R, \quad L_s = \alpha L, \quad C_s = \frac{C}{\alpha} \tag{14.69}$$

where the subscript s indicates the values after scaling. *To impedance scale by α, multiply each R and L value by α, and divide each C value by α.* The difference arises because the impedance of resistors and inductors is proportional to their element parameter R or L, while that of a capacitor is inversely proportional to C. To keep $\mathbf{V_s}$ the same, as required by (14.68), independent voltage sources are unchanged, while independent current sources must be scaled by $1/\alpha$.

Example 14.15

To impedance scale the circuit of Fig. 14.29(a) by 1000, R and L are scaled up by 1000 and C down by 1000. The voltage source is unchanged, and the current source is scaled down by 1000. All voltages in the new circuit are unchanged from their values in the old circuit; all currents are reduced by a factor of 1000.

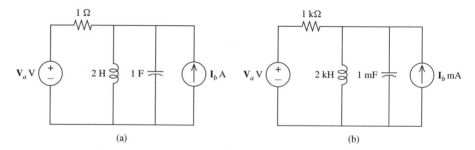

FIGURE 14.29 (a) Original circuit; (b) impedance scaled by 1000.

The gains of current-controlled current sources and voltage-controlled voltage sources in the circuit will not change when impedance scaling, since the ratio of controlled to controlling variable is the same after scaling as before. But voltage-controlled current sources will have their gain (transresistance r) scaled by α, just as the other elements whose units are ohms, and current-controlled voltage sources, with gain (transconductance g) measured in siemens, must have these gains scaled by $1/\alpha$. For instance, if a circuit with a current-controlled voltage source with source function $v(t) = 4i_c(t)$ is to be impedance scaled by 20, the new controlled source will have source function $v(t) = 80i_c(t)$. Then, with the currents in the new circuit reduced by a factor of 20, the increase in the transresistance of the controlled source by a factor of 20 will compensate and its voltage will remain unchanged as required.

A common use of impedance scaling arises in the design of op-amp circuits. Although it may be convenient to use values close to unity for R and C in the design process, this will inevitably result in a circuit whose impedance levels are too low for the use of popular low-power op amps, such as the μA741. These devices operate best at milliampere current levels when their voltage levels are an appreciable fraction of the power supply voltage, typically ±15 V. We may use convenient values like $R = 1\ \Omega$ during the design process, but if we are not to "fry" the op amp by drawing excessive current, we had best impedance scale such values to the order of kilohms before flipping the switch. This scaling also reduces capacitors to more practical values than 1 F or so. Note that since the scaled circuits have the same voltages as prescaled circuits, *there is no change to any voltage transfer function* $\mathbf{H}(s) = \mathbf{V}_{out}(s)/\mathbf{V}_{in}(s)$ *when a circuit is impedance scaled.*

Example 14.16

Consider the circuit of Fig. 14.28. To prevent excess power dissipation from the op amps resulting in circuit failure due to op amp burnout, we impedance scale by 2000. The resulting circuit, with practical values, is shown in Fig. 14.30.

Given a circuit, *frequency scaling* by β is the specification of a new set of circuit parameters such that the scaled circuit responds at frequency $\beta\omega$ exactly as the original one does at ω. There is only one way that ω enters into the calculation of any response, through the impedances $j\omega L$ and $1/j\omega C$. If we arrange that the new inductive and capacitive impedances at $\beta\omega$ are the same as the old ones at ω, all responses must be the same, as required. Now the product $j\omega L$ remains unchanged after ω is replaced by $\beta\omega$ if we also replace L by L/β. The same is true for $1/j\omega C$ if C is replaced by C/β. To

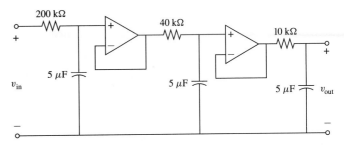

FIGURE 14.30 Circuit of Fig. 14.28 impedance scaled to practical current levels for low-power op amps.

frequency scale by β, divide each L and C by β, leaving all other elements unchanged.

$$R_s = R, \qquad L_s = \frac{L}{\beta}, \qquad C_s = \frac{C}{\beta} \qquad (14.70)$$

Example 14.17

We earlier impedance scaled the resonant circuit of Fig. 14.29(a), with result shown in Fig. 14.29(b). Let us scale the resonant frequency of this circuit upward by a factor of 10^4. This is easily accomplished by replacing the 2-H inductor by a 0.2-mH inductor and the 1-F capacitor by a 100-μF capacitor. Note that the resonant frequency of the circuits of Fig. 14.29(a) and (b) is the same, since $\omega_r = \sqrt{1/LC}$ and the product LC is the same in these circuits. Impedance scaling does not result in frequency scale changes.

Frequency scaling is often used in the design of filters. Handbooks specifying low-pass filter circuits, for instance, will need to list only the values for the low-pass whose cutoff frequency is set to $\omega_c = 1$ rad/s. The user can then frequency scale the circuit to any desired cutoff frequency.

EXERCISES

14.7.1. Suppose that a circuit is impedance scaled by 100. Consider four transfer functions: $\mathbf{H}_1(s) = \mathbf{V}_2(s)/\mathbf{V}_1(s)$, $\mathbf{H}_2(s) = \mathbf{V}_2(s)/\mathbf{I}_1(s)$, $\mathbf{H}_3(s) = \mathbf{I}_2(s)/\mathbf{I}_1(s)$, and $\mathbf{H}_4(s) = \mathbf{I}_2(s)/\mathbf{V}_1(s)$. Which will change? By what factor?
Answer $\mathbf{H}_2(s)$ will be multiplied by 100, $\mathbf{H}_4(s)$ by 1/100; others unchanged

14.7.2. If a parallel RLC passive circuit is frequency scaled by 50 and then impedance scaled by 50, what are new RLC values?
Answer $50R$, L, $C/2500$

14.7.3. A handbook lists a fourth-order Butterworth filter with $\omega_c = 1$ as

$$H(s) = \frac{\mathbf{V}_2(s)}{\mathbf{V}_1(s)} = \frac{1}{s^4 + 2.61s^3 + 3.41s^2 + 2.61s + 1}$$

If the circuit is frequency scaled by $\frac{1}{2}$ and then impedance scaled by 100, find $\mathbf{H}(s)$.

Answer Impedance scaling has no effect on a voltage transfer function:

$$\frac{0.0625}{s^4 + 1.31s^3 + 0.853s^2 + 0.326s + 0.0625}$$

14.8 SPICE AND FREQUENCY RESPONSE

SPICE may be used to determine frequency response, both gain and phase shift as functions of frequency, using the .AC control statement. The gain $|\mathbf{H}(j\omega)|$ is specified by adding the suffix M (magnitude) to the name of a current or voltage to be output, and the phase shift $\underline{/\mathbf{H}(j\omega)}$ by the suffix P. For instance, the output control statement

```
.PLOT    AC    VM(4)    VP(4)
```

will result in the plotting of both the magnitude and phase of the voltage at node 4 in the frequency range specified on the .AC control statement. If the amplitude of the ac source being swept is 1, the magnitude of the output voltage VM(4) is directly equal to the gain. If the Bode coordinates, gain in decibels versus log frequency, are desired for the gain curve, the suffix DB is used instead of M. The keyword DEC (decade) is then used in the .AC control statement as shown in Example 14.18.

SPICE Example 14.18

We will check our earlier calculation of the the Bode gain plot, Fig. 14.16(b), for the resonant circuit of Fig. 14.16(a). The SPICE input file is

```
EXAMPLE 14.18
*
I    0    1    AC    1    0
L    1    2    10
R    2    0    1
C    1    0    0.1
.AC    DEC    25    .01        1
.PLOT AC VDB(1)
.END
```

We have specified decibel units for the output that together with the logarithmic frequency axis specified by the keyword DEC and the 1-A input, make the output identical to the gain of the circuit. The output is shown in Fig. 14.31(a). Note that the resonant peak appears at $f = 0.1585$ Hz or 0.996, agreeing closely with our calculation in Example 14.11, and the value at the peak is listed as 40.01 dB, comparing favorably with our calculated 40 dB. The corner frequency at $\omega = 0.1$ rad/s ($f = 0.0159$ Hz) has a gain of +3 dB in Fig. 14.16(b); this checks well with the gain of 3.08 dB at

```
**** 04/29/94 11:25:50 ********* Evaluation PSpice (January 1991) ************

EXAMPLE 14.18


****    AC ANALYSIS                      TEMPERATURE =   27.000 DEG C


*********************************************************************************


        FREQ        VDB(1)


(*)----------      0.0000E+00   2.0000E+01   4.0000E+01   6.0000E+01   8.0000E+01
                  - - - - - - - - - - - - - - - - - - - - - - - - - - - -
        1.000E-02  1.479E+00 .*           .            .            .            .
        1.096E-02  1.728E+00 .*           .            .            .            .
        1.202E-02  2.010E+00 .*           .            .            .            .
        1.318E-02  2.328E+00 . *          .            .            .            .
        1.445E-02  2.684E+00 . *          .            .            .            .
        1.585E-02  3.078E+00 . *          .            .            .            .
        1.738E-02  3.513E+00 . *          .            .            .            .
        1.905E-02  3.987E+00 .  *         .            .            .            .
        2.089E-02  4.501E+00 .  *         .            .            .            .
        2.291E-02  5.055E+00 .  *         .            .            .            .
        2.512E-02  5.647E+00 .   *        .            .            .            .
        2.754E-02  6.278E+00 .   *        .            .            .            .
        3.020E-02  6.945E+00 .    *       .            .            .            .
        3.311E-02  7.649E+00 .    *       .            .            .            .
        3.631E-02  8.389E+00 .    *       .            .            .            .
        3.981E-02  9.166E+00 .     *      .            .            .            .
        4.365E-02  9.981E+00 .     *      .            .            .            .
        4.786E-02  1.084E+01 .      *     .            .            .            .
        5.248E-02  1.174E+01 .       *    .            .            .            .
        5.754E-02  1.269E+01 .       *    .            .            .            .
        6.310E-02  1.371E+01 .        *   .            .            .            .
        6.918E-02  1.479E+01 .         *  .            .            .            .
        7.586E-02  1.597E+01 .         *  .            .            .            .
        8.318E-02  1.727E+01 .          *..            .            .            .
        9.120E-02  1.872E+01 .           *.            .            .            .
        1.000E-01  2.039E+01 .           *            .            .            .
        1.096E-01  2.237E+01 .            . *          .            .            .
        1.202E-01  2.485E+01 .            .  *         .            .            .
        1.318E-01  2.820E+01 .            .    *       .            .            .
        1.445E-01  3.331E+01 .            .       *    .            .            .
        1.585E-01  4.001E+01 .            .          * .            .            .
        1.738E-01  3.391E+01 .            .       *    .            .            .
        1.905E-01  2.854E+01 .            .    *       .            .            .
        2.089E-01  2.506E+01 .            .  *         .            .            .
        2.291E-01  2.250E+01 .            . *          .            .            .
        2.512E-01  2.046E+01 .            *            .            .            .
        2.754E-01  1.875E+01 .           *..            .            .            .
        3.020E-01  1.725E+01 .         *  .            .            .            .
        3.311E-01  1.591E+01 .        *   .            .            .            .
        3.631E-01  1.469E+01 .        *   .            .            .            .
        3.981E-01  1.355E+01 .       *    .            .            .            .
        4.365E-01  1.247E+01 .      *     .            .            .            .
        4.786E-01  1.145E+01 .      *     .            .            .            .
        5.248E-01  1.047E+01 .      *     .            .            .            .
        5.754E-01  9.527E+00 .    *       .            .            .            .
        6.310E-01  8.607E+00 .    *       .            .            .            .
        6.918E-01  7.708E+00 .   *        .            .            .            .
        7.586E-01  6.827E+00 .  *         .            .            .            .
        8.318E-01  5.960E+00 .  *         .            .            .            .
        9.120E-01  5.105E+00 . *          .            .            .            .
        1.000E+00  4.259E+00 . *          .            .            .            .
                  - - - - - - - - - - - - - - - - - - - - - - - - - - - -

                    JOB CONCLUDED
                    TOTAL JOB TIME          .22
```

FIGURE 14.31a   SPICE outputs for Example 14.18.

EXAMPLE 14.18

```
****      AC ANALYSIS                          TEMPERATURE =   27.000 DEG

*****************************************************************************


        FREQ      VDB(1)


(*)----------     3.9000E+01   3.9500E+01   4.0000E+01   4.0500E+01   4.1000E+01
                  - - - - - - - - - - - - - - - - - - - - - - - - -
     1.570E-01  3.973E+01 .          .         *    .           .           .
     1.571E-01  3.977E+01 .          .          *   .           .           .
     1.572E-01  3.980E+01 .          .           * .           .           .
     1.574E-01  3.983E+01 .          .           *  .           .           .
     1.575E-01  3.986E+01 .          .           *  .           .           .
     1.576E-01  3.988E+01 .          .            * .           .           .
     1.577E-01  3.991E+01 .          .            * .           .           .
     1.579E-01  3.993E+01 .          .            * .           .           .
     1.580E-01  3.995E+01 .          .             *.           .           .
     1.581E-01  3.997E+01 .          .             *.           .           .
     1.582E-01  3.998E+01 .          .             *            .           .
     1.583E-01  4.000E+01 .          .             *            .           .
     1.585E-01  4.001E+01 .          .             *            .           .
     1.586E-01  4.002E+01 .          .             .*           .           .
     1.587E-01  4.003E+01 .          .             .*           .           .
     1.588E-01  4.004E+01 .          .             .*           .           .
     1.590E-01  4.004E+01 .          .             .*           .           .
     1.591E-01  4.004E+01 .          .             .*           .           .
     1.592E-01  4.004E+01 .          .             .*           .           .
     1.593E-01  4.004E+01 .          .             .*           .           .
     1.594E-01  4.004E+01 .          .             .*           .           .
     1.596E-01  4.003E+01 .          .             .*           .           .
     1.597E-01  4.002E+01 .          .             .*           .           .
     1.598E-01  4.001E+01 .          .             *            .           .
     1.599E-01  4.000E+01 .          .             *            .           .
     1.601E-01  3.999E+01 .          .             *            .           .
     1.602E-01  3.997E+01 .          .            *.            .           .
     1.603E-01  3.995E+01 .          .            *.            .           .
     1.604E-01  3.993E+01 .          .           * .            .           .
     1.606E-01  3.991E+01 .          .           * .            .           .
     1.607E-01  3.989E+01 .          .          *  .            .           .
     1.608E-01  3.986E+01 .          .         *   .            .           .
     1.609E-01  3.984E+01 .          .         *   .            .           .
     1.610E-01  3.981E+01 .          .        *    .            .           .
     1.612E-01  3.978E+01 .          .       *     .            .           .
     1.613E-01  3.975E+01 .          .       *     .            .           .
     1.614E-01  3.971E+01 .          .      *      .            .           .
     1.615E-01  3.968E+01 .          .     *       .            .           .
     1.617E-01  3.964E+01 .          .    *        .            .           .
     1.618E-01  3.960E+01 .          .   *         .            .           .
     1.619E-01  3.956E+01 .          . .*          .            .           .
     1.620E-01  3.952E+01 .          . .*          .            .           .
     1.621E-01  3.948E+01 .         *.            .            .           .
     1.623E-01  3.944E+01 .        * .            .            .           .
     1.624E-01  3.939E+01 .       *  .            .            .           .
     1.625E-01  3.934E+01 .      *   .            .            .           .
     1.626E-01  3.930E+01 .      *   .            .            .           .
     1.628E-01  3.925E+01 .    *     .            .            .           .
     1.629E-01  3.920E+01 .   *      .            .            .           .
     1.630E-01  3.915E+01 .   *      .            .            .           .
                  - - - - - - - - - - - - - - - - - - - - - - - - -


        JOB CONCLUDED
        TOTAL JOB TIME            .23
```

FIGURE 14.31b   SPICE outputs for Example 14.18.

$f = 0.1585$ in the SPICE output. In the example it was mentioned that the location of the resonant peak could only be approximated using our Bode plotting method. We may use SPICE to examine the details of a gain curve in any region by changing the sampling parameters on the $.AC$ control statement. Examining the gain near the peak by using the control statement

```
.AC LIN 25 .157 .163
```

we get the output shown in Fig. 14.31(b). To four significant digits, the peak gain is 40.04 dB, and the peak occurs somewhere between 0.1588 and 0.1594 Hz, or 0.998 and 1.002 rad/s.

EXERCISES

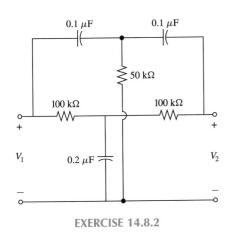

EXERCISE 14.8.2

14.8.1. Use SPICE to create plots of the gain and phase-shift curves for the circuit of Fig. 14.26 for $0 < \omega < 10$, using linear frequency and output scales with a sample every 0.1 rad/s.

14.8.2. The bandstop filter shown is known as a twin-T filter. Determine the frequency in hertz at the bottom of the notch to four significant digits.
Answer 15.92 Hz

SUMMARY

The frequency response of a circuit is the combination of gain and phase shift experienced by an input sinusoid as a function of its frequency. Design criteria for practical circuits such as filters and amplifiers are often expressed in terms of a desired frequency response, that is, gain and phase shift versus frequency.

- The frequency response function $\mathbf{H}(j\omega)$ is found from the transfer function $\mathbf{H}(s)$ by replacing s by $j\omega$. It exists for stable circuits only.

- $|\mathbf{H}(j\omega)|$ is the gain at frequency ω, $\underline{/\mathbf{H}(j\omega)}$ is the phase shift.

- The decibel equivalent of a positive real number is 20 times the log base 10 of the number. Adding 20 dB scales a number by 10, subtracting 20 dB scales by 0.1. The decibel equivalent of a product is the sum of each factor's decibel equivalents, of a ratio is the difference of decibel equivalents.

- The uncorrected Bode plot is a piecewise linear approximation to the plot of gain in dB versus log frequency.

- The uncorrected Bode plot is the sum of plots of individual factors.

- There is a 3-dB correction inside the knee of each simple factor.

- Filters are circuits with passbands, in which the gain is close to its maximum value, and stopbands, in which the gain is much lower. Filters are used to create selective responses to inputs in different frequency bands.

- The gain of all elements, including op amps, eventually declines with frequency. The frequency at which the gain of an op amp has declined to unity is called the unity-gain bandwidth.

- In the absence of loading, the transfer function of a cascade is the product of transfer functions.

- Transfer functions may be decomposed into a product of first- and second-order factors, each factor designed separately, and then cascaded.

PROBLEMS

14.1. Find the output phasor V_o if the input phasor is $V_i = 3/\underline{30°}$ and $H(s) =$

(a) $\dfrac{1}{s+1}$

(b) $\dfrac{5s}{s+5}$

(c) $\dfrac{10(s-1)}{s^3+2s^2+s+2}$

14.2. Find $\mathbf{H}(j\omega)$ for the indicated input and output.

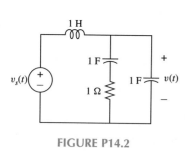

FIGURE P14.2

14.3. Find $\mathbf{H}(j\omega)$ for the indicated input and output.

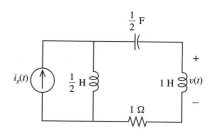

FIGURE P14.3

14.4. Find the gain $g(\omega)$ and phase shift $\phi(\omega)$ functions. Make rough sketches using linear scales.

(a) $\dfrac{1}{s+1}$

(b) $\dfrac{s}{s+1}$

(c) $\dfrac{s+1}{s+2}$

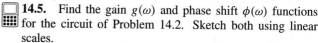 **14.5.** Find the gain $g(\omega)$ and phase shift $\phi(\omega)$ functions for the circuit of Problem 14.2. Sketch both using linear scales.

14.6. Repeat Problem 14.5 for the circuit of Problem 14.3.

14.7. Repeat Problem 14.5 for the circuit of Problem 14.4(c).

14.8. Find the gain $g(\omega)$ function for this circuit. Sketch using linear scales.

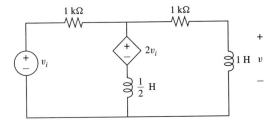

FIGURE P14.8

14.9. $\frac{d^2 v_0(t)}{dt^2} - v_0(t) = v_i(t)$. Find $H(s) = \frac{V_0(s)}{V_i(s)}$, $\mathbf{H}(j\omega)$, and the forced output $v_0(t)$ when $v_i(t) = 4\cos 2t$ V. Is this also the steady-state output?

14.10. Find the forced response $v(t)$. For what range of the controlled source gain r does this correspond to the steady-state response? Explain.

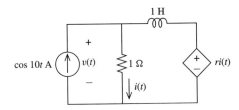

FIGURE P14.10

14.11. Find $\mathbf{H}(s) = \mathbf{I}(s)/\mathbf{V}_S(s)$, and the forced response $i(t)$ for $v_s = \sin 2t$ V. For what range of α does this correspond to the steady-state response? Explain.

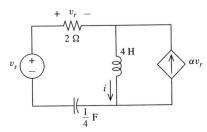

FIGURE P14.11

14.12. $1.778 = 5$ dB. Convert the following to decibels without use of a calculator.
(a) 17.78
(b) 17.78×10^3
(c) 0.001778
(d) 3.557
(e) $\sqrt{3.557}$
(f) $\sqrt[4]{3.557}$
(g) 8.89×10^{17}
(h) 8.89×10^{-15}

14.13. Convert the following to decibels without use of a calculator.
(a) $2^{3/2}$
(b) $2^{3/4}$
(c) $(5)(2)(\sqrt{1000})$
(d) $500^{1/3}$
(e) $(0.4)^9$
(f) 4×10^{30}
(g) $\frac{5}{4}$
(h) $(4/5)^9$

14.14. $1.778 = 5$ dB. Convert the following from decibels to natural numbers without using a calculator.
(a) 15 dB
(b) 18 dB
(c) 24 dB
(d) −15 dB
(e) −5 dB
(f) −105 dB

14.15. Convert the following from decibels to natural numbers without using a calculator.
(a) −26 dB
(b) +26 dB
(c) +17 dB
(d) −33 dB
(e) −39 dB
(f) −390 dB

14.16. Using a calculator, convert the following numbers to decibels.
(a) 11.74
(b) 0.00986
(c) 4.1×10^{-12}
(d) $4.1 \times 10^{+12}$
(e) 2.132
(f) $\sqrt{17}$

14.17. Using a calculator, convert the following decibel values to natural numbers.
(a) 41.7 dB
(b) −11.9 dB
(c) +17 dB
(d) +170 dB
(e) −33.3 dB
(f) 0.001 dB

14.18. Put $H(s)$ in standard form. Identify each factor as a real root or a complex root pair factor. For real root factors, specify break frequency; for complex root pair factors, specify natural frequency and damping factor ω_n, ζ.
$H(s) = :$

(a) $\dfrac{s}{(s+1)(s+2)}$

(b) $\dfrac{-10s^2}{(s^2 + s + 4)^2}$

(c) $\dfrac{1}{s} - \dfrac{s}{s-1}$

14.19. Repeat Problem 14.18 for $\mathbf{H}(s) = \mathbf{Y}(s)/\mathbf{X}(s)$, where

$$\frac{d^3 y}{dt^3} + 3\frac{d^2 y}{dt^2} + 3\frac{dy}{dt} + y(t) = 2\frac{dx}{dt} - x(t)$$

14.20. Repeat Problem 14.18 for the circuit of Problem 14.2.

14.21. Find $\mathbf{H}(s) = \mathbf{I}(s)/\mathbf{I}_s(s)$ and put in standard form. Specify all break frequencies.

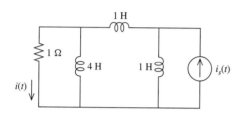

FIGURE P14.21

14.22. For what value of R is the damping factor $\zeta = 0.90$? What is the range of natural frequencies in this circuit as R is varied $0 \le R < \infty$?

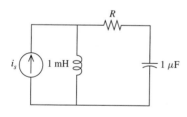

FIGURE P14.22

14.23. Consider the real root factor $(s - p)$ of multiplicity 1. Determine to three significant digits on a calculator how many decades away from the break frequency $|p|$ the difference between the exact (corrected) and the uncorrected Bode plot equals $\frac{3}{2}$ dB, 1 dB, $\frac{1}{2}$ dB, and 0.1 dB. Then express these frequencies in terms of p.

14.24. Repeat Problem 14.23 for the complex root pair factor $(s^2 + \omega_n s + \omega_n^2)$ for which the break frequency is ω_n.

14.25. Label the decibels gain at each break frequency.

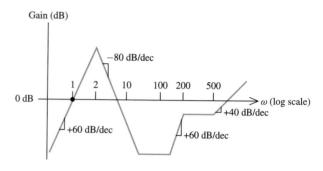

FIGURE P14.25

14.26. Specify a $\mathbf{H}(s)$ whose uncorrected Bode gain plot is shown in Problem 14.25. Is the answer unique?

14.27. Label the break frequencies. Specify a $\mathbf{H}(s)$ with no complex root pair factors with this uncorrected Bode gain plot.

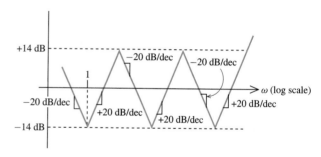

FIGURE P14.27

14.28. Sketch the Bode gain plots, corrected and uncorrected. Label each slope and label all breakpoints by their ω and decibel values. The output variable is indicated.

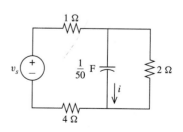

FIGURE P14.28

14.29. Repeat Problem 14.28 for this circuit.

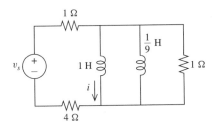

FIGURE P14.29

14.30. Repeat Problem 14.28 for this circuit.

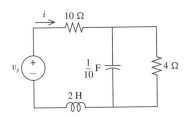

FIGURE P14.30

14.31. Repeat Problem 14.28 for this circuit.

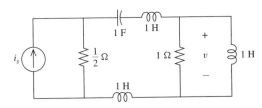

FIGURE P14.31

14.32. Repeat Problem 14.28 for this circuit.

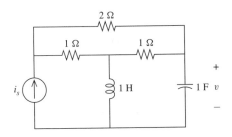

FIGURE P14.32

14.33. Sketch the Bode gain plots, corrected and uncorrected. Label each slope and all breakpoints. $\mathbf{H}(s) =$:

(a) $\dfrac{-10}{s + 10}$

(b) $\dfrac{3}{s}$

(c) $\dfrac{s + 1}{s + 2}$

14.34. Repeat Problem 14.33 for these second-order transfer functions. $\mathbf{H}(s) =$:

(a) $\dfrac{1}{(s + 1)(s + 20)}$

(b) $\dfrac{-100}{s^2 + s + 100}$

(c) $\dfrac{5}{s + 1} - \dfrac{3s}{s + 1000}$

14.35. Repeat Problem 14.33 for these higher-order transfer functions. $\mathbf{H}(s) =$:

(a) $\dfrac{500}{s^4}$

(b) $\dfrac{s^2 + s + 10^4}{s^2 + 200 + 10^4}$

(c) $\dfrac{s^2(s + 10)^2}{(s + 1)^4(s^2 + 2s + 90)}$

14.36. Draw three phasor diagrams showing Z_R, Z_L, and Z_C at $\omega = \frac{1}{10}\omega_r$, ω_r, and $10\omega_r$ for a series RLC circuit.

14.37. Draw three phasor diagrams showing Y_R, Y_L, and Y_C at $\omega = \frac{1}{2}\omega_r$, ω_r, and $2\omega_r$ for a parallel RLC circuit.

14.38. Design a series RLC circuit with $\omega_r = 50$ rad/s and $B = 4$ rad/s. Repeat for parallel RLC.

14.39. Design a series RLC circuit so that $\omega_r = 50 \times 10^3$ rad/s and $Q = 40$. Repeat for parallel RLC.

14.40. An op amp with gain–band width product GBW $= 10^6$ rad/s is used as an inverting amp with $R_F/R_A = -1$. At what radian frequency ω is the actual gain -3 dB? -20 dB? -120 dB?

14.41. Starting from (3.19b), derive the single-pole frequency response of a noninverting amp with $\mathbf{A}(j\omega) = A_o\omega_o/(j\omega + \omega_o)$. Use the approximation $\frac{1}{A_o} \ll R_A/(R_A + R_f)$. What is the corner frequency ω_C?

14.42. What is the highest gain R_F/R_A we can attain with an inverting amp and have 30,000 rad/s bandwidth? Assume the op amp has parameters $A_o = 10^5$, $\omega_o = 10$ rad/s.

14.43. Sketch the corrected and uncorrected Bode gain plots. Label all slopes and breakpoints (dB and ω). Assume ideal voltage amplifier op amp model with very high gain.

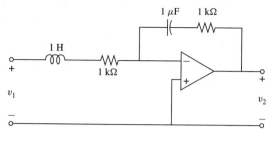

FIGURE P14.43

14.44. Repeat Problem 14.43 for this circuit.

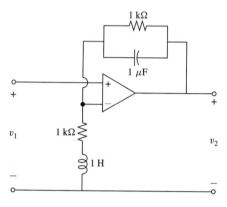

FIGURE P14.44

14.45. Design an active filter with voltage transfer function $\mathbf{H}(s) = V_2(s)/V_1(s) = -1000/(s + 1000)$.
(a) Using a buffered RC voltage divider.
(b) Using $-Z_F/Z_A$ inverting filter.

14.46. Repeat Problem 14.45 for $\mathbf{H}(s) = s/(s + 1000)$.

14.47. Design an active filter with $\mathbf{H}(s) = V_2(s)/V_s(s) = (S + 2000)/(s + 1000)$.

14.48. Design active filters with the following voltage transfer functions. Keep all resistors in the range from 5 to 500 kΩ. Draw the labeled circuit diagram.

(a) $\dfrac{s}{s+1}$

(b) $\dfrac{-3s+1}{s+50}$

(c) $\dfrac{s+1}{s}$

14.49. Repeat Problem 14.48 for the following second-order systems.

(a) $\dfrac{1}{(s+1)(s+2)}$

(b) $\dfrac{10}{s^2+24s+20}$

(c) $\dfrac{s^2}{s^2+\frac{1}{10}s+50}$

SPICE Problems

14.50. Repeat Problem 14.48 for the following higher order systems. Sketch the Bode gain plot and verify using SPICE.

(a) $\dfrac{1}{(s+1)(s+2)(s+3)}$

(b) $\dfrac{(s+1)(s+3)}{(s+2)(s+4)(s+6)}$

(c) $\dfrac{-2}{s(s^2+s+16)}$

(d) $\dfrac{10^4}{(s+10)^4}$

14.51. Use the frequency dependent op-amp model given below with $A_o = 10^5$, $R_b = 1$ Ω, $C_b = 0.1$ F, $R_o = 30$ Ω, $R_{in} = 1$ MΩ and SPICE to determine the Bode gain curves. Plot the results from 1 decade below the lowest break frequency to 1 decade above the highest decade. Compare with uncorrected hand-drawn Bode plots. These are the same $H(s)$ given in Problem 14.48.

(a) $\dfrac{s}{s+1}$

(b) $\dfrac{-3s+1}{s+50}$

(c) $\dfrac{s+1}{s}$

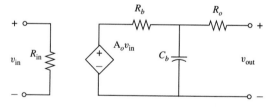

FIGURE P14.51

14.52. Repeat Problem 14.51 for the following $\mathbf{H}(s)$ of Problem 14.49.

(a) $\dfrac{1}{(s+1)(s+2)}$

(b) $\dfrac{10}{s^2+24s+20}$

(c) $\dfrac{s^2}{s^2+\frac{1}{10}s+50}$

14.53. Design a circuit with $\mathbf{H}(s) = s(s+1)/(s+2)^2$. The gain at $\omega = 1$ rad/s will be 0.283. Use cut-and-try on the leftmost capacitor to raise the gain at $\omega = 1$ to $+2.50$ dB. Show the original circuit diagram and specify the new value of the capacitance as α in the relation $C_{new} = \alpha C_{old}$. See Problem 14.51 for op-amp parameters.

14.54. Design a circuit with the given uncorrected Bode gain plot. Then add another circuit in cascade that levels the high-frequency gain at $+40$ dB by introducing a -20 dB/dec break at $\omega = 1000$ rad/s. Use SPICE to verify, using the op-amp parameters from Problem 14.51.

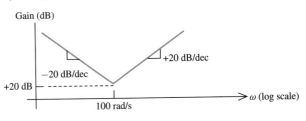

FIGURE P14.54

More Challenging Problems

14.55. Repeat Problem 14.23 for $(s - p)^r$.

14.56. Repeat Problem 14.23 for $(s^2 + \frac{1}{50}\omega_n s + \omega_n^2)^2$ (a) for $r = 1$; (b) for any integer $r > 0$.

14.57. Repeat Problem 14.21 for this circuit.

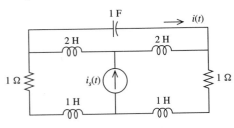

FIGURE P14.57

14.58. Find a circuit which has the given uncorrected Bode gain plot. Assume all poles and zeroes are real.

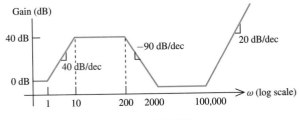

FIGURE P14.58

14.59. Repeat 14.58, but change the pair of poles with break frequency 200 rad/s to being complex with $\zeta = 0.001$.

14.60. Find an equation whose only unknown is the *exact* resonant frequency ω_r for this circuit. With the help of a calculator, using cut-and-try solve this equation to three significant digits of accuracy. Check using SPICE.

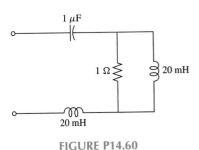

FIGURE P14.60

14.61. Estimate the resonant frequency for this circuit in two ways: first, as the natural frequency ω_n of the Bode plot; second, as the frequency ω_0 at which the phase of $\mathbf{Z}(j\omega)$ equals zero.

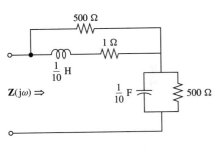

FIGURE P14.61

14.62. Suppose four identical inverting amplifier stages with gain $R_F/R_A = 10$ are put in cascade. Assuming no loading, what is the bandwidth of this overall circuit? What is the gain–bandwidth product? Compare to the gain–bandwidth product of a single stage with $R_F/R_A = 10$. Assume the op amp gain–bandwidth product is 10^6 rad/s.

15

Mutual Inductance and Transformers

James Clerk Maxwell
1831–1879

[Maxwell's revelations are] the most profound and most fruitful that physics has experienced since the time of Newton.

Albert Einstein

The first half of the nineteenth century saw an explosion of new discoveries in electricity and magnetism: the fundamental Laws of Kirchhoff and Ohm, experiments by Ampere and Faraday revealing the intimate link between these two apparently distinct forms of energy, the invention by Morse of the telegraph. But there was little agreement on the underlying physical principles beneath these phenomena, most favoring dubious ideas such as instantaneous action at a distance. It fell to James Clerk Maxwell to pull this unruly collection of new findings into a single unified theory centered around his brilliant conception of the electromagnetic field. "Maxwell's Equations" are the solid platform on which all subsequent advances in electromagnetic sciences have been anchored.

Maxwell was born in Edinburgh, Scotland on June 13, 1831. His skills as a mathematician were revealed early, publishing a paper on geometric constructions at age 14, and having two other papers published by the prestigious Royal Academy before his twentieth birthday. After graduation from Trinity College at Cambridge University, Maxwell spent nine years in full-time academic positions, first at Marishcal College, Aberdeen, then Kings College, London, before semi-retirement to his country estate in 1865 at age 34. During this brief period he set down the basic principles of his electromagnetic theory in a series of papers to the Cambridge Philosophical Society and the Royal Society, and made fundamental contributions to many other areas of physics as well, including a kinetic theory of gases, color vision, a theory of heat, and even the nature of Saturn's rings. No theoretical purist, in 1871 Maxwell accepted the post of first director of the Cavendish Laboratory at Cambridge, which under his guidance and inspiration quickly emerged as the leading experimental physical laboratory of its day.

Chapter Contents

One way in which elements can be made to interact, to influence one another's current and voltage behavior, is by using conducting wire to join their terminals. This is the only kind of interaction between circuit elements we have considered so far. Indeed, we defined an electric circuit as a set of elements and the conducting wire that connects them.

There is another way in which elements may be made to interact, one that does not require material contact. We saw in Chapter 5 that a voltage is induced across the terminals of an inductor in the presence of a time-varying magnetic field. This field may be produced by the inductor itself, which we called self-inductance, or it may be produced by a second inductor. This is called mutual inductance, which we explore in this chapter.

We begin with a basic physical description of mutual inductance, deriving the $i-v$ laws and power exchanges for sets of interacting inductors, or coupled coils. In Section 15.2 the analysis of circuits containing coupled coils is considered in detail. Useful circuits that do not themselves contain coupled coils, but are equivalent to, and may be substituted for, those which do are developed in Section 15.3. Next we turn to the special but practically important case of perfect coupling, the ideal transformer. The use of these devices to transform the levels of currents, voltages, and impedances as well as to isolate subcircuits from one another is of great practical importance and is described next. The chapter closes with a note on the use of SPICE with circuits containing mutual inductance and transformers.

15.1 MUTUAL INDUCTANCE

In Chapter 5 we found that the current i in an inductor, or *coil* as we sometimes refer to it, produces a magnetic field ϕ that threads through its N turns, producing a flux linkage

$$\lambda = N\phi \tag{15.1}$$

In a linear inductor (we shall continue to assume linearity of all inductors and magnetic media throughout), the flux produced is linear in i, and thus so is λ, which may therefore be written

$$\lambda = Li \tag{15.2}$$

The constant of proportionality L is the inductance of the inductor. Standard SI units of flux ϕ and flux linkage λ are webers (Wb), and inductance L is in henries (H). Faraday's law asserts that a voltage is induced across the terminals of the coil equal to the time rate of change of its flux linkage λ, leading to the familiar element law for an inductor,

$$v = \frac{d\lambda}{dt} = L\frac{di}{dt} \tag{15.3}$$

This previous analysis assumed that the only flux linking the coil was that due to its own current; that is, the coil is a simple or *uncoupled* coil. Sometimes the turns of a coil are threaded by the flux produced by the currents of one or more other coils, in which case we have *coupled coils*. Consider Fig. 15.1, in which two coils have fixed relative locations and orientations. The current i_1 in coil 1 produces flux ϕ_{11}, part of which threads coil 2 and part of which does not:

$$\phi_{11} = \phi_{21} + \phi_{L1} \tag{15.4}$$

Here ϕ_{21} is the component of ϕ_{11} that threads coil 2, and the remainder, ϕ_{L1}, is called coil 1's *leakage flux*. Coil 2 also has a current, i_2, producing a flux ϕ_{22}, where

$$\phi_{22} = \phi_{12} + \phi_{L2} \tag{15.5}$$

ϕ_{12} is the component of ϕ_{22} threading the turns of coil 1, and ϕ_{L2} is its leakage flux.
The total flux ϕ_1 threading coil 1 is the sum of components due to i_1 and i_2,

$$\phi_1 = \phi_{11} + \phi_{12} \tag{15.6}$$

and the net flux linkage for coil 1 is, by (15.1) and the above,

$$\lambda_1 = N_1\phi_1 = N_1\phi_{11} + N_1\phi_{12} \tag{15.7}$$

Each term on the right is a flux linkage and, by linearity, is proportional to the current producing it. The first term is due to coil 1's own current,

$$N_1\phi_{11} = L_1 i_1 \tag{15.8}$$

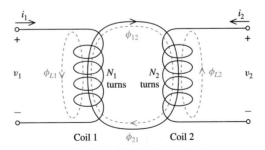

FIGURE 15.1 Pair of coupled coils. Solid flux lines are produced by the current i_1 in coil 1 and dashed flux lines by i_2 in coil 2.

where the constant of proportionality L_1 is called the *self-inductance* of coil 1, and, as in (15.2), has units of henries. The second term is produced by and is proportional to the current in the other coil:

$$N_1\phi_{12} = \pm M_{12}i_2 \qquad (15.9)$$

The positive constant M_{12} is called the *mutual inductance*. This is also measured in henries, as the dimensional equivalence of (15.8) and (15.9) requires.

The reason for the $\pm$ sign in (15.9) is that the flux component ϕ_{12} produced by the current in coil 2 may be in the *same* or *opposite* direction as that produced by coil 1 itself. If the two coils are oriented such that positive i_2 produces flux threading coil 1 in the same direction as the flux threading coil 1 due to its own positive i_1, then the plus sign must be used; otherwise, we use the minus sign.

For a given pair of coupled coils, the correct sign can always be determined from a detailed physical description of the coils, how they are wound, and their relative locations and orientations in space. To avoid the need for specifying the required physical detail, however, we shall soon introduce a shorthand system called the dot convention, which will immediately identify the proper sign.

Substituting (15.8) and (15.9) into (15.7) and differentiating, by Faraday's law

$$v_1 = \frac{d\lambda_1}{dt} = L_1\frac{di_1}{dt} \pm M_{12}\frac{di_2}{dt} \qquad (15.10)$$

Reversing the roles played by the two coils, it follows that

$$v_2 = \frac{d\lambda_2}{dt} = L_2\frac{di_2}{dt} \pm M_{21}\frac{di_1}{dt} \qquad (15.11)$$

where L_2 is the self-inductance of coil 2 and M_{21} is the mutual inductance relating current in coil 1 with the resulting voltage induced in coil 2. Note from these equations that there need be no *electric* circuit connection between two coils in order for the current in one to influence the voltage in the other; a *magnetic* circuit (flux path) is sufficient. The ability to produce electrical response yet maintain *electrical isolation*, that is, no electric circuit connection, is one primary use of coupled coils.

We next determine the energy stored in a pair of coupled coils. Along the way, we will discover the relation between the mutual inductances M_{12} and M_{21} and between the signs to be associated with the mutual inductance terms in the i–v laws (15.10) and (15.11). The instantaneous power delivered to the coils in Fig. 15.1 is

$$p_1 = v_1i_1 = L_1i_1\frac{di_1}{dt} \pm M_{12}i_1\frac{di_2}{dt} \qquad (15.12a)$$

$$p_2 = v_2i_2 = L_2i_2\frac{di_2}{dt} \pm M_{21}i_2\frac{di_1}{dt} \qquad (15.12b)$$

Assume that at $t = 0$ all currents and fluxes are zero, so there is no initial energy stored in the coupled coils; $w(0) = 0$. Since inductors are lossless elements, the total energy $w(t_1)$ stored in the system at a later time $t = t_1$ is simply the sum of all the energy delivered to coils 1 and 2 between times 0 and t_1:

$$w(t_1) = \int_0^{t_1} p_1\, dt + \int_0^{t_1} p_2\, dt \qquad (15.13)$$

If we hold i_2 at its initial value of zero during the entire time interval $[0, t_1]$, the second term in (15.12a) is zero, as are both terms in (15.12b), and

$$w(t_1) = \int_0^{t_1} L_i i_i \frac{di_1}{dt} dt$$

$$= L_1 \int_0^{i_1(t_1)} i_1 \, di_1 = \tfrac{1}{2} L_1 i_1^2(t_1) \tag{15.14}$$

This is the form derived in Chapter 5 for energy stored in a uncoupled inductor. For a subsequent time interval $[t_1, t_2]$, suppose we hold i_1 constant at its value at t_1, $i_1(t_1)$. Then $w(t_2)$, the energy stored at t_2, is given by

$$w(t_2) - w(t_1) = \int_{t_1}^{t_2} p_1 \, dt + \int_{t_1}^{t_2} p_2 \, dt$$

With i_1 constant between t_1 and t_2, $di_1/dt = 0$, and by (15.12a) and (15.12b) and the above,

$$w(t_2) - w(t_1) = \int_{t_1}^{t_2} \left(\pm M_{12} i_1 \frac{di_2}{dt} \right) dt + \int_{t_1}^{t_2} L_2 i_2 \frac{di_2}{dt} dt$$

$$= \left[\int_0^{i_2(t_2)} (\pm M_{12} i_1) \, di_2 + \int_0^{i_2(t_2)} L_2 i_2 \, di_2 \right]$$

Evaluating the integrals and using (15.14) for $w(t_1)$ with $i_1(t_1) = i_1(t_2)$, the stored energy in the pair of coupled coils at time t_2 is

$$w(t_2) = \tfrac{1}{2} L_1 i_1^2(t_2) + \tfrac{1}{2} L_2 i_2^2(t_2) \pm M_{12} i_1(t_2) i_2(t_2) \tag{15.15}$$

Now, starting back at time $t = 0$ with no stored energy, if we reverse the roles of coils 1 and 2, that is, hold i_1 at zero until t_1 and then hold i_2 constant thereafter, we would have $w(t_2)$ given by (15.15), but with the coil indexes reversed; that is,

$$w(t_2) = \tfrac{1}{2} L_2 i_2^2(t_2) + \tfrac{1}{2} L_1 i_1^2(t_2) \pm M_{21} i_2(t_2) i_1(t_2) \tag{15.16}$$

The last two equations each evaluate exactly the same quantity: the energy stored in the pair of coupled coils at a time t_2 when the currents are $i_1(t_2)$ in coil 1 and $i_2(t_2)$ in coil 2. Only the manner in which this state was reached is different. Since the energy stored does not depend on how a final state is reached, only on the value of that state, these two quantities must be identical. This implies that the last terms on the right of (15.15) and (15.16) are equal, and two important facts are revealed. The first is that

$$M_{12} = M_{21} = M \tag{15.17}$$

The mutual inductances are identical. We will use the symbol M for their common value, the *mutual inductance* of the pair of coupled coils. Rewriting (15.10) and (15.11) using

M, we have *the i–v laws for coupled coils*:

$$v_1 = L_1 \frac{di_1}{dt} \pm M \frac{di_2}{dt} \tag{15.18a}$$

$$v_2 = \pm M \frac{di_1}{dt} + L_2 \frac{di_2}{dt} \tag{15.18b}$$

Simplifying (15.15) or (15.16) by use of M and suppressing the time argument, *the energy stored in a pair of coupled coils is*

$$w = \tfrac{1}{2} L_1 i_1^2 + \tfrac{1}{2} L_2 i_2^2 \pm M i_1 i_2 \tag{15.19}$$

The second fact revealed by noting that the last terms in (15.15) and (15.16) must be identical is that *the same sign must be taken for both mutual inductance terms*. For a given pair of coupled coils, if a plus sign were taken in one equation and a minus in the other, (15.15) and (15.16) would not evaluate to identical energies, as they must. Note that the sign selected for the mutual inductance terms in the *i–v* laws (15.18) carries over to the mutual inductance term in the energy equation (15.19) as well.

Example 15.1

This example shows how the self and mutual inductances and the sign associated with the mutual inductance terms in (15.18) can be determined experimentally. For a pair of coupled coils in a fixed relative orientation, suppose that we inject a ramp test current $i_1(t) = t$ A into coil 1, while holding coil 2 open circuited, and with a voltmeter measure the responses to be dc voltages $v_1(t) = 6$ V and $v_2(t) = -4$ V. Then, by (15.18a) and (15.18b),

$$v_1 = L_1 \frac{d}{dt}(t) \pm M \frac{d}{dt}(0) = L_1 = 6 \text{ V}$$

$$v_2 = \pm M \frac{d}{dt}(t) + L_2 \frac{d}{dt}(0) = \pm M = -4 \text{ V}$$

The self-inductance L_1 of coil 1 is 6 H, and noting that M is defined to be nonnegative, the mutual inductance M is 4 H. The minus sign clearly must be taken in both equations in the *i–v* laws for these coupled coils (15.18); that is, their *i–v* laws are

$$v_1 = 6 \frac{di_1}{dt} - 4 \frac{di_2}{dt}$$

$$v_2 = -4 \frac{di_1}{dt} + L_2 \frac{di_2}{dt}$$

To determine L_2, we set $i_1 = 0$ and inject a convenient time-varying current into coil 2. Suppose that the response v_2 to the unit ramp $i_2(t) = t$ under these conditions is 5 V. Then $L_2 = 5$ H by the last equation. Finally, suppose that at some time t we have instantaneous coil currents $i_1(t) = 15$ A and $i_2(t) = -9$ A. Then the energy stored in the pair of coupled coils at t is, by (15.19),

$$w = \tfrac{1}{2}(6)(15^2) + \tfrac{1}{2}(5)(-9)^2 - 4(15)(-9) = 1417.5 \text{ J}$$

The *coefficient of coupling* k is a measure of the degree to which the flux produced by one coil threads another and is defined by

$$k = \frac{M}{\sqrt{L_1 L_2}} \qquad (15.20)$$

If there is no coupling between the coils, $M = k = 0$. A pair of coils with zero coupling coefficient k is equivalent to two simple, uncoupled coils, as can be seen by setting $M = 0$ in the i–v laws (15.18). If there is coupling, then by (15.8), (15.9), and (15.17),

$$k = \frac{\sqrt{M_{21} M_{12}}}{\sqrt{L_1 L_2}} = \sqrt{\frac{\phi_{21}}{\phi_{11}}} \sqrt{\frac{\phi_{12}}{\phi_{22}}}$$

Since both numerators on the right are components of their respective denominators, it follows that $k \leq 1$. Thus the coupling coefficient k is bounded by

$$0 \leq k \leq 1 \qquad (15.21)$$

or, equivalently, the mutual inductance M is bounded by

$$0 \leq M \leq \sqrt{L_1 L_2} \qquad (15.22)$$

A pair of coils with coupling coefficient k near its upper bound of unity is said to be *tightly coupled.* Such coils are particularly efficient in linking parts of an electrical circuit that cannot be wired together directly, a possibility that we pursue in our discussion of transformers in Section 15.4.

EXERCISES

15.1.1. When $i_1(t) = 5 \cos 2t$ A and coil 2 is held open circuited, the voltages in Fig. 15.1 are found to be $v_1(t) = 2 \sin(2t + 180°)$ V and $v_2(t) = 3 \sin(2t + 180°)$ V. When the roles are reversed, $i_2(t) = 5 \cos 2t$ A and coil 1 is held open circuited, and $v_2(t) = 10 \sin(2t + 180°)$ V. Find L_1, L_2, M, and the sign of the mutual inductance terms in the coupled coil i–v laws (15.19).

Answer $\tfrac{1}{5}$ H; 1 H; $\tfrac{3}{10}$ H; plus sign

15.1.2. The current $i_1 = +1$ A produces flux $\phi_{11} = 2$ mW, of which half threads the turns of coil 2; $\phi_{21} = 1$ mW. $i_2 = +1$ A produces flux $\phi_{22} = 10$ mW in the same direction as ϕ_{21}. Coil 1 has 100 turns and coil 2

has 900 turns. Find the fraction of ϕ_{22} that threads coil 1 and the energy stored in the pair of coupled coils when $i_1 = +1$ mA and $i_2 = +5$ mA.

Answer $\frac{9}{10}$; 117.1 μJ

15.1.3. Determine the coefficient of coupling for Exercises 15.1.1 and 15.1.2.

Answer 0.671; 0.671

15.2 CIRCUITS WITH MUTUAL INDUCTANCE

The i–v laws for a pair of coupled coils, given in (15.18) and repeated here for convenience, are

$$v_1 = L_1 \frac{di_1}{dt} \pm M \frac{di_2}{dt}$$

(15.23a)

$$v_2 = \pm M \frac{di_1}{dt} + L_2 \frac{di_2}{dt}$$

(15.23b)

To analyze a circuit containing coupled coils, we must know which signs to use. In Section 15.1 we saw that the same sign must be used for both mutual inductance terms in these equations. The shorthand notation we use for indicating which sign should be used is called the *dot convention*. Circuit diagrams will include a pair of dots, one for each inductor, and values of the two self-inductances and the mutual inductance, as shown in Fig. 15.2.

To apply the dot convention, first assign current and voltage reference directions separately for each inductor so that each satisfies the passive sign convention (arrows for i_1 and i_2 point into plus ends of v_1 and v_2, respectively). Then the *dot convention* is the following rule: *if both current reference arrows point into dotted ends or both into undotted ends of the inductors, use the plus sign for both mutual inductance terms in the i–v laws. Otherwise, use the minus sign.*

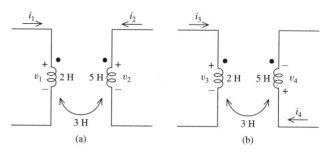

FIGURE 15.2   (a) Coupled coils; (b) same coils but different variables.

Example 15.2

Let us find the i–v laws for the coupled coils in Fig. 15.2(a) and (b). Applying the dot convention for Fig. 15.2(a), we first check that the passive sign convention is satisfied separately for (i_1, v_1) and (i_2, v_2). Then, since the current reference arrows both point into dotted ends, the plus signs in (15.23) are taken, and the i–v

laws are

$$v_1 = 2\frac{di_1}{dt} + 3\frac{di_2}{dt} \qquad (15.24a)$$

$$v_2 = 3\frac{di_1}{dt} + 5\frac{di_2}{dt} \qquad (15.24b)$$

For the same coils, suppose that we assign the current and voltage variables as in Fig. 15.2(b). Again (i_3, v_3) and (i_4, v_4) separately satisfy the passive sign convention. This time one arrow points into a dotted end and the other into an undotted end, so

$$v_3 = 2\frac{di_3}{dt} - 3\frac{di_4}{dt} \qquad (15.25a)$$

$$v_4 = -3\frac{di_3}{dt} + 5\frac{di_4}{dt} \qquad (15.25b)$$

Examining the figure, if these are the same coils, then $v_3 = v_1$, $i_3 = i_1$, $i_4 = -i_2$, and $v_4 = -v_2$. Substituting these into (15.25) recovers (15.24), so the i–v laws are completely equivalent. It does not matter which way the reference direction pairs are chosen. *Separate sets of reference directions for the two coils may be freely chosen without regard to the dots.* Of course, it will often be slightly more convenient to choose the current arrows so that minus signs in the i–v laws may be avoided, as in Fig. 15.2(a), but not Fig. 15.2(b).

The Laplace transform or s-domain version of the i–v laws (15.23) is very convenient for analyzing circuits involving initial condition responses or switching transients. Using the Laplace transform differentiation property (Table 12.1), the i–v laws for coupled coils in the s-domain are, from (15.23),

$$\mathbf{V}_1(s) = \{sL_1\mathbf{I}_1(s) - L_1i_1(0-)\} \pm \{sM\mathbf{I}_2(s) - Mi_2(0-)\} \qquad (15.26a)$$

$$\mathbf{V}_2(s) = \pm\{sM\mathbf{I}_1(s) - Mi_1(0-)\} + \{sL_2\mathbf{I}_2(s) - L_2i_2(0-)\} \qquad (15.26b)$$

In the important special case of zero initial conditions, these simplify to

$$\mathbf{V}_1(s) = sL_1\mathbf{I}_1(s) \pm sM\mathbf{I}_2(s) \qquad (15.27a)$$

$$\mathbf{V}_2(s) = \pm sM\mathbf{I}_1(s) + sL_2\mathbf{I}_2(s) \qquad (15.27b)$$

Example 15.3

Consider the circuit of Fig. 15.3(a) and the desired circuit response $v_3(t)$, $t > 0$. The initial conditions at time $t = 0-$ are all zero, since the unit step $u(t)$ does not energize the source until $t = 0$. For $t > 0$ there is only one unknown mesh current, i_m, in Fig. 15.3(b). Assign reference directions (i_1, v_1), (i_2, v_2), which separately satisfy

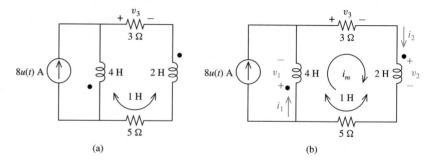

FIGURE 15.3 (a) Circuit for Example 15.2; (b) labeled for analysis.

the passive sign convention for the two inductors as shown. The mesh equation is

$$\mathbf{V}_1(s) + 3\mathbf{I}_m(s) + \mathbf{V}_2(s) + 5\mathbf{I}_m(s) = 0$$

Using the s-domain i–v laws (15.27) to eliminate the voltages,

$$[4s\mathbf{I}_1(s) + s\mathbf{I}_2(s)] + 8\mathbf{I}_m(s) + \{s\mathbf{I}_1(s) + 2s\mathbf{I}_2(s)\} = 0$$

From Fig. 15.3(b) we see that $i_m = i_2$ and $i_1 = i_m - 8u(t) = i_2 - 8u(t)$. Using these to eliminate i_m and i_1 yields

$$4s\left(\mathbf{I}_2(s) - \frac{8}{s}\right) + s\mathbf{I}_2(s) + 8\mathbf{I}_2(s) + s\left(\mathbf{I}_2(s) - \frac{8}{s}\right) + 2s\mathbf{I}_2(s) = 0$$

Solving, we obtain

$$\mathbf{I}_2(s) = \frac{5}{s+1}$$

$$v_3(t) = 3i_2(t) = 15e^{-t} \text{ V}, \qquad t > 0$$

In Example 15.3 we see that writing mesh equations for circuits with coupled coils is a straightforward two-step process. First the mesh equations are written in terms of the coil voltages; then the i–v laws (15.26) or (15.27) are used to eliminate the coil voltages and the resulting coil currents are expressed in terms of mesh currents. The process can be speeded by writing the mesh equations directly in the mesh currents in a single step, but at the peril of sign errors. It is recommended that the two steps be done explicitly until the student feels quite confident in writing the mesh equations directly in terms of the currents.

Example 15.4

We seek the current $i_2(t)$, $t > 0$ for the circuit of Fig. 15.4. At $t = 0-$ the circuit is in dc steady state, and by the coupled coil i–v laws (15.23), when the currents are constant there is no voltage across the inductors. *Coupled coils behave like short circuits in dc steady state*, just as uncoupled coils do. Then $i_2(0-) = \frac{1}{2}$ A and since the switch is open for $t < 0$, $i_1(0-) = 0$. For $t > 0$ the switch is closed, and the s-domain mesh equations are

$$4\mathbf{I}_1(s) + \mathbf{V}_1(s) = \frac{2}{s} \qquad (15.28a)$$

$$\mathbf{V}_2(s) + 2\mathbf{I}_2(s) + \left[s\mathbf{I}_2(s) - \frac{1}{2}\right] = \frac{1}{s} \qquad (15.28b)$$

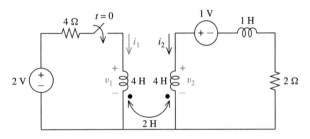

FIGURE 15.4 Circuit for Example 15.4.

Using the s-domain $i–v$ laws (15.26) to eliminate the voltages and then rewriting in vector-matrix form for convenient solution, we have

$$4\mathbf{I}_1(s) + [4s\mathbf{I}_1(s) + 2s\mathbf{I}_2(s) - 1] = \frac{2}{s} \qquad (15.29a)$$

$$[4s\mathbf{I}_2(s) - 2 + 2s\mathbf{I}_1(s)] + 2\mathbf{I}_2(s) + \left[s\mathbf{I}_2(s) - \frac{1}{2}\right] = \frac{1}{s} \qquad (15.29b)$$

$$\begin{bmatrix} 4s+4 & 2s \\ 2s & 5s+2 \end{bmatrix} \begin{bmatrix} \mathbf{I}_1(s) \\ \mathbf{I}_2(s) \end{bmatrix} = \begin{bmatrix} \dfrac{s+2}{s} \\ \dfrac{5s+2}{2s} \end{bmatrix}$$

Solving for $\mathbf{I}_2(s)$ by Cramer's rule,

$$\Delta = \begin{vmatrix} 4s+4 & 2s \\ 2s & 5s+2 \end{vmatrix} = 16s^2 + 28s + 8$$

$$\mathbf{I}_2(s) = \frac{1}{\Delta} \begin{vmatrix} 4s+4 & \dfrac{2s+4}{2s} \\ 2s & \dfrac{5s+2}{2s} \end{vmatrix} = \frac{4s^2+5s+2}{2s(4s^2+7s+2)}$$

There are poles at $s = 0$ and $s = (-7 \pm \sqrt{17})/8$ or $s = -0.360$, -1.39. The partial fraction expansion for these three simple real poles is

$$\mathbf{I}_2(s) = \frac{\frac{1}{2}}{s} + \frac{-0.242}{s+0.360} + \frac{0.242}{s+1.39}$$

$$i_2(t) = \frac{1}{2} + 0.242(e^{-1.39t} - e^{-0.36t}), \qquad t > 0$$

Since transfer functions are ratios of s-domain outputs to inputs when initial conditions are zero, the zero-initial-condition form of the $i–v$ laws (15.27) may also be used for determination of a transfer function $\mathbf{H}(s)$ and, if the circuit is stable, by replacement of s by $j\omega$, the frequency response $\mathbf{H}(j\omega)$. Recall that circuits with no dependent sources or isolated LC loops are always stable, and in any event, stability can be checked by determining whether the sign of the real part of each pole of the system is negative.

Example 15.5

We wish to find the current transfer ratio $\mathbf{H}(s) = \mathbf{I}_2(s)/\mathbf{I}_1(s)$ for the circuit shown in Fig. 15.5 and the frequency response $\mathbf{H}(j\omega)$. There

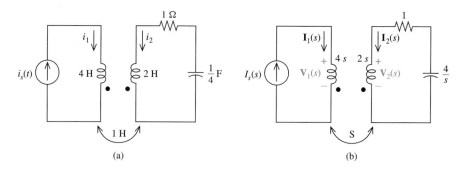

FIGURE 15.5 (a) Circuit; (b) labeled s-domain circuit.

are two meshes but only one unknown mesh current. The single mesh equation is

$$\left(\frac{4}{s} + 1\right)\mathbf{I}_2(s) + \mathbf{V}_2(s) = 0 \qquad (15.30)$$

where, since both currents flow into undotted ends of their respective inductors, the dot convention requires use of the plus sign for the mutual inductance term, and

$$\mathbf{V}_2(s) = s\mathbf{I}_1(s) + 2s\mathbf{I}_2(s)$$

Substituting the last into (15.30) and solving for the response $\mathbf{I}_2(s)$ gives

$$\mathbf{I}_2(s) = \frac{-\frac{1}{2}s^2}{s^2 + \frac{1}{2}s + 2}\mathbf{I}_1(s)$$

The desired transfer function is

$$\mathbf{H}(s) = \frac{-\frac{1}{2}s^2}{s^2 + \frac{1}{2}s + 2}$$

and the frequency response of this circuit is

$$\mathbf{H}(j\omega) = \frac{\frac{1}{2}\omega^2}{(2 - \omega^2) + j(\omega/2)}$$

Equations (15.26) and (15.27) are the v forms of the i–v laws for a pair of coupled coils, in which the voltages are found in terms of the currents. These are convenient for mesh analysis, as the examples have shown. For nodal analysis it is necessary to work with the i form of these i–v laws. Rewriting (15.27) in vector-matrix form, we have

$$\begin{bmatrix} \mathbf{V}_1(s) \\ \mathbf{V}_2(s) \end{bmatrix} = \begin{bmatrix} sL_1 & \pm sM \\ \pm sM & sL_2 \end{bmatrix} \begin{bmatrix} \mathbf{I}_1(s) \\ \mathbf{I}_2(s) \end{bmatrix} \qquad (15.31)$$

Inverting the matrix yields

$$\begin{bmatrix} \mathbf{I}_1(s) \\ \mathbf{I}_2(s) \end{bmatrix} = \frac{1}{s^2(L_1 L_2 - M^2)} \begin{bmatrix} sL_2 & -(\pm sM) \\ -(\pm sM) & sL_1 \end{bmatrix} \begin{bmatrix} \mathbf{V}_1(s) \\ \mathbf{V}_2(s) \end{bmatrix} \qquad (15.32)$$

Expanding back into coupled scalar equation form, we have *the i-form of the i–v laws for coupled coils with zero initial conditions*:

$$\mathbf{I}_1(s) = \frac{1}{sL_{e1}}V_1(s) - \left(\pm\frac{1}{sM_e}\right)V_2(s) \qquad (15.33a)$$

$$\mathbf{I}_2(s) = -\left(\pm\frac{1}{sM_e}\right)V_1(s) + \frac{1}{sL_{e2}}V_2(s) \qquad (15.33b)$$

where we have defined equivalent inductance parameters

$$L_{e1} = L_1 - \frac{M^2}{L_2}, \qquad L_{e2} = L_2 - \frac{M^2}{L_1}, \qquad M_e = \frac{L_1L_2}{M} - M \qquad (15.34)$$

Note the minus sign in front of the term $1/sM_e$. *If the dot convention requires that the positive sign be taken in the mutual inductance term, the corresponding terms* ($\pm 1/sM_e$) *in the i form of the i–v laws are negative, and vice versa.* In other words, if both current reference directions point into dotted ends or both into undotted ends of their inductors, then minus signs will be needed in this form of the *i–v* laws.

Using (15.26), when the initial conditions are not zero, (15.31) becomes

$$\begin{bmatrix} V_1(s) + L_1i_1(0-) \pm Mi_2(0-) \\ V_2(s) \pm Mi_1(0-) + L_2i_2(0-) \end{bmatrix} = \begin{bmatrix} sL_1 & \pm sM \\ \pm sM & sL_2 \end{bmatrix}\begin{bmatrix} \mathbf{I}_1(s) \\ \mathbf{I}_2(s) \end{bmatrix}$$

The same matrix is inverted as in (15.31); thus the *i* form for the *i–v* law is that shown in (15.33) but with $V_1(s)$ and $V_2(s)$ in (15.33) replaced by $V_1(s) + L_1i_1(0-) \pm Mi_2(0-)$ and $V_2(s) \pm Mi_1(0-) + L_2i_2(0-)$, respectively.

Example 15.6

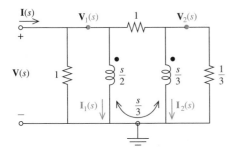

FIGURE 15.6   Circuit for Example 15.6.

We wish to determine the impedance $\mathbf{Z}(s) = \mathbf{V}(s)/\mathbf{I}(s)$ of the two-terminal subcircuit shown in Figure 15.6. The nodal equations are

$$\mathbf{I}_1(s) + 2V_1(s) - V_2(s) = \mathbf{I}(s)$$
$$\mathbf{I}_2(s) + 4V_2(s) - V_1(s) = 0 \qquad (15.35)$$

We will eliminate $\mathbf{I}_1(s)$ and $\mathbf{I}_2(s)$ using (15.33). The equivalent inductance parameters are

$$L_{e1} = \frac{1}{2} - \frac{(\frac{1}{3})^2}{\frac{1}{3}} = \frac{1}{6} \text{ H} \qquad (15.36a)$$

$$L_{e2} = \frac{1}{3} - \frac{(\frac{1}{3})^2}{\frac{1}{2}} = \frac{1}{9} \text{ H} \qquad (15.36b)$$

$$M_e = \frac{(\frac{1}{2})(\frac{1}{3})}{\frac{1}{3}} - \frac{1}{3} = \frac{1}{6} \text{ H} \qquad (15.36c)$$

and using the coil i–v laws (15.33) in (15.35), we obtain

$$\left[\frac{6}{s}\mathbf{V}_1(s) - \frac{6}{s}\mathbf{V}_2(s)\right] + 2\mathbf{V}_1(s) - \mathbf{V}_2(s) = \mathbf{I}(s) \quad (15.37a)$$

$$\left[-\frac{6}{s}\mathbf{V}_1(s) + \frac{9}{s}\mathbf{V}_2(s)\right] + 4\mathbf{V}_2(s) - \mathbf{V}_1(s) = 0 \quad (15.37b)$$

Solving the second equation for $\mathbf{V}_2(s)$ yields

$$\mathbf{V}_2(s) = \frac{s+6}{4s+9}\mathbf{V}_1(s)$$

Substituting into (15.37a) eliminates $\mathbf{V}_2(s)$, and

$$\left(\frac{6}{s}+2\right)\mathbf{V}_1(s) - \left(\frac{6}{s}+1\right)\left(\frac{s+6}{4s+9}\right)\mathbf{V}_1(s) = \mathbf{I}(s) \quad (15.38a)$$

$$\mathbf{Z}(s) = \frac{\mathbf{V}_1(s)}{\mathbf{I}(s)} = \frac{4s^2 + 9s}{7s^2 + 30s + 18} \quad (15.38b)$$

Compared to the v forms (15.26) and (15.27), the i forms of the i–v laws for coupled coils are more cumbersome. This makes mesh analysis preferable to nodal analysis for most problems. It is sometimes the case, however, as in Example 15.6, that there are significantly fewer node voltage variables than mesh current variables, and nodal analysis is warranted in these cases.

While the s-domain analysis presented in this section may be applied to any circuit problem, phasor analysis is more efficient for finding the forced or ac steady-state response to sinusoidal excitation. As shown in Chapter 14, s-domain relationships computed with zero initial conditions may be converted to the phasor domain simply by replacement of s by $j\omega$. The phasor i–v laws for coupled coils are, from (15.27),

$$\mathbf{V}_1(j\omega) = j\omega L_1 \mathbf{I}_1 \pm j\omega M \mathbf{I}_2 \quad (15.39a)$$

$$\mathbf{V}_2(j\omega) = \pm j\omega M \mathbf{I}_1 + j\omega L_2 \mathbf{I}_2 \quad (15.39b)$$

and with L_{e1}, L_{e2}, and M_e defined as in (15.34),

$$\mathbf{I}_1 = \frac{1}{j\omega L_{e1}}\mathbf{V}_1 - \left(\pm\frac{1}{j\omega M_e}\right)\mathbf{V}_2$$

$$\mathbf{I}_2 = -\left(\pm\frac{1}{j\omega M_e}\right)\mathbf{V}_1 + \frac{1}{j\omega L_{e1}}\mathbf{V}_2$$

For instance, the phasor impedance $\mathbf{Z}(j\omega)$ for the circuit of Example 15.6 is, replacing s by $j\omega$ in (15.38b),

$$\mathbf{Z}(j\omega) = \frac{4(j\omega)^2 + 9(j\omega)}{7(j\omega)^2 + 30(j\omega) + 18}$$

$$= \frac{-4\omega^2 + j9\omega}{(18 - 7\omega^2) + j30\omega} \tag{15.40}$$

From Fig. 15.6 we see that in dc steady state, the input is shorted by the leftmost inductor and the impedance is therefore zero, consistent with setting $\omega = 0$ in (15.40). In the high-frequency limit, the inductors in Fig. 15.6 behave like open circuits and the impedance is that of a 1-Ω resistor in parallel with a $\frac{4}{3}$-Ω resistor, or $\frac{4}{7}$-Ω. This checks with the limiting value in (15.40) as $\omega \to \infty$.

To sum up this section, circuits containing coupled coils may be analyzed without undue difficulty using either mesh or nodal analysis. The mesh equations are written first in terms of coil voltage variables, and then the v form of the i–v laws (15.26), or (15.27) if there are zero initial conditions, is used to eliminate these voltages. The dot convention is used to determine the proper sign to be used with the mutual inductance terms in the i–v laws. Similarly, nodal analysis equations are first written to contain coil current variables, which are replaced using the i form of the i–v laws. The i form uses equivalent inductance parameters L_{e1}, L_{e2}, and M_e defined in (15.34) and, particularly when there are zero initial conditions, is somewhat less convenient to use than the v form and mesh analysis.

EXERCISES

15.2.1. Find i_1 for $t > 0$ if $M = 1/\sqrt{2}$ H and the circuit is in steady state at $t = 0-$.
Answer $6 - 2e^{-4/3t}$ A

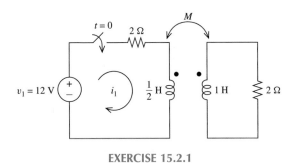

EXERCISE 15.2.1

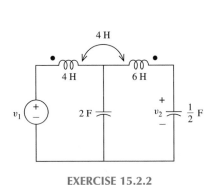

EXERCISE 15.2.2

15.2.2. Find v_2 in ac steady state if $v_1 = 8e^{-2t} \cos t$ V.
Answer $\sqrt{2}e^{-2t} \cos(t + 45°)$ V

15.2.3. Find the transfer function $\mathbf{V}_2/\mathbf{V}_1$. Note that each inductor is coupled with two others, so the i–v laws for each pair of coils should have two mutual inductance terms.

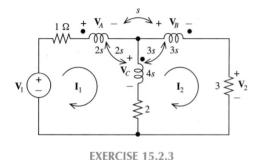

EXERCISE 15.2.3

Answer $\dfrac{3(10s+2)}{30s^2+49s+11}$

15.2.4. Write the i forms of the s-domain i–v laws for Fig. 15.2(a) with zero initial conditions and Fig. 15.2(b) with arbitrary initial conditions.

Answer (a) $\mathbf{I}_1 = \dfrac{5}{s}\mathbf{V}_1 - \dfrac{3}{s}\mathbf{V}_2$

$\mathbf{I}_2 = -\dfrac{3}{s}\mathbf{V}_1 + \dfrac{2}{s}\mathbf{V}_2$

(b) $\mathbf{I}_1 = (5/s)[\mathbf{V}_3 + 2i_3(0-) - 3i_4(0-)] + (3/s)[\mathbf{V}_4 - 3i_1(0-) + 5i_4(0-)]$

$\mathbf{I}_2 = (3/s)[\mathbf{V}_3 + 2i_3(0-) - 3i_4(0-)] + (2/s)[\mathbf{V}_4 - 3i_1(0-) + 5i_4(0-)]$

15.3 MUTUAL INDUCTANCE AND TRANSFORMERS

The most common practical use of the phenomenon of mutual inductance in electric circuits is in transformers. *A transformer is a two-port circuit containing coupled coils wound around a common core.* A typical transformer is sketched in Fig. 15.7. Each coil, or *winding*, terminates in a pair of nodes called a *port*, which is available for connection to external circuitry. One of these ports is called the *primary*. Usually, the primary is connected to an external source, as shown in the figure. The other port, called the *secondary*, is usually connected to a load. In power transformer applications, the load might be a passive impedance, while in electronics applications it might be the input to a filter, amplifier, or other active circuit.

The rectangular ring in Figure 15.7 around which both coils are wound is called the *core* of the transformer. The core is constructed largely from ferromagnetic material to facilitate production of strong magnetic fields in response to the currents flowing in the coils and to guide these fluxes through both primary and secondary windings. The core can be a simple doughnut of soft iron or a sophisticated layered construction of composite materials with lamina containing alloys of cobalt, selenium, and other materials with desirable magnetic properties.

Transformers are among the most versatile of electric devices and are widely used over the full range of circuits applications, including power, industrial, and consumer electronics. Their main use is to scale, or transform, primary circuit variables to levels better suited to drive the circuit connected to the secondary. Current, voltage, and

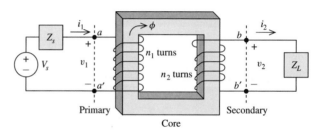

FIGURE 15.7 Transformer.

impedance levels may be transformed using this device, as we shall see. In addition, transformers are often used to electrically isolate the primary circuit from the secondary.

Consider the basic circuit shown in Fig. 15.8. Since transformers are most frequently used in ac steady-state circuits, we shall use phasor analysis. The self-inductance of the primary winding is L_1, that of the secondary is L_2, and the mutual inductance of the transformer is M. The external circuit attached to the primary has been simplified to its Thevenin equivalent, with Thevenin equivalent voltage source $\mathbf{V}_s$ and impedance $\mathbf{Z}_s$. The load has also been reduced (if necessary) to its Thevenin equivalent, and the absence of an equivalent source in series with Z_L implies that there are no independent sources in the load circuit attached to the secondary.

Note carefully the reference directions shown in Fig. 15.8. Following the standard convention for transformers, *the primary current and voltage are defined to satisfy the passive sign convention, while the secondary current and voltage* ($\mathbf{I}_2$, $\mathbf{V}_2$) *do not.* Our statement of the dot convention in Section 15.2 requires that the passive sign convention be satisfied by both coils separately, so we shall use $\mathbf{I}_2$ and $-\mathbf{V}_2$ as the current and voltage variables for the secondary when writing i–v laws.

The phasor i–v laws for the transformer are, from (15.39),

$$\mathbf{V}_1 = j\omega L_1 \mathbf{I}_1 - j\omega M \mathbf{I}_2 \tag{15.41a}$$

$$-\mathbf{V}_2 = -j\omega M \mathbf{I}_1 + j\omega L_2 \mathbf{I}_2 \tag{15.41b}$$

Since $\mathbf{I}_1$ points into a dotted end and $\mathbf{I}_2$ into an undotted end, the dot convention dictates minus signs for the mutual inductance terms. The coil voltage variables used in (15.41) are those that satisfy the passive sign convention together with $\mathbf{I}_1$ and $\mathbf{I}_2$, that is, $\mathbf{V}_1$ and $-\mathbf{V}_2$. Noting from the figure that $\mathbf{V}_2 = Z_L \mathbf{I}_2$, we may solve (15.41b) for $\mathbf{I}_2$ in terms of $\mathbf{I}_1$

$$\mathbf{I}_2 = \frac{j\omega M}{Z_L + j\omega L_2} \mathbf{I}_1 \tag{15.42}$$

Using this expression to eliminate $\mathbf{I}_2$ from (15.41a), we have

$$\mathbf{V}_1 = \left(j\omega L_1 + \frac{\omega^2 M^2}{Z_L + j\omega L_2} \right) \mathbf{I}_1 \tag{15.43}$$

This expression has an interesting interpretation. It suggests that the impedance $\mathbf{Z} = \mathbf{V}_1/\mathbf{I}_1$ looking into the primary port is equivalent to a simple (uncoupled) inductor of L_1

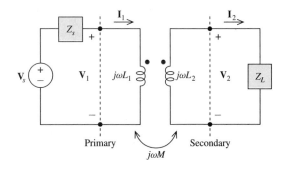

FIGURE 15.8 Basic transformer circuit (phasor domain).

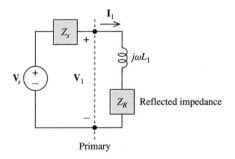

FIGURE 15.9 Secondary winding and load reflected into primary.

henries in series with a second impedance, as shown in Fig. 15.9. This second impedance shows the effect of the secondary winding and load as it appears in the primary circuit and is called the *reflected impedance* $\mathbf{Z}_R(j\omega)$.

$$\mathbf{Z}_R(j\omega) = \frac{\omega^2 M^2}{Z_L + j\omega L_2} \tag{15.44}$$

The subcircuits to the right of the primary terminals in Figs. 15.8 and 15.9 are equivalent subcircuits, and thus the one-loop circuit of Fig. 15.9 can often be used to simplify analysis of the basic transformer circuit of Fig. 15.8.

The secondary-to-primary current ratio I_2/I_1 and voltage ratio V_2/V_1 are also of interest. By (15.42),

$$\frac{I_2}{I_1} = \frac{j\omega M}{Z_L + j\omega L_2} \tag{15.45}$$

and since

$$\frac{V_2}{V_1} = \frac{Z_L I_2}{V_1} = Z_L \frac{I_2}{I_1} \frac{I_1}{V_1} \tag{15.46}$$

then, by (15.43)–(15.46),

$$\frac{V_2}{V_1} = \frac{j\omega M Z_L}{j\omega L_1(Z_L + j\omega L_2) + \omega^2 M^2} \tag{15.47}$$

Equations (15.45) and (15.47) show how the primary current and voltage are transformed in passage through the transformer, and (15.44) shows how the secondary impedance is transformed when reflected into the primary circuit.

Example 15.7

A 60-Hz household circuit that supplies $V_s = 156\underline{/0}$ V at source impedance $Z_s = 20\ \Omega$ is to be used to drive a load $Z_L = 5\ \mathrm{k}\Omega$. Determine the current, voltage, and power delivered to the load if it

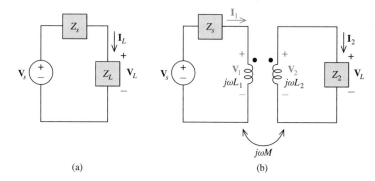

FIGURE 15.10   Two ways to connect a source and load.

is connected directly to the source as in Fig. 15.10(a), and compare with the case of connection through a transformer with $L_1 = \frac{1}{4}$ H, $L_2 = 10$ H, and $M = 1$ H, as shown in Fig. 15.10(b).

In Fig. 15.10(a) with direct connection, the load voltage is

$$\mathbf{V}_L = \frac{Z_L}{Z_L + Z_s}\mathbf{V}_s = \frac{5000}{5020}(156) = 155.4 \text{ V}$$

The load current and power dissipation are

$$\mathbf{I}_L = \frac{\mathbf{V}_s}{Z_L + Z_s} = \frac{156}{5020} = 0.031 \text{ A}$$

$$P_L = \frac{R_L|I_L|^2}{2} = 2500(0.031^2) = 2.4 \text{ W}$$

With direct connection, 2.4 W is delivered to the load at 155.4 V and 3.1 mA ac steady-state amplitude. Turning to the transformer-coupled circuit Fig. 15.10(b), the reflected impedance is

$$Z_R = \left.\frac{\omega^2 M^2}{Z_L + j\omega L_2}\right|_{\omega=377} = \frac{(377^2)(1^2)}{5000 + j3770} = 18.1 - j13.7 \text{ }\Omega$$

Examining Fig. 15.9, the primary current and voltage are

$$\mathbf{I}_1 = \frac{\mathbf{V}_2}{j\omega L_1 + Z_R + Z_s} = \frac{156\underline{/0}}{38.1 + j80.6} = 0.75 - j1.58 \text{ A}$$

$$\mathbf{V}_1 = (j\omega L_1 + Z_R)\mathbf{I}_1 = (18.1 + j80.6)(0.75 - j1.58)$$

$$= 141 + j31.9 \text{ V}$$

Then, by the primary-to-secondary ratios (15.45),

$$\mathbf{I}_2 = \frac{j\omega M}{Z_L + j\omega L_2}\mathbf{I}_1 = \frac{j377}{5000 + j3770}(0.75 - j1.58)$$

$$= 0.103 - j0.021 \text{ A}$$

$$\mathbf{V}_2 = Z_L\mathbf{I}_2 = 5000(0.103 - j0.021) = 516 - j106 \text{ V}$$

The power delivered to the load in this case is

$$P_L = \frac{R_L|I_2|^2}{2} = 2500|0.103 - j0.021|^2 = 27.6 \text{ W}$$

Comparing these values, we see that there is more than an order of magnitude of additional power delivered to the load when transformer coupled. Recall from Chapter 10 that maximum power transfer to the load would require that $\mathbf{Z}_L = \mathbf{Z}_s^* = 20 \ \Omega$. When connected directly, as in Fig. 15.10(a), the impedance mismatch is severe, with $\mathbf{Z}_L = 5000 \ \Omega$. The effect of transformer coupling is seen in Fig. 15.9, in which the secondary impedance reflected into the primary is equal to $j\omega L_1 + \mathbf{Z}_R = 18.1 + j80.6 \ \Omega$. The impedance mismatch is much less severe and more power flow to the load results. Finally, note that in the transformer-coupled case

$$\left|\frac{\mathbf{V}_2}{\mathbf{V}_1}\right| = \left|\frac{516 - j106}{141 + j31.9}\right| = 3.64$$

$$\left|\frac{\mathbf{I}_2}{\mathbf{I}_1}\right| = \left|\frac{0.103 - j0.021}{0.75 - j1.58}\right| = 0.060$$

That is, this transformer acts to step up the voltage and step down the current.

Another practical use of transformers that we shall mention is to achieve *electrical isolation*. Isolation is useful both for safety and for establishing separate reference voltage levels in primary and secondary circuits. Consider the two circuits shown in Fig. 15.10. In Fig. 15.10(a) the source is directly coupled to the load, and in 15.10(b) they are transformer coupled. Calling the negative end of the source the ground node, suppose that a person in contact with ground accidentally touches the node marked $+V_L$. The body is a relatively good conductor and forms a current path to ground. In the direct-connected case, current will flow from the source through the person to ground, but in the transformer-coupled case the secondary circuit is electrically isolated from ground; there is no return path for current flow and no danger of shock. As another use of isolation, suppose that the source and load in Fig. 15.10 are separated by some physical distance and that each must be grounded. The two local grounds may not be identical; some potential difference between them may cause unwanted current flow through what is called a *ground loop* unless source and load are electrically isolated as in the transformer-coupled case.

Finally, if the transformer is not being used for isolation, if one terminal of the primary and secondary are, or can be, tied together, the equivalent circuit shown in Fig. 15.11 may be used to simplify analysis. If one of the dots is relocated to the other end of its inductor, M must be replaced by $-M$ in the three places it appears in the equivalent circuit. The two mesh equations for the circuit of Fig. 15.11(b) simply repeat the $i–v$ laws (15.41) for a transformer, and the equivalent circuit has the advantage of containing only simple (uncoupled) inductors. The inductors in the model may be negative, which is not possible physically using passive elements, but which causes no difficulty in the mathematical analysis.

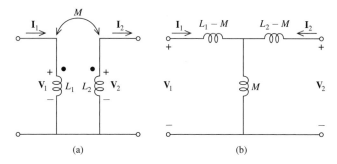

FIGURE 15.11   (a) Three-terminal transformer; (b) equivalent circuit.

EXERCISES

15.3.1.   Find the ratios $\mathbf{V}_2/\mathbf{V}_1$ and $\mathbf{I}_2/\mathbf{I}_1$ for Fig. 15.5 at $\omega = 4$ rad/s.
Answer $0.058\underline{/-126°}$; $0.565\underline{/-172°}$

15.3.2.   Find the real average power delivered to a 100-Ω load resistor across the secondary of Fig. 15.2(a) for $v_1(t) = 50\cos 377t$ V.
Answer 6.18 W

15.3.3.   Find ω for which the reflected impedance in Fig. 15.8 is purely real if $L_2 = 2$ H and Z_L is a 6-Ω resistor in series with a $\frac{1}{32}$-F capacitor.
Answer 4 rad/s

15.4   IDEAL TRANSFORMERS

Any pair of coupled coils may be used as a transformer and will scale, or transform, the voltage, current, and impedance levels of the circuit. The secondary-to-primary current and voltage ratios (15.45) and (15.47) show that, in general, this scaling varies with frequency ω and load impedance Z_L. It is often desirable that the secondary-to-primary current and voltage ratios of the transformer be fixed, independent of load and frequency. For instance, if we are using a transformer to step up the generated voltage in a power transmission circuit, we would want the transmitted line voltage to remain steady even as customers came on or off the line and the effective load Z_L varied.

To see how this can be accomplished, let us consider an N_1-turn coil with self-inductance L_1 producing flux ϕ_{11} in response to its own current i_1. Suppose that the number of turns is increased from N_1 to αN_1. With the wire assumed to be vanishingly thin, the αN_1 turns occupy the same space as the N_1 turns did, and current i_1 in each of αN_1 turns is physically indistinguishable from current αi_1 in each of N_1 turns. By linearity of the coil in i_1 (15.8), the flux produced by αi_1 is $\alpha \phi_{11}$. Increasing N_1 by a factor of α increases ϕ_{11} by the same factor; in other words, the flux produced by the current i_1 is linearly related to the number of turns N_1. Calling the constant of proportionality β_1, then, by (15.8),

$$\frac{\phi_{11}}{i_1} = \beta N_1 = \frac{L_1}{N_1}$$

or
$$L_1 = \beta_1 N_1^2 \qquad\qquad (15.48)$$

That is, the self-inductance of a given inductor is proportional to its turns squared. The constant β_1 is determined by the detailed physical properties of the magnetic path. If a pair of coupled coils 1 and 2 share the same magnetic path, in other words, if they are unity coupled, then $\beta_1 = \beta_2$ and by (15.48) *the self-inductances of unity-coupled coils are in the ratio of their turns squared,*

$$\frac{L_2}{L_1} = \frac{N_2^2}{N_1^2} = n^2$$

where $n = N_2/N_1$ is called the *turns ratio* of the transformer.

In the case of unity coupling, $M = \sqrt{L_1 L_2}$ and, by (15.47), the secondary-to-primary voltage ratio is

$$\frac{\mathbf{V}_2}{\mathbf{V}_1} = \frac{j\omega Z_2 \sqrt{L_1 L_2}}{j\omega L_1 (Z_2 + j\omega L_2) + \omega^2 L_1 L_2} = \sqrt{\frac{L_2}{L_1}} = n \qquad (15.49)$$

Thus, for any unity-coupled transformer, the secondary-to-primary voltage ratio has the property we seek: It is a constant independent of load or frequency. The value of the voltage ratio is simply the turns ratio n.

The secondary-to-primary current ratio (15.45) in the unity-coupled case is

$$\frac{\mathbf{I}_2}{\mathbf{I}_1} = \frac{j\omega \sqrt{L_1 L_2}}{Z_L + j\omega L_2} \qquad (15.50)$$

Even with unity coupling, the current ratio still generally depends on ω and $\mathbf{Z}_L$. If, however, we take L_2 to be unboundedly large, $\mathbf{Z}_L$ may be neglected in the denominator of (15.50), and

$$\frac{\mathbf{I}_2}{\mathbf{I}_1} = \frac{j\omega \sqrt{L_1 L_2}}{j\omega L_2} = \sqrt{\frac{L_1}{L_2}} = \frac{1}{n} \qquad (15.51)$$

Transformers satisfying both (15.49b) and (15.51) have current and voltage ratios that do not depend on load or frequency and are said to be ideal transformers. *An ideal transformer is a unity-coupled transformer whose self-inductances L_1 and L_2 are both unboundedly large but whose ratio L_2/L_1 is finite. Ideal transformers satisfy the i–v laws*

$$\mathbf{V}_2 = \pm n \mathbf{V}_1 \qquad\qquad (15.52a)$$

$$\mathbf{I}_2 = \pm \frac{1}{n} \mathbf{I}_1 \qquad\qquad (15.52b)$$

where the dot convention is applied as follows: Define the primary variables $(\mathbf{I}_1, \mathbf{V}_1)$ to satisfy, and the secondary variables $(\mathbf{I}_2, \mathbf{V}_2)$ to violate the passive sign convention. Then if $\mathbf{I}_1$ or $\mathbf{I}_2$ points into a dotted end while the other points into an undotted end, use plus signs in both i–v laws (15.52); otherwise, use minus signs in both. This is an awkward but completely equivalent restatement of the original dot convention, which is necessitated by the standard transformer practice of reversing the current reference direction in the secondary.

The circuit symbol for an ideal transformer, shown in Fig. 15.12(a), is similar to that for a nonideal transformer except that the turns ratio is specified rather than the inductances L_1, L_2, and M (all three of which are infinitely large), and a pair of parallel

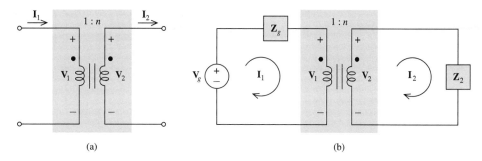

(a)   (b)

FIGURE 15.12 (a) Ideal transformer circuit symbol; (b) ideal transformer with primary source and secondary load.

lines is drawn to suggest the ring of ferromagnetic material comprising the core of this unity-coupled transformer.

Example 15.8

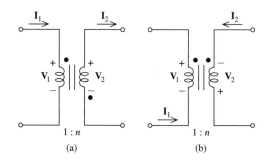

(a)   (b)

FIGURE 15.13 Two circuits for Example 15.8.

Specify the $i-v$ laws for the two ideal transformers shown. In Fig. 15.13(a), $(\mathbf{I}_1, \mathbf{V}_1)$ satisfies and $(\mathbf{I}_2, \mathbf{V}_2)$ violates the passive sign convention, as required in the statement of the dot convention for ideal transformers. Since both currents flow into dotted ends, we must take minus signs. The $i-v$ laws for Fig. 15.13(a) are consequently

$$\mathbf{V}_2 = -n\mathbf{V}_1$$

$$\mathbf{I}_2 = -\frac{1}{n}\mathbf{I}_1$$

In Fig. 15.13(b), the given primary variables $(\mathbf{I}_1, \mathbf{V}_1)$ do not satisfy the passive sign convention, but $(\mathbf{I}_1, -\mathbf{V}_1)$ do, so we will use this pair as our chosen primary variables. $(\mathbf{I}_2, \mathbf{V}_2)$ violates the passive sign convention, as required. Then, using the selected sets of port variables $(\mathbf{I}_1, -\mathbf{V}_1)$ and $(\mathbf{I}_2, \mathbf{V}_2)$, one current points into a dotted and the other into an undotted end, so plus signs will be used in the $i-v$ laws. Writing $-\mathbf{V}_1$ in place of $\mathbf{V}_1$ in (15.52), the $i-v$ laws for Fig. 15.13(b) are

$$\mathbf{V}_2 = +n(-\mathbf{V}_1)$$

$$\mathbf{I}_2 = +\frac{1}{n}\mathbf{I}_1$$

or

$$\mathbf{V}_2 = -n\mathbf{V}_1$$

$$\mathbf{I}_2 = +\frac{1}{n}\mathbf{I}_1$$

Figure 15.12(b) shows an ideal transformer connected to a source and load. The impedance looking into the primary port is

$$\frac{\mathbf{V}_1}{\mathbf{I}_1} = \frac{\mathbf{V}_2/n}{n\mathbf{I}_2} = \frac{Z_L}{n^2} \tag{15.53}$$

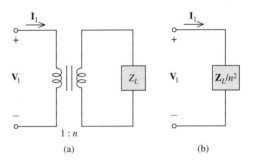

FIGURE 15.14 (a) Ideal transformer and secondary load; (b) equivalent when reflected into the primary circuit.

Thus *an ideal transformer together with its secondary load Z_L is equivalent to an impedance of value Z_L/n^2 reflected into the primary circuit*, as shown in Fig. 15.14. It is often convenient to use this reflected impedance to eliminate the transformer and one loop before analyzing a circuit containing an ideal transformer.

The simple relationships (15.52) and (15.53) reveal most of the main uses of transformers. Equation (15.52a) shows that an ideal transformer can be used to *step up* a voltage level if the secondary winding has more turns than the primary, $n > 1$, to *step down* the voltage level if $n < 1$, which results in an equivalent increase in the current level as seen in (15.52b), and to *impedance scale* a load. Each of these uses is illustrated in the next example.

Example 15.9

Figure 15.15 shows a power delivery circuit. A hydropower-driven ac generator produces the Thevenin equivalent voltage phasor $\mathbf{V}_s$ through a 0.1-Ω Thevenin equivalent impedance. Its output is increased using a 10:1 step-up transformer for more efficient transmission by long lines. The transmission lines have an equivalent impedance of 1 Ω. The power is to be delivered to a 4-Ω load through a second ideal transformer. We wish to determine the turns ratio n of the second ideal transformer to maximize the power supplied by the source. The equivalent of this circuit to the right of the terminals b–b', the primary of the right transformer, is an impedance of value

$$Z_1 = \frac{Z_L}{n^2} = \frac{4}{n^2}$$

The left transformer has an equivalent secondary load impedance of 1 Ω in series with $\mathbf{Z}_2$, or $1 + 4/n^2$ Ω. Reflecting this into its primary circuit, the impedance to the right of a–a' is

$$Z_2 = \frac{1 + (4/n^2)}{10^2}$$

For maximum power transfer, the equivalent load impedance Z_2 seen by the source should equal the complex conjugate of the source impedance, which is identical with $\mathbf{Z}_s = 0.1$ Ω, since it is purely

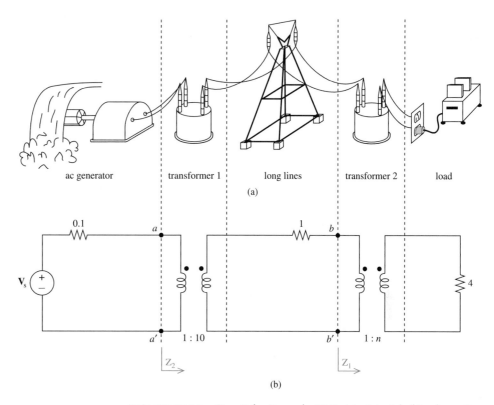

FIGURE 15.15 Circuit for Example 15.9: (a) pictorial; (b) schematic.

real. Thus, for maximum power supplied to the rest of the circuit by the source,

$$Z_2 = \frac{1 + (4/n^2)}{10^2} = \frac{1}{10}$$

Solving for the required turns ratio,

$$n = \sqrt{\frac{4}{9}} = \frac{2}{3}$$

Thus a 10:1 step-up transformer is used to generate higher voltage for efficient transmission, and then a 2:3 step-down transformer is used to impedance match to the given load.

Just as we can reflect the load into the primary, we may also reflect the source into the secondary circuit. The open-circuit voltage V_2 in Fig. 15.16(a) is

$$\mathbf{V}_T = \pm n\mathbf{V}_1 = \pm n\mathbf{V}_s \qquad (15.54)$$

with $\mathbf{V}_1 = \mathbf{V}_s$ since $\mathbf{I}_2$ and hence also $\mathbf{I}_1$ are zero. We next kill the source and determine the impedance looking into the secondary terminals. A transformer with turns ratio n,

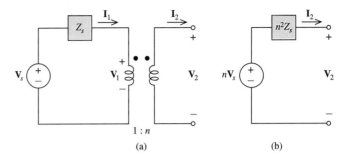

FIGURE 15.16 (a) Source and ideal transformer; (b) equivalent when re-flected into secondary circuit.

when the primary and secondary are reversed, has turns ratio $1/n$. By (15.53), the impedance looking into the secondary is

$$Z_T = \frac{Z_s}{(1/n)^2} = n^2 Z_s \qquad (15.55)$$

leading to the Thevenin equivalent circuit shown in Fig. 15.16(b). *An ideal transformer, together with its primary voltage source $\mathbf{V}_s$ of source impedance Z_s, reflects into the sec-ondary circuit as a source $\pm n\mathbf{V}_s$ of source impedance $n^2 Z_s$. The sign follows from the dot convention.* Both reflecting the load into the primary (15.53) and the source into the secondary (15.54)–(15.55) result in an equivalent circuit with one less loop and no transformer. Note that reflection to the secondary eliminates the primary subcircuit, so it is most useful when the primary variables are not required. Similarly, reflection into the primary, as in (15.53), obliterates the secondary variables, so it is best suited where primary variables are of interest.

Example 15.10

This example continues Example 15.9. For the same circuit, suppose that we change the goal to maximizing power delivered to the 4-Ω load. Reflecting the source and 10:1 step-up transformer into its secondary circuit yields the equivalent shown in Fig. 15.17(a). Then reflecting the primary circuit of the remaining transformer into its secondary yields the circuit of Fig. 15.17(b). For maximum power into the 4-Ω load, we need $11n^2 = 4$, or

$$n = \frac{2}{\sqrt{11}}$$

So the maximum power delivery to the 4-Ω load occurs if we use a step-down transformer with turns ratio 2:$\sqrt{11}$. It is interesting to note that the turns ratio for maximum power supplied by the source computed in Example 15.9, 2:3, is not the same. The 2:3

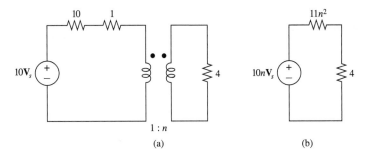

FIGURE 15.17 (a) Circuit of Fig. 15.15 reflected into secondary of left transformer; (b) reflected again into secondary of right transformer.

step-down transformer draws more power out of the source, but less of this power is dissipated by the 4-Ω load and more by the 1-Ω transmission lines than using the turns ratio $n = 2/\sqrt{11}$.

We close our discussion of the ideal transformer by observing that it is a lossless element. The complex power delivered to the primary is

$$S_1 = \tfrac{1}{2}V_1 I_1^* \tag{15.56}$$

Recalling from Chapter 10 that $S = \tfrac{1}{2}VI^*$ is interpreted as power delivered to the element if the terminal variables satisfy the passive sign convention, the complex power delivered to the secondary is

$$S_2 = -\frac{1}{2}V_2 I_2^* = -\frac{1}{2}nV_1 \left(\frac{I_1}{n}\right)^* = -S_1 \tag{15.57}$$

Thus the sum of the power delivered to the primary and secondary equals zero, and the ideal transformer is lossless. The source supplies S_1 to the primary, and the secondary dissipates $-S_1$; that is, it supplies S_1 to the rest of the circuit. Power flows from primary to secondary without losses. Of course, perfectly lossless flows of power are not possible, and the ideal transformer is just that, an idealization. But transformers with efficiencies beyond 95% (less than 5% power loss) are common, and the ideal transformer is a very useful idealization that is quite well approximated in practice.

EXERCISES

15.4.1. Find the phasors V_1 and V_2 in Fig. 15.6 if a current source $I = I_s$ is connected across the input and and the transformer is replaced by an ideal one with turns ratio $n = 10$.
Answer $I_s/382$; $10\,I_s/382$

15.4.2. Find V_1, V_2, I_1, and I_2.
Answer $10/{-36.9°}$ V; $50/143.1°$ V; $2/0$ A; $0.4/180°$ A

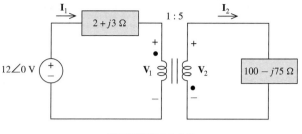

EXERCISE 15.4.2

15.4.3. Find the rms values of the primary and secondary currents in Fig. 15.12(b) if the phasor $\mathbf{V}_g = 16$ V, $Z_g = 2 + j$ Ω, $Z_2 = 8 - 4j$ Ω, and $n = 2$.

Answer $2\sqrt{2}$ A; $\sqrt{2}$ A

15.5 SPICE AND COUPLED COILS

Mutual inductance is signaled in SPICE by the inclusion of an element statement of the form

| K | LXXXXX | LYYYYY | KVAL |

LXXXXX and LYYYYY are the two inductors, and KVAL is the value of the coupling coefficient. $k = M/\sqrt{L_1 L_2}$, where L_1 and L_2 are the self-inductances and M is the mutual inductance as in (15.20). The value of k is required to be in the range $0 < k < 1$, so ideal transformers must be approximated using k set just below 1.000. For purposes of the dot convention, the dots are taken to be located at the first nodes specified on the LXXXXX and LYYYY element statements.

SPICE Example 15.11

We will use SPICE to determine the impedance at the frequencies 1 Hz, 10 Hz, 100 Hz, and 1 kHz of the circuit analyzed in Example 15.6 and redrawn as Fig. 15.18.

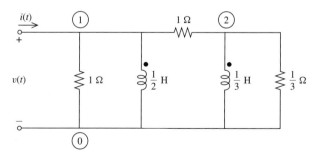

FIGURE 15.18 Circuit for Example 15.11.

The coupling coefficient for the coupled coils is

$$k = \frac{M}{\sqrt{L_1 L_2}} = \frac{\frac{1}{3}}{\sqrt{\frac{1}{6}}} = \sqrt{\frac{2}{3}}$$

which is entered in the k statement of the SPICE input file.

```
Example 15.11
*
IG      0       1       AC      1
R1      1       0       1
L1      1       0       0.5
L2      2       0       0.33333
K       L1      L2      0.8165
R2      1       2       1
R3      2       0       0.33333
.AC     DEC     1       1              1K
.PRINT AC       VM(1) VP(1)
.END
```

The output file contains the print table

```
FREQ          VM(1)         VP(1)

1.00E+00      5.24E-01      1.64E+01
1.00E+01      5.71E-01      1.85E+00
1.00E+02      5.71E-01      1.86E-01
1.00E+03      5.71E-01      1.86E-02
```

Our use of a unit current source phasor makes the terminal voltage phasor $V(1)$ identical to the impedance of this circuit. Thus $VM(1)$ and $VP(1)$ in the table are the magnitude and angle of the impedance at each frequency. From our previous result (15.38b),

$$Z(j\omega) = \frac{-4\omega^2 + j9\omega}{(18 - 7\omega^2) + j30\omega}$$

Evaluating at $\omega = 2\pi$, we obtain

$$Z(j2\pi) = \frac{-158 + j56.5}{-258 + j188} = 0.524\underline{/16.4°}$$

which agrees with the first line of the SPICE output table. The other values may be checked similarly.

SPICE does not permit "floating" nodes; that is, there must always be a dc path between every pair of nodes so that the voltage of each node relative to the reference node (node 0) is well defined. If we wish to use SPICE to analyze a circuit containing a transformer that splits the circuit into two electrically isolated subcircuits, we must take special steps to prevent SPICE from complaining. A shorting wire can be introduced between the two inductors without affecting any element current or voltage. If there is a connection between subcircuits containing the two inductors, but it is a purely capacitive connection, one whose dc impedance is infinite, a very large resistance should be introduced into the connection added in order to prevent significant current flow in this dummy connection.

Determine the ac steady-state power delivered to the 5-Ω load resistor if the transformer is unity coupled (Figure 15.19), and determine the coupling coefficient at which the power is reduced to 10% of this value. The SPICE input file for the unity-coupling evaluation is shown below. Note that the lower ends of the inductors have been tied together to add a dc path between the primary and secondary subcircuits and that the coupling coefficient has been set at 0.99999 to comply with the permissible range $0 < k < 1$.

```
Example 15.12
*
V1          1       0       AC      156
RS1         2       1
LPRI        2       0       1
LSEC        0       3       10
K           LPRI    LSEC            0.99999
C1          3       0       50U
RL          3       0       5
.AC         LIN     1       60      60
.PRINT  AC  VM(3)
.END
```

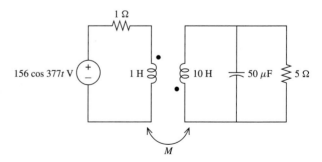

FIGURE 15.19 Circuit for Example 15.12.

The resulting SPICE output indicates that VM(3) = 164.1 V, so the power delivered to the 5-Ω load is $164.1^2/(10) = 2.693$ kW.

The load voltage at which the power is reduced by 90% is VM(3)²/ 10 = (0.10)(2693) or VM(3) = 51.9 V. Rerunning the input file with the value of k decreased by cut-and-try, we find that this condition is met for $k = 0.994$ (to three significant digits). Note that only a small leakage flux is required to sharply reduce the efficiency of a transformer-coupled power delivery circuit.

Finally, an ideal transformer may be modeled for SPICE as a pair of controlled sources satisfying the i–v laws (15.52) for this type of device. The model is shown in Fig. 15.20.

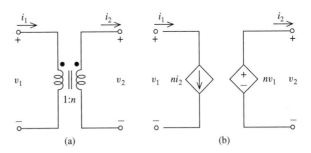

FIGURE 15.20   Ideal transformer model for SPICE: (a) circuit device; (b) dependent source model.

EXERCISES

15.5.1.   Find $v_2(t)$ for $v_1(t) = V_g \cos(1000t + \phi_g)$ using SPICE.
   Answer Use a pair of dependent sources to model the ideal transformer. $v_2(t) = 10.45V_g \cos(1000t + \phi_g - 145.3°)$.

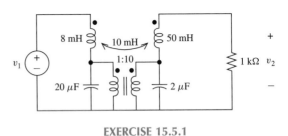

EXERCISE 15.5.1

15.5.2.   Check the transient response calculation of Example 15.4 using SPICE.
   Answer The plotted output from the SPICE input file shown agrees closely with the result $i_2(t)$, $t > 0$, of the example.

```
EXE15-5.2
*
V1          1         0      DC            2
R1          1         2      4
L1          0         2      4        IC=0
L2          0         3      4        IC=-0.5
K1          1         L2     0.5
V2          4         3      DC           -1
L3          4         5      1        IC=-0.5
R2          5         0      2
.TRAN            .1        10    UIC
.PRINT     TRAN      I(V2)
.END
```

SUMMARY

Elements may interact electrically, through the wires that connect them, or magnetically, when the magnetic field spawned by one element threads the other. Mutual inductance causes the terminal current and voltage of one inductor to depend on those of another through their electromagnetic interaction.

- The voltage induced in a coupled coil consists of a term proportional to the derivative of its own current plus terms proportional to the derivatives of the currents flowing through each of its coupled coil partners.

- The dot convention requires positive coupling terms in the i–v laws if both inductive currents flow into, or both out of, the dotted ends.

- The coupling coefficient between two coupled coils is their mutual inductance divided by the geometric mean of their self-inductances.

- A transformer is a device for changing voltage, current, or impedance levels in a circuit. It consists of a pair of coupled coils on a common core.

- An ideal transformer steps up the voltage by the turns ratio and steps down the current by the negative inverse turns ratio.

- An ideal transformer dissipates no electrical power.

PROBLEMS

15.1. Suppose $L_1 = L_2 = 0.1$ H, $M = 0.05$ H, $N_1 = N_2 = 1000$ turns. Find the leakage fluxes ϕ_{L1}, ϕ_{L2} and the total fluxes ϕ_1 and ϕ_2 threading the inductors if $i_1 = 3t + 2$ A, $i_2 = 3$ A.

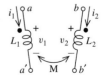

FIGURE P15.1

15.2. A pair of coupled coils have self-inductances $L_1 = 1$ H, $L_2 = 5$ H. If $i_1(t) = 5\cos 2t$ A, $i_2(t) = 2\cos 2t$ A, what is the maximum amplitude A of the sinusoidal flux linkage $\lambda_1(t) = A\cos(2t + \phi)$ threading coil 1? The minimum amplitude A?

15.3. When a' and b' in the figure for Problem 15.1 are shorted, the inductance between a and b is 1 H, and when a' is shorted to b, the inductance between a and b' is 7 H. Find the mutual inductance M.

15.4. $L_1 = L_2 = 2$ H, $M = 1$ H and positive di_1/dt leads to negative v_2. Find i_1 and i_2 for $t > 0$.

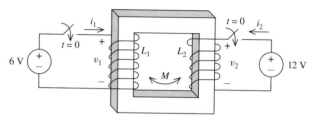

FIGURE P15.4

15.5. If $\omega(0) = 0$, determine the power $p(t)$ and energy $\omega(t)$ for the coupled coils of Problem 15.1 above, with the indicated currents.

15.6. Suppose $L_1 = 1$ H, $L_2 = 5$ H, $M = 2$ H in the figure for Problem 15.1. Find the peak instantaneous power in watts supplied to the inductors if $i_1(t) = 3\cos t$ A, $i_2 = \sin(t + 45°)$ A. Is this power dissipated in the form of heat? Explain.

15.7. When two identical coils are each made to simultaneously carry a dc current of 1 A, they each have flux linkage $i_1 = i_2 = 10$ Wb. When they are shielded from one another so no flux from one threads the other, the flux linkage rises to 12 Wb. Find M in both cases.

15.8. Write the i–v laws in the time and s domains. In both cases $i_1(0-) = 1$ A, $i_2(0-) = 0$.

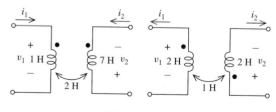

FIGURE P15.8

15.9. Find i for $t > 0$ if $i(0) = 0$ and $v(0) = 4$ V.

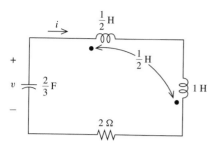

FIGURE P15.9

15.10. Determine $Z(s)$, the impedance looking into this circuit.

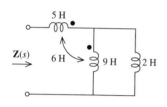

FIGURE P15.10

15.11. Find $i_2(t)$ in ac steady state.

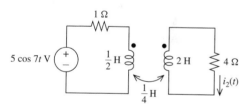

FIGURE P15.11

15.12. Find the ac steady state value of $v_R(t)$.

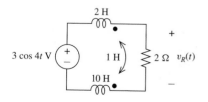

FIGURE P15.12

15.13. Find the steady-state value of v.

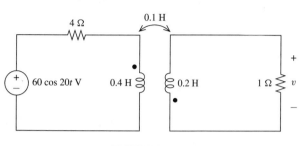

FIGURE P15.13

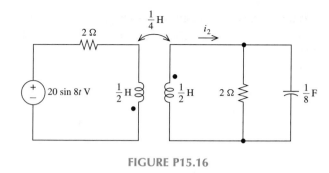

FIGURE P15.16

15.17. (a) Find v for $t > 0$ if $v_g = 4u(t)$ V and (b) find the steady-state value of v if $v_g = 4\cos 8t$ V and the output terminals are loaded with a resistor of 8 Ω.

15.14. Find the steady-state currents i_1 and i_2.

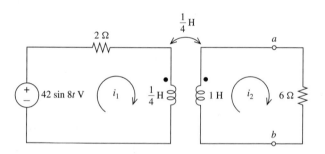

FIGURE P15.14

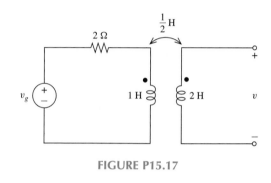

FIGURE P15.17

15.18. Find $\mathbf{I}_1$ and $\mathbf{I}_3$ using mesh analysis.

15.15. Find the power dissipated in the 1-Ω resistor in the primary circuit and the power factor seen by the source.

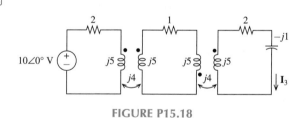

FIGURE P15.18

15.19. Find v for $t > 0$ if $i_g = 2u(t)$ A. Use mesh analysis and assume no initial stored energy at $t = 0$.

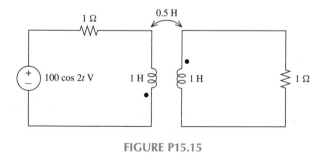

FIGURE P15.15

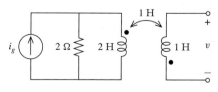

FIGURE P15.19

15.16. Find the steady-state current i_2.

15.20. Repeat 15.19 using nodal analysis.

15.21. Find the ac steady state current $i_2(t)$ using mesh analysis.

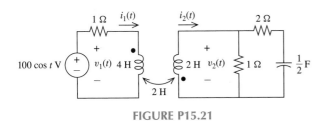

FIGURE P15.21

15.22. Find the voltage **V** across the source.

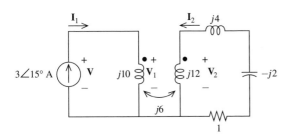

FIGURE P15.22

15.23. Determine the reflected impedance Z_R, current and voltage ratios $\mathbf{I}_2/\mathbf{I}_1$, $\mathbf{V}_2/\mathbf{V}_1$ for the circuit of Problem 15.21.

15.24. Determine the reflected impedance Z_R, current and voltage ratios $\mathbf{I}_2/\mathbf{I}_1$, $\mathbf{V}_2/\mathbf{V}_1$ for the circuit of Problem 15.22.

15.25. Solve 15.21 using the uncoupled coil model of Fig. 15.11 and nodal analysis.

15.26. Find $\mathbf{I}_L$.

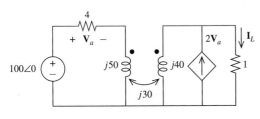

FIGURE P15.26

15.27. Find the transfer function

$$\mathbf{H}(s) = \mathbf{V}_2(s)/\mathbf{V}_1(s)|_{\mathbf{I}_2(s)=0}$$

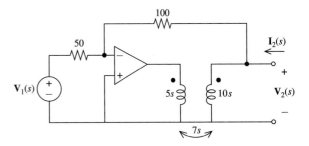

FIGURE P15.27

15.28. Find the steady-state voltage v using the equivalent circuit of Fig. 15.11(b) for the transformer.

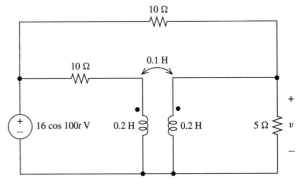

FIGURE P15.28

15.29. Find $i(t)$ for $t > 0$ if $i(0+) = 0$.

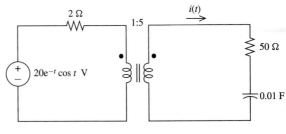

FIGURE P15.29

15.30. Find the average power delivered to the 8-Ω resistor.

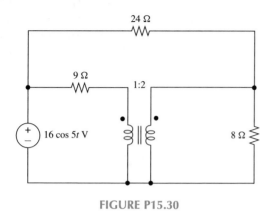

FIGURE P15.30

15.31. Write the $i–v$ laws in the time and s domains. Do not assume initial currents are zero.

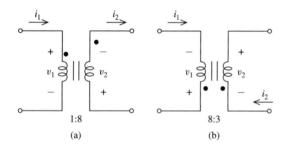

FIGURE P15.31

15.32. Find the ac steady state value of $v_2(t)$ for $v_1(t) = 50 \sin 3t$ V.

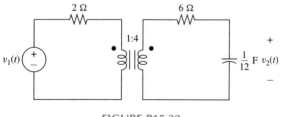

FIGURE P15.32

15.33. Find $v_2(t)$ in Problem 15.32 if all initial conditions at time $t = 0$ are zero and $v_1(t) = u(t)$.

15.34. If $\mathbf{I} = 2\underline{/0}$ A, what is the turns ratio n?

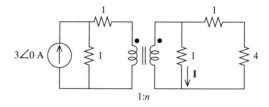

FIGURE P15.34

15.35. Find $\mathbf{I}_1$ in Problem 15.18 by reflecting impedances into primaries twice.

15.36. Find the real average power delivered by the source.

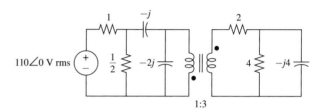

FIGURE P15.36

SPICE Problems

15.37. Use SPICE to determine the ac steady state values for the four port variables.

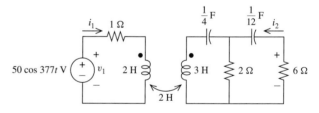

FIGURE P15.37

15.38. Repeat Problem 15.37 if the coupled coils are replaced by an ideal transformer with turn ratio $n = 20$.

15.39. Find the value for M at which the ac steady state amplitude of v_2 is reduced to $\frac{1}{10}$ of its value when the coils are unity coupled.

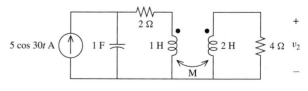

FIGURE P15.39

15.40. Plot i for $0 < t < 5$ s in the circuit Problem 15.9.

15.41. In the circuit of Problem 15.28, replace the 10-Ω resistor bridging the transformer with a series connection of a 10-Ω resistor and a 1000-mF capacitor. Use SPICE to plot the frequency response for the voltage v of the resulting circuit in the interval $1 < f < 50$ Hz.

More Challenging Problems

15.42. When b is shorted to b', the other inductor satisfies $v_1 = 4di_1/dt$. When a is shorted to a', the other inductor satisfies $v_2 = 2di_2/dt$. If $L_1 = 5$ H, find M and indicate whether the dot should be at b or b'.

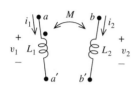

FIGURE P15.42

15.43. Show that a real transformer, with $0 \le k \le 1$, satisfies the passivity condition $\omega(t) \ge 0$ for all t.

15.44. Find $i_1(t)$ in ac steady state. The 4 H and 6 H coils are coupled with $M_a = 2$ H (filled dots). The 6 H and 5 H coils are also coupled with $M_b = 4$ H (open dots). The 4 H and 5 H coils are not coupled.

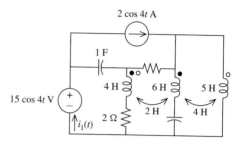

FIGURE P15.44

15.45. Find $\mathbf{I}_3$ in Problem 15.18 by reflecting sources into secondaries twice.

16

Two-Port Circuits

Alexander Graham Bell
1847–1922

Mr. Watson, come here. I want you.

Alexander Graham Bell

Undoubtedly the most common and most widely used electrical instrument is the telephone, invented by the Scottish-American scientist Alexander Graham Bell. The date was June 2, 1875, when Bell and his assistant Thomas Watson transmitted a musical note. The first intelligible telephone sentences, "Mr. Watson, come here. I want you," were spoken inadvertently by Bell himself on March 10, 1876, when he called Watson to come to an adjoining room to help with some spilled acid.

Bell was born in Edinburgh, Scotland. His father, Alexander Melville Bell, was a well-known speech teacher and his grandfather, Alexander Bell, was also a speech teacher. Young Bell, after attending the University of Edinburgh and the University of London, also became a speech teacher. In 1866 Bell became interested in trying to transmit speech electrically after reading a book describing how vowel sounds could be made with tuning forks. Shortly afterwards, Bell's two brothers died of tuberculosis and Melville Bell moved his family to Canada for health reasons. In 1873 young Graham became a professor at Boston University and began his electrical experiments in his spare time. It was there that he formed his partnership with Watson and went on to his great invention. Bell's telephone patent was the most valuable one ever issued, and the telephone opened a new age in the development of civilization.

Chapter Contents

Two-port circuits are those in which two pairs of terminals, or ports, are identified as attachment points for sources, loads, or other circuitry. Two-port analysis is most efficient not when all the currents and voltages in a circuit are of equal interest, but rather, it is the relationship between a designated input circuit variable (current or voltage) and output variable that matters. For instance, while there are many currents and voltages present within the chassis of a graphics equalizer we may purchase for audio applications, it is specifically the relationship between the voltage routed to the input port and that which emerges at the output port which determines the usefulness of this device. Presumably, the output has superior frequency response characteristics, and the other voltages and currents are of interest only to the extent that they permit the desired input–output behavior. Two-port analysis allows us to focus on this relationship, collapsing the many equations of nodal or mesh analysis to a single pair of equations governing the port variables.

We begin by defining the relationships and properties of two-port circuits. In Section 16.2 the variety of two-port equations is catalogued and the method for writing these equations from circuit diagrams is developed. Equivalent models for two-ports are described next, and the effect of interconnecting two-ports together in series, parallel, and cascade is considered in Section 16.4. The chapter concludes with a look at SPICE support for the derivation and use of two-port circuits.

16.1 TWO-PORT CIRCUITS

A *one-port circuit* [Fig. 16.1(a)] contains exactly two terminals at which connections to external elements are permitted. Since no other external connections are permitted, if current $i_1(t)$ flows into one terminal, by Kirchhoff's current law the same current must flow out the other. This pair of terminals forms a *port*, or terminal pair, to which other

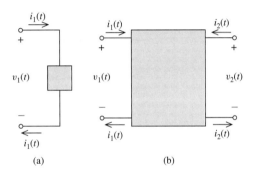

FIGURE 16.1   (a) One-port circuit; (b) two-port circuit.

circuits may be anchored. The requirement that the current into the circuit at one terminal of the port equal that flowing out the other terminal of the port is called the *port condition*.

A two-port circuit has two ports available for external connection, each of which is assumed to satisfy the port condition. Reference directions for the port variables will usually be assigned as shown in Fig. 16.1(b). Most often the currents flowing out of the lower terminals will not be explicitly labeled but are understood to be equal and opposite to the currents of the corresponding upper terminals. The left port will be referred to somewhat arbitrarily as the *input port* and the right as the *output port*. Our goal in this discussion is to characterize two-port circuits, that is, to learn how to write down the relationships linking their port variables, and explore their uses.

Why might we wish to represent a circuit as shown in Fig. 16.1(b), a featureless "black box" out of which four wires protrude? One answer touches what might be called the engineering method of problem solving. *When confronted with a challenging problem, break it up into a set of manageable subproblems, solve each separately, and then link the subproblem solutions together.* Remove the case from a radio receiver sometime and take a close look inside. There is considerable complexity in there: circuit boards, wires, and switches forming intricate patterns. No designer could have, or did, visualize that circuit in its detailed totality as a unified solution to the design problem. Most likely, the designer started by considering the first subtask needing to be done and then proceeded toward a solution by breaking the problem into several simple, manageable steps. To get started, an antenna is needed to capture the radio signal. We route the output from the antenna into an amplifier to strengthen it before proceeding. Then we must narrowband filter the signal to the station we are interested in receiving. Then . . . and so on. Each task defines a simple transformation, some desired relationship between the module's input signal and its output. The overall design (Fig. 16.2) is executed as an interconnection of two-port circuits, with the output from one two-port becoming the input to the next. To apply this general and very useful strategy in the context of linear circuits, we need to understand its basic building block, the two-port circuit, how it is described, and how these descriptions may be combined. This is the main work of the remainder of this chapter.

A second reason for our interest in two-port circuits is that they focus attention where it often matters the most. In many circumstances we are interested in the input–output behavior of a circuit, not in its detailed interior currents and voltages. Requiring that only the two sets of port variables be involved in a two-port description effectively

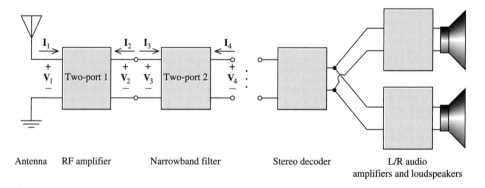

FIGURE 16.2 Modular receiver design by interconnected two-ports.

hides the details of the circuit inside an impenetrable black box where it cannot clutter up the description. Compare, for instance, the mesh equations for a 10-mesh circuit to its two-port description. If we don't care about the inner currents and voltages as long as they "get the job done," we are much better served by a set of equations with four unknowns, the port variables, than 10, the mesh currents. As an example, the power distribution circuit shown end to end in Fig. 15.15(a) contains many internal current and voltage variables. But only a few may be truly significant, the inputs and outputs of the five two-port currents shown in that figure. A two-port description is an efficient, *task-oriented* way to describe a circuit when the circuit's function is conveyed by a few input–output variables.

We limit attention to two-ports containing linear passive *(RLC)* elements and controlled sources. Independent sources are excluded since they act as additional inputs and should properly be assigned an additional port, resulting in an *n-port circuit*, which will not be discussed further. We shall also assume that there is no initial stored energy inside the two-port; that is, all initial conditions are zero.

A two-port with the external circuits to which it is connected is shown in Fig. 16.3. Since all two-port circuits will be studied in the *s*-domain in this chapter, we drop the *s*-argument for convenience, writing $\mathbf{V}_1(s)$ as $\mathbf{V}_1$, and so on. Each port is assumed to

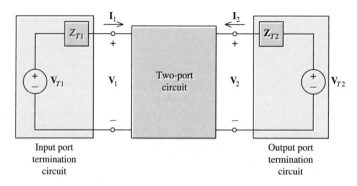

FIGURE 16.3 Two-port in *s*-domain with port terminations shown. (Dependence of all quantities on *s* has been suppressed.)

be closed by, or terminated in, a one-port circuit called a *port termination circuit*. While frequently the input port is terminated in a source and the output in a load impedance, the termination circuits for both input and output ports may in general contain independent sources. For a given pair of port termination circuits, the circuit is now completely specified, and the four port variables $\mathbf{I}_1, \mathbf{V}_1, \mathbf{I}_2$, and $\mathbf{V}_2$ will take on definite values. By what equations are these values determined? The i–v laws for the port termination circuits, one at the input and one at the output port, yield two equations among these four variables. For the Thevenin equivalents shown in Fig. 16.3, these are the equations

$$\mathbf{V}_1 + \mathbf{Z}_{T1}\mathbf{I}_1 = \mathbf{V}_{T1}, \qquad \mathbf{V}_2 + \mathbf{Z}_{T2}\mathbf{I}_2 = \mathbf{V}_{T2} \qquad (16.1)$$

Two more equations are needed, and they can only come from the two-port itself. Moreover, there cannot be more than two independent equations in the two-port description, since then the port variables will be overdefined. *Any two-port description consists of two equations in the four port variables.*

Let us see what form this pair of equations may take. To be specific, we select the port voltages $\mathbf{V}_1$ and $\mathbf{V}_2$ as the dependent variables and the currents $\mathbf{I}_1$ and $\mathbf{I}_2$ as the independent variables in the two-port equations. Since the equations describing the two-port itself must be fixed regardless of how it is terminated, we may get at these equations by applying any port termination circuits that we wish. Two "experiments" will be sufficient to reveal the form of the two-port equations. Setting the output port current $\mathbf{I}_2(s) = 0$ and terminating the input port with a current source $\mathbf{I}_1$, by linearity the voltages produced must each be of the form

$$\mathbf{V}_1|_{\mathbf{I}_2=0} = \mathbf{z}_{11}\mathbf{I}_1, \qquad \mathbf{V}_2|_{\mathbf{I}_2=0} = \mathbf{z}_{21}\mathbf{I}_1 \qquad (16.2)$$

Before proceeding, we observe that *setting a port current to zero is equivalent to open-circuiting that port,* so we have open-circuited the output port in this first experiment. For the second of our experiments, we set the input port current $\mathbf{I}_1 = 0$ (open-circuiting the input port) and terminate the output port with a current source $\mathbf{I}_1$. The response must once again be linear in the source:

$$\mathbf{V}_1|_{\mathbf{I}_1=0} = \mathbf{z}_{12}\mathbf{I}_2, \qquad \mathbf{V}_2|_{\mathbf{I}_2=0} = \mathbf{z}_{22}\mathbf{I}_2 \qquad (16.3)$$

Then, by superposition, each total voltage response $\mathbf{V}_1$ and $\mathbf{V}_2$ is the sum of responses to the two current sources when the other is killed, or

$$\mathbf{V}_1 = \mathbf{z}_{11}\mathbf{I}_1 + \mathbf{z}_{12}\mathbf{I}_2 \qquad (16.4a)$$

$$\mathbf{V}_2 = \mathbf{z}_{21}\mathbf{I}_1 + \mathbf{z}_{22}\mathbf{I}_2 \qquad (16.4b)$$

These two equations are the desired two-port description. They are not the only two-port description possible, as we shall see shortly, but they form a complete characterization of the two-port in the following sense: *given the two-port description* (16.4), *then for any specific pair of port termination circuits we can determine its input–output behavior, that is, all the port variables,* as will be illustrated in Example 16.1. The only requirement is that the two-port circuit satisfy the port condition at its input and output ports.

The two-port description (16.4) is called the *impedance*, or *z-parameter*, two-port description. The name follows from the observation that each parameter $\mathbf{z}_{11}$, $\mathbf{z}_{12}$, $\mathbf{z}_{21}$,

z_{22} in these equations has units of ohms. The impedance two-port description may be written compactly as

$$\begin{bmatrix} \mathbf{V}_1 \\ \mathbf{V}_2 \end{bmatrix} = \begin{bmatrix} \mathbf{z}_{11} & \mathbf{z}_{12} \\ \mathbf{z}_{21} & \mathbf{z}_{22} \end{bmatrix} \begin{bmatrix} \mathbf{I}_1 \\ \mathbf{I}_2 \end{bmatrix} \qquad (16.5)$$

where the square matrix is called the *impedance matrix* or the *z-parameter matrix*.

Example 16.1

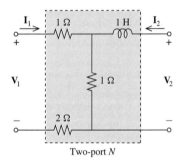

FIGURE 16.4 Two-port circuit for Example 16.1.

We wish to find the impedance two-port description of the circuit shown in Fig. 16.4 and use this description to determine the port currents and voltages under the two different port termination conditions shown in Fig. 16.6.

To find (16.5), we perform the experiment shown in Fig. 16.5(a), open-circuiting the output port to establish the port condition $\mathbf{I}_2 = 0$, applying an $\mathbf{I}_1$ current source to the input port, and determining the resulting port voltages. By (16.4)

$$\mathbf{z}_{11} = \left. \frac{\mathbf{V}_1}{\mathbf{I}_1} \right|_{\mathbf{I}_2=0}, \qquad \mathbf{z}_{21} = \left. \frac{\mathbf{V}_2}{\mathbf{I}_1} \right|_{\mathbf{I}_2=0} \qquad (16.6)$$

Examining Fig. 16.5(a), by series–parallel combinations,

$$\mathbf{V}_1 = 4\mathbf{I}_1$$
$$\mathbf{V}_2 = \mathbf{I}_1$$

Since V_1 and V_2 have been computed under the open-circuit condition $\mathbf{I}_2 = 0$, then by (16.6),

$$\mathbf{z}_{11} = 4, \qquad \mathbf{z}_{21} = 1$$

The remaining two z-parameters are specified by (16.4) to be

$$\mathbf{z}_{12} = \left. \frac{\mathbf{V}_1}{\mathbf{I}_2} \right|_{\mathbf{I}_1=0}, \qquad \mathbf{z}_{22} = \left. \frac{\mathbf{V}_2}{\mathbf{I}_2} \right|_{\mathbf{I}_1=0}$$

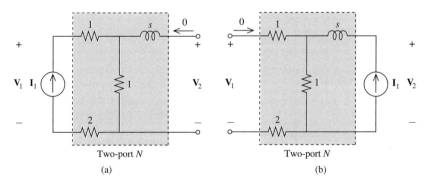

FIGURE 16.5 Two experiments for determining z-parameters.

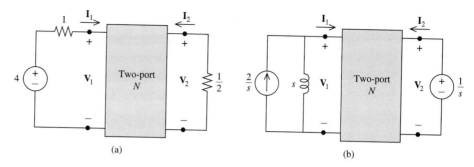

FIGURE 16.6 (a) One pair of port terminations; (b) another pair.

To find them we perform the experiment shown in Fig. 16.5(b), re-locating the open circuit to the input port ($I_1 = 0$) and the current source I_2 to the output port. There is no current through the leftmost two resistors, and the port voltages under these conditions are

$$V_1 = I_2$$
$$V_2 = (s + 1)I_2$$

Thus

$$z_{12} = 1, \qquad z_{22} = s + 1$$

and the z-parameter two-port description is

$$\begin{bmatrix} V_1 \\ V_2 \end{bmatrix} = \begin{bmatrix} 4 & 1 \\ 1 & s+1 \end{bmatrix} \begin{bmatrix} I_1 \\ I_2 \end{bmatrix} \tag{16.7}$$

To find the specific values of the port variables for the given port termination circuits in Fig. 16.6, combine the two-port description (16.7) with the two separate i–v laws for the port termination circuits. In the case of Fig. 16.6(a), these are

$$V_1 = 4 - I_1$$
$$V_2 = -\tfrac{1}{2}I_2 \tag{16.8}$$

Together, (16.7) and (16.8) comprise four equations in the four port variables. Eliminating the currents by solving (16.8) for I_1, I_2 and substituting into (16.7) gives

$$\begin{bmatrix} V_1 \\ V_2 \end{bmatrix} = \begin{bmatrix} 4 & 1 \\ 1 & s+1 \end{bmatrix} \begin{bmatrix} 4 - V_1 \\ -2V_2 \end{bmatrix}$$

Combining like terms yields

$$5V_1 + 2V_2 = 16 \tag{16.9a}$$
$$V_1 + (2s + 3)V_2 = 4 \tag{16.9b}$$

Multiplying the last by 5 and subtracting from the equation above it, we have

$$-(10s + 15)V_2 + 2V_2 = -4$$

or
$$V_2 = \frac{4}{10s + 13}$$

Back-substituting into (16.9a) gives us

$$V_1 = \frac{32s + 40}{10s + 13}$$

and using the last two in (16.8), we obtain

$$I_1 = \frac{8s + 12}{10s + 13} \tag{16.10a}$$

$$I_2 = \frac{-8}{10s + 13} \tag{16.10b}$$

which completes the evaluation of the port variables for the termination shown in Fig. 16.6(a). For different port termination circuits [Fig. 16.6(b)], we have the same two-port, so we will still use the same two-port description (16.7). The i–v laws for the new port termination circuits are

$$I_1 + \frac{V_1}{s} = \frac{2}{s} \tag{16.11a}$$

$$V_2 = \frac{1}{s} \tag{16.11b}$$

Using (16.11) to eliminate V_1 and V_2 from (16.7) and solving as above yields the port variables and completes the example.

$$V_1 = \frac{12s - 5}{s^2 + 5s + 3}, \qquad V_2 = \frac{1}{s}$$

$$I_1 = \frac{2s^2 + 2s - 1}{s(s^2 + 5s + 3)}, \qquad I_2 = \frac{-s + 4}{s(s^2 + 5s + 3)}$$

The example illustrates that the two-port description (16.5) may be used to determine the port variables for any pair of one-port termination circuits. We say that it is a *complete characterization of the input–output or port behavior of the two-port circuit*. Note that this characterization is not only complete but very concise. Regardless of the complexity of the two-port, its count of meshes, nodes, and elements, *the two-port description summarizes the port behavior in just two equations containing four parameters*. These are the z-parameters in the impedance two-port description (16.5), but may be defined in various other ways, as we shall see in Section 16.2.

If the port condition is violated, the assumptions under which the two-port description was derived are not valid and equations such as (16.4) are not valid. We must insist that the current that flows into the upper terminal of each port also flow out the lower in order that the circuit be operating as a two-port circuit. As long as each port termination circuit is a one-port, we are safe. An example of a four-terminal circuit that does not behave as a two-port is shown in Fig. 16.7(b).

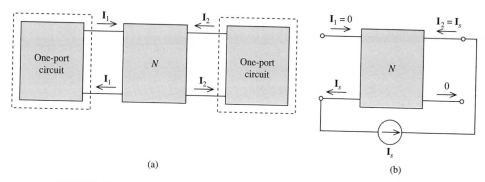

(a)   (b)

FIGURE 16.7   (a) Two-port N satisfying port conditions; (b) port conditions
violated. N is *not* being operated as a two-port circuit.

EXERCISES

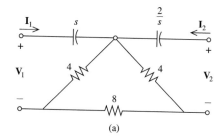

(a)

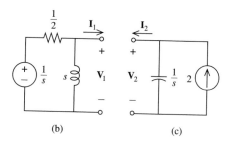

(b)   (c)

EXERCISE 16.1.1   (a) Two-port; (b) input port
termination circuit; (c) output port termination
circuit.

16.1.1.   Find the z-parameters for the two-port shown in part (a) of the
figure.

Answer $z_{11} = \dfrac{3s+1}{s}$; $z_{12} = z_{21} = 1$; $z_{22} = \dfrac{3s+2}{s}$

16.1.2.   Find the i–v laws for the one-ports shown in parts (b) and (c) of
the figure.

Answer $\mathbf{V}_1 = \dfrac{-s}{2s+1}\mathbf{I}_1 + \dfrac{2}{2s+1}$; $\mathbf{V}_2 = \dfrac{-1}{s}\mathbf{I}_2 + \dfrac{2}{s}$

16.1.3.   Find $\mathbf{I}_1$ and $\mathbf{I}_2$ if the two-port of part (a) of the figure is terminated
by the one-ports shown in parts (b) and (c).

Answer

$$\mathbf{I}_1 = \frac{2s(s+2)}{19s^3 + 35s^2 + 18s + 3},$$

$$\mathbf{I}_2 = \frac{2(6s^2 + 5s + 1)}{(s+1)(19s^3 + 35s^2 + 18s + 3)}$$

16.2   TWO-PORT PARAMETERS

Selection of the voltages as dependent variables in the two-port equations led to the
impedance two-port description

$$\begin{bmatrix} \mathbf{V}_1 \\ \mathbf{V}_2 \end{bmatrix} = \begin{bmatrix} \mathbf{z}_{11} & \mathbf{z}_{12} \\ \mathbf{z}_{21} & \mathbf{z}_{22} \end{bmatrix} \begin{bmatrix} \mathbf{I}_1 \\ \mathbf{I}_2 \end{bmatrix}$$

in which the two-port is described in terms of four z-parameters. Let us now instead choose the currents to be the dependent variables. Following the argument of Section 16.1, there will still be two linear equations, and the new two-port description will be of the form

$$\begin{bmatrix} \mathbf{I}_1 \\ \mathbf{I}_2 \end{bmatrix} = \begin{bmatrix} \mathbf{y}_{11} & \mathbf{y}_{12} \\ \mathbf{y}_{21} & \mathbf{y}_{22} \end{bmatrix} \begin{bmatrix} \mathbf{V}_1 \\ \mathbf{V}_2 \end{bmatrix} \qquad\qquad (16.12)$$

Since each parameter y_{ij} in this equation has units of siemens, we will call this the *admittance*, or *y-parameter*, two-port description.

The manner in which the y-parameters may be determined for a specific circuit is very similar to the manner in which the z-parameters were computed in Section 16.1. Two experiments are performed in which first one, then the other, independent port variable is set to zero. In the present case the independent variables are the port voltages, and *setting a port voltage to zero is equivalent to short-circuiting that port.* The other independent port variable is considered to be a source (voltage source in the present case). In each experiment both dependent variables, port currents in this case, are determined. The y-parameters in the first column of the admittance matrix in (16.12) then follow from the results of the first experiment,

$$\mathbf{y}_{11} = \left.\frac{\mathbf{I}_1}{\mathbf{V}_1}\right|_{\mathbf{V}_2=0}, \qquad \mathbf{y}_{21} = \left.\frac{\mathbf{I}_2}{\mathbf{V}_1}\right|_{\mathbf{V}_2=0} \qquad\qquad (16.13)$$

and the second column from the results of the second experiment,

$$\mathbf{y}_{12} = \left.\frac{\mathbf{I}_1}{\mathbf{V}_2}\right|_{\mathbf{V}_1=0}, \qquad \mathbf{y}_{22} = \left.\frac{\mathbf{I}_2}{\mathbf{V}_2}\right|_{\mathbf{V}_1=0} \qquad\qquad (16.14)$$

The calculation of these parameters may be simplified even a bit further. In each experiment there is a single independent source, the voltage source $\mathbf{V}_1$ in the first experiment (16.13) and $\mathbf{V}_2$ in the second (16.14). Suppose that its s-domain value is set to unity. Then by (16.13),

$$\mathbf{y}_{11} = \left.\mathbf{I}_1\right|_{\mathbf{V}_1=1,\mathbf{V}_2=0}, \qquad \mathbf{y}_{21} = \left.\mathbf{I}_2\right|_{\mathbf{V}_1=1,\mathbf{V}_2=0}$$

The first column of the y matrix is the value of the port currents with the output port short-circuited and a unit voltage source $\mathbf{V}_1(s) = 1$ *put across the input port.* Then shorting the input port and setting a unit source in the output port, by (16.14),

$$\mathbf{y}_{12} = \left.\mathbf{I}_1\right|_{\mathbf{V}_1=0,\mathbf{V}_2=1}, \qquad \mathbf{y}_{22} = \left.\mathbf{I}_2\right|_{\mathbf{V}_1=0,\mathbf{V}_2=1}$$

The second column of the y matrix is the value of the port currents with the input port short-circuited and a unit voltage source $V_2(s) = 1$ *put across the output port.*

Example 16.2

To find the y-parameters for the given two-port circuit, first short-circuit the output port and put a unit voltage source in the input port as in Fig. 16.8(a). The controlled source carries no current in this case, and the conductances $\frac{1}{2}$ S and $\frac{1}{4}$ S are in parallel,

$$\mathbf{I}_1 = \left(\tfrac{1}{2} + \tfrac{1}{4}\right)(1) = \tfrac{3}{4} = \mathbf{y}_{11}$$

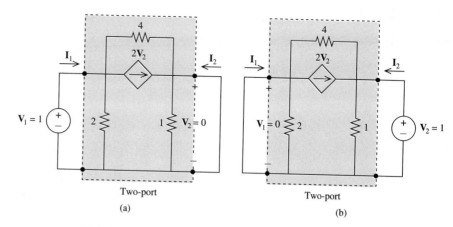

FIGURE 16.8 (a) Two-port with $V_1 = 1$, $V_2 = 0$; (b) with $V_1 = 0$, $V_2 = 1$.

and, by current division,

$$I_2 = \frac{-\frac{1}{4}}{\frac{1}{4} + \frac{1}{2}} I_1 = \left(-\frac{1}{3}\right)\left(\frac{3}{4}\right) = \frac{-1}{4} = y_{21}$$

Reversing these port termination conditions as shown in Fig. 16.8(b), by KCL, we obtain

$$I_1 = 2(1) - \tfrac{1}{4} = \tfrac{7}{4} = y_{12}$$

$$I_2 = \left(1 + \tfrac{1}{4}\right)(1) - 2 = -\tfrac{3}{4} = y_{22}$$

Thus the y-parameter two-port description is

$$\begin{bmatrix} I_1 \\ I_2 \end{bmatrix} = \begin{bmatrix} y_{11} & y_{12} \\ y_{21} & y_{22} \end{bmatrix} \begin{bmatrix} V_1 \\ V_2 \end{bmatrix}$$

$$= \begin{bmatrix} \frac{3}{4} & \frac{7}{4} \\ \frac{-1}{4} & \frac{-3}{4} \end{bmatrix} \begin{bmatrix} V_1 \\ V_2 \end{bmatrix}$$

There are six ways in which two variables can be selected from a set of four, and six corresponding two-port descriptions. Along with the impedance and admittance forms introduced thus far, they are the hybrid, inverse hybrid, transmission, and inverse transmission two-port descriptions. The definitions are shown in Table 16.1.

The similarity of form among the various two-port descriptions guarantees that determining their parameter matrices proceeds similarly. The steps are summarized as follows:

1. *Experiment 1:* Set the second independent variable to zero. Open circuit its port if it is a current and short its port if a voltage. Set the first independent variable to 1, inserting a unity voltage source in its port if it is a voltage; otherwise, insert a unity current source.

2. Determine the two dependent variables. This is the first column of the parameter matrix.

Table 16.1   Two-Port Parameter Descriptions

$$\begin{bmatrix} \mathbf{V}_1 \\ \mathbf{V}_2 \end{bmatrix} = \begin{bmatrix} \mathbf{z}_{11} & \mathbf{z}_{12} \\ \mathbf{z}_{21} & \mathbf{z}_{22} \end{bmatrix} \begin{bmatrix} \mathbf{I}_1 \\ \mathbf{I}_2 \end{bmatrix}$$

1a. Impedance (z-parameter)

$$\begin{bmatrix} \mathbf{I}_1 \\ \mathbf{I}_2 \end{bmatrix} = \begin{bmatrix} \mathbf{y}_{11} & \mathbf{y}_{12} \\ \mathbf{y}_{21} & \mathbf{y}_{22} \end{bmatrix} \begin{bmatrix} \mathbf{V}_1 \\ \mathbf{V}_2 \end{bmatrix}$$

1b. Admittance (y-parameter)

$$\begin{bmatrix} \mathbf{V}_1 \\ \mathbf{I}_2 \end{bmatrix} = \begin{bmatrix} \mathbf{h}_{11} & \mathbf{h}_{12} \\ \mathbf{h}_{21} & \mathbf{h}_{22} \end{bmatrix} \begin{bmatrix} \mathbf{I}_1 \\ \mathbf{V}_2 \end{bmatrix}$$

2a. Hybrid

$$\begin{bmatrix} \mathbf{I}_1 \\ \mathbf{V}_2 \end{bmatrix} = \begin{bmatrix} \mathbf{g}_{11} & \mathbf{g}_{12} \\ \mathbf{g}_{21} & \mathbf{g}_{22} \end{bmatrix} \begin{bmatrix} \mathbf{V}_1 \\ \mathbf{I}_2 \end{bmatrix}$$

2b. Inverse hybrid

$$\begin{bmatrix} \mathbf{V}_1 \\ \mathbf{I}_1 \end{bmatrix} = \begin{bmatrix} \mathbf{t}_{11} & \mathbf{t}_{12} \\ \mathbf{t}_{21} & \mathbf{t}_{22} \end{bmatrix} \begin{bmatrix} \mathbf{V}_2 \\ -\mathbf{I}_2 \end{bmatrix}$$

3a. Transmission (ABCD)

$$\begin{bmatrix} \mathbf{V}_2 \\ -\mathbf{I}_2 \end{bmatrix} = \begin{bmatrix} \mathbf{s}_{11} & \mathbf{s}_{12} \\ \mathbf{s}_{21} & \mathbf{s}_{22} \end{bmatrix} \begin{bmatrix} \mathbf{V}_1 \\ \mathbf{I}_1 \end{bmatrix}$$

3b. Inverse transmission

3. *Experiment 2:* As in experiment 1, but set the first independent variable to zero and the second to unity.

4. Solve for the two dependent variables. This is the second column of the parameter matrix.

Example 16.3

Let us find the hybrid and transmission parameters for the two-port circuit shown in Fig. 16.9. The independent variables in the hybrid description are $\mathbf{I}_1$ and $\mathbf{V}_2$; the dependent variables are $\mathbf{V}_1$ and $\mathbf{I}_2$. The experiments for finding the hybrid description are shown in Fig. 16.9.

The impedance $\mathbf{Z}_{in}$ looking into the circuit of experiment 1 is

$$\mathbf{Z}_{in} = \frac{2s}{s+2} + \frac{1[1+(4/s)]}{1+1+(4/s)} = \frac{\frac{5}{2}s+2}{s+2}$$

$$\mathbf{V}_1 = \mathbf{Z}_{in}\mathbf{I}_1 = \frac{\frac{5}{2}s-2}{s+2}$$

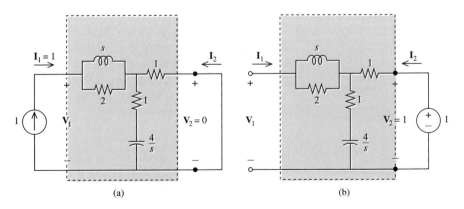

(a)   (b)

FIGURE 16.9   (a) First experiment for determining hybrid parameters; (b) second experiment.

and by current division

$$I_2 = \frac{-1}{1 + [s/(s+4)]}(1) = \frac{-\frac{1}{2}s - 2}{s + 2}$$

V_1 and I_2 form the first column of the hybrid matrix (entry 2a in Table 16.1). To get the second column, we open port 1 and put a unity voltage source across port 2, as shown in Fig. 16.9(b). Since $V = I = 0$ across the parallel RL,

$$I_2 = \frac{V_2}{2 + (4/s)} = \frac{s}{2s + 4}$$

$$V_1 = \left(1 + \frac{4}{s}\right)I_2 = \frac{s + 4}{2(s + 2)}$$

Thus the hybrid matrix is

$$\begin{bmatrix} \mathbf{h}_{11} & \mathbf{h}_{12} \\ \mathbf{h}_{21} & \mathbf{h}_{22} \end{bmatrix} = \begin{bmatrix} \dfrac{5s + 4}{2(s + 2)} & \dfrac{s + 4}{2(s + 2)} \\ \dfrac{-(s + 4)}{2(s + 2)} & \dfrac{s}{2(s + 2)} \end{bmatrix}$$

Turning to the transmission parameters, in Fig. 16.10(a) we have set $-I_2 = 0$ and $V_2 = 1$. By KVL around the right loop, $V_a = 1$ and

$$I_1 = I_a = \frac{V_a}{a + (4/s)} = \frac{s}{s + 4}$$

$$V_1 = \left(\frac{2s}{s + 2} + 1 + \frac{4}{s}\right)I_1 = \frac{3s^2 + 6s + 8}{(s + 2)(s + 4)}$$

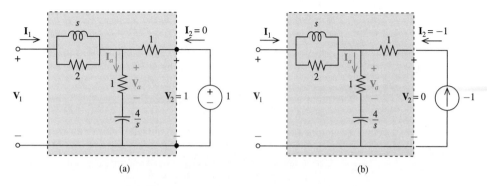

(a) (b)

FIGURE 16.10 (a) First experiment for determining transmission parameters; (b) second experiment.

In the second experiment of Fig. 16.10(b), the conditions set above are reversed; $-\mathbf{I}_2 = 1$ and $\mathbf{V}_2 = 0$. The right loop equation is $V_a = 1$, so

$$\mathbf{I}_a = \frac{\mathbf{V}_a}{1 + (4/s)} = \frac{s}{s+4}$$

$$\mathbf{I}_1 = \mathbf{I}_a + 1 = \frac{2s+4}{2+4}$$

$$\mathbf{V}_1 = \frac{2s}{s+2}\mathbf{I}_1 + \mathbf{V}_a = \frac{5s+4}{s+4}$$

and the transmission, or *ABCD*, parameter matrix is

$$\begin{bmatrix} \mathbf{t}_{11} & \mathbf{t}_{12} \\ \mathbf{t}_{21} & \mathbf{t}_{22} \end{bmatrix} = \begin{bmatrix} \dfrac{3s^2 + 6s + 8}{(s+2)(s+4)} & \dfrac{5s+4}{s+4} \\ \dfrac{s}{s+2} & \dfrac{2s+4}{s+4} \end{bmatrix}$$

which completes the example.

For a given two-port circuit, we may compute six distinct two-port descriptions. What are the relationships among them? They can hardly be independent of one another, since each claims to be a complete representation of the same behavior (input–output behavior) of the same circuit. Consider, for instance, the relationship between the impedance and hybrid parameters. If we know the impedance parameters, we must be able to specify the outcomes of the two experiments that define the columns of the hybrid matrix, since each experiment simply sets the port termination circuits, and any two-port description gives us the port variables for any specified port termination conditions.

The completeness of any one of the two-port descriptions suggests that we must be able to convert among them. *In the case of two descriptions occupying the same row of Table 16.1, the parameter matrices are related by being inverses of each other.* This follows from the observation that the dependent and independent variables are reversed in the two representations across a row. If $\mathbf{M}_1$ and $\mathbf{M}_2$ are two square matrices such that for all vectors $\mathbf{A}$ and $\mathbf{B}$ we have $\mathbf{M}_1\mathbf{A} = \mathbf{B}$ and $\mathbf{M}_2\mathbf{B} = \mathbf{A}$, as we do across a row of Table 16.1, then inverting the second, $\mathbf{B} = \mathbf{M}_2^{-1}\mathbf{A}$, and from the first, $\mathbf{M}_1\mathbf{A} = \mathbf{M}_2^{-1}\mathbf{A}$; that is, the matrices are inverses of one another. The impedance and admittance matrices are a pair of matrix inverses, as are the hybrid and inverse hybrid matrices and the transmission and inverse transmission matrices (which explains the use of *inverse* in their names).

Rules for conversion between all pairs of two-port descriptions are gathered in Table 16.2. Each entry shows the conversion in terms of a matrix divided by a scalar that is common to the four elements of the matrix. Examples of the use of this table follow.

Table 16.2 Two-Port Parameter Conversions

From: To:	z	y	h	g	t	s
z	$\begin{bmatrix} z_{11} & z_{12} \\ z_{21} & z_{22} \end{bmatrix}$	$\dfrac{\begin{bmatrix} y_{22} & -y_{12} \\ -y_{21} & y_{22} \end{bmatrix}}{Y}$	$\dfrac{\begin{bmatrix} H & h_{12} \\ -h_{21} & 1 \end{bmatrix}}{h_{22}}$	$\dfrac{\begin{bmatrix} 1 & -g_{12} \\ g_{21} & G \end{bmatrix}}{g_{11}}$	$\dfrac{\begin{bmatrix} t_{11} & T \\ 1 & t_{22} \end{bmatrix}}{t_{21}}$	$\dfrac{\begin{bmatrix} -s_{22} & -1 \\ -S & -s_{11} \end{bmatrix}}{s_{21}}$
y	$\dfrac{\begin{bmatrix} z_{22} & -z_{12} \\ -z_{21} & z_{11} \end{bmatrix}}{Z}$	$\begin{bmatrix} y_{11} & y_{12} \\ y_{21} & y_{22} \end{bmatrix}$	$\dfrac{\begin{bmatrix} 1 & -h_{12} \\ h_{21} & H \end{bmatrix}}{h_{11}}$	$\dfrac{\begin{bmatrix} G & g_{12} \\ -g_{21} & 1 \end{bmatrix}}{g_{22}}$	$\dfrac{\begin{bmatrix} t_{22} & -T \\ -1 & t_{11} \end{bmatrix}}{t_{12}}$	$\dfrac{\begin{bmatrix} -s_{11} & 1 \\ S & -s_{22} \end{bmatrix}}{s_{12}}$
h	$\dfrac{\begin{bmatrix} Z & z_{12} \\ -z_{21} & 1 \end{bmatrix}}{z_{22}}$	$\dfrac{\begin{bmatrix} 1 & -y_{12} \\ y_{22} & Y \end{bmatrix}}{y_{11}}$	$\begin{bmatrix} h_{11} & h_{12} \\ h_{21} & h_{22} \end{bmatrix}$	$\dfrac{\begin{bmatrix} g_{22} & -g_{12} \\ -g_{21} & g_{22} \end{bmatrix}}{G}$	$\dfrac{\begin{bmatrix} t_{12} & T \\ -1 & t_{21} \end{bmatrix}}{t_{22}}$	$\dfrac{\begin{bmatrix} -s_{12} & 1 \\ -S & -s_{21} \end{bmatrix}}{s_{11}}$
g	$\dfrac{\begin{bmatrix} 1 & -z_{12} \\ z_{21} & Z \end{bmatrix}}{z_{11}}$	$\dfrac{\begin{bmatrix} Y & y_{12} \\ -y_{21} & 1 \end{bmatrix}}{y_{22}}$	$\dfrac{\begin{bmatrix} h_{22} & -h_{12} \\ -h_{21} & h_{11} \end{bmatrix}}{H}$	$\begin{bmatrix} g_{11} & g_{12} \\ g_{21} & g_{22} \end{bmatrix}$	$\dfrac{\begin{bmatrix} t_{21} & -T \\ 1 & t_{12} \end{bmatrix}}{t_{11}}$	$\dfrac{\begin{bmatrix} -s_{21} & -1 \\ S & -s_{12} \end{bmatrix}}{s_{22}}$
t	$\dfrac{\begin{bmatrix} z_{11} & Z \\ 1 & z_{22} \end{bmatrix}}{z_{21}}$	$\dfrac{\begin{bmatrix} -y_{22} & -1 \\ -Y & -y_{11} \end{bmatrix}}{y_{21}}$	$\dfrac{\begin{bmatrix} -H & -h_{11} \\ -h_{22} & -1 \end{bmatrix}}{h_{21}}$	$\dfrac{\begin{bmatrix} 1 & g_{22} \\ g_{11} & G \end{bmatrix}}{g_{21}}$	$\begin{bmatrix} t_{11} & t_{12} \\ t_{21} & t_{22} \end{bmatrix}$	$\dfrac{\begin{bmatrix} s_{22} & -s_{12} \\ -s_{21} & s_{22} \end{bmatrix}}{S}$
s	$\dfrac{\begin{bmatrix} z_{22} & -Z \\ -1 & z_{11} \end{bmatrix}}{z_{12}}$	$\dfrac{\begin{bmatrix} -y_{11} & 1 \\ Y & y_{22} \end{bmatrix}}{y_{12}}$	$\dfrac{\begin{bmatrix} 1 & -h_{11} \\ -h_{22} & H \end{bmatrix}}{h_{12}}$	$\dfrac{\begin{bmatrix} -G & g_{22} \\ -g_{11} & -1 \end{bmatrix}}{g_{12}}$	$\dfrac{\begin{bmatrix} t_{22} & -t_{12} \\ -t_{21} & t_{11} \end{bmatrix}}{T}$	$\begin{bmatrix} s_{11} & s_{12} \\ s_{21} & s_{22} \end{bmatrix}$

$$Z = z_{11}z_{22} - z_{12}z_{21} \qquad\qquad Y = y_{11}y_{22} - y_{12}y_{21}$$

$$H = h_{11}h_{22} - h_{12}h_{21} \qquad\qquad G = g_{11}g_{22} - g_{12}g_{21}$$

$$T = t_{11}t_{22} - t_{12}t_{21} \qquad\qquad S = s_{11}s_{22} - s_{12}s_{21}$$

Example 16.4

The z-parameters for a given two-port are

$$\begin{bmatrix} V_1 \\ V_2 \end{bmatrix} = \begin{bmatrix} 4 & 3s \\ -s & 1 \end{bmatrix} \begin{bmatrix} I_1 \\ I_2 \end{bmatrix}$$

To convert to the hybrid two-port description we use the third row, first column of Table 16.2, with $z_{11} = 4$, $z_{12} = 3s$, $z_{21} = -s$, $z_{22} = 2$, and $Z = z_{11}z_{22} - z_{12}z_{21} = 3s^2 + 8$. Applying these conversion factors, we obtain

$$\begin{bmatrix} h_{11} & h_{12} \\ h_{21} & h_{22} \end{bmatrix} = \dfrac{\begin{bmatrix} 3s^2 + 8 & 3s \\ s & 1 \end{bmatrix}}{2} \begin{bmatrix} \frac{3}{2}s^2 + 4 & \frac{3}{2}s \\ \frac{s}{2} & \frac{1}{2} \end{bmatrix}$$

The derivation of each of these conversion rules consists simply of solving the original two-port description for the dependent variables of the description we are converting to. For instance, in converting from impedance to hybrid, solve the second impedance equation (16.4b) for I_2, substitute into the first (16.4a), and gather terms. This yields V_1 and I_2 as the dependent variables and I_1 and V_2 as the independent variables, which is the desired hybrid description. The formulas in the corresponding Table 16.2 entry are the hybrid parameters in terms of the original (z) parameters.

Example 16.5

In Example 16.3 we computed the hybrid and transmission parameters for the circuit shown in Fig. 16.9. The transmission parameters were found to be

$$
\begin{bmatrix} \mathbf{t}_{11} & \mathbf{t}_{12} \\ \mathbf{t}_{21} & \mathbf{t}_{22} \end{bmatrix} = \begin{bmatrix} \dfrac{3s^2 + 6s + 8}{(s+2)(s+4)} & \dfrac{5s+4}{s+4} \\ \dfrac{s}{s+4} & \dfrac{2s+4}{s+4} \end{bmatrix}
$$

Applying the transmission to the hybrid conversion cell from Table 16.2 with these transmission parameters and

$$
\mathbf{T} = \frac{3s^2 + 6s + 8}{(s+2)(s+4)} \frac{2s+4}{s+4} - \frac{5s+4}{s+4} \frac{s}{s+4} = 1
$$

yields

$$
\begin{bmatrix} \mathbf{h}_{11} & \mathbf{h}_{12} \\ \mathbf{h}_{21} & \mathbf{h}_{22} \end{bmatrix} = \frac{\begin{bmatrix} \dfrac{5s+4}{s+4} & 1 \\ -1 & \dfrac{s}{s+4} \end{bmatrix}}{(2s+4)/(s+4)} = \begin{bmatrix} \dfrac{5s+4}{2(s+2)} & \dfrac{s+4}{2(s+2)} \\ \dfrac{-(s+4)}{2(s+2)} & \dfrac{s}{2s+4} \end{bmatrix}
$$

which agrees with the hybrid matrix calculated independently in Example 16.3.

Being able to convert from one to another shows vividly that there is much redundancy in the information conveyed by the six separate two-port descriptions. What is the point of having six when one will do? One reason is that *not every two-port description exists for every two-port circuit.* Examining Table 16.2, note that to convert between descriptions, division by a parameter is required. If that parameter happens to be zero, it is not possible to perform the conversion because the corresponding two-port description does not exist.

A second practical reason for having several different two-port descriptions is that it is sometimes easier to determine one set of two-port parameters than another. For instance, if the two-port description of a transistor circuit is to be determined by bench measurement, the four parameters comprising the hybrid description are easily measurable for this type of circuit. Others, requiring experiments with an open-circuited output port, are more difficult to arrange without driving the circuit out of linear operation. Or we may be calculating the two-port description of a circuit using mesh analysis, and the open-circuited port conditions of the impedance description may lead to fewer equations than the short-circuited admittance description.

Finally, when two-ports are interconnected, it is far easier to determine the overall two-port description of the interconnected circuit if we are using the right two-port parameters. Which is right depends on the manner in which the two-ports are interconnected, as we shall see in Section 16.3.

To summarize, in this section we have defined the six two-port descriptions: impedance, admittance, hybrid and inverse hybrid, and transmission and inverse transmission.

The general procedure for determining any of these descriptions is to perform two experiments, in each case establishing a zero value for one independent variable by open- or short-circuiting its port, putting a source in the other port to reflect the other independent variable, and computing the dependent variables in the resulting circuit. Each experiment produces one column of the parameter matrix. The six descriptions are related by the conversion factors listed in Table 16.2.

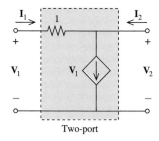

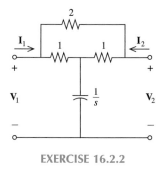

EXERCISE 16.2.1

EXERCISE 16.2.2

16.2.1. Find the y-parameters. Which, if any, of the two-port descriptions fail to exist for this circuit?

Answer $y_{11} = y_{22} = 1$, $y_{12} = 0$, $y_{21} = -1$; inverse transmission does not exist

16.2.2. Determine all six of the two-port descriptions.

Answer

$$\mathbf{Z} = \frac{1}{4s}\begin{bmatrix} 3s+4 & s+4 \\ s+4 & 3s+4 \end{bmatrix}; \mathbf{Y} = \frac{1}{2s+4}\begin{bmatrix} 3s+4 & -(s+4) \\ -(s+4) & 3s+4 \end{bmatrix};$$

$$\mathbf{h} = \frac{1}{3s+4}\begin{bmatrix} 2s+4 & s+4 \\ -(s+4) & 4s \end{bmatrix}; \mathbf{g} = \frac{1}{3s+4}\begin{bmatrix} 4s & -(s+4) \\ s+4 & 2s+4 \end{bmatrix};$$

$$\mathbf{t} = \frac{1}{s+4}\begin{bmatrix} 3s+4 & 2s+4 \\ 4s & 3s+4 \end{bmatrix}; \mathbf{s} = \frac{1}{s+4}\begin{bmatrix} 3s+4 & -2(s+2) \\ -4s & 3s+4 \end{bmatrix}$$

16.2.3. A *bilateral* circuit has $z_{12} = z_{21}$. What equalities does a bilateral circuit satisfy in terms of the other two-port parameters?

Answer $y_{12} = y_{21}$, $h_{12} = -h_{21}$, $g_{12} = -g_{21}$; $S = T = 1$ (S and T are the determinants of the **s** and **t** matrices as in Table 16.2)

16.3 TWO-PORT MODELS

Thus far in our discussion of two-port circuits we have focused on the uses and the procedures for converting a circuit diagram to a set of four parameters, the process of two-port *analysis*. The reverse process, in which we find a circuit that agrees with a given set of two-port parameters, is called two-port *synthesis* or *modeling*.

Two-port models are useful in several contexts. They permit us to visualize a circuit described by a set of two-port parameters, to consider the effects of interconnecting them with other elements or subcircuits, and to use SPICE to analyze them. In the case of coupled coils, they permit us to model a circuit with mutual inductance by one that does not contain mutual inductance.

To see how a class of two-port models may be easily derived, consider the z-parameter description

$$V_1 = z_{11}I_1 + z_{12}I_2 \tag{16.15a}$$

$$V_2 = z_{21}I_1 + z_{22}I_2 \tag{16.15b}$$

Any circuit that satisfies these two equations is called a circuit model for this two-port description. We may use any circuit configuration that agrees with these equations, although it is usually to our advantage to select a simple one. With each term having units of volts, we are led to the model shown in Fig. 16.11(a). On the input side, by (16.15a), V_1 is the sum of two voltage drops. One is proportional to the current through it, so it must be an impedance, and the other is proportional to a current elsewhere, so it must be a current-controlled voltage source. The output port model circuit follows similarly.

Most of the other two-port descriptions yield models in the same natural way. For the hybrid case, for instance, the equations are

$$V_1 = h_{11}I_1 + h_{12}V_2 \tag{16.16a}$$

$$I_2 = h_{21}I_1 + h_{22}V_2 \tag{16.16b}$$

The first equation suggests a KVL input port model, like the previous model, and this is shown in Fig. 16.11(b). The second has units of amperes and is taken to be a KCL equation. The port current I_2 is apparently the sum of two currents, one flowing through an impedance of value $1/h_{12}$ and the other through a current-controlled current source as shown in Fig. 16.11(b).

Circuit models for the admittance and inverse hybrid descriptions follow similarly. Only the transmission and inverse transmission cases do not so naturally evoke circuit models of this type, and if a circuit model is needed first, it is recommended that a conversion to impedance, admittance, hybrid, or inverse hybrid parameters be done.

A pair of coupled coils is an important example of a two-port circuit. The i–v laws for coupled coils (15.27), repeated here, are

$$V_1 = sL_1I_1 \pm sMI_2 \tag{16.17a}$$

$$V_2 = \pm sMI_1 + sL_2I_2 \tag{16.17b}$$

This pair of equations may also be recognized as an impedance two-port description of the coupled coils. Consequently, a model of the form shown in Fig. 16.11(a) with

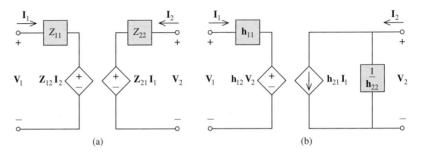

(a)   (b)

FIGURE 16.11 Two-port circuit models: (a) z-parameter; (b) hybrid parameter.

$\mathbf{z}_{11} = sL_1$, $\mathbf{z}_{22} = sL_2$, and $\mathbf{z}_{12} = \mathbf{z}_{21} = \pm sM$ may always be used to replace the pair of coupled coils. The model replaces the complication of mutual inductance with that of a pair of controlled sources and thus may not always be of benefit. A less general but simpler model, for a pair of coupled coils with a common terminal, was shown in Fig. 15.11.

The forms for these circuit models are not unique. For instance, Thevenin–Norton transformations may be applied to the circuit model for either port, yielding a new model with a different source. Since many different circuits have the same two-port description, it is to be expected that many circuit models will be available for the same set of two-port parameters. Indeed, the essence of the two-port description is that it preserves the port behavior of the circuit while obscuring the internal detail. When we resupply the internal circuit detail by creating an explicit circuit model, there will always be many different ways we can do this.

The models described so far have no wires connecting input and output ports; that is, their two ports are electrically isolated. Many two-port circuits, such as multistage op-amp circuits, have a common node linking input and output. Since the ports in the models above are electrically isolated, a single wire added between the ports will not carry any current, and no current or voltage drop will be changed. So the models above, with the addition of a shorting wire, can be used to model common-node (common-ground) as well as electrically isolated two-ports. Circuits in which there is a common node between input and output ports are also called *three-terminal two-ports*. A three-terminal two-port and a model for it of the type introduced above are shown in Fig. 16.12.

In addition to the standard two-port model containing two controlled sources shown in Fig. 16.12(b), three-terminal two-ports have another type of model that takes advantage of the common-node configuration. Consider a general two-port with the z-parameters given in (16.15). Suppose that we add the term $(\mathbf{z}_{12} - \mathbf{z}_{21})\mathbf{I}_1$ to both sides of (16.15b), resulting in

$$\mathbf{V}_1 = \mathbf{z}_{11}\mathbf{I}_1 + \mathbf{z}_{12}\mathbf{I}_2 \tag{16.18a}$$

$$\mathbf{V}_2' = \mathbf{V}_2 + (\mathbf{z}_{12} - \mathbf{z}_{21})\mathbf{I}_1 = \mathbf{z}_{12}\mathbf{I}_1 + \mathbf{z}_{22}\mathbf{I}_2 \tag{16.18b}$$

Let us find the z-parameter description of the circuit of Fig. 16.13, using terminals $a–a'$ as the output port and $\mathbf{V}_2'$ as the output port voltage. This circuit contains three impedances, labeled $\mathbf{z}_{11} - \mathbf{z}_{12}$, $\mathbf{z}_{22} - \mathbf{z}_{21}$, and $\mathbf{z}_{12}$. Computing the z parameters in the usual way, it is

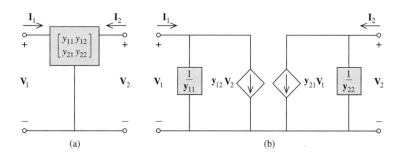

(a)   (b)

FIGURE 16.12   (a) Three-terminal two-port; (b) model.

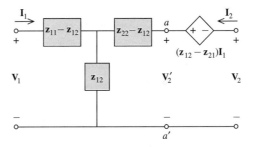

FIGURE 16.13 Circuit model for three-terminal two-port.

easily verified that this circuit obeys the two-port description

$$\mathbf{V}_1 = \mathbf{z}_{11}\mathbf{I}_1 + \mathbf{z}_{12}\mathbf{I}_2 \tag{16.19a}$$

$$\mathbf{V}_2' = \mathbf{z}_{12}\mathbf{I}_1 + \mathbf{z}_{22}\mathbf{I}_2 \tag{16.19b}$$

Next observe from the figure that $\mathbf{V}_2' = \mathbf{V}_2 + (\mathbf{z}_{12} - \mathbf{z}_{21})\mathbf{I}_1$. Combining this with (16.19), we are back with the general form (16.15). Thus *Fig. 16.13 is a general circuit model for a three-terminal two-port with impedance parameters* $\mathbf{z}_{11}$, $\mathbf{z}_{12}$, $\mathbf{z}_{21}$, and $\mathbf{z}_{22}$. In contrast to the previous models, it contains a single controlled source.

In the important special case $\mathbf{z}_{12} = \mathbf{z}_{21}$, even this one remaining controlled source disappears, resulting in an impedance-only model. Two-port circuits for which $\mathbf{z}_{12} = \mathbf{z}_{21}$ or, equivalently, in terms of y parameters $\mathbf{y}_{12} = \mathbf{y}_{21}$, are called *bilateral*. It can be shown that all circuits containing only impedances are bilateral and hence can be modeled by three impedances in the Y configuration shown in Fig. 16.13.

Example 16.6	Let us find a compact two-port model for the circuit of Fig. 16.14(a), a model with no internal loops. Begin by finding the z-parameters for this circuit. Open-circuiting the output and placing a unit current source in the input port, the loop equation is

$$1(\mathbf{I}_a - 1) + (s + 3)\mathbf{I}_a = 0$$

whence $\mathbf{I}_a = 1/(s + 4)$.

$$\mathbf{V}_1 = 2\mathbf{I}_1 + 1(1 - \mathbf{I}_a) = 2 + \frac{s + 3}{s + 4} = \frac{3s + 11}{s + 4}$$

and

$$\mathbf{V}_2 = 3\mathbf{I}_a = \frac{3}{s + 4}$$

$\mathbf{V}_1$ and $\mathbf{V}_2$ form the first column of the z-matrix. Next, open-circuiting the input and putting the unit current source on the output yields

$$3(\mathbf{I}_a + 1) + (s + 1)\mathbf{I}_a = 0$$

Solving for $\mathbf{I}_a$ and then $\mathbf{V}_1$ and $\mathbf{V}_2$ as in the first experiment, the

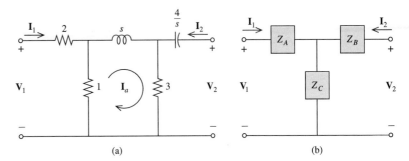

FIGURE 16.14 (a) Circuit for Example 16.6; (b) two-port model with no internal loops.

z-matrix is

$$
\begin{bmatrix} \mathbf{z}_{11} & \mathbf{z}_{12} \\ \mathbf{z}_{21} & \mathbf{z}_{22} \end{bmatrix} = \begin{bmatrix} \dfrac{3s + 11}{s + 4} & \dfrac{3}{s + 4} \\ \dfrac{3}{s + 4} & \dfrac{3s^2 + 7s + 16}{s(s + 4)} \end{bmatrix}
$$

Comparing this with Fig. 16.13, a two-port model for this three-terminal bilateral circuit is shown in Fig. 16.14(b) with

$$
\mathbf{Z}_A = \mathbf{z}_{11} - \mathbf{z}_{12} = \frac{3s + 8}{s + 4}
$$

$$
\mathbf{Z}_B = \mathbf{z}_{22} - \mathbf{z}_{12} = \frac{3s^2 + 4s + 16}{s(s + 4)}
$$

$$
\mathbf{Z}_C = \mathbf{z}_{12} = \frac{3}{s + 4}
$$

Any pair of coupled coils forms a bilateral two-port circuit, as can be seen from (16.17). If the two coils can be tied together to form a three-terminal two-port circuit, the impedance-only model of Fig. 16.13 with $\mathbf{z}_{12} = \mathbf{z}_{21}$ is a very useful circuit model for the coupled coils. The model, previously introduced in Fig. 15.11(b), replaces the pair of coupled coils, with their magnetic coupling, by three simple (uncoupled) coils. Note that even if the pair of coupled coils does not have a common reference node, as long as there is no external electrical connection between the coils, a shorting wire can always be installed between the coils without affecting any element's current or voltage drop. This is due to the generalized KCL, which guarantees that a single wire between otherwise electrically isolated subcircuits carries no current.

Example 16.7

Find an equivalent circuit for Fig. 16.15(a) that has no coupled coils or controlled sources. The i–v laws for the coupled coils are

$$
\mathbf{V}_1 = 2s\mathbf{I}_1 - 2s\mathbf{I}_2 \tag{16.20a}
$$

$$
\mathbf{V}_2 = -2s\mathbf{I}_1 + 3s\mathbf{I}_2 \tag{16.20b}
$$

Connecting a and a', we may use the three-terminal model shown in Fig. 16.13, which, for the bilateral case $\mathbf{z}_{12} = \mathbf{z}_{21}$, is impedance-only. With $\mathbf{z}_{11} - \mathbf{z}_{12} = 4s$, $\mathbf{z}_{22} - \mathbf{z}_{12} = 5s$, and $\mathbf{z}_{12} = -2s$, the result is shown in Fig. 16.15(b).

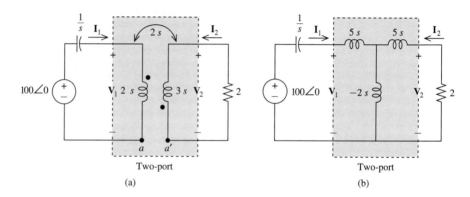

FIGURE 16.15 (a) Circuit for Example 16.7; (b) equivalent.

Finally, recall from Chapter 11 that a three-terminal Y-configured set of impedances such as those in Fig. 16.15(b) or (16.13) has a Δ-configured equivalent. Thus the Y–Δ transformation may be used to convert these models to equivalent models having the same number of elements, but fewer loops (Y configuration) or fewer nodes (Δ configuration). The Y–Δ and Δ–Y conversion formulas are given in Section 11.5 as (11.31) and (11.32).

Example 16.8

A certain three-terminal two-port circuit has z-parameters,

$$\begin{bmatrix} \mathbf{z}_{11} & \mathbf{z}_{12} \\ \mathbf{z}_{21} & \mathbf{z}_{22} \end{bmatrix} = \begin{bmatrix} \dfrac{2(4s+1)}{4s+3} & \dfrac{4s}{4s+3} \\ \dfrac{-2(2s+3)}{4s+3} & \dfrac{4s(s+2)}{4s+3} \end{bmatrix}$$

Following Fig. 16.13, a circuit model for this circuit is shown in Fig. 16.16(a). Using the Y–Δ conversion formulas (11.31) with

$$\mathbf{Y}_a = \frac{4s+3}{2(2s+1)}, \qquad \mathbf{Y}_b = \frac{4s+3}{4s(s+1)}, \qquad \mathbf{Y}_c = \frac{4s+3}{4s}$$

we arrive at the equivalent two-port model shown in Fig. 16.16(b). In both models, the values inside the boxes are impedances.

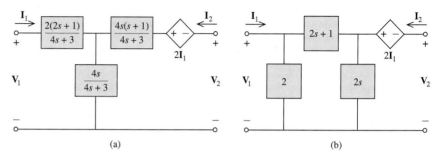

FIGURE 16.16 (a) Two-port model for Example 16.8; (b) equivalent two-port model using Y–Δ transformation.

Now (16.21) is in the form of a transmission two-port description

$$\begin{bmatrix} \mathbf{V}_1 \\ \mathbf{I}_1 \end{bmatrix} = \begin{bmatrix} \mathbf{t}_{11} & \mathbf{t}_{12} \\ \mathbf{t}_{21} & \mathbf{t}_{22} \end{bmatrix} \begin{bmatrix} \mathbf{V}_4 \\ -\mathbf{I}_4 \end{bmatrix} \tag{16.22}$$

of the overall interconnected circuit, where the t_{ij}'s are the overall transmission parameters for the cascaded pair of of circuits. Comparing (16.21) and (16.22), it must be the case that

CASCADE

$$\begin{bmatrix} \mathbf{t}_{11} & \mathbf{t}_{12} \\ \mathbf{t}_{21} & \mathbf{t}_{22} \end{bmatrix} = \begin{bmatrix} \mathbf{t}_{11}^a & \mathbf{t}_{12}^a \\ \mathbf{t}_{21}^a & \mathbf{t}_{22}^a \end{bmatrix} \begin{bmatrix} \mathbf{t}_{11}^b & \mathbf{t}_{12}^b \\ \mathbf{t}_{21}^b & \mathbf{t}_{22}^b \end{bmatrix} \tag{16.23}$$

The overall transmission matrix for a cascaded pair of two-ports is the product of their transmission matrices, with the input two-port premultiplying.

To find the overall transmission description, we need only multiply the individual transmission matrices together in the natural order (input two-port matrix first). Recall that matrices do not in general commute; that is, the matrix product AB is not in general equal to BA, so order matters.

The simplicity with which the overall transmission parameters of the cascaded interconnection can be derived from the individual transmission parameters does not carry over to other two-port descriptions. For instance, there is no simple relationship between the impedance parameters of two circuits and the impedance parameters of their cascade interconnection. For this reason it is useful to convert to transmission parameters before computing the two-port description of a cascade.

Example 16.9

Two circuits A and B are cascaded as in Fig. 16.17. We wish to know the hybrid description of the resulting two-port, given the individual hybrid descriptions

$$A: \quad \begin{bmatrix} \mathbf{h}_{11}^a & \mathbf{h}_{12}^a \\ \mathbf{h}_{21}^a & \mathbf{h}_{22}^a \end{bmatrix} = \begin{bmatrix} s & -3 \\ -s & s+3 \end{bmatrix}$$

$$B: \quad \begin{bmatrix} \mathbf{h}_{11}^a & \mathbf{h}_{12}^a \\ \mathbf{h}_{21}^a & \mathbf{h}_{22}^a \end{bmatrix} = \begin{bmatrix} 4 & 1 \\ -1 & s+1 \end{bmatrix}$$

We first convert the given descriptions to transmission parameters. For A, $s(s+3) - 3s = s^2$ and the conversion is, by Table 16.2,

$$\begin{bmatrix} \mathbf{t}_{11}^a & \mathbf{t}_{12}^a \\ \mathbf{t}_{21}^a & \mathbf{t}_{22}^a \end{bmatrix} = \frac{\begin{bmatrix} -\mathbf{H}^a & -\mathbf{h}_{11}^a \\ -\mathbf{h}_{22}^a & -1 \end{bmatrix}}{\mathbf{h}_{21}^a}$$

$$= \begin{bmatrix} s & 1 \\ \dfrac{s+3}{s} & \dfrac{1}{s} \end{bmatrix}$$

Doing the same for B, $\mathbf{H}^b = 4(s+1) + 1 = 4s + 5$ and

$$\begin{bmatrix} \mathbf{t}_{11}^b & \mathbf{t}_{12}^b \\ \mathbf{t}_{21}^b & \mathbf{t}_{22}^b \end{bmatrix} = \begin{bmatrix} 4s+s & 4 \\ s+1 & 1 \end{bmatrix}$$

The overall transmission matrix is their product, with the input two-port A's transmission matrix written first (premultiplying).

$$\begin{bmatrix} \mathbf{t}_{11} & \mathbf{t}_{12} \\ \mathbf{t}_{21} & \mathbf{t}_{22} \end{bmatrix} = \begin{bmatrix} \mathbf{t}_{11}^a & \mathbf{t}_{12}^a \\ \mathbf{t}_{21}^a & \mathbf{t}_{22}^a \end{bmatrix} \begin{bmatrix} \mathbf{t}_{11}^b & \mathbf{t}_{12}^b \\ \mathbf{t}_{21}^b & \mathbf{t}_{22}^b \end{bmatrix}$$

$$= \begin{bmatrix} s & 1 \\ \dfrac{s+3}{s} & \dfrac{1}{s} \end{bmatrix} \begin{bmatrix} 4s+5 & 4 \\ s+1 & 1 \end{bmatrix}$$

Carrying out the matrix multiplication, the transmission parameters of the cascade of A and B are

$$\mathbf{t}_{11} = s(4s+5) + 1(s+1) = 4s^2 + 6s + 1$$

$$\mathbf{t}_{12} = s(4) + 1(1) = 4s + 1$$

$$\mathbf{t}_{21} = \frac{s+3}{s}(4s+5) + \frac{1}{s}(s+1) = \frac{4s^2 + 18s + 16}{s}$$

$$\mathbf{t}_{22} = \frac{s+3}{s}(4) + \frac{1}{s}(1) = \frac{4s+13}{s}$$

Table 16.1 cites the conversion from transmission to hybrid parameters as

$$\begin{bmatrix} \mathbf{h}_{11} & \mathbf{h}_{12} \\ \mathbf{h}_{21} & \mathbf{h}_{22} \end{bmatrix} = \frac{\begin{bmatrix} \mathbf{t}_{12} & \mathbf{T} \\ -1 & \mathbf{t}_{21} \end{bmatrix}}{t_{22}}$$

where $\mathbf{T} = \mathbf{t}_{11}\mathbf{t}_{22} - \mathbf{t}_{12}\mathbf{t}_{21}$. The final result is

$$\begin{bmatrix} \mathbf{h}_{11} & \mathbf{h}_{12} \\ \mathbf{h}_{21} & \mathbf{h}_{22} \end{bmatrix} = \frac{\begin{bmatrix} 4s+1 & \dfrac{-3}{s} \\ -1 & \dfrac{4s^2 + 18s + 16}{s} \end{bmatrix}}{(4s+13)/s}$$

$$= \begin{bmatrix} \dfrac{s(4s+1)}{4s+13} & \dfrac{-3}{4s+13} \\ \dfrac{-s}{4s+13} & \dfrac{4s^2 + 18s + 16}{4s+13} \end{bmatrix}$$

Equation (16.23) is easily generalized to a chain of three or more in cascade. Suppose that the cascade order is ABC, with A the input two-port and C the output two-port. Then $\mathbf{T}_{ab} = \mathbf{T}_a\mathbf{T}_b$, where $\mathbf{T}_a$ and $\mathbf{T}_b$ are the transmission matrices for A and B and $\mathbf{T}_{ab}$ is the transmission matrix of A cascaded with B. When this is cascaded with C, we have our three-way cascade: $\mathbf{T}_{abc} = \mathbf{T}_{ab}\mathbf{T}_c = \mathbf{T}_a\mathbf{T}_b\mathbf{T}_c$. *Transmission matrices in a chain of circuits in cascade multiply in natural order.*

Another common interconnection is the *parallel* interconnection shown in Fig. 16.18. The plus terminals of each port are tied together, and that node is defined as the plus terminal of the interconnected set. Since

$$\begin{bmatrix} \mathbf{I}_1 \\ \mathbf{I}_2 \end{bmatrix} = \begin{bmatrix} \mathbf{I}_1^a \\ \mathbf{I}_2^a \end{bmatrix} + \begin{bmatrix} \mathbf{I}_1^b \\ \mathbf{I}_2^b \end{bmatrix}$$

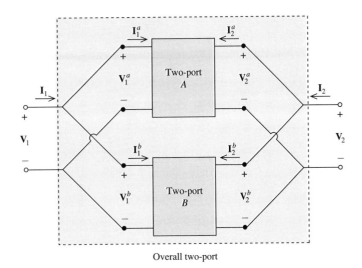

Overall two-port

FIGURE 16.18   Parallel interconnection of two-ports.

then, using the admittance matrices for the two circuits,

$$\begin{bmatrix} I_1 \\ I_2 \end{bmatrix} = \begin{bmatrix} y_{11}^a & y_{12}^a \\ y_{21}^a & y_{22}^a \end{bmatrix} \begin{bmatrix} V_1^a \\ V_2^a \end{bmatrix} + \begin{bmatrix} y_{11}^b & y_{12}^b \\ y_{21}^b & y_{22}^b \end{bmatrix} \begin{bmatrix} V_1^b \\ V_2^b \end{bmatrix}$$ (16.24)

But the figure also makes clear that the voltages are equal:

$$\begin{bmatrix} V_1 \\ V_2 \end{bmatrix} = \begin{bmatrix} V_1^a \\ V_2^a \end{bmatrix} = \begin{bmatrix} V_1^b \\ V_2^b \end{bmatrix}$$ (16.25)

And combining the last two equations yields

$$\begin{bmatrix} I_1 \\ I_2 \end{bmatrix} = \left\{ \begin{bmatrix} y_{11}^a & y_{12}^a \\ y_{21}^a & y_{22}^a \end{bmatrix} + \begin{bmatrix} y_{11}^b & y_{12}^b \\ y_{21}^b & y_{22}^b \end{bmatrix} \right\} \begin{bmatrix} V_1 \\ V_2 \end{bmatrix}$$

This is an equation of the form

$$\begin{bmatrix} I_1 \\ I_2 \end{bmatrix} = \begin{bmatrix} y_{11} & y_{12} \\ y_{21} & y_{22} \end{bmatrix} \begin{bmatrix} V_1 \\ V_2 \end{bmatrix}$$

so it must be the admittance (y-parameter) description of the parallel combination.

PARALLEL

$$\begin{bmatrix} y_{11} & y_{12} \\ y_{21} & y_{22} \end{bmatrix} = \begin{bmatrix} y_{11}^a + y_{11}^b & y_{12}^a + y_{12}^b \\ y_{21}^a + y_{21}^b & y_{22}^a + y_{22}^b \end{bmatrix}$$ (16.26)

The overall admittance matrix for parallel two-ports is the sum of their individual admittance matrices.

Just as the transmission parameters are most compatible with cascade interconnects, it is the y-parameters that are best suited for the parallel case. Note that while parallel means something different for one-ports and two-ports, in both cases we may summarize the result by saying that parallel admittances add. In one case we are referring to

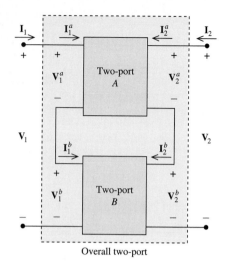

FIGURE 16.19 Series interconnection of two-ports.

scalar admittances and in the other to admittance matrices, but nonetheless, the rule is satisfyingly the same.

The third and final type of interconnection we shall consider is the series interconnection shown in Fig. 16.19. Applying KVL at each port, we have

$$\begin{bmatrix} \mathbf{V}_1 \\ \mathbf{V}_2 \end{bmatrix} = \begin{bmatrix} \mathbf{V}_1^a \\ \mathbf{V}_2^a \end{bmatrix} + \begin{bmatrix} \mathbf{V}_1^b \\ \mathbf{V}_2^b \end{bmatrix}$$

Substituting the impedance descriptions for A and B gives

$$\begin{bmatrix} \mathbf{V}_1 \\ \mathbf{V}_2 \end{bmatrix} = \begin{bmatrix} \mathbf{z}_{11}^a & \mathbf{z}_{12}^a \\ \mathbf{z}_{21}^a & \mathbf{z}_{22}^a \end{bmatrix} \begin{bmatrix} \mathbf{I}_1^a \\ \mathbf{I}_2^a \end{bmatrix} + \begin{bmatrix} \mathbf{z}_{11}^b & \mathbf{z}_{12}^b \\ \mathbf{z}_{21}^b & \mathbf{z}_{22}^b \end{bmatrix} \begin{bmatrix} \mathbf{I}_1^b \\ \mathbf{I}_2^b \end{bmatrix} \qquad (16.27)$$

But the input port currents are all the same, as are the output port currents:

$$\begin{bmatrix} \mathbf{I}_1 \\ \mathbf{I}_2 \end{bmatrix} = \begin{bmatrix} \mathbf{I}_1^a \\ \mathbf{I}_2^a \end{bmatrix} = \begin{bmatrix} \mathbf{I}_1^b \\ \mathbf{I}_2^b \end{bmatrix}$$

When substituted into (16.27), this is

$$\begin{bmatrix} \mathbf{V}_1 \\ \mathbf{V}_2 \end{bmatrix} = \left\{ \begin{bmatrix} \mathbf{z}_{11}^a & \mathbf{z}_{12}^a \\ \mathbf{z}_{21}^a & \mathbf{z}_{22}^a \end{bmatrix} + \begin{bmatrix} \mathbf{z}_{11}^b & \mathbf{z}_{12}^b \\ \mathbf{z}_{21}^b & \mathbf{z}_{22}^b \end{bmatrix} \right\} \begin{bmatrix} \mathbf{I}_1 \\ \mathbf{I}_2 \end{bmatrix}$$

SERIES

$$\begin{bmatrix} \mathbf{z}_{11} & \mathbf{z}_{12} \\ \mathbf{z}_{21} & \mathbf{z}_{22} \end{bmatrix} = \begin{bmatrix} \mathbf{z}_{11}^a + \mathbf{z}_{11}^b & \mathbf{z}_{12}^a + \mathbf{z}_{12}^b \\ \mathbf{z}_{21}^a + \mathbf{z}_{21}^b & \mathbf{z}_{22}^a + \mathbf{z}_{22}^b \end{bmatrix}$$

The overall impedance matrix for series two-ports is the sum of their individual impedance matrices.

Example 16.10

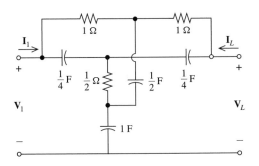

FIGURE 16.20 Circuit for Example 16.10.

We wish to know the transfer function $\mathbf{H}(s) = \mathbf{V}_2(s)/\mathbf{V}_1(s)|_{\mathbf{I}_2(s)=0}$, called the open-circuit voltage transfer ratio, for the circuit in Fig. 16.20. Since any two-port description is a complete characterization of the port behavior of a circuit, a transfer function involving only port variables such as this $\mathbf{H}(s)$ should be easily computable once a two-port description is known. Examining Fig. 16.20, the overall circuit consist of two T-shaped two-ports in parallel, all in series with a two-port consisting of just a capacitor, as shown in Fig. 16.20. Since parallel admittances add, let us begin by finding the admittance matrices for N_1 and N_2. For N_1 we first set $\mathbf{V}_1 = 1$ and $\mathbf{V}_2 = 0$. Then

$$\mathbf{I}_1 = \frac{1}{1 + (2/s)/(1 + 2/s)} = \frac{s+2}{s+4}$$

and by current division

$$-\mathbf{I}_2 = \frac{1}{1 + (s/2)} \frac{s+2}{s+4} = \frac{2}{s+4}$$

Reversing the locations of the short and the unit voltage source, by the symmetry of the circuit we see that we will get the same values as above with indexes reversed. Thus, for N_1,

$$N_1: \quad \begin{bmatrix} \mathbf{y}_{11} & \mathbf{y}_{12} \\ \mathbf{y}_{21} & \mathbf{y}_{22} \end{bmatrix} = \frac{1}{s+4} \begin{bmatrix} s+2 & -2 \\ -2 & s+2 \end{bmatrix} \quad (16.28)$$

N_2 is analyzed in the same way. With $\mathbf{V}_1 = 1$ and $\mathbf{V}_2 = 0$, we have

$$\mathbf{I}_1 = \frac{1}{(4/s) + [(1/2)(4/s)]/(\frac{1}{2} + 4/s)} = \frac{s(s+8)}{8(s+4)}$$

and by current division

$$-\mathbf{I}_2 = \frac{s/4}{(s/4) + 2} \frac{s(s+8)}{8(s+4)} = \frac{s^2}{8(s+4)}$$

N_2 has the same symmetry as N_1 did, so

$$N_2: \quad \begin{bmatrix} \mathbf{y}_{11} & \mathbf{y}_{12} \\ \mathbf{y}_{21} & \mathbf{y}_{22} \end{bmatrix} = \frac{1}{8(s+4)} \begin{bmatrix} s^2 + 8s & -s^2 \\ -s^2 & s^2 + 8s \end{bmatrix} \quad (16.29)$$

Adding (16.28) and (16.29), we arrive at the admittance description of N_{12}, which is the parallel interconnection of N_1 and N_2:

$$N_{12}: \quad \begin{bmatrix} \mathbf{y}_{11} & \mathbf{y}_{12} \\ \mathbf{y}_{21} & \mathbf{y}_{22} \end{bmatrix} = \frac{1}{8(s+4)} \begin{bmatrix} s^2 + 16s + 16 & -(s^2 + 16) \\ -(s^2 + 16) & s^2 + 16s + 16 \end{bmatrix}$$

Next we need to combine N_{12} in series with N_3. Since series impedance matrices add, we will convert our description of N_{12} to the impedance form. But the z and y matrices are inverses of one another, so

$$N_{12}: \quad \begin{bmatrix} \mathbf{z}_{11} & \mathbf{z}_{12} \\ \mathbf{z}_{21} & \mathbf{z}_{22} \end{bmatrix} = \begin{bmatrix} \mathbf{y}_{11} & \mathbf{y}_{12} \\ \mathbf{y}_{21} & \mathbf{y}_{22} \end{bmatrix}^{-1}$$

$$= 8(s+4) \begin{bmatrix} s^2 + 16s + 16 & -(s^2 + 16) \\ -(s^2 + 16) & s^2 + 16s + 16 \end{bmatrix}^{-1}$$

where we have used the fact that if $P = aQ$, where P and Q are square matrices and a is a scalar, then $P^{-1} = (1/a)Q^{-1}$. The determinant is

$$(s^2 + 16s + 16)^2 - (s^2 + 16)^2 = 32s^3 + 256s^2 + 512s$$

$$= 32s(s+4)^2 \qquad (16.30)$$

and

$$N_{12}: \quad \begin{bmatrix} \mathbf{z}_{11} & \mathbf{z}_{12} \\ \mathbf{z}_{21} & \mathbf{z}_{22} \end{bmatrix} = \frac{1}{4s(s+4)} \begin{bmatrix} s^2 + 16s + 16 & s^2 + 16 \\ s^2 + 16 & s^2 + 16s + 16 \end{bmatrix}$$

We need the z-matrix for N_3 as well. Inspection of Fig. 16.21(b) shows that with a unit current source in either port and the other

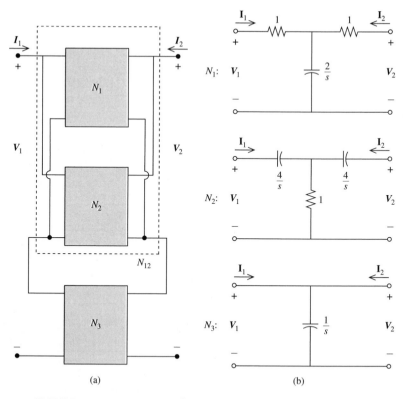

FIGURE 16.21 (a) Circuit of Fig. 16.20 as the interconnection of three two-ports; (b) N_1, N_2, N_3.

open-circuited the port voltages are $\mathbf{V}_1 = \mathbf{V}_2 = 1/s$, so

$$N_3: \quad \begin{bmatrix} \mathbf{z}_{11} & \mathbf{z}_{12} \\ \mathbf{z}_{21} & \mathbf{z}_{22} \end{bmatrix} = \begin{bmatrix} \dfrac{1}{s} & \dfrac{1}{s} \\ \dfrac{1}{s} & \dfrac{1}{s} \end{bmatrix}$$

Adding the z-parameters for the series two-ports N_{12} and N_3, the z-parameters for the overall circuit are

$$\begin{bmatrix} \mathbf{z}_{11} & \mathbf{z}_{12} \\ \mathbf{z}_{21} & \mathbf{z}_{22} \end{bmatrix}$$

$$= \frac{1}{4s(s+4)} \begin{bmatrix} s^2 + 20s + 32 & s^2 + 4s + 32 \\ s^2 + 4s + 32 & s^2 + 20s + 32 \end{bmatrix} \tag{16.31}$$

We stated earlier that any transfer function, such as $\mathbf{V}_2/\mathbf{V}_1|_{\mathbf{I}_2=0}$ required here, can easily be determined from a two-port description. The z-parameters satisfy

$$\mathbf{V}_1 = \mathbf{z}_{11}\mathbf{I}_1 + \mathbf{z}_{12}\mathbf{I}_2$$

$$\mathbf{V}_2 = \mathbf{z}_{21}\mathbf{I}_1 + \mathbf{z}_{22}\mathbf{I}_2$$

Setting $\mathbf{I}_2 = 0$ and dividing the second equation by the first gives

$$\mathbf{H}(s) = \left. \frac{\mathbf{V}_2(s)}{\mathbf{V}_1(s)} \right|_{\mathbf{I}_2(s)=0} = \frac{\mathbf{z}_{21}}{\mathbf{z}_{11}} \tag{16.32}$$

and in light of (16.31),

$$\mathbf{H}(s) = \frac{s^2 + 4s + 32}{s^2 + 20s + 32}$$

which completes the example.

A final cautionary note on the use of these results is required. We saw earlier that if we are not careful, it is possible to terminate a four-terminal circuit in such a way that the port condition is violated and the circuit is not being operated as a two-port circuit. If the port condition fails to be true, our basic two-port representations and all results that flow from them fail to apply. It is also possible, even likely, that certain interconnections between two-ports will cause port condition violations for one or more of the interconnected two-ports. To use the results of this chapter, a port condition violation must be prevented.

One way to guarantee that the port condition is satisfied everywhere is to introduce a 1:1 ideal isolation transformer at either one of the ports of each two-port being interconnected, as shown in Fig. 16.22(a). Since the primary is just a solid length of wire connecting upper and lower terminals, the port condition will clearly be satisfied there. And since it is satisfied there, by KCL applied to the two-port as a whole, it is also satisfied at the other port as well. If, say, five two-ports are placed in parallel, four transformers will be sufficient, since the fifth two-port must satisfy the port condition if all the rest do.

If we do not wish to add isolation transformers, we must take care that the port conditions are satisfied at the overall input and output port of the interconnection, that is, no cross connections as in Fig. 16.22(b). Even so, the individual circuits making up

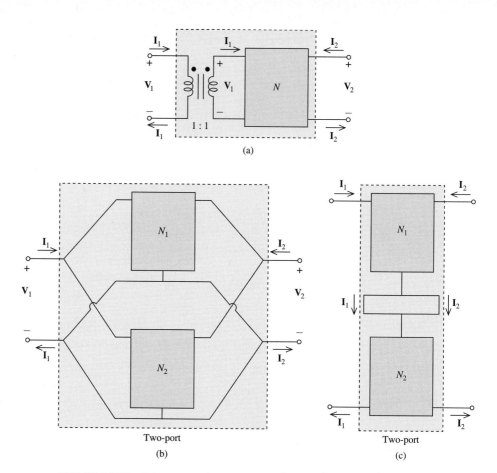

FIGURE 16.22 Interconnections guaranteed to satisfy port condition (a) by transformer coupling; (b) parallel; (c) series.

the interconnection may be in violation of the port condition. Only in cascade is the port condition of each circuit satisfied as long as the overall circuit's port condition is satisfied. For series–parallel circuits there is a useful special case. The port condition is always trivially satisfied for two-ports in parallel or in series if each two-port is of the three-terminal type, as shown in Fig. 16.22(b) and (c), and the common terminals of both circuits are tied together as shown. For instance, in Fig. 16.22(c), since $I_a + I_b = I_1 + I_2$ and I_a and I_b flow through parallel shorts, we may assign $I_a = I_1$ and $I_b = I_1$, thus satisfying the port condition for each two-port in series. Other series and parallel interconnections should be considered on a case basis. If there is no common terminal connection, they may well violate the port condition, and two-port analysis may not be applicable.

In this section we have studied cascade, parallel, and series interconnections of two-ports. We found that cascade transmission matrices multiply, parallel admittance matrices add, and series impedance matrices add. The simplicity of these results encourages us to convert, with the help of Table 16.2, to the appropriate two-port matrix before attempting to determine the overall two-port representation. Each interconnected circuit must continue to satisfy the port condition for this procedure to be applicable, for only then do we truly have an interconnection of two-port circuits as they have been defined and studied. If we are not working with three-terminal two-ports, this may necessitate adding a 1:1 isolation transformer.

EXERCISES

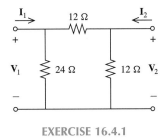

EXERCISE 16.4.1

16.4.1. Find the t matrix and determine the voltage across the input port if four of these circuits are put in cascade, the output is open-circuited, and a +2-A dc source placed in the input port.

Answer $t_{11} = 2$, $t_{12} = 12$, $t_{21} = \frac{1}{6}$, $t_{22} = \frac{3}{2}$; 20.2 V

16.4.2. Show that when the two-ports (a) and (b) are put in parallel and a unit current source is put in the input with shorted output as shown in (c), the port condition is violated for both two-ports.

Answer For port condition to be satisfied, need $I_{a1} = I'_{a1}$, $I_{a2} = I'_{a2}$, $I_{b1} = I'_{b1}$, $I_{b2} = I'_{b2}$. But $I_{a1} = I_{a2} = I'_{b2} = I'_{b1} = 0$, while $I_{b1} = I'_{a1} = 1$, $I_{b2} = I'_{a2} = -1$, so none of the port conditions is satisfied.

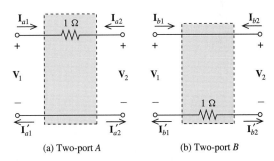

(a) Two-port A (b) Two-port B

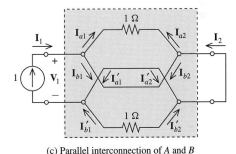

(c) Parallel interconnection of A and B

EXERCISE 16.4.2

16.4.3. Find the y-parameters for the parallel interconnection of the two circuits of Exercise 16.4.2 with and without placing an ideal 1:1 isolation transformer in either of the ports of either of the two-ports.

Answer With an isolation transformer, $y_{11} = y_{22} = -y_{12} = -y_{21} = 2$. Without, y-parameters don't exist (infinite admittance, input, and output are shorted together)

SPICE may be used to determine two-port descriptions, but only at specific frequencies. Replacing s by $j\omega$ in any of the two-port descriptions results in phasor-domain equations, which can be evaluated using SPICE in .AC mode.

SPICE Example 16.11

Let us check the results of Example 16.10, in which the z-parameters for the circuit of Fig. 16.20 were found to be

$$\begin{bmatrix} \mathbf{z}_{11} & \mathbf{z}_{12} \\ \mathbf{z}_{21} & \mathbf{z}_{22} \end{bmatrix} = \frac{1}{4s(s+4)} \begin{bmatrix} s^2 + 20s + 32 & s^2 + 4s + 32 \\ s^2 + 4s + 32 & s^2 + 20s + 32 \end{bmatrix}$$

Replacing s by $j\omega$, the z-parameters at the angular frequency ω are

$$\begin{bmatrix} \mathbf{z}_{11}(j\omega) & \mathbf{z}_{12}(j\omega) \\ \mathbf{z}_{21}(j\omega) & \mathbf{z}_{22}(j\omega) \end{bmatrix}$$

$$= \frac{1}{4j\omega(j\omega+4)} \begin{bmatrix} (32-\omega^2)+20j\omega & (32\omega^2)+4j\omega \\ (32-\omega^2+4j\omega) & (32-\omega^2)+20j\omega \end{bmatrix}$$

We will use SPICE to check the value of the z-parameters at the frequencies of 1 and 10 Hz. Two experiments are required to find the z-parameters for a two-port circuit. The SPICE input file for experiment 1, in which a unit current source is located in the input port and the output port is open-circuited, is shown next. Note that 100-MΩ resistors have been put in parallel with the capacitors to eliminate floating nodes because SPICE insists on dc paths between all nodes. Since the impedance of each capacitor is many orders of magnitude less, the error introduced is negligible.

```
EXAMPLE 16.11
*
I1        1        0        AC            1
C1        1        2            0.25
RD1       1        2     100MEG
R1        2        3            0.50
C2        2        5            0.25
RD2       2        5     100MEG
C3        3        0            1
RD3       3        0     100MEG
R2        1        4            1
R3        4        5            1
C4        4        3            0.50
.AC     LIN        2            1            10
.PRINT   AC     VR(1)     VI(1)       VR(5)    VI(5)
.END
```

A portion of the output follows:

FREQ	VR(1)	VI(1)	VR(5)	VI(5)
1.00E+00	-3.94E-01	5.44E-01	-1.05E-01	9.18E-02
1.00E+01	-2.52E-01	6.35E-02	-2.48E-01	1.28E-04

These values correspond to the first column of the z-matrix for $\omega = 2\pi$ and $\omega = 20\pi$. For experiment 2 the input file is amended by editing the I1 source statement to reference node 5 rather than node 1, thus moving the unit source to the output port and the open circuit to the input port. The corresponding SPICE output is the second column of the z-matrix for these two frequencies.

FREQ	VR(1)	VI(1)	VR(5)	VI(5)
1.00E+00	-1.05E-01	9.18E-02	-3.94E-01	5.44E-01
1.00E+01	-2.48E-01	1.28E-04	-2.52E-01	6.35E-02

The values do indeed check. For instance, at $f = 1$ Hz or $\omega = 2\pi$ rad/s,

$$\mathbf{z}_{11}(j2\pi) = \left.\frac{(32 - \omega^2) + 20j\omega}{4j\omega(j\omega + 4)}\right|_{\omega=2\pi} = -0.394 + j0.0545$$

which agrees with the results of the first experiment for V(1). At $f = 10$ Hz or $\omega = 20\pi$ rad/s,

$$\mathbf{z}_{22}(j20\pi) = \left.\frac{(32 - \omega^2) + 20j\omega}{4j\omega(j\omega + 4)}\right|_{\omega=20\pi} = -0.252 + j0.0635$$

in accord with the second experiment's value for V(5).

SPICE may also be used to determine circuit variables for circuits containing two-ports. In this case the two-port descriptions must first be converted to circuit models as described in Section 16.3.

SPICE Example 16.12

Let us determine the Bode gain plot (gain in dB versus log frequency) in the range 1 Hz to 1 kHz for the circuit of Fig. 16.23. Replacing the two-port by its hybrid model of Fig. 15.28(b), we have the circuit of Fig. 16.24.

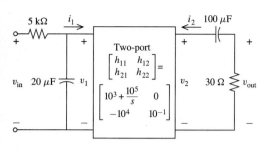

FIGURE 16.23

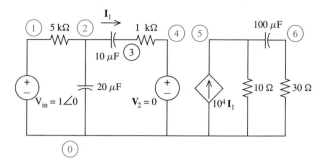

FIGURE 16.24

Note that a dummy voltage source has been installed so that the controlling current I_1 can be referenced. The resulting output for the following input file is shown in Fig. 16.25.

```
             FREQ          VDB(6)


    (*)----------    -3.0000E+01  -2.0000E+01  -1.0000E+01  0.0000E+00  1.0000E+01
                     -  -  -  -  -  -  -  -  -  -  -  -  -  -  -  -  -  -  -  -  -
     1.000E+00  -2.140E+01  .            *       .           .          .       .
     1.259E+00  -1.848E+01  .          .  *      .           .          .       .
     1.585E+00  -1.579E+01  .          .       * .           .          .       .
     1.995E+00  -1.329E+01  .          .        .  *         .          .       .
     2.512E+00  -1.096E+01  .          .        .      *.    .          .       .
     3.162E+00  -8.758E+00  .          .        .         *  .          .       .
     3.981E+00  -6.657E+00  .          .        .         .   *         .       .
     5.012E+00  -4.645E+00  .          .        .         .      *      .       .
     6.310E+00  -2.719E+00  .          .        .         .         *   .       .
     7.943E+00  -8.949E-01  .          .        .         .           *.        .
     1.000E+01   7.994E-01  .          .        .         .           . *       .
     1.259E+01   2.318E+00  .          .        .         .           .   *     .
     1.585E+01   3.601E+00  .          .        .         .           .     *   .
     1.995E+01   4.579E+00  .          .        .         .           .       * .
     2.512E+01   5.187E+00  .          .        .         .           .        *.
     3.162E+01   5.379E+00  .          .        .         .           .        *.
     3.981E+01   5.139E+00  .          .        .         .           .        *.
     5.012E+01   4.488E+00  .          .        .         .           .       * .
     6.310E+01   3.476E+00  .          .        .         .           .     * . .
     7.943E+01   2.172E+00  .          .        .         .           .   *     .
     1.000E+02   6.472E-01  .          .        .         .           . *       .
     1.259E+02  -1.037E+00  .          .        .         .          *.         .
     1.585E+02  -2.830E+00  .          .        .         .       *   .         .
     1.995E+02  -4.696E+00  .          .        .         .    *      .         .
     2.512E+02  -6.611E+00  .          .        .         . *         .         .
     3.162E+02  -8.556E+00  .          .        .        * .          .         .
     3.981E+02  -1.052E+01  .          .        .     *.    .          .         .
     5.012E+02  -1.250E+01  .          .        .   *     .           .         .
     6.310E+02  -1.449E+01  .          .      *  .         .          .         .
     7.943E+02  -1.648E+01  .          .   *     .         .          .         .
     1.000E+03  -1.847E+01  .          . *       .         .          .         .
                     -  -  -  -  -  -  -  -  -  -  -  -  -  -  -  -  -  -  -  -  -
```

FIGURE 16.25   SPICE plot for Example 16.12.

```
EXAMPLE 16.12
*
VIN      1     0     AC       1
R1       1     2     5K
C1       2     0     20U
C2       2     3     10U
R2       3     4     1K
VD       0     4     AC       0
F1       0     5     VD       10K
R3       5     0     10
C3       5     6     100U
RL       6     0     30
.AC      DEC   10    1        1K
.PLOT    AC    VDB(6)
.END
```

EXERCISES

16.5.1. Find the admittance matrix Y for this circuit. Note that there are no storage elements, so Y is independent of frequency.

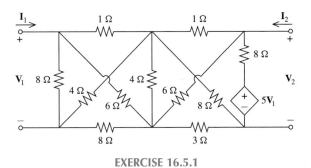

EXERCISE 16.5.1

$$\text{Answer } y = \begin{bmatrix} 0.353 & -0.368 \\ -0.641 & 1.05 \end{bmatrix}$$

16.5.2. Find the frequency f at which the magnitude of the look-in impedance is minimized.

Answer 79.2 Hz

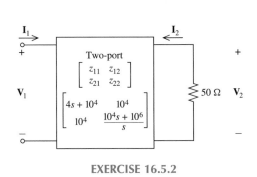

EXERCISE 16.5.2

SUMMARY

The current and voltage at any pair of wires, or input port, affect those at a second pair, or output port, with which it is connected electrically or magnetically. Two-port circuit representations are sets of equations that show the input–output relationships at the two ports.

- A circuit with two pairs of accessible terminals, each satisfying the port condition, is called a two-port circuit. Such circuits are described by pairs of equations involving the four port variables.

- There are six distinct sets of two-port parameters: impedance or z-parameter, admittance or y-parameter, hybrid, inverse hybrid, transmission, and inverse transmission. They each result by specific choice of two of the four port variables to act as independent variables in the two-port equations.

- The transmission matrices of two-ports in cascade multiply, the admittance matrices of two-ports in parallel add, and the impedance matrices of two-ports in series add.

- Placing an ideal transformer in either port of a two-port guarantees that the port condition is satisfied.

PROBLEMS

16.1. Find the port variables i_1, i_2, v_1, v_2.

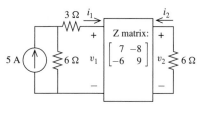

FIGURE P16.1

16.2. Find the transmission and inverse transmission parameters.

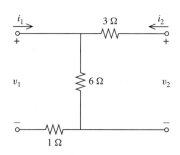

FIGURE P16.2

16.3. The y-parameters for a certain two-port are

$$\begin{bmatrix} y_{11} & y_{12} \\ y_{21} & y_{22} \end{bmatrix} = \begin{bmatrix} 1 + \frac{1}{s} & -\frac{2}{s} \\ -\frac{2}{s} & 3 + \frac{3}{s} \end{bmatrix}$$

Find the hybrid parameters by using Table 16.2.

16.4. Find the y-parameters.

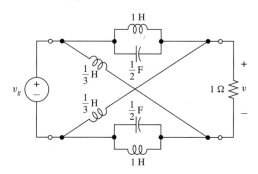

FIGURE P16.4

16.5. Show that the hybrid parameters may be obtained from the z-parameters by

$$g_{11} = \frac{1}{z_{11}}, \qquad g_{12} = -\frac{z_{12}}{z_{11}}$$

$$g_{21} = \frac{z_{21}}{z_{11}}, \qquad g_{22} = \frac{\Delta_z}{z_{11}}$$

16.6. Find the h- and g-parameters.

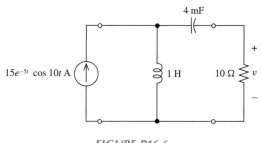

FIGURE P16.6

16.7. Show that the transmission parameters, **A**, **B**, **C**, and **D**, where

$$\mathbf{V}_1 = \mathbf{A}\mathbf{V}_2 - \mathbf{B}\mathbf{I}_2$$

$$\mathbf{I}_1 = \mathbf{C}\mathbf{V}_2 - \mathbf{D}\mathbf{I}_2$$

are given by

$$\mathbf{A} = \frac{\mathbf{z}_{11}}{\mathbf{z}_{21}}, \qquad \mathbf{B} = \frac{\Delta_z}{\mathbf{z}_{21}}$$

$$\mathbf{C} = \frac{1}{\mathbf{z}_{21}}, \qquad \mathbf{D} = \frac{\mathbf{z}_{22}}{\mathbf{z}_{21}}$$

16.8. Find the transmission parameters of the two-port network in Fig. P16.6.

16.9. Find the y-parameters of the network shown. Terminate the output port with a 1-Ω resistor, and find the resulting network function $\mathbf{V}_2/\mathbf{V}_1$.

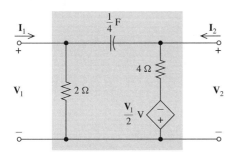

FIGURE P16.9

16.10. Let $\mathbf{h}_{11} = 1 \text{ k}\Omega$, $\mathbf{h}_{12} = 10^{-4}$, $\mathbf{h}_{21} = 100$, and $\mathbf{h}_{22} = 10^{-4}$ S in Fig. 16.11(b) and find the network function $\mathbf{V}_2/\mathbf{V}_1$ if port 2 is open-circuited.

16.11. Find the z-parameters for this two-port circuit.

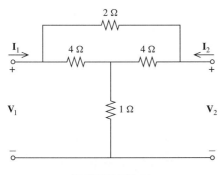

FIGURE P16.11

16.12. Find the z-parameters for this two-port circuit.

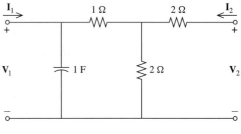

FIGURE P16.12

16.13. A circuit has admittance description

$$\begin{bmatrix} \mathbf{I}_1(s) \\ \mathbf{I}_2(s) \end{bmatrix} = \begin{bmatrix} s+1 & -s \\ -s & s+2+\frac{1}{s} \end{bmatrix} \begin{bmatrix} \mathbf{V}_1(s) \\ \mathbf{V}_2(s) \end{bmatrix}$$

Find the s-domain port variables $\mathbf{I}_1(s)$, $\mathbf{I}_2(s)$, $\mathbf{V}_1(s)$, $\mathbf{V}_2(s)$ if $\mathbf{V}_1(s) = 1/s$ and the output port is terminated by a parallel RLC circuit with $R = 2 \ \Omega$, $L = 4$ H, and $C = 1$ F.

16.14. Find the z-parameters for this two-port. Repeat for the n-parameters (do not use Table 16.2).

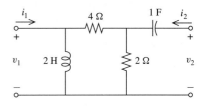

FIGURE P16.14

16.15. Find the inverse hybrid and inverse transmission parameters for the circuit of Problem 16.14. (Do not use Table 16.2.)

16.16. Find all six of the two-port parameter descriptions for the circuit of Problem 16.11, computing the g-parameter descriptions from two experiments and the rest from Table 16.2.

16.17. Find the z-parameters, and find the impedance looking into the input port with the output port open-circuited $\mathbf{V}_1/\mathbf{I}_1|_{\mathbf{I}_2=0}$ and the impedance looking into the output port with the input open-circuited $\mathbf{V}_2/\mathbf{I}_2|_{i_1=0}$.

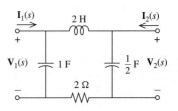

FIGURE P16.17

16.18. Determine the transmission parameters, and the transfer function $\mathbf{I}_1/\mathbf{I}_2|_{\mathbf{V}_1=0}$.

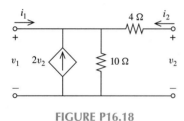

FIGURE P16.18

16.19. Determine a two-port's open-circuit voltage gain transfer function, defined as $\mathbf{H}(s) = \mathbf{V}_2(s)/\mathbf{V}_1(s)|_{I_2(s)=0}$, in terms of its z-parameters $\mathbf{z}_{11}(s)$, $\mathbf{z}_{12}(s)$, $\mathbf{z}_{21}(s)$, $\mathbf{z}_{22}(s)$. Repeat for the hybrid parameters.

16.20. Find the transconductance

$$\mathbf{H}(s) = \mathbf{I}_2(s)/\mathbf{V}_1(s)|_{\mathbf{I}_1(s)=0}$$

for a general two-port in terms of its inverse hybrid parameters and in terms of its transmission parameters.

16.21. Sketch a circuit for which the hybrid parameters do not exist, but all other parameters exist. Compute the values of all other 5 two-port descriptions for your circuit.

16.22. The symmetrical lattice is terminated in 1 Ω and $\mathbf{Z}_a$ and $\mathbf{Z}_b$ are as shown. Find the voltage ratio transfer function.

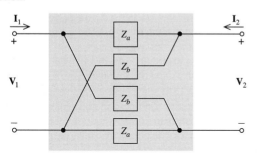

FIGURE P16.22

16.23. A certain two-port has y-parameter descriptions

$$\begin{bmatrix} y_{11} & y_{12} \\ y_{21} & y_{22} \end{bmatrix} = \begin{bmatrix} s+1 & -s \\ -1 & s+2 \end{bmatrix}$$

Sketch a three-terminal model which includes no controlled current sources.

16.24. Find a model which has no internal loops for the circuit of a Problem 16.12.

16.25. Find a model which has no internal nodes (nodes not accessible at the input or output ports) for the circuit of Problem 16.11.

16.26. Show that the given circuit is equivalent to the general two-port network. Note how it differs from the equivalent circuit of Fig. 16.13.

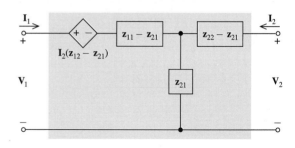

FIGURE P16.26

16.27. Show that the given circuits are equivalent to the general two-port network. Note how the two circuits differ.

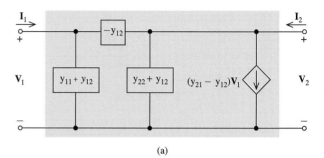

(a)

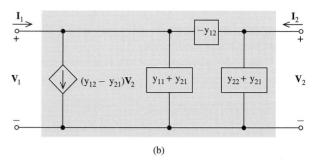

(b)

FIGURE P16.27

16.28. A circuit with inverse hybrid description

$$\begin{bmatrix} \mathbf{I}_1 \\ \mathbf{V}_2 \end{bmatrix} = \frac{1}{(2s+3)} \begin{bmatrix} 1 & -(s+1) \\ (s+1) & s^2+7s+8 \end{bmatrix} \begin{bmatrix} \mathbf{V}_1 \\ \mathbf{I}_2 \end{bmatrix}$$

is terminated by a 3-Ω load resistor across the output port. What is its admittance $Y(j\omega)$ at $\omega = 10$ rad/s?

16.29. Find the hybrid parameters of the two-port consisting of two copies of the circuit of Problem 16.11 in parallel.

16.30. Find the y-parameters of the two-port consisting of two copies of the circuit of Problem 16.11 in cascade.

16.31. A circuit has transmission matrix

$$\begin{bmatrix} V_1(s) \\ I_1(s) \end{bmatrix} = \begin{bmatrix} 12 & 1 \\ -3 & 2s \end{bmatrix} \begin{bmatrix} V_2(s) \\ -I_2(s) \end{bmatrix}$$

Find the four port variables $V_1(s)$, $I_1(s)$, $V_2(s)$, $I_2(s)$ if both ports are terminated with copies of the one-port shown, aligned so that terminal a is connected to the plus ends of both ports.

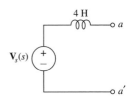

FIGURE P16.31

☐ **16.32.** Find the z-parameters of three copies of the circuit of Problem 16.18 in cascade.

16.33. Find the g-parameters of three copies of the circuit of Problem 16.18 in parallel.

☐ **16.34.** Find the inverse transmission parameters of three copies of the circuit of Problem 16.18 in series.

☐ **16.35.** Find the y-parameter description.

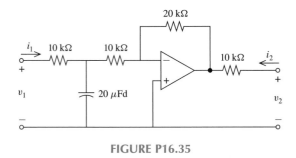

FIGURE P16.35

SPICE Problems

16.36. Find the z-parameters $\mathbf{z}_{11}(j\omega)$, $\mathbf{z}_{12}(j\omega)$, $\mathbf{z}_{21}(j\omega)$, and $\mathbf{z}_{22}(j\omega)$ at $\omega = 1$ and $\omega = 100$ rad/s. Use the ideal

voltage amplifier model for the op amp with open loop gain 10^5.

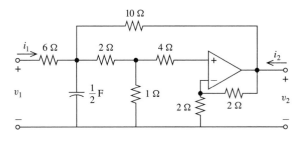

FIGURE P16.36

16.37. Determine the y-parameters $\mathbf{y}_{11}(j\omega)$, $\mathbf{y}_{12}(j\omega)$, $\mathbf{y}_{21}(j\omega)$, $\mathbf{y}_{22}(j\omega)$ at $\omega = 377$ rad/s if 10 copies of the circuit of Problem 16.39 are put in cascade. Solve using SPICE and subcircuits .SUBCKT.

16.38. Find the ac steady-state value of $v_2(t)$ using SPICE.

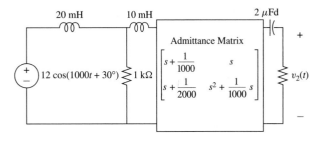

FIGURE P16.38

More Challenging Problems

16.39. Find the y-parameters for this circuit. Repeat for the transmission parameters (do not use Table 16.2).

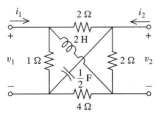

FIGURE P16.39

16.40. Find the hybrid parameters for this two-port with a–a' the input port and b–b' the output port. Then find $v_{aa'} = v_1$, $v_{bb'} = v_2$ in terms of $\mathbf{I}_g(s)$.

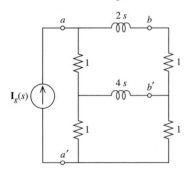

FIGURE P16.40

16.41. Find the h-matrix for the two-port.

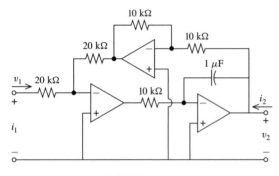

FIGURE P16.41

16.42. Show that for the circuit of Problem 16.4, in terms of the transmission parameters we have

$$\frac{V_2}{V_1} = \frac{1}{A + B}$$

16.43. The circuit of Problem 16.22 (no termination) is a *lattice* with *series arms* both equal to $\mathbf{Z}_a$ and *cross arms*

both equal to $\mathbf{Z}_b$. It is called a *symmetrical* lattice because the series arms are equal and the cross arms are equal. Show that the z- and y-parameters are given by

$$\mathbf{z}_{11} = \mathbf{z}_{22} = \frac{1}{2}(\mathbf{Z}_b + \mathbf{Z}_a)$$

$$\mathbf{z}_{12} = \mathbf{z}_{21} = \frac{1}{2}(\mathbf{Z}_b - \mathbf{Z}_a)$$

and

$$\mathbf{y}_{11} = \mathbf{y}_{22} = \frac{1}{2}(\mathbf{Y}_b + \mathbf{Y}_a)$$

$$\mathbf{y}_{12} = \mathbf{y}_{21} = \frac{1}{2}(\mathbf{Y}_b - \mathbf{Y}_a)$$

where $\mathbf{Y}_a = 1/\mathbf{Z}_a$ and $\mathbf{Y}_b = 1/\mathbf{Z}_b$.

16.44. Find the hybrid and inverse hybrid descriptions for this circuit.

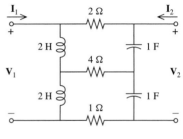

FIGURE P16.44

16.45. Use superposition to show that if a two-port contains independent sources, its two-port equations will have an extra source term. For instance, the impedance description is

$$\begin{bmatrix} \mathbf{V}_1 \\ \mathbf{V}_2 \end{bmatrix} = \begin{bmatrix} \mathbf{Z}_{11} & \mathbf{Z}_{12} \\ \mathbf{Z}_{21} & \mathbf{Z}_{21} \end{bmatrix} \begin{bmatrix} \mathbf{I}_1 \\ \mathbf{I}_2 \end{bmatrix} + \begin{bmatrix} \mathbf{V}_{1s} \\ \mathbf{V}_{2s} \end{bmatrix}$$

How would the new term $\begin{bmatrix} \mathbf{V}_{1s} \\ \mathbf{V}_{2s} \end{bmatrix}$ be found?

17

Fourier Series and Transform

Jean Baptiste Joseph Fourier
1768–1830

A great mathematical poem.

Lord Kelvin on the Fourier Series

In 1822 a greatly influential pioneer work on the mathematical theory of heat conduction was published by the great French mathematician, Egyptologist, and administrator Jean Baptiste Joseph Fourier. It was a masterpiece not only because of the new field of heat conduction that it explored, but also because of the infinite series of sinusoids that it developed; the latter became famous as the Fourier series. With the Fourier series, we are no longer restricted in the shortcut phasor methods to circuits whose inputs are sinusoids.

Fourier was born in Auxerre, France, the son of a tailor. He attended a local military school conducted by Benedictine monks and showed such proficiency in mathematics that he later became a mathematics teacher in the school. Like most Frenchmen his age he was swept into the politics of the French Revolution and its aftermath and more than once came near to losing his life. He was one of the first teachers in the newly formed Ecole Polytechnique and later became its professor of mathematical analysis. At age 30 Fourier was appointed scientific advisor by Napoleon on an expedition to Egypt and for four years was secretary of the Institute d'Egypte, the work of which marked Egyptology as a separate discipline. He was prefect of the department of Isere from 1801 to 1814, where he wrote his famous treatise on heat conduction. He completed a book on algebraic equations just before his death in 1830.

Chapter Contents

Almost every mathematical function of practical interest can be expressed as a superposition of sinusoids. This insight, bequeathed to us by the great French mathematician and physicist Jean Baptiste Joseph Fourier (1768–1830), has important implications for circuit analysis. Since linear circuits satisfy the superposition principle, it means that we can write down the response to any input if we know the response merely to sinusoids. The response to sinusoids is the frequency response, which was introduced in Chapter 14. In the clear light of Fourier's observation, the idea of frequency response takes on expanded meaning. It is a tool for predicting the response of a linear circuit to any input whatsoever, not just sinusoidal inputs. The ability to express an arbitrary signal as a superposition of sinusoids also implies that phasors are a far more general tool than we may have suspected thus far. Upon Fourier's expansion of an arbitrary input into sinusoids, phasor analysis may be applied to each component sinusoid in the circuit. Superposition may then be invoked to determine the overall response. Thus in linear circuits phasors may be used quite generally to determine outputs driven by nonsinusoidal inputs as well as sinusoidal ones.

The form taken by the superposition into sinusoidal components hinges on whether the signal is periodic. For periodic signals, we will get a straightforward sum of sinusoids called a *Fourier series*. Otherwise, the superposition takes the form of an integral rather than a sum, and the resulting representation is called a *Fourier transform*. We first discuss some of the basic properties of periodic signals and then the process by which the Fourier series for periodic functions may be found. Two forms for the Fourier series are introduced, the trigonometric and complex exponential forms, which are closely related through the Euler identity. We use this result and the principle of superposition to find the response of a linear circuit to any periodic signal. Next we turn to nonperiodic signals, defining the Fourier transform and showing its relation to both nonsinusoidal signals and to the Laplace transform introduced in Chapter 12. Some useful properties of the Fourier transform are developed, and the chapter concludes with a brief discussion of SPICE and the Fourier domain.

As first noted in our study of the properties of sinusoids in Chapter 8, a function $f(t)$ is said to be *periodic* if there is a real constant $T > 0$ such that

$$f(t + T) = f(t) \qquad \text{for all } -\infty < t < +\infty \tag{17.1}$$

The smallest T satisfying (17.1) is called the *period* of $f(t)$ and has units of time. Equation (17.1) implies that

$$f(t + nT) = f(t) \quad \text{for all } -\infty < t < +\infty, \text{ all integers } n \tag{17.2}$$

The period T is the smallest amount we may shift the graph of $f(t)$ and have it coincide with its unshifted form, while (17.2) identifies other values of shift that have the same effect: $2T$, $3T$, $-T$, and so on.

Periodic functions are completely specified by their values over any period, as can be seen from (17.2). Let

$$f_T(t) = f(t)[u(t - t_0) - u(t - t_0 - T)] \tag{17.3}$$

be a function that may be nonzero only over a finite interval of duration T starting at t_0. Then, adding together shifted copies of $f_T(t)$, we define

$$f(t) = \sum_{n=-\infty}^{\infty} f_T(t + nT) \tag{17.4}$$

as the *periodic extension* of $f_T(t)$. $f_T(t)$ is, in turn, called the *generator* of its periodic extension $f(t)$. t_0 and T are the *initial time* and *duration* of the generator $f_T(t)$. $f(t)$ is clearly periodic, since by (17.4)

$$f(t + mT) = \sum_{n=-\infty}^{\infty} f_T\left(t + (n + m)T\right)$$

and with $i = n + m$,

$$= \sum_{i=-\infty}^{\infty} f_T(t + iT) = f(t)$$

While this verifies that $f(t)$ is periodic, note that the duration T of the generator $f_T(t)$ may not always coincide with the period of $f(t)$, since there may be a smaller number satisfying (17.1) for $f(t)$. For instance, if $f(t)$ is a periodic function of period T and we consider any $f_T(t)$ in (17.3) with duration $2T$, then its periodic extension $f(t)$ just reproduces $f(t)$ itself. The period of $f(t)$ is still T, not the duration $2T$ of its generator $f_T(t)$. These ideas are illustrated in Example 17.1.

Example 17.1

Consider the function $f(t) = e^{-|t|}$. It is clearly not a periodic function. The periodic extensions of two generator functions of different duration and initial time, both taken from $f(t)$, are shown in Fig. 17.1. One period of the periodic extension has been tinted green to aid visualization.

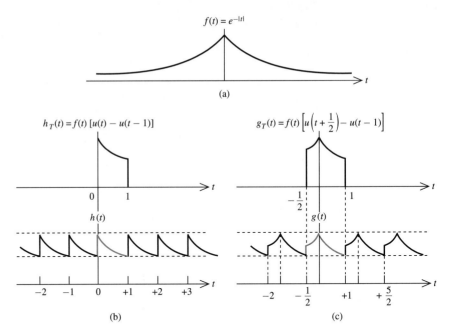

$f(t) = e^{-|t|}$

(a)

$h_T(t) = f(t)\,[u(t) - u(t-1)]$

$g_T(t) = f(t)\left[u\left(t + \dfrac{1}{2}\right) - u(t-1)\right]$

$h(t)$

$g(t)$

(b)

(c)

FIGURE 17.1 (a) $f(t)$; (b) generator $h_T(t)$ and its periodic extension; (c) different generator $g_T(t)$ and its periodic extension.

The minimum shift of the function $h(t)$ in Fig. 17.1(b) for which the graph coincides with itself is clearly $T = 1$, and that of Fig. 17.1(c) is $T = \frac{3}{2}$. Consider next a generator $h'_T(t)$ of duration 3 taken from the function $h(t)$ of Fig. 17.1(b) starting at $t = 0$:

$$h'_T(t) = h(t)[u(t) - u(t-3)]$$

This generator contains exactly three periods of $h(t)$. Shifting by 3 and adding as in (17.4), we see that the periodic extension $h'(t)$ of $h'_T(t)$ simply replicates $h(t)$ three periods at a time. So $h'(t) = h(t)$, and the period of this function is $T = 1$, as noted above. This is smaller than the duration 3 of the generator $h'_T(t)$.

Periodic functions may possess *symmetry properties* that facilitate their analysis. *A function $f(-t) = f(t)$ for all t is said to be an even function, and $f(t) = -f(-t)$ for all t is called an odd function.* Periodic functions need only be tested for evenness or oddness over the symmetrically located single period $-T/2 \leq t < +T/2$. Given any $t' = t + nT$ with $-T/2 \leq t < +T/2$, then by periodicity $f(t') = f(t)$, and if $f(t) = f(-t)$, then $f(-t') = f(-t - nT) = f(-t) = f(t')$ and the function is even. A similar argument holds for $f(t)$ periodic and odd.

The cosine function $A \cos \omega t$ is even since $A \cos \omega(-t) = A \cos \omega t$, and the sine function is odd, since $A \sin \omega(-t) = -A \sin \omega t$. The complex exponential function $e^{j\omega t}$ is composed of the sum of an even real part, $\cos \omega t$, and an odd imaginary part, $\sin \omega t$.

If $f_e(t)$ and $g_e(t)$ are both even functions, the general linear combination of them,

$$h(t) = K_1 f_e(t) + K_2 g_e(t)$$

is also even, since

$$h(-t) = K_1 f_e(-t) + K_2 g_e(-t)$$

$$= K_1 f_e(t) + K_2 g_e(t) = h(t)$$

while by similar reasoning any linear combination of odd functions is odd. Thus the *sums and differences of even functions are even and of odd functions are odd.* Products of even and odd functions inherit symmetry properties in the same fashion as products of even and odd integers do. For $f_e(t)$ and $g_e(t)$ even,

$$h(t) = f_e(t) g_e(t)$$

implies that

$$h(-t) = f_e(-t) g_e(-t) = f_e(t) g_e(t) = h(t)$$

so $h(t)$ is even. The product of an even function $f_e(t)$ and an odd function $g_o(t)$ is

$$h(t) = f_e(t) g_o(t)$$

so

$$h(-t) = f_e(-t) g_o(-t) = -f_e(t) g_o(t) = -h(t)$$

and the product is odd. *The product of even functions is even, of odd functions is also even, and of an even and an odd function is odd.*

While arbitrary functions are neither even nor odd, they can be rewritten as the sum of two components each possessing symmetry. *Any $f(t)$ may be expressed as the sum of an even function and an odd function.* Define

$$f_e(t) = \frac{f(t) + f(-t)}{2} \tag{17.5a}$$

$$f_o(t) = \frac{f(t) - f(-t)}{2} \tag{17.5b}$$

as the *even part* $f_e(t)$ and the *odd part* $f_o(t)$ of $f(t)$. Clearly, $f_e(t) = f_e(-t)$ and $f_o(t) = -f_o(-t)$ as required of even and odd functions. Adding (17.5a) and (17.5b) gives

$$f_e(t) + f_o(t) = f(t)$$

verifies that their sum is $f(t)$.

Example 17.2

Find the even and odd parts of the periodic function $f(t) = 2\cos(3t + 30°)$. Because of the 30° phase shift, this function has neither even nor odd symmetry. By (17.5a) its even part is

$$f_e(t) = \tfrac{1}{2}[2\cos(3t + 30°) + 2\cos(-3t + 30°)]$$

$$= \cos(3t + 30°) + \cos(3t - 30°)$$

Using the identities for cosine of sum and difference angles,

$$= \left(\sqrt{\tfrac{3}{2}} \cos 3t - \tfrac{1}{2} \sin 3t \right) + \left(\sqrt{\tfrac{3}{2}} \cos 3t + \tfrac{1}{2} \sin 3t \right) \tag{17.6}$$

$$f_e(t) = \sqrt{3} \cos 3t$$

and, by (17.5b),

$$f_o(t) = \tfrac{1}{2}[2\cos(3t + 30°) - 2\cos(-3t + 30°)]$$

$$= \left(\sqrt{\tfrac{3}{2}}\cos 3t - \tfrac{1}{2}\sin 3t\right) - \left(\sqrt{\tfrac{3}{2}}\cos 3t + \tfrac{1}{2}\sin 3t\right)$$

or $\quad f_o(t) = -\sin 3t$ $\hfill$ (17.7)

$f_e(t)$ and $f_o(t)$ are clearly even and odd, respectively. The result $f(t) = f_e(t) + f_o(t)$ with $f_e(t)$ in (17.6) and $f_o(t)$ in (17.7) in fact follows directly from the identity for the cosine of a sum angle (these trigonometric identities are tabulated on the inside front cover). *The even part of a sinusoid is a cosine function, and the odd part is a sine function.*

Finally, we recall certain averages of periodic functions first defined in Chapter 7 in the context of ac steady state. Functions of period T possess *mean* or *average value*

$$f_{av} = \frac{1}{T}\int_{t_0}^{t_0+T} f(t)\,dt \hfill (17.8)$$

and *mean square value*

$$|f|_{av}^2 = \frac{1}{T}\int_{t_0}^{t_0+T} |f(t)|^2\,dt \hfill (17.9)$$

where the integrals range over one period T, and t_0 is arbitrary. The mean square value is also called the *average power* in $f(t)$. The square root of the mean square value is the rms value of $f(t)$:

$$f_{rms} = \sqrt{|f|_{av}^2} = \sqrt{\frac{1}{T}\int_{t_0}^{t_0+T} |f(t)|^2\,dt} \hfill (17.10)$$

For instance, the average value of the function $f(t)$ shown in Fig. 17.1(b) is

$$f_{av} = \int_0^1 e^{-t}\,dt = 1 - e^{-1}$$

and the mean square and rms values are

$$|f|_{av}^2 = \int_0^1 e^{-2t}\,dt = \tfrac{1}{2}(1 - e^{-2})$$

$$f_{rms} = \sqrt{|f|_{av}^2} = \sqrt{\frac{e^2 - 1}{2e^2}}$$

The calculation of these averages is simplified somewhat for even or odd functions. For an even function $f_e(t)$,

$$f_{e\ av} = \frac{1}{T}\int_{-T/2}^{T/2} f_e(t)\,dt = \frac{1}{T}\int_{-T/2}^{0} f_e(t)\,dt + \frac{1}{T}\int_0^{T/2} f_e(t)\,dt$$

Changing the variable of integration from t to $-x$ in the first integral on the far right, we have

$$\frac{1}{T} \int_{-T/2}^{T/2} f_e(t)\, dt = \frac{-1}{T} \int_{T/2}^{0} f_e(-x)\, dx + \frac{1}{T} \int_{0}^{T/2} f_e(t)\, dt = \frac{2}{T} \int_{0}^{T/2} f_e(t)\, dt \qquad (17.11)$$

Thus *the average value of an even periodic function is its average over the half-period [0, T/2].* If $f(t)$ is odd, the two integrals in the sum above cancel,

$$\frac{1}{T} \int_{-T/2}^{T/2} f_o(t)\, dt = \frac{-1}{T} \int_{T/2}^{0} f_o(-x)\, dx + \frac{1}{T} \int_{0}^{T/2} f_o(t)\, dt = 0 \qquad (17.12)$$

and *the average value of an odd periodic function is zero.*

Each of these averages is of interest in the decomposition of periodic functions into their sinusoidal components, to which we turn next.

EXERCISES

17.1.1. Show that the general linear combination of odd functions is odd and that the product of odd functions is even.

17.1.2. Find and sketch the even and odd parts of $h(t)$ shown in Fig. 17.1(b).
Answer For $0 < t < 1$, $f_e(t) = \frac{1}{2}(e^{-t} + e^{t-1})$, $f_o(t) = \frac{1}{2}(e^{-t} - e^{t-1})$.

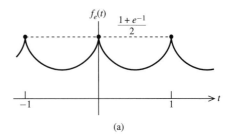

(a)

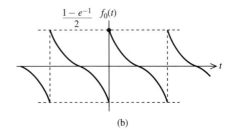

(b)

EXERCISE 17.1.2

17.1.3. Is $f(t) = \sin t + \cos \omega t$ periodic for $\omega = \frac{2}{3}$? If so, what is the period of $f(t)$? Repeat for $\omega = \sqrt{2}$.
Answer Yes; $T = 6\pi$; no

17.2   TRIGONOMETRIC FOURIER SERIES

Our earlier work on ac steady state emphasized the fact that many practical sources are sinusoidal. By no means, however, are all periodic source functions that are of practical interest sinusoidal. Consider, for instance, the four periodic functions shown in Fig. 17.2. The first, or sawtooth, wave is used to sweep a beam of electrons across the face of a cathode ray tube to generate an image. Its generator is a ramp with initial time $t_0 = 0$ and duration T. The second, fully rectified sine wave, is common in communications

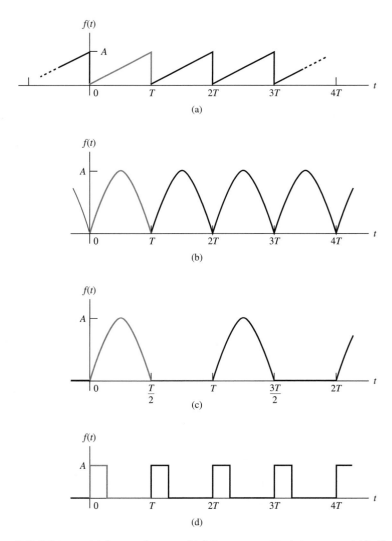

FIGURE 17.2   (a) Sawtooth wave; (b) full-wave rectified sine wave; (c) half-wave rectified sine wave; (d) square wave. One period has been tinted green.

receivers and in power supplies. Its generator is a half-period from a sine wave. The next, or half-rectified sine wave, is also found in many power supplies, and the square wave in the clock circuits of digital computers. None of these periodic functions is sinusoidal. Here we consider how functions such as these may be expressed as sums of sinusoids.

As Fourier showed, if a periodic function $f(t)$ with period T satisfies a set of rather general conditions, it may be represented by the infinite series of sinusoids

$$f(t) = \frac{a_0}{2} + a_1 \cos \omega_0 t + a_2 \cos 2\omega_0 t + \cdots$$
$$+ b_1 \sin \omega_0 t + b_2 \sin 2\omega_0 t + \cdots$$

or, more compactly,

$$f(t) = \frac{a_0}{2} + \sum_{n=1}^{\infty} (a_n \cos n\omega_0 t + b_n \sin n\omega_0 t) \qquad (17.13)$$

where $\omega_0 = 2\pi/T$. This series is called the *trigonometric Fourier series*, or simply the *Fourier series*, of $f(t)$. The a_n's and b_n's are called the *Fourier coefficients* and depend on $f(t)$.

The Fourier coefficients may be determined rather easily. Let us begin by obtaining a_0, which may be done by integrating both sides of (17.13) over a full period; that is,

$$\int_0^T f(t)\, dt = \int_0^T \frac{a_0}{2}\, dt$$
$$+ \sum_{n=1}^{\infty} \int_0^T (a_n \cos n\omega_0 t + b_n \sin n\omega_0 t)\, dt$$

Each term in the last summation is an integral of a sinusoid over n of its periods $T_n = T/n$. Since the average value of a sinusoid over a period is zero, the summation is zero, and therefore we have simply

$$a_0 = \frac{2}{T} \int_0^T f(t)\, dt \qquad (17.14)$$

Next, let us multiply (17.13) through by $\cos m\omega_0 t$, where m is a nonzero integer, and integrate. This yields

$$\int_0^T f(t) \cos m\omega_0 t\, dt = \int_0^T \frac{a_0}{2} \cos m\omega_0 t\, dt$$
$$+ \sum_{n=1}^{\infty} a_n \int_0^T \cos m\omega_0 t \cos n\omega_0 t\, dt \qquad (17.15)$$
$$+ \sum_{n=1}^{\infty} b_n \int_0^T \cos m\omega_0 t \sin n\omega_0 t\, dt$$

Using trigonometric identities found on the front cover, we have

$$\cos m\omega_0 t \cos n\omega_0 t = \tfrac{1}{2}\left(\cos(n+m)\omega_0 t + \cos(n-m)\omega_0 t\right)$$

$$\cos m\omega_0 t \sin n\omega_0 t = \tfrac{1}{2}\left(\sin(n+m)\omega_0 t + \sin(n-m)\omega_0 t\right)$$

Applying these to (17.15), each term in the sum is once again zero, since it is the integral of a sinusoid over an integral of periods, except the term where $n = m$ in the first summation. This term is given by

$$a_m \int_0^T \cos^2 m\omega_0 t \, dt = \frac{\pi}{\omega_0} a_m = \frac{T}{2} a_m$$

so that

$$a_m = \frac{2}{T} \int_0^T f(t) \cos m\omega_0 t \, dt, \qquad m = 1, 2, 3, \ldots \qquad (17.16a)$$

Finally, multiplying (17.13) by $\sin m\omega_0 t$, integrating, and applying the same argument, we have

$$b_m = \frac{2}{T} \int_0^T f(t) \sin m\omega_0 t \, dt, \qquad m = 1, 2, 3, \ldots \qquad (17.16b)$$

We note that (17.14) is the special case, $m = 0$, of (17.16a) (which is why we used $a_0/2$ instead of a_0 for the constant term). Also, since $f(t)$ and $\cos m\omega_0 t$ have periods of T and T/m respectively, we may integrate their product over any interval of length T, such as t_0 to $t_0 + T$, for arbitrary t_0, and the results will be the same. The same argument applies to $f(t)$ and $\sin m\omega_0 t$ as in (17.16b). Therefore, we may summarize by giving the Fourier coefficients in the form

$$\boxed{\begin{aligned} a_n &= \frac{2}{T} \int_{t_0}^{t_0+T} f(t) \cos n\omega_0 t \, dt, &\quad n = 0, 1, 2, \ldots \\[2mm] b_n &= \frac{2}{T} \int_{t_0}^{t_0+T} f(t) \sin n\omega_0 t \, dt, &\quad n = 1, 2, 3, \ldots \end{aligned}} \qquad (17.17)$$

We have replaced the subscript m by n to correspond to the notation of (17.13).

The term $a_n \cos n\omega_0 t + b_n \sin n\omega_0 t$ in (17.13) is sometimes called the nth *harmonic*, by analogy with music theory. The case $n = 1$ is the first harmonic, or *fundamental*, with fundamental frequency ω_0. The case $n = 2$ is the second harmonic with frequency $2\omega_0$, and so on. Note that frequency $2\omega_0$ of the second harmonic is one *octave* (a factor of 2) above the fundamental, and the fourth harmonic is an octave above the second and two octaves above the fundamental. The term $a_0/2$ is the constant, or dc, component and by (17.14) may be seen to be the average value of $f(t)$ over a period. It may be found quite often by inspection of the graph of $f(t)$.

The conditions that (17.13) is the Fourier series representing $f(t)$, where the Fourier coefficients are given by (17.17), are, as we have said, quite general and hold for almost any function we are likely to encounter in engineering. Sufficient conditions called the *Dirichlet conditions* are:

1. $f(t)$ is piecewise continuous.
2. $f(t)$ has isolated maxima and minima.
3. $f(t)$ is absolutely integrable over a period,

$$\int_0^T |f(t)| \, dt < \infty \tag{17.18}$$

Recall that piecewise continuity was defined in Section 12.1, and isolated maxima and minima means every finite time interval contains only a finite number of such points. Note that in the Dirichlet conditions for Laplace transformability in Section 12.1 condition 3 is replaced by $f(t)$ being of exponential order. Functions satisfying the Dirichlet conditions for Fourier series (17.18) are equal everywhere to their Fourier series (17.13), (17.17), except perhaps at points of discontinuity $t = \tau$ where the series converges to the average value $\frac{1}{2}[f(\tau^-) + f(\tau^+)]$ on either side of the discontinuity.

Example 17.3

As an example, let us find the Fourier series for the sawtooth wave of Fig. 17.3, given by

$$f(t) = t, \qquad -\pi < t < \pi$$
$$f(t + 2\pi) = f(t) \tag{17.19}$$

Since $T = 2\pi$, we have $\omega_0 = 2\pi/T = 1$. If we choose $t_0 = -\pi$, the first equation of (17.17) for $n = 0$ yields

$$a_0 = \frac{1}{\pi} \int_{-\pi}^{\pi} t \, dt = 0$$

For $n = 1, 2, 3, \ldots$, we have

$$a_n = \frac{1}{\pi} \int_{-\pi}^{\pi} t \cos nt \, dt$$

$$= \frac{1}{n^2 \pi} (\cos nt + nt \sin nt) \Big|_{-\pi}^{\pi}$$

$$= 0$$

$$b_n = \frac{1}{\pi} \int_{-\pi}^{\pi} t \sin nt \, dt$$

$$= \frac{1}{n^2 \pi} (\sin nt + nt \cos nt) \Big|_{-\pi}^{\pi}$$

$$= -\frac{2 \cos n\pi}{n}$$

$$= \frac{2(-1)^{n+1}}{n}$$

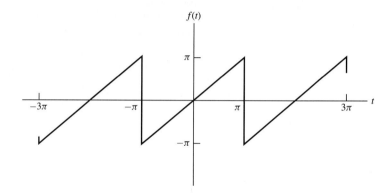

FIGURE 17.3 Sawtooth wave.

The case $n = 0$ had to be considered separately because of the appearance of n^2 in the denominator in the general case. Also, since $a_0/2$ is the average value of the sawtooth wave over a period, by inspection of Fig. 17.3 we see that $a_0 = 0$.

From our results, the Fourier series for the sawtooth wave is

$$f(t) = 2\left(\frac{\sin t}{1} - \frac{\sin 2t}{2} + \frac{\sin 3t}{3} - \cdots\right) \qquad (17.20)$$

The fundamental and the second, third, and fifth harmonics are sketched for one period in Fig. 17.4. If a sufficient number of terms in (17.20) are taken, the series can be made to approximate $f(t)$ very nearly. For example, the first 10 harmonics are added and the result sketched in Fig. 17.5.

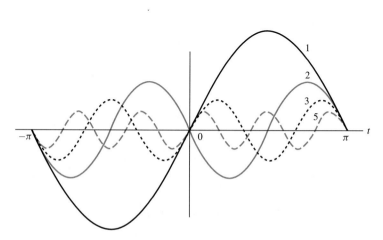

FIGURE 17.4 Four harmonics of (17.20).

Chapter 17   Fourier Series and Transform

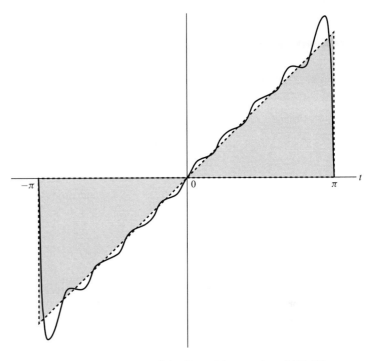

FIGURE 17.5 Sum of the first 10 harmonics of (17.20).

Example 17.4

As another example, suppose that we have the generator function

$$f_T(t) = \begin{cases} 0, & -2 < t < -1 \\ 6, & -1 < t < 1 \\ 0, & 1 < t < 2 \end{cases}$$

with duration $T = 4$. Its periodic extension is $f(t)$, the square wave shown in Fig. 17.6. Evidently, the period of $f(t)$ is $T = 4$ and $\omega_0 = 2\pi/T = \pi/2$. If we take $t_0 = 0$ in (17.17), we must break each integral into three parts since, on the interval from 0 to 4, $f(t)$ has 0, 6, and 0 values. If $t_0 = -1$, we only have to divide the integral into two parts since $f(t) = 6$ on -1 to 1 and $f(t) = 0$ on

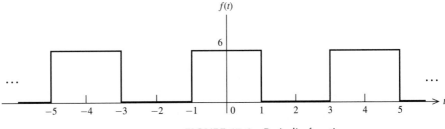

FIGURE 17.6 Periodic function.

1 to 3. Therefore, let us choose this value of t_0 and obtain

$$a_0 = \frac{2}{4} \int_{-1}^{1} 6 \, dt + \frac{2}{4} \int_{1}^{3} 0 \, dt = 6$$

Also, we have

$$a_n = \frac{2}{4} \int_{-1}^{1} 6 \cos \frac{n\pi t}{2} \, dt + \frac{2}{4} \int_{1}^{3} 0 \cos \frac{n\pi t}{2} \, dt$$

$$= \frac{12}{n\pi} \sin \frac{n\pi}{2}$$

and finally,

$$b_n = \frac{2}{4} \int_{-1}^{1} 6 \sin \frac{n\pi t}{2} \, dt + \frac{2}{4} \int_{1}^{3} 0 \sin \frac{n\pi t}{2} \, dt$$

$$= 0$$

Thus the Fourier series is

$$f(t) = 3 + \frac{12}{\pi} \left(\cos \frac{\pi t}{2} - \frac{1}{3} \cos \frac{3\pi t}{2} + \frac{1}{5} \cos \frac{5\pi t}{2} - \cdots \right)$$

Since there are no even harmonics, we may put the result in the more compact form

$$f(t) = 3 + \frac{12}{\pi} \sum_{n=1}^{\infty} \frac{(-1)^{n+1} \cos\{[(2n-1)\pi t]/2\}}{2n-1}$$

Anyone doubting the importance of selecting a good value of t_0 in (17.17) may wish to take $t_0 = 0$. The results will be the same, but not the effort.

Now let us see how calculation of the Fourier coefficients a_n and b_n may be simplified if $f(t)$ has symmetry. In (17.17) we need to integrate the functions

$$g(t) = f(t) \cos n\omega_0 t$$

$$h(t) = f(t) \sin n\omega_0 t$$

If $f(t)$ is even, then since $\cos n\omega_0 t$ is even and $\sin n\omega_0 t$ is odd, $g(t)$ is even and $h(t)$ is odd. Therefore, by (17.11) and (17.12), for $f(t)$ even

$$a_n = \frac{4}{T} \int_{0}^{T/2} f(t) \cos n\omega_0 t \, dt, \quad n = 0, 1, 2, \ldots$$

$$b_n = 0, \qquad\qquad\qquad\qquad\quad n = 1, 2, 3, \ldots$$

(17.21)

Chapter 17 Fourier Series and Transform

and for $f(t)$ odd,

$$a_n = 0, \qquad\qquad n = 0, 1, 2, \ldots$$

$$b_n = \frac{4}{T} \int_0^{T/2} f(t) \sin n\omega_0 t \, dt, \quad n = 1, 2, 3, \ldots$$

(17.22)

In either case one entire set of coefficients is zero, and the other set is obtained by taking twice the integral over half the period.

In summary, *an even function has no sine terms, and an odd function has no constant or cosine terms in its Fourier series.* For instance, the sawtooth function of Example 17.3 is odd, and its Fourier series has no cosine terms, while the square wave of Example 17.4 is even and there are no sine terms in its Fourier series.

Example 17.5

To illustrate the use of symmetry, let us find the Fourier coefficients of the square wave of period $T = 2$ defined by

$$f(t) = \begin{cases} 4, & 0 < t < 1 \\ -4, & 1 < t < 2 \end{cases}$$

Since $T = 2$, $\omega_0 = \pi$. The function is odd and, therefore, by (17.22) we have

$$a_n = 0, \qquad n = 0, 1, 2, \ldots$$

$$b_n = \frac{4}{2} \int_0^1 4 \sin n\pi t \, dt$$

$$= \frac{8}{n\pi}[1 - (-1)^n]$$

Therefore,

$$b_n = \begin{cases} 0, & n \text{ even} \\ \dfrac{16}{n\pi}, & n \text{ odd} \end{cases}$$

EXERCISES

17.2.1. Find the Fourier series representation of the square wave

$$f(t) = \begin{cases} 4, & 0 < t < 1 \\ -4, & 1 < t < 2 \end{cases}$$

$$f(t + 2) = f(t)$$

Answer $\dfrac{16}{\pi} \displaystyle\sum_{n=1}^{\infty} \dfrac{\sin(2n - 1)\pi t}{2n - 1}$

17.2.2. Find the Fourier coefficients for

$$f(t) = \begin{cases} 3, & 0 < t < 1 \\ -1, & 1 < t < 4 \end{cases}$$

$$f(t + 4) = f(t)$$

Answer $a_0 = 0$, $a_n = \dfrac{4}{n\pi} \sin \dfrac{n\pi}{2}$, $b_n = \dfrac{4}{n\pi} \left(1 - \cos \dfrac{n\pi}{2}\right)$, $n = 1, 2, 3, \ldots$

17.2.3. Find the Fourier series for

$$f(t) = \begin{cases} 2, & 0 < t < \dfrac{\pi}{2} \\ 0, & \dfrac{\pi}{2} < t < \pi \end{cases}$$

$$f(t + \pi) = f(t)$$

Answer $1 + \dfrac{4}{\pi} \displaystyle\sum_{n=1}^{\infty} \dfrac{\sin(2n - 1)}{2n - 1}$

17.2.4. Find the Fourier series for the sawtooth wave of Example 17.3 using symmetry properties.

17.2.5. Find the Fourier series for the half-wave rectified sinusoid with generator

$$f_T(t) = 4 \cos 2t \left[u\left(t + \dfrac{\pi}{2}\right) - u\left(t - \dfrac{\pi}{2}\right) \right]$$

of duration π and initial time $t_0 = -\pi/2$ shown in the figure.

Answer $a_0 = \dfrac{8}{\pi}$, $a_1 = 2$, $a_n = \dfrac{8}{\pi(1 - n^2)} \cos \dfrac{n\pi}{2}$, $n = 2, 3, 4, \ldots$; $b_n = 0$, $n = 1, 2, 3, \ldots$

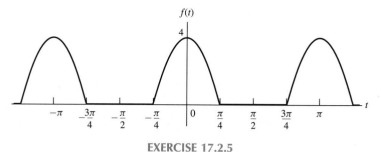

EXERCISE 17.2.5

17.2.6. Find the Fourier series for the full-wave rectified sinusoid

$$f(t) = |4 \sin 2t|$$

shown in the accompanying figure.

Answer $\dfrac{16}{\pi} \left(\dfrac{1}{2} + \displaystyle\sum_{n=1}^{\infty} \dfrac{1}{1 - 4n^2} \cos 4nt \right)$

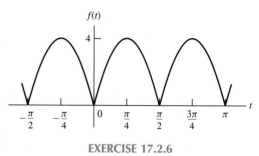

EXERCISE 17.2.6

17.3 THE EXPONENTIAL FOURIER SERIES

We may obtain yet another form of the Fourier series by replacing the sine and cosine functions by their exponential equivalents, using Euler's formula. This form is called the *exponential* Fourier series and is extremely useful, particularly in considering frequency responses, one of the most important applications of Fourier series. In this section we consider the exponential Fourier series, and in the following section we investigate the frequency responses or frequency spectra.

By the Euler identity,

$$\cos n\omega_0 t = \tfrac{1}{2}\left(e^{jn\omega_0 t} + e^{-jn\omega_0 t}\right)$$

$$\sin n\omega_0 t = \frac{1}{j2}(e^{jn\omega_0 t} - e^{-jn\omega_0 t})$$

Substituting into (17.13) and gathering terms yields

$$f(t) = \frac{a_0}{2} + \sum_{n=1}^{\infty}\left[\left(\frac{a_n - jb_n}{2}\right)e^{jn\omega_0 t} + \left(\frac{a_n + jb_n}{2}\right)e^{-jn\omega_0 t}\right] \qquad (17.23)$$

If we define a new coefficient c_n by

$$c_n = \frac{a_n - jb_n}{2} \qquad (17.24)$$

and substitute for a_n and b_n from (17.17) with $t_0 = -T/2$, we have

$$c_n = \frac{1}{T}\int_{-T/2}^{T/2} f(t)(\cos n\omega_0 t - j\sin n\omega_0 t)\, dt$$

or, by Euler's formula,

$$c_n = \frac{1}{T}\int_{-T/2}^{T/2} f(t)e^{-jn\omega_0 t}\, dt \qquad (17.25)$$

We observe also that for real $f(t)$, a_n and b_n are both real, so c_n^* (the conjugate of c_n) is given by

$$c_n^* = \frac{a_n + jb_n}{2}$$

$$= \frac{1}{T}\int_{-T/2}^{T/2} f(t)(\cos n\omega_0 t + j\sin n\omega_0 t)\, dt$$

which is evidently c_{-n} (c_n with n replaced by $-n$). That is,

$$c_{-n} = \frac{a_n + jb_n}{2} = c_n^* \qquad (17.26)$$

Finally, let us observe that

$$\frac{a_0}{2} = \frac{1}{T} \int_{-T/2}^{T/2} f(t) \, dt$$

which by (17.25) is

$$\frac{a_0}{2} = c_0 \qquad\qquad (17.27)$$

Summing up, (17.24), (17.26), and (17.27) enable us to write (17.23) in the form

$$f(t) = c_0 + \sum_{n=1}^{\infty} c_n e^{jn\omega_0 t} + \sum_{n=1}^{\infty} c_{-n} e^{-jn\omega_0 t}$$

$$= \sum_{n=0}^{\infty} c_n e^{jn\omega_0 t} + \sum_{n=-1}^{-\infty} c_n e^{jn\omega_0 t}$$

We have combined c_0 with the first summation and replaced the dummy summation index n by $-n$ in the second summation. The result is more compactly written as

$$f(t) = \sum_{n=-\infty}^{\infty} c_n e^{jn\omega_0 t} \qquad\qquad (17.28)$$

where c_n is given by (17.25). This version of the Fourier series is called the *exponential Fourier series*.

Example 17.6

As an example, let us find the exponential series for the square wave of Example 17.5 given by

$$f(t) = \begin{cases} 4, & 0 < t < 1 \\ -4, & 1 < t < 2 \end{cases}$$

with $T = 2$. We have $\omega_0 = 2\pi/T = \pi$, and thus by (17.25)

$$c_n = \frac{1}{2} \int_{-1}^{1} f(t) e^{-jn\pi t} \, dt$$

ior $n \neq 0$, this is

$$c_n = \frac{1}{2} \int_{-1}^{0} (-4) e^{-jn\pi t} \, dt + \frac{1}{2} \int_{0}^{1} 4 e^{-jn\pi t} \, dt$$

$$- \frac{4}{jn\pi} [1 - (-1)^n]$$

Also, we have

$$c_0 = \frac{1}{2} \int_{-1}^{1} f(t)\,dt$$

$$= -\frac{1}{2} \int_{-1}^{0} 4\,dt + \frac{1}{2} \int_{0}^{1} 4\,dt$$

$$= 0$$

Since $c_n = 0$ for n even and $c_n = 8/jn\pi$ for n odd, we may write the exponential series in the form

$$f(t) = \frac{8}{j\pi} \sum_{n=-\infty}^{\infty} \frac{1}{2n-1} e^{j(2n-1)\pi t} \qquad (17.29)$$

The reader may verify that this result is equivalent to that obtained in Example 17.5.

It is straightforward to convert between the trigonometric and exponential Fourier series. By (17.24), the exponential series coefficients may be derived from the trigonometric by

$$c_n = \frac{a_n - jb_n}{2}$$

and by (17.24) and (17.26), the trigonometric coefficients are

$$a_n = c_{-n} + c_n = 2\,\text{Re}(c_n) \qquad (17.30a)$$

$$b_n = c_{-n} - c_n = -2\,\text{Im}(c_n) \qquad (17.30b)$$

EXERCISES

17.3.1. Find the exponential Fourier series for $f(t)$ periodic with period $T = 4$.

$$f(t) = \begin{cases} 1, & -1 < t < 1 \\ 0, & 1 < |t| < 2 \end{cases}$$

Answer $\dfrac{1}{2} + \dfrac{1}{\pi} \displaystyle\sum_{\substack{n=-\infty \\ n\neq 0}}^{\infty} \dfrac{\sin(n\pi/2)}{n} e^{jn\omega t/2}$

17.3.2. The function

$$\text{Sa}(x) = \frac{\sin x}{x}$$

called the *sampling function*, occurs frequently in communication theory. [A closely related function is the *sinc function*, defined by sinc $x = (\sin \pi x)/\pi x = \text{Sa}(\pi x)$.] Show that if we define

$$c_0 = \lim_{n \to 0} c_n$$

then the result of Exercise 17.3.1 may be given by

$$f(t) = \frac{1}{2} \sum_{n=-\infty}^{\infty} \text{Sa}\left(\frac{n\pi}{2}\right) e^{jn\pi t/2}$$

17.3.3. Generalize the results of Exercise 17.3.2 by finding the exponential series of a train of pulses of width $\delta < T$, described by

$$f(t) = \begin{cases} 1, & -\dfrac{\delta}{2} < t < \dfrac{\delta}{2} \\ 0, & \dfrac{\delta}{2} < |t| < \dfrac{T}{2} \end{cases}$$

$$f(t + T) = f(t)$$

Answer $\dfrac{\delta}{T} \displaystyle\sum_{n=-\infty}^{\infty} \text{Sa}\left(\dfrac{n\pi\delta}{T}\right) e^{j2n\pi t/T}$

17.3.4. Show that if $f(t)$ is even, the exponential Fourier coefficients are

$$c_n = \frac{a_n}{2} = \frac{2}{T} \int_0^{T/2} f(t) \cos n\omega_0 t \, dt$$

and if $f(t)$ is odd, they are

$$c_n = \frac{b_n}{2j} = \frac{2}{jT} \int_0^{T/2} f(t) \sin n\omega_0 t \, dt$$

Use the results to obtain (17.29).

17.3.5. Find the exponential Fourier series for the function of Exercise 17.2.3.

Answer $1 + \displaystyle\sum_{n=-\infty}^{+\infty} \left[\frac{-j2}{(2n-1)\pi} e^{-j(2n-1)2t} \right]$

17.4 RESPONSE TO PERIODIC INPUTS

The principle of superposition asserts that in any linear circuit the forced response to an independent source with source function

$$v_s(t) = v_o(t) + v_1(t) + v_2(t) + \cdots$$

is the sum of responses to each $v_i(t)$. Here $v_s(t)$ may be either a current or a voltage. Superposition may be applied to find the response to any periodic excitation $v_s(t)$ that has been represented by a Fourier series. The response to each term of the Fourier series may be found using phasor analysis and these component responses added to arrive at the overall response.

Example 17.7

Let us compute the forced response $i(t)$ in the circuit shown in Fig. 17.7, where the input $v(t)$ is the square wave $f(t)$ of Example 17.4. Since the terms in the Fourier series expansion of the source are at distinct frequencies, the phasor circuit of Fig. 17.7(b) has been labeled so that it can be applied for each component

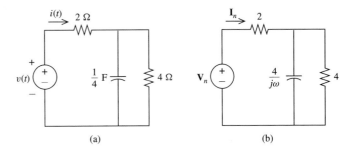

(a)   (b)

FIGURE 17.7   (a) Circuit for Example 17.7; (b) phasor circuit.

frequency ω. The source phasor has been labeled as $\mathbf{V}_n$, the phasor associated with the nth term $v_n(t)$, which will depend on the frequency $\omega = n\omega_0$ of that term. The output when this phasor source is used is labeled $\mathbf{I}_n$. The impedance of the circuit is

$$Z = 2 + \frac{(4/j\omega)4}{(4/j\omega) + 4} = \frac{6 + j\omega 2}{1 + j\omega}$$

Then the current $\mathbf{I}_n$ is

$$\mathbf{I}_n = \frac{\mathbf{V}_n}{Z} = \frac{1 + j\omega}{2(3 + j\omega)}\mathbf{V}_n \qquad (17.31)$$

In Example 17.4 the Fourier series expansion of the square wave $v(t)$ of period $T = 4$ ($\omega_o = \pi/2$) was found to be of the pure cosine type since $v(t)$ is an even function.

$$v(t) = \frac{a_0}{2} + \sum_{n=1}^{\infty} a_n \cos \frac{n\pi}{2}t$$

Here we have the source expressed as a series:

$$v(t) = v_o(t) + v_1(t) + v_2(t) + \cdots \qquad (17.32)$$

where

$$v_o(t) = \frac{a_0}{2} \qquad (17.33a)$$

$$v_n(t) = a_n \cos\left(\frac{n\pi t}{2}\right), \qquad n = 1, 2, \ldots \qquad (17.33b)$$

It was found in the example that $a_0 = 6$, and for $n = 1, 2, \ldots$,

$$a_n = \frac{12}{n\pi} \sin \frac{n\pi}{2} \qquad (17.34)$$

Then the source phasors corresponding to these source terms are $\mathbf{V}_0 = a_0/2 = 3$ and $\mathbf{V}_n = a_n$. The response to $\mathbf{V}_0$ is

$$\mathbf{I}_0 = \left.\frac{1 + j\omega}{2(3 + j\omega)}\right|_{\omega=0} \mathbf{V}_0 = \left(\tfrac{1}{6}\right) 3 = \tfrac{1}{2} \qquad (17.35)$$

and to the nth term in the Fourier series of the excitation ($n = 1, 2, \ldots$)

$$\mathbf{I}_n = \left.\frac{1+j\omega}{2(3+j\omega)}\right|_{\omega=(n\pi)/2} \mathbf{V}_n \tag{17.36a}$$

$$= \frac{6}{n\pi}\left(\frac{2+jn\pi}{6+jn\pi}\right) \sin\frac{n\pi}{2} \tag{17.36b}$$

These are the phasor responses to each input component. The overall response is

$$i(t) = i_0(t) + i_1(t) + i_2(t) + \cdots \tag{17.37a}$$

$$= \mathbf{I}_0 + \sum_{n=1}^{\infty} |\mathbf{I}_n| \cos\left(\frac{n\pi}{2}t + \theta_n\right) \tag{17.37b}$$

where, as usual in phasor analysis, the amplitude of the sinusoidal response is the magnitude of the phasor, and the phase shift is the angle of the phasor:

$$\theta_n = \underline{/\mathbf{I}_n} \tag{17.38}$$

Equation (17.37b) is the trigonometric Fourier series for the forced response $i(t)$ in which sine and cosine terms at the same frequency have been combined. Using (17.36b), we can evaluate the Fourier coefficients in this combined form,

$$|\mathbf{I}_n| = \begin{cases} \dfrac{6}{n\pi}\sqrt{\dfrac{4+n^2\pi^2}{36+n^2\pi^2}}, & n = 1, 3, 5, \ldots \\ 0, & n = 0, 2, 4, \ldots \end{cases}$$

Since the even terms are missing, we need only calculate the phase angles for n odd:

$$\theta_n = \begin{cases} \tan^{-1}\dfrac{n\pi}{2} - \tan^{-1}\dfrac{n\pi}{6}, & n = 1, 5, 9, \ldots \\ \tan^{-1}\dfrac{n\pi}{2} - \tan^{-1}\dfrac{n\pi}{6} + \pi, & n = 3, 7, 11, \ldots \end{cases}$$

This completes the example.

The reader may notice a close similarly between this analysis and use of the frequency response function $\mathbf{H}(j\omega)$ introduced in Chapter 14. Indeed, the response to periodic inputs may be computed directly in terms of $\mathbf{H}(j\omega)$. Recall that the forced response $\mathbf{V}_o(\omega)$ to the phasor input $\mathbf{V}_i(\omega)$ at frequency ω is

$$\mathbf{V}_o(\omega) = \mathbf{H}(j\omega)\mathbf{V}_i(\omega) \tag{17.39}$$

Given the trigonometric Fourier series for a general periodic input

$$v_i(t) = \frac{a_o}{2} + \sum_{n=1}^{\infty}(a_n \cos n\omega_0 t + b_n \sin n\omega_0 t)$$

we have input phasors at each $\omega = n\omega_0$ given by

$$\mathbf{V}_i(n\omega_0) = |\mathbf{V}_i(n\omega_0)|\underline{/\theta_i(n\omega_0)} = \sqrt{a_n^2 + b_n^2}\underline{/-\tan^{-1}(b_n/a_n)} \qquad (17.40)$$

for $n = 1, 2, \ldots$ and at $n = 0$ (dc), by

$$\mathbf{V}_i(0) = |\mathbf{V}_0|\underline{/\theta_0} = \frac{a_0}{2} \qquad (17.41)$$

Then the output phasor at $\omega = n\omega_0$ is the input scaled by the frequency response evaluated at that frequency,

$$\mathbf{V}_o(n\omega_0) = \mathbf{H}(jn\omega_0)\mathbf{V}_i(n\omega_0) \qquad (17.42)$$

and the overall sinusoidal output is the sum of the individual sinusoidal output components:

$$v_o(t) = \sum_{n=0}^{\infty} |\mathbf{V}_o(n\omega_0)| \cos[n\omega_0 t + \theta_0(n\omega_0)]$$

where, by (17.40),

$$|\mathbf{V}_o(n\omega_0)| = |\mathbf{H}(n\omega_0)||\mathbf{V}_i(n\omega_0)| \qquad (17.43)$$

$$\theta_0(n\omega_0) = \underline{/\mathbf{H}(n\omega_0)} + \theta_i(n\omega_0) \qquad (17.44)$$

These last remind us that the gain experienced by the component at frequency $n\omega_0$ is the magnitude of the frequency response at that frequency, and its phase shift is the angle of the frequency response. *Each component of the Fourier series expansion of the input has gain and phase shift given by* $\mathbf{H}(jn\omega_0)$, *the frequency response function of the circuit evaluated at the component's frequency* $n\omega_0$.

Example 17.8

We wish to determine the steady-state output $v(t)$ when the input $i(t)$ is the periodic impulse train shown in Fig. 17.8. The circuit is clearly stable (there are no controlled sources or isolated LC loops), so the steady-state response exists and can be computed from the frequency response $\mathbf{H}(j\omega)$. The frequency response is derived from the transfer function $\mathbf{H}(s)$ by replacing s by $j\omega$ as shown in Chapter 14.

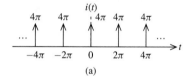

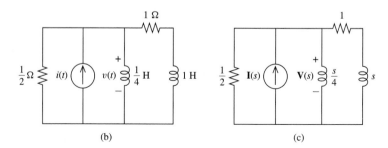

FIGURE 17.8 (a) $i(t)$; (b) circuit for Example 17.8; (c) s-domain circuit.

The s-domain node equation using zero initial conditions is

$$\left(2 + \frac{4}{s} + \frac{1}{s+1}\right) \mathbf{V}(s) = \mathbf{I}(s)$$

so the transfer function is

$$\mathbf{H}(s) = \frac{\mathbf{V}(s)}{\mathbf{I}(s)} = \frac{1}{2 + (4/s) + [1/(s+1)]} = \frac{s^2 + s}{2s^2 + 7s + 4} \quad (17.45)$$

Replacing s by $j\omega$, we have the frequency response function

$$\mathbf{H}(j\omega) = \frac{-\omega^2 + j\omega}{(4 - 2\omega^2) + j\omega 7}$$

The magnitude and angle of this function at $\omega = n\omega_0$ will give the gain and phase shift to associate with each frequency component $n = 0, 1, 2, \ldots$. To determine these components, we need a Fourier series representation of the input $i(t)$. Computing the coefficients in the usual fashion and noting ihat $T = 2\pi$, so $\omega_o = 1$ rad/s, we obtain

$$a_n = 4 \int_{-\pi}^{\pi} \delta(t) \cos(nt)\, dt = 4, \qquad n = 0, 1, 2, \ldots \quad (17.46a)$$

$$b_n = 4 \int_{-\pi}^{\pi} \delta(t) \sin(nt)\, dt = 0, \qquad n = 0, 1, 2, \ldots \quad (17.46b)$$

The phasors for each component of the input are therefore simply

$$\mathbf{I}(n\omega_0) = \mathbf{I}_n = 4\underline{/0}, \qquad n = 0, 1, 2, \ldots \quad (17.47)$$

so the output phasors are

$$\mathbf{V}_n = \mathbf{H}(jn\omega_0)\mathbf{I}_n \qquad (17.48)$$

or

$$\mathbf{V}_n = \left[\frac{-\omega^2 - j\omega}{(4 - 2\omega^2) + j\omega 7} \bigg|_{\omega=n} \right] 4 = 4 \left[\frac{-n^2 + jn}{(4 - 2n^2) + j7n} \right]$$

Note that the dc component will be zero, since $\mathbf{V}_o = 0$. The Fourier series for the output is then

$$v(t) = \sum_{n=1}^{\infty} |\mathbf{V}_n| \cos(nt + \underline{/\mathbf{V}_n}) \qquad (17.49)$$

or, after carrying out the algebra,

$$|\mathbf{V}_n| = 2\sqrt{\frac{n^4 + n^2}{n^4 + \frac{33}{4}n^2 + 4}}, \qquad n = 1, 2, 3, \ldots \qquad (17.50a)$$

$$\underline{/\mathbf{V}_n} = -\tan^{-1}\frac{1}{n} - \tan^{-1}\frac{7n}{4 - 2n^2}, \qquad n = 1, 2, 3, \ldots \qquad (17.50b)$$

This completes the required calculation. Note that the strength of the input (17.47) is the same at each frequency $n\omega_0$, while the output components vary in strength due to the variation of gain $|\mathbf{H}(j\omega)|$ with frequency. For instance, the dc component of the output is zero. This is as we might expect, since the steady-state voltage response of an inductor to a dc current is zero, and we are superposing steady-state responses at each frequency. At very high frequencies, the response amplitudes all go to 2, reflecting the input amplitudes of 4 and a gain that approaches $\frac{1}{2}$ as ω goes to infinity [at high frequencies almost all the current $i(t)$ flows through the $\frac{1}{2}$-Ω resistor].

In the preceding examples we used the trigonometric Fourier series to describe the input. Equivalently, the exponential Fourier series may be used:

$$v_i(t) = \sum_{n=-\infty}^{\infty} c_n e^{jn\omega_0 t} \qquad (17.51)$$

Since the forced response to each complex exponential component $e^{j\omega t}$ of the input is multiplied by $H(j\omega)$, using superposition the forced response is

$$v_o(t) = \sum_{n=-\infty}^{\infty} H(jn\omega_0)c_n e^{jn\omega_0 t} \qquad (17.52)$$

While this is an appealingly direct route to $v_o(t)$ compared to using the formulas (17.17) and (17.40) required for the trigonometric Fourier series, in most cases it is desirable to express the output $v_o(t)$ in terms of real functions, that is, sinusoids. Thus the terms at $\pm n$ in (17.52) still need to be combined by the Euler identity, and some part of the apparent computational advantage of the exponential form is lost.

In this section we have computed the response to a periodic input from the output phasors due to each Fourier component of the input. These phasors may be determined directly from the phasor circuit or, with greater efficiency, from the frequency response function $\mathbf{H}(j\omega)$. Either the trigonometric or exponential Fourier series may be used to describe the input.

EXERCISES

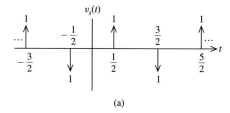

(a)

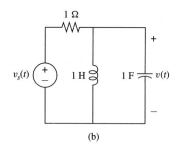

(b)

EXERCISE 17.4.3 (a) $v_s(t)$; (b) circuit for Exercise 17.4.3.

17.4.1. The periodic signal $f(t)$ of period $T = 2$ which equals $\cosh t$ [$\cosh t = \frac{1}{2}(e^{+t} + e^{-t})$] for $-1 < t < +1$, is passed through an ideal low-pass filter $\mathbf{H}(j\omega) = u(\omega + 4) - u(\omega - 4)$. Find the forced output $g(t)$.

$$Answer\ (e^1 - e^{-1})\left[1 + \frac{1}{1 + \pi^2}\cos(\pi t + 180°)\right]$$

17.4.2. The current into a parallel combination of a 4-Ω resistor and a 1-H inductor is the square wave $f(t)$ of Exercise 17.2.3. What is the current into the 1-H inductor [defined to satisfy the passive sign convention with $f(t)$]. Express in terms of exponential Fourier series.

$$Answer\ w_o = 2; c_n = 0 \text{ for } n \text{ even}; c_n = 4/([jn\pi][2 + jn]) \text{ for}$$
n odd

17.4.3. Find the steady-state response $v(t)$ in the circuit to the periodic impulse train $v(t)$ shown. Express as a trigonometric Fourier series.

Answer

$$v(t) = \sum_{n=1}^{\infty}\left[\frac{2n\pi(1 - n^2\pi^2)\sin(n\pi/2)}{(1 - n^2\pi^2)^2 + n^2\pi^2}\right]\cos n\pi t$$

$$+ \left[\frac{2n^2\pi^2\sin(n\pi/2)}{(1 - n^2\pi^2)^2 + n^2\pi^2}\right]\sin n\pi t$$

17.5 DISCRETE SPECTRA AND PHASE PLOTS

The Fourier series reveals that a given periodic signal is actually the sum of sinusoidal or complex exponential components at harmonically related frequencies $n\omega_0$, each having its own strength and phase. It is useful to be able to visualize the relative strength and phase of these components and how these vary with the frequency $n\omega_0$ of the component. Such graphs are called *discrete spectra* and *discrete phase* plots.

From (17.13) we see that combining sine and cosine components into one phase-shifted component at each frequency $n\omega_0$ the Fourier series

$$f(t) = \frac{a_0}{2} + \sum_{n=1}^{\infty} A_n \cos(n\omega_0 t + \phi_n) \qquad (17.53)$$

has amplitude

$$A_n = \sqrt{a_n^2 + b_n^2}, \qquad n = 1, 2, \ldots$$

and phase

$$\phi_n = \tan^{-1} \frac{-b_n}{a_n} = -\tan^{-1} \frac{b_n}{a_n} \qquad (17.54)$$

With $A_o = a_o/2$ and A_n for $n = 1, 2, \ldots$ as above, the variation of A_n with $n, n = 0, 1, 2, \ldots$, reveals how the amplitude of the harmonically related frequency components in $f(t)$ changes with frequency, and ϕ_n versus n shows how the phase changes with frequency. This same information may be extracted from the exponential Fourier series. In light of (17.24),

$$|c_n| = \frac{1}{2}\sqrt{a_n^2 + b_n^2} = \frac{1}{2} A_n \qquad (17.55)$$

$$\underline{/c_n} = \tan^{-1} \frac{-b_n}{a_n} = \phi_n \qquad (17.56)$$

Thus a plot of $|c_n|$ versus n reveals the variation of amplitude with frequency and that of $\underline{/c_n}$, the variation of phase with frequency. The plot of $|c_n|$ versus n for all n is called the *discrete amplitude spectrum* or *line spectrum* of $f(t)$ and that of $\underline{/c_n}$ versus n, the *discrete phase spectrum*.

Example 17.9

Suppose that we apply the square wave of Fig. 17.9(b) as a voltage $v(t)$ across a 1-F capacitor. The current through the capacitor $i(t)$ will satisfy

$$i(t) = \frac{dv}{dt} \qquad (17.57)$$

Since over one period, say $-\frac{3}{2} \leq t < \frac{1}{2}$,

$$v(t) = 4 - 8u(t+1) + 8u(t), \qquad -\frac{3}{2} \leq t < \frac{1}{2} \qquad (17.58)$$

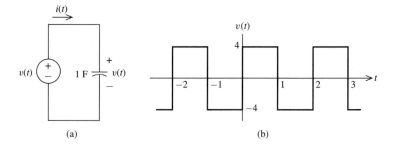

(a)   (b)

FIGURE 17.9   Circuit for Example 17.9.

then, differentiating, over one period the value of $i(t)$ will be

$$i(t) = -8\delta(t+1) + 8\delta(t), \qquad -\tfrac{3}{2} \le t < \tfrac{1}{2} \qquad (17.59)$$

We wish to compare the discrete amplitude and phase spectra of the periodic signals $v(t)$ and $i(t)$. By the results of Example 17.5,

$$v(t): \quad c_n = \begin{cases} \dfrac{8}{jn\pi}, & n \text{ odd} \\[2mm] 0, & n \text{ even} \end{cases} \qquad (17.60)$$

Thus the amplitude spectrum is

$$v(t): \quad |c_n| = \begin{cases} \dfrac{8}{n\pi}, & n \text{ odd} \\[2mm] 0, & n \text{ even} \end{cases} \qquad (17.61)$$

and the phase spectrum is

$$v(t): \quad \angle c_n = -90°, \qquad n \text{ odd} \qquad (17.62)$$

With $\omega_0 = \pi$ the exponential Fourier series for $i(t)$ has coefficients

$$i(t): \quad c_n = \frac{1}{2}\int_{-3/2}^{1/2} [-8\delta(t+1) + 8\delta(t)]e^{-jn\pi t}\,dt = 4(1 - e^{jn\pi})$$

Here we have selected $-\tfrac{3}{2}$ as the initial time for the integration so that the impulses do not coincide with either of the limits of integration. Then

$$i(t): \quad |c_n| = \begin{cases} 8, & n \text{ odd} \\ 0, & n \text{ even} \end{cases} \qquad (17.63)$$

$$\angle c_n = 0°, \qquad n \text{ odd} \qquad (17.64)$$

The discrete amplitude and phase spectra of the periodic signals $v(t)$ and $i(t)$ are shown in Fig. 17.10. Note that, while $v(t)$ has frequency components whose strength rolls off with large ω (large n), the effect of the differentiation of $v(t)$ is to strengthen the high-frequency components. Indeed, we know that the derivative of a sinusoid is

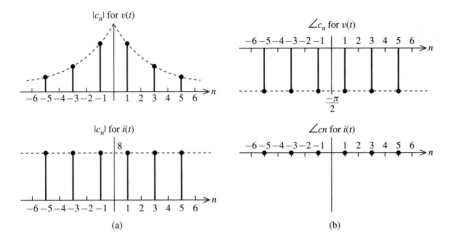

FIGURE 17.10 (a) Discrete amplitude spectra for input $v(t)$ and output $i(t)$; (b) discrete phase spectra.

scaled by ω, so the high-frequency components are relatively more prominent. Also, the differentiation results in a $+90°$ phase shift in all frequency components.

Finally, it is often of interest to know how the power in a periodic signal is divided among its frequency components. For instance, the purpose of rectifying a sinusoid as in Fig. 17.2(b) may be to produce dc power. The resulting Fourier series exhibits a dc component with power, but has power at many other frequencies as well. Or we may pass a signal through a filter and need to know how much of the power in the output lies within the passband of the filter and how much in the stopband.

Multiplying $f(t)$ by its conjugate $f^*(t)$ in (17.28) and integrating over one period yields

$$\int_{t_0}^{t_0+T} |f(t)|^2\,dt = \int_{t_0}^{t_0+T} \sum_{n=-\infty}^{\infty} \sum_{m=-\infty}^{\infty} c_n c_m^* e^{j(n-m)\omega_o t}\,dt$$

Integrating term by term, we have zero for all terms for which $n - m$ is not zero, since we are integrating the complex exponential over an integral number of periods, and both its real part $\cos(n - m)\omega_0 t$ and imaginary part $\sin(n - m)\omega_0 t$ have zero average value. Thus, with the remaining terms $n = m$ having constant integrands,

$$\int_{t_0}^{t_0+T} |f(t)|^2\,dt = T \sum_{n=-\infty}^{\infty} |c_n|^2$$

Dividing by T, the average power or mean-square value of $f(t)$ is

$$|f|_{\text{av}}^2 = \sum_{n=-\infty}^{\infty} |c_n|^2 \qquad (17.65)$$

We may identify each term on the right as the power of a single frequency component of $f(t)$. Equation (17.65) is the *Parseval relation,* which asserts that *the average power in a periodic signal is the sum over all components n of the power* $|c_n|^2$ *in its exponential Fourier series components.* A plot of $|c_n|^2$ versus n is called the *discrete power spectrum* of the signal $f(t)$. It consists of a set of equally spaced *spectral lines* that show us how the power in a periodic signal is distributed in the frequency domain.

Example 17.10

Determine the amplitude and power spectra of the periodic extension $f(t)$ of the generator $f_T(t) = e^{-t}$ with duration $T = 1$ and initial time $t_0 = 0$ first used in Example 17.1. A sketch of $f(t)$ is shown in Fig. 17.1(b) and repeated as Fig. 17.11 for convenience.

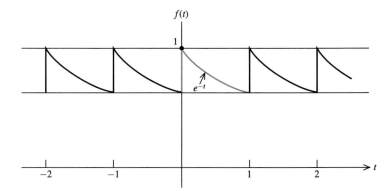

FIGURE 17.11 $f(t)$ for Example 17.10. One period has been tinted green.

With $\omega_0 = 2\pi$, c_n is given by

$$c_n = \int_0^1 e^{-(1+j2\pi n)t}\, dt = \frac{1}{1 + j2\pi n}[1 - e^{-(1+j2\pi n)}]$$

$$c_n = \frac{1 - e^{-1}}{1 + j2\pi n}$$

The amplitude spectrum is given by

$$|c_n| = \frac{1 - e^{-1}}{\sqrt{(2\pi n)^2 + 1}}$$

and the power spectrum by

$$|c_n|^2 = \frac{(1 - e^{-1})^2}{(2\pi n)^2 + 1}$$

So, for instance, the dc power $(n = 0)$ in $f(t)$ is $(1 - e^{-1})^2$. Since the total power in $f(t)$ is

$$|f|_{av}^2 = \frac{1}{T} \int_0^T f^2(t)\,dt = \int_0^1 e^{-2t}\,dt$$

$$= \tfrac{1}{2}(1 - e^{-2})$$

then the fraction of the total power in this waveform that is at dc is

$$\frac{2(1 - e^{-1})^2}{1 - e^{-2}} = 0.924$$

If used to develop dc power, this waveform is 92.4% efficient, with 7.6% "ripple" to be low-pass filtered out. The amplitude and power spectra are plotted in Fig. 17.12.

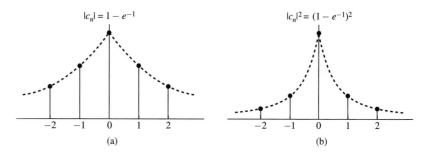

FIGURE 17.12   (a) Amplitude spectrum of $f(t)$; (b) power spectrum.

EXERCISES

17.5.1.   Suppose that the waveform $f(t)$ of Example 17.10 is passed through a high-pass filter with frequency response $\mathbf{H}(j\omega) = j\omega/(1 + j\omega)$. Sketch the discrete line spectrum of the output $g(t)$.

Answer $|c_n| = \dfrac{(1 - e^{-1})2\pi|n|}{1 + (2\pi n)^2}$

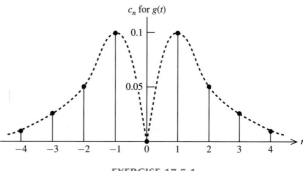

EXERCISE 17.5.1

17.5.2. Rewrite the Parseval relation (17.65) in terms of the trigonometric Fourier series coefficients.

Answer $|f|^2_{av} = \dfrac{1}{2}\left[\dfrac{a_o^2}{2} + \displaystyle\sum_{n=1}^{\infty}(a_n^2 + b_n^2)\right]$

17.5.3. How are the discrete amplitude spectrum $|c_n|$ and discrete phase spectrum θ_n of $f(t)$ related to the discrete amplitude spectrum $|c'_n|$ and discrete phase spectrum θ'_n of its derivative $df(t)/dt$?

Answer $|c'_n| = |n||\omega_o||c_n|$; $\theta'_n = \theta_n + 90°$

17.6 THE FOURIER TRANSFORM

The discrete spectra show us that the energy in periodic functions is highly concentrated in the frequency domain over just a discrete set of frequencies. These functions have no energy whatsoever at the vast majority of frequencies, those that happen not to satisfy $\omega = n\omega_0$ for integer n. If we were to pass a periodic signal through any filter whose gain at the frequencies $n\omega_0$ was zero, then regardless of its gain elsewhere in between these frequencies, nothing would come out at all.

Nonperiodic functions are more even handed, scattering their energy continuously throughout the frequency domain. The *Fourier transform* is a function of continuous ω, which reveals how the energy in a nonperiodic function is distributed over the continuous frequency domain, just as the Fourier series showed how periodic functions distribute their energy only over their discrete harmonic frequencies $n\omega_0$.

We begin the development of the Fourier transform by considering a function $f(t)$ defined on an infinite interval, but that is not periodic and therefore cannot be represented by a Fourier series. It may be possible, however, to consider the function to be periodic with an *infinite* period and extend our previous results to include this case. The development we will give is nonrigorous, but the results may be obtained rigorously if $f(t)$ satisfies the Dirichlet conditions of Section 17.2 with the absolute integrability condition over a period replaced by full-scale absolute integrability.

$$\int_{-\infty}^{\infty} |f(t)|\,dt < \infty \tag{17.66}$$

Our strategy is to consider the periodic extension $f_p(t)$ of the generator $f_T(t)$ given by

$$f_T(t) = f(t)\left[u\left(t + \frac{T}{2}\right) - u\left(t - \frac{T}{2}\right)\right]$$

as illustrated in Fig. 17.13. The exponential Fourier series for $f_p(t)$ is

$$f_p(t) = \sum_{n=-\infty}^{\infty} c_n e^{j2\pi nt/T} \tag{17.67}$$

where

$$c_n = \frac{1}{T}\int_{-T/2}^{T/2} f_T(x)e^{-j2\pi nx/T}\,dx \tag{17.68}$$

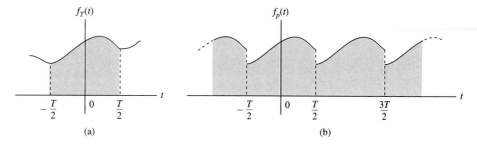

FIGURE 17.13   (a) Function $f_T(t)$; (b) its periodic extension $f_p(t)$.

We have replaced ω_0 by $2\pi/T$ and are using the dummy variable x instead of t in the coefficient expression. Our intention is to let $T \to \infty$, in which case $f_p(t) \to f(t)$. The energy distribution over frequency in the periodic $f_p(t)$ (17.67) will, in the limit, reveal how the energy in the nonperiodic $f(t)$ is distributed.

Since the limiting process requires that $\omega_0 = 2\pi/T \to 0$, for emphasis we replace $2\pi/T$ by $\Delta\omega$. Therefore, substituting (17.68) into (17.67), we have

$$
\begin{aligned}
f_p(t) &= \sum_{n=-\infty}^{\infty} \left[\frac{\Delta\omega}{2\pi} \int_{-T/2}^{T/2} f_T(x) e^{-jxn\Delta\omega}\, dx \right] e^{jtn\Delta\omega} \\
&= \sum_{n=-\infty}^{\infty} \left[\frac{1}{2\pi} \int_{-T/2}^{T/2} f_T(x) e^{-j(x-t)n\Delta\omega}\, dx \right] \Delta\omega
\end{aligned}
\tag{17.69}
$$

If we define the function

$$
g(\omega, t) = \frac{1}{2\pi} \int_{-T/2}^{T/2} f_T(x) e^{-j\omega(x-t)}\, dx
\tag{17.70}
$$

then clearly the limit of (17.69) is given by

$$
f(t) = \lim_{\substack{T \to \infty \\ \Delta\omega \to 0}} \sum_{n=-\infty}^{\infty} g(n\Delta\omega, t)\, \Delta\omega
\tag{17.71}
$$

By the fundamental theorem of integral calculus, the last result appears to be

$$
f(t) = \int_{n=-\infty}^{\infty} g(\omega, t)\, d\omega
\tag{17.72}
$$

But in the limit, $f_T \to f$ and $T \to \infty$ in (17.69), so that what appears to be $g(\omega, t)$ in (17.69) is really its limit, which by (17.70) is

$$
\lim_{T \to \infty} g(\omega, t) = \frac{1}{2\pi} \int_{-\infty}^{\infty} f(x) e^{-j\omega(x-t)}\, dx
$$

Therefore, (17.71) is actually

$$f(t) = \frac{1}{2\pi} \int_{-\infty}^{\infty} \left[\int_{-\infty}^{\infty} f(x) e^{-j\omega(x-t)} \, dx \right] d\omega \qquad (17.73)$$

We may write this result in the form

$$f(t) = \frac{1}{2\pi} \int_{-\infty}^{\infty} \left[\int_{-\infty}^{\infty} f(x) e^{-j\omega x} \, dx \right] e^{j\omega t} \, d\omega \qquad (17.74)$$

Defining the expression in brackets to be the function

$$\mathbf{F}(j\omega) = \int_{-\infty}^{\infty} f(t) e^{-j\omega t} \, dt \qquad (17.75)$$

where we have changed the variable of integration from x to t, we have

$$f(t) = \frac{1}{2\pi} \int_{-\infty}^{\infty} \mathbf{F}(j\omega) e^{j\omega t} \, d\omega \qquad (17.76)$$

The function $\mathbf{F}(j\omega)$ is called the *Fourier transform* of $f(t)$, and $f(t)$ is called the *inverse Fourier transform* of $\mathbf{F}(j\omega)$. These facts are often stated symbolically as

$$\mathbf{F}(j\omega) = \mathcal{F}[f(t)]$$
$$f(t) = \mathcal{F}^{-1}[\mathbf{F}(j\omega)] \qquad (17.77)$$

where $\mathcal{F}$ and $\mathcal{F}^{-1}$ denote the operations of taking the Fourier transform and inverse transform (just as $\mathcal{L}$ and $\mathcal{L}^{-1}$ denoted Laplace transformation and inverse transformation). Also, (17.75) and (17.76) are together called the *Fourier transform pair*.

Writing (17.67) and (17.76) together, the Fourier series and transform are quite similar.

$$\text{Fourier series:} \quad f_p(t) = \sum_{\substack{\omega=n\omega_0 \\ n=-\infty}}^{+\infty} c_n e^{j\omega t} \qquad (17.78a)$$

$$\text{Fourier transform:} \quad f(t) = \int_{-\infty}^{+\infty} \mathbf{F}(j\omega) e^{j\omega t} \, d\omega \qquad (17.78b)$$

Both express a time function as a superposition of frequency components. In the series case the time function is periodic and only a discrete set of frequencies is needed, so the superposition is a sum. In the transform case every frequency ω contributes an incremental amplitude $\mathbf{F}(j\omega)d\omega$, so an integral over all ω is needed.

Example 17.11

As an example, let us find the transform of

$$f(t) = e^{-at}u(t)$$

where $a > 0$. By definition (17.75), we have

$$\mathcal{F}[e^{-at}u(t)] = \int_{-\infty}^{\infty} e^{-at}u(t)e^{-j\omega t}\,dt$$

$$= \int_{0}^{\infty} e^{-(a-j\omega)t}\,dt$$

or

$$\mathcal{F}[e^{-at}u(t)] = \frac{1}{-(a+j\omega)}e^{-(a-j\omega)t}\Big|_{0}^{\infty}$$

Because $a > 0$, the upper limit is

$$\lim_{t\to\infty} e^{-at}(\cos\omega t - j\sin\omega t) = 0$$

since the expression in parentheses is bounded while the exponential goes to zero. Thus we have

$$\mathcal{F}[e^{-at}u(t)] = \frac{1}{a+j\omega} \qquad (17.79)$$

Example 17.12

As another example, let us find the transform of the single rectangular pulse

$$f(t) = \begin{cases} A, & -\dfrac{\delta}{2} < t < \dfrac{\delta}{2} \\[2mm] 0, & |t| > \dfrac{\delta}{2} \end{cases} \qquad (17.80a)$$

which is shown in Fig. 17.14. By definition we have

$$\mathbf{F}(j\omega) = \int_{-\infty}^{\infty} f(t)e^{-j\omega t}\,dt$$

$$= \int_{-\delta/2}^{\delta/2} Ae^{-j\omega t}\,dt$$

$$= \frac{2A}{\omega}\frac{e^{j\omega\delta/2} - e^{j\omega\delta/2}}{j2}$$

or

$$\mathbf{F}(j\omega) = \frac{2A}{\omega}\sin\frac{\omega\delta}{2} \qquad (17.80b)$$

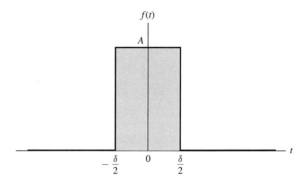

<figure>
FIGURE 17.14   Finite pulse of width δ.
</figure>

The reader may have noticed a close similarity between the first example, which led to the Fourier transform

$$\mathcal{F}(e^{-at}u(t)) = \frac{1}{j\omega + a} \tag{17.81}$$

and the Laplace transform of the same time function from Chapter 12,

$$\mathcal{L}(e^{-at}u(t)) = \frac{1}{s + a} \tag{17.82}$$

Even the steps in the derivation of these results, (12.3) and (12.4) for the Laplace transform and Example 17.11 for the Fourier transform, echo one another almost exactly.

This is no coincidence. *For any Fourier-transformable function $f(t)u(t)$ which is zero for $t < 0$, the two transforms are simply related by*

$$\mathcal{F}[f(t)u(t)] = \mathcal{L}[f(t)u(t)]\Big|_{s=j\omega} \tag{17.83}$$

Let us call a function $f(t)$ that is zero for $t < 0$ a *one-sided $t \geq 0$ function*. Then (17.83) asserts that *whenever it exists, the Fourier transform of any one-sided $t \geq 0$ function is identical to the Laplace transform of the same function with s evaluated at $j\omega$.* This is certainly the case for the one-sided $t \geq 0$ function $e^{-at}u(t)(a > 0)$, which was the subject of Example 17.11, as can be seen from (17.79).

To see that it is true in general, we first note that the above claim implies that as long as $f(t)u(t)$ is Fourier transformable, it will also be Laplace transformable. Looking back at the requirements for transformability, any one-sided $t \geq 0$ function satisfying the Dirichlet conditions must be absolutely integrable and is therefore of exponential order (with $\sigma = 0$) and hence Laplace transformable. Indeed, *the set of Fourier transformable functions is a subset of the Laplace transformable functions.* Conditions for Fourier transformability are more stringent. To complete the justification of (17.83), write the definitions of the transforms together:

$$\text{Fourier transform: } \quad \mathbf{F}(j\omega) = \int_{-\infty}^{\infty} f(t) e^{-j\omega t}\, dt$$

$$\text{Laplace transform: } \quad \mathbf{F}(s) = \int_{0-}^{\infty} f(t) e^{-st}\, dt$$

Clearly, if both exist and $f(t)$ is zero for $t < 0$, these integrals are identical after identification of s with $j\omega$.

The equality of Laplace and Fourier transforms at $s = j\omega$ for one-sided $t \geq 0$ functions will prove useful in deriving Fourier transform pairs from known Laplace transform pairs, such as those gathered in Table 12.2.

Example 17.13

Find the Fourier transform of $f(t) = e^{-t}\sin 2t u(t)$. This is a one-sided $t \geq 0$ function. From Table 12.2,

$$\mathcal{L}[e^{-t}\sin 2t u(t)] = \frac{2}{(s+1)^2 + 4}$$

By (17.68), we have the desired Fourier transform simply by replacing s by $j\omega$.

$$\mathbf{F}(j\omega) = \frac{2}{(j\omega + 1)^2 + 4}$$

Even the Fourier transforms of general two-sided functions $f(t)$, those not zero for $t < 0$ or for $t > 0$, may be found from the Laplace transforms of one-sided $t \geq 0$ functions such as those listed in the Table 12.2. This technique is demonstrated in the next example.

Example 17.14

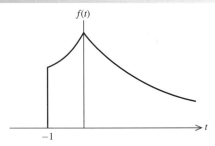

FIGURE 17.15 $f(t)$ for Example 17.14.

We wish to find the Fourier transform of $f(t) = e^{-|t|} u(t+1)$ shown in Fig. 17.15. Write $f(t)$ as the sum of a one-sided $t \geq 0$ function and a one-sided $t \leq 0$ function:

$$f(t) = e^{-t}u(t) + e^{+t}(u(t+1) - u(t)) \qquad (17.84)$$

Then

$$\mathbf{F}(j\omega) = \mathcal{F}(e^{-t}u(t)) + \mathcal{F}[e^{+t}(u(t+1) - u(t))] \qquad (17.85)$$

The first term, the transform of the one-sided $t \geq 0$ function $e^{-t}u(t)$ is given by (17.79) with $a = 1$.

$$\mathcal{F}(e^{-t}(u(t+1) - u(t))) = \int_{-1}^{0} e^{t}e^{-j\omega t}\, dt = -\int_{1}^{0} e^{-\tau}e^{j\omega \tau}\, d\tau$$

The second term may be evaluated by changing variables $\tau = -t$ in the integral. Now if we replace $j\omega$ by $-s$, this integral is identified as the Laplace transform of the time function $e^{-t}(u(t) - u(t-1))$.

Then we can compute this Laplace transform using the techniques of Chapter 12:

$$\mathcal{L}[e^{-t}u(t) - e^{-t}u(t-1)] = \frac{1}{s+1} - \frac{e^{-1}e^{-s}}{s+1} = \frac{1 - e^{-(s+1)}}{s+1} \quad (17.86)$$

Reversing the substitution of $j\omega$ by s (replacing s by $-j\omega$), we have the Fourier transform of the one-sided $t \leq 0$ function in (17.84):

$$\mathcal{F}[e^{+t}(u(t+1) - u(t))] = \left. \frac{1 - e^{-(s+1)}}{s+1} \right|_{s=-j\omega}$$

$$= \frac{1 - e^{-(-j\omega+1)}}{-j\omega + 1} \quad (17.87)$$

Then, combining the two terms in (17.84), we obtain

$$\mathbf{F}(j\omega) = \frac{1}{j\omega + 1} + \frac{1 - e^{j\omega - 1}}{-j\omega + 1} = \frac{2 - (j\omega + 1)e^{j\omega - 1}}{1 + \omega^2}$$

This is the desired Fourier transform.

EXERCISES

17.6.1. Confirm the result of Example 17.14. Find $\mathbf{F}(j\omega)$ by evaluating the integral definition of the Fourier transform (17.75) directly.

17.6.2. Find $\mathcal{F}[e^{-a|t|}]$, where $a > 0$.

Answer $\dfrac{2a}{\omega^2 + a^2}$

17.6.3. Find $\mathcal{F}[-\frac{1}{4}(t-3)e^{-(t-3)}\sin 2(t-3)u(t-3)]$.

Answer $\dfrac{e^{-3j\omega}(j\omega + 1)}{[j\omega + 1]^2 + 4^2}$

17.6.4. Which of these functions satisfies the Dirichlet conditions? $\sin t$, $1/t$, $u(t)$, $u(t+1) - u(t-1)$, $\cos\frac{1}{t}(u(t+1) - u(t-1))$.

Answer No; no; no; yes; no

17.6.5. Find the Fourier transform of $f(t) = e^{-2t}$, $t > 0$, e^{+3t}, $t < 0$.

Answer $\dfrac{5}{\omega^2 + 6 + j\omega}$

17.7 FOURIER TRANSFORM PROPERTIES

Just as Laplace transform properties summarized in Table 12.1 proved useful in working in the s-domain, we will derive and apply Fourier transform properties in this section. Indeed, the close relation between these transforms suggests that Fourier properties may be derived from Laplace properties. Here we derive the desired Fourier properties from the integral definition (17.75) of this transform.

One of the most often used properties is linearity, which is a consequence of the fact that the Fourier transform is an integral transform, and thus it is a *linear* operation. That is, the transform of the combination

$$v(t) = c_1 f_1(t) + c_2 f_2(t)$$

is the combination of the transforms

$$\mathcal{F}(c_1 f_1(t) + c_2 f_2(t)) = c_1 \mathbf{F}_1(j\omega) + c_2 \mathbf{F}_2(j\omega) \qquad (17.88)$$

where $\mathbf{F}_1$ and $\mathbf{F}_2$ are, respectively, the Fourier transforms of f_1 and f_2, and c_1 and c_2 are constants. This concept of linearity enables us to readily find transforms of relatively complicated functions from transforms of simpler functions.

Example 17.15

For example, the transform of the function

$$f(t) = 2(e^{-2t} - e^{-3t})u(t)$$

is, by linearity and (17.79),

$$\mathbf{F}(j\omega) = \frac{2}{2 + j\omega} - \frac{2}{3 + j\omega}$$

$$= \frac{2}{(2 + j\omega)(3 + j\omega)}$$

Another operation involving Fourier transforms that we shall find useful is that of time differentiation. Suppose that we wish to find the Fourier transform of the derivative of a function $f(t)$. By definition

$$f(t) = \frac{1}{2\pi} \int_{-\infty}^{\infty} \mathbf{F}(j\omega)e^{j\omega t}\, d\omega$$

from which we obtain

$$\frac{df(t)}{dt} = \frac{1}{2\pi} \int_{-\infty}^{\infty} \frac{d}{dt}[\mathbf{F}(j\omega)e^{j\omega t}]\, d\omega$$

$$= \frac{1}{2\pi} \int_{-\infty}^{\infty} [j\omega \mathbf{F}(j\omega)]e^{j\omega t}\, d\omega$$

Therefore, we have

DIFFERENTIATION

$$\mathcal{F}\left(\frac{df(t)}{dt}\right) = j\omega \mathbf{F}(j\omega) \qquad (17.89)$$

That is, the transform of the derivative of f is found by simply multiplying the transform of f by $j\omega$. This result may be readily extended to the general case:

$$\mathcal{F}\left(\frac{d^n f(t)}{dt^n}\right) = (j\omega)^n \mathbf{F}(j\omega) \qquad (17.90)$$

where $n = 0, 1, 2, \ldots$. We are assuming that the derivatives involved exist and that the interchange of operations of differentiation and integration is valid. The conditions under which this is true are quite general and hold for almost any function we are likely to encounter.

Example 17.16

As another example, let us find the transform of $f(t - \tau)$, where τ is a constant. Replacing t by $t - \tau$ in (17.76), we have

$$f(t - \tau) = \frac{1}{2\pi} \int_{-\infty}^{\infty} \mathbf{F}(j\omega) e^{j\omega(t-\tau)} \, d\omega$$

$$= \frac{1}{2\pi} \int_{-\infty}^{\infty} [\mathbf{F}(j\omega) e^{-j\omega\tau}] e^{j\omega t} \, d\omega$$

from which we conclude, by (17.76), that

TIME SHIFT

$$\mathcal{F}(f(t - \tau)) = \mathbf{F}(j\omega) e^{-j\omega\tau} \qquad (17.91)$$

The physical significance of this result is that a *delay* in the time domain [the function $f(t-\tau)$ is $f(t)$ delayed τ seconds] corresponds to a phase shift (an addition of $-\omega\tau$ to the phase) in the frequency domain. Note that τ may be positive or negative.

The similarity of the integrals for $f(t)$ and its transform in (17.75) and (17.76) suggests that interchanging t and ω in some way would lead to a new set of transforms. Indeed, if in (17.75), we replace ω by x and t by $-\omega$ and multiply both sides of the equation by 2π, we have

$$2\pi f(-\omega) = \int_{-\infty}^{\infty} \mathbf{F}(jx) e^{-jx\omega} \, dx$$

The right member, as we see from (17.75), is the Fourier transform of $\mathbf{F}(jt)$. Thus we have

DUALITY PROPERTY

$$\mathcal{F}(\mathbf{F}(jt)) = 2\pi f(-\omega) \qquad (17.92)$$

That is, the *Fourier transform* $\mathbf{F}(j\omega)$ *with* ω *replaced by* t *is a time function whose Fourier transform is* $2\pi f(t)$, *with* t *replaced by* $-\omega$.

Example 17.17

As examples, if we apply (17.92) to (17.79) and (17.80b), respectively, we have

$$\mathcal{F}\left(\frac{1}{a + jt}\right) = 2\pi e^{a\omega} u(-\omega), \qquad a > 0$$

and

$$\mathcal{F}\left(\frac{2A}{t}\sin\frac{t\delta}{2}\right) = \begin{cases} 2\pi A, & |\omega| < \dfrac{\delta}{2} \\[2ex] 0, & |\omega| > \dfrac{\delta}{2} \end{cases}$$

We have tabulated the Fourier transform operations derived in this section, along with others, in Table 17.1. The reader is asked to derive the others as exercises and problems.

Table 17.1 Fourier Transform Operations

	$f(t)$	$\mathbf{F}(j\omega)$
1. Linearity	$c_1 f_1(t) + c_2 f_2(t)$	$c_1 \mathbf{F}_1(j\omega) + c_2 \mathbf{F}_2(j\omega)$
2. Time/frequency scaling	$f\left(\dfrac{t}{a}\right)$	$|a|\mathbf{F}(j\omega a)$
3. Time shift	$f(t - \tau)$	$e^{-j\omega\tau}\mathbf{F}(j\omega)$
4. Frequency shift	$e^{j\omega_0 t} f(t)$	$\mathbf{F}(j\omega - j\omega_0)$
5. Duality property	$\mathbf{F}(jt)$	$2\pi f(-\omega)$
6. Time/frequency reversal	$f(-t)$	$\mathbf{F}(-j\omega)$
7. n-Fold differentiation	$\dfrac{d^n}{dt^n} f(t)$	$(j\omega)^n \mathbf{F}(j\omega), n = 0, 1, 2, \ldots$
8. n-Fold t-multiplication	$t^n f(t)$	$(-1)^n \dfrac{d^n}{d(j\omega)^n}\mathbf{F}(j\omega), n = 0, 1, 2, \ldots$

We may also use Fourier transforms to extend the concept of frequency response, considered in Chapter 14, to circuits with nonsinusoidal excitations. As we will see, the response functions are exactly the same as in the case of circuits with sinusoidal excitations. Therefore, all the properties of transfer functions that we have considered are valid in this more general case. The only difference is that here the inputs and outputs are Fourier transforms rather than phasors.

We begin by considering the general circuit of Fig. 17.16, where, to be specific, we have taken the input and output as the voltages $v_i(t)$ and $v_o(t)$, respectively. We could as well let one or both of these functions be currents. We will take the describing equation

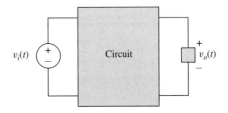

FIGURE 17.16 General circuit.

of the circuit to be

$$a_n \frac{d^n v_o}{dt^n} + a_{n-1} \frac{d^{n-1} v_o}{dt^{n-1}} + \cdots + a_1 \frac{dv_o}{dt} + a_0 v_o$$
$$= b_m \frac{d^m v_i}{dt^m} + b_{m-1} \frac{d^{m-1} v_i}{dt^{m-1}} + \cdots + b_1 \frac{dv_i}{dt} + b_0 v_i$$

Taking the Fourier transform of both sides, making use of entry 7 in Table 17.1 and linearity, we have

$$[a_n(j\omega)^n + a_{n-1}(j\omega)^{n-1} + \cdots + a_1 j\omega + a_0]\mathbf{V}_o(j\omega)$$
$$= [b_m(j\omega)^m + b_{m-1}(j\omega)^{m-1} + \cdots + b_i j\omega + b_0]\mathbf{V}_i(j\omega)$$

The functions $\mathbf{V}_o$ and $\mathbf{V}_i$ are the Fourier transforms of the output and input functions, v_o and v_i. From this result we may write

$$\frac{\mathbf{V}_o(j\omega)}{\mathbf{V}_i(j\omega)} = \mathbf{H}(j\omega)$$

$$= \frac{b_m(j\omega)^m + b_{m-1}(j\omega)^{m-1} + \cdots + b_0}{a_n(j\omega)^n + a_{n-1}(j\omega)^{n-1} + \cdots + a_0} \qquad (17.93)$$

where $\mathbf{H}(j\omega)$ is identical to the transfer function of Chapter 13 evaluated at $s = j\omega$.

This development indicates that the transfer function, defined previously as the ratio of the output Laplace transform to the input Laplace transform, is precisely the same as the ratio of the output Fourier transform to the input Fourier transform. Of course, both input and output must satisfy the Fourier transform Dirichlet conditions including full-scale integrability (17.66) for (17.93) to be valid.

Example 17.18

For example, suppose that the input $v_i(t)$ is given by

$$v_i(t) = e^{-3t} u(t)$$

and is related to the output $v_o(t)$ by the equation

$$\frac{dv_o}{dt} + 2v_o = 2v_i$$

Transforming the equation, we have

$$(j\omega + 2)\mathbf{V}_o(j\omega) = 2\mathbf{V}_i(j\omega)$$

from which the transfer function is

$$\mathbf{H}(j\omega) = \frac{\mathbf{V}_o(j\omega)}{\mathbf{V}_i(j\omega)} = \frac{2}{j\omega + 2}$$

Also by (17.79), we have

$$\mathbf{V}_i(j\omega) = \frac{1}{3 + j\omega}$$

so that the transform of the output is

$$\mathbf{V}_o(j\omega) = \mathbf{H}(j\omega)\mathbf{V}_i(j\omega)$$

$$= \frac{2}{(2 + j\omega)(3 + j\omega)}$$

This can be rewritten, after partial fraction expansion, as

$$\mathbf{V}_o(j\omega) = 2\frac{1}{2 + j\omega} - 2\frac{1}{3 + j\omega}$$

so that, by linearity, we have

$$v_o(t) = 2(e^{-2t} - e^{-3t})u(t)$$

In Example 17.18 we used partial fraction expansion to facilitate inverse Fourier transformation, just as we did previously with Laplace transforms. *All the same rules for partial fraction expansion apply, with $j\omega$ replacing s in the rules and procedures.*

the time-domain responses obtained in this manner are responses of initially *relaxed* circuits. (No initial energy is stored.) We could develop general methods for using Fourier transforms to find time-domain solutions, but the Fourier transform is better suited for other applications. We prefer, therefore, to use the *Laplace transform*, which unlike the Fourier transform takes into account nonzero initial conditions for such problems.

Since the transfer functions of this chapter are identical to those considered earlier, we will not go into the details of their frequency responses, poles and zeros, natural frequencies, and so on. These topics have all been considered in Chapters 12 and 13.

EXERCISES

17.7.1. Derive operations 2, 4, 6, and 8 of Table 17.1.

17.7.2. Find the inverse transform of $2e^{-j\omega}/(\omega^2 + 1)$.
Answer $e^{-|t-1|}$

17.7.3. If $x(t)$ is the input and $y(t)$ is the output, find $\mathbf{H}(j\omega)$ using Fourier transforms, where $y'' + 4y' + 3y = 4x$.
Answer $4/(3 - \omega^2 + j4\omega)$

17.7.4. If in Exercise 17.7.3 $x = e^{-2t}u(t)$, find $\mathcal{F}[y(t)]$.
Answer $4/[(j\omega + 1)(j\omega + 2)(j\omega + 3)]$

17.7.5. Verify in Exercise 17.7.4 that

$$\mathcal{F}[y(t)] = \frac{2}{j\omega + 1} - \frac{4}{j\omega + 2} + \frac{2}{j\omega + 3}$$

and find $y(t)$ for the case $y(0) = y'(0) = 0$.
Answer $(2e^{-t} - 4e^{-2t} + 2e^{-3t})u(t)$

17.8 SPICE AND FOURIER ANALYSIS

SPICE can be used to find the Fourier series for a periodic function $f(t)$. The form directly supported by SPICE is

$$f(t) = d_o + \sum_{n=1}^{\infty} D_n \sin(n\omega_0 t + \psi_n) \tag{17.94}$$

Comparing (17.53) and (17.94), the latter can be easily converted to the more familiar phase-shifted cosine form (17.53) by the change of variables

$$d_o = \frac{a_o}{2} \tag{17.95a}$$

$$D_n = A_n \tag{17.95b}$$

$$\psi_n = \phi_n - \frac{\pi}{2} \tag{17.95c}$$

To use SPICE for determining a Fourier series, we first produce the generator function $f_T(t)$ of the periodic function $f(t)$ of interest. This is done by performing a transient analysis, as explained in Section 6.8, on a circuit that produces $f_T(t)$ as part of one of its transient currents or voltages. Recall that the syntax of the `.TRAN` control statement is

`.TRAN       TSTEP       TSTOP       <UIC>`

where `TSTEP` is the time step between values to be printed or plotted, `TSTOP` is the end time of the transient analysis, which always begins at $t = 0$, and the optional key word `UIC` directs SPICE to use initial conditions specified on L and C element cards, rather than initial conditions derived from a dc steady-state analysis.

732   Chapter 17   Fourier Series and Transform

Assuming that a transient analysis has been specified, the control statement that directs SPICE to subsequently perform a Fourier series analysis is the .FOUR statement, whose format is

```
.FOUR    TINV    F1 < F2 F3... >
```

The first parameter TINV is the inverse of the duration T of $f_T(t)$, the generator for the Fourier series. TINV is therefore also the fundamental frequency $f_o = \omega_o/2\pi$ measured in hertz. The initial time for the generator is taken to be TSTOP-T, where TSTOP is the end time of the .TRAN transient analysis and T the duration of $f_T(t)$, that is, the inverse of TINV. F1 is the current or voltage containing $f_T(t)$, and F2, F3, ... are optional additional currents or voltages whose Fourier series is desired. If more than one current or voltage is specified, the same generator duration and initial time are assumed to apply to all. Only the first 10 components of the Fourier series are computed and output by SPICE, the dc component, the fundamental at $f_o =$ TINV, and the next eight harmonic components.

In Section 17.1 it was pointed out that if a generator $f_T(t)$ consists of an exact integer number of repetitions of a waveform, the period of its periodic extension will not coincide with the specified duration of the generator (it will be shorter). SPICE simply assumes that the generator does not have this property, usually a good assumption since SPICE computes approximate sample values, not exact continuous values, for the generator and could not tell if there is an exact repetition in any event. Therefore, TINV will always be output as the fundamental frequency f_o of the Fourier series that SPICE computes and be listed as component 1 (fundamental component) of the output, as shown next.

SPICE Example 17.19

As an example, let us find the Fourier coefficients of the waveform of Fig. 17.3. To perform a transient analysis, we need to define a circuit to which the Fourier decomposition can be applied to the response. Suppose that we use the simple circuit composed of a current source with an output flowing from the reference node to node 1 equal to that of Fig. 17.3 connected to a 1-Ω resistor. The Fourier coefficients for this voltage can then be calculated using the .FOUR command. A circuit file that will produce this result using the PWL (piecewise linear waveform transient specification in I) is

```
EXAMPLE FOR THE FOURIER SERIES OF FIG. 17.3
*DATA STATEMENTS
I 0 1 PWL(0 0A 3.1416 3.1416A 3.1417 -3.1416A 6.2832 0A)
R 1 0 1
.TRAN 0.1 6.2832
.FOUR 0.159 V(1)
.END
```

In this program, the generator duration is defined as the interval from 0 to 6.2832 s, the transient response interval. Therefore, a fundamental frequency of 0.159 Hz is used in the $\cdot$ FOUR command. The resulting solution for the decomposition is

```
FOURIER COMPONENTS OF TRANSIENT RESPONSE V(1)
DC COMPONENT = -2.497780E-05
HARMONIC  FREQUENCY  FOURIER    NORMALIZED   PHASE       NORMALIZED
NO        (HZ)       COMPONENT  COMPONENT    (DEG)       PHASE (DEG)
   1      1.590E-01  2.000E+00  1.000E-00    5.805E-04   0.000E+00
   2      3.180E-01  1.001E+00  5.002E-01   -1.800E-02  -1.800E+02
   3      4.770E-01  6.676E-01  3.338E-01    1.936E-04  -3.869E-04
   4      6.360E-01  5.013E-01  2.506E-01    1.800E-02   1.800E+02
   5      7.950E-01  4.016E-01  2.008E-01    1.156E-04  -4.649E-04
   6      9.540E-01  3.353E-01  1.676E-01    1.800E-02   1.800E+02
   7      1.113E+00  2.880E-01  1.440E-01    8.192E-05  -4.986E-04
   8      1.272E+00  2.526E-01  1.263E-01   -1.800E-02  -1.800E+02
   9      1.431E+00  2.252E-01  1.126E-01    6.606E-05  -5.144E-04
```

Once the Fourier coefficients are determined, the series can be written using (17.94) to describe the original function. The response of a network to a periodic function can now be found by applying the $\cdot$ DC command for the dc component d_0 to the circuit and the $\cdot$ AC command for each harmonic component (amplitude D_n, frequency nf_0, and phase ψ_n). The solution, by superposition, is then the sum of the individual components in the time domain.

SPICE may also be used to find 19 samples from the Fourier transform $\mathbf{F}(j\omega)$ of certain time functions $f(t)$, those that are zero outside a finite interval of time. Suppose that $f(t)$ is zero outside the time interval $[t_0, t_0 + T]$. Let $\omega_1 = 2\pi/T$. Evaluating (17.75), the integral defining the Fourier transform $\mathbf{F}(j\omega)$, at $\omega = n\omega_1$, is

$$\mathbf{F}(jn\omega_1) = \int_{-\infty}^{\infty} f(t)e^{-jn\omega_1 t}\, dt$$

Since $f(t)$ is zero outside the time interval $[t_o, t_o + T]$, we may reduce the limits of integration and then note that the resulting integral is identical to (17.25), the equation for the exponential Fourier series coefficient c_n when scaled by T:

$$\mathbf{F}(jn\omega_1) = \int_{t_0+T}^{t_0} f(t)e^{-jn\omega_1 t}\, dt = Tc_n \qquad (17.96)$$

Thus SPICE may be used to compute a set of samples from the Fourier transform, $\mathbf{F}(jn\omega_1)$, $n = 0, \pm 1, \ldots, \pm 9$, of a function $f(t)$ zero outside a finite interval $[t_o, t_o + T]$. $f(t)$ is treated as the generator $f_T(t)$ for a Fourier series analysis as described above and the resulting $\cdot$ FOUR Fourier series coefficients are converted to exponential form. These are in turn scaled by T, the duration of $f_T(t)$, and we have 19 samples of the Fourier transform of $f(t)$, that is, $\mathbf{F}(j\omega)$ for $\omega = n\omega_1$, $n = 0, \pm 1, \ldots, \pm 9$.

Let us find some samples from the Fourier transform $\mathbf{F}(j\omega)$ of the time function $f(t)(u(t + \pi) - u(t - \pi))$, for the $f(t)$ sketched in Fig. 17.3. For $T = 2\pi$, we have $\omega_1 = 1$, and the c_n in (17.95) are the exponential Fourier series coefficients corresponding to the coefficients found in Example 17.19. For instance, the fundamental, or harmonic component 1, found in that example was found to be $D_1 = 2.00$ and $\psi_1 = 0.000584°$. Let us round this result to $2\sin t$. In terms of the trigonometric Fourier series coefficients, this implies $a_1 = 0$ and $b_1 = 2$. Thus

$$c_1 = \tfrac{1}{2}(a_1 - jb_1) = -j$$

By (17.95),

$$\mathbf{F}(j\omega)\Big|_{\omega=1} = Tc_n = -j2\pi$$

In other words, the Fourier transform of the time-limited function $f(t)(u(t + \pi) - u(t - \pi))$ has the value $-j2\pi$ at $\omega = 1$. Using the remaining coefficients from Example 17.19, we may compute $\mathbf{F}(j\omega)$ at $\omega = n\omega_1 = n$, $n = 0, \pm 1, \ldots, \pm 9$.

Finally, certain dialects of SPICE permit more complete information on the Fourier transform to be gathered. Using the postprocessor PROBE available with PSpice, the magnitude and phase, or if preferred the real and imaginary parts, of $F(j\omega)$ may be displayed as high-resolution graphics plots. In this book we have chosen to concentrate on those core features available using "plain-vanilla" SPICE and thus leave the reader with access to PSpice with PROBE to explore this useful feature on her or his own.

EXERCISES

17.8.1. Using SPICE, find the Fourier coefficients for the waveform of Fig. 17.2(b).

17.8.2. Using SPICE, find the Fourier coefficients for the waveform shown if $T = 10 \ \mu s$ and $A = 1$.

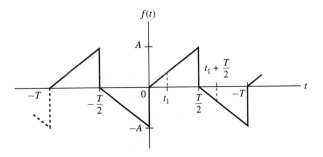

EXERCISE 17.8.2

Fourier analysis is the decomposition of a waveform into sinusoids of different frequencies, amplitudes, and phases. Suppose that the input to a linear circuit is analyzed this way. Since the gain and phase shift of each sinusoidal component may easily be determined from the frequency response, and the output is simply their sum, Fourier analysis is often a convenient path to determining a circuit's response.

- Periodic functions are represented by Fourier series, nonperiodic functions by Fourier transforms.

- The trigonometric Fourier series is an infinite sum of sines and cosines at integer multiple, or harmonic, frequencies. Its Fourier coefficients are real numbers.

- The exponential Fourier series is an infinite sum of complex exponentials of integer multiple (harmonic) frequencies. Its Fourier coefficients are complex numbers.

- The two forms of the Fourier series are related by $c_n^* = (a_n + jb_n)/2$, where c_n is the exponential Fourier coefficient, a_n the cosine coefficient, and b_n the sine coefficient.

- $|c_n|^2$ equals the power in the nth frequency component $n\omega_o$.

- The Fourier transform is defined as the integral transform

$$\mathbf{F}(j\omega) = \int_{-\infty}^{+\infty} f(t)e^{-j\omega t}\, dt$$

- $|\mathbf{F}(j\omega)|$ corresponds to the strength of the sinusoidal component at frequency ω, $\underline{/\mathbf{F}(j\omega)}$ to the phase shift at frequency ω.

- The Fourier transform of a sum is the sum of Fourier transforms; scaling a time function scales its transform equally. Fourier transformation is a linear operation.

- Table 17.1 of Fourier transform operations may be used to derive new transforms from known ones.

- The Fourier transform of the impulse response $h(t)$ is the frequency response $\mathbf{H}(j\omega)$.

PROBLEMS

17.1. Let $f_1(t) = u(t+1) - u(t-1)$. Sketch the periodic extensions of each of the following generators taken from $f_1(t)$ and indicate the period. T is the duration and t_0 the initial time for the generator.
(a) $T = 4, t_0 = -2$
(b) $T = 10, t_0 = -2$
(c) $T = 2, t_0 = 0$

17.2. Let $f_2(t) = r(t+1) - 2r(t) + 2r(t-1) - 2r(t-2) + r(t-3)$. Sketch the periodic extension of the following generators taken from $f_2(t)$ and indicate their period. T is the duration and t_0 the initial time for the generator.
(a) $T = 2, t_0 = -1$
(b) $T = 4, t_0 = -1$
(c) $T = 8, t_0 = -4$

17.3. Show that if the even part of $f(t)$ is identically zero, then $f(t)$ has odd symmetry and if the odd part is zero it has even symmetry.

17.4. Suppose $f_1(t)$ and $f_2(t)$ are periodic with periods nT_1 and mT_2, repeatedly, where n and m are non-negative integers.
(a) Is $\alpha f_1(t) + \beta f_2(t)$ periodic? If so, what is its period?
(b) Is $f_1(t)f_2(t)$ periodic? If so, what is its period?
(c) Is $(1 + f_1(t))f_2(t)$ periodic? If so, what is its period?

17.5. Find the even and odd parts $f_e(t)$ and $f_o(t)$ if $f(t) = :$
(a) $f_2(t)$ from Problem 17.2
(b) $\sin^2 2\pi t$

(c) $4\cos(2t + 17°)$

(d) $4\cos(2t + 17°)(u(t - 1) - u(t + 1))$

17.6. If the t, $f(t)$ axes are translated to the new τ, $g(\tau)$ axes shown so that the origin for the new axes is the point (t_0, f_0), the relations between the variables are $g = f - f_0$ and $\tau = t - t_0$. Show that

$$f(t) = f_0 + g(t - t_0)$$

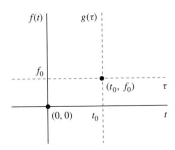

FIGURE P17.6

17.7. Determine f_{av}, $|f|^2_{av}$, and f_{rms} for the periodic extensions of each of the generators described in Problem 17.1.

(a) $T = 4, t_1 = -2$

(b) $T = 10, t_0 = -2$

(c) $T = 2, t_0 = 0$

17.8. Determine f_{av}, $|f_{av}|^2$, and f_{rms} for the following $f(t)$'s:

(a) Fig. 17.2(a)

(b) Fig. 17.2(b)

(c) Fig. 17.2(c)

(d) Fig. 17.2(d)

17.9. For $f(t)$ complex, if Re $f(t)$ is periodic with period T_1 and Im $f(t)$ is periodic with period T_2, is $f(t)$ necessarily periodic? Justify.

17.10. Find the trigonometric Fourier series for the function

$$f(t) = 4|\cos 2t|$$

17.11. Find the trigonometric Fourier series for the function

$$f(t) = 4\sin(2t + \pi/3), \quad -\frac{\pi}{6} < t < \frac{\pi}{3}$$

$$= 0, \quad \frac{\pi}{3} < t < \frac{5\pi}{6}$$

$$f(t + \pi) = f(t)$$

17.12. Obtain the trigonometric series of $f(t)$ defined by

$$f(t) = 1, \quad -1 < t < 1$$

$$= 0, \quad 1 < |t| < 2$$

$$f(t + 4) = f(t)$$

17.13. Find the trigonometric Fourier series for the periodic waveform shown.

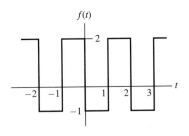

FIGURE P17.13

17.14. Find the trigonometric Fourier series for the periodic extensions of each of the generators taken from $f_1(t)$ in Problem 17.1:

(a) $T = 4, t_0 = -2$

(b) $T = 10, t_0 = -2$

(c) $T = 2, t_0 = 0$

17.15. Find the trigonometric Fourier series for $g(t) = df/dt$, with $f(t)$ shown in the figure of Problem 17.13.

17.16. (a) Find the trigonometric Fourier series for $f(t)$ shown.

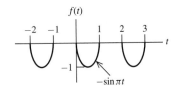

FIGURE P17.16

(b) Find the trigonometric Fourier series for $f(t - 1)$.

(c) Use linearity to find the trigonometric Fourier series for $f(t - 1) - f(t)$.

17.17. Fnd the trigonometric Fourier series for the $f(t)$ shown ($T = 4$). For $-1 \leq t \leq 1$, $f(t) = -t^2 + 1$.

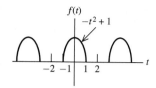

FIGURE P17.17

17.18. Find the trigonometric Fourier series for the $f(t)$ shown. The equation is

$$f_T(t) = \begin{cases} \sin t, & -\pi \le t < +\pi \\ -\sin t, & -2\pi \le t < -\pi \text{ and } \pi \le t < 2\pi \end{cases}$$

with $T = 4\pi$ and $t_0 = -2\pi$.

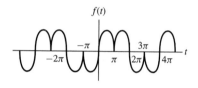

FIGURE P17.18

17.19. Find the trigonometric Fourier series for the functions

(a) $f(t) = 2$, $0 < t < \pi$
      $= 1$, $\pi < t < 2\pi$
     $f(t + 2\pi) = f(t)$
(b) $f(t) = e^t$, $-\pi < t < \pi$
     $f(t + 2\pi) = f(t)$
(c) $f(t) = t^2$, $-1 < t < 1$
     $f(t + 2) = f(t)$

17.20. Find the trigonometric Fourier series for the functions

(a) $f(t) = 2t + 1$, $0 < t < 1$
      $= 2t - 1$, $-1 < t < 0$
     $f(t + 2) = f(t)$
(b) $f(t) = |t|$, $-1 < t < 1$
     $f(t + 2) = f(t)$
(c) $f(t) = \dfrac{A}{T}t$, $0 < t < T$
     $f(t + T) = f(t)$
(d) $f(t) = 1 - |t|$, $-1 < t < 1$
     $f(t + 2) = f(t)$

17.21.

(a) Show that for $f(t)$ with half-wave symmetry, if $f(t)$ is even, all the Fourier coefficients are zero except

$$a_{2n-1} = \frac{8}{T} \int_0^{T/4} f(t) \cos(2n - 1)\omega_0 t \, dt$$

and if $f(t)$ is odd, all the coefficients are zero except

$$b_{2n-1} = \frac{8}{T} \int_0^{T/4} f(t) \sin(2n - 1)\omega_0 t \, dt$$

(b) Find the trigonometric Fourier series for the function $f(t)$, one period of which is shown. (Note that this function is even and half-wave-symmetric.)

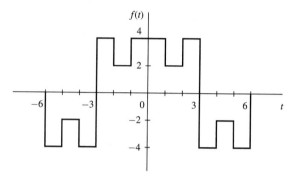

FIGURE P17.21

17.22. Let a_n, b_n be the Fourier series coefficients for $f(t)$. Define $g(t) = f(t - \tau)$ to be $f(t)$ time shifted to the right by τ. For each of the following τ's, determine a_n', b_n', the coefficients for $g(t)$, in terms of a_n and b_n. T is the period of $f(t)$.

(a) $\tau = T/2$
(b) $\tau = T$
(c) $\tau = +T/4$

17.23. Suppose a time function $f(t)$ is time scaled to $g(t) = f(ct)$. Relate the coefficients a_n, b_n and the ω_0 of the Fourier series for the two functions $f(t)$ and $g(t)$.

17.24. Find the trigonometric Fourier series for df/dt if $f(t)$ is the sawtooth waveform shown in Fig. 17.3.

17.25. Let $f(t)$ be the periodic extension of the generator $f_T(t) = te^{-t}$ with $t_0 = 0$ and $T = 2$. Find the trigonometric Fourier series.

17.26. Find the exponential Fourier series for $f(t)$ of Problem 17.13.

17.27. Find the exponential Fourier series for the periodic extension of each of these generators taken from $f_1(t)$ in Problem 17.1.

(a) $T = 4, t_0 = -2$
(b) $T = 10, t_0 = -2$
(c) $T = 2, t_0 = 0$

17.28. Find the exponential Fourier series for the periodic extension $f(t)$ of the generator $f_T(t) = u(t-1) + \delta(t)$ with $t_0 = -1$, $T = 4$.

17.29. Let $g(t) = f(ct)$ so that $g(t)$ is a time scaled version of $f(t)$ with time compression factor c. Relate c_n, the exponential Fourier series coefficients of $f(t)$, to c_n', the coefficients for $g(t)$.

17.30. Let $g(t) = f(t-\tau)$. Relate the exponential Fourier series coefficients c_n of $f(t)$ and c_n' of $g(t)$.

17.31. (a) What does it signify about $f(t)$ if all its exponential Fourier series coefficients are purely real?
(b) Repeat for purely imaginary.

17.32. Find the exponential Fourier series for $f(t) = 2\left|\cos t - \frac{1}{2}\right|$.

17.33. Show that if $f(t)$ has half-wave symmetry, that is, $f[t + T/2] = -f(t)$, then $c_n = 0$ for all n even.

17.34. Find the amplitudes A_n as in (17.53) of the dc, fundamental, and second harmonic components of $i(t)$ in Fig. 17.7 if $v_1(t) = f(t)$ in Fig. 17.3.

17.35. Find the complex Fourier series for $v(t)$ if $i_{s_1}(t)$ is the waveform illustrated in the following figures. Take $T = 1$ in each case, and $v_{s_1}(t) = 12$ V dc.
(a) Fig. 17.2(a)
(b) Fig. 17.2(b)
(c) Fig. 17.2(c)

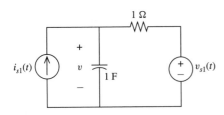

FIGURE P17.35

17.36. Find the forced response $v(t)$ if v_g is the function of Problem 17.12.

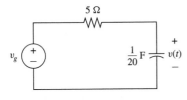

FIGURE P17.36

17.37. Find the first three terms of the forced response $i(t)$ if $v_g(t)$ is the function of Problem 17.12.

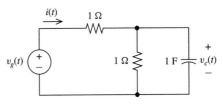

FIGURE P17.37

17.38. Determine the forced output $i(t)$ if the input $i_s(t)$ is the periodic extension of the generator $i_{s_T}(t) = \delta(t)$ with $t_0 = -1$ and $T = 2$. Express as in (17.53).

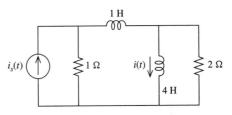

FIGURE P17.38

17.39. Determine the value of the amplitudes A_n of the dc component and the first three harmonics of the forced response $v(t)$ if $v_s(t)$ is the square wave $f(t)$ of Example 17.5.

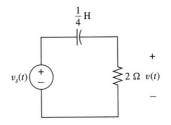

FIGURE P17.39

17.40. Find the power delivered by the source in the circuit of Fig. P17.37 if v_g is the function of Problem 17.19(a).

17.41. Find the discrete amplitude, phase, and power spectra for $i(t)$ in Problem 17.39. Sketch.

17.42. Find the discrete amplitude, phase, and power spectra for $v(t)$ in Problem 17.38. Sketch.

17.43. What fraction of the total power in the full-wave rectified waveform of Fig. 17.2(b) is in its dc component? In the dc plus fundamental components?

17.44. What is the strongest frequency component in $f(t)$? If $f(t) = v_s(t)$ is high-pass filtered by the circuit of Problem 17.39, creating $v(t)$, what is its strongest frequency component then?

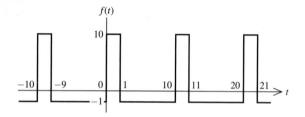

FIGURE P17.44

17.45. Find the Fourier transforms of
(a) $f(t) = u(t-a) - u(t-b), a < b$
(b) $f(t) = e^{-at} \cos bt \, u(t), a > 0$
(c) $f(t) = e^{-at} \sin bt \, u(t), a > 0$
(d) $f(t) = e^{-at}[u(t) - u(t-2)]$

17.46. Determine the Fourier transform.

FIGURE P17.46

17.47. Find the Fourier transform of $f(t)$.

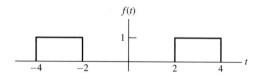

FIGURE P17.47

17.48. Find the Fourier transforms of $e^{-\alpha t} \cos \beta t u(t)$ and $e^{+\alpha t} \cos \beta t u(-t)$ ($\alpha > 0$).

17.49. Find the transform $D(j\omega)$ of

$$d(t) = 1/\Delta \left[u\left(t + \frac{\Delta}{2}\right) - u\left(t - \frac{\Delta}{2}\right) \right]$$

Noting that $d(t) \to \delta(t)$ as $\Delta \to 0$, show that the limit of $D(j\omega)$ as $\Delta \to 0$ coincides with the Fourier transform of $\delta(t)$ computed directly from the integral definition 17.75.

17.50. Solve the following differential equations by Fourier transforms:
(a) $\dfrac{dy}{dt} + 2y = 0$

(b) $\dfrac{dy}{dt} + 2y = e^{-t}u(t)$

(c) $\dfrac{d_2 y_2}{dt^2} + 3\dfrac{dy}{dt} + 2y = e^{-|t|}$

17.51. Find the inverse Fourier transforms.
(a) $\dfrac{6}{j\omega - 3}$

(b) $\dfrac{6}{\omega^2 + 9}$

(c) $\dfrac{2}{(j\omega - 1)(j\omega - 3)}$

17.52. A circuit has input $v_s(t) = e^{-t}u(t)$ and output $v(t) = (3te^{-t} + 4e^{-t})u(t)$. Find the frequency response function $H(j\omega)$ and the output $v(t)$ if the input is $v_s(t) = e^{-|t|} \sin t$.

17.53. Use linearity and a trigonometric identity to find the Fourier transform of

$$f(t) = e^{-t} \sin\left(2t + \frac{\pi}{4}\right) u(t)$$

SPICE Problems

17.54. Use SPICE to determine the values of the first 10 coefficients and phase angles for the periodic extension with period $T = 6$ of the $f(t)$ of Figure 17.3.

17.55. Find the central 19 values of the amplitude, phase, and power spectra of the periodic extension of the generator $f(t) = te^{-1000t}$ with initial time $t_0 = 0$ and $T = 3$ ms using SPICE.

17.56. Compute the exponential Fourier series coefficients $c_n, n = 0, \pm1, \pm2$ for the impulse train $f(t)$ with period $T = 2\pi$ if $f(t) < \delta(t)$ in the period $-\pi \le t < \pi$. Convert to the form (17.88) used by SPICE. Now use SPICE to determine how small Δ must be in the approximation $\delta(t) \approx d(t) = 1/\Delta \, [u(t) - u(t - \Delta)]$ in order that the dc and first two harmonic frequency components reported by SPICE agree in both amplitude (d_0, D_n) and phase (ψ_n) to within two significant figures. Any Δ satisfying this requirement is acceptable.

17.57. Use SPICE to help sketch $|F(j\omega)|$ for $f(t) = (1 - e^{-t})(u(t) - u(t - 2))$. First, find a simple single-loop circuit containing a dc source whose forced response is $v(t) = 1 - e^{-t}$. Then use $v(t), 0 \le t < 2$ and SPICE to find $|F(j\omega)|, \omega = n\pi$ for $n = 0, \pm1, \pm2, \dots, \pm9$. Sketch $|F(j\omega)|$ based on these sample values.

More Challenging Problems

17.58. Let $f_1(t) = A_1 \cos(\omega_1 t + \theta_1)$, $f_2 = A_2 \cos(\omega_2 t + \theta_2)$. Find necessary and sufficient conditions of ω_1 and ω_2 so that $f_1(t) + f_2(t)$ is periodic, repeat for $f_1(t) \cdot f_2(t)$.

17.59. Let $f_T(t)$ be the generator for its periodic extension $f(t)$, and let $g_T(t) = \int_{t_0}^{t} f_T(\tau) \, d\tau$ be the generator for g(t). Relate the coefficients a_n, b_n and the ω_0's for the trigonometric Fourier series for $f(t)$ and $g(t)$.

17.60. Repeat Problem 17.35 but with $v_{s1}(t) = h(t)$ shown in Fig. 17.1(b), and $i_s(t)$ the $f(t)$, $T = 1$, in:
(a) Fig. 17.2(a)
(b) Fig. 17.2(b)
(c) Fig. 17.2(c)
(d) Fig. 17.2(d)

17.61. Consider a system where an input signal $x(t)$ is multiplied by a function $m(t) = 2 + \cos \omega_0 t$. The response of the system is $y(t) = m(t)x(t)$ and the signal is said to have been *amplitude modulated*. (a) Find $\mathbf{Y}(j\omega)$ in terms of $\mathbf{X}(jw)$. (b) If

$$\mathbf{X}(j\omega) = 1, \qquad |\omega| < \omega_c$$

$$= 0, \qquad \text{elsewhere}$$

and $\omega_0 > 2\omega_c$, sketch the graph of $\mathbf{Y}(j\omega)$. (c) If $y(t)$ is passed through an ideal low-pass filter with cutoff frequency ω_{LP} and unity gain such that $\omega_c < \omega_{LP} < \omega_0 - \omega_c$, what is the filter output?

17.62. Show that if $f(t)$ is periodic, df/dt is also periodic and its period is less than or equal to T. Give examples of both cases. Is $\int_{-\infty}^{t} f(\tau)d\tau$ periodic? Under what conditions?

17.63. Find the trigonometric Fourier series for the $f(t)$ shown. The generator $(t_0 = -1, T = 2)$ is $f_T(t) = \frac{1}{\Delta}(\delta(t - \Delta) - \delta(t + \Delta))$. Then find the limiting Fourier series as $\Delta \to 0$.

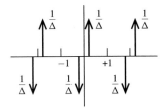

FIGURE P17.63

17.64. Use SPICE to determine 10 Fourier series coefficients for the periodic extension of the generator $f_T(t) = (t + 1)^2(u(t) - u(t - 2))$. Use initial time $t_0 = -1$ and duration $T = 2$.

18

Design of Linear Filters

Benjamin Franklin
1706–1790

He snatched the lightning from the skies and the sceptre from tyrants.

Anne Turgot

The energetic, optimistic spirit with which Benjamin Franklin tackled problems as diverse as understanding electricity, seeing his country through the treacherous war which gave it birth, and even the practical problem of securing a handsome living, came to exemplify the can-do American temperament. Equally well-beloved in Europe for his daring experiments with lightning and for the witty Poor Richard's Almanac, Franklin entered into secret negotiations in Paris shortly after the Declaration of Independence in 1776, which led to France's fateful, perhaps pivotal, entry into the Revolutionary War two years later on the American side.

Benjamin Franklin was born in Boston, Massachusetts in 1706. A childhood apprenticeship to his brother, a printer, gave him an early love for the written word and a self-taught mastery of language. In his early twenties he produced the most respected colonial newspaper, composed the widely popular Poor Richard's Almanac filled with his beloved aphorisms, and after convincing the colonies of the need, even printed their first paper currency. Fascinated upon seeing sparks emanating from a Leyden jar, Franklin reasoned that lightning has much the same character, and that the earth and sky must form a massive electric cell. In 1752 he performed his famous experiment with kite and key, confirming the common nature of lightning and electricity. It is well for the colonies that luck was with him; the next two who tried this experiment both died of electrical shock. Based on this and other experiments, Franklin developed the single-fluid theory of electrical flow, inventing terms such as battery, positive and negative charge. His ingenuity and scope as an inventor is demonstrated in the Franklin stove, bifocal glasses and the lightning rod, each still in common use today.

Chapter Contents

The approaches we have explored in the course of our study have been wide ranging, beginning with real algebraic and differential equations in the time domain, through phasors in the static complex domain, to the transform domains of Laplace and Fourier. Whatever the approach, our focus has remained the same: to advance our growing ability to formulate and solve the most general circuit analysis problems. Excepting only those few sections entitled "Design of ...," we have not looked beyond strengthening our scope and ability at circuit analysis.

Why have we so adamantly preferred analysis to design? While being able to analyze a given circuit certainly has some satisfactions, to most students of electric circuits, analysis is a means to an end rather than an end in itself. The real goal is to be able to create interesting and useful circuits, that is, to do circuit design. With perhaps a hint of regret, it was observed in the initial chapter of this book that it is vital to give analysis primacy over design because meaningful design requires and presupposes effective analysis. In retrospect this was certainly reinforced by our experience in the design examples and exercises throughout this book, in which analysis always played a prominent role. The payoff for studying circuit analysis comes when we confront nontrivial design challenges armed with the appropriate tools to address them.

In this chapter we conclude by addressing one very important class of circuit design problems, the design of linear filters. First we consider the design of passive filters, those consisting exclusively of passive RLC elements. Adding op amps and their required power supplies results in active filters, which can exhibit frequency response characteristics difficult or impossible for passive designs, and in addition permit inductor-free circuits. A general method for designing active filters of any order based on transfer function realization is presented. Classical filters, those with optimal properties such as the Butterworth, Chebyshev, and Bessel families, are next defined and their design considered. The chapter concludes with a look at filter transformation and scaling, operations that greatly expand the flexibility of a single basic design.

Although this chapter only samples the field of filter design, it clearly demonstrates the power and versatility of the analysis-based approach to circuit design. At the end of our journey, the careful reader is rewarded with a perspective from which development of real skill at circuit design may confidently be undertaken.

18.1 PASSIVE FILTERS

Filters, introduced in Chapter 14, are circuits whose input–output behavior depends strongly upon frequency. For instance, lowpass filters respond to low-frequency inputs, those below a selected cutoff frequency, by passing them to the output relatively unchanged while attenuating higher-frequency inputs. Filters are perhaps the most ubiquitous of electric circuits, embedded in practically every electronic system of any complexity.

Since inductors and capacitors have impedances that vary with frequency, filter action can be induced from passive *RLC* circuits. Passive filters should be considered where the power supply requirement of the alternative, active filters, is a significant disadvantage (e.g., in battery-operated devices). In addition, they often have a low component count, are inexpensive and reliable, and generally tolerate higher power levels than do most active filters. Passive filters have their limitations and disadvantages as well, including the requirement that inductors be included in most designs. Inductors tend to be bulky, nonideal, and poorly suited to miniaturization and integrated circuits.

Our first design example is a crossover filter for a two-loudspeaker system. Since the range of frequencies easily reproduced by a loudspeaker depends strongly upon its size and materials, it is usually more cost-effective, for high-quality reproduction, to combine large "woofers" for low frequencies and much smaller "tweeters" for higher frequencies. It would be inefficient to present the entire input signal to each speaker, since each will burn power for inputs falling outside its frequency range without producing much audible output. Crossover filters match the signal components with the speakers, routing lower frequencies to the woofer and higher frequencies to the tweeter. This is a favorable setting for the use of passive filters since relatively high power must be handled and the size of the speakers themselves can easily accommodate the bulk of the inductors used in the design.

Design Example 18.1

We wish to design a crossover network for a two-loudspeaker system. The woofer and tweeter are both modeled as 8-Ω resistors, and the power amplifier feeding the loudspeaker system has 8-Ω resistive output impedance. Assume that the woofer transduces electric to acoustic energy relatively efficiently for frequencies below 1 kHz and the tweeter for those above 1 kHz. Our design goals are to pass the appropriate frequency ranges to each speaker while maintaining a good impedance match at all frequencies to the power amplifier driving the system.

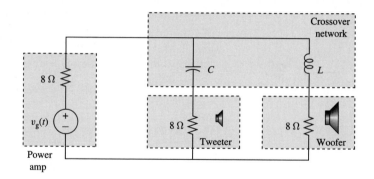

FIGURE 18.1 Two-way speaker system with crossover circuit.

A simple but effective approach is to put the woofer in series with an inductor L, the tweeter with a capacitor C, and place these two circuits in parallel across the input from the power amplifier as shown in Fig. 18.1. The inductor will present high impedance to high frequencies, blocking undesired current flow into the woofer in that frequency range, and the capacitor will do likewise for the low-frequency range.

In order that the crossover frequency (at which the magnitude of the impedance, and thus the power delivered, to the two arms of the circuit are equal) be set to a desired value ω_r, it is required that

$$|j\omega_r L| = \left| \frac{1}{j\omega_r C} \right| \qquad (18.1)$$

or

$$\omega_r = \sqrt{\frac{1}{LC}} \qquad (18.2)$$

The cutoff frequency is just the resonant frequency of the LC components. Since

$$\omega_r = 2\pi(1000) = 6283 \text{ rad/s}$$

this yields the condition

$$LC = \frac{1}{\omega_r^2} = 2.53 \times 10^{-8} \qquad (18.3)$$

While any pair of LC values satisfying (18.1) will result in higher power dissipation in the woofer at input frequencies below 1 kHz and in the tweeter above 1 kHz as we desire, the overall impedance needs to be considered. The equivalent impedance of the system at frequency ω is, by the parallel combination law,

$$Z(j\omega) = \frac{(8 + j\omega L)[8 - j(1/\omega C)]}{16 + j(\omega L - 1/\omega C)} \qquad (18.4a)$$

At very low frequencies the capacitor blocks current to the tweeter and the inductor is a very low impedance, so $Z(j\omega)$ goes to 8 Ω. Similarly, at high frequencies the impedance goes to 8 Ω, as can be seen from (18.4a). Thus for very low and very high frequencies we have an exact impedance match with the 8-Ω source, yielding maximum power transfer. At the cutoff frequency ω_r, substituting

$$\omega_r L = \frac{1}{\omega_r C} = \sqrt{\frac{L}{C}}$$

into (18.4a) yields

$$Z(j\omega_r) = \frac{(8 + j\sqrt{L/C})(8 - j\sqrt{L/C})}{16} = 4 + \frac{L}{16C} \qquad (18.4b)$$

To have an exact impedance match at the crossover frequency we need

$$Z(j\omega_r) = 4 + \frac{L}{16C} = 8$$

or
$$L = 64C \qquad (18.5)$$

Substituting the last into (18.3) we have $C^2 = 2.53 \times 10^{-8}/64$ or

$$C = 19.9 \ \mu\text{F}, \qquad L = 1.27 \ \text{mH} \qquad (18.6)$$

For these values, the impedance of the loudspeaker system is exactly matched to the 8-Ω power amplifier output impedance for very low frequencies, very high frequencies, and at the 1-kHz crossover. In between the impedance remains very close to the value desired, as can be verified with the help of SPICE.

```
Design Example 18.1: overall loudspeaker impedance
*
IS         1          0          AC         1
L          1          2          1.27M
C          1          3          19.9U
RWOOF      2          0          8
RTWEET     3          0          8
.AC    DEC          10           1          1MEG
.PRINT AC      VM(1)      VM(2)      VM(3)
.END
```

The output VM(1) from this run is the magnitude of the impedance of the loudspeaker, since we are using a $1\underline{/0}$ A current source and measuring the resulting voltage magnitude. The values printed all remain within 0.1% of 8 Ω. Following are two lines of output selected from this run:

```
FREQ        VM(1)      VM(2)     VM(3)
1.00E+02   8.00E+00  7.96E+00 7.88E-01
1.00E+05   8.02E+00  6.37E-02 8.00E+00
```

The ratio of the power delivered to the speakers at a given frequency is the ratio of their respective voltage phasor magnitudes squared. Thus at 100 Hz there is $(7.96/0.788)^2 = 102$ times as much power delivered to the woofer as the tweeter, and at 100 kHz there is $(8.00/0.0637)^2 = 15{,}770$ times as much delivered to the tweeter. The crossover network is doing its job.

Design Example 18.2

Let us expand the previous two-way design to accommodate a three-way speaker system. The added midrange speaker, of intermediate size, is most efficient at reproducing frequencies between the bass and treble extremes at which the large woofer and small tweeter operate best. All three of the speakers are modeled as 8-Ω resistors, and the center frequency of the midrange speaker is taken to be 1 kHz. Our goal, as before, is to pass the appropriate frequency range to each speaker while maintaining the overall impedance reasonably close to 8 Ω.

In the two-way design, we added storage elements in series with the woofer and tweeter which were selected to have high impedance, thus blocking current, in the frequency range better suited to the other speaker. Following this idea, we will block current to the midrange speaker at both high and low frequencies. The series LC circuit has this characteristic, and the resulting three-way speaker system is shown in Fig. 18.2.

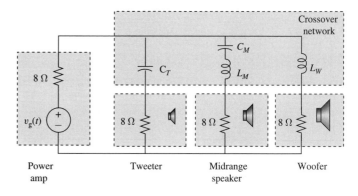

FIGURE 18.2 Three-way speaker system with crossover circuit.

Consider its behavior at very low frequencies. For any values of the storage elements L_B, C_T, L_M, and C_M making up the crossover circuit, the resulting circuit will indeed block current flow through the two less-efficient speakers while maintaining an overall impedance very close to 8 Ω. The same is true at very high frequencies. At 1 kHz we wish the impedances of the woofer and tweeter circuits to be high and that of the midrange circuit to be near 8 Ω. This will route power to the proper speaker while maintaining the overall impedance near its nominal value of 8 Ω.

By (18.4b), the equivalent impedance of the parallel woofer and tweeter circuits at their crossover frequency of 1000 Hz increases with L_W, the inductor in the bass circuit. We want this to be as large as possible, blocking current to both these speakers in the midrange frequencies. Assume that a 15-mH inductor is the largest available whose series resistance is sufficiently low to ensure that the inductor will not overheat in use. Then by (18.2),

$$\sqrt{\frac{1}{(0.015)C_T}} = 2\pi(1000)$$

yields $C_T = 1.69 \ \mu F$. Note that we can also use these LC values for the midrange crossover elements, $L_M = L_W = 15$ mH and $C_M = C_T = 1.69 \ \mu F$, as we wish the same resonant frequency for the $L_M C_M$ series resonant circuit, 1000 Hz. When this circuit is used in a SPICE simulation as described for the two-way design example above, it is found that while all other criteria are satisfied, the overall impedance of the three-way system is excessively high at frequencies near the midrange-woofer transition frequency and the midrange-tweeter transition frequency (frequencies of equal power dissipation). For instance, the impedance rises to over 270 Ω at 600 Hz and 1.66 kHz. The Q value of the midrange series resonant circuit is too high, causing its impedance to increase too quickly off-resonance. This is remedied by adjusting L_M downward while incrementing C_M upward to keep the resonant frequency fixed. After several SPICE runs while modifying these two elements, it is found that with $L_M = 180 \ \mu H$ and $C_M = 141 \ \mu F$, performance is satisfactory. The impedance is in the range 8 to 10.6 Ω everywhere, and the power allocation is as desired: at 25 Hz over 90% of the power is delivered to the woofer, at 1000 Hz over 90% to the midrange, and at 25 kHz over 90% to the tweeter.

Design Example 18.3

We wish to supply a 300-Ω resistive load with 60-Hz power drawn from the inverter power supply shown in Fig. 18.3(a). Its Thevenin equivalent impedance at 60 Hz is $300\underline{/45°}$ Ω and the open-circuit (unloaded) power supply voltage is a square wave with period $\frac{1}{60}$ s.

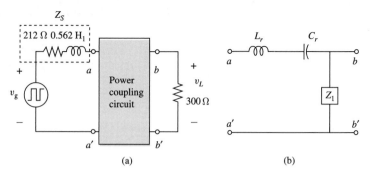

FIGURE 18.3 (a) Overall circuit; (b) power coupling filter.

This waveform contains power at 60 Hz and harmonic components at 120 Hz, 180 Hz, and so on. Our goal is to design a power coupling filter to maximize power to the load at 60 Hz while minimizing the power delivered at all harmonic frequencies. By placing a series LC resonant circuit tuned to 60 Hz as one arm of a voltage divider as in Fig. 18.3(b), we will have the full power supply terminal voltage across the load at 60 Hz, and less at all other frequencies. This will suppress the harmonic frequencies present. We may then determine the nature of an impedance Z_1 placed across the load which maximizes power to R_L at 60 Hz without considering the LC resonant circuit, which will have zero impedance at 60 Hz.

Let Z_p be the parallel combination of Z_1 and the 300-Ω resistive load. The power to the load at 60 Hz is given by $|\mathbf{V}_L|^2/300$, where, by voltage division,

$$|\mathbf{V}_L| = \left| \frac{Z_p}{Z_p + Z_s} \right| |\mathbf{V}_g|$$

and $\mathbf{V}_g$ is the open-circuit source rms voltage phasor at 60 Hz. Since $\mathbf{V}_g$ is fixed, power at 60 Hz is maximized by maximizing $|Z_p|/|Z_p + Z_s|$. But

$$\left| \frac{Z_p}{Z_p + Z_s} \right| = \left| \frac{1/Y_p}{1/Y_p + 1/Y_s} \right| = \left| \frac{Y_s}{Y_s + Y_p} \right|$$

and with Y_s fixed at $1/(300\underline{/45^\circ}) = 2.36(1 - j)$ mS, we must select Y_p to minimize $|Y_p + 0.00236(1 - j)|$. Now Y_p is the parallel combination of impedances Z_1 and $R_L = 300\ \Omega$, so the real part of $Y_p = \text{Re}\{Y_1\} + 1/300$ and the imaginary part of Y_p is just $\text{Im}\{Y_1\}$. Thus the quantity to be minimized is

$$|Y_p + 0.00236(1-j)| = |(\text{Re}\{Y_1\} + 0.00569) + j(\text{Im}\{Y_1\} - 0.00236)|$$

Chapter 18   Design of Linear Filters

Re{Y_1} must be nonnegative, since we are limited to passive filters consisting of passive elements, so the magnitude of the real part is minimized by setting Re{Y_1} = 0. The imaginary part can be set to zero by selecting Im{Y_1} = 0.00236 S. Thus the best Y_1 = j0.00236 S, or Z_1 = $-j$423.7 Ω. This is a capacitor of value

$$C_1 = \frac{1}{(2\pi)(1000)(423.7)} = 0.376 \ \mu F$$

To complete the design, we must specify L_r and C_r, the elements of the 60-Hz series resonant subcircuit. While any pair of values that resonate at 60 Hz will yield zero impedance at 60 Hz as required, the impedance at other frequencies will rise more steeply with increasing value of L_r. This can be seen by noting that the Q value of a series RLC circuit increases with L for any fixed R. The steeper the increase, the less power will be delivered to the load at the harmonic frequencies. Thus the largest practical value of L_r should be selected. Factors that limit the practical value may include weight, size, cost, and the increase of series resistance, which accompanies the increase in inductance as the number of turns is increased. Assuming that the largest practical inductor available is L_r = 15 mH, the matching capacitor is

$$C_r = \frac{1}{(0.015)(120\pi)^2} = 469 \ \mu F$$

This large value of capacitance is most conveniently attained by connecting several smaller capacitors in parallel. The power coupling filter is shown in Fig. 18.4.

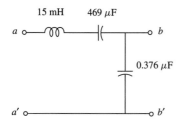

FIGURE 18.4 Power coupling filter for the example.

15 mH   469 μF

0.376 μF

EXERCISES

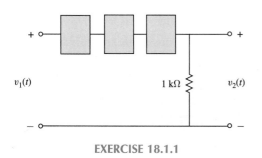

EXERCISE 18.1.1

$v_1(t)$ 1 kΩ $v_2(t)$

18.1.1. Using LC subcircuits with 1-H inductors, complete the filter design shown. $H(s) = \mathbf{V}_2/\mathbf{V}_1$ should be a filter with three stopbands centered at $\omega = 10^4$, 10^6, and 10^8 rad/s. What is the gain at these three frequencies? Which notch has the greatest bandwidth?

Answer Parallel LC circuits with C = 0.01 μF, 1 pF, 0.1 fF. Gain is zero at the center of each notch. The notch at 10^8 rad/s has the greatest bandwidth.

18.1.2. Design a transformerless coupling circuit so that the load voltage $\mathbf{V}_L$ at 1 kHz has magnitude 100 times that of the source $\mathbf{V}_S$.

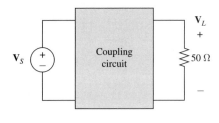

$\mathbf{V}_S$ Coupling circuit $\mathbf{V}_L$ + $50\,\Omega$ −

EXERCISE 18.1.2

Answer

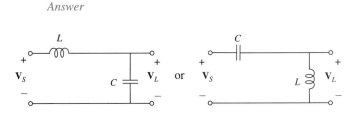

with $L = 79.6\ \mu\mathrm{H}$, $C = 318\ \mu\mathrm{F}$

18.2   ACTIVE FILTERS

The transfer function $\mathbf{H}(s)$ of a stable linear circuit, or equivalently its frequency response function $\mathbf{H}(j\omega)$, tells us what the circuit "does." It describes how the circuit will process any frequency component or combination of frequency components present in its input. For a given $\mathbf{H}(s)$ we may sketch a Bode gain plot and see if the transfer function yields the desired circuit behavior for the filter task at hand. The size and location of the pass- and stopbands, the gains in these bands, and the slopes between them are determined by $\mathbf{H}(s)$. These filter properties can be modified by adjustment of the poles, zeros, and constant factor in $\mathbf{H}(s)$.

Once we have selected a transfer function $\mathbf{H}(s)$ that meets the needs of our design task, we are faced with the *realization* problem: find a circuit whose transfer function is $\mathbf{H}(s)$. In this section we develop a general design procedure for solving the realization problem using only resistors, capacitors, and op amps, that is, a procedure leading to an inductorless active filter realization. Inductors have disadvantages as circuit elements within modern miniaturized electronic designs, and we use op amps because the ability to multiply transfer functions of op amp two-ports connected in cascade makes active filter design generally simpler than passive design.

We will have need for the building-block circuits introduced in Chapter 3. Their circuit diagrams and voltage transfer equations are included in Table 18.1. The two summers shown may, of course, have more than two inputs, in which case the voltage transfer equation has the corresponding additional terms.

Table 18.1 Building Block Circuits

Building Block	Circuit Diagram	Voltage Transfer Equation
Inverting amplifier		$V_o = \dfrac{-R_F}{R_A} V_1$
Noninverting amplifier		$V_o = \left(1 + \dfrac{R_F}{R_A}\right) V_1$
Inverting summer		$V_o = \dfrac{-R_F}{R_1} V_1 + \dfrac{-R_F}{R_2} V_2$
Noninverting summer		$V_o = \left(1 + \dfrac{R_F}{R_A}\right)\left(\dfrac{R_T}{R_1} V_1 + \dfrac{R_T}{R_2} V_2\right)$ where $R_T = R_1 \| R_2$

The voltages in Table 18.1 may be interpreted in terms of time functions $v_i(t)$ as in Chapter 3, or as phasors $\mathbf{V}_i$ or s-domain voltages $\mathbf{V}_i(s)$ where convenient. In the latter two cases, a moment's review of the derivations in Chapter 3 confirms that *the same equations apply if the resistances in these building blocks are replaced by impedances.* The resulting circuits are filters rather than amplifiers, and summing filters rather than summers.

Design Example 18.4

Let us analyze the circuit of Fig. 18.5. It is an inverting filter with impedances $Z_F(s) = R_F$ and $Z_A(s) = R_A + 1/sC$. The voltage transfer equation for this circuit is that for the inverting amp with

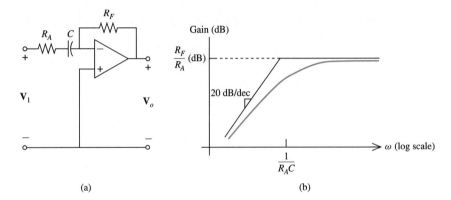

(a)

(b)

FIGURE 18.5 (a) Inverting filter $\mathbf{H}(s) = -Z_F(s)/Z_A(s)$; (b) Bode gain plot.

impedances replacing resistances,

$$\mathbf{V}_0(s) = \frac{-Z_F(s)}{Z_A(s)}\mathbf{V}_1(s)$$

$$\mathbf{H}(s) = \frac{-R_F}{R_A + 1/sC} = \frac{-(R_F/R_A)s}{s + 1/R_AC}$$

There is a pole at $s = -1/R_AC$ and a zero at $s = 0$. The Bode gain plot for this circuit is shown in Fig. 18.5(b). Note that it is a high-pass filter whose passband gain can be fixed by the ratio R_F/R_A, and whose corner frequency is $\omega_c = 1/(R_AC)$.

One inverting filter will be of particular interest to us, the *inverting integrator* shown in Fig. 18.6. Its transfer function is given by

$$\mathbf{H}(s) = \frac{Z_F(s)}{Z_A(s)} = \frac{-1/RC}{s} \qquad (18.8)$$

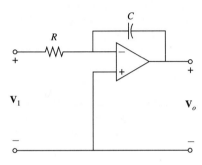

FIGURE 18.6 Inverting integrator.

The integration property of Laplace transforms listed in Table 12.1 shows us that

$$\mathbf{V}_2(s) = -\frac{1}{RCs}\mathbf{V}_1(s) \tag{18.9}$$

which implies that in the time domain,

$$v_2(t) = -\frac{1}{RC}\int v_1(t)\,dt \tag{18.10}$$

The output is an inverted and scaled integral of the input.

We are now ready to develop a procedure for realizing any rational transfer function $\mathbf{H}(s)$ whose denominator polynomial is at least the same order as its numerator. This includes all proper rational functions (numerator order less than denominator) and those of equal order. Factor $\mathbf{H}(s)$ into a product of transfer functions

$$\mathbf{H}(s) = K\mathbf{H}_1(s)\mathbf{H}_2(s)\cdots\mathbf{H}_q(s) \tag{18.11}$$

where each $\mathbf{H}_i(s)$ is either of the form

$$\mathbf{H}(s) = \frac{b_1 s + b_0}{s + a_0} \tag{18.12}$$

corresponding to a real pole $s = -a_0$ in $\mathbf{H}(s)$, or of the form

$$\mathbf{H}(s) = \frac{b_2 s^2 + b_1 s + b_0}{s^2 + a_1 s + a_0} \tag{18.13}$$

corresponding to complex conjugate pole pairs in $\mathbf{H}(s)$. Equation (18.12) is called a *bilinear* transfer function, since both numerator and denominator are linear polynomials, and (18.13) a *biquadratic (biquad)* transfer function. Our strategy will be to realize each bilinear and biquadratic factor separately, then cascade the resulting circuits. In the absence of loading, this cascade will have the desired product (18.11) as its overall transfer function. We will guarantee against loading by use of op amp circuits with high input impedance and low output impedance. The constant-gain factor K in (18.11) will be divided among the bilinear and biquad circuits.

First we shall realize the general bilinear transfer function (18.12). The backbone of the realization is the inverting integrator of Fig. 18.6, shown in block diagram form in Fig. 18.7(a). If we label its output $\mathbf{V}_0(s)$, then by (18.9) its input must be $-RCs\mathbf{V}_0(s)$. Setting $RC = 1$ for convenience, suppose that we scale $\mathbf{V}_0(s)$ by a_0, sum it with $-\mathbf{V}_1(s)$, and connect the output of the summer block to the input of the inverting integrator block as shown in Fig. 18.7(b). Then we have forced the summer output to equal $-s\mathbf{V}_0(s)$, or, by the diagram,

$$a_0\mathbf{V}_0(s) - \mathbf{V}_1(s) = -s\mathbf{V}_0(s) \tag{18.14}$$

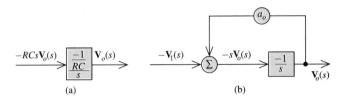

(a)   (b)

FIGURE 18.7   (a) Inverting integrator block; (b) $\mathbf{H}(s) = 1/(s + a_0)$.

Solving for $\mathbf{V}_0(s)$ gives

$$\mathbf{V}_0(s) = \frac{1}{s + a_0}\mathbf{V}_1(s) \tag{18.15}$$

Since our goal is the general bilinear $\mathbf{H}(s)$ in (18.12), (18.15) is a transfer function with the right denominator but wrong numerator. Scale and add the signals at the input and output of the inverting integrator block as shown in Fig. 18.8, labeling the output $\mathbf{V}_2(s)$. By this figure we have

$$\mathbf{V}_2(s) = (-b_1)[-s\mathbf{V}_0(s)] + b_0\mathbf{V}_0(s) = (b_1 s + b_0)\mathbf{V}_0(s) \tag{18.16}$$

which by (18.15) is

$$\mathbf{V}_2(s) = \frac{b_1 s + b_0}{s + a_0}\mathbf{V}_1(s) \tag{18.17}$$

This is the general bilinear transfer function (18.12) that we seek.

Examining Fig. 18.8, to produce this circuit requires an inverting integrator and two summers, one with coefficients -1 and a_0, the other with b_0 and $-b_1$. The summers listed in Table 18.1 both require all coefficients to be the same sign, so we shall need a new circuit, the *general summer* shown in Fig. 18.9. To determine the voltage transfer equation of this circuit, first kill all the noninverting inputs $\mathbf{V}_1, \mathbf{V}_2, \ldots, \mathbf{V}_q$. This results in an inverting summer, and by Table 18.1 the component of the output due to the remaining inputs is

$$V_1 = -\frac{R_F}{R_a}V_a - \frac{R_F}{R_b}V_b - \cdots - \frac{R_F}{R_p}V_p \tag{18.18}$$

With the noninverting inputs back in place, next kill the inverting inputs $\mathbf{V}_a, \mathbf{V}_b, \ldots, \mathbf{V}_p$. We are left with a noninverting summer which by the same table yields the output

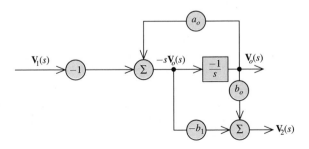

FIGURE 18.8   Bilinear block diagram $\mathbf{H}(s) = (b_1 s + b_0)/(s + a_0)$.

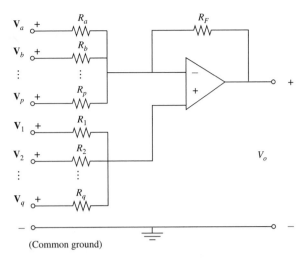

FIGURE 18.9 General summer.

component

$$V_2 = \left(1 + \frac{R_F}{R_A}\right)\left(\frac{R_T}{R_1}V_1 + \frac{R_T}{R_2}V_2 + \cdots + \frac{R_T}{R_q}V_q\right) \tag{18.19}$$

where R_T is the parallel equivalent of the noninverting input resistors $R_1, R_2, \ldots, R_q$ and R_A is the parallel equivalent of the inverting input resistors $R_a, R_b, \ldots, R_p$. Then by superposition $\mathbf{V}_0 = \mathbf{V}_1 + \mathbf{V}_2$ and the overall voltage transfer equation for the general summer of Fig. 18.9 is

$$
\begin{aligned}
V_0 = &-\frac{R_F}{R_a}V_a - \frac{R_F}{R_b}V_b - \cdots - \frac{R_F}{R_p}V_p \\
&+ \left(1 + \frac{R_F}{R_A}\right)\left(\frac{R_T}{R_1}V_1 + \frac{R_T}{R_2}V_2 + \cdots + \frac{R_T}{R_q}V_q\right)
\end{aligned}
\tag{18.20}
$$

where $R_A = R_a||R_b|| \cdots ||R_p$ and $R_T = R_1||R_2|| \cdots ||R_q$.

We will develop the procedure for specifying the resistors in (18.20) with reference to a specific example. Suppose that we seek to realize the voltage transfer equation

$$\mathbf{V}_0 = -2\mathbf{V}_a - 6\mathbf{V}_b + 3\mathbf{V}_1 + \tfrac{3}{2}\mathbf{V}_2 \tag{18.21}$$

Begin by selecting resistors to realize the inverting terms $-2\mathbf{V}_a - 6\mathbf{V}_b$. By (18.20) this can be done by selecting, for instance, $R_F = 60$ kΩ, $R_a = 30$ kΩ, and $R_b = 10$ kΩ. Next, choose values for the noninverting input resistors $R_1, R_2, \ldots, R_q$ in inverse ratio to the coefficients in their terms $3\mathbf{V}_1 + \tfrac{3}{2}\mathbf{V}_2$ (since by (18.20) they are denominator factors in these coefficients). For instance, we may select $R_1 = 30$ kΩ and $R_2 = 60$ kΩ in the present example. Then $R_A = 30$ k$\Omega||10$ k$\Omega = 7.5$ kΩ, $R_T = 30$ k$\Omega||60$ k$\Omega = 20$ kΩ, and evaluating the voltage transfer equation (18.20) yields

$$\mathbf{V}_0 = -2\mathbf{V}_a - 6\mathbf{V}_b + (1+8)(\tfrac{2}{3}\mathbf{V}_1 + \tfrac{1}{3}\mathbf{V}_2) = -2\mathbf{V}_a - 6\mathbf{V}_b + 6\mathbf{V}_1 + 3\mathbf{V}_2 \tag{18.22}$$

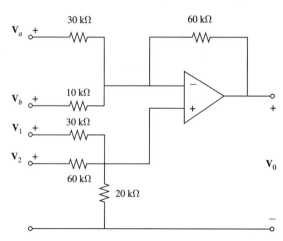

FIGURE 18.10 General summer realizing $\mathbf{V}_0 = -2\mathbf{V}_a - 6\mathbf{V}_b + 3\mathbf{V}_1 + \frac{3}{2}\mathbf{V}_2$.

If all terms match the desired voltage transfer equation, the design is complete. The inverting terms will always be correct, but more frequently the noninverting terms will each be off by a common factor, necessitating one final step to correct this discrepancy. That is the case in our example. When we compare (18.21) and (18.22), we find that each noninverting coefficient in (18.22) is too high by a factor of 2.

Suppose that we add a *dummy input* to the noninverting terminal, that is, an input $\mathbf{V}_3$ whose value is zero, connected to the noninverting terminal through a resistor R_3. Since we now have one additional resistor in the parallel equivalent defining R_T, the value of R_T will be reduced. Examining (18.20), the only effect of the dummy input is to decrease R_T, which scales all the noninverting input coefficients downward proportionally. In the present case we wish to reduce all the noninverting coefficients by a factor of 2, so we require our corrected R_T to equal half its original value of 20 kΩ or 20 k$\Omega || R_3 = 10$ kΩ. In terms of conductance,

$$50 \times 10^{-6} + G_3 = 100 \times 10^{-6} \qquad (18.23)$$

which yields $G_3 = 50 \times 10^{-6}$ S or $R_3 = 20$ kΩ. This completes the design, and the resulting circuit is shown in Fig. 18.10.

In our example the noninverting coefficients were initially too high, leading to a dummy input at the noninverting terminal. If the noninverting coefficients are, instead, initially too low, a dummy input must be placed at the inverting input terminal instead. R_A in (18.20) will then be reduced, not R_T, and the effect is to increase all noninverting coefficients. The value for this dummy input resistor is calculated above as R_3. This is demonstrated in Design Example 18.6.

Design Example 18.5

Design an all-pass filter with the transfer function

$$\mathbf{H}(s) = \frac{b_1 s + b_0}{s + a_0} = \frac{10(s-1)}{s+1}$$

Examining Fig. 18.8, we find that we need an inverting integrator, an input summer with coefficients -1 and $a_0 = +1$, and an output summer with coefficients $b_0 = -b_1 = -10$. The input summer is

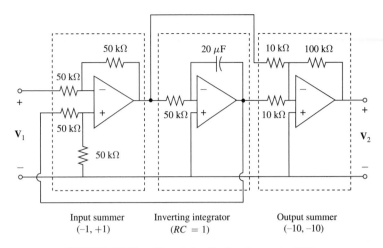

Input summer
(−1, +1)

Inverting integrator
(RC = 1)

Output summer
(−10, −10)

FIGURE 18.11 Circuit for Design Example 18.5.

a general summer, as in Fig. 18.9, with one inverting input, one noninverting input, and perhaps a dummy input as well. Following the procedure just outlined for designing a general summer, first select $R_F = R_a = 50$ kΩ to set the inverting coefficient at the desired value of −1. Then for any noninverting input resistor R_1, in the absence of a dummy input we have a general summer with $R_T = R_1$, $R_A = 50$ kΩ, and overall voltage transfer equation, by (18.20), $\mathbf{V}_0 = -\mathbf{V}_a + 2\mathbf{V}_1$. Since the noninverting coefficient is too high by a factor of 2, a dummy input at the noninverting terminal with $R_2 = R_1$ will have the desired effect of reducing R_T, and thus the noninverting coefficient in the voltage transfer equation, by a factor of 2. The final circuit is shown in Fig. 18.11.

The other type of transfer function in the cascade (18.11) is the biquad given in (18.13). As with the bilinear circuit, we first consider the related transfer function with unity numerator,

$$\mathbf{H}(s) = \frac{\mathbf{V}_0(s)}{\mathbf{V}_1(s)} = \frac{1}{s^2 + a_1 s + a_0} \tag{18.24}$$

Since this is a second-order system we need two integrators, as shown in Fig. 18.12(a). Cross-multiplying (18.24), we have

$$s^2\mathbf{V}_0(s) = -a_1 s\mathbf{V}_0(s) - a_0\mathbf{V}_0(s) + \mathbf{V}_1(s) \tag{18.25}$$

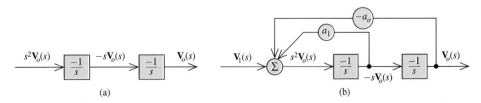

(a)

(b)

FIGURE 18.12 (a) Double integrator; (b) $\mathbf{H}(s) = 1/(s^2 + a_1 s + a_0)$.

This equation is true if the input to the leftmost integrator is the sum of the three terms on the right, leading to the block diagram shown in Fig. 18.12(b). Scaling by the required numerator coefficients and summing the outputs of the integrators and the input summer, we arrive at the desired biquad realization. By Fig. 18.12 and (18.25),

$$V_2(s) = (b_2 s^2 + b_1 s + b_0)V_0(s) = \frac{b_2 s^2 + b_1 s + b_0}{s^2 + a_1 s + a_0} V_1(s)$$

The resulting general biquadratic circuit block diagram is shown in Fig. 18.13.

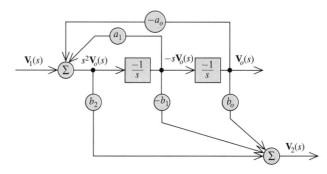

FIGURE 18.13 Biquadratic block diagram: $\mathbf{H}(s) = (b_2 s^2 + b_1 s + b_0)/(s^2 + a_1 s + a_0)$.

Design Example 18.6

We seek a biquad realization of the transfer function of a series RLC circuit given in (14.43) for the values $R = 14\ \Omega$, $L = 2$ H, $C = \frac{1}{4}$ F, namely,

$$\mathbf{H}(s) = \frac{\frac{1}{2}s}{s^2 + 7s + 2} \qquad (18.26)$$

By Fig. 18.13, we will need two inverting integrators, an input summer with coefficients $+1$, $a_1 = +7$, and $-a_0 = -2$, and at the output, since only one b_i is nonzero, the usual output summer is just an inverting amplifier, with gain $-b_1 = -\frac{1}{2}$.

Working on the input summer, we first set the inverting coefficient to -2 by selecting $R_F = 70$ kΩ and $R_a = 35$ kΩ (see Fig. 18.9). Then to get the desired ratio 1:7 for the two noninverting coefficients, we choose the inverse ratio $R_1 = 70$ kΩ, $R_2 = 10$ kΩ. With no dummy input we thus have $R_A = R_a = 35$ kΩ, $R_T = 70$ k$\Omega||10$ k$\Omega = 8.75$ kΩ, and by (18.20), the voltage transfer equation for our input summer is

$$V_0 = -2V_a + \tfrac{3}{8}V_1 + \tfrac{21}{8}V_2$$

We need to adjust the noninverting coefficients upward to achieve the desired values of $+1$ and $+7$. Applying a dummy input $\mathbf{V}_b = 0$ to the inverting terminal through an input resistor R_b will reduce R_A and thus increase these coefficients. To achieve the correct coefficient $+1$ for input $\mathbf{V}_1$ (and thus be correct for the other noninverting input since their ratio is fixed at 1:7) requires that

$$\left(1 + \frac{R_F}{R_A}\right)\frac{R_T}{R_1} = \left(1 + \frac{70 \times 10^3}{R_A}\right)\left(\frac{1}{8}\right) = +1$$

or $R_A = 10$ kΩ. The dummy input resistor R_b must therefore satisfy $R_A = 35$ k$\Omega || R_b = 10$ kΩ or, in terms of conductances,

$$\frac{1}{35 \times 10^3} + G_b = \frac{1}{10^4}$$

Thus $G_b = \frac{1}{14}$ mS, $R_b = 14$ kΩ. This completes the design of the input summer. The inverting integrators and output amplifier are straightforward, and the resulting realization is shown in Fig. 18.14.

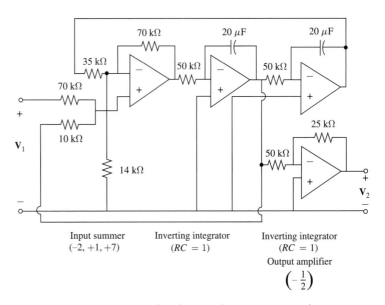

FIGURE 18.14 Biquad realization for Design Example 18.6.

The technique developed in this section can be used to realize any $\mathbf{H}(s)$ whose denominator polynomial is of the same degree or higher than its numerator polynomial. *Factor $\mathbf{H}(s)$ into bilinear and biquadratic transfer functions, realize each as above, and connect in cascade.* As long as input resistors in the range 5 kΩ to 500 kΩ are used, the input impedance of each stage will be high enough, and the output impedance low enough, that there will be negligible loading and resulting transfer function will be the desired product of these factors.

18.2.1. Design a circuit with bilinear transfer function $\mathbf{H}(s) = (2s + 1)/(s + 5)$. Make the largest resistance 100 kΩ.

Answer

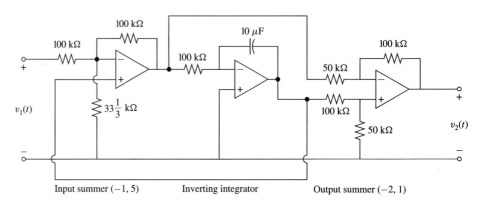

Input summer $(-1, 5)$ Inverting integrator   Output summer $(-2, 1)$

EXERCISE 18.2.1

18.2.2. Design a circuit with biquadratic transfer function $\mathbf{H}(s) = s^2/(s^2 + s + 1)$. Make 10 k$\Omega$ the smallest resistance. What kind of circuit is this?

Answer Unity gain high-pass

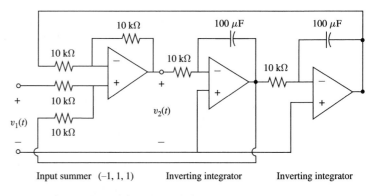

Input summer $(-1, 1, 1)$ Inverting integrator   Inverting integrator

EXERCISE 18.2.2

18.2.3. Design a circuit with the third-order transfer function $\mathbf{H}(s) = -100/s^3$. If the input to this circuit was a unit step and all initial conditions were zero, what would the output be?

Answer Output would be $v_2(t) = \frac{1}{6}t^3 u(t)$, the unit cubic singularity function.

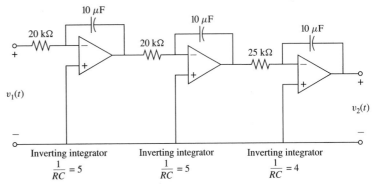

Inverting integrator
$\frac{1}{RC} = 5$

Inverting integrator
$\frac{1}{RC} = 5$

Inverting integrator
$\frac{1}{RC} = 4$

EXERCISE 18.2.3

18.3 CLASSICAL FILTERS

Ideal filters are characterized by a constant gain of unity in their passband and discontinuous jump to zero gain everywhere outside their passband. Although a useful idealization, such pure binary frequency-domain behavior cannot be realized in practice. Filters designed with linear lumped elements such as those described in this book (RLC elements, op amps, and linear controlled sources), for instance, have rational transfer functions $\mathbf{H}(s)$, and any rational transfer function with p poles has a maximum rolloff of $-20p$ dB/decade. The gain for real filters with rational transfer functions varies smoothly and continuously with frequency and is neither exactly constant in the passband nor exactly equal to zero everywhere else.

Since real filters cannot behave ideally, a performance metric by which real filters may be judged and compared is needed. In practice, there are several useful measures by which the deviation of a real filter $\mathbf{H}(s)$ from the ideal can be assessed. Each measures a different feature of the filter's deviation from the ideal, and the relative importance of each depends upon the application at hand. Consider the gain plot of a typical real lowpass filter shown in Fig. 18.15. If this were an ideal filter, the deviation of gain $|\mathbf{H}(j\omega)|_{\max} - |\mathbf{H}(j\omega)|_{\min}$ in the passband, or *passband ripple*, would be zero since the gain is everywhere constant at unity in the ideal filter passband.

For an ideal filter the *passband slope* would be 0 dB/decade throughout the passband. The rolloff, or maximum transition band slope, would be infinite and there would be no stopband ripple. In contrast to ideal filters, real filters $\mathbf{H}(s)$ take on finite nonzero value for each of these measures. In addition, the behavior of the filter's phase response $\phi(\omega) = /\mathbf{H}(j\omega)$ is important in assessing filter performance. A phase shift $\phi(\omega)$ rad corresponds to $\phi(\omega)/2\pi$ periods, or a time shift (time advance) of

$$\frac{\phi(\omega)}{2\pi} T = \frac{\phi(\omega)}{2\pi} \frac{2\pi}{\omega} = \frac{\phi(\omega)}{\omega}$$

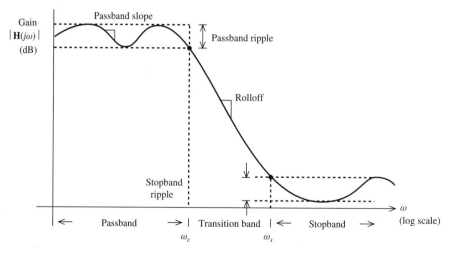

FIGURE 18.15 Filter parameters.

The *time delay* $\tau(\omega)$ at frequency ω is the negative of this time advance:

$$\tau(\omega) = \frac{-\phi(\omega)}{\omega} \tag{18.27}$$

If every frequency component within the passband has the same time delay, the output will have all its passband components in the same relative phase as they were in the input, and will be a distortionless, time-delayed version of the input. If the time delay is not constant in the passband, some frequency components will be phase shifted relative to others, and the shape of the output will not match the shape of the input even if the gain in the passband is constant. Thus constancy of time delay, or by (18.27) the degree to which $\phi(\omega)$ represents a *linear phase shift*, (phase shift $\phi(\omega)$ linear in ω) is a useful criterion of the degree to which passage through a filter causes distortion in the time domain.

For a fixed filter order [fixed number of poles in $\mathbf{H}(s)$] these criteria generally trade off. That is, adjusting filter coefficients for improved constancy in the passband may create increased passband ripple and decreased roll-off. The filter designer must consider which of these criteria are most important in a given application and to what degree. For instance, a filter for use in a medical diagnostic instrument such as an electrocardiogram (ECG) must preserve the shape of its input signal, for it is the shape that the physician studies. Thus linearity of phase is often paramount in this class of applications. For a circuit operating against strong unpredictable signals originating outside the passband, such as a doppler radar system tracking a target among clutter, the level of passband ripple may be the dominant consideration. A filter for a communications receiver seeking to tune to a single transmitter in the midst of a crowded band depends on sharp roll-off to prevent crosstalk with other transmitters nearby, and the roll-off rate would be of highest priority in the evaluation of filter performance.

Families of filters that are optimal with respect to specific selected measures of performance have long been defined and tabulated. Although it is seldom the case that only one measure of performance is relevant, these filters represent extreme performance in explicit senses and are thus of great value to the filter designer.

We describe three such families: the Butterworth, Chebyshev, and Bessel filters. We focus attention on the *normalized lowpass filters* of various order for each family. As we shall see in the next section, filters of different bandwidth and of different type (high-pass, bandpass, bandstop) can easily be derived from the normalized lowpass filter transfer function $\mathbf{H}(s)$ once that has been determined. All three of these "classical" lowpass filters are of the all-pole type; that is, their transfer functions $\mathbf{H}(s)$ possess constant numerators, so each is of the form

$$\mathbf{H}(s) = \frac{b_0}{s^n + a_{n-1}s^{n-1} + \cdots + a_1 s + a_0} = \frac{b_0}{(s - p_1) \cdots (s - p_n)} \tag{18.28}$$

Thus each may be specified by the value of its numerator b_0, and either the locations of its poles $p_k, k = 1, \ldots, n$ or its denominator polynomial coefficients $a_k, k = 0, \ldots, n-1$.

Butterworth filters are optimal in the sense of being maximally flat in the passband. No other normalized lowpass filter of a given order has a lower rate of change of gain at $\omega = 0$ than the Butterworth filter of that order. The poles of the Butterworth filter of order n are the n left-half plane solutions of the equation

$$s^{2n} = (-1)^{n+1} \tag{18.29}$$

For instance, the poles of the order $n = 3$ Butterworth are found from the equation $s^6 = 1$. The six solutions to this equation are equally spaced around the unit circle, $p_i = 1/\underline{i\pi/3}, i = 0, 1, \ldots, 5$, as can easily be verified. The three poles are thus located at $1/\underline{2\pi/3}, 1/\underline{\pi}$, and $1/\underline{4\pi/3}$. The numerator of each order Butterworth filter is $b_0 = a_0$, so the dc gain is unity (0 dB). The gain $|\mathbf{H}_{\mathrm{BUT}}(j\omega)|$ of this filter is

$$|\mathbf{H}_{\mathrm{BUT}}(j\omega)| = \frac{1}{\sqrt{1 + \omega^{2n}}} \tag{18.30}$$

The gain plot for the normalized lowpass Butterworth filter of order 3 is shown in Fig. 18.16. Note from (18.30) that the dc gain is indeed unity for all orders, and that the gain at the normalized cutoff frequency $\omega_c = 1$ rad/s is -3 dB.

Chebyshev filters are optimal in the sense of minimum passband ripple. For this to be a meaningful statement, we must carefully consider what is meant by passband and passband ripple. If the passband is defined, as is often the case, to be bounded by the first half-power (or -3 dB) frequency, then by definition all filters have the same passband ripple (3 dB) and this performance measure makes no sense. For the present purpose, define the ε-passband as *the set of frequencies for which the gain exceeds* $1/\sqrt{1 + \varepsilon^2}$. For fixed ε, a filter with larger ε-bandwidth varies less within that range of frequencies than any with a smaller ε-bandwidth. The filter of a given order with largest ε-bandwidth is the Chebyshev filter. The gain of the Chebyshev lowpass filter normalized to unity ε-bandwidth is given by

$$|\mathbf{H}_{\mathrm{CHE}}(j\omega)| = \frac{1}{\sqrt{1 + \varepsilon C_n^2(\omega)}}$$

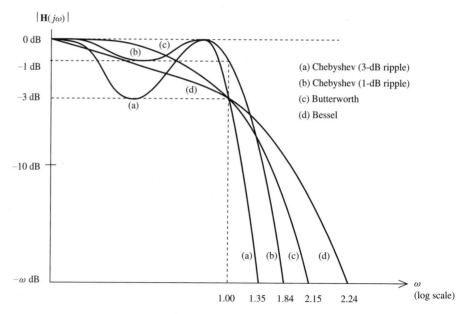

FIGURE 18.16   Gains of selected normalized lowpass filters of order 3.

where $C_n(\omega)$ is the Chebyshev polynomial of order n. The Chebyshev polynomials are in turn defined by the recursion

$$C_{n+1}(\omega) = 2\omega C_n(\omega) - C_{n-1}(\omega) \tag{18.31}$$

where $C_0(\omega) = 1$ and $C_1(\omega) = \omega$. The gain plots for the order 3 Chebyshev filters are shown in Fig. 18.16 for $\varepsilon = 0.51$ (1-dB passband ripple) and $\varepsilon = 1.0$ (3-dB ripple). Note that both Chebyshev filters have steeper roll-off than the other filters, at the cost of more rapid passband gain variation.

The *Bessel filters* are optimal in the sense of maximum constancy of time delay in the passband. The coefficients a_k, $k = 0, \ldots, n - 1$, of the denominator of $\mathbf{H}_{\text{BES}}(s)$, the lowpass Bessel filter or order n normalized to 1-s delay, are given by

$$a_k = \frac{(2n - k)!}{(2^{n-k})(n - k)!k!}$$

and the dc gain is unity; that is, $b_0 = a_0$ in (18.28) for each order. The gain for a Bessel filter of order 3 is shown in Fig. 18.16. For purposes of comparison it has been renormalized to -3-dB gain at $\omega = 1$ rad/s. Note that its performance is inferior to any of the other filters shown. Figure 18.17 shows the time delays for these filters, and it is here that the Bessel filter has its advantage. In the passband, the time delay of the Bessel filter is clearly less variable than the others, and this will lead to lower time-domain waveform distortion, as we shall see in an example to come.

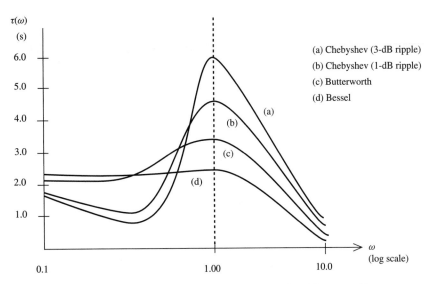

$\tau(\omega)$
(s)

6.0

5.0

4.0

3.0

2.0

1.0

(a) Chebyshev (3-dB ripple)
(b) Chebyshev (1-dB ripple)
(c) Butterworth
(d) Bessel

(a)
(b)
(c)
(d)

0.1   1.00   10.0

ω
(log scale)

FIGURE 18.17   Time delays of selected normalized lowpass filters.

Design Example 18.7

We wish to design a fourth-order Butterworth lowpass filter with +6-dB dc gain and unity −3-dB bandwidth. By (18.29), the poles of the filter will be the four left half-plane solutions of $s^8 = -1$. These are $p_1 = 1/\underline{(5\pi/8)}$, $p_2 = 1/\underline{(7\pi/8)}$, and their complex conjugates $p_1^* = 1/\underline{(-5\pi/8)}$, $p_2^* = 1/\underline{(-7\pi/8)}$. The desired transfer function is

$$\mathbf{H}(s) = \frac{2}{(s - p_1)(s - p_1^*)(s - p_2)(s - p_2^*)}$$

Multiplying conjugated terms in the denominator, this transfer function can be expressed as the product of two biquadratics:

$$\mathbf{H}(s) = \mathbf{H}_1(s)\mathbf{H}_2(s) = \left(\frac{2}{s^2 + 0.765s + 1}\right)\left(\frac{1}{s^2 + 1.85s + 1}\right)$$

We will use the cascade realization of Section 18.2. With reference to Fig. 18.13, to realize $\mathbf{H}_1(s)$ we require an input summer with coefficients −1, +1, and +0.765, two inverting integrators with $RC = 1$ and an output amplifier with gain +2. The input summer will have one inverting input with $R_F = R_a$ which we select at 50 kΩ. The two noninverting inputs have coefficients +1 and +0.765, so referring to the first as $\mathbf{V}_1$ and the second as $\mathbf{V}_2$, select input resistors in inverse ratio $R_1 = 76.5$ kΩ and $R_2 = 100$ kΩ. Then $R_A = R_a = 50$ kΩ, $R_T = R_1||R_2 = 43.3$ kΩ and the voltage transfer equation is

$$V_0 = -\frac{50}{50}V_a + \left(1 + \frac{50}{50}\right)\left(\frac{43.3}{76.5}V_1 + \frac{43.3}{100}V_2\right)$$

$$= -V_a + 1.13V_1 + 0.866V_2$$

Since V_1 should have coefficient 1, the noninverting coefficients need to be reduced by the factor $1/1.13$. This is accomplished by adding a dummy input V_3 to the noninverting terminal through a resistor R_3. The new value for R_T, which we designate as R_τ, must be $R_\tau = (1/1.13)R_T = 38.3$ kΩ. Thus 43.3 kΩ$||R_3 = 38.3$ kΩ, which is easily solved for $R_3 = 326$ kΩ. The full realization for this biquadratic is seen on the left in Fig. 18.18.

The realization procedure for $H_2(s)$ proceeds similarly. The input summer coefficients this time are -1, $+1$, and 1.85. For the input summer we select $R_F = 50$ kΩ, $R_a = 50$ kΩ, $R_1 = 185$ kΩ, and $R_2 = 100$ kΩ. Then $R_A = R_a = 50$ kΩ, $R_T = R_1||R_2 = 64.9$ kΩ, and the voltage transfer equation before dummy inputs is $V_0 = -V_a + 0.702V_1 + 1.30V_2$. This time the noninverting coefficients need to be raised, so we add a dummy input at the inverting terminal, calling the resistor R_b. The coefficient we require for V_1 is 1 rather than its current value of 0.702, so defining the new $R_A = 50$ kΩ$||R_b$, we have the requirement that

$$1 + \frac{50 \text{ k}\Omega}{R_A} = \frac{1}{0.702}\left(1 + \frac{50}{50}\right) = 2.85$$

Solving yields $R_A = 27.0$ kΩ = 50 kΩ$||R_b$, so $R_b = 58.8$ kΩ. This completes the design of the input summer. The rest is straightforward, and the realization of $H_2(s)$ is shown on the right in Fig. 18.18, completing the cascade realization of the desired fourth-order Butterworth filter. Note that since $b_0 = 1$ and the other b_i are zero, no output summer or amplifier is needed.

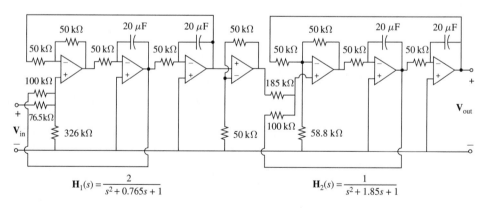

$$H_1(s) = \frac{2}{s^2 + 0.765s + 1}$$

$$H_2(s) = \frac{1}{s^2 + 1.85s + 1}$$

FIGURE 18.18 Fourth-order Butterworth realization.

Table 18.2 lists the coefficients of the transfer functions of selected classical normalized lowpass filters. It should be emphasized that the normalizations are somewhat different for each filter shown, so comparisons must be made with care. The Bessel filters are normalized to a 1-s time delay at dc, the Butterworth filters to -3-dB gain at $\omega = 1$ rad/s, and the $\frac{1}{2}$- and 1-dB Chebyshev filters to ε-passbands of $\omega = 1$ rad/s for their corresponding ε (0.350 and 0.509, respectively).

$$H(s) = \frac{b_0}{s^n + a_{n-1}s^{n-1} + \cdots + a_1 s + a_0}$$

	n	b_0	a_0	a_1	a_2	a_3	a_4
Bessel	1	1	1				
	2	3	3	3			
	3	15	15	15	6		
	4	105	105	105	45	10	
	5	945	945	945	420	105	15
Butterworth	1	1	1				
	2	1	1	1.41			
	3	1	1	2	2		
	4	1	1	2.61	3.41	2.61	
	5	1	1	3.24	5.24	5.24	3.24
Chebyshev $\left(\frac{1}{2}\ \text{dB}\right)$	1	2.86	2.86				
	2	1.43	1.52	1.43			
	3	0.716	0.716	1.53	1.26		
	4	0.358	0.379	1.03	1.72	1.20	
	5	0.179	0.179	0.753	1.31	1.94	1.17
Chebyshev (1 dB)	1	1.97	1.97				
	2	0.983	1.10	1.10			
	3	0.491	0.491	1.24	0.988		
	4	0.246	0.276	0.743	1.45	0.953	
	5	0.123	0.123	0.580	0.974	1.69	0.937

To use this table most effectively, we need to generalize the cascade method for realization developed in Section 18.2. We seek to avoid factoring third- or higher-order polynomials for filter orders greater than or equal to 3. Such factorization is required to reduce a higher-order filter realization to a cascade of biquadratic and bilinear blocks.

Following the logic of Section 18.2 to realize the general transfer function cited in Table 18.2 and (18.28) in the form of a single block rather than a cascade, we begin with a chain of n inverting integrators. Labeling the output of the rightmost $V(s)$, the other voltages are labeled as shown in Fig. 18.19. Designating the input to our filter as $V_1(s)$ and the output as $V_2(s)$ gives

$$H(s) = \frac{V_2(s)}{V_1(s)} = \frac{b_0}{s^n + a_{n-1}s^{n-1} + \cdots + a_1 s + a_0} \tag{18.32}$$

FIGURE 18.19 Chain of n inverting integrators with output $V_0(s)$.

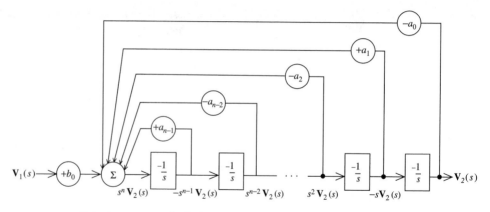

FIGURE 18.20 Block diagram for all-pole filter of even order n.

Cross-multiplying and solving for the term with the highest power of s, we obtain

$$s^n \mathbf{V}_2(s) = -a_{n-1}s^{n-1}\mathbf{V}_2(s) - a_{n-2}s^{n-2}\mathbf{V}_2(s) - \cdots - a_0\mathbf{V}_2(s) + b_0\mathbf{V}_1(s) \quad (18.33)$$

If n is even, the input to the chain of integrators shown in Fig. 18.19 is identical to the sum of terms in (18.33), and the resulting block diagram is shown as Fig. 18.20. If n is odd, the input to the chain of integrators of Fig. 18.19 is the negative of the sum of terms in (18.33). Reversing all the coefficient signs in this sum, we arrive at the block diagram for odd-order filters. Thus our all-pole filter realizations based on Fig. 18.20 for even-order filters and Fig. 18.21 for odd orders consists of n inverting integrators and a single input summer. These forms may be used to realize any all-pole filter, such as the filters listed in Table 18.2. $n+1$ op amps will always be required, one for each integrator and one for the input summer. All-pole transfer functions have constant numerators, so no output summer is required.

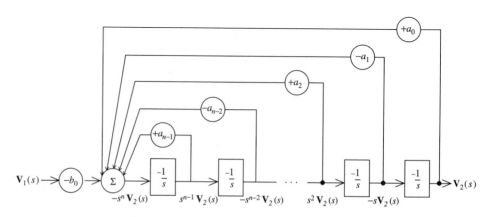

FIGURE 18.21 Block diagram for all-pole filter of odd order n.

<table>
<tr><td>

Design Example 18.8

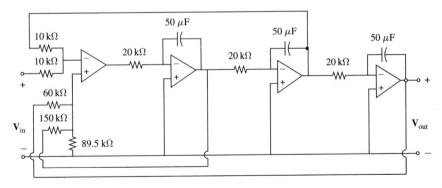

</td><td>

We wish to design the third-order Bessel lowpass filter of Table 18.2. The desired transfer function, from Table 18.2, is

</td></tr>
</table>

We wish to design the third-order Bessel lowpass filter of Table 18.2. The desired transfer function, from Table 18.2, is

$$\mathbf{H}(s) = \frac{15}{s^3 + 6s^2 + 15s + 15}$$

This transfer function is of odd order $n = 3$, so following Fig. 18.21, we require an input summer with coefficients $-b_0$, $+a_0$, $-a_1$, and $+a_2$, which are -15, $+15$, -15, and $+6$, respectively. Design of this general summer begins by fixing inverting coefficients to -15 and -15, selecting, for instance, $R_F = 150$ kΩ, $R_a = R_b = 10$ kΩ. The required noninverting terms in the desired voltage transfer equation are $+15\mathbf{V}_1 + 6\mathbf{V}_2$, where we are labeling the voltages $\mathbf{V}_1$ and $\mathbf{V}_2$ to conform with the standard general summer notation in (18.20). Taking the inverse ratio, we set $R_1 = 60$ kΩ and $R_2 = 150$ kΩ. Absent a dummy input, this fixes the design. $R_A = 10$ k$\Omega \| 10$ k$\Omega = 5$ kΩ, $R_T = 60$ k$\Omega \| 150$ k$\Omega = 42.9$ kΩ and the input summer has voltage transfer equation

$$\mathbf{V}_0 = -15\mathbf{V}_a - 15\mathbf{V}_b + \left(1 + \frac{150}{5}\right)\left(\frac{42.9}{60}\mathbf{V}_1 + \frac{42.9}{150}\mathbf{V}_2\right)$$

or $\mathbf{V}_0 = -15\mathbf{V}_a - 15\mathbf{V}_b + 22.2\mathbf{V}_1 + 8.87\mathbf{V}_2$ \hfill (18.34)

To reduce the noninverting coefficients 22.2 and 8.87 to their required values of 15 and 6, we need to reduce R_T by the addition of a dummy input resistor R_3 to the noninverting input terminal. The new value of R_T is required to be $(15/22.2)(42.9 \times 10^3) = 29$ kΩ. Then R_3 satisfies 42.9 k$\Omega \| R_3 = 29$ kΩ or $R_3 = 89.5$ kΩ. This completes the design of the input summer. The realization of the filter requires only the addition of the chain of three inverting integrators. The full realization is shown in Fig. 18.22. A SPICE input file for this circuit is shown on the following page.

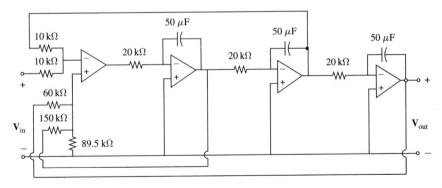

FIGURE 18.22 Realization of third-order normalized Bessel lowpass filter.

```
Design Example 18.8 3rd order Bessel LPF
*
*Main circuit:
VIN      1       0       AC        1
R1       1       2       10K
R2       2       8       10K
R3       3       10      60K
R4       3       6       150K
R5       3       0       89.5K
R6       2       4       150K
XOA13    2       4       OPAMP
XII14    6       INVINT
XII26    8       INVINT
XII38    10      INVINT
*Here is the op amp subcircuit:
.SUBCKT OPAMP    1       2         3
RIN      1       2       1MEG
EVCVS    4       0       1         2         100K
ROUT     4       3       30
.ENDS
*And the inverting integrator subckt.
*Note subckts may reference previous ones.
.SUBCKT INVINT   1       2
RII1     3       20K
CII3     2       50U
XOAII    0       3       2         OPAMP
.ENDS
*Finally the mode of analysis and print cards
.AC DEC 20       .01     1
.PLOT AC VDB(10)
.END
```

The resulting gain plot is shown in Fig. 18.23. Note that the gain of this filter at $\omega = 1$ rad/s, which is 0.159 Hz, is not normalized to -3 dB. The Bessel table entries have been normalized to a 1-s delay, which for this filter corresponds roughly to a -1-dB gain at $\omega = 1$ rad/s. A second run with the mode and output statements changed to

```
.AC LIN 50      1U      1
.PLOT AC VP(10)
```

was run to observe the phase shift of this filter. Note from the resulting plot shown in Fig. 18.24 that the filter exhibits accurately linear phase shift in its passband. The 1U entry is used to circumvent the restriction in SPICE that a frequency of zero not be used for ac analysis.

```
**** 06/12/65  15:18:43 ********* Evaluation PSpice (January 1991) *************
Design Example 18.7 3rd order Bessel LPF

****     AC ANALYSIS                      TEMPERATURE =   27.000 DEG C

********************************************************************************


      FREQ        VDB(10)


 (*)----------    -3.0000E+01  -2.0000E+01  -1.0000E+01   0.0000E+00   1.0000E+01
                 - - - - - - - - - - - - - - - - - - - - - - - - - - - - - - - -
    1.000E-02  1.206E-02 .           .            .            *            .
    1.122E-02  1.115E-02 .           .            .            *            .
    1.259E-02  1.002E-02 .           .            .            *            .
    1.413E-02  8.583E-03 .           .            .            *            .
    1.585E-02  6.779E-03 .           .            .            *            .
    1.778E-02  4.507E-03 .           .            .            *            .
    1.995E-02  1.647E-03 .           .            .            *            .
    2.239E-02 -1.956E-03 .           .            .            *            .
    2.512E-02 -6.491E-03 .           .            .            *            .
    2.818E-02 -1.220E-02 .           .            .            *            .
    3.162E-02 -1.940E-02 .           .            .            *            .
    3.548E-02 -2.846E-02 .           .            .            *            .
    3.981E-02 -3.987E-02 .           .            .            *            .
    4.467E-02 -5.425E-02 .           .            .            *            .
    5.012E-02 -7.238E-02 .           .            .            *            .
    5.623E-02 -9.523E-02 .           .            .            *            .
    6.310E-02 -1.240E-01 .           .            .            *            .
    7.079E-02 -1.604E-01 .           .            .            *            .
    7.943E-02 -2.063E-01 .           .            .            *            .
    8.913E-02 -2.643E-01 .           .            .            *            .
    1.000E-01 -3.377E-01 .           .            .            *            .
    1.122E-01 -4.306E-01 .           .            .           *.            .
    1.259E-01 -5.486E-01 .           .            .           *.            .
    1.413E-01 -6.988E-01 .           .            .           *.            .
    1.585E-01 -8.905E-01 .           .            .           *.            .
    1.778E-01 -1.136E+00 .           .            .           *.            .
    1.995E-01 -1.452E+00 .           .            .          *  .           .
    2.239E-01 -1.859E+00 .           .            .          *  .           .
    2.512E-01 -2.386E+00 .           .            .         *   .           .
    2.818E-01 -3.066E+00 .           .            .        *    .           .
    3.162E-01 -3.939E+00 .           .            .       *     .           .
    3.548E-01 -5.044E+00 .           .            .     *       .           .
    3.981E-01 -6.410E+00 .           .            .   *         .           .
    4.467E-01 -8.051E+00 .           .          *  .            .           .
    5.012E-01 -9.960E+00 .           .       *     .            .           .
    5.623E-01 -1.211E+01 .           .    *  .                  .           .
    6.310E-01 -1.445E+01 .          *  .            .           .           .
    7.079E-01 -1.696E+01 .       *     .            .           .           .
    7.943E-01 -1.959E+01 .    .*       .            .           .           .
    8.913E-01 -2.231E+01 .   *  .      .            .           .           .
    1.000E+00 -2.510E+01 .  *          .            .           .           .
                 - - - - - - - - - - - - - - - - - - - - - - - - - - - - - - - -


      JOB CONCLUDED

      TOTAL JOB TIME            .40
```

FIGURE 18.23   SPICE gain plot for a Bessel lowpass filter.

```
**** 06/12/95  15:39:35 ********* Evaluation PSpice (January 1991) *************
Design Example 18.7 3rd order Bessel LPF
****      AC ANALYSIS                        TEMPERATURE =    27.000 DEG C
*****************************************************************************

     FREQ       VP(10)
 (*)----------    -8.0000E+01  -6.0000E+01  -4.0000E+01  -2.0000E+01   0.0000E+00
              - - - - - - - - - - - - - - - - - - - - - - - -
  1.000E-06 -3.607E-04 .            .            .            .                  *
  4.083E-03 -1.472E+00 .            .            .            .                *.
  8.164E-03 -2.994E+00 .            .            .            .               *  .
  1.225E-02 -4.416E+00 .            .            .            .              *   .
  1.633E-02 -5.888E+00 .            .            .            .             *    .
  2.041E-02 -7.360E+00 .            .            .            .            *     .
  2.449E-02 -8.832E+00 .            .            .            .           *      .
  2.857E-02 -1.030E+01 .            .            .            .          *       .
  3.265E-02 -1.178E+01 .            .            .            .         *        .
  3.674E-02 -1.325E+01 .            .            .            .        *         .
  4.082E-02 -1.472E+01 .            .            .            .       *          .
  4.490E-02 -1.619E+01 .            .            .            .       *          .
  4.898E-02 -1.766E+01 .            .            .            .      *           .
  5.306E-02 -1.913E+01 .            .            .            .    *             .
  5.714E-02 -2.061E+01 .            .            .            .    *             .
  6.123E-02 -2.208E+01 .            .            .            .  *.              .
  6.531E-02 -2.355E+01 .            .            .            . *                .
  6.939E-02 -2.502E+01 .            .            .            .*                 .
  7.347E-02 -2.649E+01 .            .            .           *                   .
  7.755E-02 -2.796E+01 .            .            .          *                    .
  8.163E-02 -2.943E+01 .            .            .         *                     .
  8.571E-02 -3.090E+01 .            .            .        *                      .
  8.980E-02 -3.237E+01 .            .            .       *                       .
  9.388E-02 -3.384E+01 .            .            .      *                        .
  9.796E-02 -3.531E+01 .            .            .      *                        .
  1.020E-01 -3.678E+01 .            .            .     *                         .
  1.061E-01 -3.825E+01 .            .            .    *                          .
  1.102E-01 -3.972E+01 .            .            .   *                           .
  1.143E-01 -4.119E+01 .            .            .  *.                           .
  1.184E-01 -4.266E+01 .            .            . *                             .
  1.224E-01 -4.413E+01 .            .            .*                              .
  1.265E-01 -4.560E+01 .            .           *                               .
  1.306E-01 -4.706E+01 .            .          *                                .
  1.347E-01 -4.853E+01 .            .         *                                 .
  1.388E-01 -5.000E+01 .            .         *                                 .
  1.429E-01 -5.146E+01 .            .        *                                  .
  1.469E-01 -5.293E+01 .            .       *                                   .
  1.510E-01 -5.439E+01 .            .      *                                    .
  1.551E-01 -5.586E+01 .            .     *                                     .
  1.592E-01 -5.732E+01 .            .    *                                      .
  1.633E-01 -5.878E+01 .            .   .*                                      .
  1.673E-01 -6.025E+01 .            .  *                                        .
  1.714E-01 -6.171E+01 .            . *                                         .
  1.755E-01 -6.317E+01 .            .*.                                         .
  1.796E-01 -6.462E+01 .           *.                                           .
  1.837E-01 -6.608E+01 .          * .                                           .
  1.878E-01 -6.754E+01 .         *  .                                           .
  1.918E-01 -6.899E+01 .       *    .                                           .
  1.959E-01 -7.044E+01 .      *     .                                           .
  2.000E-01 -7.189E+01 .     *      .                                           .
              - - - - - - - - - - - - - - - - - - - - - - - -

       JOB CONCLUDED
       TOTAL JOB TIME            .52
```

774

FIGURE 18.24 SPICE phase plot for a Bessel lowpass filter.

18.3.1. For some radio direction-finding systems it is necessary to determine which of two input sinusoids, $v_1(t)$ at frequency ω_1 or $v_2(t)$ at ω_2, is stronger. If the frequencies are close, $|\omega_1 - \omega_2|$ is small. What filter family would you use to filter out high-frequency noise?

Answer The Butterworth filters have the least passband slope, so it is least likely that the stronger will be attenuated enough to have a lower output amplitude.

18.3.2. Determine the poles, transfer function, and realization for a third-order normalized Butterworth lowpass filter with passband gain +6 dB. Use as many 50-kΩ resistors as possible.

Answer Poles -1, $1/\underline{\pm120°}$;

$$\mathbf{H}(s) = \frac{2}{(s^2+s+1)(s+1)} \cdot \text{Biquad}\left(\frac{1}{s^2+s+1}\right) \text{ plus bilinear }\left(\frac{2}{s+1}\right) \text{ realization:}$$

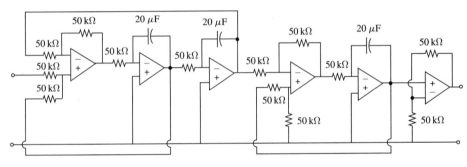

EXERCISE 18.3.2

18.3.3. Find a realization of a second-order Chebyshev $\left(\frac{1}{2} \text{ dB}\right)$ normalized lowpass filter.

Answer

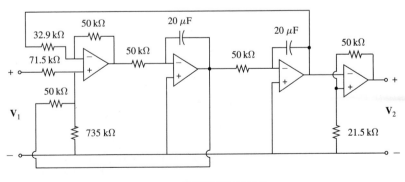

EXERCISE 18.3.3

18.4 FILTER TRANSFORMATION AND SCALING

In this section we consider ways in which one filter design may be mapped into many. We will first determine how to transform a filter from one type to another: lowpass to high-pass, high-pass to bandpass, and so on. This technique will permit us to use lowpass filter transfer functions, such as those specified in Table 18.2, to design classical filters of many types, not just lowpass filters. We will also determine how to impedance scale and frequency scale these circuits. Impedance scaling permits easy modification of a circuit designed to drive a high-impedance load to work effectively with a low-impedance load or to reduce the power dissipated by a circuit by uniform reduction of the current levels of all elements. Frequency scaling allows us to dilate or contract the frequency axis for a given frequency response function $\mathbf{H}(j\omega)$. This permits us to modify a normalized filter, such as those normalized to 1 rad/s listed in Table 18.2, to any bandwidth desired. The flexibility permitted by filter transformation and scaling permits the filter designer to work with basic normalized transfer functions, yet in the end create circuits to meet almost any application need without extensive redesign.

Given the transfer function $\mathbf{H}(s)$, let us consider the effect of creating a new transfer function $\mathbf{H}'(s) = \mathbf{H}(1/s)$, that is, of replacing s by its inverse $1/s$. The gain associated with the new transfer function is

$$|\mathbf{H}'(j\omega)| = \left| \mathbf{H}\left(-j\frac{1}{\omega} \right) \right|$$

But for any transfer function $\mathbf{H}(s)$ with real coefficients, $\mathbf{H}(-j\omega) = \mathbf{H}^*(j\omega)$, and noting that the magnitude of a complex number and its complex conjugate are equal, the expression above may be simplified slightly:

$$|\mathbf{H}'(j\omega)| = \left| \mathbf{H}\left(j\frac{1}{\omega} \right) \right| \tag{18.35}$$

After replacement of s by its inverse $1/s$, the new transfer function $\mathbf{H}'(s) = \mathbf{H}(1/s)$ has, for each ω, the same gain that $\mathbf{H}(s)$ does at the frequency $1/\omega$. As shown in Fig. 18.25, the gain curve is reflected around $\omega = 1$. Lowpass filters are mapped to high-pass, bandpass filters with passbands below $\omega = 1$ to bandpass filters with passbands above $\omega = 1$, and so on. A common use of this transformation is to map a normalized lowpass filter such as the entries of Table 18.2 into a normalized high-pass filter. The new filter will have the same gain at $\omega = 1$ rad/s as the old one, but its passband and stopband will be reversed.

The poles of the transformed filter $\mathbf{H}(1/s)$ will clearly be the inverses of the poles of the original. A pole of $\mathbf{H}(s)$ in the second quadrant, for instance, will become a pole in the third quadrant of $\mathbf{H}'(s) = \mathbf{H}(1/s)$ (if ϕ is an angle in the second quadrant, $-\phi$ will be in the third quadrant), and vice versa. Thus left-half plane poles of $\mathbf{H}(s)$ become left-half plane poles of the transformed filter, and stability is preserved. It would not be useful to define a transformation that did not preserve the stability of the circuit.

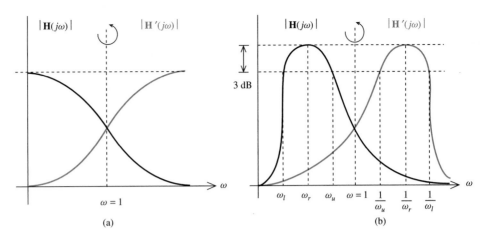

$\omega = 1$

(a)

ω_l ω_r ω_u $\omega = 1$ $\dfrac{1}{\omega_u}$ $\dfrac{1}{\omega_r}$ $\dfrac{1}{\omega_l}$

3 dB

(b)

FIGURE 18.25 Transformation $\mathbf{H}'(s) = \mathbf{H}(1/s)$: (a) lowpass transformed to high-pass; (b) bandpass with passband below $\omega = 1$ transformed to one with passband above $\omega = 1$.

Design Example 18.9

Design a normalized second-order Chebyshev high-pass filter with 1-dB passband ripple. The table entry for the normalized lowpass with 1-dB passband ripple is, by Table 18.2,

$$\mathbf{H}(s) = \frac{0.983}{s^2 + 1.10s + 1.10}$$

Replacing s by $1/s$, we have our desired high-pass filter transfer function:

$$\mathbf{H}'(s) = \frac{0.983}{(1/s)^2 + 1.10\,(1/s) + 1.10} = \frac{0.894s^2}{s^2 + s + 0.909}$$

$\mathbf{H}'(s)$ is a biquad transfer function that may be realized based on the block diagram of Fig. 18.13, as several previous biquad examples were. Here we modify the method slightly to save one op amp. Since there is only a single numerator term, the usual biquad output summer is just an output amplifier, in this case a noninverting amplifier with gain 0.894. However, since this is a linear circuit, if we scale the input to this circuit by a factor α, all responses will scale by the same factor α. So if we choose $\alpha = 0.894$, we need not apply any amplification at all at the output. The resulting block diagram is shown in Fig. 18.26(a), and corresponds to the general block diagram for all-pole filters of even order in Fig. 18.20.

From Fig. 18.26(a) the input summer will have three terms with coefficients $+0.894$, $+1$, and -0.909. Following the general summer design procedure and using the notation of (18.29), $R_F = 50$ kΩ, $R_a = 55$ kΩ, $R_1 = 100$ kΩ, and $R_2 = 89.4$ kΩ. Then $R_A = R_a = 55$ kΩ and $R_T = 100$ k$\Omega\|89.4$ k$\Omega = 47.2$ kΩ. The noninverting coefficient, which is required to be 0.894, would be,

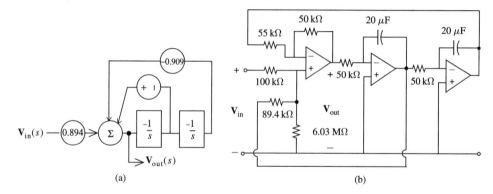

(a)   (b)

FIGURE 18.26 (a) Block diagram and (b) circuit for Design Example 18.9.

absent any dummy input,

$$\left(1 + \frac{R_F}{R_A}\right)\frac{R_T}{R_1} = 0.901$$

This is reduced to the required value of 0.894 by adding a dummy input to the noninverting terminal though a resistor R_3. With $R_\tau = R_T \| R_3$ we require that

$$\frac{R_\tau}{R_T} = \frac{0.894}{0.901}$$

Using $R_T = 47.2$ kΩ the last is solved for $R_\tau = 46.8$ kΩ, and since $R_\tau = R_T \| R_3$, the value of R_3 is found to be 6.03 MΩ. Adding the inverting integrators with $RC = 1$ as shown in Fig. 18.26(a), the realization is complete and is shown in Fig. 18.26(b).

The gain curve for the circuit of Fig. 18.26(b), as computed in an ac SPICE run, is shown as Fig. 18.27. Note the -1-dB gain at $\omega = 1$ ($f = 0.159$ Hz), as expected, and the 1-dB passband ripple characteristic of 1-dB Chebyshev filters.

A second transformation of interest is to replace s by $(s^2 + 1)/s$,

$$\mathbf{H}'(s) = \mathbf{H}\left(\frac{s^2 + 1}{s}\right) \tag{18.36}$$

This has the effect of transforming a lowpass filter $\mathbf{H}(s)$ to a bandpass $\mathbf{H}'(s)$ with center frequency at $\omega = 1$. To verify this, note that the frequency response of the transformed filter at $\omega = 1$ is that of the original at $\omega = 0$,

$$\overline{\mathbf{H}}'(j1) = \mathbf{H}\left(\frac{-1+1}{j}\right) = \mathbf{H}(j0)$$

and by similar reasoning both the dc and high-frequency gains $|\mathbf{H}'(j0)|$ and $|\mathbf{H}'(j\infty)|$ of $\mathbf{H}'(s)$ correspond to the high-frequency gain $|\mathbf{H}(j\infty)|$ of $\mathbf{H}(s)$. For instance, we may transform the second-order normalized Chebyshev lowpass of Design Example 18.9 to a

```
**** 04/06/96  15:27:34 ********* Evaluation PSpice (January 1991) *************
Design Example 18.9 2nd order Chebyshev HPF

****      AC ANALYSIS                       TEMPERATURE =   27.000 DEG C

********************************************************************************


    FREQ        VDB(4)


 (*)----------    -6.0000E+01  -4.0000E+01  -2.0000E+01   0.0000E+00   2.0000E+01
              - - - - - - - - - - - - - - - - - - - - - - - -
  1.590E-02 -4.012E+01 .              *            .            .            .
  2.002E-02 -3.609E+01 .            .   *          .            .            .
  2.250E-02 -3.205E+01 .            .       *      .            .            .
  3.172E-02 -2.799E+01 .            .          *  .            .            .
  3.994E-02 -3.390E+01 .            .            *.            .            .
  5.028E-02 -1.977E+01 .            .            * .            .            .
  6.330E-02 -1.557E+01 .            .            .   *          .            .
  7.969E-02 -1.134E+01 .            .            .      *       .            .
  1.003E-01 -7.179E+00 .            .            .         *    .            .
  1.263E-01 -3.488E+00 .            .            .            * .            .
  1.590E-01 -1.015E+00 .            .            .            *.            .
  2.002E-01 -7.349E-02 .            .            .            . *            .
  2.520E-01 -3.080E-02 .            .            .            . *            .
  3.172E-01 -2.479E-01 .            .            .            . *            .
  3.994E-01 -4.702E-01 .            .            .            . *            .
  5.028E-01 -6.394E-01 .            .            .            . *            .
  6.330E-01 -7.562E-01 .            .            .            . *            .
  7.969E-01 -8.335E-01 .            .            .            *.            .
  1.003E+00 -8.837E-01 .            .            .            *.            .
  1.263E+00 -9.158E-01 .            .            .            *.            .
  1.590E+00 -9.363E-01 .            .            .            *.            .
  2.002E+00 -9.493E-01 .            .            .            *.            .
  2.520E+00 -9.576E-01 .            .            .            *.            .
  3.172E+00 -9.628E-01 .            .            .            *.            .
  3.994E+00 -9.661E-01 .            .            .            *.            .
  5.028E+00 -9.682E-01 .            .            .            *.            .
  6.330E+00 -9.695E-01 .            .            .            *.            .
  7.969E+00 -9.703E-01 .            .            .            *.            .
  1.003E+01 -9.708E-01 .            .            .            *.            .
  1.263E+01 -9.711E-01 .            .            .            *.            .
  1.590E+01 -9.714E-01 .            .            .            *.            .
  2.002E+01 -9.715E-01 .            .            .            *.            .
  2.520E+01 -9.716E-01 .            .            .            *.            .
  3.172E+01 -9.716E-01 .            .            .            *.            .
  3.994E+01 -9.717E-01 .            .            .            *.            .
  5.028E+01 -9.717E-01 .            .            .            *.            .
  6.330E+01 -9.717E-01 .            .            .            *.            .
  7.969E+01 -9.717E-01 .            .            .            *.            .
  1.003E+02 -9.717E-01 .            .            .            *.            .
  1.263E+02 -9.717E-01 .            .            .            *.            .
  1.590E+02 -9.717E-01 .            .            .            *.            .
              - - - - - - - - - - - - - - - - - - - - - - - -


       JOB CONCLUDED

       TOTAL JOB TIME              .40
```

FIGURE 18.27

Section 18.4 Filter Transformation and Scaling

779

normalized Chebyshev bandpass filter via the transformation

$$\mathbf{H}'(s) = \frac{0.983}{\left(\dfrac{s^2+1}{s}\right)^2 + 1.10\left(\dfrac{s^2+1}{s}\right) + 1.10}$$

which, after simplification, can be written

$$\mathbf{H}'(s) = \frac{0.983s^2}{s^4 + 1.10s^3 + 3.10s^2 + 1.10s + 1}$$

This transformation doubles the order of the filter, as may be predicted from the replacement of s by an expression including s^2. Order-doubling is a by-product of the two-sided roll-off of the bandpass filter compared to the one-sided roll-off of the lowpass. The frequency range $0 \le \omega \le 1$ in the original filter is mapped into $\omega_l \le \omega \le \omega_u$, where $\omega_u - \omega_l = 1$, so the bandpass has unit passband if the original lowpass did.

A third transformation, which maps lowpass filters to bandstop filters with center frequency $\omega = 1$, is replacement of s by $s/(s^2 + 1)$, or

$$\mathbf{H}'(s) = \mathbf{H}\left(\frac{s}{s^2+1}\right) \tag{18.37}$$

Since this transformation is just the inverse of the previous, it is easy to see from the previous discussion that by exchanging the dc and high-frequency gains the gain at $\omega = 1$ will be the high-frequency gain of $\mathbf{H}(s)$ and that the gain in the new filter at both extremes of ω will be the dc gain of the original $\mathbf{H}(s)$. If $\mathbf{H}(s)$ is a lowpass, therefore, this transformation produces a bandstop filter.

Each of these transformations, when applied to a normalized filter with corner frequency $\omega = 1$, produces another normalized filter with corner or center frequency $\omega = 1$. To adjust the passbands of these filters to desired locations and sizes, we must *frequency scale* the normalized filter. Frequency scaling in the s-domain is the transfer function transformation

$$\mathbf{H}'(s) = \mathbf{H}\left(\frac{s}{\beta}\right) \tag{18.38}$$

where β is a real positive constant. Frequency scaling does not transform the type of filter, it simply expands or contracts the frequency axis. If $\beta > 1$, the passband size and location are expanded, while $\beta < 1$ corresponds to contraction of the passband parameters. For instance, if we wish a Chebyshev high-pass filter of order 2, as in Design Example 18.9, but whose passband begins at $f_c = 100$ Hz rather than 1 rad/s, we must expand the frequency axis by a factor of $\beta = 2\pi(100)$. This corresponds to the scaling

$$\mathbf{H}'(s) = \mathbf{H}\left(\frac{s}{2\pi(100)}\right)$$

Note that by substituting $s = j\omega_c = j2\pi(100)$, the scaled filter behaves at $f = 100$ Hz just as the original one did at $\omega = 1$ rad/s.

Viewed in the circuit domain, frequency scaling by the factor β is the specification of a new set of circuit parameters such that the scaled circuit responds at frequency $\beta\omega$ exactly as the original one does at ω. There is only one way ω enters into the calculation of any circuit response, through the impedances $j\omega L$ and $1/j\omega C$. If we arrange that the new inductive and capacitive impedances at $\beta\omega$ are the same as the old ones at ω, all circuit responses must be the same and the circuit is frequency scaled by β as desired. Now the product $j\omega L$ remains unchanged after ω is replaced by $\beta\omega$ if we also replace L by L/β. The same is true for $1/j\omega C$ if C is replaced by C/β. *To frequency scale a circuit by β, divide each L and C by β, leaving all other elements unchanged.*

$$R_s = R, \qquad L_s = \frac{L}{\beta}, \qquad C_s = \frac{C}{\beta} \qquad\qquad (18.39)$$

Design Example 18.10

In Design Example 18.7 we realized a fourth-order normalized Butterworth lowpass filter. Let us use this to create a fourth-order Butterworth lowpass filter with corner (-3 dB) frequency 10 kHz. Let $\beta = 2\pi(10^4)$. The transfer function $\mathbf{H}'(s)$ of the desired filter is

$$\mathbf{H}'(s) = \mathbf{H}\left(\frac{s}{2\pi(10^4)}\right)$$

where $\mathbf{H}(s)$ is the entry in Table 18.2 for the fourth-order Butterworth lowpass.

$$\mathbf{H}'(s) =$$

$$\frac{1}{\left(\frac{s}{2\pi(10^4)}\right)^4 + 2.61\left(\frac{s}{2\pi(10^4)}\right)^3 + 3.41\left(\frac{s}{2\pi(10^4)}\right)^2 + 2.61\left(\frac{s}{2\pi(10^4)}\right) + 1}$$

Note that after reduction to the standard form, the coefficients of $\mathbf{H}'(s)$ will be very large [e.g., $b_0 = a_0 = (2\pi)^4(10^{16})$]. It will not be possible to use practical values of resistances to achieve such large coefficients. On the other hand, having realized $\mathbf{H}(s)$, the normalized version of $\mathbf{H}'(s)$ (see Fig. 18.18), it is a simple matter to apply frequency scaling in the circuit domain. Since there are no inductors, by (18.39) the desired circuit is identical to that of Fig. 18.18 with the four capacitors in that figure replaced by capacitors of value

$$\frac{C}{\beta} = \frac{20 \times 10^{-6}}{2\pi \times 10^4} = 0.318 \text{ nF}$$

As this example demonstrates, it is often simpler to realize a normalized version of the desired filter, then frequency scale the circuit, than it is to frequency scale the transfer function in the s-domain before realization.

The generation of bandpass and bandstop filters from normalized lowpass filters should be done with care. Suppose that we require a bandpass filter with passband B and center frequency ω_c. Since the lowpass-to-bandpass transformation $\mathbf{H}'(s) = \mathbf{H}\left(\frac{s^2+1}{s}\right)$ maps the unit frequency interval $0 \le \omega \le 1$ into a unit interval of frequency for the bandpass, the passband of both filters will be the same size. Frequency scaling by ω_c will move the center frequency to the desired location, but also increase the bandwidth by

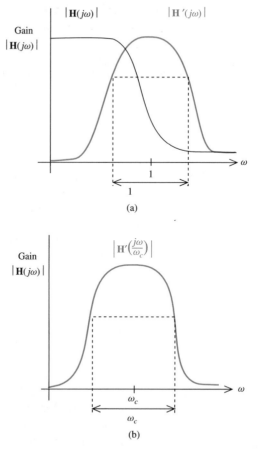

FIGURE 18.28 (a) Lowpass-to-bandpass transformation $\mathbf{H}'(s) = \mathbf{H}\!\left(\frac{s^2+1}{s}\right)$; (b) subsequent frequency scaling by ω_c.

the same factor as in Fig. 18.28. We will not have the desired bandwidth B. To remedy this we must anticipate the final scaling by ω_c by producing a bandpass filter with center frequency $\omega = 1$ rad/s but bandwidth B/ω_c. This may be done by frequency scaling the normalized filter by B/ω_c before applying the bandpass transformation. If, for instance, we frequency scale the lowpass by $\frac{1}{2}$, the bandpass bandwidth will be $\frac{1}{2}$ rad/s. After scaling by ω_c, its bandwidth will be $\omega_c/2$.

Finally, we note that throughout this book numerical values such as $R = 1\ \Omega$, $C = 2$ F, and $\omega = 1$ rad/s have been used in almost all examples, exercises, and problems. Although convenient for pedagogical purposes, and easing the labor of hand calculation, use of these values may concern the reader since they are not practical circuit parameters for most realistic design applications.

Fortunately, we can have the best of both worlds. We can frame a problem in a simple numerical setting, solve it, then convert the solution to a more practical range of circuit operation. We saw above how frequency scaling may aid in designing classical filters. We conclude with a second type of useful scaling, impedance scaling. *Impedance scaling* by α is the specification of a new set of circuit parameters for the circuit such that all voltages are unchanged but all currents are scaled by the constant $1/\alpha$. To see how a circuit may be impedance scaled, consider its mesh equations in vector matrix form,

$$\mathbf{Z}\mathbf{I} = \mathbf{V}_s \tag{18.40}$$

Here $\mathbf{I}$ is the vector of independent mesh currents, $\mathbf{V}_s$ the independent source vector, and Z a square connection matrix. If we can find a new circuit with the same $\mathbf{V}_s$ whose connection matrix is scaled to αZ but whose source vector $\mathbf{V}_s$ is unchanged, then since (18.40) implies that

$$(\alpha Z)\frac{\mathbf{I}}{\alpha} = \mathbf{V}_s \tag{18.41}$$

the currents will each be scaled by $1/\alpha$. Having studied mesh analysis, we know that the on-diagonal (i, i) element of Z contains the sum of all impedances around ith mesh, and the off-diagonal elements the negative sum of impedances on the boundary between the ith and jth meshes. If there are no controlled sources in the circuit, Z consists entirely of these impedances and will be scaled as in (18.41) if each impedance is scaled by α. That is, we must replace each RLC in the original circuit by

$$R_s = \alpha R, \qquad L_s = \alpha L, \qquad C_s = \frac{C}{\alpha} \tag{18.42}$$

where the subscript s indicates the values after scaling. *To impedance scale by α, multiply each R and L value by α, and divide each C value by α.* The difference arises, of course, because the impedance of resistors and inductors is proportional to their element parameter R or L, while that of a capacitor is inversely proportional to C. To keep $\mathbf{V}_s$ the same, as required by (18.41), independent voltage sources are left unchanged, while independent current sources must be scaled by $1/\alpha$.

The gains of current-controlled current sources and voltage-controlled voltage sources in the circuit will not change when impedance scaling, since the ratio of controlled to controlling variable is the same after scaling as before. But current-controlled voltage sources will have their gain (transresistance r) scaled by α, just as the other elements whose units are ohms, and voltage-controlled current sources, with gain (transconductance g) measured in siemens, must have these gains scaled by $1/\alpha$. For instance, if a circuit with a current-controlled voltage source with source function $v(t) = 4i_c(t)$ is to be impedance scaled by 20, the new controlled source will have source function $v(t) = 80i_c(t)$. Then, with the currents in the new circuit reduced by a factor of 20, the increase in the transresistance of the controlled source by a factor of 20 will compensate and its voltage will remain unchanged, as required.

A common use of impedance scaling arises in the design of op amp circuits. While it may be convenient to use values close to unity for R and C in the design process, this will inevitably result in a circuit whose impedance levels are too low for the use of popular low-power op amps such as the μA741. These devices operate best at milliampere current levels when their voltage levels are an appreciable fraction of the power supply voltage, typically ± 15 V. We may use convenient values such as $R = 1\ \Omega$ during the design process, but if we are not to "fry" the op amp by drawing excessive current, we had best impedance scale such values to the order of kilohms before flipping the switch. This scaling also reduces capacitors to more practical values than 1 F or so. Note that since the scaled circuits have the same voltages as prescaled, *there is no change to any voltage transfer function $\mathbf{H}(s) = V_{out}(s)/V_{in}(s)$ when a circuit is impedance scaled.*

18.4.1. Generalize the block diagram of Fig. 18.21 by adding an output summer to realize the transfer function

$$\mathbf{H}(s) = \frac{3s^2 - 4s + 1}{s^3 + 5s^2 + 7s + 1}$$

Answer

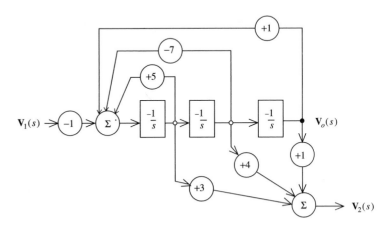

EXERCISE 18.4.1

18.4.2. Compare the op amp count for the cascade realization of the **H**(s) of Exercise 18.3.2 with that for the realization suggested in Exercise 18.4.1.
　　　　Answer Cascade realization: six or seven op amps, depending on how the numerators of the biquad and bilinear circuits are defined. The realization above has five op amps: two summers and three inverting integrators.

18.4.3. Design a third-order Butterworth high-pass filter with a −3-dB frequency of 5 kHz.
　　　　Answer First realize the normalized table entry, then scale down the capacitors by $10^4\pi$. Factor into biquad plus buffered *RC* highpass filter.

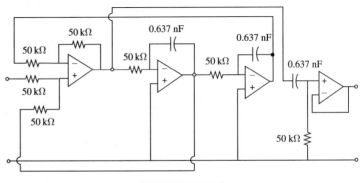

EXERCISE 18.4.3

18.4.4. Find the transfer function and a realization for a fourth-order Bessel bandpass filter with center frequency $\omega = 4$. Using either successive evaluations of $\mathbf{H}(s)$ or SPICE, find the -3-dB bandwidth.

$$\text{Answer } \mathbf{H}(s) = \frac{48s^2}{s^4 + 12s^3 + 80s^2 + 192s + 256}$$

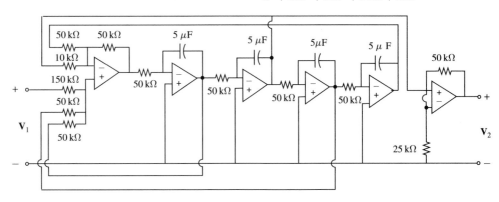

EXERCISE 18.4.4

Bandwidth is 4.00 rad/s or 1.27 Hz.

SUMMARY

Filters, modules whose input-output behavior depends upon the input frequencies, are found in almost all practical electrical and electronic circuits. Filter design is the process of specifying a circuit whose input-output behavior complies with prescribed design specifications, including the desired frequency response. In this concluding chapter a general method for schematic design of linear filters agreeing with a given frequency response $\mathbf{H}(j\omega)$ or transfer function $\mathbf{H}(s)$ is given. The result is a schematic or circuit diagram of the desired filter. Physical design, that is, selection of appropriate physical elements, their physical environment and packaging, is not considered here.

- Passive filters, those containing only passive elements, are practical solutions to filter design problems in which power supplies or other energy sources are undesirable and where discrete rather than integrated circuit implementation is planned.

- Inductorless active filters consisting of cascades of bilinear and biquadratic circuits may be used to realize any stable linear filter. The transfer functions of the several modules are found by factoring the desired transfer function.

- The general bilinear circuit consists of the interconnection of two summers and an inverting integrator.

- The general biquadratic circuit consists of the interconnection of two summers and two inverting integrators.

- Each summer and inverting integrator may be realized using a building block circuit with a single op amp.

- Butterworth filters have maximally flat response at the center of the passband, Chebyshev filters have minimal passband ripple, and Bessel filters have maximally linear phase shift in the passband.

- The coefficients of normalized classical lowpass filter transfer functions of all orders are widely found in circuit design reference books. Designs of normalized filters may be easily converted to filters of other types and passbands using filter conversion, frequency and impedance scaling.

- A normalized lowpass filter may be converted to a high-pass by replacing s by $1/s$, to a bandpass by replacing s by $(s^2 + 1)/s$, and to a bandstop filter by replacing s by $s/(s^2 + 1)$.

- Frequency scaling consists in dividing all capacitances and inductances by the frequency scale factor.

- Impedance scaling consists of multiplying all resistances, inductances and transresistances in the circuit by the desired impedance scale factor, while dividing all capacitances and transconductances by this value. To impedance scale without changing any voltages in the circuit, independent voltage sources are left unchanged but independent current sources are divided by the impedance scale factor.

PROBLEMS

18.1. (a) Design a passive filter whose voltage gain has the uncorrected Bode plot shown in the figure.

    (b) Using the filter designed in (a) above, add a load resistance R_L placed across the output port. For your filter, what value for R_L will produce −3 dB loading (that is, the largest R_L so that the loaded and unloaded gains at any given frequency differ by 3 dB)? This R_L may be specified as the smallest permissible load resistance for the circuit.

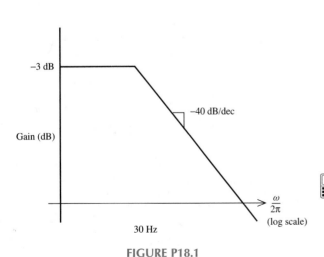

FIGURE P18.1

18.2. Repeat 18.1(a) for the uncorrected Bode plot given. Use complex conjugate poles with zero dB correction, that is, damping factor of 1/2.

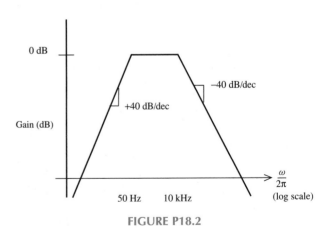

FIGURE P18.2

18.3. (a) Design an RC coupling circuit Z which will maximize the power delivered to the load Z_L at 1 kHz.

    (b) Now make the design in (a) adjustable so it can be set to maximize power delivered to Z_L at any frequency in the range 500 Hz $\leq f \leq$ 2 kHz. Which circuit parameters of Z need to be made adjustable, and over what ranges?

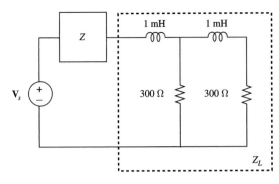

FIGURE P18.3

18.4. (a) Show how an ideal transformer may be used to make a small capacitor behave like one of greater capacitance (except at dc).

(b) Use (a) to design a passive two-terminal circuit whose impedance attains its minimum value of 1 Ω at 1 Hz. Use only one each of an ideal transformer, resistor, inductor with inductance less than or equal to 100 mH, and capacitor with capacitance less than or equal to 10 mF.

18.5. A three-phase 60-Hz generator developing 4.4 kV RMS is to be connected via lengthy power lines to a load. One phase of this balanced wye-wye system is shown in the figure. Design a passive delivery circuit by specifying its two subcircuits 1 and 2 which will deliver maximum power to the load under the condition that the power line current $\mathbf{I}_T$ cannot exceed 1 kA RMS and voltage $\mathbf{V}_T$ cannot exceed 44 kV RMS.

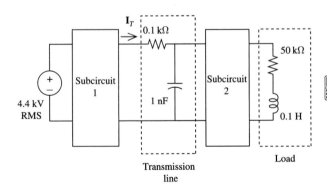

FIGURE P18.5

18.6. After filtering a periodic voltage signal detected by an ultrasonic transducer, the two principal signal components are a fundamental frequency component at 90 kHz and its harmonic at 270 kHz. The filter has introduced a phase shift of −20° in the fundamental and −30° in the re-

sponse to the harmonic component. Design a circuit which will restore the original shape of the received voltage signal waveform by adding a phase shift of +20° at 90 kHz and +30° at 270 kHz. The phase compensation circuit should not change the gain at either of these frequencies by more than ±3 dB.

18.7. A bridge circuit can be used to detect a small change in resistance. In the circuit below, light falling on the photoresistor will cause its resistance to drop, signaling presence of the light beam.

(a) Show that if $R_1 R_P = R_2 R_3$ then $I = 0$. This is the bridge's "null."

(b) The dark resistance for $R_p = 1$ kΩ, and this value drops to 800 Ω when the photoresistor is illuminated. Specify values for R_1, R_2, and R_3 so that the bridge is nulled in dark conditions, and the microammeter reaches its full-scale 50 μA reading when the photoresistor is illuminated. Such a circuit can be used to detect an interrupted light beam, for instance with elevator door safety circuits.

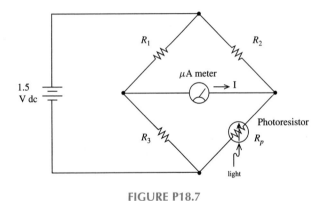

FIGURE P18.7

18.8. Design a passive *RLC* circuit whose voltage gain is given by

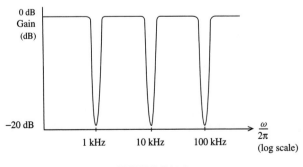

FIGURE P18.8

The gain at the bottom of each notch should be less than -20 dB and the quality factor $Q = 10$ for all three notches. The gains at all frequencies at least 1/2 decade away from the notch minima 1 kHz, 10 kHz, and 100 kHz should be within 1 dB of 0 dB.

18.9. (a) Show that this circuit, called *twin-T filter*, acts as a bandstop (notch) filter satisfying $|V_2(j\omega_0)/V_1(j\omega_0)| = 0$ at $\omega_0 = 1/RC$.

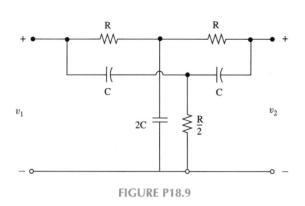

FIGURE P18.9

(b) Note that this circuit assumes exactly matched elements. Suppose the $R/2$ Ω resistor in the figure is replaced by one with resistance $(1 + \rho)(R/2)$ Ω. Sketch a graph showing the depth of the notch in dB as a function of the mismatch parameter ρ. Note that for $\rho = 0$ there is no mismatch and the notch is infinitely deep (in dB), and the larger $|\rho|$ the greater the mismatch. Use SPICE to help generate this curve.

18.10. Design a passive circuit which will drive a display containing three light emitting diodes (LEDs) which show the dominant frequency range of the input signal. The input to the circuit is an ac steady state 1 mA RMS current. If more than half the power in the input is in frequencies in the range 20 Hz–500 Hz the LED marked "bass" should light up, if more than half is in the range 500 Hz–2 kHz the LED marked "midrange" should light up, and if in the range 2 kHz–20 kHz the LED denoted "treble" should light up. Assume that the LED's light up when the voltage across them reaches 5 V RMS.

> **Except where otherwise noted, the following design rules apply to all active circuit designs:**
> 1. **All resistors should be in the range 1 kΩ to 1 MΩ;**
> 2. **Do not use any inductors;**
> 3. **Use the op amp model of Fig. 3.4 with $R_i = 1$ MΩ, $R_0 = 30$ Ω, and $A = 100{,}000$ for all SPICE runs.**

18.11. Replace the resistors in the noninverting amplifier of Table 18.1 by impedances, and use the resulting noninverting filter to design a circuit with uncorrected Bode plot shown.

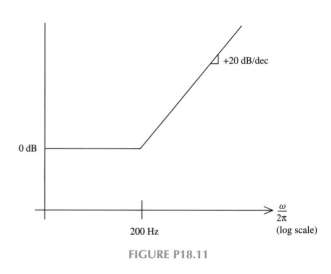

FIGURE P18.11

18.12. Repeat 18.11 for the voltage transfer function

$$H(s) = \frac{s^2 + 6s + 4}{s^2 + 4s + 4}$$

18.13. Replace the resistors of the noninverting summer of Table 18.1 by impedances, and use the resulting building block circuit to design a circuit with voltage transfer equation $V_0(s) = 1/3V_1(s) + 2/(s + 1)V_2(s)$. Note that a dummy input may be necessary.

18.14. Replace the resistors of the circuits of Table 18.1 by impedances and use the resulting building block circuits to design to the desired triple-pole voltage transfer function

$$H(s) = \frac{10}{(s + 1)^3}$$

18.15. Replace the resistors of the circuits of Table 18.1 by impedances and use the resulting building block circuits to design a circuit with the uncorrected Bode plot shown

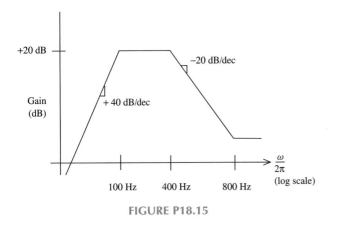

FIGURE P18.15

18.16. Design a circuit with voltage transfer function

$$H(s) = \frac{100s^2 + s + 4 \times 10^7}{(s + 200)^2}$$

What kind of filter is this? Justify.

18.17. Design a circuit with voltage transfer function

$$H(s) = \frac{s^2 - 30s + 200}{s^2 + 30s + 200}$$

What kind of filter is this? Justify.

18.18. Design a circuit with voltage transfer function

$$H(s) = \frac{3.77 \times 10^6}{s + 377}$$

Do not use any building block circuit with absolute value of gain exceeding 10 on any of its inputs.

18.19. Design an audio amplifier with 40 dB/decade roll-off in the uncorrected Bode plot in both the high and low frequencies, lower break frequency 100 Hz and upper break frequency 10 kHz, damping factor $\zeta = 1/4$ for all poles, and +100 dB gain in the center of the passband.

18.20. Design a bandpass circuit whose passband is surrounded by guard bands in which the gain is required to be very low, as shown in the accompanying figure. The gain must be +10 dB or higher at 4 kHz, the gains one octave

away at 2 kHz and 8 kHz must be −10 dB or lower, and the gain in the very low and very high frequencies must tend to 0 dB ±1 dB.

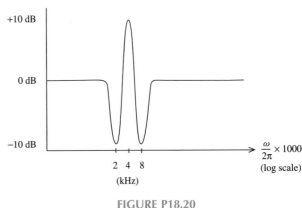

FIGURE P18.20

18.21. If the harmonics comprising a periodic signal are arbitrarily phase-shifted, the shape of the signal will be distorted. This may be caused by frequency-dependent time delays as the signal propagates through air, water or some other medium. Design a circuit called an *adjustable phase-shifter* which introduces a selectable phase shift at a selectable frequency while leaving all other phases, and all gains, relatively unchanged. The phase shift should be adjustable ±90° in the frequency range 1 kHz–10 kHz. The gain should be 0 dB ±1.5 dB at all frequencies, and the phase shift one octave or more away from the selected frequency should be less than 6°. Using SPICE, show the gain and phase shift curves when phase-shifting +30° at 2 kHz and −45° at 8 kHz. What circuit elements need to be made adjustable? (Hint: consider complex poles and zeros with the same break frequency but whose real parts have opposite signs.)

18.22. A limb-lead electrocardiogram (ECG) has three inputs: body surface voltages measured on the left arm (La), right arm (Ra), and left leg (Ll) all relative to the right leg (Rl) voltage common ground as shown. Design an ECG amp so that its output v_0 is proportional to $v_{La} - 0.5(v_{Ra} + v_{Ll})$. v_0 is referred to as aVL, the "augmented left arm lead voltage." The constant of proportionality should be such that when $v_{La} = 1$ mV and $v_{Ra} = v_{Ll} = 0$, we have $v_0 = 1$ V within the passband (0.1 Hz to 20 Hz). The output should roll off at 40 dB/decade on either side of these frequency limits. Use a multi-op amp design so that no gain on any input exceeds 10 in magnitude for any single op amp.

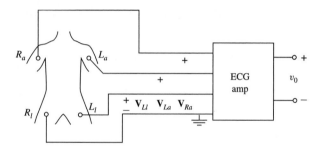

FIGURE P18.22

18.23. We need a baseband filter for a digital communications receiver. This system encodes a binary zero as a 1-kHz sinusoid and a binary one as a 2-kHz sinusoid. Design a filter which will have at least +20 dB gain at these two frequencies, and at most 0 dB gain for all frequencies at least one octave away from either. This receiver is intended for portable operation so use as few op amps as you can.

18.24. An *equalizer* is a circuit which will allow the gain in different frequency ranges to be adjusted independently. Design a four section equalizer with section ranges 10 Hz–100 Hz, 100 Hz–1000 Hz, 1 kHz–10 kHz, and 10 kHz–100 kHz so that when each slider (variable resistor control) is set in its nominal position the gain of the equalizer is 0 dB ± 1.5 dB across the entire band 10 Hz–100 kHz, and the gain in each section can be set by its slider within the range ±20 dB. Specify the resistance range of the variable resistor needed for each section. Demonstrate the circuit using SPICE when the low frequency section is set for +20 dB gain and the other sections are set at their nominal slider values, also when the four sections are set to −20 dB, −5 dB, +5 dB, and +20 dB (lowest to highest frequency sections).

18.25. The circuit shown in the figure is called a *transconductance amplifier*.

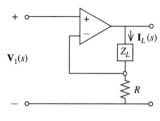

FIGURE P18.25

(a) Show that the output, the current $I_L(s)$, is equal to a constant times the input, the voltage $V_1(s)$, and find the constant of proportionality (called the *transconductance g* of the amplifier, units siemens).

(b) We wish to supply a variable load with a fixed current which is user-selectable. Design a circuit so that for a fixed setting of its input voltage, the current through its load remains constant. The load is an electric motor which we model as a series combination of a resistor R_L and a 1 mH inductor. As the load torque of the motor changes during operation, the equivalent load resistance R_L varies in the range 1 kΩ ≤ R_L ≤ 2 kΩ. As the input voltage to the circuit is adjusted in the range 0–1 V, the current delivered to the motor should vary between 0 and 5 mA RMS. For any fixed input voltage, the current should not vary by more than 5% as R_L varies over its full range as specified. Demonstrate the behavior of this design using SPICE, showing the output for input voltages of 0, 0.5, and 1.0 V and in each case setting $R_L = 1$ kΩ and repeating with $R_L = 2$ kΩ.

18.26. The circuit shown in the figure is called a *transresistance amplifier*.

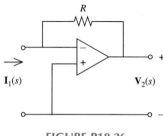

FIGURE P18.26

(a) Show that the output, the voltage $V_2(s)$, is simply equal to a constant times the input, the current $I_1(s)$, and find the constant of proportionality (the *transresistance r* of the amplifier, units ohms).

(b) A certain thermocouple produces a temperature-dependent current we wish to use to trigger an overheating alarm. Design a circuit so that when the temperature drives the thermocouple current to 50 μA, your circuit will deliver an open-circuit output voltage of 5 V at a pair of terminals marked "To *Caution* LED." When it reaches 100 μA, 5 V (open-circuit) will first become available at another pair of terminals marked "To *Danger* LED."

18.27. Design a circuit which solves the coupled differential equations

$$\frac{d^2x}{dt^2} + 3x + \frac{dy}{dt} = 5u$$

$$\frac{d^2}{dt^2}y - 2\frac{d}{dt}(x+y) = -2u$$

where $u(t)$ is the input voltage and $x(t)$ and $y(t)$ the output voltages. (Hint: transform the equations, determine the transfer functions.)

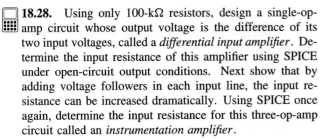

18.28. Using only 100-kΩ resistors, design a single-op-amp circuit whose output voltage is the difference of its two input voltages, called a *differential input amplifier*. Determine the input resistance of this amplifier using SPICE under open-circuit output conditions. Next show that by adding voltage followers in each input line, the input resistance can be increased dramatically. Using SPICE once again, determine the input resistance for this three-op-amp circuit called an *instrumentation amplifier*.

18.29. Design a third-order Chebyshev (1 dB) bandstop filter with center frequency 10 kHz.

18.30. Consider the design of a circuit with transfer function

$$\mathbf{H}(s) = \left(\frac{s^2 + 5s + 100}{s^2 + 40s + 400} \right)^2$$

This circuit can, of course, be designed by cascading two identical biquad circuits. Each would use four op amps, for a total of eight. Show that the op amp count can be reduced by adding a single output summer op amp to the block diagram of Fig. 18.20. Use this approach to design the circuit.

Appendix A

Matrix Methods

A set of simultaneous linear algebraic equations

$$a_{11}x_1 + a_{12}x_2 + \cdots + a_{1n}x_n = y_1$$

$$a_{21}x_1 + a_{22}x_2 + \cdots + a_{2n}x_n = y_2 \qquad \text{(A.1)}$$

$$a_{m1}x_1 + a_{m2}x_2 + \cdots + a_{mn}x_n = y_m$$

may be concisely written as a single vector-matrix equation, and matrix methods are then used for its solution. Sets of equations of this type (A.1) arise naturally in applying systematic methods of circuit analysis, such as nodal or mesh analysis, to linear circuits. Here we summarize the elements of linear algebra leading to the vector-matrix form and its solution by matrix inversion, Cramer's rule, and Gauss elimination.

A.1 MATRIX FUNDAMENTALS

A *matrix* $\mathbf{A}$ of dimension m rows by n columns is an ordered array of values a_{ij}, $i = 1, \ldots, m$; $j = 1, \ldots, n$. Matrix addition $\mathbf{A} + \mathbf{B}$ is defined for matrices of the same dimension $m \times n$ as elementwise addition:

$$\begin{bmatrix} a_{11} & a_{12} & a_{1n} \\ a_{21} & a_{22} & a_{2n} \\ a_{m1} & a_{m2} & a_{mn} \end{bmatrix} + \begin{bmatrix} b_{11} & b_{12} & b_{1n} \\ b_{21} & b_{22} & b_{2n} \\ b_{m1} & b_{m2} & b_{mn} \end{bmatrix} = \begin{bmatrix} a_{11}+b_{11} & a_{12}+b_{12} & a_{1n}+b_{1n} \\ a_{21}+b_{21} & a_{22}+b_{22} & a_{2n}+b_{2n} \\ a_{m1}+b_{m1} & a_{m2}+b_{m2} & a_{mn}+b_{mn} \end{bmatrix}$$

Scalar multiplication $k\mathbf{A}$, the operation of multiplication of a matrix $\mathbf{A}$ by a scalar k, is defined as elementwise multiplication of $\mathbf{A}$ by k:

$$k\begin{bmatrix} a_{11} & a_{12} & a_{1n} \\ a_{21} & a_{22} & a_{2n} \\ a_{m1} & a_{m2} & a_{mn} \end{bmatrix} = \begin{bmatrix} ka_{11} & ka_{12} & ka_{1n} \\ ka_{21} & ka_{22} & ka_{2n} \\ ka_{m1} & ka_{m2} & ka_{mn} \end{bmatrix} \qquad \text{(A.2)}$$

Matrix multiplication $\mathbf{AB}$ is defined between ordered pairs of matrices such that the column dimension of the left matrix $\mathbf{A}$ equals the row dimension of the right matrix $\mathbf{B}$, in which case the matrices are called *conformable*. Given conformable matrices

A of dimension $m_A \times n_A$ and **B** of dimension $m_B \times n_B$ $(n_A = m_B)$, their matrix product

$$\mathbf{C} = \mathbf{AB}$$

is defined to be the $n_A \times m_B$ matrix **C** whose elements are

$$c_{ij} = \sum_{k=1}^{n_A} a_{ik} b_{kj}$$

Matrix multiplication $\mathbf{C} = \mathbf{AB}$ can be carried out by writing **A** below and to the left of **B** and then extending the rows of **A** and columns of **B** into the space thus formed, as shown in Fig. A.1. Each intersection corresponds to an element of the matrix **C**. To determine its value, move from a given intersection to the farthest elements in **A** and **B** along the intersecting "tracks," and multiply these scalars together. Move one element closer to the intersection along both tracks, multiply these elements together, and add to the previous. Repeat until the intersection is reached and enter the accumulated value there. If the number of elements along the tracks do not pair exactly, the matrices are not conformable.

An $m \times 1$ matrix **x** with a single column is called a *column vector*, or simply a *vector*. The elements x_i, $i = 1, \ldots, n$ of a vector **x** need only one index, indicating the row location of the element. Following the rules of matrix multiplication, the product of an $m \times n$ matrix **A** and a n-vector **x** is an $m \times 1$ matrix, or m-vector **y**,

$$\mathbf{Ax} = \mathbf{y} \tag{A.3}$$

whose elements are, for $i = 1, \ldots, m$,

$$y_i = \sum_{k=1}^{n} a_{ik} x_k \tag{A.4}$$

We shall have need of the operation of matrix multiplication in converting equations to our desired form and in solving the resulting equation.

Finally, the *transpose* $\mathbf{A}^{\mathbf{T}}$ of an $m \times n$ matrix **A** is defined as an $n \times m$ matrix whose (i, j) element is the (j, i) element of **A**. That is, $\mathbf{A}^{\mathbf{T}}$ is the mirror image of **A** flipped around the main diagonal. The transpose matrix will be helpful in computing the inverse of a matrix, which we shall do shortly.

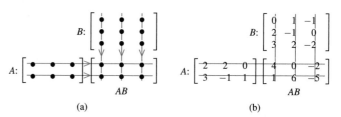

FIGURE A.1 Matrix multiplication.

A.2   CONVERSION TO VECTOR-MATRIX FORM

A set of m equations in n unknowns, such as (A.1), may be easily converted to a single equation of the form (A.3), said to be in *vector-matrix form*, with **x** referred to as the unknown vector, **y** the known vector, and **A** the connection matrix. Begin by assigning the n unknowns in the set of scalar equations as the n elements of the vector $\mathbf{x}$. Then following (A.4), consider the first given equation as the first row of the vector-matrix equation (A.3). This requires that y_i with $i = 1$ in (A.4) match y_1 in (A.1) and that the elements of the first row of the connection matrix $\mathbf{A}$ match the coefficients of the first equation in (A.1). Continuing in this manner for each row, the vector-matrix form (A.3) may be written by inspection of the set of scalar equations (A.1).

For instance, consider the three equations in three unknowns

$$3v_1 + v_2 - 2v_3 = 0$$
$$v_1 - 2v_2 - v_3 = -6 \tag{A.5}$$
$$-v_1 + v_2 - 2v_3 = 2$$

We will convert to the vector-matrix form $\mathbf{Ax} = \mathbf{y}$ with

$$\mathbf{x} = \begin{bmatrix} v_1 \\ v_2 \\ v_3 \end{bmatrix}, \qquad \mathbf{y} = \begin{bmatrix} y_1 \\ y_2 \\ y_3 \end{bmatrix} = \begin{bmatrix} 0 \\ -6 \\ 2 \end{bmatrix}$$

which will give

$$\begin{bmatrix} a_{11} & a_{12} & a_{13} \\ a_{21} & a_{22} & a_{23} \\ a_{31} & a_{32} & a_{33} \end{bmatrix} \begin{bmatrix} v_1 \\ v_2 \\ v_3 \end{bmatrix} = \begin{bmatrix} 0 \\ -6 \\ 2 \end{bmatrix}$$

According to (A.4) with $i = 1$, we must have

$$a_{11}v_1 + a_{12}v_2 + a_{13}v_3 = 0$$

Comparing the first equation in (A.5), it follows that $a_{11} = 3$, $a_{12} = 1$, and $a_{13} = -2$. Repeating for the second and third rows, the desired vector-matrix form of the equations (A.5) is

$$\begin{bmatrix} 3 & 1 & -2 \\ 1 & -2 & -1 \\ -1 & 1 & -2 \end{bmatrix} \begin{bmatrix} v_1 \\ v_2 \\ v_3 \end{bmatrix} = \begin{bmatrix} 0 \\ -6 \\ 2 \end{bmatrix}$$

In the remainder of this appendix we will summarize methods for solving the vector-matrix equation $\mathbf{Ax} = \mathbf{y}$ in the case that the connection matrix $\mathbf{A}$ is square, in other words, that there are as many unknowns as equations.

A.3   DETERMINANTS

Both principal methods for solving $\mathbf{Ax} = \mathbf{y}$, Cramer's rule and matrix inversion, depend on calculation of determinants. Let $\mathbf{A}$ be a square matrix of dimension $n \times n$. The determinant Δ of $\mathbf{A}$, denoted $|\mathbf{A}|$, is a scalar function of $\mathbf{A}$ computed as follows. For $n = 1$, $\mathbf{A}$ is the 1×1 matrix a_{11}, and it is its own determinant.

$$\Delta = a_{11} \tag{A.6}$$

For $n = 2$,

$$\Delta = \begin{vmatrix} a_{11} & a_{12} \\ a_{21} & a_{22} \end{vmatrix} = a_{11}a_{22} - a_{12}a_{21} \tag{A.7}$$

Schematically, this might be thought of as a *diagonal rule*. That is,

$$\Delta = \begin{vmatrix} a_{11} & a_{12} \\ a_{21} & a_{22} \end{vmatrix} = a_{11}a_{22} - a_{12}a_{21} \tag{A.8}$$

or Δ is a difference of the product $a_{11}a_{22}$ of elements down the diagonal to the right and the product $a_{12}a_{21}$ of the elements down the diagonal to the left.

As an example, the 2×2 determinant

$$\Delta = \begin{vmatrix} 1 & 2 \\ -3 & 4 \end{vmatrix}$$

given by

$$\Delta = (1)(4) - (2)(-3) = 10$$

A third-order, or 3×3, determinant, such as

$$\Delta = \begin{vmatrix} a_{11} & a_{12} & a_{13} \\ a_{21} & a_{22} & a_{23} \\ a_{31} & a_{32} & a_{33} \end{vmatrix} \tag{A.9}$$

may also be evaluated by a diagonal rule, given by

$$\Delta = (a_{11}a_{22}a_{33} + a_{12}a_{23}a_{31} + a_{13}a_{32}a_{21})$$
$$- (a_{13}a_{22}a_{31} + a_{23}a_{32}a_{11} + a_{33}a_{21}a_{12})$$

$$\tag{A.10}$$

The value of the determinant is a difference of products of elements down the three diagonals to the right and products of elements down the three diagonals to the left.

An example of a third-order determinant and its evaluation is given by

$$\Delta = \begin{vmatrix} 1 & 1 & 1 \\ 2 & -1 & 1 \\ -1 & 1 & 2 \end{vmatrix}$$

$$= [(1)(-1)(2) + (1)(1)(-1) + (1)(1)(2)] \tag{A.11}$$

$$[(1)(-1)(-1) + (1)(1)(1) + (2)(2)(1)]$$

$$= -7$$

A general definition of determinants may be given and used to derive a number of evaluation procedures. This is the technique usually given in elementary algebra books. However, for our purposes we shall use the diagonal rules that we have considered for second- and third-order determinants and evaluate higher-order determinants by the method of expansion by *minors*, or *cofactors*.

The *minor* A_{ij} of the element a_{ij} is the determinant of the matrix left after the ith row and the jth column are removed. For example, in (A.11) the minor A_{21} of the element $a_{21} = 2$ (second row, first column) is

$$A_{21} = \begin{vmatrix} 1 & 1 \\ 1 & 2 \end{vmatrix} = 2 - 1 = 1$$

The *cofactor* C_{ij} of the element a_{ij} is given by

$$C_{ij} = (-1)^{i+j} A_{ij} \qquad \text{(A.12)}$$

In other words, the cofactor is the *signed* minor, the minor multiplied by ± 1 with the sign depending on whether the sum of the row number and column number is even or odd.

The value of a determinant is the sum of products of the elements in any row or column and their cofactors. For example, let us *expand* the determinant of (A.9) by cofactors of the first row. The result is

$$\Delta = a_{11}C_{11} + a_{12}C_{12} + a_{13}C_{13}$$

or, by (A.12),

$$\Delta = a_{11}(-1)^{1+1}A_{11} + a_{12}(-1)^{1+2}A_{12} + a_{13}(-1)^{1+3}A_{13}$$

$$= a_{11}A_{11} - a_{12}A_{12} + a_{13}A_{13}$$

Writing out the minors explicitly, we have

$$\Delta = a_{11} \begin{vmatrix} a_{22} & a_{23} \\ a_{32} & a_{33} \end{vmatrix} - a_{12} \begin{vmatrix} a_{21} & a_{23} \\ a_{31} & a_{33} \end{vmatrix} + a_{13} \begin{vmatrix} a_{21} & a_{22} \\ a_{31} & a_{32} \end{vmatrix}$$

By the diagonal rule, this may be written as

$$\Delta = a_{11}(a_{22}a_{33} - a_{23}a_{32}) - a_{12}(a_{21}a_{33} - a_{23}a_{31}) + a_{13}(a_{21}a_{32} - a_{22}a_{31})$$

which may be simplified to (A.10).

To illustrate expansion by minors, let us evaluate the determinant of (A.11) by applying the technique to the third column. We have

$$\Delta = 1(-1)^{1+3} \begin{vmatrix} 2 & -1 \\ -1 & 1 \end{vmatrix} + 1(-1)^{2+3} \begin{vmatrix} 1 & 1 \\ -1 & 1 \end{vmatrix} + 2(-1)^{3+3} \begin{vmatrix} 1 & 1 \\ 2 & -1 \end{vmatrix}$$

$$= (2 - 1) - (1 + 1) + 2(-1 - 2) = -7$$

A.4 CRAMER'S RULE

The first method we shall summarize for solving the vector-matrix equation

$$\mathbf{Ax} = \mathbf{y}$$

where A is a square $n \times n$ connection matrix and $\mathbf{x}$ and $\mathbf{y}$ are n-vectors is called Cramer's rule. This rule states that the ith component x_i, $i = 1, \ldots, n$ of the solution is

$$x_i = \frac{\Delta_i}{\Delta}$$

where Δ is the determinant of $\mathbf{A}$ and Δ_i is the determinant of the matrix formed by replacing the ith column of $\mathbf{A}$ by the known vector $\mathbf{y}$. If $\Delta = 0$, this rule clearly cannot be applied, and $\mathbf{A}$ is said to be *noninvertible*. Sets of equations with noninvertible connection matrices do not have unique solutions and will not be discussed further.

The case we shall solve to illustrate Cramer's rule is the equation

$$\begin{bmatrix} 1 & -2 \\ 6 & 1 \end{bmatrix} \begin{bmatrix} x_1 \\ x_2 \end{bmatrix} = \begin{bmatrix} 5 \\ 4 \end{bmatrix}$$

which, evidently, is the vector-matrix form corresponding to the scalar equations

$$x_1 - 2x_2 = 5$$

$$6x_1 + x_2 = 4$$

Cramer's rule states that

$$x_1 = \frac{\Delta_1}{\Delta} = \frac{\begin{vmatrix} 5 & -2 \\ 4 & 1 \end{vmatrix}}{\begin{vmatrix} 1 & -2 \\ 6 & 1 \end{vmatrix}} = \frac{5(1) - 4(-2)}{1(1) - 6(-2)} = 1$$

and

$$x_2 = \frac{\Delta_2}{\Delta} = \frac{\begin{vmatrix} 1 & 5 \\ 6 & 4 \end{vmatrix}}{\begin{vmatrix} 1 & -2 \\ 6 & 1 \end{vmatrix}} = \frac{1(4) - 5(6)}{1(1) - 6(-2)} = -2$$

where the determinants are computed by (A.7). Cramer's rule is particularly efficient when only one or a small number of elements of the vector of unknowns are needed.

A.5   MATRIX INVERSION

The second method we shall summarize for solving the vector-matrix equation

$$\mathbf{Ax} = \mathbf{y} \tag{A.13}$$

where $\mathbf{A}$ is a square $n \times n$ connection matrix and $\mathbf{x}$ and $\mathbf{y}$ are n-vectors, is called matrix inversion. Suppose we can find an $n \times n$ matrix $\mathbf{A}^{-1}$, which has the property that

$$\mathbf{A}^{-1}\mathbf{A} = \mathbf{I} \tag{A.14}$$

where $\mathbf{I}$ is the $n \times n$ identity matrix [each (i, i) element is 1, each (i, j) element is 0 for i not equal to j]. Then, premultiplying both sides of (A.13) by $\mathbf{A}^{-1}$,

$$\mathbf{A}^{-1}\mathbf{Ax} = \mathbf{Ix} = \mathbf{A}^{-1}\mathbf{y} \tag{A.15}$$

or, since $\mathbf{Ix} = \mathbf{x}$ for any conformable matrix $\mathbf{x}$,

$$\mathbf{x} = \mathbf{A}^{-1}\mathbf{y} \tag{A.16}$$

which is the desired solution. The matrix $\mathbf{A}^{-1}$ is called the *inverse* of $\mathbf{A}$.

To determine $\mathbf{A}^{-1}$, begin by computing the $n \times n$ matrix whose (i,j)th element is the cofactor C_{ij} defined in (A.12). The transpose of this matrix of cofactors is denoted adj $\mathbf{A}$ (the *adjoint* of $\mathbf{A}$). Then

$$\mathbf{A}^{-1} = \frac{\text{adj}\mathbf{A}}{\Delta} \tag{A.17}$$

where, as usual, Δ is the determinant of $\mathbf{A}$. Equation (A.17) is an instance of scalar multiplication, that is, multiplication of a matrix (adj $\mathbf{A}$) by a scalar $(1/\Delta)$, which was defined in Sec. A.3.

To illustrate, we shall resolve the vector-matrix equation solved using Cramer's rule in the previous subsection:

$$\begin{bmatrix} 1 & -2 \\ 6 & 1 \end{bmatrix} \begin{bmatrix} x_1 \\ x_2 \end{bmatrix} = \begin{bmatrix} 5 \\ 4 \end{bmatrix}$$

The determinant of $\mathbf{A}$ is

$$\Delta = (1)(1) - (-2)(6) = 13$$

Each cofactor C_{ij} is found by striking out the ith row and jth column of $\mathbf{A}$, computing the determinant of what remains, and multiplying by $(-1)^{i+j}$. After striking out the first row and column, C_{11} is just the determinant of the remaining 1×1 matrix $+1$, multiplied by $(-1)1 + 1 = 1$, or $+1$. C_{12} is found by striking out the first row and second column, computing the determinant of the remaining matrix, 6, and multiplying by $(-1)^{1+2} = -1$. The result is $C_{12} = -6$. Proceeding in this fashion, the matrix of cofactors is

$$\begin{bmatrix} 1 & -6 \\ 2 & 1 \end{bmatrix}$$

and after transposing, to get the adjoint, the inverse is found as

$$A^{-1} = \frac{\text{adj}A}{\Delta} = \frac{1}{13} \begin{bmatrix} 1 & 2 \\ -6 & 1 \end{bmatrix} = \begin{bmatrix} \dfrac{1}{13} & \dfrac{2}{13} \\ \dfrac{-6}{13} & \dfrac{1}{13} \end{bmatrix}$$

Then applying (A.16), the solution is

$$\begin{bmatrix} x_1 \\ x_2 \end{bmatrix} = \begin{bmatrix} \dfrac{1}{13} & \dfrac{2}{13} \\ -\dfrac{6}{13} & \dfrac{1}{13} \end{bmatrix} \begin{bmatrix} 5 \\ 4 \end{bmatrix} = \begin{bmatrix} 1 \\ -2 \end{bmatrix}$$

which agrees with our result using Cramer's rule. Matrix inversion is frequently more efficient than repeated use of Cramer's rule when most or all of the elements of the unknown vector are required.

A.6 GAUSS ELIMINATION

A final method for solving $\mathbf{Ax} = \mathbf{y}$ for square matrices $\mathbf{A}$, based on successive elimination of unknowns, is called Gauss elimination. Unlike Cramer's rule and matrix inversion, it can be developed without resort to the matrix formalism. Here we present it as a matrix method, since it is useful in that context also.

We shall illustrate the method for a 3×3 case, the generalization to other dimensions is evident. Consider the vector-matrix equation

$$\begin{bmatrix} 1 & 1 & 1 \\ 2 & -1 & 1 \\ -1 & 1 & 2 \end{bmatrix} \begin{bmatrix} x_1 \\ x_2 \\ x_3 \end{bmatrix} = \begin{bmatrix} 6 \\ 3 \\ 7 \end{bmatrix} \tag{A.18}$$

Our first goal will be to *triangularize* the matrix, that is, replace the elements below the main diagonal by 0's. We may, of course, consider (A.18) as the set of three equations in three unknowns:

$$x_1 + x_2 + x_3 = 6$$

$$2x_1 - x_2 + x_3 = 3$$

$$-x_1 + x_2 + 2x_3 = 7$$

Suppose we subtract twice the first equation from the second and use this as a new second equation. Similarly, add the first equation to the third, and replace the old third equation by this one. Back in vector-matrix form,

$$\begin{bmatrix} 1 & 1 & 1 \\ 0 & -3 & -1 \\ 0 & 2 & 3 \end{bmatrix} \begin{bmatrix} x_1 \\ x_2 \\ x_3 \end{bmatrix} = \begin{bmatrix} 6 \\ -9 \\ 13 \end{bmatrix} \tag{A.19}$$

Note the set of 0's in the first column below the main diagonal. Next we add two-thirds times the second row equation to the third, which yields

$$\begin{bmatrix} 1 & 1 & 1 \\ 0 & -3 & -1 \\ 0 & 0 & \frac{7}{3} \end{bmatrix} \begin{bmatrix} x_1 \\ x_2 \\ x_3 \end{bmatrix} = \begin{bmatrix} 6 \\ -9 \\ 7 \end{bmatrix}$$

The matrix is in *upper-triangular* form. The final row equation may now be easily solved for x_3, because no other unknowns enter this equation. The solution is $x_3 = 3$. Moving up one row, after back-substitution of the value for x_3 the second equation has only the single unknown x_2. Upon division of this equation by -3, we have

$$x_2 = 3 - \frac{x_3}{3} = 3 - 1 = 2$$

and back-substituting into the first row equation,

$$x_1 = 6 - x_2 - x_3 = 6 - 2 - 3 = 1$$

which completes the solution.

The method of Gauss elimination can be viewed as the formation of an augmented matrix containing both **A** and **y**. In the case of the example just completed, this is the augmented matrix

$$\begin{bmatrix} 1 & 1 & 1 & 6 \\ 2 & -1 & 1 & 3 \\ -1 & 1 & 2 & 7 \end{bmatrix}$$

Subtracting twice the first row from the second and adding the first row to the third and the augmented matrix becomes

$$\begin{bmatrix} 1 & 1 & 1 & 6 \\ 0 & -3 & -1 & -9 \\ 0 & 2 & 3 & 13 \end{bmatrix}$$

(compare to A.19). Adding two-thirds of the second row to the third and then dividing the second row by -3 and the third by $\frac{7}{3}$ yields

$$\begin{bmatrix} 1 & 1 & 1 & 6 \\ 0 & 1 & \frac{1}{3} & 3 \\ 0 & 0 & 1 & 3 \end{bmatrix}$$

800

The last row reveals that $x_3 = 3$. The process of back-substitution may be implemented in the augmented matrix by subtracting one-third times the last row from the second,

$$\begin{bmatrix} 1 & 1 & 1 & 6 \\ 0 & 1 & 0 & 2 \\ 0 & 0 & 1 & 3 \end{bmatrix}$$

and then subtracting the third and the second rows from the first:

$$\begin{bmatrix} 1 & 0 & 0 & 1 \\ 0 & 1 & 0 & 2 \\ 0 & 0 & 1 & 3 \end{bmatrix}$$

The last element in the top row is $x_1 = 1$, the second row $x_2 = 2$, and the bottom row, $x_3 = 3$, as was previously calculated. The process of transforming the first n columns to the identity matrix results in the desired solution as the remaining column.

Appendix B

···

Complex Numbers and the Complex Exponential

B.1   COMPLEX NUMBERS

From our earliest training in arithmetic we have dealt with *real* numbers, such as 3, −5, $\frac{4}{7}$, π, and so on, which may be used to measure distances along the *real line* from a fixed point. A number such as x that satisfies

$$x^2 = -4 \qquad (B.1)$$

is not a real number and is customarily, and unfortunately, called an *imaginary* number. To deal with imaginary numbers, an *imaginary unit*, denoted by j, is defined by

$$j = \sqrt{-1} \qquad (B.2)$$

Thus we have $j^2 = -1$, $j^3 = -j$, $j^4 = 1$, and so on. (We might note that mathematicians use the symbol i for the imaginary unit, but in electrical engineering this might be confused with current.) An imaginary number is defined as the product of j with a real number, such as $x = j2$. In this case $x^2 = (j2)^2 = -4$, and thus x is a solution of (B.1).

A *complex* number is the sum of a real number and an imaginary number, such as

$$A = a + jb \qquad (B.3)$$

where a and b are real. The complex number A has a *real part*, a, and an *imaginary part*, b, which are sometimes expressed as

$$a = \text{Re}A$$

$$b = \text{Im}A$$

It is important to note that both parts are real, in spite of their names.

The complex number $a + jb$ may be represented on a rectangular coordinate plane, or a *complex plane*, by interpreting it as a point (a, b). That is, the horizontal coordinate is a and the vertical coordinate is b, as shown in Fig. B.1, for the case $4 + j3$. Because of

803

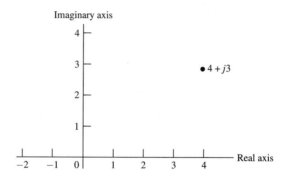

Imaginary axis

• $4 + j3$

Real axis

FIGURE B.1   Graphical representation of a complex number.

this analogy with points plotted on a rectangular coordinate system, (B.3) is sometimes called the *rectangular form* of the complex number A.

The complex number $A = a + jb$ may also be uniquely located in the complex plane by specifying its distance r along a straight line from the origin and the angle θ that this line makes with the real axis, as shown in Fig. B.2. From the right triangle thus formed, we see that

$$r = \sqrt{a^2 + b^2}$$

$$\theta = \tan^{-1}\frac{b}{a}$$

(B.4)

and that

$$a = r\cos\theta$$

$$b = r\sin\theta$$

(B.5)

We denote this representation of the complex number by

$$A = r\underline{/\theta}$$

(B.6)

which is called the *polar form*. The number r is called the *magnitude* and is sometimes denoted by

$$r = |A|$$

The number θ is the *angle* or *argument* and is often denoted by

$$\theta = \text{ang } A = \text{arg } A$$

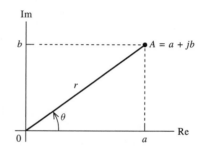

FIGURE B.2   Two forms of a complex number.

App. B   Complex Numbers and the Complex Exponential

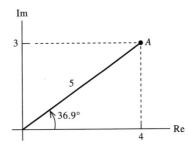

Im

3 - - - - - - - - - - - - - - - - - • A

5

36.9°

4

Re

FIGURE B.3   Two forms of a complex number A.

We may easily convert from rectangular to polar form, or vice versa, by means of
(B.4) and (B.5). For example, the number A, shown in Fig. B.3, is given by

$$A = 4 + j3 = 5\underline{/36.9°}$$

since by (B.4)

$$r = \sqrt{4^2 + 3^2} = 5$$

$$\theta = \tan^{-1}\frac{3}{4} = 36.9°$$

The *conjugate* of the complex number $A = a + jb$ is defined to be

$$A^* = a - jb \tag{B.7}$$

That is, j is replaced by $-j$. Since we have

$$|A^*| = \sqrt{a^2 + (-b)^2} = \sqrt{a^2 + b^2} = |A|$$

and

$$\arg A^* = \tan^{-1}\left(\frac{-b}{a}\right) = -\tan^{-1}\frac{b}{a} = -\arg A$$

we may write, in polar form,

$$(r\underline{/\theta})^* = r\underline{/-\theta} \tag{B.8}$$

We may note from the definition that if A^* is the conjugate of A then A is the
conjugate of A^*. That is, $(A^*)^* = A$.

The operations of addition, subtraction, multiplication, and division apply to complex numbers exactly as they do to real numbers. In the case of addition and subtraction,
we may write, in general,

$$(a + jb) + (c + jd) = (a + c) + j(b + d) \tag{B.9}$$

and

$$(a + jb) - (c + jd) = (a - c) + j(b - d) \tag{B.10}$$

That is, to add (or subtract) two complex numbers, we simply add (or subtract) their real
parts and their imaginary parts.

As an example, let $A = 3 + j4$ and $B = 4 - j1$. Then

$$A + B = (3 + 4) + j(4 - 1) = 7 + j3$$

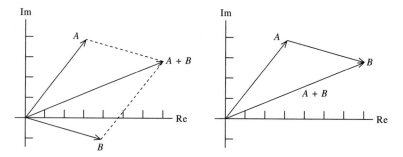

FIGURE B.4 Two methods of graphical addition.

This may also be done graphically, as shown in Fig. B.4(a), where the number A and B are represented as vectors from the origin. The result is equivalent to completing the parallelogram or to connecting the vectors A and B in head-to-tail manner, as shown in Fig. B.4(b), as the reader may check by comparing the numbers. For this reason, complex number addition is sometimes called vector addition.

In the case of multiplication of numbers A and B given by

$$A = a + jb = r_1 \cos\theta_1 + jr_1 \sin\theta_1$$
$$B = c + jd = r_2 \cos\theta_2 + jr_2 \sin\theta_2$$

(B.11)

we have

$$AB = (a + jb)(c + jd) = ac + jad + jbc + j^2bd$$
$$= (ac - bd) + j(ad + bc)$$

(B.12)

Alternatively we have, from (B.11),

$$AB = (r_1 \cos\theta_1 + jr_1 \sin\theta_1)(r_2 \cos\theta_2 + jr_2 \sin\theta_2)$$
$$= r_1 r_2 [(\cos\theta_1 \cos\theta_2 - \sin\theta_1 \sin\theta_2) + j(\sin\theta_1 \cos\theta_2 + \cos\theta_1 \sin\theta_2)]$$
$$= r_1 r_2 [\cos(\theta_1 + \theta_2) + j \sin(\theta_1 + \theta_2)]$$

Therefore, in polar form we have

$$(r_1 \underline{/\theta_1})(r_2 \underline{/\theta_2}) = r_1 r_2 \underline{/\theta_1 + \theta_2}$$

(B.13)

and hence we may multiply two numbers by multiplying their magnitudes and adding their angles.

From this result we see that

$$AA^* = (r \underline{/\theta})(r \underline{/-\theta}) = r^2 \underline{/0} = |A|^2 \underline{/0}$$

Since $|A|^2 \underline{/0}$ is the real number $|A|^2$, we have

$$|A|^2 = AA^*$$

(B.14)

Division of a complex number by another, such as

$$N = \frac{A}{B} = \frac{a + jb}{c + jd}$$

App. B   Complex Numbers and the Complex Exponential

results in an irrational denominator, since $j = \sqrt{-1}$. We may rationalize the denominator and display the real and imaginary parts of N by writing

$$N = \frac{AB^*}{BB^*} = \frac{a + jb}{c + jd} \cdot \frac{c - jd}{c - jd}$$

which is

$$N = \frac{(ac + bd) + j(bc - ad)}{c^2 + d^2} \tag{B.15}$$

We may also show by the method used to obtain (B.13) that

$$\frac{r_1/\theta_1}{r_2/\theta_2} = \frac{r_1}{r_2}/\theta_1 - \theta_2 \tag{B.16}$$

As examples, let $A = 4 + j3 = 5\underline{/36.9°}$ and $B = 5 + j12 = 13\underline{/67.4°}$. Then we have

$$AB = (5)(13)\underline{/36.9° + 67.4°} = 65\underline{/104.3°}$$

and

$$\frac{A}{B} = \frac{5}{13}\underline{/36.9° - 67.4°} = 0.385\underline{/-30.5°}$$

Evidently, it is easier to add and subtract complex numbers in rectangular form and to multiply and divide them in polar form.

B.2   COMPLEX EXPONENTIAL FUNCTION

Complex numbers are widely used in electrical engineering because of the intimate link between real sines and cosines and the *complex exponential function* $e^{j\theta}$. To expose this relationship, define

$$g = \cos\theta + j\sin\theta \tag{B.17}$$

Differentiating,

$$\frac{dg}{d\theta} = j(\cos\theta + j\sin\theta) = jg$$

This is a first-order unforced linear differential equation with general solution $ke^{s\theta}$, where s satisfies the characteristic equation

$$s - j = 0$$

and, evaluating (B.17) at $\theta = 0$, $k = 1$. Thus $g = e^{j\theta}$ and, by (B.17),

$$e^{j\theta} = \cos\theta + j\sin\theta \tag{B.18a}$$

Converting to polar form and noting that $\cos^2\theta + \sin^2\theta = 1$,

$$e^{j\theta} = 1\underline{/\theta} \tag{B.18b}$$

Equation (B.18a,b) are the rectangular and polar forms of *Euler's formula*, or the *Euler identity*. Replacing θ by $-\theta$ in (B.18a),

$$e^{-j\theta} = \cos\theta - j\sin\theta \tag{B.19}$$

Combining (B.18a) and (B.19),

$$\cos\theta = \frac{1}{2}\left(e^{j\theta} + e^{-j\theta}\right) \tag{B.20a}$$

$$\sin\theta = \frac{1}{2j}\left(e^{j\theta} - e^{-j\theta}\right) \tag{B.20b}$$

The polar form of the Euler formula (B.18b) suggests that any complex number may be represented as a complex exponential. Let x have magnitude $r \geq 0$ and angle θ. Then

$$x = r\underline{/\theta}$$
$$= (r\underline{/0})(1\underline{/\theta})$$
$$= re^{j\theta}$$

Thus, along with the rectangular and polar forms to represent complex numbers, we have the *exponential form*

$$x = a + jb \quad \text{(rectangular form)} \tag{B.21a}$$
$$= r\underline{/\theta} \quad \text{(polar form)} \tag{B.21b}$$
$$= re^{j\theta} \quad \text{(exponential form)} \tag{B.21c}$$

For instance, the complex number in rectangular form $x = 1 + j$ has polar representation $x = \sqrt{2}\underline{/45°}$ and exponential representation $x = \sqrt{2}e^{j45°}$.

A principal use of the exponential form is in forming phase-shifted complex exponentials. For instance, suppose

$$i(t) = xe^{j\omega t} \tag{B.22}$$

where x is a complex number. Then, using the exponential form (B.21),

$$i(t) = re^{j\theta}e^{j\omega t} = re^{j(\omega t + \theta)}$$

By the Euler formula (B.18a), the last is easily recognized to have real part $r\cos(\omega t + \theta)$ and imaginary part $r\sin(\omega t + \theta)$. Extracting real and imaginary parts of expressions such as (B.22) is an integral part of phasor analysis.

The exponential form can be used to concisely justify the multiplication and division formulas presented earlier. Let

$$x = r_1\underline{/\theta_1}, \qquad y = r_2\underline{/\theta_2}$$

Then

$$xy = (r_1e^{j\theta_1})(r_2e^{j\theta_2}) = (r_1r_2)e^{j(\theta_1 + \theta_2)}$$
$$= r_1r_2\underline{/(\theta_1 + \theta_2)}$$

as asserted in (B.13), and

$$\frac{x}{y} = \frac{r_1e^{j\theta_1}}{r_2e^{j\theta_2}} = \frac{r_1}{r_2}e^{j(\theta_1 - \theta_2)}$$
$$= \frac{r_1}{r_2}\underline{/(\theta_1 - \theta_2)}$$

as in (B.16).

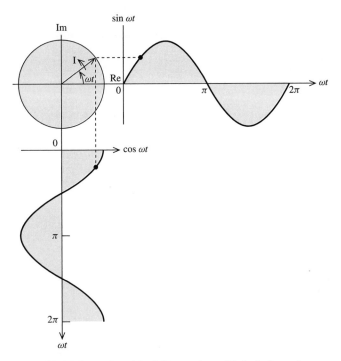

FIGURE B.5   Graphical illustration of Euler's formula.

Euler's formula is illustrated graphically in Fig. B.5. A unit vector is rotating around a circle in the direction shown, with an angular velocity of ω rad/s. Therefore, in t seconds it has moved through an angle ωt as shown, and thus the vector may be specified by $1\underline{/\omega t}$ or $e^{j\omega t}$. Its real part is the projection on the horizontal axis, given by $\cos \omega t$, and its imaginary part is the projection on the vertical axis, given by $\sin \omega t$. That is,

$$e^{j\omega t} = \cos \omega t + j \sin \omega t$$

which is Euler's formula in rectangular form (B.18a). The projections trace out the cosine and sine waves, as shown, as the vector rotates with period $\omega T = 2\pi$ or $T = 2\pi/\omega$.

Appendix C

....

Circuit Topology

The general methods of nodal and mesh analysis emphasized in this text result in an equal number of equations as unknowns. This in itself is not enough to guarantee that we can use these equations to "break" the circuit, that is, evaluate their unknowns, the node voltages or mesh currents, and from them all other voltages and currents. It is easy to write k equations in k unknowns for which there are no solutions or for which an infinite number of equally valid solutions exist. In neither case could we go on to complete the circuit analysis from these equations, that is, find unique values for all currents and voltages.

In this appendix we shall present some of the key results from a study of the circuit *topology*, or the manner in which the elements are interconnected. This proves an effective way to determine how many equations are required to analyze a given circuit, which ones are independent, and the most efficient set of equations for analysis. We seek the smallest sets of equations with equal numbers of unknowns, while, also seeking to avoid the dual pitfalls of an infinite number of solutions or no solutions at all.

C.1 NETWORK GRAPH

In studying the topology of a circuit, we will not distinguish individual elements by *what* they are, resistors, current sources, and so on, but by *where* they are. The configuration of lines and nodes formed by replacing elements of a circuit by featureless lines is called its *circuit graph* or *network graph*. To emphasize the great generality of this graph, its capacity to represent networks of all kinds, not just electric circuits, we will refer to it as a network graph hereafter. The lines in a network graph are called its *branches*.

A network graph is *connected* if there is a path between any two nodes. Our discussion will assume the graph is connected, since, if it is not, each disjoint part may be analyzed separately as a connected graph.

A *tree* is a subgraph with all nodes of the full network graph but no loops. An example of a simple circuit, its network graph, and two distinct trees are shown in Fig. C.1. While a given network graph with B branches and N nodes may have many

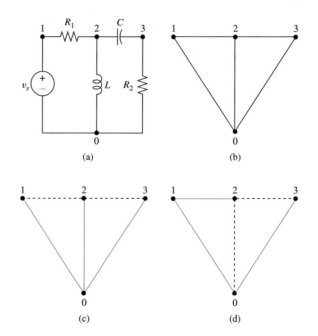

FIGURE C.1 (a) Circuit; (b) network graph; (c) one tree; (d) another tree.

trees, each tree has exactly $N-1$ branches. This follows from the fact that it must connect all N nodes, but cannot have any closed loops. The branches in a tree are called *tree branches* and the remaining branches are *links*. For instance, the branch b_{01} connecting nodes 0 and 1 in Fig. C.1 is a tree branch for both trees shown, while b_{02} is a tree branch for Fig. C.1(c) and a link for the tree of Fig. C.1(d). Since there are $N - 1$ tree branches, there must be $B - (N - 1) = B - N + 1$ links in any tree. The set of these $B - N + 1$ links is called the *cotree* associated with a given tree. The union of the tree and cotree is the entire network graph.

C.2 NODAL ANALYSIS

Consider a network graph with N nodes, B branches, and a given tree. Adding a single link to the tree results in a closed loop, since there is a tree path between the nodes at each end of the link, and the second path formed by the link itself must close a loop. Thus, by KVL, the link voltage is fixed by the tree branch voltages. On the other hand, the set of tree branch voltages themselves are completely unconstrained by KVL, since they contain no closed loops. The tree branch voltages may take on any set of values independently of one another. Thus the $N - 1$ tree branch voltages are an independent set of voltages, and the remaining $B - N + 1$ link voltages are fixed by the tree branch voltages. All branch voltages may be expressed in terms of the $N - 1$ tree branch voltages.

It is well known that a set of k independent equations in k unknowns possesses a unique solution. Thus, if we can find $N - 1$ independent equations in the $N - 1$ tree branch voltages, we can find their values and, since they fix the link voltages, all remaining voltages as well. From the resulting complete set of voltages we may recover the currents by the element i–v laws. Thus, if we can find $N - 1$ independent equations in the tree branch voltages, these will suffice for complete network analysis.

Let us write KCL at any one node in the network graph. Moving to any other node, the KCL equation at this second node must be independent of the first if $N > 2$. This follows from the fact that the second equation contains at least one branch not referenced in the first equation. For if the second equation contained all the same branches as the first, every branch connected to one node would terminate at the other, in other words, $N = 2$. Moving to a third node, its KCL contains a new branch not referenced in either of the first two as long as $N > 3$, or, if it does not, then $N = 3$. We conclude that any set of $N - 1$ KCL node equations is independent.

For instance, referring to Fig. C.1(c), let i_{nm} refer to the current flowing in branch b_{nm} from node n to node m. Then KCL at node 0 is

$$i_{01} + i_{02} + i_{03} = 0$$

and KCL at node 1 is

$$i_{10} + i_{12} = 0$$

This second equation has a new branch current, i_{12}, not in the first, so is independent. KCL at node 2 is

$$i_{21} + i_{20} + i_{23} = 0$$

Once again the new KCL is independent of the others, since it contains a new branch current, i_{23}. The final KCL equation, at node 3, is

$$i_{32} + i_{30} = 0$$

Neither term is new to the set of four node equations, and this equation is not independent of the first three. In fact, it is simply the negative sum of the first three equations. In this network graph there are four nodes, $N = 4$, there are $N - 1 = 3$ independent KCL nodal equations, and the fourth is not independent of the first three.

Now suppose we choose to write $N - 1$ KCL node equations in terms of $N - 1$ tree branch voltages. Each branch voltage, whether tree branch or link, may be expressed in terms of the $N - 1$ tree branch voltages, and, with the help of the i–v laws, so can the branch currents. Since the resulting $N - 1$ equations in $N - 1$ unknowns are independent, it follows that we can find unique solutions for the tree branch voltages, and hence all voltages and currents. We conclude that any $N - 1$ KCL node equations in terms of the tree branch voltages for any tree are independent and will suffice for complete circuit analysis.

It is usually more convenient to work with node voltages than tree branch voltages, since then we are not required to construct a tree. $N - 1$ given node voltages fix the full set of $N - 1$ tree branch voltages, since the tree branch voltages are differences of node voltages. Equally, the $N - 1$ tree branch voltages fix the $N - 1$ node voltages, as can be seen by starting with a tree branch connected to the reference node, labeling the voltage at the reference node zero, and at the other node voltage equal to the tree branch voltage. Since the two sets of $N - 1$ voltages are equivalent to one another, and one is independent, so must the other be.

We conclude that KCL equations at any $N - 1$ nodes written in terms of the node voltages are independent and will suffice for complete circuit analysis. This is what we refer to as nodal analysis in Chapter 4 and elsewhere, and we see that it is guaranteed to work for every circuit.[1]

C.3   BASIC LOOP ANALYSIS AND MESH ANALYSIS

Turning attention from voltages to currents, for any network graph with a given tree, the set of tree branch currents is fixed by the values of the $B - N + 1$ link currents. For if we set all link currents to zero, since there are no loops in the tree, there are no closed loops anywhere and all tree branch currents are fixed at zero. Any individual link current can be assigned independently of the others, as can be seen by setting all but one to zero and noting the last may take on any value, since it is part of a closed loop with no other links in the loop. We conclude that the set of $B - N + 1$ link currents is independent, and the tree branch currents are fixed by the link currents.

If all but one link is removed from a network graph, there will remain one closed loop. Call the set of these $B - N + 1$ loops, one per link, the *basic loops* associated with the given tree. The set of KVL equations written around these loops is independent, since each equation contains a branch not contained in any of the others, that is, its defining link. For instance, referring to Fig. C.1(d), with $B = 5$ and $N = 4$, the $B - N + 1 = 2$ links are b_{20} and b_{23}. The two basic loops are closed by first inserting b_{20} into the tree, and then replacing with b_{23}. These basic loops are 2012 and 23012, enumerated by the order that nodes are traversed.

Thus the $B - N + 1$ independent KVL equations around the basic loops, written in terms of the $B - N + 1$ link currents, also referred to as *basic loop currents*, can be solved for the link currents. The remaining tree branch currents are fixed by them, and, given the full set of branch currents, the branch voltages follow from the $i-v$ laws. We conclude that *for any given tree, KVL written around the $B - N + 1$ basic loops in terms of the $B - N + 1$ basic loop currents are independent and are sufficient for complete circuit analysis.* This is referred to as *basic loop analysis* and, just like nodal analysis, is guaranteed to work for all circuits.

Just as we found it convenient to convert in the end from tree branch voltages to node voltages in our previous discussion of nodal analysis, it is preferable wherever possible to work with meshes rather than basic loops. Meshes are defined only for *planar networks*, those networks whose graphs can be drawn on a plane without any branches crossing. A *mesh* is defined as an empty closed loop in a planar network graph, that is, a loop containing no branches inside it. Mesh current paths can be identified by inspection, while basic loops must be constructed from a tree.

The set of KVL mesh equations is independent, since each added mesh equation brings in at least one new branch (if the network graph contains meshes fully enclosed by sets of other meshes, add them before the meshes that enclose them). By their definitions,

---

[1] With the understanding that each voltage source defines a supernode that should be counted as a single node (voltage sources are topologically equivalent to short circuits). Thus a circuit with five nodes and one voltage source will have three nodal analysis equations.

there are as many KVL mesh equations as mesh currents. Moreover, each branch current may be expressed in terms of mesh currents, and, with the help of the i–v laws, each branch voltage. It follows that the KVL mesh equations written in terms of mesh currents constitute an equal number of independent equations as unknowns and thus may be solved uniquely for the mesh currents. All branch currents, and hence branch voltages, may be determined from the resulting mesh currents. Mesh analysis is thus guaranteed to suffice for complete analysis of any planar circuit.[2]

The relative efficiency of mesh analysis depends on how many equations it requires (compared to $N-1$ nodal equations and $B-N+1$ basic loop equations). Suppose we reconstruct a given planar network one mesh at a time. The first mesh has as many nodes as branches, say k_1 of each. The next, and all subsequent meshes added, will add one less node than branch to the growing reconstruction. Each branch in a new mesh will also add a new node, except the last branch in the mesh, which returns to the previously reconstructed subcircuit. Thus, if k_2 branches are added with the second mesh, k_3 with the third, and so on, upon full reconstruction we have M meshes, where

$$B = k_1 + k_2 + \cdots + k_M$$

$$N = k_1 + (k_2 - 1) + (k_3 - 1) + \cdots + (k_M - 1)$$

$$= B - (M - 1)$$

so the number of meshes M is

$$M = B - N + 1 \tag{C.1}$$

which is the same as the number of basic loops. *We conclude that KVL equations around the $B-N+1$ basic loops associated with any tree, expressed in terms of basic loop currents, or KVL equations around the $B-N+1$ meshes in a planar circuit, expressed in terms of the mesh currents, are linearly independent and sufficient for complete circuit analysis.*

To illustrate, consider the circuit of Fig. C.2(a). The network graph is shown in Fig. C.2(b), with a tree colored in green. Associated with this tree are four basic loops: 014530, 03520, 0120, and 3453. The current source I_{s1} reduces the number of required loop equations by one, since the basic loop current marked i_4 is in fact equal to I_{s1}, a known value. The three basic loop equations, associated with the first three basic loops cited above, are

Basic loop 1: $(R_1 + R_2 + R_6 + R_7)i_1 - R_7 i_2 + R_1 i_3 = V_{s2} - R_6 I_{s1}$

Basic loop 2: $-R_7 i_1 + (R_3 + R_4 + R_7)i_2 + R_4 i_3 = -V_{s2}$

Basic loop 3: $R_1 i_1 + R_4 i_2 + (R_1 + R_5 + R_4)i_3 = 0$

The four meshes and their mesh currents are shown for this planar circuit in Fig. C.3. There are three unknown mesh currents marked i_5, i_6, and i_7. Temporary removal of the current source collapses the two meshes 01430 and 3453 into the single supermesh

---

[2]With the understanding that each current source decreases the number of meshes by one (current sources are topologically equivalent to open circuits). Thus a circuit with four meshes and one current source will have three KCL mesh analysis equations.

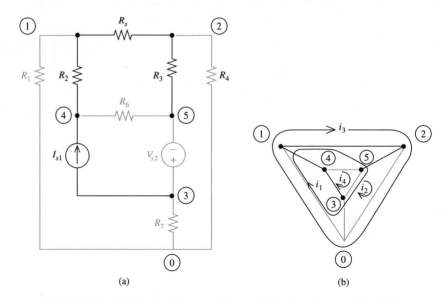

FIGURE C.2 (a) Circuit; (b) network graph showing tree and basic loop currents.

014530. The mesh equations are

Supermesh: $\quad R_1 i_5 + R_2(i_5 - i_6) + R_6(I_{s1} - i_5 - i_6) + R_7(i_5 - i_7) \quad = V_{s2}$

Mesh 6: $\quad R_5 i_6 + R_3(i_6 - i_7) + R_6(i_6 - I_{s1} + i_5) + R_2(i_6 - i_5) \quad = 0$

Mesh 7: $\qquad\qquad\qquad R_4 i_7 + R_7(i_7 - i_5) + R_3(i_7 - i_6) \quad = -V_{s2}$

Which of these methods, nodal, mesh, or basic loop analysis, should be used to study a given circuit? If $N - 1 \ll B - N + 1$, as is the case for circuits with many elements in parallel (many branches per node), nodal analysis will be more efficient. If $B - N + 1 \ll N + 1$, as in circuits with many elements in series, and consequently few elements per node, either basic loop or mesh analysis would be a better choice. For planar circuits, mesh analysis has the edge over basic loop analysis, since no tree is needed. Since most circuits of low and medium complexity, such as we work with in this book, are planar, nodal and mesh analysis are the methods we have emphasized in this book.

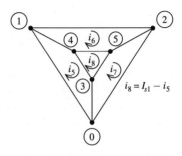

FIGURE C.3 Meshes and mesh currents for the circuit of Fig. C.2(a).

Appendix **D**

• •

SPICE Reference Guide

This appendix lists the basic input statement formats and usage rules for the family of circuit simulators referred to throughout this book as SPICE. This family embraces many variations, or "dialects," some public domain (SPICE1, SPICE2, SPICE3) and some proprietary (PSpice, HSPICE). The dialects differ somewhat both with respect to restrictions and enhancements. Where dialects differ in their restrictions, the most restrictive form is listed here. For instance, some dialects permit the first character of the title statement to be white space and others do not, here you are warned not to use white space. The dialects also differ widely in enhancements of various kinds; some for instance offer support for high resolution graphics and printing. For capabilities beyond the universal common core discussed here, the user is referred to the documentation of the specific software program. *The syntax summarized here is a common core applicable to all dialects of SPICE.*

SPICE1, which stands for Simulation Program with Integrated Circuit Emphasis (Version 1), was written by faculty and students at the University of California, Berkeley, largely with public grant support, and released into the public domain in 1972. The first major rewrite, SPICE2, released in 1975, improved accuracy and robustness and expanded the set of supported circuit devices. SPICE2 still remains the most widely used member of the SPICE family in computing environments favoring multiuser batch processing, such as large university mainframes. The mainframe model of digital computing prevalent in 1975 gradually gave way to a networked server-workstation model where each user had a high resolution screen and printer available. Recognizing this opportunity, SPICE3 appeared in 1985 complete with a graphical user interface, point-and-click controls, and high resolution plots. The popularity of the C programing language in network-based computing led SPICE's authors to abandon FORTRAN, the language of SPICE1 and SPICE2, in favor of recoding in C. While designed for interactive use and incorporating menus and a graphical user interface, SPICE3 also supports ASCII batch mode operation, making it backward compatible with earlier releases.

The first SPICE dialect designed for use on personal computers, PSpice, was released in 1984 by MicroSim, Inc. It remains the most widely used of the proprietary dialects, in part because of the availability of a free Student Evaluation Version. PSpice has been ported to most popular PC and workstation operating systems: DOS (Microsoft, Inc.), Windows (Microsoft, Inc.), MacintoshOS (Apple, Inc.), Unix, and others. As with other proprietary dialects, PSpice supports many enhancements, including advanced graphics, simulation of digital and mixed-mode circuits, and stimulus editing for non-standard input signals.

D.1 SPICE INPUT FILE

All descriptive and control information is passed to SPICE in an ordinary text (ASCII) file called the SPICE input file. This file is a set of statements separated by *Return* keystrokes, one statement per line. It begins with the title statement and ends with the .END statement. In between are three kinds of statements: netlist statements, control statements and comment statements. Netlist statements define the circuit: element types, values, parameters, and interconnections. Control statements specify the mode of analysis, output variables and type of output that are desired and other control information. Statements may appear in the input file in any order, with the exceptions that the file must begin and end with the title and .END statements and that all statements between a .SUBCKT statement and the next .ENDS (END Subcircuit) statement are interpreted to be part of the subcircuit definition.

Netlist and control statements consist of fields separated by any amount of white space (space and/or tab keystrokes). These fields are of three types: keyword, text and numeric. Keyword fields must be typed exactly as specified, such as the keyword .END. Text fields may consist of any ASCII characters. Numbers may be entered in any of the following formats:

Format	Examples
INTEGER	77, −56
FLOATING-POINT	7.54, −33.65
EXPONENT	100E-02, 1.755E3

Suffixes in numerical fields that scale by powers of 10 are

Suffix	Factor	Suffix	Factor
K	10^{+3}	M	10^{-3}
MEG	10^{+6}	U	10^{-6}
G	10^{+9}	N	10^{-9}
T	10^{+12}	P	10^{-12}
		F	10^{-15}

For instance, the following are all equivalent numerical fields:

1.05E6   1.05MEG   1.05E3K   0.00105G

The location of all elements in the circuit and the location and reference directions of voltages and currents are indicated by specifying their nodes. One node of the circuit must be declared the reference node, which is given the node number 0. Other nodes are identified by unsigned integers.

The voltage drop between nodes N and M is specified as V(N, M), with positive sign of the reference direction of V(N, M) at node N and negative sign at M. V(N, 0) may be abbreviated V(N).

Currents are specified in the form I(VNAME), where VNAME is an independent voltage source declared in the netlist (see section D.3 for netlist statement format). The reference direction arrow for the current points from the minus end of VNAME (the second node in its netlist statement) to the plus end (first node on its netlist statement). If current through an element other than an independent voltage source is required, a dummy voltage source, with +0 V value, must be inserted in series with this element.[1] Any statement may be continued to the next line by including a plus sign (+) in the first column of the new line to indicate continuation.

D.2 TITLE AND COMMENT STATEMENTS

A *title statement* is the first statement of the SPICE input file. It is restricted to a single line, which should not start with white space (space or tab), but is otherwise free text format. The title statement is used to identify the circuit under study. An example is

```
DC analysis of circuit 6.2
```

Comment statements are used for embedding descriptive or explanatory free text information within the SPICE input file and for separating parts of the file. Any line beginning with an asterisk (*) is a comment statement and is ignored by the SPICE compiler. Examples are

```
*This circuit becomes unstable for R2<15K
*
```

D.3 NETLIST STATEMENTS

The circuit to be investigated is specified by a netlist. A netlist is a list of all the elements in a network, their types, values, parameters and locations. The circuit cannot contain a closed loop of voltage sources or a closed region into which only current sources flow. Each element in the circuit is completely specified by a single *netlist statement*. The formats for netlist statements by element type follow.

---

[1] The restriction that only currents through voltage sources may be declared is not enforced in all SPICE dialects. Some permit forms such as I(R1). Consult your documentation.

Resistor

```
R<NAME> <(+) NODE> <(-) NODE> <VALUE>
```

⟨NAME⟩ denotes any 1- to 7-chapter alphanumeric string for labeling the element.

(+) and (−) NODES define the polarity of the resistor connection. Positive current flows from (+) NODE through the resistor to (−) NODE.

⟨VALUE⟩ is the nonzero resistance value (positive or negative) in ohms.

Example D.1:
```
RLOAD 12 3 10K
```

denotes a 10-kΩ resistor labeled RLOAD connected between nodes 12 and 3 having current flowing from positive node 12 through the resistor to negative node 3.

Capacitor

```
C<NAME> <(+) NODE> <(-) NODE> <VALUE> [IC = <INITIAL VALUE>]
```

⟨NAME⟩ denotes any 1- to 7-character alphanumeric string.

(+) and (−) NODES define the polarity of the capacitor connection. Positive current flows from (+) NODE through the capacitor to (−) NODE.

⟨VALUE⟩ is the nonzero value (positive or negative) in farads.

⟨INITIAL VALUE⟩ is optional and denotes the initial capacitor voltage at time $t = 0$ of (+) NODE with respect to (−) NODE for a transient response analysis.

Example D.2:
```
CEXT 2 3 10U
CEXT 2 3 10U IC=4
```

denotes a 10-μF capacitor labeled EXT connected between nodes 2 and 3. If IC = 4 is included, then an initial voltage of 4 V for node 2 with respect to 3 exists at $t = 0$. This is used only in the transient response.

Inductor

```
L<NAME> <(+) NODE> <(-) NODE> <VALUE> [IC = <INITIAL VALUE>]
```

⟨NAME⟩ denotes any 1- to 7-character alphanumeric string.

(+) and (−) NODES define the polarity of the inductor connection. Positive current flows from (+) NODE through the inductor to (−) NODE.

⟨VALUE⟩ is the nonzero value (positive or negative) in henrys.

⟨INITIAL VALUE⟩ is optional and denotes the initial inductor current at time $t = 0$ form (+) NODE through the inductor to (−) NODE for a transient response analysis.

Example D.3:
```
L12 100 0 10M
L12 100 0 10M IC=-0.5
```

denotes a 10-mH inductor labeled 12 connected between nodes 100 and 0 (reference node). If IC $= -0.5$ is included, then an initial current of -0.5 A from node 100 through the inductor to node 0 exists at $t = 0$. This is used only in the transient response.

Linear Transformer

```
K<NAME> L<INDUCTOR NAME A> L<INDUCTOR NAME B> <COUPLING VALUE>
```

K⟨NAME⟩ couples two inductors A and B using a dot convention that is determined by the node assignments of inductors A and B. The polarity is determined by the order of the nodes in the L devices and not by the order of the inductors in the K statement. The dotted terminals of inductors A and B are those connected to the positive (first labeled) nodes in their defining statements.

⟨NAME⟩ denotes any 1- to 7-character alphanumeric string.

⟨COUPLING VALUE⟩ is the coefficient of mutual coupling in the range $0 \leq k < 1$.

Example D.4:

```
LPRI 2 3 500M
LSEC 5 4 400M
KXFRM LPRI LSEC 0.98
```

denotes a linear transformer having a mutual coupling of 0.98 between LPRI and LSEC. Dotted terminals for the polarity of the coupling are the terminal of LPRI connected to node 2 and the terminal of LSEC connected to node 5.

Independent Source

```
I<NAME> <(+) NODE> <(-) NODE> [TYPE <VALUE>] [TRANSIENT SPEC.]
V<NAME> <(+) NODE> <(-) NODE> [TYPE <VALUE>] [TRANSIENT SPEC.]
```

I denotes independent current source.

V denotes independent voltage source.

⟨NAME⟩ denotes any 1- to 7-character alphanumeric string.

(+) and (−) NODES define the polarity of the source. Positive current flows from (+) NODE through the source to (−) NODE (the reference arrow points to the (−) NODE).

TYPE is DC (default) for a dc source and ac for a sinusoidal ac source.

⟨VALUE⟩ is a dc value for dc or a magnitude and phase (in degrees) for ac. Default values are zero.

[TRANSIENT SPEC.] is used in transient analysis only and can be one of the following:

```
EXP(<x1> <x2> <td1> <tc1> <td2> <tc2>)
```

The EXP form causes the output current or voltage to be ⟨x1⟩ for the first ⟨td1⟩ seconds. Then, the output decays exponentially from ⟨x1⟩ to ⟨x2⟩ with a time constant of ⟨tc1⟩. The decay lasts ⟨td2⟩ seconds. Then, the output decays from ⟨x2⟩ back to ⟨x1⟩ with a time constant of ⟨tc2⟩.

```
PULSE(<x1> <x2> <td> <tr> <tf> <pw> <per>)
```

The PULSE form causes the output to start at ⟨x1⟩ and remain for ⟨td⟩ seconds. The output then goes linearly from ⟨x1⟩ to ⟨x2⟩ during the next ⟨tr⟩ seconds. The output remains ⟨x2⟩ for ⟨pw⟩ seconds. It then returns linearly to ⟨x1⟩ during the next ⟨tf⟩ seconds. It remains at ⟨x1⟩ for ⟨per⟩ − (⟨tr⟩ + ⟨pw⟩ + ⟨tf⟩ seconds and the cycle repeats, excluding the initial delay of ⟨td⟩ seconds.

```
PWL(<t1> <x1> <t2> <x2> .... <tn> <xn>)
```

The PWL form describes a piecewise linear waveform. Each pair of time–output values specifies a corner of the waveform. The output at times between corners is the linear interpolation of the current at the corners.

```
SIN(<xoff> <xampl> <freq> <td> <itc>)
```

The SIN form causes the output to start at ⟨xoff⟩ and remain for ⟨td⟩ seconds. Then, the output becomes an exponentially damped sine wave described by the equation

$$xoff + xampl \cdot \sin\{2\pi \cdot [freq \cdot (TIME-td)]\} \cdot e^{-(TIME- td) * itc}$$

Example D.5:   `IG1 2 3 0.2A   or   IG1 2 3 DC 0.2A`

denotes a dc current source labeled G1 supplying 0.2 A from node 2 to 3 through the source.

```
ISOURCE 0 5  AC 10 64
```

denotes a $10/64°$ A ac source labeled SOURCE connected between nodes 0 and 5 with the positive terminal to node 0.

```
I2 0 2 EXP(10M 0M 0 0.1 1)
```

denotes an exponential current of $10e^{-10t}$ mA flowing from node 0 to 2 through the source in the interval $0 < t < 0.1$ s for transient analysis.

Dependent Sources

Voltage-controlled Voltage Source (VCVS)

```
E<NAME> <(+) NODE> <(-) NODE> <(+ CONTROLLING) NODE>
+        <(-CONTROLLING) NODE> <GAIN>
```

⟨NAME⟩ denotes any 1- to 7-character alphanumeric string.

(+) and (−) NODES define the polarity of the source. Positive current flows from (+) NODE through the source to (−) NODE.

(+ CONTROLLING) and (− CONTROLLING) are in pairs and define a set of controlling voltages that are multiplied by ⟨GAIN⟩. A particular node may appear more than once, and the output and controlling nodes need not be different.

The plus sign on the second line of this statement indicates continuation.

Example D.6:   `EBUFF 1 3 11 9 2.4`

denotes a VCVS labeled BUFF connected between nodes 1 and 3, with the voltage of node 1 with respect to 3 equal to 2.4 times the voltage of node 11 with respect to 9.

Current-controlled Current Source (CCCS)

`F<NAME> <(+) NODE> <(-) NODE> <((CONTROLLING V DEVICE) NAME> <GAIN>`

⟨NAME⟩ denotes any 1- to 7-character alphanumeric string.

(+) and (−) NODES define the polarity of the source. Positive current flows from (+) NODE through the source to (−) NODE. The current through ⟨(CONTROLLING V DEVICE) NAME⟩ multiplied by ⟨GAIN⟩ determines the output current. Positive controlling current flows from the (+) NODE of (CONTROLLING V DEVICE) to the (−) node.

(CONTROLLING V DEVICE) is an independent voltage source.

Example D.7:   `F23 3 7 VOUT 1.2`

denotes a CCCS labeled 23, with current flowing from node 3 through the source to node 7 having a value equal to 1.2 times that of the current flowing in independent voltage source VOUT positive current (from + terminal to − terminal).

Voltage-controlled Current Source (VCCS)

```
G<NAME> <(+) NODE> <(-) NODE> <(+ CONTROLLING) NODE>
+        <(- CONTROLLING) NODE> <TRANSCONDUCTANCE>
```

⟨NAME⟩ denotes any 1- to 7-character alphanumeric string.

(+) and (−) NODES define the polarity of the source. Positive controlling current flows from (+) NODE of the (CONTROLLING V DEVICE) to (−) NODE.

(+ CONTROLLING) and (− CONTROLLING) are in pairs and define a set of controlling voltages that are multiplied by ⟨TRANSCONDUCTANCE⟩.

Example D.8:   `GAMP 4 3 1 9 1.7`

denotes a VCCS labeled AMP connected between nodes 4 and 3, with the current flowing from node 4 to 3 through the source equal to 1.7 times the voltage of node 1 with respect to 9.

Current-controlled Voltage Source (CCVS)

```
H<NAME> <(+) NODE> <(-) NODE> <(CONTROLLING V DEVICE) NAME>
+       <TRANSRESISTANCE>
```

⟨NAME⟩ denotes any 1- to 7-character alphanumeric string.

(+) and (−) NODES define the polarity of the source. Positive controlling current flows from (+) NODE of the (CONTROLLING DEVICE) to (−) NODE. The current through ⟨(CONTROLLING V DEVICE) NAME⟩ multiplied by ⟨TRANSRESISTANCE⟩ determines the output voltage.

(CONTROLLING V DEVICE) is an independent voltage source.

Example D.9:   `HIN 4 8 VDUMMY 7.7`

denotes a CCVS labeled IN with voltage between nodes 4 and 8 (node 4 positive) having a value equal to 7.7 times that of the current flowing in independent voltage source VDUMMY (from + terminal to − terminal).

Subcircuit Call Statement

```
X<NAME> [NODE]* <(SUBCIRCUIT) NAME>
```

⟨NAME⟩ denotes any 1- to 7-character alphanumeric string.

[NODE]* denotes a list of nodes required by the subcircuit definition.

⟨(SUBCIRCUIT) NAME⟩ is the name of the subcircuit's definition (see .SUBCKT statement below). There must be the same number of nodes in the call as in the subcircuit's definition.

This statement causes the referenced subcircuit to be inserted into the circuit, with the given nodes replacing the nodes in the .SUBCKT definition in order. It allows defining a block of circuitry once and then the use of that block in several places.

Example D.10:   `XBUFF 4 1 7 9 UNITAMP`

denotes a call to a subcircuit that replaces the call statement by the content of the file UNITAMP for SPICE analysis.

D.4 SOLUTION CONTROL STATEMENTS

All statements beginning with a period (.) are control statements. The commands to be discussed in this section are the .AC, .DC, .FOUR, .IC, .LIB, .SUBCKT, .TF, and .TRAN statements. The output control statements .PLOT and .PRINT will be described in the next section.

.AC

AC analysis: The .AC statement is used to calculate the frequency response of a circuit over a range of frequencies. It has the form

```
.AC [LIN][OCT][DEC] <(POINTS) VALUE> <(START FREQUENCY) VALUE>
+                   <(END FREQUENCY) VALUE>
```

LIN, OCT, or DEC are keywords that specify the type of sweep as follows:

LIN: Linear sweep. The frequency varies linearly from START FREQUENCY to END FREQUENCY. ⟨(POINTS) VALUE⟩ is the number of points in the sweep.

OCT: Sweep by octaves. The frequency is swept logarithmically by octaves. ⟨(POINTS) VALUE⟩ is the number of points per octave.

DEC: Sweep by decades. The frequency is swept logarithmically by decades. ⟨(POINTS) VALUE⟩ is the number of points per decade.

Exactly one of LIN, OCT, or DEC must be specified.

⟨(END FREQUENCY) VALUE⟩ must not be less than ⟨(START FREQUENCY) VALUE⟩, and both must be greater than zero. The entire sweep may specify only one point if desired.

Example D.11:
```
.AC LIN 1 100HZ 100HZ
```

denotes an ac steady-state solution for a network having a frequency of 100 Hz.

```
.AC LIN 101 100KHZ 200KHZ
```

denotes a linear frequency response having 101 points evenly distributed in the range from 100 to 200 kHz.

.DC

DC analysis: The .DC statement causes a DC sweep analysis to be performed for the circuit. It has the form

```
.DC <(SWEEP VARIABLE) NAME> <(START) VALUE> <(END) VALUE>
+   <(INCREMENT) VALUE>
```

⟨(SWEEP VARIABLE) NAME⟩ is a name of an independent current or voltage source. It is swept linearly from ⟨(START) VALUE⟩ to ⟨(END) VALUE⟩. The increment size is ⟨(INCREMENT) VALUE⟩. ⟨(START) VALUE⟩ may be greater or less than ⟨(END) VALUE⟩; that is, the sweep may go in either direction. ⟨(INCREMENT) VALUE⟩ must be greater than zero. The entire sweep may specify only one point if desired.

Example D.12:   `.DC VIN 10V 10V 1V`

denotes a dc solution for a circuit with independent voltage source VIN = 10 V.

`.DC IGEN 1M 10M 1M`

denotes a dc sweep for independent current source IGEN being swept from 1 to 10 mA in 1-mA steps.

.FOUR

Fourier analysis: Fourier analysis performs a decomposition into Fourier components as the result of a transient analysis. A .FOUR statement requires a .TRAN statement (described below). It has the form

`.FOUR <(FREQUENCY) VALUE> <(OUTPUT VARIABLE)>*`

⟨(OUTPUT VARIABLE)⟩ is a list of one or more variables for which the Fourier components are desired. The Fourier analysis is done by starting with the results of the transient analysis for the specified output variables. From these voltages or currents, the dc component, the fundamental frequency, and the second through ninth harmonics are calculated. The fundamental frequency is ⟨(FREQUENCY) VALUE⟩, which specifies the period for the analysis. The transient analysis must be at least 1/⟨(FREQUENCY) VALUE⟩ seconds long.

Example D.13:   `.FOUR 10KHZ V(5) V(6,7) I(VSENS3)`

yields the Fourier components for variables V(5), V(6, 7), and I(VSENS3). The fundamental frequency for the decomposition is set equal to 10 kHz.

.IC

Initial transient conditions: The .IC statement is used to set initial conditions for transient analysis. It has the form

`.IC <V(<NODE>) = <VALUE>>*`

Each ⟨VALUE⟩ is a voltage that is assigned to ⟨NODE⟩ for the initial node voltage at time $t = 0$ for the transient analysis.

Example D.14:   `.IC V(2)=5 V(5)=-4V V(101)=10`

denotes setting the initial node voltages of nodes 2, 5, and 101 to 5, −4, and 10 V, respectively, at $t = 0$.

Library file: The .LIB statement is used to reference a subcircuit library in another file. It has the form

```
.LIB <(FILE) NAME>
```

⟨(FILE) NAME⟩ is the name of the subcircuit library file.

Example D.15:
```
.LIB OPAMP.LIB
```

denotes that the subcircuit library file is named OPAMP.LIB.

Operating (bias) point dc analysis: The .OP statement calculates all the dc node voltages and the currents in all voltage sources. It has the simple form

```
.OP
```

Subcircuit definition: The .SUBCKT statement is used to define a subcircuit that is called using the X statement described previously. It has the form

```
.SUBCKT <NAME> [NODE]*
```

The .SUBCKT statement begins the definition of a subcircuit. The definition is ended with a .ENDS statement. All statements between .SUBCKT and .ENDS are included in the definition. Whenever the subcircuit is called by an X (subcircuit call) statement, all the statements in the definition replace the calling statement.

Transfer function: The .TF statement produces the dc small-signal transfer function. It has the form

```
.TF <(OUTPUT VARIABLE)> <(INPUT SOURCE) NAME>
```

The gain from ⟨(INPUT SOURCE) NAME⟩ to ⟨(OUTPUT VARIABLE)⟩ is calculated along with the input and output resistances. ⟨(OUTPUT VARIABLE)⟩ may be a current or a voltage: however, in the case of a current it is restricted to be the current through a voltage source.

Example D.16:
```
.TF V(3) IIN
```

produces the dc small-signal transfer function for V(3)/IIN, the input resistance seen from the terminals of independent current source IIN, and the output resistance seen at node V(3).

.TRAN

Transient analysis: The .TRAN statement causes a transient analysis to be performed for the circuit. It has the form

```
.TRAN <(PRINT STEP) VALUE> <(FINAL TIME) VALUE> [UIC]
```

The transient analysis calculates the behavior of the circuit over time, starting at $t = 0$ and going to ⟨(FINAL TIME) VALUE⟩. ⟨(PRINT STEP) VALUE⟩ is the time interval used for plotting or printing the results of the analysis. The keyword UIC (use initial conditions) causes the initial conditions set for capacitors and inductors with the IC specification to be used.

Example D.17:
```
.TRAN 1NS 100NS UIC
```

produces a transient analysis in the interval from 0 to 100 ns, with plotted or printed output on 1-ns intervals.

D.5 OUTPUT CONTROL STATEMENTS

The output control statements for plotting and printing are the .PLOT, .PRINT, and .WIDTH statements.

.PLOT

The .PLOT statements allows results from dc, ac, and transient analysis to be output in the form of *line printer* plots [for example, Fig. 14.44(a) or (b)]. It has the form

```
.PLOT [DC][AC][TRAN][OUTPUT VARIABLE]*
+      ([<(LOWER LIMIT) VALUE>, <(UPPER LIMIT) VALUE>])
```

DC, AC, and TRAN are the analysis types that can be output. Exactly one analysis type must be specified. [OUTPUT VARIABLE]* is a list of the output variables desired for plotting. Currents are specified in the form I(VNAME), where VNAME is an independent voltage source and I(VNAME) flows through the voltage source from its + terminal to its − terminal. For ac analysis, the suffix M, P, R, or I can be added to the variable specifiers V or I. These indicate magnitude, phase, real, and imaginary parts. A maximum of eight output variables are allowed in one .PLOT statement.

The range and increment of the x-axis are fixed by the analysis being plotted. The range of the y-axis can be set by adding (⟨⟨LOWER LIMIT⟩ VALUE⟩, ⟨⟨UPPER LIMIT⟩ VALUE⟩) to output variables with the same y-axis range. Each occurrence defines one y-axis with the specified range. All output variables that come between it and the next range to the left are put on its corresponding y-axis. If no y-axis limits are specified, the program automatically determines the plot limits.

Example D.18:　　　`.PLOT DC V(2) V(3,5)`

plots the dc response for V(2) and V(3, 5),

`.PLOT AC VM(3) VP(3) IR(V1) II(V1)`

plots the magnitude and phase of V(3) and the real and imaginary parts of the current through V1.

`.PLOT TRAN V(5) V(2, 3) (0, 5V) I(VCC) (-5MA,5MA)`

plots the transient response of V(5) and V(2, 3) between the limits of 0 and 5 V and I(VCC) between the limits of −5 and 5 mA.

.PRINT

Print: The .PRINT statement allows results from dc, ac, and transient analysis to be output in the form of tables. It has the form

`.PRINT [DC][AC][TRAN][(OUTPUT VARIABLE)]*`

DC, AC, and TRAN are the analysis types that can be output. Exactly one analysis type must be specified. [(OUTPUT VARIABLE)]* is a list of output variables desired (see .PLOT statement above). There is no limit to the number of output variables. The output format (characters per line) is determined be the specification of the .WIDTH command.

Example D.19:　　　`.PRINT DC V(1) I(V12)`

prints the dc values for V(1) and I(V12).

`.PRINT AC VM(1,5) VP(1,5) IR(V2) II(V2)`

prints the magnitude and phase of V(1, 5) and the real and imaginary parts of I(L2).

`.PRINT TRAN V(7) I(VCC) V(3,1)`

prints the transient response for V(7), I(VCC), and V(3, 1).

Width: The .WIDTH statement sets the width of the output. It has the form

```
.WIDTH OUT = <VALUE>
```

⟨VALUE⟩ is the number of columns and must be either 80 or 132.

Example D.20:
```
.WIDTH OUT = 132
```

D.6   END STATEMENTS

End statements for subcircuits and circuit files are .ENDS and .END, respectively.

End of subcircuit definition: The .ENDS statement marks the end of a subcircuit definition (started by a .SUBCKT statement). It has the form

```
.ENDS [(SUBCIRCUIT) NAME]
```

It is good practice to repeat the subcircuit name, although this is not required.

End of circuit: The .END statement marks the end of the circuit file. It has the form

```
.END
```

Appendix E

Answers to Selected Odd-Numbered Problems

CHAPTER 1

1.1. 456 MJ
1.3. 3529 km/hr
1.5. 1116 s
1.7. -4 C; 2 A; -3 A
1.9. 96 mW; 92 μC; 736 μJ
1.11. -11.8 W
1.13. $-4(t^3 - 15t^2 + 50t)$ W; 0 J
1.15. $\frac{1}{2}e^t$ A
1.17. $-\frac{1}{8}$ J
1.19. V-source supplying; $\frac{-1}{36}$ V
1.21. (a) 14.4 kJ; (b) 1.2 kC
1.23. 10 S; 6 J
1.25. p odd
1.27. Never
1.29. 100; 43
1.31. passive
1.33. passive

CHAPTER 2

2.1. $i_1 - i_4 - i_6 = 0$; $i_2 + i_4 + i_5 = 0$; $-i_3 - i_5 + i_6 = 0$
2.5. (a) $-v_5 - v_2 - v_1 + v_3 + v_4 = 0$; (b) $v_5 + v_2 + v_1 - v_3 - v_4 = 0$; (c) $v_3 + v_4 = v_1 + v_2 + v_5$
2.11. 12 kW
2.13. 24 V; 12 V; -20 V; -32 V
2.15. 10 V; -10 V; -10 V; -10 V
2.17. $v_0 = 5 \sin 2t$ V; $v_1 = \frac{5}{3} \sin 2t$ V;
$v_2 = v_3 = \frac{10}{3} \sin 2t$ V; $i_0 = i_1 = \frac{1}{6} \sin 2t$ A;
$i_2 = \frac{1}{9} \sin 2t$ A; $i_3 = \frac{1}{18} \sin 2t$ A

2.19. $\sqrt{2}$
2.21. $v = -4i$; $i = -\frac{1}{4}v$
2.23. (a) 800 Ω, 780 Ω; (b) 720 Ω, 702 Ω
2.25. 40 V
2.27. 5 Ω; 10 Ω
2.29. 4 resistors; 240 resistors
2.31. $\frac{205}{36}$ A; $\frac{205}{9}$ W
2.35. 10 kΩ; 60 mA
2.37. 4 resistors; 200 resistors
2.39. $v_T = -6$ V, $R_{TH} = 6$ Ω
2.41. Voltage across 1-Ω resistor, current through it.
2.43. $v_T = v_s$; $R_{TH} = 0$ Ω
2.45. 96 kΩ
2.47. $i_n = \frac{-2}{15}$ A; $R_N = \frac{5}{2}$ Ω
2.49. $v_T = \frac{4}{9}$ V, $R_{TH} = R_N = \frac{8}{9}$ Ω, $i_N = \frac{1}{2}$ A
2.51. $R_N = \infty$
2.53. All in series: 14, 4 copies of 10 in parallel, 7 copies of 100 in parallel.
2.55. $v_T = 28$ V, $R_{TH} = R_N = 8$ Ω; $i_N = \frac{7}{2}$ A
2.57. (a) $6i_2 + 24$; (b) $\frac{1}{9} \sin t + \frac{8}{3}$ A, $-\frac{2}{3} \sin t + 8$ V
2.59. $v_T = 11$ V, $R_{TH} = R_N = \frac{77}{6}$ Ω, $i_N = \frac{6}{11}$ A

CHAPTER 3

3.3. -1 S
3.5. 3 A
3.7. $-\frac{2}{3} \sin 2t$ V
3.9. $v_T = 8$ V, $R_T = \frac{4}{7}$ Ω
3.11. VCVS -1; CCVS -10
3.13. $-\frac{2}{5}$ A
3.15. 1 A

3.17. .331 V

3.19. $R_F = 45 \text{ k}\Omega$, $R_A = 5 \text{ k}\Omega$*

3.21. $R_F = 30 \text{ k}\Omega$, $R_A = 5 \text{ k}\Omega$*

3.23. 5 mW, $\frac{20}{9}$ mW

3.25. Yes

3.27. -2

3.29. $R_F = 100 \text{ k}\Omega$, $R_1 = 50 \text{ k}\Omega$, $R_2 = 20 \text{ k}\Omega$*

3.31. $v_3 = \frac{1}{2}v_1 - 2v_2$

3.33. $v_3 = \frac{1}{2}v_1 - 2v_2$

3.37. $v_3 = \frac{2}{3}v_1 - 2v_2$ (both)

3.39. 101 MΩ

CHAPTER 4

4.1. $\frac{-3}{25}$

4.5. 63 V

4.7. $\frac{3}{10}$ V

4.9. $-\frac{5}{4}$ A

4.11. $-\frac{3}{2}$ A, $\frac{22}{3}$ A, $\frac{2}{3}$ A

4.13. 1 A

4.15. 18 mA

4.17. 20 V, 12 V

4.19. 600 V, -3 kV

4.21. 20 V

4.23. 21 mA

4.25. $\frac{1}{4}$ A

4.27. $(i_1 - i_1) + (i_2 - i_2) + \cdots + (i_n - i_n) = 0$

4.29. $\frac{11}{4}$ V

4.31. $\frac{-23}{54}$ V, $\frac{-5}{6}$ V, $\frac{47}{216}$ A

4.33. $\frac{-27}{50}$ V, $\frac{-93}{50}$ V, $\frac{3}{40}$ A

4.35. 8 W

4.37. 14 mA

4.39. 8 V

4.41. 21 mA

4.43. $v_2 = \frac{5}{2}v_{g1} - \frac{3}{2}v_{g2}$

4.45. -2

4.47. -2 A, 3 A

4.49. For $v_{si} = 0$ V, $i_1 = -.24$ V, $i_2 = -.48$ V, etc.

4.53. -2 A, 3 A

4.55. $\frac{3}{7}$ V

4.57. $\frac{13}{7}$ A, $\frac{44}{7}$ V

4.59. $\frac{5}{4}$

CHAPTER 5

5.1. $40 \times 10^{-6} [\cos 6t - 6t \sin 6t]$ A

---

*Many possible solutions

5.3. 5 F, 7 F

5.7. 10 mA

5.9. $5e^{-10t}$ V

5.11. $\frac{1}{20}$ μF, 160 μJ

5.13. $\frac{1}{2} \sin^2 t$ J, $\frac{1}{2}t + \frac{1}{4}\sin 2t$ J, R

5.15. $\frac{126}{55}$ μF

5.17. 14×10^6 copies of $41\frac{1}{3}$ μF caps in parallel all in series.*

5.19. $-\frac{8}{3}\sin 2t$ A, $\frac{4}{3}(\cos 2t - 1)$ V

5.21. $-2(t - 1)$ V

5.23. $\frac{1000}{3}$ rad/s

5.25. piecewise constant ("staircase")

5.27. (a) $500t^2 - 5t + 10$ for $0 < t < 10$ ms, periodic.

5.29. (a) 0; (b) 0.2 V; (c) 10 $\cos 100t$ V; (d) $0.5e^{-5t}$ V

5.31. (a) $-10\sin 10t$ mV; (b) $-50\sin 20t$ μW;
(c) $5\cos^2 10t$ μJ; (d) $\frac{3\pi}{40}$ S

5.33. $\frac{1}{2}t^4 - 2t^3 + 2t^2$ J

5.35. $\frac{L_2}{L_1}$

5.37. 10 mH

5.39. 15 mH

5.41. 4 mA, 4 mA

5.43. V-source shorted

5.45. 2 A, -2 A

5.47. $\frac{4}{R}$ A, 0 A

5.51. 0 W all

CHAPTER 6

6.1. $2e^{-t}$ V, $\frac{4}{5}e^{-t}$ A

6.3. $-12e^{-(R/L)t}$ V

6.5. $\frac{2R}{L}$

6.7. 1 s, $100e^{-t/\tau}$ V, $100e^{-t/\tau}$ μA

6.9. $\tau = \frac{1}{2}$ s.

6.11. 1 μF and 1 Ω, 0.865 J, 1 J

6.13. $C_1 L_1 \frac{d^2 i_2}{dt^2} + \left(C_1 L_1 + \frac{L_1}{R_2}\right)\frac{di_2}{dt} + \left(1 + \frac{R_1}{R_2}\right)i_2 = 0$

6.15. 1 s

6.17. $\frac{-24}{5}e^{-1400t}$ mA

6.19. $\frac{1}{4}$ s

6.21. Downward capacitive current: $-\frac{3}{80}e^{-(9/4)t}$, etc.

6.23. $4e^{-4t}$ V

6.25. $4 - 2e^{-8000t}$ mA

6.27. $-2 + 4e^{-(t/3)}$ A

6.29. $-16 + 28e^{-t}$ V

6.31. 0 V

6.33. 0 A

6.35. $8 - 8e^{-(4/7)t}$ V

6.37. 3 A

6.39. $2u(t+4) + 2u(t+3) - 6u(t+2) + 12u(t-2) - 10u(t-\frac{5}{2})$

6.41. $3r(t) - 4r(t-4) - 2r(t-8) + 3r(t-10) - 2u(t-15)$

6.43. $\frac{1}{3}\left(1 - e^{-(12/5)t}\right)(u(t) - u(t-1)) +$
$\frac{1}{3}\left(e^{12/5} - 1\right)e^{-(12/5)t} \times u(t-1)$ V

6.45. (a) 0; (b) $-\frac{5}{18}e^{-2t}$ A; (c) $-\frac{5}{18}\sin 2t$ A

6.47. (a) $4\left(e^{-2t} - 1\right)u(t)$ V;
(b) $-4\left(1 - e^{-2t}\right)u(t) + 4\left[1 - e^{-2(t-1)}\right]u(t-1)$ V

6.49. $1 - 2e^{-1000t}$ V

6.53. $\frac{L}{R}$ s

6.55. $\frac{11}{3}\left(-1 + e^{-[(15/2)\times 10^6)]t}\right)$ mA, etc.

6.57. $v_2 = -\frac{1}{2}\frac{d}{dt}v_1$

6.59. $(1 - e^{10^5 t})u(t)$

7.1. $\frac{di}{dt} + \frac{4}{15}i = -\frac{1}{3}\frac{di_g}{dt}$, $\frac{di_1}{dt} + 4i_1 = -4i_g$, $\frac{di_2}{dt} + 2i_2 = -2i_g$, no, no

7.3. $\frac{d^2 v_2}{dt^2} + 1150\frac{dv_2}{dt} + 300{,}000v_2 = 0$

7.5. $5\frac{d^2 v_2}{dt^2} + 11\frac{dv_2}{dt} + 4v_2 = 4\frac{dv_g}{dt} + 4v_g$

7.7. $\frac{d^2 v_{out}}{dt^2} = \frac{1}{R_1 R_2 C_1 C_2}v_g$, 2nd order

7.9. $K_1 + K_2 = 1$

7.11. $2te^{-7t}$

7.13. $\frac{5}{3}e^{-1000t} - \frac{2}{3}e^{-7000t}$ mA

7.15. $5e^{-t} - 2e^{-7t}$ A

7.17. $e^{-2t}(4\cos t + 2\sin t)$ A

7.19. $6(1 + 4t)e^{-2t}$ A

7.21. $-2te^{-4t}$

7.25. $e^{-2t}\left[\frac{8}{13}\cos 3t + \frac{12}{13}\sin 3t\right]$ V,
$e^{-2t}\left[\frac{2}{5}\cos 3t + \frac{3}{5}\sin 3t\right]$ V

7.27. (a) $5 - 5e^{-t} - e^{-6t}$ A; (b) $5e^{-t} - e^{-6t} - 5e^{-2t}$ A;
(c) $\left(\frac{3}{5} + 2t\right)e^{-t} - \frac{8}{5}e^{-6t}$ A

7.29. $8e^{-t} - 2e^{-4t} + 2$ A

7.31. (a) $3\left[1 - e^{-t}(\cos t + \sin t)\right]$ A;
(b) $\frac{3}{2}\left[1 - 2e^{-t} + e^{-2t}\right]$ A;
(c) $3\left[1 - (1 + 2t)e^{-2t}\right]$ A

7.33. 0 A

7.35. $e^{-t}\left[\frac{8}{3}\cos\sqrt{3}t + \frac{16\sqrt{3}}{9}\sin\sqrt{3}t\right]$ A

7.37. 0 A

7.39. 0 V

7.41. $\frac{1}{3}$ kΩ

7.43. $\left[4 - (4 + 8t)e^{-2t}\right]u(t)$ V

7.45. $v_0 = -35000e^{-50000t}u(t)$ V

7.47. $i_1 = \frac{18}{5} + e^{-(11/38)t}[-.423\cos 2.0t - .13\sin 2.0t]$ A, etc.

7.49. $\frac{d^2 x_1}{dt^2} + 5\frac{dx_1}{dt} + 4x_1 = 8t + 14$, $3\frac{d^2 x_1}{dt^2} + 10\frac{dx_2}{dt} + 8x_2 = 16t + 28$, $x_1 = e^{-4t} - 2e^{-t} + 2t + 1$, $x_2 = -6e^{-(4/3)t} + 5e^{-2t} + 2t + 1$*

8.1. V_m: 4, 3; ω: 2 rad/s, 2 rad/s, ϕ: $107°$, $-139.3°$; T: π s, π s

8.3. $\sin(\omega t + n\pi) = (-1)^n \sin\omega t$, $\cos(\omega t + n\pi) = (-1)^n \cos\omega t$

8.5. Yes, yes.

8.7. $i_f = \frac{V_m}{\sqrt{R^2 + \omega^2 L^2}}\cos(\omega t + \tan^{-1}\frac{R}{\omega L})$ A

8.9. $6\frac{d^3 v_1}{dt^3} + 14\frac{d^2 v_1}{dt^2} + 6\frac{dv_1}{dt} = \cos\frac{t}{2} - 4\sin\frac{t}{2}$, $0.991\cos(\frac{t}{2} - 71.3°)$ V

8.11. 4 s, 2, $\pi/2$ rad/s

8.13. (a) $\frac{4\sqrt{5}}{37}\underline{/63.5°}$; (b) $1.55\underline{/-55°}$; (c) $48.8\underline{/1.5°}$;
(d) $45.3\underline{/-2.0°}$

8.15. $-i_1$, $3i_2 + j(i_1 - i_2)$

8.17. $\frac{4\sqrt{53}}{53}\cos(7t - 74.1°)$

8.19. $2V_m\angle\phi_v$, $V_m\angle(\phi_v - 45°)$

8.21. $\frac{d^2 i}{dt^2} + \frac{2di}{dt} + i = -24\sin 2t$, (t: mS, i: mA), $\frac{24}{5}\sin(12t + 53.1°)$ mA

8.23. (a) $5\sqrt{2}\cos(20t + 135°)$; (b) $5\cos(20t + 216.9°)$;
(c) $13\cos(20t - 67.4°)$; (d) $10\cos 2t$; (e) $5\sin 20t$

8.25. 0.346 μF

8.27. $\frac{1}{300}$ H

8.31. $5\cos 10t$ A

8.33. 98.04 rad/s

8.35. $11.1\underline{/-144°}$ V, $11.1\cos(2t - 144°)$ V

8.37. $4.56\underline{/115°}$, $4.56\cos(10t + 115°)$ A

8.39. $12\cos(t - 53.1°)$ V, $3\cos(t - 53.1°)$ V

8.41. $4\cos(10t - 53.1°)$ A

8.43. (a) $4\cos(t + 36.9°)$ A; (b) $5\cos 2t$ A;
(c) $4\cos(4t - 36.9°)$ A

8.45. $.55\cos(4t - 164°)$ A, $0.27\cos(4t - 164°)$ V

8.47. $150\cos(100t + 91.1°)$ V

8.49. $0.14\sin(3t + 109°)$ V

8.51. $8.07\cos(1000t + 58.4°)$ mA

8.53. $-w\omega^2$, same as for R or L.

9.1. $\frac{Z_2}{Z_1 + Z_2}110\underline{/0}$ V

9.3. $\frac{1}{2} + j\frac{\sqrt{3}}{2}$ Ω, 1 Ω, $\frac{1}{2} - j\frac{\sqrt{3}}{2}$ Ω

9.5. $2\cos t$ mA

9.7. $V_T = 24.7\underline{/-134°}$, $Z_T = Z_N = 3 + j12\ \Omega$, $I_N = 2\underline{/150°}$ A

9.9. $-.062 - j.012\ \Omega$

9.11. $2\sqrt{2}\cos(2t + 45°)$ V

9.13. $4\sqrt{5}\cos(300t + 63.4°)$ mA

9.15. $4.54\cos(2000t - 93.9°)$ V

9.17. $92.6\underline{/180°}$ A

9.19. $-2\sin 2000t$ V

9.23. $2\cos(8t + 53.1°)$ V

9.27. $19.4\cos(2000t - 40.0°)$ V

9.29. $1.13\cos(10t - 73.6°)$ V

9.31. $.645\underline{/97.4°}$ A, $.62\underline{/-82.6°}$ A

9.33. $2\sqrt{2}\cos(6t - 135°) + 6$ V

9.35. $3.16\cos(1000t + 61.6°) +$
$2220\cos(2000t - 124°)$ V

9.37. $6.57\underline{/38.8°}$ V, $6.57\cos(t + 38.8°)$ V

9.41. $4.54\cos(2000t - 93.9°)$ V

9.43. $3.01\underline{/87.4°}$ mA

9.45. $2.85\cos(t - 0.57°)$ mV,
$5.71\cos(2t - 1.15°)$ mV, etc.

9.47. 0

9.49. $v_2(t) = V_m\cos(\omega t + \phi)$ where $V_m = \left|\frac{\omega^2 - 4j\omega - 4}{\omega^2 - 4j\omega - 2}\right|$,
$\phi = \tan^{-1}\left[\frac{\omega^2 - 4j\omega - 4}{-\omega^2 + 4j\omega + 2}\right]$

CHAPTER 10

10.1. $5\frac{2}{3}$ kW

10.3. 500 W

10.5. $\frac{3}{2}RI_m^2$

10.7. $\frac{1}{2}$ W, independent of ϕ

10.9. $-90°$

10.11. 138 W

10.13. (a) $2\sqrt{2}$ A; (b) $2\sqrt{3}$ A; (c) $\frac{I_m}{2}$ A; (d) $\frac{1}{\sqrt{2}}I_m$ A

10.15. $P_R = 13.7$ W, etc.

10.19. $63.2\ \mu$F

10.21. $Q_{L1} = 2.5 \times 10^{-4}$ VA, $Q_{L2} = 2.5 \times 10^{-4}$ VA, etc.

10.23. $P_{1k\Omega} = \frac{5}{26}$ mW, $P_{2k\Omega} = \frac{1}{13}$ mW, $P_L = P_C = 0$,
$P_V = -\frac{7}{26}$ mW

10.25. 4 V

10.27. $S_I = -2 - j\frac{2}{3}$; $S_V = 2 - j120$

10.29. 1.92 W, 15.4 W, 9.6 W

10.31. $P_{Idc} = 8$ kW, $P_{Iac} = 479$ W, $P_{Vac} = -465$ W

10.33. (a) Set $\frac{\partial P_L}{\partial X_L} = 0$, $\frac{\partial P_L}{\partial R_L} = 0$; (b) Set $\frac{\partial P_L}{\partial R_L} = 0$

10.35. -1 s

10.37. $Z_L = 4.19\underline{/24.8°}$, RL circuit

10.39. $\Sigma S_i = 0 + j0$

10.41. $\underline{/Z} > 0$ each case

10.43. $\frac{1}{3}$ H

10.45. $\frac{-100}{7}\ \Omega$

10.47. (a) 0.943 lagging; (b) $93.7\ \mu$F

10.49. 1.62 MΩ

10.51. ∞

10.53. $\omega < 1$ rad/s

CHAPTER 11

11.1. $10\sqrt{3}\underline{/90°}$ A

11.3. 0 A

11.5. $111\underline{/1.5°}$ A, $111\underline{/-178.5°}$ A, 0 A

11.7. $4\sqrt{3}\underline{/30°}\ \Omega$, 1.8 kW

11.9. 240 V rms, $\frac{80}{3}$ A rms

11.11. $115\underline{/(30°, -90°, +150°)}$ V,
$31.9\underline{/(-56.3°, -176.3°, 63.7°)}$ A

11.13. $|I_{aA}| = 4400/\sqrt{(20 + .01l)^2 + 16}$

11.15. 20.96 V

11.17. 35.35 A rms, 11.3 kW, 3.75 kW

11.19. $20\underline{/15°}$, $1280 + j341$, $1920 + j512$

11.21. $\cos^2\omega t$ has period half that of $\cos\omega t$

11.23. $\frac{100}{\sqrt{3}}$ A rms, 15 kW

11.25. 60 A rms, 10.8 kW

11.27. Advantage: better spatial balancing of motor torque, disadvantage: more wires, etc.

11.29. $255\underline{/(-8°, -128°, +112°)}$, 18 kW

11.31. $2530\underline{/23°}\ \Omega$, 4.7 kW

11.33. $15\sqrt{2}$ A rms

11.35. $I_{aA} = 7.97 - j0.48$ A rms, $I_{bB} = 10.8 - j1.83$ A rms, $I_{cC} = 2.86 + j2.31$ A rms, 2280 W

11.37. $14 + 14\sqrt{3}j\ \Omega$, $-\sqrt{3}$ kVAR

11.39. (a) $6 + 3j\ \Omega$, (b) $\frac{22 + j21}{5}\ \Omega$, $6 - j\ \Omega$, $6 - j\ \Omega$

11.41. $111\underline{/1.5°}$ A, $111\underline{/-178.5°}$ A, 0 A

11.43. $\frac{14}{3}\ \Omega$

11.45. $111\underline{/1.5°}$ A, $111\underline{/-178.5°}$ A, 0 A

11.47. $|I_{cC}| = 100$ A ($50\sqrt{2}$ A rms) for $R = 6.2\ \Omega$

11.49. No such $R_{p'} < 100\ \Omega$ either case.

11.51. Standard mesh equations.

11.53. $188\ \Omega$

CHAPTER 12

12.1. (a) $\frac{5}{s+9} - \frac{2}{s+1} + \frac{1}{s-1}$, (b) same as (a), (c) $\frac{j4}{s^2+4}$

12.3. No, not of exponential order.

12.5. $\frac{1}{t-1}$ not $p\omega$ continuous

12.7. $t^{10}e^{-10t} \cdot t^* = 35.8$ s

12.9. $\frac{4.89s - 3.11}{s^2 + 9}$

12.11. (a) $\frac{1}{s}(1 - e^{-2s}) + \frac{e^{-2s}}{s+1}$, $Re(s) > -1$;

(b) $\frac{3e^{-4s}}{s^2}$, $Re(s) > 0$; (c) $\frac{10}{s+1} - \frac{2\pi}{s^2+(2\pi)^2}$, $Re(s) > -1$

12.15. (a) $\left[\delta^{(n)}(t) - \delta^{(n)}(t-1)\right] + \frac{1}{2}\left[\delta^{(n)}(t-2) - \delta^{(n)}(t-3)\right] + \frac{1}{4}\left[\delta^{(n)}(t-4) - \delta^{(n)}(t-5)\right]$,

$n = 0, 1$; (b) $2\left[\delta^{(n)}(t-2) - 2\delta^{(n)}(t-3) + \delta^{(n)}(t-4)\right]$, $n = -1, 0$; (c) $2\left[\delta^{(n)}(t) - 4\delta^{(n)}(t-1) + \delta^{(n-1)}(t-1) - \delta^{(n-1)}(t-3)\right]$, $n = 0, 1$ where $\delta^{(0)}(t) = \delta(t)$, $\delta^{(-1)}(t) = u(t)$.

12.17. (a) $2 - e^{-s} - e^{-2s} + e^{-3s} + e^{-(7/2)s} - 2e^{-4s}$;

(b) $\frac{5}{s} - \frac{5e^{-4s}}{s^2} + \frac{5e^{-6s}}{2s^2}$; (c) $5\left[1 - \frac{e^{-4s}}{s} + \frac{e^{-6s}}{2s}\right]$

12.19. 0

12.21. (a) $\frac{s}{s+1}$; (b) $\frac{-1}{s+1}$; (c) $\frac{2s}{(s+1)^2+4}$; (d) $\frac{2s}{(s+1)^2+4}$

12.23. (a) $\frac{1}{(s+2)^2} + \frac{2}{s+2}$; (b) $\frac{2}{(s+4)^2+4}$; (c) $\frac{s+3}{(s+3)^2-4}$;

(d) $\frac{s+1}{(s+1)^2+16}$

12.25. (a) $\frac{1}{(s/a)+1}$; (b) $\frac{1}{(s+a-1)+1}$

12.27. (a) $\delta(t) + e^{-t}u(t)$; (b) $\delta^{(1)}(t) + \delta(t) - e^{-t}u(t)$; (c) $\delta^{(3)}(t) + \delta^{(1)}(t) + \frac{1}{2}\left[e^t - e^{-t}\right]u(t)$

12.29. $1 - \cos t + \sin t$

12.31. (a) $\left[2e^{-t} - e^{-t}(\cos 2t + 2\sin 2t)\right]u(t)$;

(b) $e^{-t}(1 + t - \cos t - \sin t)u(t)$;

(c) $\left[-2\sin 3t + 2e^{-2t}(2\cos 3t - \sin 3t)\right]u(t)$;

(d) $\left[4e^{-2t} + e^{-t}(-4\cos 2t + \frac{13}{4}\sin 2t - 5\sin 2t - \frac{5}{2}t\cos 2t)\right]u(t)$

12.33. (a) $-\cos t + \sin t$; (b) $e^{-t}\sin t$; (c) $1 + e^t\sin t$; (d) $3e^{-t} + 2\sin t$; (e) $e^{-t} - e^{-3t} - 2te^{-3t}$; (f) $e^{-t}(1 + 2t) - e^{-3t}(1 + 4t)$

12.35. (a) $2e^{-t} - e^{-2t}$; (b) $e^t - 2e^{-t} - 1$; (c) $6\left[\delta^{(1)}(t) + \delta(t) + e^{-t}\right]$

12.37. (a) $-\frac{9}{4}te^{-(3/2)t} + \frac{3}{2}e^{-(3/2)t}$; (b) $[1 - 7t]e^{-3t}$; (c) $\left[\frac{7}{2}t^2 - 14t + 7\right]e^{-t}$

12.39. (a) $-6e^{-9t}$ A

12.41. (a) $\cos t$; (b) $\sin t$; (c) $(3 - t)e^{-t} - 3e^{-2t}$

12.43. $1 + 3e^{-4t}$

12.45. $t - \frac{6}{5}\left[1 - e^{-(5/6)t}\right]$, $t + \frac{3}{5}\left[1 - e^{-(5/6)t}\right]$

12.47. (a) $\frac{4s}{s^4+6s^2+25}$; (b) $\frac{s^3-3s}{s^4-6s^2+25}$; (c) $\frac{s^2-3}{(s^2+9)(s^2+1)}$

12.49. $\frac{1}{4}\left[e^t - e^{-t}\right] + \frac{1}{2}\left[\cos t + e^{\alpha t}\cos(\beta t + 45°) + e^{-\alpha t}\cos(\beta t - 45°)\right]$, $\alpha = \beta = \frac{\sqrt{2}}{2}$

CHAPTER 13

13.1. $\frac{2s^4+16s^2+8}{s^3+6s}$ Ω

13.3. $\frac{12s^3+6s^2+30s+5}{40s^2+20s}$ S

13.5. $\frac{-(2s^2+3s+2)}{2s^3+3s^2+6s}$

13.7. $V_T = \frac{s^2+s+2}{3s^2}$, $Z_T = \frac{10}{3}(3s + 1)$

13.9. 1 Ω in series with ($\frac{1}{2}$ H in parallel with 1 F)

13.11. Series LC: Z has zeroes at $\pm j\frac{1}{\sqrt{LC}}$ and pole at 0, Y has poles at $\pm j\frac{1}{\sqrt{LC}}$ and zero at 0, etc.

13.13. (a) $e^{-t}(\cos 2t + \sin 2t)$ A; (b) $\frac{2}{5}\cos t - \frac{4}{5}\sin t + e^{-t}(\frac{8}{5}\cos 2t - \frac{9}{5}\sin 2t)$ A

13.15. $6 - 2e^{-9t}$ V

13.17. Yes, any circuits with same Thevenin equivalent.

13.19. $\frac{16}{s+2}$ V, $\frac{4}{s+2}$ V

13.21. $\frac{61s^3+132s^2+84s+12}{s(s+1)^2(3s^2+5s+1)}$ V

13.23. 3 V, 4 A

13.25. $(.715e^{-2.66t} + .285e^{+1.41t})u(t)$ V

13.27. $\frac{4(s^2+5)}{s^2+2s+5}$

13.29. $\frac{20}{s^2+2s+5}$

13.31. $\frac{s(4s+1)}{16s^2+12s+1}$

13.33. $\left[2.71e^{-.655t} + 1.04e^{-.095t}\right]u(t)$ A

13.35. $\frac{2s^2}{3s+4}$

13.37. $\left[-26te^{-2t} - 18e^{-2t} + 15.2\sin(t - 62.6°)\right]u(t)$ V

13.39. $\left[te^{-t} + 84.4e^{-t}\cos(.707t - 166°)\right]u(t)$ V

13.41. $k < 6$

13.43. $-\frac{1}{8} \pm j\frac{\sqrt{15}}{8}$, yes

13.45. $0.21\cos(t - 143°)$ A, $0.052\cos(t + 129°)$ V

13.49. $-1, 0$

13.51. Let (n_1, d_1) be orders of numerator, denominator in $Z(s)$, (n_2, d_2) in $V_s(s)$. Then $d_1 - n_1 < d_2 - n_2 - 1$

13.53. $2(e^{-t} - e^{-1})u(t - 1)$

13.55. Transforms of both are $sF(s)G(s)$.

13.57. $\frac{2s(2s+1)}{s+2}$, $4\delta'(t) - 6\delta(t) + 12e^{-2t}u(t)$ V, $(-4e^{-t} + 12e^{-2t})u(t)$ V

13.59. $e^{-2t}(1 - \cos t)u(t)$

13.61. $e^{-t}(2 + t - \frac{1}{2}t^2)$ A

13.63. $\left[.318e^{-.18st}\cos(1.56t - 57.9°) + .388e^{-.44t}\cos(.093t + 48.6°)\right]$

CHAPTER 14

14.1. (a) $\frac{3}{\sqrt{\omega^2+1}}\underline{/30° - \tan^{-1}\omega}$; (b) $\frac{15\omega}{\sqrt{\omega^2+25}}\underline{/120° - \tan^{-1}\frac{\omega}{5}}$

(c) $\left[\left(30\sqrt{1+\omega^2}\right) / \left(\sqrt{4+\omega^2}|1-\omega^2|\right)\right]$

$\underline{/30° - \tan^{-1}\omega\frac{\omega}{2} - \tan^{-1}\frac{\omega}{2}} - \underline{/(1 - \omega^2)}$

14.3. $\frac{-j\omega^3}{(4-3\omega^2)+j2\omega}$

14.5. $\sqrt{\frac{\omega^2+1}{\omega^6+2\omega^4-3\omega^2+1}}$, $\tan^{-1}\omega - \tan^{-1}\left(\frac{\omega-\omega^3}{1-2\omega^2}\right)$

14.7. $\sqrt{\frac{\omega^2+1}{\omega^2+2}}$, $\tan^{-1}\omega - \tan^{-1}\frac{\omega}{2}$

14.9. $\frac{1}{s^2-1}$, $\frac{-1}{\omega^2+1}$, $-\frac{4}{5}\cos 2t$ V, no.

14.13. (a) 9 dB; (b) $\frac{9}{2}$ dB; (c) 50 dB; (d) 18 dB;
(e) −72 dB; (f) 612 dB; (g) 2 dB; (h) −18 dB

14.15. (a) .05, (b) 20; (c) $5\sqrt{2}$ (d) $.01\sqrt{5}$; (e) $.005\sqrt{5}$;
(f) $\sqrt{10} \times 10^{-20}$

14.17. (a) 121.62; (b) .254; (c) 7.08; (d) 3.16×10^8,
(e) .0216, (f) 1.0001

14.19. $\frac{2(s-1/2)}{(s+1)^3}$, triple real pole with break frequency 1 rad/s, zero with break frequency 1/2 rad/s.

14.21. $\frac{\frac{1}{2}s}{s+3/4}$, 3/4 rad/s

14.25. 18 dB, −38 dB, −38 dB, −20 dB, −20 dB

14.27. $\omega_1 = 25.1$ rad/s, $\omega_2 = 631$ rad/s, $\omega_3 = 15.8$ krad/s, $\omega_4 = 398$krad/s, $\frac{(5\times10^{-7})(s+1)^2(s+w_2)^2(s+w_4)^2}{s(s+\omega_1)^2(s+\omega_3)^2}$

14.29. $\frac{1/6}{(s+25/3)}$

14.31. $\frac{\frac{1}{4}s^2}{(s+1.47)(s^2+.279s+.34)}$

14.39. Series: 1 H, 400 pF, 12.5 Ω; parallel: 1 H, 400 pF, 2 MΩ

14.41. $\frac{A_0 W_0}{jw + \frac{A_0\omega_0 R_A}{R_A+R_F}}$, $\frac{A_0\omega_0 R_A}{R_A+R_F}$ rad/s

14.61. $\frac{-1}{s(s+1)(s+3)}$

15.1. $(\frac{3}{2}t + 1) \times 10^{-4}$ W, $\frac{3}{2} \times 10^{-4}$ W,
$(3t + \frac{7}{2}) \times 10^{-4}$ W, $(\frac{3}{2}t + 4) \times 10^{-4}$ W

15.3. $\frac{3}{2}$ H

15.5. $0.9t + 1.05$ W, $0.45t^2 + 1.05t$ J

15.7. 2 H, 0 H

15.9. $4(e^{-t} - e^{-3t})$ A

15.11. $\left[-4e^{-(t/5)} + 75\cos(4t - 87.1°)\right]$ mA

15.13. $\frac{5\sqrt{2}}{2}\sin(2t + 135°)$ V

15.15. 1250 W, 0.6 lagging

15.17. (a) $2e^{-2t}$ V; (b) $\frac{2\sqrt{2}}{3}\cos(8t - 45°)$ V

15.19. $-2e^{-t}$ V

15.21. $35.9\cos(t + 152°)$ A

15.23. $2\underline{/76.4°}$, $1\underline{/22.6°}$, $.343\underline{/-59.0°}$

15.25. $35.9\cos(t + 152°)$ A

15.27. −2

15.29. $\sqrt{2}e^{-t}\cos(t + 45°) - e^{-t}$ A

15.31. (a) $V_2 = -8V_1$, $I_2 = \frac{1}{8}I_1$, $v_2 = -8v_1$, $i_2 = \frac{1}{8}i_1$;
(b) $V_2 = -\frac{3}{8}V_1$, $I_2 = \frac{8}{3}I_1$, $v_2 = -\frac{3}{8}v_1$, $i_2 = \frac{8}{3}i_1$

15.33. $4(1 - e^{-(6/19)t})u(t)$ V

15.35. $2.67\underline{/-47.7°}$

15.37. $v_1 = 50\cos 377t$ V, $i_1 = 0.2\cos(377t - 89.6°)$ A,
$v_2 = 0.2\cos(377t - 89.3°)$ V, $i_2 = .033\cos(377t + 90.7°)$ A

15.39. .761 H

15.41. $\frac{t_{21}}{t_{11}t_{22}-t_{12}t_{21}}$, $\frac{g_{11}}{g_{12}}$

15.43. $w \geq \frac{1}{2}L_1 l_1^2 + \frac{1}{2}L_2 l_1^2 - \sqrt{L_1 L_2}l_1 l_2 = \frac{1}{2}(\sqrt{L_1}l_1 + \sqrt{L_2}l_2)^2 \geq 0$

15.45. $1.27\underline{/-158°}$ A

16.1. 2.34 A, .938 A, 8.91 V, −5.62 V

16.3. $\begin{bmatrix} \frac{s}{s+1} & \frac{2}{s+1} \\ 3 & \frac{2s^2+6s-1}{s(s+1)} \end{bmatrix}$

16.5. Solve z-parameter equations for I_1 and V_2.

16.7. Solve z-parameter equations for V_1 and I_2.

16.9. $\frac{1}{4}\begin{bmatrix} s+2 & -s \\ 2-4s & s+1 \end{bmatrix}$, $\frac{2s-1}{2s+10}$

16.11. $\begin{bmatrix} 3.4 & 2.6 \\ 2.6 & 3.4 \end{bmatrix}$

16.13. $V_1 = \frac{1}{s}$, $V_2 = \frac{4s(8s^2+10s+5)}{(4s^2+2s+1)^2}$,
$I_1 = \frac{s+1}{s} - \frac{4s^2(8s^2+10s+5)}{(4s^2+2s+1)^2}$, $I_2 = \frac{-(8s^2+10s+5)}{4s^2+2s+1}$

16.15. $\begin{bmatrix} \frac{s+3}{6s} & -\frac{1}{3} \\ +\frac{1}{3} & \frac{4s+3}{3s} \end{bmatrix}$, $\begin{bmatrix} \frac{(2s+3)(s+1)}{2s^2} & -\frac{(4s+3)}{s} \\ -\frac{(s+3)}{2s} & 3 \end{bmatrix}$

16.17. $\frac{1}{s(2s^2+2s+3)}\begin{bmatrix} 2(s^2+s+1) & 2 \\ 2 & 2(2s^2+2s+1) \end{bmatrix}$,
$\frac{2(s^2+s+1)}{s(2s^2+2s+3)}$, $\frac{2(2^2+2s+1)}{s(2s^2+2s+3)}$

16.19. $\frac{z_{21}}{z_{11}}$, $\frac{-h_{21}}{h_{11}h_{22}-h_{12}h_{21}}$

16.23. Fig. 16.30 with $z_{11} - z_{12} = \frac{2}{s^2+2s+2}$, $z_{22} - z_{12} = \frac{1}{s^2+2s+2}$, $z_{22} - z_{21} = \frac{s}{s^22s+2}$, $z_{12} = s$

16.25. Symmetric π-circuit with 6 Ω across each port, connected by 1.85 Ω.

16.27. (a) Apply KCL at input, output ports, (b) apply KCL at input, output ports.

16.29. $\begin{bmatrix} .706 & .765 \\ 1.00 & .588 \end{bmatrix}$

16.33. $\begin{bmatrix} -5.70 & -9.00 \\ 1.00 & 1.33 \end{bmatrix}$

16.35. $10^{-4}\begin{bmatrix} \frac{s+5}{s+10} & 0 \\ \frac{10}{s+10} & 1 \end{bmatrix}$

16.37. $\begin{bmatrix} 2.52 & 1.64 \times 10^7 \\ -1.58 \times 10^{-7} & 1.26 \end{bmatrix}$

16.43. Nodal analysis with negative terminal of v_1 as reference yields $y_{11} = y_{12}$ and $y_{12} = y_{21}$. Inverting yields the z-parameters.

836 App. E Answers to Selected Odd-Numbered Problems

16.45. Dependent variable pair is the superposition of a dependent variable pair plus new term due to sources. To find new term, set both independent variables to zero (open or short the ports), leave internal sources in, compute dependent variables. They equal the new terms.

CHAPTER 17

17.1. (a) $T_p = 4$, (b) $T_p = 10$, (c) $T_p = 2$

17.3. $f_e(t) = \frac{1}{2}[f(t) + f(-t)] = 0$ implies $f(t) = -f(t)$ (odd symmetry). Similarily $f_0(t) = 0$ implies $f(t) = +f(-t)$.

17.5. (a) $\frac{1}{2}[(r(t+1) + r(-t+1)) - 2(r(t) + r(-t)) + 2(r(t-1) + r(-t-1)) - 2(r(t-2) + r(-t-2)) + (r(t-3) + r(-t-3))]$, $\frac{1}{2}[(r(t+1) - r(-t+1)) - 2(r(t) - r(-t)) + 2(r(t-1) - r(-t-1)) - 2(r(t-2) - r(-t-2)) + (r(t-3) - r(-t-3))]$;
(b) $\sin^2 2\pi$, 0;
(c) $3.83\cos 2t - 1.17\sin 2t$ (d) $3.83\cos 2t\,[u(t-1) - u(t+1)]$, $-1.17\sin 2t\,[u(t-1) - u(t+1)]$

17.7. (a) $\frac{1}{2}, \frac{1}{2}, \frac{\sqrt{2}}{2}$; (b) $\frac{1}{5}, \frac{1}{5}, \frac{\sqrt{5}}{5}$; (c) $\frac{1}{2}, \frac{1}{2}, \frac{\sqrt{2}}{2}$

17.9. No. $f(t)$ possible if and only if T_2/T_1 rational number.

17.11. $\frac{4}{\pi} + 2\sin 2(t + \frac{\pi}{6}) - \frac{8}{\pi}\sum_{n=1}^{\infty} \frac{1}{4n^2-1}\cos 4n(t + \frac{\pi}{6})$

17.13. $\frac{1}{2} - \frac{6}{\pi}\sum_{n=1}^{\infty} \frac{\sin(2n-1)\pi t}{(2n-1)}$

17.15. $-6\sum_{n=1}^{\infty} \cos(2n-1)\pi t$

17.17. $\omega_0 = \frac{\pi}{2}, b_n = 0, a_n = \left[\frac{(n\omega_0)^3 - (n\omega_0)^2 - 2}{(n\omega_0)^3}\right]\sin n\omega_0 - \left[\frac{2}{(n\omega_0)^2}\right]\cos n\omega_0$

17.19. (a) $\frac{3}{2} + \frac{2}{\pi}\sum_{n=1}^{\infty} \frac{\sin(2n-1)t}{(2n-1)}$, (b) $\omega_0 = 1$, $a_n = \frac{(-1)^n(e^{\pi} - e^{-\pi})}{\pi(1+n^2)}, b_n = \frac{(-1)^{n+1}(e^{\pi} - e^{\pi})}{\pi(1+n^2)}$, (c) $\omega_0 = \pi, a_n = \frac{4(-1)^n}{(n\pi)^2}, b_n = 0$

17.21. (a) Break into two integrals, use symmetry; (b) $\omega_0 = \frac{\pi}{6}, b_n = 0, a_{2n-1} = \frac{4}{(2n-1)\pi}\left[2\sin\frac{(2n-1)\pi}{6} - 2\sin\frac{(2n-1)\pi}{3} + 4(-1)^{n+1}\right]$

17.23. a_n, b_n are all the same; ω_0 for $g(t)$ is c times ω_0 for $f(t)$.

17.25. $\omega_0 = \pi, a_n = \frac{e^{-2}}{1+(n\pi)^2}\left[(n\pi)^2 - 3\right] + \frac{1-(n\pi)^2}{(1+(n\pi)^2)^2}$, $b_n = \frac{-2n\pi e^{-2}}{1+(n\pi)^2} + \frac{2n\pi}{(1+(n\pi)^2)^2}(1 - 2e^{-2})$

17.27. (a) $\omega_0 = \frac{\pi}{2}, c_n = \frac{1}{n\pi}\sin\frac{n\pi}{2}$; (b) $\omega_0 = \frac{\pi}{5}, c_n = \frac{1}{n\pi}\sin\frac{n\pi}{2}$; (c) $\omega_0 = \pi, c_n = \frac{1}{jn\pi}$ (n even), $c_n = 0$ (n odd)

17.29. Same for all n.

17.31. (a) $f(t)$ even; (b) $f(t)$ odd

17.33. Evaluate (16.25).

17.35. (a) $c_0 = A + 24, c_n = \frac{jA}{(2n\pi)(1+jn\omega_0)}$; (b) $c_0 = \frac{2A}{\pi} + 24, c_n = \frac{2A}{\pi(1-4n^2)(1+jn\omega_0)}$ ($n \neq 0$); (c) $c_0 = \frac{2A}{\pi} + 24, c_1 = \frac{-jA}{4(j\omega_0+1)}, c_{2n} = \frac{-A}{\pi(4n^2-1)(j2n\omega_0+1)}, c_{2n-1} = 0$ for $n > 1$

17.37. $\frac{1}{4} + \frac{2}{\pi}\sqrt{\frac{\pi^2+4}{\pi^2+16}}\cos\left(\frac{\pi t}{2}\tan^{-1}\frac{\pi}{2} - \tan^{-1}\frac{\pi}{4}\right) - \frac{2}{3\pi}\sqrt{\frac{9\pi^2+4}{9\pi^2+16}}\cos\left(\frac{3\pi t}{2} + \tan^{-1}\frac{3\pi}{2} - \tan^{-1}\frac{3\pi}{4}\right)$

17.39. n odd: $A_n = \frac{16}{(n\pi)^2+4}$; n even: $A_n = 0$

17.41. $\frac{8}{\sqrt{4+(n\pi)^2}}, -\tan^{-1}\frac{n\pi}{2}, \frac{64}{4+(n\pi)^2}$

17.43. $\frac{8}{\pi^2}, \frac{88}{9\pi^2}$

17.45. (a) $\frac{1}{j\omega}\left[e^{ja\omega} - e^{-jb\omega}\right]$; (b) $\frac{a+j\omega}{(a+j\omega)^2+b^2}$; (c) $\frac{b}{(a+j\omega)^2+b^2}$; (d) $\frac{1-e^{-2(a+j\omega)}}{a+j\omega}$

17.47. $\frac{2}{\omega}[\sin 4\omega - \sin 2\omega]$

17.49. $\lim_{\Delta \to 0}\left[\frac{2}{\Delta}\sin\frac{\Delta}{2}\omega\right] = 1$

17.51. (a) $-6e^{+3t}u(-t)$; (b) $-6e^{-3|t|}$; (c) $(e^t - e^{3t})u(-t)$

17.53. $\frac{1}{\sqrt{2}}\frac{3+j\omega}{5-\omega^2+j2\omega}$

17.57. $|F(jn\pi)| = 1.135(n = 0), 0.262(n = \pm 1), 0.136(n = \pm 2)$, etc.

17.61. (a) $2\mathbf{X}(j\omega) + \frac{1}{2}[\mathbf{X}(j[\omega - \omega_0]) + \mathbf{X}(j[\omega + \omega_0])]$; (b) $2\mathbf{X}(j\omega)$

Index

direct, 8
divider, 44
division, 44, 361
electric, 1, 6
full-scale, 61
gain, 81
leakages, 196
mesh, 139
phase, 454
reference direction, 7, 11
sensitivity, 63
short-circuit, 51
sinusoidal, 8
terminal, 33
current–divider principle, 44, 45
current sources, 13, 142
   controlled, 80
   current-controlled (CCCS), 81
   dependent, 80
      element statements of, in SPICE, 159
   ideal, 56
   independent, 13–14, 34, 53, 128
   Norton equivalent, 49
   in parallel, 46
   practical, 55–56
   in series, 41
   voltage-controlled (VCCS), 81

D

d'Arsonval meter, 61
damping, 280
   critical, 279
   factor, 567
   ratio, 278
dc. *See* Current, direct
dc steady state, 193
dc sweep, 161
decibel scale, 563
delta connection, 461, 463–464, 466
   phasor diagram, 463
describing equation, 272
determinant(s), 795
dielectric. *See* Insulator(s)
differential equation(s), 209
   characteristic equation of, 213

first-order, 209
   forcing term in, 222
Dirac, Paul, 490
Dirichlet conditions, 699
discrete phase plots, 714
discrete spectra, 714
dot convention, 616
double-subscript notation, 447
dual equations, 146
duality, 145, 147–148
   of capacitor and inductor, 187
   property, 728

E

Edison, Thomas Alva, 359
electric field(s), 26, 178
electric circuit, 2
electrical isolation, 628
electricity
   animal, 1
   metallic, 1
electrocardiogram (ECG), 764
electromagnetic (EM) field theory, 2
electromotive force, 9
electrons, 6
electrophorus, 1
element(s)
   active, 12–14
   ideal, 15
   linear, 119
   lossless, 408
   in parallel, 42
   passive, 12, 32, 179, 189
   two-terminal, 13, 34
element law(s), 13, 23, 34
   and the s-domain, 521–522
energy, 10, 11
equation(s)
   characteristic, 213
   dual, 146
   of phasor, 347
equivalence, 35
equivalent circuits, 35–36
Euler identity (Euler's formula), 324, 705, 807, 809
excitations. *See* Sources, independent

impedance, 338
    equivalents, 342
    input, 591
    matrices, 652, 674
    output, 591
    phase, 453
    reflected, 626
    scaling, 595, 632, 782
    z-parameter, 658
improved op amp model, 86–87, 93, 100
impulse product rule, 490
impulse response, 541
inductance,
    circuits with, 616
    mutual, 612, 624
    self-, 612
    and transformers, 624
induction, electromagnetic, 186
inductor(s), 2, 185
    energy storage in, 188–189
    practical, 196
    series and parallel, 190, 191
    voltage–current relations for, 335
initial condition generator, 526
initial value theorem, 540
input, 15
    dummy, 758
    inverting, 84
    noninverting, 84
input–output,
    analysis, 530
    relationship, 16
insulator, 174
    perfect, 32
integrated circuits, 60
integration, 494
integrodifferential equations, 509
interconnection laws. *See under* Kirchhoff's laws
International System of Units, 3, 211
inverse–inverse law, 44
integrator, inverting, 754

J

joule (J), 4
Joule, James Prescott, 4, 401

K

KCL. *See* Kirchhoff's laws, of current
kelvin, 3
kilogram, 3
Kirchhoff, Gustav Robert, 22, 117
Kirchhoff's laws, 23
    of current (KCL), 23–25, 29
    and impedance equivalents, 342
    and phasors, 342
    and the s-domain, 521–525
    of voltage (KVL), 23, 25–27
        sum of drops form, 26
        sum of rises form, 26
KVL. *See* Kirchhoff's laws, of voltage

L

ladder network, 121
Laplace, Pierre Simon, 479
Laplace transform(s), 481–485
    differentiation, 493
    integration, 494
    inversion integral, 485
    i–v laws of, 617
    pairs, 501, 541
    poles of, 483
    properties, 500
        frequency shift, 496
        time-frequency sealing, 498
        time shift, 496
        t-multiplication, 499
    zeros of, 483
Laplace transformation, 481
    conditions for, 481
leakage flux, 611
Leibniz's rule, 492
linear circuit, 119
linear combination, 120
linearity, 119, 121, 484, 727
loading, 100, 591
loop, 138
    ground, 628
loop analysis, generalized, 138
loudspeaker system, 745
    three-way, 748
lumped-parameter circuit, 23

LAPLACE TRANSFORM PAIRS

	$f(t)$	$\mathbf{F}(s)$

Singularity functions

1. Unit impulse $\delta(t)$ 1
2. Unit step $u(t)$ $1/s$
3. Unit ramp $r(t) = tu(t)$ $1/s^2$
4. Unit parabola $p(t) = 1/2t^2 u(t)$ $1/s^3$
5. n^{th} integral of impulse $\delta^{(-n)}(t)$ $1/s^n$
6. Unit doublet $\delta'(t)$ s
7. n^{th} deriv. of impulse $\delta^{(n)}(t)$ s^n

Ordinary functions

8. Constant 1 $1/s$
9. t t $1/s^2$
10. Power of t $t^{n-1}/(n-1)!$ $1/s^n$
11. Exponential e^{-at} $1/(s+a)$
12. t-mult. exponential te^{-at} $1/(s+a)^2$
13. repeated t-mult. exp. $t^{n-1}e^{-at}/(n-1)!$ $1/(s+a)^n$
14. sine $\sin \omega t$ $\omega/(s^2 + \omega^2)$
15. cosine $\cos \omega t$ $s/(s^2 + \omega^2)$
16. damped sine $e^{-at}\sin \omega t$ $\omega/([s+a]^2 + \omega^2)$
17. damped cosine $e^{-at}\cos \omega t$ $(s+a)/([s+a]^2 + \omega^2)$
18. t-mult. sine $t \sin \omega t$ $2\omega s/(s^2 + \omega^2)^2$
19. t-mult. cosine $t \cos \omega t$ $(s^2 - \omega^2)/(s^2 + \omega^2)^2$

20. rectified sine $|\sin \omega t|$ $\dfrac{\omega}{s^2 + \omega^2}\coth\dfrac{\pi s}{2\omega}$

21. quartic I $\dfrac{1}{2\omega^3}[\sinh \omega t - \sin \omega t]$ $1/(s^4 - a^4)$

22. quartic II $\dfrac{1}{2\omega^2}[\cosh \omega t - \cos \omega t]$ $s/(s^4 - a^4)$

FORGIVE US
OUR TRESPASSES

BY

LLOYD C. DOUGLAS

Author of
'*Magnificent Obsession*'

BOSTON AND NEW YORK
HOUGHTON MIFFLIN COMPANY
The Riverside Press Cambridge
1932